Chemistry as a Diagnostic of Star Formation

Conference Sponsors

CITA ICAT

COM DEV
SPACE

CSA ASC

University of
Waterloo

Chemistry as a Diagnostic of Star Formation

Proceedings of the Conference held in Waterloo, Canada, August 21–23, 2002

Edited By
Charles L. Curry and Michel Fich

NRC·CNRC
NRC Research Press
Ottawa 2003

Printed in Canada on acid-free paper.

Publisher's Note: This book has been prepared from camera ready copy provided by the editors.

ISBN 0–660–19089–3
NRC No. 46321

National Library of Canada cataloguing in publication data

Main entry under title :

Chemistry as a diagnostic of star formation : proceedings of the conference held at the University of Waterloo, Waterloo, Ontario, Canada, August 21-23, 2002

Issued by the National Research Council of Canada.
ISBN 0-660-19089-3

1. Stars -- Formation - Congresses.
2. Cosmochemistry - Congresses.
I. Curry, Charles L. (Charles Leo), 1966- .
II. Fich, Michel.
III. National Research Council Canada.

QB806.C43 2003 523.8 C2003-980255-8

Inquiries: Proceedings Program, NRC Research Press, National Research Council of Canada, Ottawa, Ontario K1A 0R6, Canada.

Correct citation for this publication: Chemistry as a Diagnostic of Star Formation (2003). Edited by Charles L. Curry and Michel Fich. Published in 2003 by NRC Research Press, Ottawa ON, K1A 0R6, Canada, 479 Pages.

Cover Figures:

Spectrum: M-band, R = 25; 000 Keck/NIRSPEC spectrum of the low-mass protostar L1489 IRS, showing narrow ^{12}CO absorption lines and a broad multi-component band of solid CO. (From Boogert et al., this volume, page 175).

Polycyclic aromatic hydrocarbon (PAH) molecule. Reprinted with permission of David Jansen, Leiden Observatory.

Artist's conception of the APEX telescope. (From Schilke et al., this volume, page 20).

TABLE OF CONTENTS

PREFACE

It has often been remarked that our knowledge of how stars form is sensitive to, dependent upon, or subject to our current understanding of the chemical composition and state of the interstellar medium. However, this is a vast understatement of the true importance of astrochemistry for interstellar medium studies. Star formation, in particular, a process that was still poorly understood only thirty years ago, has progressed essentially in lock-step with our knowledge of the chemical composition and state of the gaseous and particulate material in these highly visibly obscured regions. One need only revisit the rough conclusions arrived at from broadband continuum observations in the early 1970's to see that the field has been utterly transformed by use of the more sophisticated spectral techniques now at astronomers' disposal.

Plans for future instrumentation indicate at least as rapid an evolution in the coming years. With several large, international observatories and instrument programs now in their design or early construction phases, it is more important than ever that all of those working in the field get up to speed on the current state of astrochemical knowledge. Thus the conference was conceived as an opportunity to bring together those working on myriad aspects of astrochemistry as it relates to star formation: including those whose work focussed essentially on theory, laboratory experiment, observations, and instrumentation. The idea also bore a sense of moderate urgency: within the next decade, several international collaborations already well past the design phase (ALMA, *Herschel, SIRTF*, and *NGST*) promise to produce a wealth of data that perhaps very few workers in the area today are well prepared to interpret.

As in other areas of science, major developments in star formation in past decades have been opportunistic in two important ways: first, by exploiting technological advances in telescope design, spectroscopy, and other instrumentation to obtain the highest resolution and most penetrating views of dense molecular clouds and protostellar regions; and second, by taking full advantage of the rapid pace of improvement in computer hardware and design, permitting the simultaneous calculation of increasingly detailed dynamical and chemical processes. That both approaches are not just necessary but, one hopes, symbiotic, was a view that seemed to be shared by the majority of the participants.

Ultimately, roughly 130 scientists and students from 15 countries converged on Waterloo in the dog days of August, for what turned out to be a lively discussion and, for many of the participants, a reunion long overdue. The scientific sessions were structured with the above motivations in mind. The conference opened with a preview of the next generation of telescopes and detectors, their projected capabilities, and the ambitious science goals motivating their design. This was followed by a unit on fundamental processes, which highlighted important work going on both in the lab and "in the field," where much of the discussion surrounded the high observed abundance of multiply deuterated molecules. The often indirect connections between chemistry and physics in star-forming regions were explored in depth in Sessions III and V, with the intervening unit dedicated to the burgeoning area of dust continuum observations and their interpretation. The final session took on issues in high-mass star formation, which presents its own unique conundrums and observational challenges.

The detailed scientific highlights of the conference may be found in the review presented by George Mitchell. Four separate indexes are provided at the back of the book to aid in locating topics of particular interest—we hope that readers find this a useful feature. There are certainly a number of interesting scientific surprises in these pages; on a lighter note, some may be intrigued to learn that "observations of seven white dwarfs ..." have been conducted, in a location that will be left for the curious reader to discover. The prize for perspicuity goes to Geoff Blake and Di Li for their conscientious use of the term "isotopologue" (with no hits in an ADS abstract search of pre-2002 astrophysics papers), while the misapplied "isotopomer" continues to hold sway in the astrophysical literature[1] (150 abstract-only hits on ADS since 1993). Out of respect for author autonomy, the editors resisted the urge to right this semantical injustice; but clearly the path has been struck.

The editors would like to extend our thanks to the other members of the SOC (Ted Bergin, Ewine van Dishoeck, Neal Evans, Xander Tielens and David Williams) who provided valuable advice in putting together the scientific program, which by all accounts was the "right" one, given the high-quality mix of contributed and invited talks placed before us. We are grateful to the several organizations who provided financial support. Donations from COM DEV, the Canadian Space Agency, the Canadian Institute for Theoretical Astrophysics, and the Faculty of Science at the University of Waterloo helped make possible the attendance of not only invited speakers, but also a large number of graduate students, and also helped defray various other costs incurred by the conference.

The support, advice, and judgment provided by other members of the LOC (Peter Bernath and Walt Duley) is gratefully acknowledged. Conference computing issues were deftly handled by Scott Nicoll, as were the financial aspects by Colleen Buchanan. The generosity and flexibility of Waterloo Co-operative Residence Incorporated enabled us to accommodate nearly 70 of the conferees in rooms just a few minutes' walk from the conference venue, and all for a donation that was certainly out of proportion to their efforts. In the hectic days leading up to the meeting, Heather Anderson, Laura Parker, and Heather Mann gave generously of their time and patience and helped manage a sometimes hectic registration process.

As manuscripts began to flood in near the submission deadline, the editors received help from Kristin Woodley, who helped to extract and process papers as they arrived by email. CC is grateful to the CCCma at the University of Victoria for the use of its computing facilities in the process of preparing the final manuscript. Few, if any, publications emerge error-free from the editing process, and we are aware that this volume will be no different. However, we have endeavoured to excise the most conspicuous of these from the text, with help from several authors who promptly resubmitted materials when necessary. We extend our apologies for any errors that remain. If the product will be useful to researchers for a reasonable time into the future, then we feel that our aim as editors, and one of the many goals of the conference, will have been attained.

Charles Curry
Michel Fich

August, 2003

Reference

Goldsmith, P.F., Li, D., Bergin, E.A., Melnick, G.J., Tolls, V., Howe, J.E., Snell, R.L., & Neufeld, D.A. 2002, ApJ, 576, 814

[1]In an example used by Goldsmith et al. (2002), $H^{13}CCCN$, $HC^{13}CCN$, and $HCC^{13}CN$ are isotopomers, while CO and ^{13}CO are isotopologues.

CONFERENCE PARTICIPANTS

Aikawa, Yuri	Kobe University
Allers, Katelyn	University of Texas, Austin
Andre, Philippe	CEA, Paris
Bacmann, Aurore	European Southern Observatory
Basu, Shantanu	University of Western Ontario
Beck, Tracy	Gemini Observatory
Benedettini, Milena	CNR–IFSI, France
Bensch, Frank	CfA, Harvard
Bergin, Ted	CfA, Harvard
Bernard, Jean-Philippe	CESR–CNRS, France
Bernath, Peter	University of Waterloo
Blake, Geoffrey	Caltech
Boogert, Adwin	Caltech
Boudet, Nathalie	CESR–CNRS, France
van Broekhuizen, Fleur	Leiden University
Carey, Sean	SSC/Caltech
Castets, Alain	Bordeaux Observatory
Caux, Emmanuel	CESR, Toulouse
Chihara, Hiroki	Osaka University
Choi, Minho	ASIAA
Ciolek, Glenn	Rensselaer Polytechnic Institute
Comito, Claudia	MPIfR, Bonn
Cunningham, Cliff	Mercury Magazine, ASP
Curry, Charles	University of Waterloo
De Buizer, James	Cerro Tololo Observatory
Decamp, Nicolas	University of Paris
Di Francesco, James	National Research Council Canada
van Dishoeck, Ewine	Leiden Observatory
Doty, Steven	Denison University
Douglas, Kevin	University of Calgary
Duley, Walt	University of Waterloo
Dunbar, Robert	Case Western Reserve University
Evans, Neal	University of Texas, Austin
Feldman, Paul	National Research Council Canada
Fich, Mike	University of Waterloo
Fuller, Gary	UMIST, Manchester
Geballe, Thomas	Gemini Observatory
Gerin, Maryvonne	ENS, Paris
Gibb, Andy	University of Maryland
Glinski, Robert	Tennessee Tech
Glover, Simon	American Museum of Natural History
Gregersen, Erik	McMaster University
Habart, Emilie	Arcetri Observatory
Harju, Jorma	University of Helsinki
Hassel, George	Rensselaer Polytechnic Institute

Henning, Thomas	MPIfA, Heidelberg
Herczeg, Gregory	JILA/University of Colorado
Hily-Blant, Pierre	Paris Observatory
Hirota, Tomoya	Kagoshima University
Hogerheijde, Michiel	Steward Observatory
Honda, Mitsuhiko	Subaru Telescope
Houde, Martin	Caltech
Jeyakumar, Solai	University of Cologne
Joblin, Christine	CESR-CNRS, France
Johnstone, Doug	National Research Council Canada
Jonkheid, Bastiaan	Leiden University
Jørgensen, Jes	Leiden University
Kalenski, Sergei	Astro Space Centre, Moscow
Keane, Jacqueline	NASA Ames Reserach Centre
Kessler, Jacqueline	Caltech
Kim, Sunguen	Smithsonian Astrophysical Observatory, Harvard
Knez, Claudia	University of Texas, Austin
Koike, Chiyoe	Kyoto Pharmaceutical University
Kontinen, Samu	University of Helsinki
Kulesa, Craig	University of Arizona
Kwok, Sun	University of Calgary
Lacy, John	University of Texas
Lahuis, Fred	SRON, Groningen
Lai, Shih-Ping	Jet Propulsion Laboratory, Pasedena
Lee, Jeong-Eun	University of Texas, Austin
Li, Di	Smithsonian Astrophysical Observatory, Harvard
Liu, Sheng-Yuan	Caltech
Loinard, Laurent	UNAM, Mexico City
Mandy, Margot	University of Northern British Columbia
Maret, Sebastien	CESR-CNRS, France
Markwick, Andrew	UMIST, Manchester
Matsumura, Soko	McMaster University
Matthews, Brenda	University of California at Berkeley
Meier, David	University of Illinois
Millar, Tom	UMIST, Manchester
Mitalas, Romas	University of Western Ontario
Mitchell, George	Saint Mary's University
Miville-Deschenes, Marc-Antoine	Canadian Institute for Theoretical Astrophysics
Nassar, Ray	University of Waterloo
Nayral, Celine	CESR-CNRS, France
Nehme, Cyrine	University of Paris
Nikolic, Silvana	Onsala Space Observatory
Noel, Benjamin	CESR-CNRS, France

Oka, Takeshi	University of Chicago
Oka, Tomoharu	University of Tokyo
Olofsson, Henrik	Onsala Space Observatory
Osamura, Yoshihiro	Rikkyo University
Pagani, Laurent	Paris Observatory
Peng, Ruisheng	Caltech
Petrie, Simon	Australian National University
Phillips, Tom	Caltech
Plume, René	University of Calgary
Poidevin, Frederick	University of Montreal
Pontoppidan, Klaus	Leiden Observatory
Qi, Chunhua	Center for Astrophysics, Harvard
Reid, Michael	McMaster University
Ristorcelli, Isabelle	CESR–CNRS, France
Roberge, Wayne	Rensselaer Polytechnic Institute
Roberts, Helen	Ohio State University
Roser, Joseph	Syracuse University
Saito, Tomoki	University of Tokyo
Schilke, Peter	MPIfR, Bonn
Schoier, Fredrik	Leiden Observatory
Schutte, Willem	Leiden Observatory
Scott, Alan	EMS Technologies, Ottawa
Shematovich, Valery	RAS, Moscow
Stantcheva, Tatiana	Ohio State University
van der Tak, Floris	MPIfR, Bonn
Takahashi, Junko	Meiji Gakuin University
Teyssier, David	LERMA, Paris Observatory
Thi, Wing-Fai	University College London
Tielens, Xander	Kapteyn Institute
Tilley, David	McMaster University
Tothill, Nick	Saint Mary's University
Vastel, Charlotte	Caltech
Velusamy, Thangasamy	Caltech
Volgenau, Nikolaus	University of Maryland
Wakelam, Valentine	University of Bordeaux
Walmsley, Malcolm	Arcetri Observatory
Walsh, Andrew	Smithsonian Astrophysical Observatory, Harvard
Widicus, Susanna	Caltech
Williams, David	University College London
Wilson, Christine	McMaster University
Wong, Tony	Australia Telescope National Facility, Sydney
Yate, Christopher	UMIST, Manchester
Yoshida, Hiroshige	Caltech

CHEMISTRY AS A DIAGNOSTIC OF STAR FORMATION
http://astro.uwaterloo.ca/sfchem2002
21 – 23 August 2002
University of Waterloo, Canada
Program Outline
1. Fundamental processes
2. Chemistry, physical conditions, and structure of dark clouds
3. Chemical diagnostics of infall, outflow, and disks
4. High-mass star formation
5. The role of dust continuum observations
6. Issues of ongoing and future interest (Primordial star formation, spectroscopic signatures of brown dwarfs and extrasolar planets, observatories, telescopes, instrumentation, and future missions)
Invited Speakers
Phillipe Andre
Ted Bergin
Geoff Blake
Neil Evans
Maryvonne Gerin
Thomas Henning
Tom Millar
Tom Phillips
Xander Tielens
Ewine van Dishoeck
Malcolm Walmsley
David Williams
Scientific Organizing Committee
Ted Bergin
Charles Curry
Neil Evans
Michel Fich (Chair)
Xander Tielens
Ewine van Dishoeck
David Williams
Local Organizing Committee
Peter Bernath
Charles Curry (Chair)
Walt Duley
Michel Fich
Image of S106 ©Suburu Telescope, National Astronomy Observatory of Japan. 2001

George Mitchell and Mike Fich at the reception.

Geoff Blake and Peter Bernath at the reception.

Nathalie Boudet confers during a poster session.

Phillippe Andre answers a question while the session chair, Neal Evans, stands by and Doug Johnstone readies his talk.

Steve Doty asks the speaker a question, that Sean Carey (to his right) and David Teyssier (in foreground) are the first to hear.

Thomas Henning is slightly amused at what he is hearing.

The audience pays close attention.

Tanglefoot: the Canadian folk music band at the conference dinner.

PART I:
THE FUTURE OF ASTROCHEMISTRY INSTRUMENTATION

FUTURE MILLIMETER/SUBMILLIMETER INSTRUMENTATION AND SCIENCE OPPORTUNITIES: THE EXAMPLE OF DEUTERATED MOLECULES

Thomas G. Phillips and Charlotte Vastel

California Institute of Technology, 320-47, Pasadena, CA 91125, U.S.A.

Abstract

During the next decade a tremendous advance will take place in instrumentation for spectroscopy of the interstellar medium. Major new facilities (ALMA, SOFIA, APEX, LMT, *Herschel* and others) will be constructed and commissioned, so that the science opportunities, in the field of astrochemistry, will increase by a huge factor. This will be enhanced by the new receivers with greater bandwidth and sensitivity. The new opportunities will be in the area of astrochemistry of distant objects, through greater sensitivity, or new spectral ranges due to the platforms above the Earth's atmosphere.

Various aspects of new spectral ranges are discussed, with emphasis on H_2O lines, features previously hidden under H_2O or O_2 lines, light hydrides and particularly on deuterium in molecules. Recently, multiply deuterated species have been detected (e.g., ND_3) in cold dense regions of the interstellar medium. It is argued here that it is possible that so much deuterium could be trapped, by the fractionation process, into heavy molecules such as ND_3, and species such as H_2D^+ and possibly D_2H^+, that D and HD might be depleted. This would be the mechanism for the large dispersion of [D]/[H] values found in the interstellar medium. Light molecules (hydrides and deuterides) generally have large fundamental rotation frequencies, often lying in the HIFI bands. The deuterides are an especially suitable case, because the species exist mainly in cold dense regions, where molecules are in their ground states and THz observations will best be carried out by absorption spectroscopy against background dust continuum sources such as Sgr B2 and W49N.

Key words: ISM: molecules – astrochemistry – deuterium – submillimeter

1. Introduction

A great deal is already known about the nature of the gas and dust in the interstellar medium, obtained from a wide variety of telescopes and facilities. Most of these have been ground-based, but some were airborne, balloon-borne, or space missions. In particular, high resolution, heterodyne ($R \geq 10^6$) spectrometers have provided information on the molecular and (surprisingly) atomic species in the various types of interstellar clouds. It has indicated the physical nature of the gas, mostly from the velocity structure and tested the chemical models—e.g., the ion-molecule reaction scheme, which was of course developed to explain the observed molecular content of the clouds. Figure 1 (Phillips & Keene 1992) shows a submillimeter and far-infrared spectrum of an interstellar cloud, as it might be observed from a space platform. Features include the line-forest of heavy molecules at the longer wavelengths, light hydrides and deuterides at shorter wavelengths, important atoms and ions, such as C^0, C^+, and O^0, and dust continuum and resonances. Also large molecules may possibly be seen via their bending modes.

Compared to other branches of astronomy, the submillimeter has been slow to develop. This is in part due to the low photon energy (compared to optical) and the small number of arriving photons (compared to the radio), making it difficult to construct detection equipment; but it is also due to the poor overall transmission of the Earth's atmosphere. As is seen in Figure 2 (*top*), even from a high mountain site such as Mauna Kea, Hawaii, the transmission is only partial up to about 900 GHz (330 μm) and very small above that. Of course, from airborne altitudes—e.g., 12 km for SOFIA—transmission is much better (Figure 2, *bottom*) but is still badly contaminated by H_2O and O_2 lines. There is a clearly indicated need for space-based telescopes.

A third, and possibly dominant reason for the underdeveloped nature of the field is the lack of major instruments. Because the millimeter/submillimeter observers

Chemistry as a Diagnostic of Star Formation (2003). *Edited by Charles L. Curry and Michel Fich.* Published in 2003 by NRC Research Press, Ottawa ON, K1A 0R6, Canada. ISBN 0-660-19089-3.

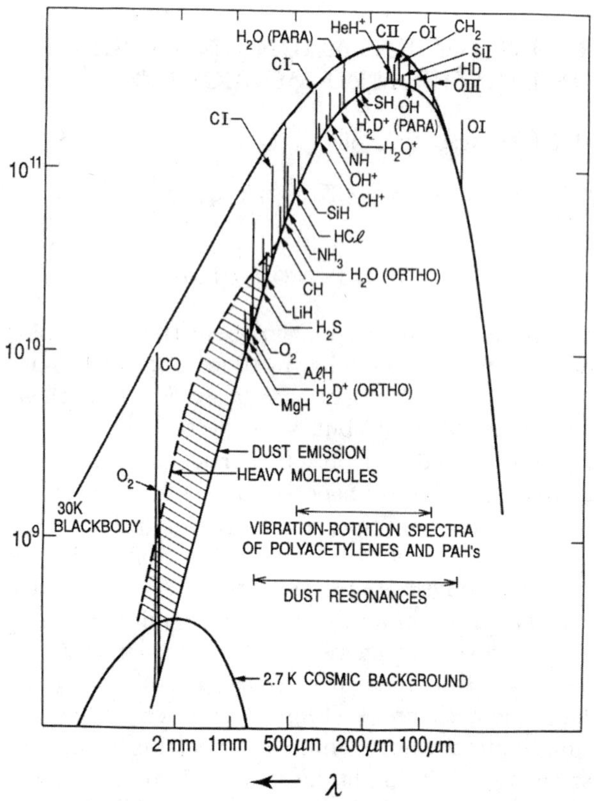

Figure 1. Illustration of the principal spectral lines expected at submillimeter wavelengths, superimposed on the dust continuum emission of a $T = 30\,K$ interstellar cloud.

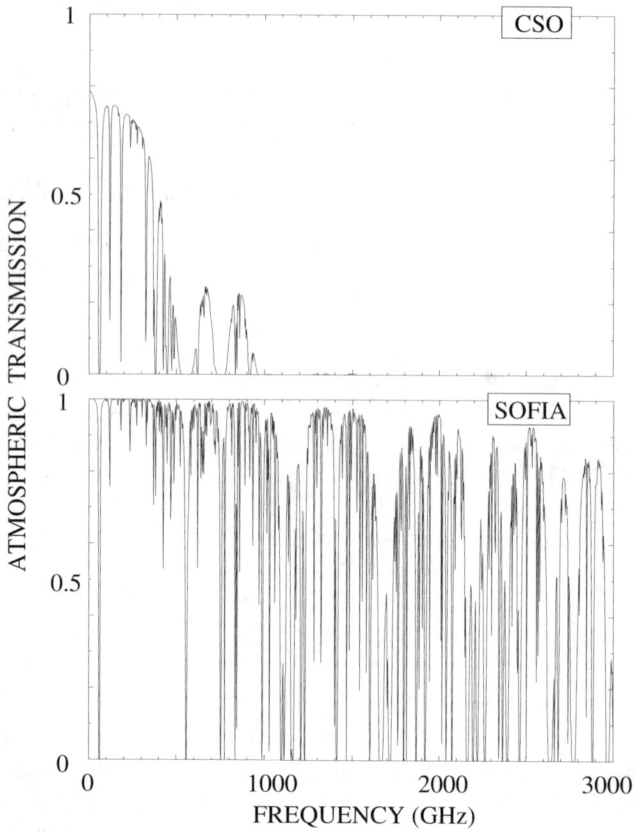

Figure 2. Comparison between the atmospheric transmission at CSO (with a standard precipitable water column of 1 mm) and for SOFIA (with a standard precipitable water column of 10 μm), both at a latitude of 45°.

are developing their field later than the optical, radio and X-ray communities, we have had to wait our turn for major funding. The wait is almost over and the full potential of this exciting field is about to be realized in the coming decade!

2. New Instruments

We are concerned with the new opportunities for interstellar chemistry, which include investigation of molecules and atoms in the gas phase, dust grains and surface chemistry, and the interactions between the gas and dust phases—including accretion of molecules onto grains and the opposite process, where the accretion material is returned to the gas phase. Here we discuss this only from the point of view of gas spectroscopy.

Table 2 lists many of the major new telescopes and interferometers which will (hopefully) be available within the next decade. Compared to what we have today, this will be a staggering new capability and the field will lead astronomy in that it will discover and define the new objects and new concepts to be investigated. One might worry that there will not be enough trained people in the field to fully utilize such a cornucopia of facilities, so university-based facilities should expand their efforts in this field now!

ALMA represents the biggest step forward, providing an unprecedented spectroscopic capability, through all the atmospheric windows, for nearby and distant objects. This huge sensitivity will probably manifest itself most dramatically for nearby objects in the chemistry of prestellar disks around Young Stellar Objects (YSOs), and, at the other end of the distance scale, for the chemistry of the interstellar medium of primordial galaxies. One caveat might be mentioned: high spectral resolution is often required where there is high spatial resolution.

Table 1. New facilities for interstellar chemistry

Name	Aperture	Platform	Available
SOFIA	2.5 m	Airborne (747)	2005
Herschel	3.5 m	Space (L2)	2007
ALMA	64 × 12 m	Atacama (Chile)	2010
CARMA	6 × 10 m	White Mts. (California, USA)	2005
SMA	8 × 6 m	Mauna Kea (Hawaii, USA)	2003
e-SMA	(8 × 6 m) + 15 m + 10 m	Mauna Kea (Hawaii, USA)	2005
LMT	50 m	Sierra Negra (Mexico)	2004
APEX	12 m	Atacama (Chile)	2005
South Pole	8 m	South Pole	2006
GBT	100 m	West Virginia (USA)	2002

Even with single-dish beams of 30″, local galaxies such as Cen A show disk features of only $10\,\mathrm{km\,s^{-1}}$ in width. Thus, ALMA must provide correlator capability for high spectral resolution ($\sim 1\,\mathrm{km\,s^{-1}}$) at full galaxy bandwidths ($\geq 1000\,\mathrm{km\,s^{-1}}$) on as many baselines as possible.

ALMA will make images, at 1 mm wavelength, with the same (0.1″) or better resolution than achieved by the Hubble Space Telescope (*HST*) at visible wavelengths. This capability will enable many new imaging programs, including:

1. Studies of the evolution of dust and gas from molecular clouds, to circumstellar disks, to planetary systems;
2. Studies of the enigmatic hot cores across the Galaxy;
3. Studies of YSOs across the Galaxy;
4. Investigation of the chemistry of circumstellar disks, using multiple lines of many species, possibly by means of line surveys; and
5. Measurements of cloud temperatures across the Galaxy, for studies of depletion in cold cores and chemical fractionation.

SOFIA, the new NASA airborne observatory, will give an improved view of much of the region of the spectrum blocked by the atmosphere. With a 2.5 m telescope and much improved spectrometers compared to the KAO, many specific new spectroscopic features can be searched for, with the exception of H_2O and O_2 lines, and features still hidden by the strong atmospheric lines. Ultimately the spectroscopy of the interstellar medium and nearby galaxies will be best studied from HIFI, the heterodyne instrument on *Herschel*, the ESA cornerstone mission. From the L2 point, an ideal satellite location, studies such as line surveys can be carried out continuously and without contamination from H_2O atmospheric lines. Both SOFIA and *Herschel* will extend the high-quality spectroscopy range to about 2 THz or possibly higher.

Actually, in spite of the impressive size and cost of the new instruments, much of the new capability is provided by the smallest component: the detector element. Since their inception (in ∼1979), SIS devices have increased sensitivity by a factor approaching 100. Now, essentially quantum noise-limited, the route to further improvement is in more bandwidth and more sophisticated device structure. For instance, IF bandwidths are at about 4 GHz and are planned for up to 20 GHz. For single-dish work, the single junction element is being replaced by devices detecting both polarizations, having instantaneous on-off subtraction, sideband separation, and possibly balanced mixing. Each of these features implies the use of two junctions; thus up to 16 detectors could be used for spectroscopy of a distant point source, resulting in a sensitivity gain of about 5 or so. Similarly, the increased frequency range (∼2 THz) is being supported by new beam-lead, broadband, high output power diode multipliers for the local oscillators.

3. New Spectral Opportunities

While ALMA provides sensitivity and angular resolution breakthroughs, SOFIA and *Herschel* (HIFI) generate opportunities for observing in new spectral ranges, due to the avoidance of some or all of the atmospheric H_2O. HIFI will provide a novel view of much of the submillimeter spectrum, totally unperturbed by atmospheric H_2O. It was intended that HIFI would have total frequency coverage from about 480 GHz to 2.7 THz, but the technical

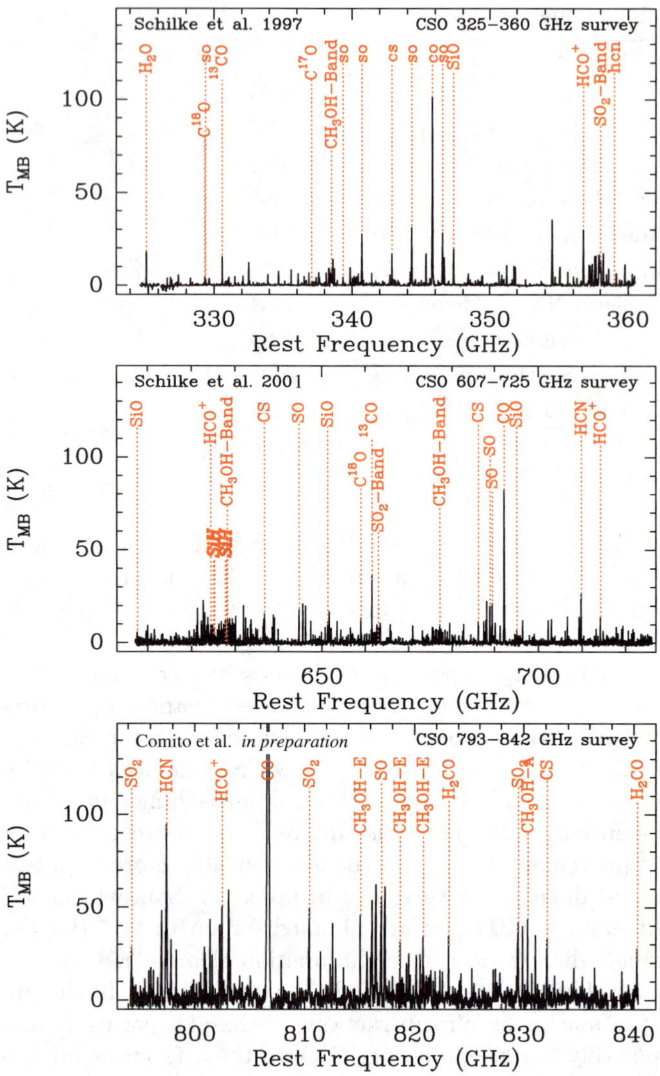

Figure 3. Line surveys of Orion with the CSO telescope.

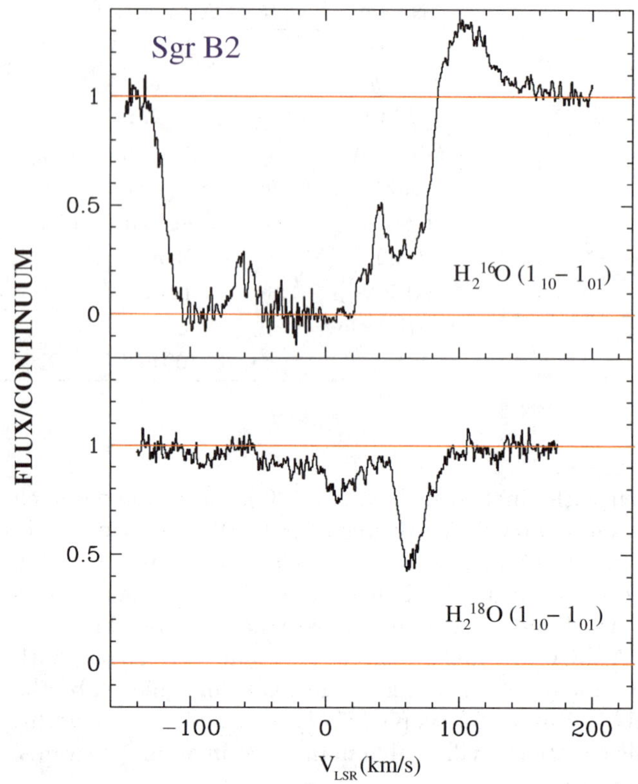

Figure 4. $H_2^{16}O$ and $H_2^{18}O$ line detections by the *SWAS* satellite toward Sagittarius B2.

difficulty of production of the local oscillators, combined with the limited budget and fixed launch schedule, has forced the abandonment of the highest frequency channel, so the highest frequency is now 1.9 THz (the frequency of the C^+ line).

3.1. Line Surveys

Much of the HIFI data will be taken in the form of line surveys. These have been spectacularly successful on ground-based telescopes (Fig. 3), and the methodology has been refined by means of simulations (Comito & Schilke 2001), to make it suitable for rapid frequency scanning such that a full spectral survey of THz bandwidth can be completed, by HIFI, for strong sources in a day or less. This compares with the many weeks taken on ground-based telescopes to obtain only 100 GHz of data. The factor ~ 100 of improvement in time is mostly due to the automation necessary for space and the vastly improved designs of mixers and local oscillators.

3.2. H_2O

Some H_2O lines have been observed from the ground and from the KAO, but the bulk of observations are of the 557 GHz line from *SWAS* and also *Odin*. Figures 4 and 5 show some amazing *SWAS* H_2O spectra, indicating a complex structure to the line shape, often including self-absorption. The three sources presented in these figures are Sgr B2 (Neufeld et al. 2000), W51 (Neufeld et al. 2002) and W49N (Plume et al., in preparation). These are distant compact H II regions, where absorption line obser-

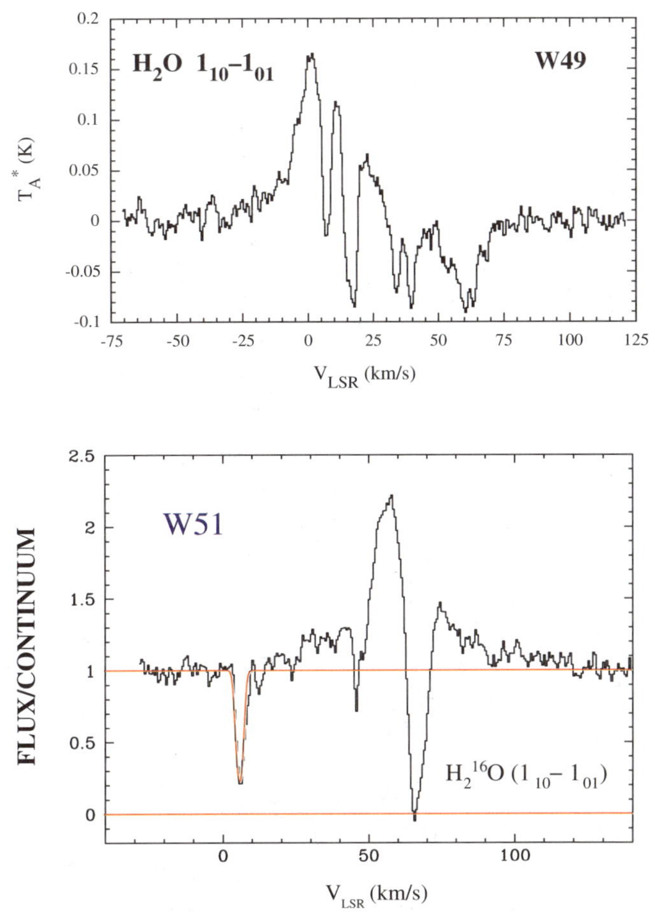

Figure 5. Two spectacular examples with W49N, a distant ($\sim$ 11.4 kpc) compact H II region whose line of sight twice crosses the Sagittarius spiral arm, and W51 located at $\sim$ 7 kpc.

vations can trace the water vapor content in clouds along the line of sight.

HIFI will have many transitions of H_2O and $H_2{}^{18}O$ available, most of which have not been observed previously, and including those in the important excitation path to the ground-state (Fig. 6). These many available transitions will probe the gas cooling properties of H_2O in objects with a wide range of densities, such as collapsing protostellar envelopes (Ceccarelli et al. 1996), where H_2O becomes the primary coolant as the density increases in the inner regions.

3.3. Hidden Species

Besides the H_2O lines themselves, there exists a body of spectral features never observed because of their proximity in frequency to the H_2O and O_2 lines. Some examples are lines of H_2D^+ (e.g., $2_{1,1} \rightarrow 2_{1,2}$ at 1.112 THz), CH^+ (e.g., $1 \rightarrow 0$ at 836 GHz), NH (e.g., $N = 1 \rightarrow 0$, $J = 0 \rightarrow 1$ at 946 GHz), and LiH (e.g., $1 \rightarrow 0$ at 444 GHz).

Molecular lines such as HCl ($1 \rightarrow 0$) at 626 GHz and SiH at 627 GHz are hard to observe from the ground, being in the wings of H_2O lines. Many of the "hard to detect from the ground" species are light, fast rotators, typified by the diatomic hydrides. Some of these, e.g., NaH, CaH, and MgH, have been searched for in $J = 0 \rightarrow 1$ transitions using atmospheric windows for ground-based telescopes—with negative results—but can be searched for now with increased sensitivity, in two ways. The first method (which can be carried out from the ground in favorable cases) makes use of cold clouds along the line of sight to powerful continuum sources, such as Sgr B2. There, in the cold dark clouds, the molecules will be in the ground state and would be detected in absorption. This method avoids the confusion in the spectra of GMCs, since the line forest is largely made up of highly excited molecular lines, which, of course, is not the case for the hydrides in cold clouds. A good example is seen in the *SWAS* detection of the ground state H_2O line in Sgr B2 (Fig. 4). The second method does suffer from GMC line confusion, but makes use of the new high frequency capabilities of HIFI and SOFIA, where the high–J lines of the various hydrides will have much higher line strengths than $1 \rightarrow 0$; hopefully this will overcome the line forest effect in the high temperature and high density sources.

Generally, above 1 THz the near-zero atmospheric transmission has prevented studies so far, but several windows of limited size and strength have recently been detected from high-altitude sites. Space platforms will allow studies of HD ($1 \rightarrow 0$) at 2.67 THz, HeH^+ ($1 \rightarrow 0$) at 2.01 THz, N^+ at 1.46 and 2.46 THz, HF at 1.23 THz, FeH at 1.41 THz, OH at 1.83 THz, OH^+ at 2.46 THz, and others. Some species have been detected by *ISO*: e.g., CH^+ in the high–J transitions (Cernicharo et al. 1997), HF (Neufeld et al. 1997) and HD (Wright et al. 1999; Caux et al. 2002). HIFI will be a great improvement compared to what used to be available on *ISO*.

4. Deuterium

One of the most important aspects of molecular spectroscopy of the interstellar medium is the study of the abundance of deuterium. The new facilities will provide a much wider range of frequencies and sensitivities, so that the deuterium studies can be extended to new species and new objects. The fundamental aspect of deuterium

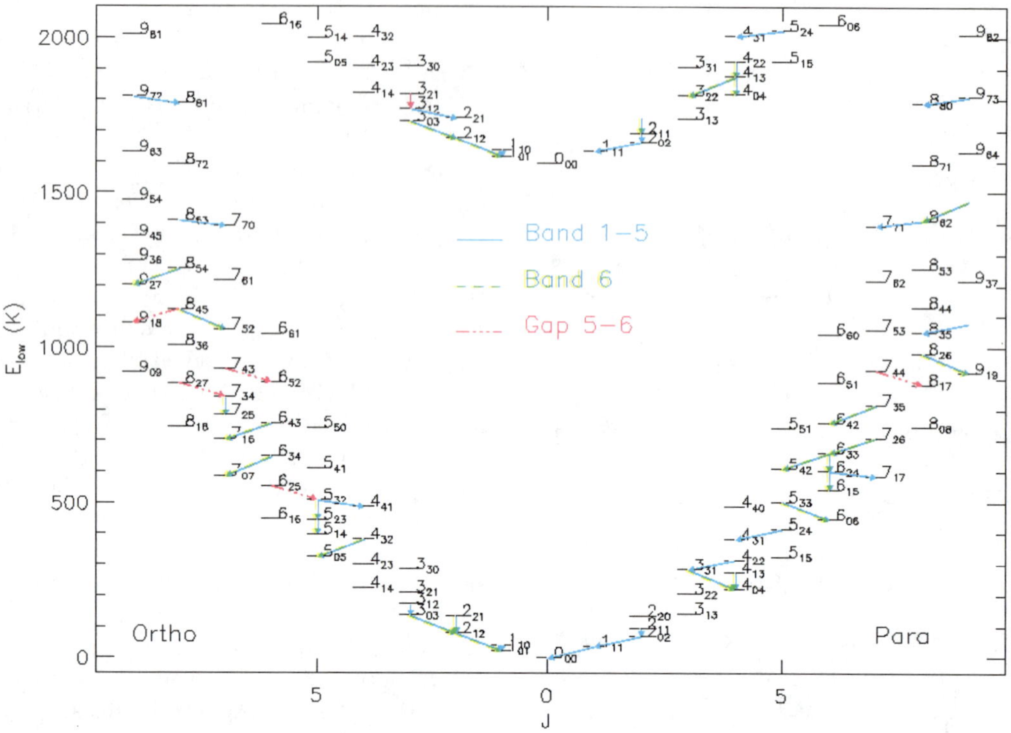

Figure 6. Water transitions available to HIFI. Bands 1–5 cover the frequency range 472–1280 Gz, while Band 6 covers 1400-1904 GHz. Gap 5–6 is the frequency range between Band 5 and Band 6, which may not be covered.

chemistry is that the chemical fractionation process (see below) forces deuterium atoms into heavy molecules, at the expense of hydrogen, whenever the medium is cold, with amazing results.

Deuterium-bearing molecules have become the target of many observations in recent years and several models have been developed to account for them (Tielens 1983; Roberts & Millar 2000; Roberts & Millar 2000b). More than 20 such molecules have been detected to date in interstellar clouds with abundances, relative to the non-deuterated counterpart, ranging from 10^{-4} to 10^{-1}. It is accepted that deuterium is produced during the Big Bang and it is generally believed that since the Big Bang, deuterium has been destroyed but not created in the nuclear reactions occurring inside stars.

The most reliable determinations of the D/H ratio are based on spectroscopic measurements of Lyman series ultraviolet absorption lines from foreground interstellar gas. In our Galaxy, this has been obtained (via D I and H I observations) with satellites such as the International Ultraviolet Explorer (*IUE*), *Copernicus*, the Extreme UltraViolet Explorer (*EUVE*), the Hubble Space Telescope (*HST*) and, most recently, the Far Ultraviolet Spectroscopic Explorer (*FUSE*). Measurements of the D/H ratio toward high-redshift systems like quasars (Tytler et al. 1999) seemed to show more dispersion than expected and an inverse correlation of this abundance with H I column density. If this is real, it would suggest that in these high H I column density systems, some processing of D/H must have occurred (Vidal-Madjar 2001; Fields et al. 2001).

It is interesting to note that abundances of deuterium measured in the interstellar medium also appear to show considerable dispersion. From published values, D/H ranges from $\sim 5 \times 10^{-6}$ to $\sim 4 \times 10^{-5}$. *FUSE* observations of seven white dwarfs and sub-dwarfs lead to a D/H ratio of $(1.52 \pm 0.08) \times 10^{-5}$ (Moos et al. 2002), to be compared with the value of $(1.5 \pm 0.1) \times 10^{-5}$ (Linsky 1998) determined from *HST* observations of late-type stars. Both measurements refer to warm interstellar gas, located within 100 pc of the Sun. However, differences by a factor of two have been derived from *Copernicus* and

HST data toward stars located between 100 and 500 pc from the Sun. The ratio seems constant within 100 pc, but seems to vary at considerably greater distances.

There is no identified process which can explain such large variability, and without a better understanding it is not justified to use an average D/H ratio to represent the primordial deuterium abundance. We will argue below that the chemistry of the interstellar medium could be responsible, in that it can extract large amounts of deuterium which becomes trapped in molecules and on grains.

4.1. Basic Chemistry

Deuterium-bearing species are good probes of the cold phases of molecular clouds prior to star formation and many recent observations point to the fact that their abundances relative to their hydrogenated analogues are larger, by a factor up to 10^4, than the solar neighborhood value of $\sim 1.5 \times 10^{-5}$ (see references above). Therefore, the relative abundance of isotopomers does not measure the relative abundances of the isotopes themselves. The chemical fractionation process arises from differences in the molecular binding energies caused by different zero-point vibration energies. Somewhat surprisingly, this can lead to a detectable quantity of the triply deuterated ammonia molecule (see ff. Sec. 4.2).

In molecular clouds, hydrogen and deuterium are predominantly in the form of H_2 and HD, respectively. So the HD/H_2 ratio should closely follow the D/H ratio. Since the zero-point energies of HD and H_2 differ greatly (Fig. 7), chemical fractionation will favor the production of HD compared to H_2.

Deuterium is initially removed from the atomic phase through charge exchange with H^+, followed by reaction with the abundant H_2. HD could further interact with D^+ again to give D_2:

$$H^+ + D \longleftrightarrow H + D^+ \quad (1)$$

$$D^+ + H_2 \longleftrightarrow HD + H^+ + \Delta E_1 \quad (2)$$

$$D^+ + HD \longleftrightarrow D_2 + H^+ + \Delta E_2 \quad (3)$$

Reactions (2) and (3) are exothermic, as substituting an H atom for a D atom in a polyatomic molecule generally leads to a gain in energy. These energies may be computed (at 0 K) by the differences between the zero-point energies of the products and reactants. The energies ΔE_1 and ΔE_2 are given in Figure 7.

In dense, cold regions of the interstellar medium, D will be initially nearly all absorbed into HD. The abundant ion available for interaction is H_3^+, which gives H_2D^+:

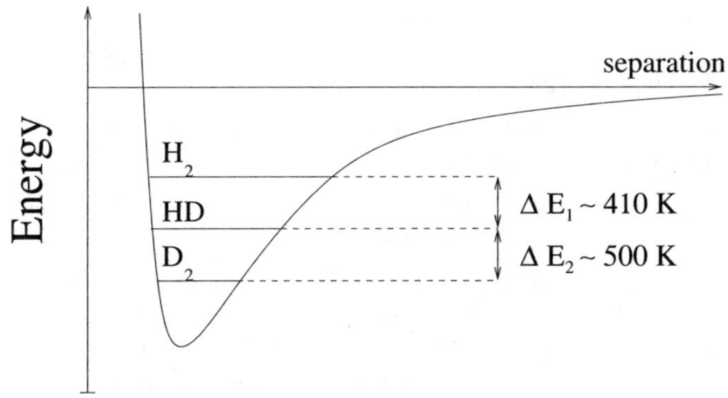

Figure 7. H_2, HD and D_2 potential energy diagram. ΔE_i is the difference between the zero-point energies relative to the minimum of the molecular potential curve.

$$H_3^+ + HD \longleftrightarrow H_2D^+ + H_2 + \Delta E_3, \quad (4)$$

where $\Delta E_3/k \sim 230$ K for a typical temperature of a dark cloud of about 10 K (Millar et al. 1989). The reverse reaction does not occur efficiently in cold, dense clouds where the temperature is obviously much lower than ΔE_3. Therefore, the degree of fractionation of H_2D^+ becomes non-negligible. This primary fractionation can then give rise to a second fractionation:

$$H_2D^+ + CO \longleftrightarrow DCO^+ + H_2$$
$$\longleftrightarrow HCO^+ + HD.$$

In dark clouds, H_3^+ gives rise to HCO^+ via the reaction:

$$H_3^+ + CO \longleftrightarrow HCO^+ + H_2.$$

4.2. Multiply Deuterated Molecules

The study of doubly deuterated interstellar molecules has been booming in the last few years since the surprising discovery of a large amount (~ 5 percent) of doubly deuterated formaldehyde in the low-mass protostar IRAS 16293-2422 (Ceccarelli et al. 1998; Loinard et al. 2000). This is more than an order of magnitude higher than in Orion KL where D_2CO was first detected by 1990. This first discovery was followed by many other studies which confirmed the presence of very large amount of doubly deuterated formaldehyde (D_2CO), as well as ammonia (ND_2H) (Roueff et al. 2000; Loinard et al. 2001). Gas phase chemical models account relatively well for the observations of the abundances of singly deuterated molecules, but may not be able to completely reproduce the large deuterations observed for multiply deuterated molecules. The large deuterations could also be a product of active chemistry on the grain surfaces as predicted by Tielens (1983) with two processes:

1. Deuteration during the mantle formation phase;
2. Evaporation of the mantle ices resulting from heating by the newly-formed star, with injection into the gas phase of the deuterated species.

The year 2002 appears to be a turning point in the study of deuteration processes with the first detection of a triply deuterated molecule, ND_3. Until now, the possibility for detecting triply deuterated molecules was so remote that their lines were omitted in the spectroscopic catalogs for astrophysics. The ground state rotational transition at 309.91 GHz of ND_3 has been detected with the CSO towards the Barnard 1 cloud (Lis et al. 2002; see Fig. 8) and the NGC 1333 IRAS 4A region (Van der Tak et al. 2002), with abundance ratios $[ND_3]/[NH_3] \sim 10^{-3}$ and $[ND_3/H_2] \sim 10^{-11}$, respectively. These authors conclude that reactions in the gas phase, rather than grain surface chemistry, are more likely to produce these high degrees of deuteration. However, they cannot totally rule out the possibility that surface processes also contribute to the formation of ND_3 (cf. Rodgers & Charnley 2001). Moreover, the recent discovery of doubly deuterated methanol towards IRAS 16293-2422 ($CHD_2OH/CH_3OH \sim 0.2$; Parise et al. 2002) cannot be accounted for in gas phase models. The most promising route for such methanol deuteration seems to be by surface chemistry. Recent observations of the extended D_2CO emission towards L1689N (Ceccarelli et al. 2002) pointed out the difficulty in explaining the D_2CO abundance, either with gas phase models or with grain surface chemistry. The many routes to deuteration are not fully understood and more observations are necessary to elucidate theses processes.

Very recently, doubly deuterated hydrogen sulfide has been detected by Vastel et al. (in preparation) with the CSO towards the Barnard 1 cloud and the DCO^+ ($3 \rightarrow 2$) emission peak of NGC 1333 IRAS 4A. Figure 9 shows, as an example, the first detection of D_2S ($1_{1,1} \rightarrow 0_{0,0}$), together with HDS ($1_{0,1} \rightarrow 0_{0,0}$).

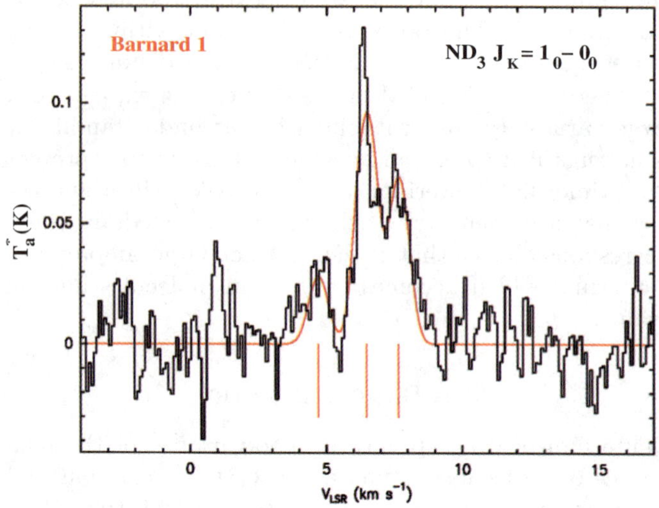

Figure 8. Spectrum of the 309.91 GHz ND_3 line towards Barnard 1 observed at the CSO. The red curves show the fits to the position and strength of the hyperfine components.

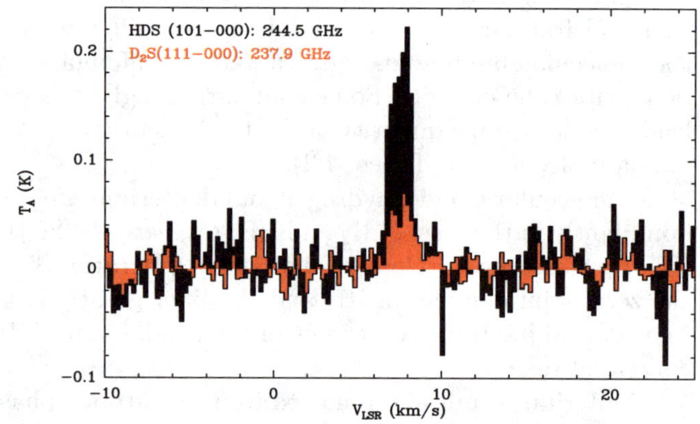

Figure 9. HDS ($1_{0,1} \rightarrow 0_{0,0}$) and D_2S ($1_{1,1} \rightarrow 0_{0,0}$) detections with the CSO towards the DCO^+ ($3 \rightarrow 2$) emission peak of NGC 1333 IRAS 4A (Vastel et al., in preparation).

4.3. Depletion of CO

Gas phase species are expected to be depleted at the centers of cold, dark clouds, since they tend to stick to the dust grains. A series of recent observations has shown that, in some cases, the abundance of molecules like CO decreases toward the center of cold, dense clouds (L1498: Willacy et al. 1998; IC 5146: Kramer et al. 1999, Bergin et al. 2001; L977: Alves et al. 1999; L1544: Caselli et al. 1999; L1689B: Jessop & ward-Thompson 2001; Bacmann et al. 2002). These decreases in abundance have been interpreted as resulting from the depletion of molecules onto dust grains (Bergin & Langer 1997; Charnley 1997). It is now clear that these drops in abundance are typical of all dense cores. The removal of these reactive species affects the gas phase chemistry and particularly the deuterium fractionation within the cloud. Indeed, the removal of species which would normally destroy H_3^+ (e.g.,

CO; Roberts & Millar 2000) means that the H_3^+ is more likely to react with HD and produce H_2D^+ (reaction [4] above).

H_2D^+ has been detected toward the low luminosity protostars NGC 1333 IRAS 4A (Stark et al. 1999) and IRAS 16293-2422 (Stark et al., in preparation) with a H_2D^+/H_2 ratio of 3×10^{-12} in the former. Recently, Caselli et al. (in preparation) detected the H_2D^+ 372 GHz line towards two prestellar cores, L1544 and L1521F, both probably on the verge of collapse.

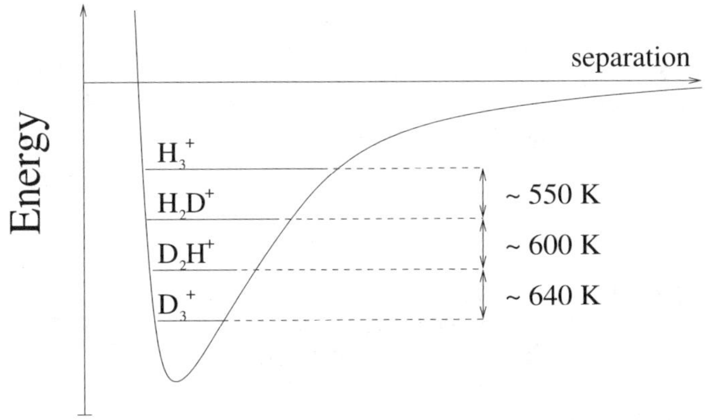

Figure 10. H_3^+, H_2D^+, D_2H^+ and D_3^+ potential energy diagram.

4.4. Modification of the Deuterium Chemistry

The large differences found in the values of the deuterium abundance in the interstellar medium poses the question as to the mechanisms responsible for these variations. A possible answer is the chemical fractionation process, which, in cold regions of the interstellar medium, steadily forces the deuterium into the heavy molecules. The trend is to minimize the free energy, which implies forming the heaviest species, with the smallest uncertainty energy (see Fig. 10). As shown in the previous section, in the very cold (~ 10 K) regions where, e.g., ND_3 is detected, CO is in fact heavily depleted by accretion onto grains. For example, $[CO/H_2] \sim 5 \times 10^{-6}$ (Bacmann et al. 2002), leaving HD at $[HD/H_2] \sim 5 \times 10^{-5}$ as the most abundant molecule available for reaction with H_3^+ and H_2D^+. Another example is the case of HD 112 μm absorption towards a cold molecular cloud in the line of sight of W49 where Caux et al. (2002) found that HD is 12 times more abundant than CO. The result may be the production of more high deuterium content molecules:

$$H_3^+ + HD \longleftrightarrow H_{2D}^+ + H_2 + \Delta E_a$$

$$H_2D^+ + HD \longleftrightarrow D_2H^+ + H_2 + \Delta E_b$$

$$D_2H^+ + HD \longleftrightarrow D_3^+ + H_2 + \Delta E_c$$

where ΔE_a, ΔE_b and ΔE_c are the released energies of the exothermic reactions. Using the zero-point energies computed by Carney (1980), and the energy of the first allowed rotational state of the H_3^+ molecule permitted by the Pauli exclusion principle (~ 92 K), these values are: $\Delta E_a \simeq 230$ K, $\Delta E_b \simeq 180$ K, and $\Delta E_c \simeq 230$ K.

The species D_2H^+ may indeed be detectable in sufficiently cold, CO-depleted regions. Because of the low temperature of the highly-deuterated regions (~ 10 K), the lowest-lying transition of H_2D^+ and D_2H^+ will be most appropriate and can possibly be searched for in absorption against Sagittarius B2:

- H_2D^+ $(1_{01} - 0_{00})$ at 1370.15 GHz
- H_2D^+ $(2_{11} - 2_{12})$ at 1111.74 GHz
- D_2H^+ $(1_{11} - 0_{00})$ at 1476.60 GHz
- D_2H^+ $(2_{20} - 2_{11})$ at 1370.05 GHz.

(*Note*: It appears to be purely by chance that the first and last frequencies are so nearly equal.)

4.5. D, HD Depleted by Fractionation?

It would be a difficult task to make an inventory of all D-substituted molecular species throughout the Galaxy, to see if the measured values of the deuterium abundance could be affected. The fractionation might work in favor of metal species (e.g., ND_3) in high metallicity galaxies, or in favor of purely hydrogenic species (e.g., H_2D^+ and D_2H^+) in low metallicity objects. At any rate, we can ask if it is physically possible for the former case to occur. If we take the local values of C, N and O abundances and count the total number of bonds available for substitution of H by D, we get 3×10^{-3} [H]. But [D]/[H] is only $\sim 10^{-5}$, so there does exist the possibility of such an effect. In fact, if the product of the fraction of available bonds actually occupied by D (f_D) times the fraction of the interstellar medium which is cold enough for strong fractionation and CO depletion (f_C) approaches 10^{-3}, then there will be a significant effect:

$$f_D \times f_C \geq 10^{-3}.$$

We know that $f_D \sim 10^{-1}$ or more in some cases, so if $f_C \geq 10^{-2}$, then measured values of the deuterium abundance may be significantly affected.

Acknowledgements

We acknowledge support from the NSF under contract AST-9980846. We thank Eric Herbst, John Pearson, Tom Millar, and Dieter Gerlich for helpful conversations. We are also grateful to Paul Goldsmith for providing us with the figures of *SWAS* observations towards Sgr B2, W49N, and W51.

References

Alves, J., Lada, C.J., & Lada, E.A. 1999, ApJ, 515, 265
Bacmann, A., et al. 2002, A&A, 389, 6
Bergin, E.A., & Langer, W.D. 1997, ApJ, 486, 316
Bergin, E., et al. 2001, ApJ, 557, 209
Carney, G.D. 1980, Chem. Phys., 54, 103
Caselli, P., et al. 1999, ApJ, 523, 165
Caux, E., et al. 2002, A&A, 383, 9
Ceccarelli, C., Hollenbach, D., & Tielens, A.G.G.M. 1996, ApJ, 471, 400
Ceccarelli, C., et al. 1998, A&A, 338, 43
Ceccarelli, C., et al. 2002, A&A, 381, 17
Cernicharo, J., et al. 1997, ApJ, 483, L65
Charnley, S.B. 1997, MNRAS, 291, 455
Comito, C., & Schilke, P. 2001, ESASP, 460, 389
Fields, B.D., et al. 2001, ApJ, 563, 653
Jessop, N.E., & Ward-Thompson, D. 2001, MNRAS, 323, 1025
Kramer, C., et al. 1999, A&A, 342, 257
Linsky, J.L. 1998, Sp. Sci. Rev., 84, 285
Lis, D.C., et al. 2002, ApJ, 571, 55
Loinard, L., et al. 2000, A&A, 359, 1169
Loinard, L., et al. 2001, ApJ, 552, 163
Millar, T.J., Bennett, A., & Herbst, E. 1989, ApJ, 340, 906
Moos, H.W., et al. 2002, ApJS, 140, 3
Neufeld, D., Zmuidzinas, J., Schilke, P., & Phillips, T.G. 1997, ApJ, 488, L141
Neufeld, D.A., et al. 2000, ApJ, 539, 111
Neufeld, D.A., et al. 2002, ApJ, 580, 278
Parise, B., Ceccarelli, C., Tielens, A.G.G.M., & Herbst, E. 2002, A&A, 393, L49
Phillips, T.G., & Keene, J. 1992, IEEEP, 80, 1662
Roberts, H., & Millar, T.J. 2000a, A&A, 361, 388
Roberts, H., & Millar, T.J. 2000b, A&A, 364, 780
Rodgers, S.D., & Charnley, S.B. 2001, ApJ, 553, 613
Roueff, E., et al. 2000, A&A, 354, L63
Schilke, P., Groesbeck, T.D., Blake, G.A., & Phillips, T.G. 1997, ApJS, 108, 301
Schilke, P., et al. 2001, ApJS, 132, 281
Stark, R., Van der Tak, F., & van Dishoeck, E. 1999, ApJ, 521, 67
Tielens, A.G.G.M. 1983, ApJ, 119, 177
Tytler, D., et al. 1999, AJ, 117, 63
Turner, B.E. 1990, ApJ, 362, 29
Van der Tak, F., et al. 2002, A&A, 388, L53
Vidal-Madjar, A. 2001, in *Cosmic Evolution*, Paris Symposium Proc., ed. E. Vangioni-Flam, R. Ferlet, & M. Lemoine (New Jersey: World Scientific), 49
Willacy, K., Langer, W., & Velusamy, T. 1998, ApJ, 507, 171
Wright, C.M., et al. 1999, ApJ, 515, 29

ELISA: A SMALL BALLOON-BORNE EXPERIMENT FOR A LARGE SUBMILLIMETER SURVEY OF THE GALAXY

J.P. Bernard[1,2], I. Ristorcelli[1], M. Giard[1], G. Lagache[2], C. Meny[1], J.P. Torre[3], M. Armengaud[1], J.P. Crussaire[2], B. Leriche[2], and Y. Longval[2]

[1]CESR, 9 av. du Colonel Roche, BP 4346-31028, Toulouse, Cedex 4, France
[2]IAS, UMR-CNRS, Université Paris Sud, 91405 Orsay, Cedex, France
[3]SA, UPR-CNRS, Verrieres-le-Buisson, France

Abstract

This paper briefly presents the technical aspects and scientific objectives of the balloon-borne Experiment for Large Infrared Survey Astronomy (ELISA). The emphasis is upon the impact on star formation studies and the synergies existing between ELISA and future space missions, with respect to both technical and scientific aspects.

ELISA is a small balloon project for an experiment dedicated to measuring the far-infrared to submillimeter continuum emission of dust over a large fraction of the sky with unprecedented sensitivity and angular resolution. The primary mirror of the telescope, similar to the one used for the *Top-Hat* mission, will have a diameter of 1 m, ensuring an angular resolution of about 3.5′. PACS-type bolometer arrays will be used in four photometric bands centered at 170, 240, 400, and 650 μm, providing a $22' \times 45'$ instantaneous field of view per channel. A liquid He cryostat will host the cold optics, including the secondary mirror of the telescope, as well as the detectors, which will be cooled to 0.3 K using an ^{3}He closed-cycle refrigerator. Mapping of the sky will be accomplished by rotating the gondola over a large azimuth range (up to 60 degrees). The pointing of the experiment will be maintained to a constant elevation during the azimuth scans through a feedback loop using the signal from a large format, fast stellar sensor, operating day and night. The total mass of the experiment will be lower than 500 kg.

The scientific goal of the experiment is to map the diffuse submillimeter emission along a large fraction of the Milky Way. The data obtained will be used to derive the emission properties of the dust grains in the interstellar medium, such as their temperature and emissivity. The measurements will be particularly suited to detect and study the cold component of the Galaxy, and therefore the first steps toward star formation. It will also systematically measure the polarization of the dust emission and should lead to the detection of a few thousand point sources such as newly formed stars and distant galaxies. In addition to these goals, the ELISA project will serve as a test bed for the detector technology to be used for the *Herschel* space mission to be launched in 2007. The ELISA data will also be used to help calibrate the observations of the *Herschel* and *Planck* space missions and to plan the large scale surveys to be undertaken with *Herschel*. Owing to these objectives, three flights of the ELISA experiment, including one from Southern hemisphere, are foreseen in the period from 2005 to 2007.

The ELISA project is an international collaboration including France (CESR, IAS, CEA, CNES), Netherlands (SSD/ESTEC), Denmark (DSRI), England (QMW), USA (JPL/Caltech), and Italy (ASI). Flight operations will be managed by the balloon division at CNES.

Key words: ISM: dust grains, cold clouds – Stars: formation

1. Introduction

Interstellar dust plays a key role in the process of star formation and in the energetic equilibrium of the Galaxy. It therefore seems essential to understand its physical and chemical properties, in particular in the densest phases of the interstellar medium (ISM). Current dust models consider three main components, differing in composition, size, structure and emission mechanism (e.g., Desert et al. 1990). The smallest particles (PAHs) and the very small grains (sizes $\leq$ 15 nm) are transiently heated and emit in the infrared range, which has been extensively studied with *ISO*. The largest grains are in thermal equilibrium, dominating the far-infrared and submillimeter spectrum with a continuum emission following

$$I_\nu = \kappa(\nu/\nu_0)^\beta B_\nu(T), \qquad (1)$$

where T is the average temperature, β the dust emissivity spectral index, and κ is proportional to the column density.

Chemistry as a Diagnostic of Star Formation (2003). *Edited by Charles L. Curry and Michel Fich.* Published in 2003 by NRC Research Press, Ottawa ON, K1A 0R6, Canada. ISBN 0-660-19089-3.

In order to constrain both T and β, submillimeter multi-band observations are needed, including the range $100 - 300\,\mu$m (inaccessible from the ground), since it contains the peak of the emission spectrum for dust at $T \leq 30$ K. These measurements are particularly well-suited to characterize the interstellar cold component, and trace the morphology and structure of the clouds, especially during the very early steps toward star formation, when the cloud opacity plays a dominant role. The emission from large grains is so far poorly constrained by observations since the far-infrared and submillimeter domain is still almost unexplored, excepted at the large angular scale of *COBE* (7°) (e.g., Boulanger et al. 1996; Lagache et al. 1998).

The PRONAOS balloon-borne experiment (Lamarre et al. 1994) mapped the dust emission in four photometric channels from 200 to 650 μm, toward a few selected regions of the ISM during its three flights in 1994, 1996, and 1999. These observations have brought new insights about the nature of large grains in the ISM, and raise a number of questions about the nature and evolution of dust. PRONAOS not only revealed the existence of cold condensations ($T \sim 12$ K) in star-forming regions (Ristorcelli et al. 1998; Dupac et al. 2001), but also in translucent and optically thin dust clouds at high galactic latitude. This cannot be explained by the standard dust models currently in use, but can be interpreted as implying the existence of porous dust aggregates (Bernard et al. 1999; Stepnik et al. 2001). The observations also showed a significant correlation between the dust equilibrium temperature T and the spectral index β, which may reflect new quantum processes within the grains specific to low temperatures (see Boudet et al., this volume). However, the PRONAOS observations were limited to a very small fraction of the sky (a few square degrees), and the characterization of dust emission clearly calls for more statistics and a larger survey. Due to a limited spectral range ($\lambda \geq 350\mu$m) and the subtraction of low spatial frequencies, ground observations cannot constrain both temperature and spectral index of dust, nor can they measure the faint extended emission.

2. Main Objectives

In this context, the science objective of the ELISA experiment is to survey the galactic emission from different sites of the ISM. For that purpose, two specific observing modes will be used. A large survey will enable mapping of the galactic plane emission and a deep survey at high galactic latitudes will be performed to study the emission from more diffuse clouds. The expected sensitivities for each observing mode are given in Table 1. A large census of the dust emission will be deduced from those surveys. From the temperature and spectral index maps, we will study statistically the correlation between these two parameters. Spectral and spatial variations of the large grain emissivity will be analysed in relation to the physical conditions (e.g., density and turbulence), the other dust components (VSGs and PAHs), and the molecular abundances. A specific study of the galactic cold component will be performed. The grain properties in cold cores are expected to be very different from those in the diffuse ISM, and ELISA observations should enable the characterization of both spatial and spectral changes of the dust properties due to molecular ice mantles and molecular and/or grain coagulation leading to aggregates.

2.1. Galactic plane: Large-scale survey

An understanding of the very early phase of star formation requires a better knowledge of the composition and physical and chemical properties of prestellar cold and dense cores. In particular, the opacity, which is dominated by dust grains, must play a dominant role in the phase preceding the collapse, and may influence the fragments' sizes. It is therefore expected that global quantities such as the initial mass function and star formation efficiency should be strongly linked to the dust absorption and emission properties. In addition, since dust grains participate directly in most of the physical processes influencing the gas ionization stage, their properties could significantly impact the ambipolar diffusion efficiency, which is thought to be one of the major regulating processes of star formation.

With ELISA, we plan to conduct a large-scale survey along the galactic plane, at $b \leq 20^o$, down to 2 MJy/sr (3σ) at 650 μm. The famous nearby star forming regions (e.g., ρ Ophiuchus, Taurus, Orion, and Serpens) will be mapped, as well as more distant large molecular complexes (e.g., Cygnus). Such a survey will also cover a large fraction of the galactic ridge and therefore allow the derivation of dust properties as a function of distance to the galactic center, using the galactic rotation curve and correlation with velocity information in the H I and molecular large-scale surveys. This survey will allow a systematic search and study of very cold condensations. A large number of detections is expected, including the thousands of cold cores detected in extinction against the

Table 1. Main instrument characteristics

Wavelength Band	1	2	3	4
$\lambda_{\rm center}(\mu{\rm m})^a$	170	240	400	650
NEB[b] (MJy sr^{-1} Hz$^{-1/2}$)/bol	1.23	0.93	0.63	0.49
Large-scale survey:[c]				
Det. limit at 3σ (MJy sr^{-1})	12.6	9.5	2.1	1.7
N(H)(cm^{-2}), A_V (mag, 3σ)			1.5×10^{21}	0.9
Deep survey:[d]				
Det. limit at 3σ (MJy sr^{-1})	1.5	1.1	0.25	0.2
N(H)(cm^{-2}), A_V (mag, 3σ)			1.8×10^{20}	0.1
Payload instrument mass	$\leq$150 kg			
Primary mirror diameter	1100 mm, F/1			
Azimuthal scanning	1.2 deg s^{-1}, $\Delta Az = \pm 30°$			
Pointing accuracy	$\pm 15''$ in elevation			
Beam; field of view	$3.5'$; $22' \times 45'$			

[a] $\Delta\lambda/\lambda \sim 30\%$.
[b] Noise Equivalent Brightness per bolometer.
[c] 700 deg^2/h.
[d] 10 deg^2/h.

infrared background in the *MSX* (Egan et al. 1998) and ISOGAL (Hennebelle et al. 2000) surveys.

2.2. High-latitude clouds: Deep survey

ELISA will aim at limited observations (typically a few tenths of a square degree per cloud, due to the low-level emission) of a representative sample of cirrus clouds. These will allow, for the first time, a complete census of dust temperature and emissivity properties of dust toward diffuse regions (typically $A_V \leq 0.1$) at the angular scale of a few arcminutes. The size of the maps, the number of different cirrus and their careful selection should guarantee minimum observational bias and the statistical relevance of the results. For these observations, comparison to H I and molecular line observations measurements will be crucial, in particular to derive the dust emissivity. We plan to include several intermediate and high-velocity H I clouds (IVCs and HVCs, respectively) in the cirrus sample, in order to search for their dust emission—if any—in the submillimeter. Detection of cold dust and derivation of dust temperature in these clouds may be the only way to derive their distance to the galactic plane and help elucidate their origin. The results obtained toward diffuse clouds by ELISA should be very useful to anticipate the methods to be used to subtract galactic foreground contribution in the *Planck* data. They will be particularly relevant to the problem of the separation between thermal dust emission and grain rotation, which is likely to dominate the spectrum at longer wavelengths.

2.3. Impact on Star Formation Studies

The *IRAS* Point Source Catalog (PSC) remains the reference database for statistical study of star formation in our Galaxy. The *IRAS* wavelength coverage has enabled the study of the evolutionary sequence from cold and deeply embedded young stellar objects to progressively hotter and older stars (Class I, II, and III objects). However, the coldest and youngest sources (Class –I and 0), the protostars in the isothermal contraction phase and in the phase preceding gravitational collapse, can only be detected in the submillimeter, because their emission peak is in the 100–300 μm region. Similarly, the study of the early stage of massive star formation suffers from a lack of good candidates, expected to be rare due to the very short timescales involved and cannot be readily identified in the *IRAS* PSC. Submillimeter measurements seem to be the best way to distinguish them from the far more common solar-mass YSOs. Yet no extensive PSC similar to *IRAS* exists in the FIR and submillimeter. Generally speaking, the short characteristic lifetime of YSOs and the necessity of unbiased studies calls for large surveys. An angular resolution better than that of DIRBE or FIRAS is necessary to separate individual sources efficiently. From simple extrapolations of the *IRAS* PSC, we estimate that the

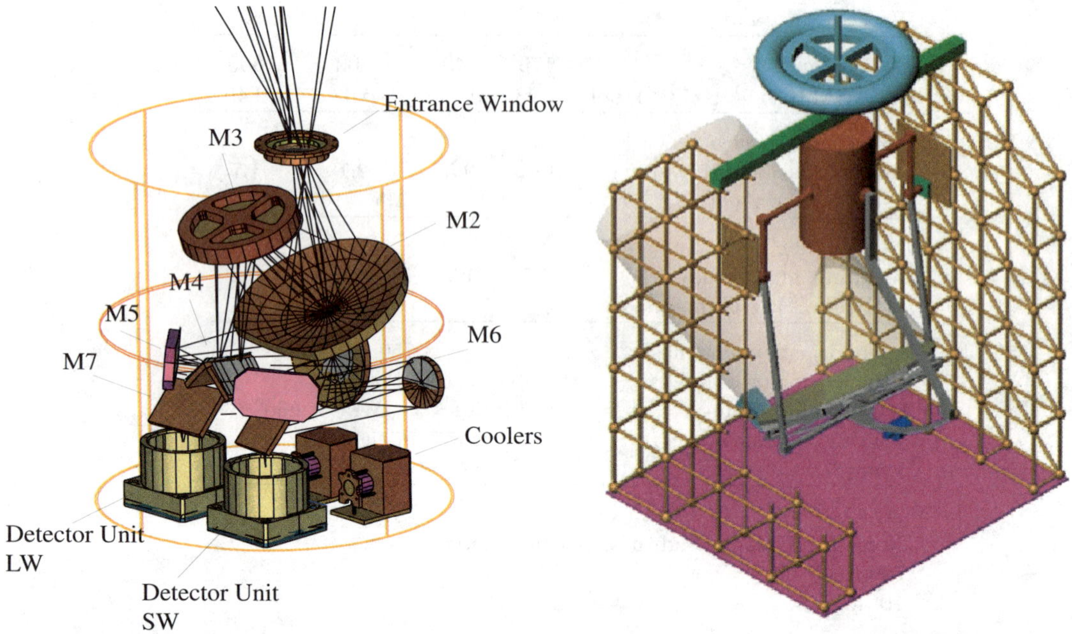

Figure 1. Schematic view of the ELISA experiment. *Left*, cryost

ELISA PSC will contain between 3×10^4 (at 650 μm) and 6×10^4 (at 200 μm) entries, mainly consisting of YSOs in the galactic plane.

Our current plan is to equip each photometric channel of ELISA with a set of polarizers, which will allow measurement of both the degree and the direction of polarization of the radiation. To date, very few measurements of the polarization have been obtained on large scales in the submillimeter at an angular resolution of a few arcminutes. We expect that most of the polarization in the ELISA channels will be coming from polarized dust emission. Although polarization from dust is known to be fairly low (< 10 percent), the high sensitivity of ELISA should allow detection in clouds with A_V=25 and A_V=3 magnitudes in the galactic and deep survey modes, respectively. We expect that these measurements will help strongly constrain the nature of the dust and the efficiency of its alignment with respect to the magnetic field. In particular, fractal grain aggregates should have different polarization and alignment properties than classical dust grains. The correlation between the dust temperature and the observed polarization should therefore shed new light on the nature of cold dust. In dense star-forming regions, such observations should reveal the orientation of the magnetic field, which is of prime importance in understanding star-forming mechanisms.

3. Main instrument characteristics

The ELISA experiment is designed to fly at a ceiling altitude of around 4 mbar (37 km) in the stratosphere, for a large galactic survey to be performed in three flights of 10 to 20 hours each, including one flight in the southern hemisphere. The institutes taking part in the main project developments are: CESR-Toulouse, IAS-Orsay, CEA-Saclay, CNES-Toulouse, DSRI-Copenhagen, SSD/ESTEC-Noordwijk, QMW-London, LAT-Toulouse, and La Sapienza-U.-Roma, together with the scientific collaboration of Toronto, Nagoya and Tokyo Universities. An angular resolution of 3.5′ was chosen as a compromise between the need of having a resolution better or similar to *IRAS* and the aim of surveying a large galactic fraction per flight, with a small project carrying a 1-m diameter telescope and a payload lighter than 500 kg.

The telescope is a 1-m diameter off-axis Gregorian, whose primary is the carbon-fiber mirror of the TOPHAT experiment, provided by DSRI. The secondary mirror is integrated within the cryostat, and the equivalent focal distance is about 2 m (see Fig. 1).

We propose to have one measurement channel coinciding with the *COBE*/DIRBE band at 240 μm in order to monitor and correct for possible drifts due to remaining atmospheric emission or parasitic signals of in-

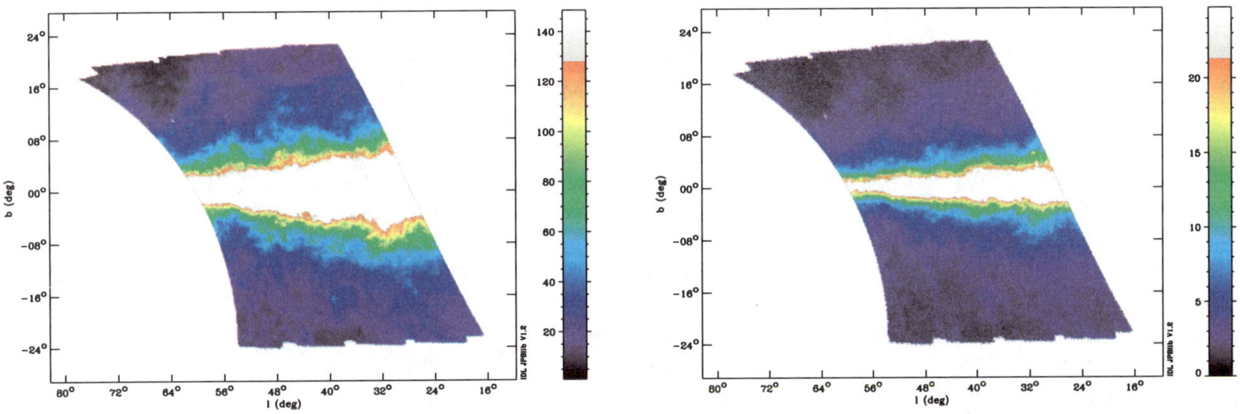

Figure 2. Simulated ELISA observations at 200 (*left*) and 650 μm (*right*) corresponding to 2 hours of observation in the large survey scanning mode (700 deg^2/h). The detection of a typical cold condensation (13 K) is shown at (l,b)=(64°, 11°)(see Sec. 4).

strumental nature. A preliminary study to optimize both T and β leads to four submillimeter wide-band channels ($\delta\lambda/\lambda \sim 30\%$) centered at 170, 240, 400 and 650 μm.

In the nominal configuration, we will use the PACS-type bolometer arrays for the four channels, developed by CEA and LETI. An adequate cold optics scheme allows spatial splitting of the beam in order to have two 16×16 bolometer arrays per channel. Specific attention was given to the radiation thermalisation and the minimization of internal straylight. The large focal plane to be used with the arrays (50 mm for two bands), and its flatness produce a sophisticated cold optics layout (see Fig. 1).

The liquid He cryostat holding the cold optics and the detectors will be cooled down to 0.3 K using a ^{3}He closed-cycle fridge. The expected sensitivity corresponds to the photon noise limit for each band, an estimate of which is given in Table 1.

Mapping of the sky will be accomplished by rotating the gondola over a large azimuth range (±30°) at constant elevation, in order to reduce the residual atmospheric contribution. A servo-control loop ensures the stabilisation in elevation with a 15″ accuracy, by means of a reaction wheel, a magnetometer and a fast and large-field (15°) stellar sensor, operating day and night. The elevation ranges from 15° to 60°. The selected 1.2 deg s^{-1} rotation speed for the scanning is a compromise between on the one hand, the need to cover a large amplitude (integrating the sky rotation) and reduce instrument drifts and, on the other hand, the need to distinguish point sources from parasitic "spikes," given both the detectors' and stellar sensor response times.

4. Observing Strategy

The scientific objectives previously described have opposite requirements in terms of observing strategy: the need to reach a sufficient integration time per beam for the (T, β) determination even for low level brightness emission versus the need to survey a large fraction of the galactic plane. These constraints suggest the use of two observing modes: a large survey mode corresponding to a sky coverage of 700 square degrees per hour, and a deep survey for faint region mapping, with a rate of a few tens of square degrees per hour. These different coverage rates will be reached by adapting the elevation step between scans and/or repeating observations in order to increase the signal-to-noise ratio. Detailed modeling will allow optimization of the observing strategy, along with parameters such as scan speed, length of the scans, and amplitude of the elevation corrections. A simulation of the ELISA observations in large survey mode is shown in Figure 2. The submillimeter galactic emission has been extrapolated from *IRAS*, with an average dust spectrum corresponding to $T = 17.5$ K, $\beta = 2$. A cold condensation with a typical spectrum ($T = 13$ K, $\beta = 2$ has been superimposed on the 100 μm emission near $l = 64°$, $b = 11°$. Such a cold component, undetected at 100 μm and too diluted in the DIRBE beam, could be revealed by ELISA during the large galactic survey.

5. Synergy with *Planck* and *Herschel*

Herschel instruments will allow analysis of the dust grain emission and gas-grain interaction with an unprecedented sensitivity, spectral coverage and angular resolution. However, as for *ISO*, the constraints of an observatory mission will lead to a limited number of observed regions, which will be selected in advance, on the basis of the current datasets and models. The PRONAOS experiment confirmed that a submillimeter extrapolation from *IRAS* emission is very uncertain. Most of the cold condensations discovered with PRONAOS are undetected at $100\,\mu$m, and a significant grain emissivity variation has been shown in the submillimeter range, at the angular scale of a few arcminutes. Both are very difficult to study with DIRBE and FIRAS measurements due to insufficient angular resolution (1° and 7°, respectively). The ELISA submillimeter survey will be very useful to prepare and optimize observations with *Herschel*, and should identify new targets (cold condensations in particular). It will play a role similar to that of the *IRAS* survey for *ISO*, since the *Planck* "early release" PSC will contain exclusively point sources and galaxy clusters and the extended emission maps of *Planck* will not be delivered in time for follow-up observations with *Herschel*.

The FIRAS instrument concept has provided the best absolute calibration in the far-infrared and submillimeter range. It will be a reference for the flight calibration of SPIRE, in particular on extended sources. However, the angular resolution of the two instruments is very different ($30''$ and 7°) which makes direct and systematic comparisons very difficult. ELISA will provide data at $3.5'$ resolution cross-calibrated with FIRAS, which it will be possible to use as a secondary calibrator for SPIRE.

In addition, the results obtained with ELISA will be very useful to anticipate methods to be used to subtract the galactic foreground contribution in the *Planck* data. It will be particularly relevant to the problem of separation between dust emission and grain rotation, which is likely to dominate the spectrum at longer wavelength. Finally, the ELISA polarized large scale maps will allow a global study of the polarized galactic foregrounds, which will have to be carefully taken into account when measuring CMB polarization with future space missions such as *Planck*.

ACKNOWLEDGEMENTS

We are deeply indebted to Guy Serra who initiated the concept of the ELISA experiment. Guy died in August 2000. He was scientifically renowned as a pioneer of infrared and submillimeter astronomy. He will remain in our mind both as a great scientist and a generous humanist.

We are very grateful to the ELISA participants from the institutes of CEA-Saclay, CNES-Toulouse, DSRI-Copenhagen, SSD/ESTEC-Noordwijk, QMW-London, LAT-Toulouse, La Sapienza-U-Roma, Toronto, Nagoya and Tokyo Universities.

REFERENCES

Bernard, J.P., et al. 1999, A&A, 347, 640
Boulanger, F., et al. 1996, A&A, 312, 256
Desert, F.X., Boulanger, F., & Puget, J.L. 1990, A&A, 237, 215
Dupac, X., et al. 2001, ApJ, 553, 604
Egan, M.P., et al. 1998, ApJ, 494, L199
Hennebelle, P., et al. 2000, ESA-SP 455, 125
Lagache, G., et al. 1998, A&A, 333, 709
Lamarre, J.M., et al., 1994, IR Phys. Technol., 35, 277
Ristorcelli, I., et al. 1998, ApJ, 496, 267
Stepnik, B., et al. 2003, A&A, 398, 551

APEX: A NEW SUBMILLIMETER TELESCOPE IN CHILE

Peter Schilke

Max-Planck-Institut für Radioastronomie, Auf dem Hügel 69, 53121 Bonn, Germany

Abstract

Chemical studies of active star formation regions are best conducted in the submillimeter region of the spectrum: it is here that the more excited lines linked directly to the forming star can be found, and there is less confusion due to cold foreground gas. Unfortunately, ground-based observing of the submillimeter region of the spectrum is hampered by the atmospheric transmission, which is severely limited by water line absorption. A solution is to go to high mountain sites. Indeed, the JCMT and CSO observatories on Mauna Kea, Hawaii, have been very successful. An even better site is envisioned for the Atacama Large Millimeter Array (ALMA): the Chajnantor Plain in northern Chile, at an altitude of 5000 m (as compared to 4200 m for Mauna Kea).

The Max-Planck-Institute for Radioastronomy, together with the University of Bochum, the Onsala Space Observatory, and the European Southern Observatory, is constructing a 12-m single dish telescope on the Chajnantor site, the Atacama Pathfinder EXperiment (APEX) to explore the southern submillimeter sky. This instrument is scheduled to start operating by the end of 2003, and will eventually have a full set of heterodyne receivers covering the range 200-1500 GHz, plus large bolometer arrays.

Apart from being at a site with better transmission than Mauna Kea, APEX will have access to the virtually unexplored—in the submillimeter—southern sky. The latter contains unique objects, such as the Magellanic Clouds and the galactic center, which are not (or only with difficulty) observable at the highest frequencies from the northern hemisphere.

Key words: ISM: molecules – telescopes – submillimeter

1. Scientific Rationale

The submillimeter region of the electromagnetic spectrum is a window to the cool universe: objects in the temperature range 30–200 K radiate a substantial amount of their line and continuum emission in this frequency range. At the same time, the submillimeter sky is transparent enough to look through large column densities of dust, which hamper observations in the infrared. In other words, science in the submillimeter is mostly a science of origins: stars being formed, galaxies being assembled. For line studies of galactic star forming regions, there are other advantages: while many lines, notably CO, in the traditional millimeter range are ill-suited to pinpoint or study centers of activity, due to confusion with the ambient molecular clouds, the usually more highly excited submillimeter lines are better probes, since the ambient gas does not radiate at these frequencies. The submillimeter lines are also important tools in excitation studies of star forming regions. For astrochemists, the submillimeter regime also has the attraction of containing several low lying states of light hydrides: e.g., HCl, SiH, NH_2, and many others.

Very cold objects, such as dark clouds, do not show prominent molecular line emission in the submillimeter range, with the exception of the atomic carbon fine structure line at 492 GHz. But such objects are prominent in submillimeter dust emission, which can be used, together with millimeter dust maps, to set tight constraints on the mass and temperature of these regions.

Even many intrinsically hot regions are best observed in the millimeter/submillimeter region, since they are so deeply embedded (extinctions of $A_V \approx 1000$ mag are not uncommon) that they are completely invisible at optical to mid-IR wavelengths.

While the interstellar medium is largely transparent in the submillimeter regime, the Earth's atmosphere unfortunately is not (Fig. 1). The absorption, particularly of the low excitation water lines at 557, 752, 1113 and 1670 GHz, limit the transmission in the atmosphere particularly at submillimeter wavelengths. Therefore, all existing ground-based submillimeter telescopes are built on high mountains, the best currently developed site being Mauna Kea on Hawaii at 4200 m, with its CSO and

Chemistry as a Diagnostic of Star Formation (2003). *Edited by Charles L. Curry and Michel Fich.* Published in 2003 by NRC Research Press, Ottawa ON, K1A 0R6, Canada. ISBN 0-660-19089-3.

JCMT telescopes, and the Submillimeter Array under construction.

An even better site has been found on Llano de Chajnantor, in Chile, at an altitude of 5000 m. Here, the Atacama Large Millimeter Array (ALMA)[1] will erect its 64 telescopes scheduled to be finished in 2011. This site not only has significantly better atmospheric transmission than Mauna Kea, it also opens the southern sky which, so far, has been studied in the submillimeter regime only with a small telescope on the South Pole, AST/RO. Close to Chajnantor, on Pampa la Bola, at 4800 m, Japan has built its 10 m ASTE submillimeter telescope, in operation since 2002.

The southern location, apart from allowing access to unique objects such as the Magellanic Clouds, the nearest AGN Centaurus A and the peculiar star η Carinae, also is much better situated for studies of the Galactic center. The site characteristics are such that not only the classical submillimeter windows below 1 THz are accessible, but also some atmospheric windows beyond open up.

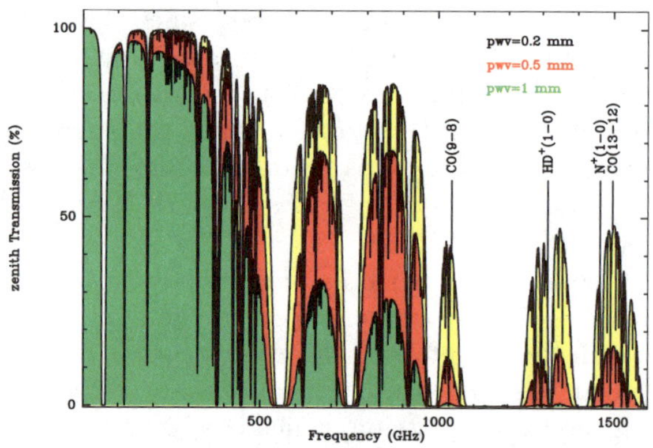

Figure 1. Average atmospheric transmission in Hawaii *(green)*, Chajnantor *(red)* and best possible transmission in Chajnantor *(yellow)*.

2. Project Description

For all the reasons mentioned in the previous section, the Max-Planck-Institut für Radioastronomie (MPIfR), in collaboration with the Universität Bochum, the European Southern Observatory (ESO), and the Onsala Space Observatory (OSO), decided to erect and operate a large single-dish submillimeter telescope on the Llano de Chajnantor. The telescope[2] is a modified copy of the NRAO ALMA prototype, which is built by VERTEX Antennentechnik, Germany (see Fig. 2). The antenna diameter is 12 m, and the anticipated surface accuracy is 18 μm, which should allow measurements up to the THz range. While solar observations are not foreseen, the surface has been treated to scatter diffusely in the optical and IR, so that daytime observations, even close to the Sun, are possible. The blind pointing accuracy is 2″, the tracking accuracy 0.6″. A small optical pointing telescope will be available.

Figure 2. Artist's conception of the APEX telescope from the front and back. The instrument containers are clearly visible.

[1] See the ESO and NRAO web pages at http://www.eso.org/projects/alma and http://www.alma.nrao.edu.

[2] http://www.mpifr-bonn.mpg.de/div/mm/apex.html

The modifications to the ALMA antenna design consist of two Nasmyth cabins, which allow operation of receivers in both Nasmyth foci, and an instrument container at the back of the antenna which accommodates the correlators and the IF section. An overview of the instruments is given in Sec. 4.

The antenna erection and commissioning will benefit from the fact that this is the second antenna of its type, and it is hoped that many problems will already be recognized and eliminated by the experience with the US ALMA prototype. To go one step further in compatibility, it has been decided to use the same antenna control system as the ALMA prototype antennas, TICS/ACS. This, of course, has to be adapted to the different kinds of instruments APEX will possess.

The antenna will arrive in Chile in April 2003, and the rest of the year will be used for commissioning of the antenna and some instruments. The surface setting is initially being done via photogrammetric measurements; subsequently, holographic methods will be used. Operational readiness is foreseen by the end of 2003.

3. Operation

Due to the extreme altitude (5000 m), safety considerations play an important role in defining the operations model. Eventually, the antenna will be operated fully remotely from San Pedro, where a base camp is being constructed. The telescope will be connected to the base camp by a 34 Mbit s^{-1} radio link. In the initial phase, fully remote operation will probably not be possible, because the system will not yet be in a stable state. During this time, the telescope will be operated by staff at the site, where oxygenated control rooms and some other facilities will be available. The observations will be conducted either by the staff (service observing with flexible scheduling), or by observers from San Pedro. Under regular circumstances, no observer should be at the telescope site during observations. This restriction may be lifted for PI campaigns, but special safety measures will have to be taken in this case. Remote observing from Europe is possible, but limited by the available bandwidth out of San Pedro, which currently is only 64 kbit s^{-1}.

The observing time distribution is planned as follows: MPIfR/Bochum, 45%; ESO, 24%; OSO, 21%; and Chile, 10%. Each group will have its own methods for time allocation, although collaboration is encouraged. Regular operation is foreseen for 10 months per year and 14 hours per day. The telescope will not operate in the Bolivian winter from mid-January to mid-March, since the weather in this period is much worse than during the rest of the year. These restrictions exist because of limited manpower, and can be lifted during PI campaigns, when extra manpower is available.

4. Instrumentation

The telescope has been equipped with Nasmyth cabins to have more space for instruments. It is foreseen to equip the Cassegrain cabin with large bolometer arrays, and the Nasmyth cabins with heterodyne receivers covering all atmospheric bands from 200 to 950 GHz, and selected bands at higher frequencies. Bolometer arrays at various wavelenghts are also planned (Kreysa et al. 2002). A summary of the instruments is given in Table 1.

Table 1. APEX instrumentation

Instrument	λ, ν range	Pixels	Avail.	Status[a]
		Heterodyne		
APEX-1	210-275 GHz	1	2003	f
APEX-2	275-370 GHz	1	2004	f
APEX-3	375-500 GHz	2	2004	f
CHAMP+	600-720 GHz	7	2005	p
CHAMP+	780-960 GHz	7	2005	p
APEX-T2	1240-1400 GHz	1	2005	f
		Bolometers		
LABOCA	870 μm	300	2003	f
350 μm	350 μm	37	2004	f
S-Z Array[b]	2 mm	300	2005	p

[a] *f* is a facility instrument, while *p* denotes a PI instrument, available only in collaboration with the instrument providers.
[b] The S-Z Array is an external PI instrument by the University of California, Berkeley, in collaboration with the MPIfR.

The heterodyne receivers will use correlator backends. For the single pixel receivers, the ATNF/SEST correlator will be available, providing 2 times 1 GHz bandwidth. CHAMP+ will use the MACS correlator (Güsten et al. 1998), which has 32 times 1 GHz.

Acknowledgements

APEX would never have happened without the initiative and persistence of Karl Menten. The Max-Planck-Gesellschaft provided funds for the antenna purchase, the partners Catherine Cesarsky (ESO), Roy Booth (OSO) and Rolf Chini (Bochum) made the operation feasible. Lars-Åke Nyman, the future site manager, was heavily involved in planning the details of operation, and Dirk Muders, head of APEX software development, made antenna control possible. Special thanks goes to the instrument teams in Bonn, Bochum and Onsala.

References

Güsten, R. et al., 1998, Proc. SPIE, 3357, *Advanced Technology MMW, Radio, and Terahertz Telescopes*, ed. T.G. Phillips, 167

Kreysa, E., Gemünd, H-P. Raccanelli, A., Reichertz, L.A., & Siringo, G., 2002, AIP Conference Proceedings, No. 616, 262

PART II:
FUNDAMENTAL PROCESSES

THE MOLECULAR UNIVERSE

A.G.G.M. Tielens

SRON/Kapteyn Institute, The Netherlands

Abstract

During the last decade it has become increasingly clear that the molecular universe is widespread and very diverse. The near- and mid-IR spectra of the diffuse interstellar medium of the Milky Way and other galaxies is dominated by the vibrational emission of large (50 C-atoms) polycyclic aromatic hydrocarbon (PAH) molecules. These emission features are also bright in many regions of massive star formation as well as in the ashes expelled by stars in their death throes. These ubiquitous molecules lock up about 1–10 percent of all the available elemental carbon. It is also now clear that the Diffuse Interstellar Bands in the visible spectral range—which have presented a spectroscopic enigma since their discovery 75 years ago—are due to electronic transitions of interstellar molecules. Besides these PAHs, likely candidates include acetylenic C-chains and fullerenes. Similarly, spectroscopic studies of the rotational, millimeter and submillimeter spectra of molecular clouds—the birth sites of new stars—have revealed a rich inventory of molecules which has increased at a pace of about 5 new molecules per year over the last 25 years. Molecules have now been discovered as far back as $z = 4.7$, when the first burst of star formation accompanied galaxy formation in the universe. Indeed, cooling by molecules probably played as key a role in star formation at that time as it does in the present universe. In the solar system, the chemical composition of comets still attests to the heritage of the interstellar cloud from which the solar system formed 4.5 billion years ago, while carbonaceous meteorites display a very diverse pattern of organic species including a wide variety of amino acids many of which are unknown in nature on Earth. This talk will review the richness of the molecular universe and the processes that play a role in the origin and evolution of these species.

Key words: astrochemistry – ISM: molecules

Chemistry as a Diagnostic of Star Formation (2003). *Edited by Charles L. Curry and Michel Fich.* Published in 2003 by NRC Research Press, Ottawa ON, K1A 0R6, Canada. ISBN 0-660-19089-3.

ESTIMATES OF THE H_2 FORMATION RATE IN PDRs

Emilie Habart[1], Francois Boulanger[2], L. Verstraete[2], C.M. Walmsley[1], and G. Pineau des Forêts[2]

[1]Osservatorio Astrofisico di Arcetri, Florence, Italy
[2]Institut d'Astrophysique Spatiale, Orsay, France

Abstract

The main aim of this study is to provide estimates of the molecular hydrogen formation rate in a sample of nearby photon-dominated regions (PDRs) using results from both *ISO* and ground–based telescopes. The physical conditions in the PDR layers from which H_2 emission is observed ($n_H \simeq 10^4 - 10^5$ cm^{-3}, $T_{\rm gas} > 300$ K) differ considerably from those in the diffuse clouds where one can observe the UV lines of H_2 ($n_H \simeq 100 - 1000$ cm^{-3}, $T_{\rm gas} \simeq 50 - 100$ K) and upon which most estimates of the molecular hydrogen formation rate to date have been based. Thus the results from PDRs allow important constraints to be placed upon the mechanisms for forming molecular hydrogen on grain surfaces. Our results show that a high H_2 formation rate $R_f \sim 0.3 - 1.5 \times 10^{-16}$ cm^3 s^{-1} (or $1 - 5$ times the "standard" rate estimated by Jura [1975] for diffuse clouds) is required to explain the observed H_2 line intensity ratios. Thus, the higher grain and gas temperatures in PDRs do not seem to impede the formation of H_2. This result and the correlation between the H_2 and aromatic dust emission found in our PDR sample suggest that H_2 forms efficiently (at least in PDRs) via recombination with chemically attached H-atoms on small carbonaceous grains.

Key words: ISM: molecules – ISM: structure – ISM: clouds

1. Introduction

The formation of the molecular hydrogen is a key process affecting the thermal and density structure and the chemical evolution of the interstellar medium (Combes & Pineau des Forêts 2000). Although there is a consensus that H_2 forms on the surface of dust grains (Hollenbach & Salpeter 1971; Jura 1975; Duley & Williams 1984), the mechanism is not yet understood. Numerous theoretical and experimental works have thus been dedicated to the study of the H_2 formation process. Another approach to the question is to study the efficiency of the H_2 formation in different regions of the ISM as well as to compare the H_2 emission with those of dust grains in order to probe by which process and on which kind of grains H_2 forms. *ISO*, due to its ability to observe with the Short Wavelength Spectrometer (SWS; Kessler et al. 1996) a series of H_2 pure rotational lines, provide new insight on the H_2 excitation and formation rate in various photon-dominated regions (PDRs) of the ISM. In fact, the gas temperature probed by the population distribution of low H_2 rotational levels, as well as the H_2 lines intensity, were towards several PDRs higher than expected by models (Bertoldi 1997; Draine & Bertoldi 1999; Thi et al. 1999; Habart et al. 2002a; Li et al. 2002). One explanation of this is that the H_2 formation is large at high gas temperatures, moving the H^0/H_2 transition zone closer to the edge of the PDR (Habart et al. 2002a). This would cause the higher temperatures and lines intensities observed.

In this study, we consider a sample of PDRs with various excitation conditions and investigate the H_2 formation process using observations of H_2 emission obtained by *ISO*-SWS and ground-based telescopes. PDRs, where stellar radiation plays a dominant role in determining the chemical and thermal state of the gas (for a recent review see Hollenbach & Tielens 1999), are privileged objects for the study of physical and chemical processes of the ISM. In this paper, we first describe the sample of PDRs studied (Sec. 2) and the model used to analyse the H_2 lines emission (Sec. 3). Thus, we estimate the H_2 formation rate under PDR conditions using the H_2 line intensity ratios as diagnostic (Sec. 4). We also compare the intensities measured in H_2 fluorescent line with those in the aromatic bands (Sec. 5). Finally, we discuss which kind of H_2 formation mechanisms could explain our estimated rates (Sec. 6).

2. Sample of PDRs observed by *ISO*

In the *ISO* data base, we have selected five nearby PDRs which sample well the range of excitation condition cov-

Chemistry as a Diagnostic of Star Formation (2003). *Edited by Charles L. Curry and Michel Fich.* Published in 2003 by NRC Research Press, Ottawa ON, K1A 0R6, Canada. ISBN 0-660-19089-3.

ered by the SWS observations. At the low excitation end, we have the ρ Ophiuchi West interface (hereafter called OphW) and the S140 PDR. Then, we have the PDR IC 63 and the reflection nebula NGC 2023 where we focus on the filament at 60″ south of the central star. Finally, at the high excitation end, we have the Orion Bar taken at the position of the peak of the fluorescent H_2 emission. These nearby ($d \sim$100-500 pc), well-observed PDRs are ideal targets to study the formation of H_2 in excited regions of the ISM.

In Table 2, we summarize the physical conditions—e.g., FUV (6 eV< $h\nu$ <13.6 eV) incident radiation field and gas density critical parameters for determining the thermochemical structure of PDRs—prevailing in each region determined from previous studies (see references in Habart et al. 2002b). χ is the incident FUV field in units of the equivalent Draine (1978) flux of 2.6×10^{-3} erg s^{-1} cm^{-2} appropriate to the average interstellar medium. It is determined from the projected distance to the exciting star and its expected FUV luminosity. Estimations based on observed atomic lines of C II, C I or O I are also reported. The proton gas density n_H has been derived using various observational constraints: measurements of fine-structure line emission as well as C radio recombination lines and CO, CN, CS, HCO^+ or HCN line ratios, H_2 fluorescent lines emission or observed stratification of different tracers of PDRs.

In Table 2, we report the observed intensities for the H_2 0-0 S(3) and 1-0 S(1) lines, obtained respectively with *ISO*-SWS and from ground based observations. High spatial resolution observations ($\sim$1″) of the 1-0 S(1) line emission have been smoothed to the $\sim$20″ beam of the SWS. For measurements made with beams larger than the SWS observations, we have scaled the flux according to the beam ratio. For each region, we also give in Table 2 the excitation temperature characterizing H_2 pure rotational levels with $J \leq 7$. Finally, based on ISOCAM observations[1], we give in Table 2 the aromatic dust emission diluted in the beam of the SWS. In the following, aromatic dust particles will be hereafter referred to as polycyclic aromatic hydrocarbons (PAHs). This is a generic term which encompasses large aromatic molecules and tiny carbonaceous dust grains containing up to a few 100 atoms and with radii of a few Å to a few tens of Å.

[1] From the brightness in the LW2 filter (5-8.5 μm), dominated by the aromatic dust emission (Boulanger et al. 1998), we can estimate the emission using the following relationship: $I_{\rm PAH}(2-15\mu m) \simeq 2\nu I_\nu(5-8.5\mu m)$ based on ISOCAM-CVF spectrum (corrected from the dust continuum emission) taken in PDRs.

3. PDR MODEL

In order to analyse the H_2 emission observations from the PDRs, we use an updated version of the stationary PDR model described in Le Bourlot et al. (1993). In this model, a PDR is represented by a semi-infinite plane-parallel slab with an incident radiation field on the interface. The input parameters are (i) the incident FUV field χ, and (ii) the proton gas density n_H. We will consider constant density models with $n_H = 10^4 - 10^5$ cm^{-3} and $\chi = 10^2 - 10^5$ in the range of values prevailing in our PDR sample (see Table 2). With the inputs χ and n_H the model solves the chemical and thermal balances starting from the slab edge at each A_v-step in the cloud. The H_2 abundance results from a balance between the formation of H_2 on dust grains and the photodissociation of H_2 by FUV flux, which is attenuated by dust extinction and self-shielding in the H_2 lines. The adopted H_2 formation rate, R_f (cm^3 s^{-1}), is a semi-empirical rate coefficient:

$$R_f = \gamma \left(\frac{T_{\rm gas}}{100K}\right)^{0.5} S(T_{\rm gas}), \quad (1)$$

with γ a numerical factor in cm^3 s^{-1}, $T_{\rm gas}$ the gas temperature, and S the sticking coefficient of H_2 onto grains. The coefficient S, which is uncertain, is approximated by $1/[1 + T_{\rm gas}/400\ {\rm K} + (T_{\rm gas}/400\ {\rm K})^2]$ (Burke & Hollenbach 1983; Bertoldi 1997). The standard value inferred from observations of H_2 UV absorption lines in interstellar diffuse clouds, $R_f \sim 3 \times 10^{-17}$ cm^3 s^{-1} (Jura 1975), corresponds to $\gamma^0 = 4.3 \times 10^{-17}$ cm^3 s^{-1}.

4. H_2 LINE INTENSITY RATIOS

The (v, J) excited states of H_2 can be populated by inelastic collisions with gas phase species, UV pumping and by the formation process. The population distribution of H_2 levels is function of the gas density, the gas temperature and the UV flux. Then, the H_2 line intensity ratios—indicators of the physical conditions in the photodissociation front where hydrogen becomes molecular—are expected to probe the H_2 formation rate which controls the location of H^0/H_2 transition zone towards the PDRs.

In Figure 1, we show the H_2 [0-0 S(3)]/[1-0 S(1)] line intensity ratio predicted by the model (edge-on, as are most of our PDRs except IC 63) in the H_2 emission regions as a function of χ for different H_2 formation rate. Higher H_2 formation rate implies a higher fraction of H_2 which allows hydrogen to compete more effectively with the dust for UV photons and causes the photodissociation front to move closer to the edge of the cloud. Then, the 1-0 S(1) line, proportional to the fraction of the UV photon flux pumping H_2, increases roughly linearly with the

Table 1. Sample of PDRs observed by *ISO*

Object	χ	n_H (cm^{-3})	A_V[a] (mag)	T_{rot}[b] (K)	I [0-0 S(3)][c]	I [1-0 S(1)][d]	PAH[d]/10^3	Ref.
OphW	250	10^4	10	330	13.7	3.1	3.6	(1)
S140	$100-250$	$10^4 - 5\times10^4$	10	500	16.8	4.4	-	(2a,b)
IC 63	600	$5\times10^4 - 5\times10^5$	7	650	10	1.84	0.8	(3a,b,c)
NGC 2023	$500-3000$	$10^4 - 10^5$	10	330	16.5	7	11	(4a,b,c)
Orion Bar	$5000 - 2.5\times10^4$	$5\times10^4 - 3\times10^5$	10	400	59.7	13	62	(5a,b,c)

[a] Visual extinction within the PDR inferred from submm dust emission or CO observations.
[b] Excitation temperature of H_2 pure rotational levels with $J \leq 7$.
[c] Intensities in 10^{-5} erg s^{-1} cm^{-2} sr^{-1} not corrected for dust attenuation.
[d] Intensities diluted in the $\sim$20″ beam of SWS or multiplied by a beam factor.

REFERENCES—(1) Data corresponding to the peak of the H_2 emission in Habart et al. (2002a); (2a) Timmermann et al. (1996); (2b) Draine & Bertoldi (1999); (3a) Thi et al. (1999) (note that the observations cannot be fit by a single excitation temperature; rather, two components of thermalized emission at temperatures of $\sim$ 100 K and $\sim$ 600 K are needed); (3b) Luhman et al. (1997); (3c) data from Cesarsky; (4a) Moutou et al. (1999); (4b) Field et al. (1998); (4c) data from Abergel; (5a) data from Bertoldi; (5b) van Der Werf et al. (1996); (5c) Cesarsky et al. (2000).

H_2 formation rate (see Eq. (2) in Sec. 5 and Habart et al. 2002b). The low J rotational $v = 0-0$ H_2 lines have a somewhat different dependence upon R_f because at densities of interest to us ($> 10^4$ cm^{-3}), their population distribution is essentially in LTE and the line intensities depend mainly on the temperature in the photodissociation front where hydrogen becomes molecular. In the range of χ and n_H studied here, the temperature profile as a function of depth in the PDR is fairly insensitive to molecular hydrogen (heating is mainly due to photoelectric emission from dust grains and cooling due to fine structure line emission of O I and C II), but higher H_2 formation rates cause the photodissociation front to shift closer to the surface where the temperature is higher. Thus, the 0-0 S(3) line intensity increases as R_f increases with a consequent increase of temperature at the dissociation front. This effect also causes a sharper increase in 0-0 S(3) than in 1-0 S(1) and thus the ratio [0-0 S(3)]/[1-0 S(1)] increases with increasing R_f as shown in Figure 1. However, for high values of R_f, high densities, and low radiation fields, saturation sets in as the fraction of incident UV photons absorbed by H_2 approaches unity.

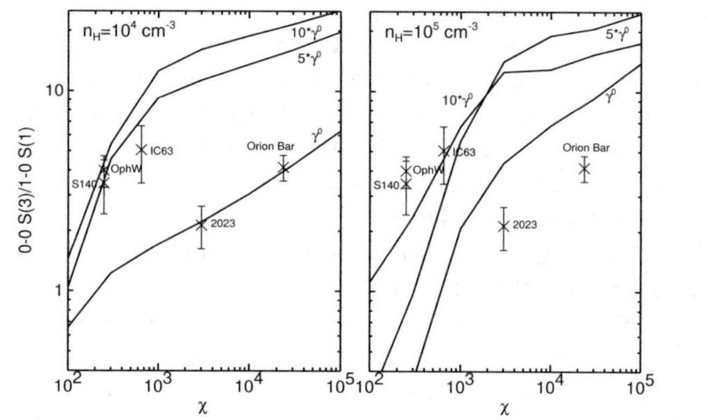

Figure 1. H_2 [0-0 S(3)]/[1-0 S(1)] line intensity ratio as predicted by model (edge-on) and as observed for our PDR sample (crosses with error bars) versus the FUV incident radiation field, χ. Model results, presented for two gas densities, are shown for three different H_2 formation numerical factors $\gamma=\gamma^0$, $5\gamma^0$, and $10\gamma^0$ (defined in Eq. [1]). χ is determined from the projected distance to the exciting star and its expected FUV luminosity (see Sec. 2).

In summary, we find that the [0-0 S(3)]/[1-0 S(1)] line intensity ratios depend strongly on the H_2 formation rate via the variation of both the gas temperature and the UV flux in the H^0/H_2 transition zone. Thus, by comparing this H_2 line intensity ratio predicted by the model (essentially independent of the geometry and the total column density) with observations from PDRs where χ and n_H have been determined from other observations, we can expect to probe the H_2 formation rate in PDRs.

We now compare these predictions with observational data from our PDRs sample described in Sec. 2. The data have been corrected for dust attenuation using the extinction curve of Draine (1989) and the visual extinction reported in Table 2. We find that for OphW ($n_H \sim 10^4$ cm^{-3}, $\chi \sim$250) and S140 ($n_H \sim 1-5\times10^4$ cm^{-3}, $\chi \sim 100-250$), as well as for IC 63 ($n_H \sim 5-15\times10^4$

cm^{-3}, $\chi \sim 650$), models with a high H_2 formation rate (e.g., $\gamma \geq 5\gamma^0$) roughly reproduce the observed [0-0 S(3)]/[1-0 S(1)] line ratio. For the standard H_2 formation rate, the [0-0 S(3)]/[1-0 S(1)] line ratio is in fact underestimated by a factor of about 4. For the highly excited PDRs, i.e., NGC 2023 ($n_H \sim 10^4 - 10^5$ cm^{-3}, $\chi \sim 500-5000$) and the Orion Bar ($n_H \sim 5\times10^4-3\times10^4$ cm^{-3}, $\chi \sim 0.5 - 2.5 \times 10^4$), we find, on the contrary, that models with the standard H_2 formation rate roughly match the data. In conclusion, for PDRs exposed to a moderate radiation field ($100 \leq \chi \leq 1000$), a relatively large H_2 formation rate (e.g., $\gamma \sim 5\gamma^0 \sim 2 \times 10^{-16}$ cm^3 s^{-1}) in warm gas seems required to account for the observed H_2 line intensity ratios, while for highly excited PDRs ($\chi > 1000$), the standard H_2 formation rate matches the data better.

It must be emphasized that these results are based on three fundamental assumptions. Firstly, we assume a uniform homogeneous gas density whereas PDRs are generally expected to be present at the edge of a density gradient and could be clumpy. Nevertheless, because the modelling of the H_2 excitation mainly depends upon the average gas density in the H_2 emission zone, density structure effects are probably minor. Secondly, we assumed a static, equilibrium PDR. In reality, the propagation of the ionization and photodissociation fronts will bring fresh H_2 into the zone emitting line radiation. However, we note that models of non-equilibrium PDRs by Storzer & Hollenbach (1998) predict that for an advection velocity of the order of 1 km s^{-1}, the H_2 [0-0 S(3)]/[1-0 S(1)] line intensity ratio varies by less than a factor ~2. Thirdly, we assume the dust grain abundance to be constant towards the PDRs, while former studies have shown that there are abundance variations of the small carbonaceous grains in interstellar clouds (Boulanger et al. 1988; Boulanger et al. 1990; Bernard et al.1993) and this might affect—via the photoelectric effect—the gas temperature (Habart et al. 2001). In the next section, we discuss the coupling between the aromatic hydrocarbon grains and H_2.

5. H_2 FLUORESCENCE TO PAH EMISSION RATIO

In PDRs, dust grains are intimately coupled to the evolution of the gas. Firstly, the grain photoelectric heating is a strong component of the gas heat budget. Secondly, the formation of H_2 necessarily occurs on dust grains. These processes are expected to be sensitive to the size distribution of the dust grains and, especially, to the abundance of small grains. In fact, recent theoretical (Bakes & Tielens 1994; Weingartner & Draine 2001) and observational (Habart et al. 2001) work has shown that small grains (radius ≤ 100 Å) dominate the photoelectric heating. Furthermore, given that small grains make a dominant contribution to the total grain surface, it is plausible that they play a dominant role in H_2 formation. Therefore, if H_2 forms on small grains, we can expect that both the grain photoelectric heating rate and the H_2 formation rate scale with their abundance. In particular, increasing their abundance would lead to higher gas temperature and would bring the H^0/H_2 transition closer to the edge; this could significantly enhance the rotational and fluorescent emission of H_2. Habart et al. (2002a) have shown that, assuming that R_f scales with the PAH abundance, the enhancement in R_f required to account for the observed H_2 emission from OphW may also result from an increase of PAH abundance. Based on this, we investigate what we can predict concerning the influence of PAHs on the H_2 formation process.

To free oneself from the influence of PAHs in the thermal balance, we study here the ratio between the H_2 fluorescent and PAH emission. For $N_H \leq 10^{21}$ cm^{-2}, we can write (see Habart et al. 2002b)

$$\frac{I_{H_2(\text{fluorescent})}}{I_{\text{PAH}}} \propto \frac{R_f}{[C/H]_{\text{PAH}}} \left(\frac{n_H}{\chi}\right), \qquad (2)$$

with $[C/H]_{\text{PAH}}$ the carbon abundance locked up in PAHs. Then, by studying the observed H_2 fluorescence to PAH emission ratio from PDRs covering a wide range of n_H/χ and PAH abundance, we should be able to probe the H_2 formation on PAHs; in the case of H_2 forms on PAHs, we expect that the $I_{\text{H2}}/I_{\text{PAH}}$ ratio scales with n_H/χ. This will be true if the $R_f/[C/H]_{\text{PAH}}$ ratio is constant and does not vary from one region to another.

We now compare the ratio of the H_2 1-0 S(1) line to PAH emission observed from the PDRs in our sample to model results. For the modelling, we take a constant $R_f/[C/H]_{\text{PAH}}$ ratio corresponding to the values derived in the OphW PDR, e.g., $\gamma = 5\gamma^0$ and $[C/H]_{\text{PAH}} = 0.5 \times 10^{-4}$ (Habart et al. 2002a). The PAH emission has been calculated using the PAH absorption cross-section of Verstraete & Léger (1992) and assuming an MRN size distribution with $\alpha = 3.5$. In Figure 2, we show the H_2 [1-0 S(1)]/PAH emission ratio predicted by the model in the H_2 emission zone as a function of n_H/χ. As expected from equation (2), the ratio increases proportionally with n_H/χ.

The observed ratios for our PDRs sample are now compared. From one PDR to another, the change of the ratio seems to result mainly from the physical conditions and n_H/χ variations. Then, the $R_f/[C/H]_{\text{PAH}}$ ratio appears to be roughly constant. In our PDR sample, because

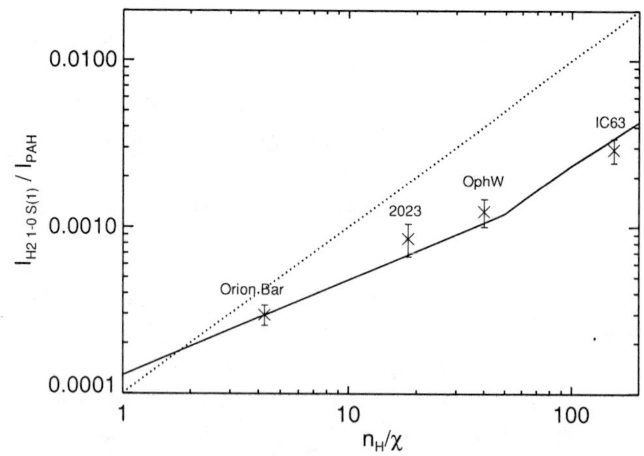

Figure 2. Ratio between the H_2 1-0 S(1) line and PAH emission as predicted by the edge-on model (solid line) and as observed (crosses with error bars) as a function of n_H/χ. The dotted straight line shows the linear dependence of the emission ratio with n_H/χ expected from equation (2). For the PDRs in our sample, n_H/χ is taken equal to the ratio between the average value of the n_H determinations given in Table 2 and the χ estimated from the projected distance between the exciting star and its FUV luminosity.

of the various illumination conditions, PAH abundance variations are expected: from regions exposed to moderate to high radiation field, the PAH abundance should decrease. In fact, former studies based on *ISO* observations of aromatic dust emission suggest that the PAH carbon abundance is $[C/H]_{\mathrm{PAH}} \sim 4 \times 10^{-5}$ in OphW (Habart et al. 2002a), $\sim 2.5 \times 10^{-5}$ in NGC 2023 (Verstraete et al. 2001), and $\sim 1 \times 10^{-5}$ in the Orion Bar (from Verstraete, private communication). Further, these abundance variations seem correlated to the H_2 formation rate estimated in Sec. 4: we find low values of both $[C/H]_{\mathrm{PAH}}$ and γ ($\sim \gamma^0$) for the Orion Bar, while for OphW higher values are found.

In conclusion, the H_2 formation rate seems related to the PAH abundance and consequently PAHs may play a relevant role in the H_2 formation process. However, this comparison needs first to be extended to more objects with a variety of different conditions. We are currently comparing the H_2 fluorescent emission with those of the PAHs within the harsh environment of the 30 Doradus star-forming region, where the different conditions (metallicity, radiation field) may potentially influence both the available grain surface area and the efficiency of H_2 formation (Boulanger et al. in preparation). In addition, in future work this comparison should be extended to the larger small grains; e.g., the Very Small Grains.

6. Discussion

The high H_2 formation rate found here in several PDRs is incompatible with the formation process by physisorbed H atoms (Langmuir-Hinshelwood) whereby two adsorbed and mobile physisorbed H atoms interact to recombine and desorb as H_2. Such a mechanism is in fact efficient only in moderately irradiated regions ($\chi < 100$), where the equilibrium temperature of big grains drops below 30 K (Pirronello et al. 1997; Pirronello et al. 1999). This suggests that H_2 forms, at least in PDRs, via a mechanism involving strongly bound H atoms. This process will be efficient at higher dust surface temperatures than the pure Langmuir-Hinshelwood mechanism.

In Habart et al. (2002b), we have attempted to interpret these observationally determined rates with simple empirical models of H_2 formation involving an H atom bound to the surface in a chemisorbed site. The second H atom from the gas phase reachs this site directly (*direct* chemisorption or Eley-Rideal, [Duley 1996; Parneix & Brechignac 1998], or after diffusion on the grain surface (*indirect* chemisorption). From these, we conclude that an "indirect chemisorption" model analogous to that discussed by Cazaux & Tielens (2002a,b) is capable of explaining most of our PDR data. This requires a fraction of occupied chemisorbed sites of around ten percent and an activation energy barrier against the recombination of a physisorbed H atom and a neighbouring chemisorbed H atom of $\leq$0.1 eV. Moreover, small grains, which make a dominant contribution to the total grain surface, have numerous chemically-bonded hydrogen atoms (because of their small size, they are disordered) and spend most of their time at low temperature (Guhathakurta & Draine 1989), may be the most promising surface for forming H_2 in PDRs. This is also favored by the correlation between the H_2 and PAH emission found in our PDR sample, which suggests that R_f scales with the PAH abundance (see Sec. 5). However, a better knowledge of the small grains' properties (temperature, coverage of absorbed H atom, abundance, etc.) and of the mobilities of H atoms on realistic grain surfaces are critical.

Given our ignorance concerning the interstellar grain properties and also our lack of understanding of surface reactions in the interstellar context, one is forced to some extent to use the astrophysical data at different ISM locations to guide our estimates of the formation of H_2. The agreement between the H_2 line and dust infrared emission obtained by *ISO* with the H_2 UV absorption line observations obtained recently by *FUSE* (Rachford et al. 2002; Tumlinson et al. 2002; Sonnentrucker et al. 2002; Gry et al., in preparation) and the UV dust extinction curve gives us the possibility to probe the dependence of the H_2 formation with the physical parameters from the diffuse ISM to star formation regions. Studying regions such as the Magellanic clouds where the different metallicity, extinction curve, and radiation field potentially may influence both the available grain surface area and the efficiency of H_2 formation is also clearly useful (Tumlinson et al. 2002). Finally, this kind of approach should be expanded with future instruments such as *SIRTF* and, in the longer term, *NGST* (MIRI, 5-29 μm) to various astrophysical environments.

References

Bakes, E. L. O., & Tielens, A. G. G. M. 1994, ApJ, 427, 822

Bernard, J. P, Boulanger, F., & Puget, J. L. 1993, A&A, 277, 609

Bertoldi, F. 1997, in *First ISO Workshop on Analytical Spectroscopy*, 67

Boulanger, F., Beichman, C., Desert, F. X., Helou, G., Perault, M., & Ryter, C. 1988, ApJ, 332, 328

Boulanger, F., Falgarone, E., Puget, J. L., & Helou, G. 1990, ApJ, 364, 136

Boulanger, F., Boisssel, P., Cesarsky, D., & Ryter, C. 1998, A&A, 339, 194

Burke, J. R., & Hollenbach, D. J. 1983, ApJ, 265, 223

Cazaux, S., & Tielens, A. G. G. M. 2002a, ApJ, 575, L29

Cazaux, S., & Tielens, A. G. G. M. 2002b, ApJ, 577, L127

Cesarsky, D., Jones, A. P., Lequeux, J., & Verstraete, L. 2000, A&A, 358, 708

Combes, F., & Pineau des Forêts, G. 2000, in *Molecular Hydrogen in Space*, ed. F. Combes & G. Pineau des Forêts (Cambridge: Cambridge University Press)

Draine, B. T., & Bertoldi, F. 1999, in *The Universe as Seen by ISO*, ed. P. Cox & M. Kessler, 533

Draine, B. T. 1978, ApJ Supp., 36, 595

Draine, B. T. 1989, in *Infrared Spectroscopy in Astronomy*, 93

Duley, W. W., & Williams, D. A. 1984, in *Interstellar Chemistry*, ed. W.W. Duley & D.A. Williams

Duley, W. W. 1996, MNRAS, 279, 591

Field, D., Lemaire, J. L., Pineau Des Forêts, G., Gerin, M., Leach, S., Rostas, F., & Rouan, D. 1998, A&A, 333, 280

Guhathakurta, P., & Draine, B. T. 1989, ApJ, 345, 230

Habart, E., Verstraete, L., Boulanger, F., Pineau Des Forêts, G., Le Peintre, F., & Bernard, J. P. 2001, A&A, 373, 702

Habart, E., Boulanger, F., Verstraete, L., Abergel, A., Falgarone, E., & Pineau des Forêts, G. 2002a, A&A, in press

Habart, E., Boulanger, F., Verstraete, L., Walmsley, C. M., & Pineau des Forêts, G. 2002b, in preparation

Hollenbach, D., & Salpeter, E. E. 1971, ApJ, 163, 155

Hollenbach, D. J., & Tielens, A. G. G. M. 1999, Rev. Mod. Phys., 71, 173

Jura, M. 1975, ApJ, 197, 575

Kessler, M. F., Steinz, J. A., Anderegg, M. E., Clavel, J., Drechsel, G. 1996, A&A, 315, L27

Le Bourlot, J., Pineau des Forêts, G., Roueff, E., & Flower, D. R. 1993, A&A, 267, L233

Li, W., Evans, N. J. I., Jaffe, D. T., van Dishoeck, E. F., & Thi, W. F. 2002, ApJ, 568, 242

Luhman, M. L., Luhman, K. L., Benedict, T., Jaffe, D. T., & Fischer, J. 1997, ApJ, 480, L133

Moutou, C., Verstraete, L., Sellgren, K., & Léger, A. 1999, in *The Universe as Seen by ISO*, 727

Parneix, P., & Brechignac, P. 1998, A&A, 334, 363

Pirronello, V., Liu, C., Shen, L., & Vidali, G. 1997, ApJ, 475, L69

Pirronello, V., Liu, C., Roser, J.E., & Vidali, G. 1999, A&A, 344, 681

Rachford, B. L., Snow, T. P., Tumlinson, J., Shull, J. M., Blair, W. P. 2002, ApJ, 577, 221

Sonnentrucker, P., Friedman, S. D., Welty, D. E., York, D. G., & Snow, T. P. 2002, ApJ, 576, 241

Storzer, H., & Hollenbach, D. 1998, ApJ, 495, 853

Thi, W. F., van Dishoeck, E. F., Black, J. H., Jansen, D. J., Evans, N. J., & Jaffe, D. T. 1999, in *The Universe as Seen by ISO*, 529

Timmermann, R., Bertoldi, F., Wright, C. M., Drapatz, S., Draine, B. T., Haser, L., & Sternberg, A. 1996, A&A, 315, L281

Tumlinson, J., Shull, J. M., Rachford, B. L., Browning, M. K., Snow, T. P. 2002, ApJ, 566, 857

van Der Werf, P. P., Stutzki, J., Sternberg, A., & Krabbe, A. 1996, A&A, 313, 633

Verstraete, L., & Léger, A. 1992, A&A, 266, 513

Verstraete, L., Pech, C., Moutou, C., Sellgren, K., Wright, C. M. 2001, A&A, 372, 981

Weingartner, J. C., & Draine, B. T. 2001, ApJ Supp., 134, 263

FORMATION OF MOLECULAR HYDROGEN ON AMORPHOUS WATER ICE

Joseph E. Roser[1], Gianfranco Vidali[1], Giulio Manicò[2], and Valerio Pirronello[2]

[1]Physics Department, Syracuse University, 201 Physics Build., Syracuse N.Y., 13244-1130, USA
[2]Dipartimento di Metodologie Fisiche e Chimiche per l'Ingegneria, Università di Catania, Viale Doria 6, 95125 Catania, Sicily, Italy

Abstract

In this work, we address the question of quantifying the efficiency of formation of hydrogen molecules on ice mantles primarily due to surface processes. The recombination of hydrogen was studied on amorphous water ice surfaces prepared using different methods and treated thermally and with UV photons. We used an apparatus dedicated to the study of physical/chemical processes on surfaces at low temperature, consisting of two atomic beam lines, an angle-resolved detection system, a time-of-flight section and spectroscopic instrumentation to measure the internal excitation of molecules departing from the sample. The experimental results were analyzed to obtain information on the energetics of the $H_{surface} + D_{surface} \rightarrow HD_{surface}$ reaction. The application of these results to the problem of H_2 formation in actual astrophysical environments is presented.

Key words: astrochemistry – molecular processes – ISM: molecules

1. Introduction

The problem of understanding how H_2 is formed in the interstellar medium has been with astrophysicists since the detection of H_2 in diffuse interstellar clouds. The importance in knowing more about H_2 rests on the fact that molecular hydrogen is a key aide in star formation and in the synthesis of a large number of molecules found in the interstellar medium (ISM). As is well known, the abundance of H_2 observed in the ISM cannot be explained by two-body reactions of H atoms since there is no third body to carry away the excess energy before the molecule dissociates again (Duley and Williams 1984). Alternative gas phase routes to H_2, such as through ion-neutral reactions, have too small a reaction rate to explain the observed abundances (Stancil & Dalgarno 1998). Grain surfaces in the interstellar medium have been proposed as the sites of H_2 formation. Dust grains present in the interstellar medium are believed to consist of core particles composed of amorphous silicates or carbonaceous material with a characteristic grain size $\sim 0.1\mu$m. In regions with large ultraviolet (UV) fluxes, the bare core particles are the active surfaces for chemistry to take place, with only simple chemical species being able to survive dissociation by UV light. In regions where the UV flux has been sufficiently attenuated, chemical species formed on the grains or in the gas phase can "freeze-out" and form an ice layer coating the grain. The primary constituent of the mantles of these grains is amorphous water ice, while other molecules such as CO, CO_2, CH_3OH, and more complex species have also been identified in the ice mantles.

In astrophysical environments, at least three different phases of water ice have been identified. The primary water ice component in ISM grain mantles is a high density amorphous ice phase. In laboratory experiments, this high density ice can be produced by water vapor deposition onto a sufficiently thermally conducting substrate cooled to ~10-20 K. The other two water ice components are low-density amorphous water ice (prepared by heating a high density deposit to ~80 K, or by vacuum deposition at ~77 K), and a cubic crystalline phase produced by heating a vacuum ice deposit to ~150 K. The presence of multiple water ice phases in the same ISM environment has been interpreted as evidence for thermal processing of the grain mantles during the lifetime of a molecular cloud.

Initial calculations of hydrogen recombination on crystalline surfaces by Hollenbach et al. (1971) lead to the assessment that quantum tunneling on the surface is rapid and would allow any two adsorbed H atoms on a grain to quickly recombine. Calculations by Smoluchowski (1983) indicated that H atom mobility on amorphous or polycrystalline grains could be up to two orders of magnitude slower than for crystalline grains. In the case of amorphous water ice, it was suggested by Mayer & Pletzer (1986) that the porous nature of the amorphous

Chemistry as a Diagnostic of Star Formation (2003). *Edited by Charles L. Curry and Michel Fich.* Published in 2003 by NRC Research Press, Ottawa ON, K1A 0R6, Canada. ISBN 0-660-19089-3.

ice would lead to increased hydrogen formation within the ice pores.

Except for early attempts (Govers et al. 1980) performed on surfaces and under conditions far from the ones encountered in the interstellar medium (Schutte 1986), our laboratory experiments (Pirronello et al. 1997a; Pirronello et al. 1997b; Pirronello et al. 1999; Manicò et al. 2001) were the first systematic experiments to yield recombination efficiencies and desorption kinetics of H_2 formation on realistic grain analogues. Our work led to the conclusion that H mobility on a terrestrial sample of polycrystalline olivine and on an amorphous carbon sample prepared in an inert atmosphere was dominated by thermal diffusion at surface temperatures above ~7 K; quantum tunneling played a secondary role. In this communication, we are primarily interested in H_2 formation on polar ices (i.e., H_2O-rich). We decided to study the $H_{surface} + D_{surface} \rightarrow HD_{surface}$ reaction on pure water ice rather than on an ice analogue (which contains CO, CO_2 and hydrocarbons among other molecules) in order to remove chains of hydrogenation reactions involving CO.

2. Experiment

The experimental apparatus used for molecular hydrogen studies is discussed in detail in Vidali et al. (1998) and Roser et al. (2002) and is summarized here. The apparatus consists of a stainless-steel ultra-high vacuum (UHV) chamber with a base pressure of 10^{-10} torr (see Fig. 1). The sample is placed at the end of a UHV-compatible cold finger that can be cooled with liquid helium down to ~5 K. The cold finger is mounted on a rotatable flange to allow the sample to be presented toward different probes in the system. The sample holder is surrounded by a copper shield for additional thermal radiation shielding and to prevent reactions taking place on the upper sections of the sample assembly. The temperature of the sample can be monitored by a calibrated silicon diode located directly behind it and the temperature can be controlled by a constantan heating coil placed behind the sample and a resistive heater wound higher up on the sample holder. The temperature of the sample can be set between 5 to 30 K for the experiments described here. Higher temperatures can be achieved by throttling the liquid helium flow. The sample used for the amorphous ice experiments is a copper blank that can be used as a substrate for amorphous ice deposits from water vapor introduced into the vacuum chamber through a UHV leak valve and capillary. For a high density amorphous ice deposit the ~10 K sample is lowered directly into the path of the water vapor streaming from the nozzle, while for a highly porous, low density deposit the sample is kept out of the stream of direct vapor flow and allowed to adsorb water molecules effusing onto the sample from the background pressure of H_2O in the chamber. Alternatively, a low density ice can be formed by heating a high density deposit; in our experiments we heated the high density ice to ~90 K for at least 5 minutes to convert it to low density ice. The two types of ice, amorphous high density ice and amorphous low density ice, could be easily distinguished by the distinctive features they gave in the thermal desorption of HD, as shown below.

The ice-coated sample was then irradiated by atomic hydrogen from two atomic beam lines, one operating with hydrogen and the other with deuterium. In each beam line, the gas flowed into a water-cooled Pyrex source where 50 W of RF energy was used to dissociate the molecular gas with approximately 90% efficiency. The dissociated gas was sprayed with a nozzle down a series of differentially pumped vacuum chambers with collimating slits to produce a beam of gas with a diameter of ~3 mm on the 10 mm diameter sample. The two beam lines were both aligned to irradiate with gas exactly the same spot on the sample, and the intensity of each beam can be measured with a rotatable quadrupole mass spectrometer (QMS) mounted in the main chamber. The use of beams to deliver H to the sample, rather than dissociating hydrogen in the sample chamber, as is often done in other studies, is important for three reasons. First, it allows us to finely calibrate the amount of H that goes onto the sample. There is only a small increase in the background pressure of H or H_2 due to the scattering of H atoms from the sample. Using a beam in our studies we can keep the estimated coverage (obtained from a Langmuir adsorption curve [Vidali et al. 1998]) well below monolayer completion. Second, the technique allows control of the temperature of the beams by passing the atoms through a short aluminum nozzle cooled to the desired temperature by a copper braid connected to a liquid nitrogen reservoir. Third, the beams deliver H onto the sample without flooding the chamber with electrons or photons produced in the dissociation of molecular hydrogen using hot filaments.

The same QMS is used to measure the desorbing products of the recombination reaction on the surface both during the atomic H and D irradiation and during a heat pulse applied to the sample as part of a thermal desorption experiment. The advantage of using both a H and a D atomic beam line is that the detector's signal is due primarily to HD formed on the sample, thus eliminating the major sources of spurious contributions, such as the undissociated fraction of molecular hydrogen from the beam lines and molecular hydrogen desorbing from the walls of the vacuum chamber. Consequently, the detection of HD gives a very clean signal that can be corrected for the detector's instrument response and for the different routes of molecule formation. The purpose of the TPD experiment is to accelerate the time of reactions, by desorbing from the sample already formed molecules or setting atoms in motion, and to learn about the reaction kinetics (Vidali et al. 1998; Pirronello et al. 1997b).

Additional ice processing was performed by rotating the sample holder in front of a UHV MgF_2 viewport and exposing it to UV radiation from a PerkinElmer model 1155 flashlamp. The flashes were focused on the sample with a MgF_2 plano-convex lens to give a uniform illumination to the entire ice surface. Since the threshold wavelength for UV transmission in MgF_2 is ~120 nm, relatively high energy photons that are implicated in photochemistry on the ice surface are transmitted to the sample. We estimate an average total flux of radiation of ~5 milliwatts for wavelengths below 350 nm (corresponding to a flux of 6×10^{15} photons/sec at 250 nm). In experiments exploring the effect of ultraviolet light on the ice layer, we first exposed the freshly prepared ice deposit to 15 minutes of UV before proceeding to the atomic hydrogen exposure followed by thermal desorption. The ratio of the flux of photons to the sample surface to the flux of H atoms was approximately the same as that in a typical ISM region. Obviously, the flux of H in the laboratory experiment had to be much higher than the actual flux in a dense cloud.

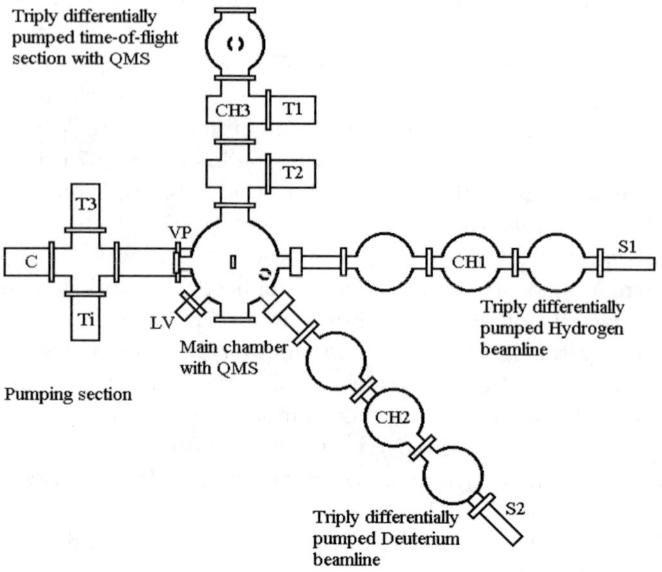

Figure 1. Schematic of the apparatus: top view. S1 and S2 denote H_2 and D_2 RF dissociation sources; CH1, CH2 and CH3 the positions of mechanical choppers; T1, T2, and T3 are turbopumps while Ti is a titanium sublimation pump and C a cryopump; LV is the leak valve for introducing water vapor into the system through a capillary; and VP is the MgF_2 viewport for admitting UV light into the system. From Roser et al. (2002).

3. Results

By comparing the desorption yield of HD to the surface exposure to H and D and correcting for the instrument response and the probabilities for forming H_2 and D_2 instead of HD, we can determine the recombination efficiency, which is the product of the probability that a H atom will stick and accommodate itself to the surface and the probability of recombination for atoms on the surface. In Figure 2 we present a graph of the recombination efficiency as a function of temperature for the high density amorphous and low density amorphous water ice surfaces. The recombination efficiency is thus somewhat larger for the low density amorphous surface, and by comparison larger than previous results for olivine and amorphous carbon (Pirronello et al. 1997b; Pirronello et al. 1999).

Further information about the H_2 recombination reaction can be derived from the desorption kinetics of the molecular species. The desorption rate as a function of time can be analyzed with the Polanyi-Wigner equation, $dN_{H_2}/dt = k_m N_H^m$ where N is the coverage of a chemical species on the surface, m is the integral order of the kinetics, and k_m is an energy and temperature dependent prefactor. In Figure 3 we illustrate desorption traces for thermal desorption from high density ice and low density ice formed by heating a high density ice to produce a phase change to low density.

Here we see different features in the kinetics of the reaction $H_{surface} + D_{surface} \rightarrow HD_{surface}$ taking place on the low density and high density phases. Instead, we observe similar desorption traces (not shown) in the two realizations of low density ice phases (heat treated and deposited from the gas phase). The heterogeneity of each amorphous surface produces a distribution of binding energy sites (and corresponding activation energies for atom migration) for adsorbed species that leads to two desorption peaks. In Roser et al. (2002), we present a simple desorption model in which the prefactor k_m of each peak

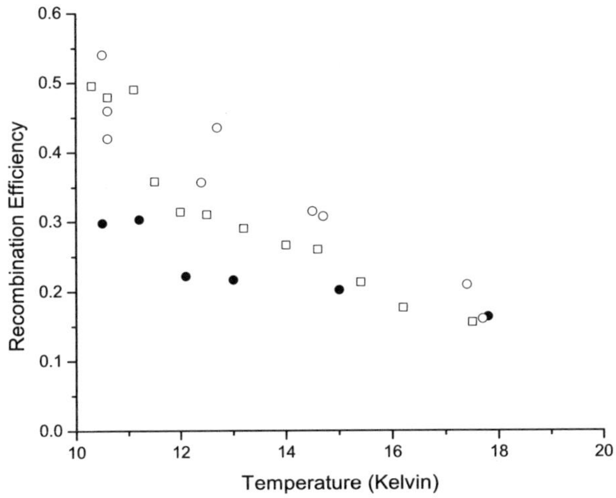

Figure 2. Recombination efficiency of hydrogen versus temperature for high density amorphous ice (filled circles), low density amorphous ice formed by heating a high density ice (open circles), and vapor deposited low density ice (open squares) (from Roser et al. 2002).

depends upon a Gaussian distribution of activation energies for recombination and desorption, and fit this model to desorption traces from high density and low density ices.

For thermal desorption of H_2 from water ices with UV processing, we observed different changes in the spectra depending whether the UV–treated surface was low density amorphous ice or high density amorphous ice. For low density amorphous ice, UV treatment shifts the leftmost peak to lower temperature, while the effect on the high density amorphous ice is to enhance the leftmost peak and suppress the high temperature peak. There is also a slightly larger desorption yield for the processed versus the unprocessed high density amorphous ice. Thermal desorption of HD after repeated applications of UV light onto the same ice produced little to no additional change in the desorption spectra. An interpretation of these changes, and in general of the features of the desorption kinetics from the untreated ice, requires a microscopic model of the surface and of the mechanisms of diffusion and desorption. We are currently working on a model that can explain these features, similar to what we did for the data from olivine and amorphous carbon (Katz et al. 1999).

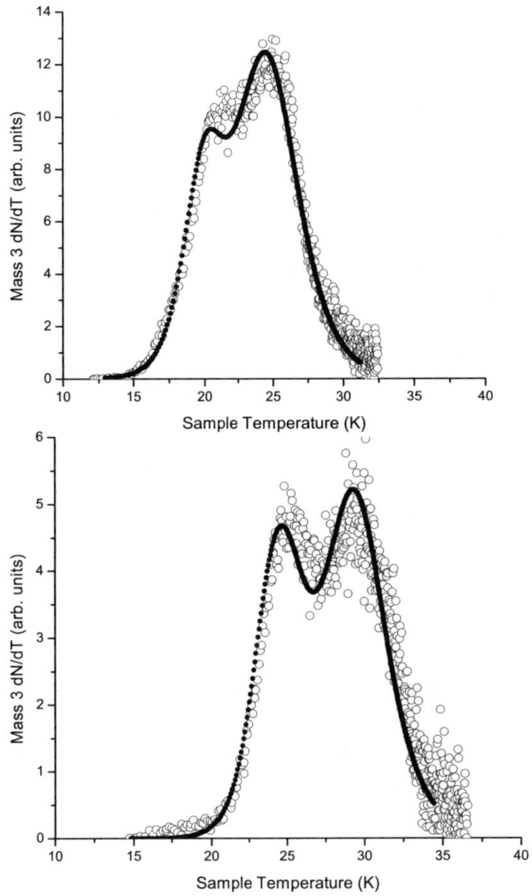

Figure 3. Desorption traces of HD for low density amorphous water ice (top) and high density amorphous water ice (bottom) formed by heating a high density ice. Open circles are data and solid curves are a fit using the kinetics model of Roser et al. (2002).

4. Astrophysical Implications

Water ice in the ISM can be detected and analyzed using the 3.07 μm absorption band. The best fit to the 3.07 μm feature is typically given by a mixture of high density amorphous, low density amorphous, and cubic crystalline ice dominated by the high density amorphous phase (Jenniskens et al. 1995). H_2O is observed in dark cloud regions above a threshold visual extinction A_v, with a linear relationship of optical depth to A_v above the threshold (O'Neill & Williams 1999). Our results show that the hydrogen recombination efficiency in dark cloud environments where water ice mantles persist on the grains is in agreement with what is required by observations.

In a further experiment planned for the immediate future, the energetics of the 4.5 eV of energy released by the H_2 recombination reaction will be investigated with a time-of-flight mass spectrometer. The measurement of the kinetic energy of the departing molecule that has just formed on the surface of the dust grain analogue should give information both on the amount of energy that these reactions dump into the ISM and on the microscopic processes leading to the formation of H_2. In this experiment, the flux of H_2 generated by the TPD will be chopped for time tagging; the chopped beam will then be detected by another high performance quadrupole mass spectrometer located at the end of a triply differentially pumped time-of-flight section attached to the main scattering chamber. This experiment will measure the translational energy of the desorbing H_2 molecules from the thermal desorption experiment—a key piece of information in determining the full product energy distribution of the system.

5. Acknowledgments

G.V. was supported by NASA through grants NAG5-4998 NAG5-6822 and NAG5-9093. V.P. was supported by the Italian Ministry for University and Scientific Research through grant 21043088.

References

Duley, W.W., & Williams, D.A. 1984, *Interstellar Chemistry* (Academic Press).

Govers, T.R., Mattera, L. & Scoles, G. 1980, J. Chem. Phys., 72, 5446

Hollenbach, D.J., Werner, M.W., & Salpeter, E.E. 1971, ApJ, 163, 155

Jenniskens, P., Blake, D.F., Wilson, M. A., & Pohorille, A. 1995, ApJ, 455, 389

Katz, N., Furman,I., Biham,O., Pirronello, V., & Vidali, G. 1999, ApJ, 522, 305

Manicò, G., Ragunì, G., Pirronello, V., Roser, J.E., & Vidali, G. 2001, ApJ, 548, L253

Mayer, E., & Pletzer, R. 1986, Nature, 319, 298

O'Neill, P.T., & Williams, D.A. 1999, Ap&SS, 266, 539

Pirronello, V., Liu, C., Shen, L., & Vidali, G. 1997a, ApJ, 475, L69

Pirronello, V., Biham, O., Liu, C., Shen, L., & Vidali, G. 1997b, ApJ, 483, L131

Pirronello, V., Liu, C., Roser, J.E., & Vidali, G. 1999, A&A, 344, 681

Roser, J. E., Manicò, G., Pirronello, V., & Vidali, G. 2002, ApJ, in press

Schutte, W., in *The Cosmic Dust Connection*, M. Greenberg ed. (Dordrecht: Kluwer), 1

Smoluchowski, R. 1983, J. Phys. Chem., 87, 4229

Stancil, P.C. & Dalgarno, A. 1998, *Faraday Discussion*, 109, 61

Vidali, G., Pirronello, V., Liu, C., & Shen, L. 1998, Astrophys. Lett. Comm., 35, 423

Vidali, G., Roser, J. Liu, C., Pirronello, V., Biham, O., 1998b, in *Proceedings of the NASA Laboratory Space Science Workshop*

THE IMPORTANCE OF THE GAS-GRAIN INTERACTION

T.J. Millar

Department of Physics, UMIST, P. O. Box 88, Manchester M60 1QD, U. K.

Abstract

It has been known for about 30 years that the gas-grain interaction in interstellar clouds is responsible for the formation of the simplest molecule in the Universe; namely, H_2. It has also been known for about 30 years that the abundance of HD observed in diffuse clouds is so large that it must be formed in gas phase reactions. As a result of this, and the success of gas phase chemistry in understanding the abundances and presence of molecules in interstellar clouds, it has been a paradigm that gas phase processes determine gas phase composition and abundances. However, in the last decade or so, it has become apparent that the gas-grain interaction can have a significant effect on the physical and chemical evolution of dense interstellar clouds. In this article, I review the evidence for this and show that in regions of low-mass star formation, the interaction can have significant observational consequences. I also discuss how observations of gas in high-mass star-forming regions can be used to probe grain surface processes.

Key words: astrochemistry – molecular processes – ISM: clouds – ISM: dust – ISM: molecules

1. Introduction

The presence of interstellar dust in molecular clouds has been known for about 100 years, initially through its absorption and scattering of starlight. Detailed modelling of the extinction and polarization observed in the interstellar medium has led to the broad identification of the composition and size distribution associated with these solid particles. Interstellar grains are produced primarily in the outflows from late-type stars and fall into two broad compositional classes: silicates, produced in oxygen-rich stars; and carbons, produced in carbon-rich stars. Both classes have been found in amorphous and crystalline forms. The size distribution of interstellar grains can be described by a power law: $n(a)da \propto a^{-3.5}da$, and combining the optical properties of likely interstellar materials with this size distribution leads to the dust cross sections needed to explain the absolute amount of extinction in both the visible and UV parts of the spectrum. The typical integrated cross section of the particles causing extinction in the visible is about $10^{-22}\,n$ cm^2 per unit volume, where n is the number density of hydrogen nuclei per unit volume. The larger extinction observed in the UV implies that these dust grains are smaller and have an integrated surface area about 10 times larger than those dust grains causing the visible extinction.

In addition to these species, there is also much evidence, primarily through the detection of emission bands in the infrared, for the presence of a population of large molecules or small grains, each perhaps consisting of about 50–100 carbon atoms and taking up 10 percent of the total carbon budget. These molecules, called polycyclic aromatic hydrocarbons (PAHs), are thought to be formed in the circumstellar envelopes of carbon-rich objects and ejected into the interstellar medium. If their properties are such that they are photo-resistant to the interstellar UV radiation field, and in this respect some self-selection may be occurring, they can populate interstellar clouds.

Observations of interstellar dust show that it comprises about 1 percent of the mass of interstellar gas, with which it is well mixed. In diffuse clouds, UV absorption line measurements show that many of the elements have abundances which are depleted below their solar or cosmic abundances at levels which are in agreement with the sizes and compositions derived from extinction. In dense clouds, a new form of dust extinction is seen—absorption of infrared radiation by ice-like mantles which have been accreted onto the surfaces of interstellar grains. These mantles are primarily composed of water, but contain substantial components of methanol, carbon monoxide, and carbon dioxide, among others. These mantles can grow on a time scale shorter than the dynamical time scale of the cloud and thus lead to a substantial

Chemistry as a Diagnostic of Star Formation (2003). *Edited by Charles L. Curry and Michel Fich.* Published in 2003 by NRC Research Press, Ottawa ON, K1A 0R6, Canada. ISBN 0-660-19089-3.

reduction in the gas phase abundances of the C, N, O elements. The typical accretion time scale for species X is $t_{\rm ac} = (S_X \pi a^2 n_{\rm d} v_X)^{-1}$, where S_X is the sticking coefficient and is close to unity for most species at very low temperatures, v_X is the relative velocity of species X with respect to the grain, and $\pi a^2 n_{\rm d}$ is the grain cross section per unit volume. For simplicity we take the grains to be spheres of radius a and number density $n_{\rm d}$. In fact, the analysis is independent of the precise geometric form of the grains since the grain cross section can be derived directly from the observations, and has the typical value given above. Thus, we find $t_{\rm ac} = 3 \times 10^9 \, n^{-1}$ yr, where n is the hydrogen number density. In dense clouds, with $n > 10^5$ cm^{-3}, the accretion time scale becomes much shorter than the collapse time scale. Thus, for an interstellar cloud which is collapsing to high density to form a star, we predict that all species which can freeze out onto the surfaces of interstellar grains will indeed do so. The only materials which do not are those which have very low sticking coefficients: H_2 and related ions, and the inert elements, including He. Thus we see that in star-forming regions, there may be clumps of high density gas which are collapsing to form a star but in which the normal molecular tracers used to follow the dynamics are missing. In this article, I discuss molecular line tracers of this high density, cold gas in star forming regions. In particular, I shall argue that deuterium-bearing molecules offer the best possibility to trace the dynamics and history of gas in dense clumps.

2. Grain mantles

Infrared observations of absorption bands clearly show that thick molecular ice mantles grow on grain cores in dense, cold gas. In many lines of sight, and in particular those towards embedded protostars, the column density of a molecule in solid form usually exceeds its column density in the gas phase. This is true even for CO which shows that not only can absolute gas phase abundances be affected by accretion but also overall elemental abundances. If CO, which has a relatively small binding energy, can be heavily depleted from the gas phase, it is likely that almost all other gas phase species which have higher binding energies are also depleted heavily. The presence of particular molecules in grain mantles can be explained either by accretion of the molecule from the gas phase or by synthesis on the surface. In general, the fractional abundances of molecules in ices are larger than those seen in cold, dense gas so that surface formation seems preferred. Given the abundance of atomic hydrogen and the mobility of this atom on a cold surface, synthesis appears to involve the hydrogenation of accreted atoms and radicals. This hydrogenation can, however, be hindered by the presence of activation barriers, such as those which prevent the complete conversion of CO to CH_3OH. Models of surface chemistry remain very uncertain because of the sensitivity of the final abundances to these poorly known barriers. Some of this sensitivity may be removed if the surfaces are 'activated' by interaction with UV photons or cosmic ray particles. Nonetheless, the presence and composition of ice mantles in cold clouds indicate that the gas freezes out onto the dust on relatively short time scales and that some surface processing of the accreted gas does take place.

The interaction of the gas and dust not only leads to depletion of the gas phase but it also affects the remaining gas phase chemistry in ways that may give us observational probes of the depletion phase. The most sensitive means of detecting high density gas involved in the earliest stages of star formation may be to observe deuterium-bearing molecules. I shall show that this is most easily achieved in regions of low-mass star formation. In high-mass star formation regions, the observation of the protostellar phase is more difficult because of the short time scales involved. Here it turns out that one method of studying the protostellar phase is to observe the evaporated products of the ice mantles—that is, grain materials heated by the new star. In this article, I shall discuss how observations of these different regions give information on the importance of the gas-grain interaction.

3. Low-mass star formation

At low temperatures, molecules can be heavily fractionated in deuterium by a number of processes. The most important of the primary fractionation reactions is that between H_3^+ and HD which forms H_2D^+ in a reaction which is exothermic by about 220 K:

$$H_3^+ + HD \longleftrightarrow H_2D^+ + H_2 + \Delta E. \qquad (1)$$

Thus, the back reaction is unimportant at low temperatures; although given the large abundance of H_2 in molecular clouds, the back reaction begins to dominate loss of H_2D^+ once the temperature is larger than about 20 K. One can write the fractionation in H_2D^+ relative to that in the reservoir HD by equating the formation and loss rates of H_2D^+. One finds

$$\frac{[H_2D^+]}{[H_3^+]} = \frac{k_{1\rm f}}{k_{1\rm r} + \alpha_1 f(e) + \Sigma k_M f(M)} \frac{[HD]}{[H_2]},$$

where $f(X)$ is the fractional abundance of X relative to H_2, α_1 is the dissociative recombination coefficient of H_2D^+ with electrons, and k_M is the rate coefficient of H_2D^+ with species M. One can re-write this equation as

$$R(H_2D^+) = S_{\mathrm{H_2D^+}}(T)\, f(HD).$$

Since the rate coefficients, particularly k_{1r} and α_1, are temperature dependent—$k_{1r} = k_{1f}e^{-\Delta E/T}$ and $\alpha_1 \propto T^{-0.5}$—the enhancement factor is a function of temperature. Reaction (1) is a more complex system than discussed here and effects such as rotational excitation and the ortho-para ratio of H_2 can also affect the rate constants. These have been discussed more fully by Roberts et al. (2002).

At the low temperatures and low fractional ionizations of dense clouds, the enhancement factor reduces to

$$S_{\mathrm{H_2D^+}} = \frac{k_{1\mathrm{f}}}{\Sigma k_M f(M)},$$

which is now independent of temperature. With the usual ion-molecule rate coefficients of 10^{-9} cm^3 s^{-1}, one finds $S_{\mathrm{H_2D^+}} = [f(M)]^{-1}$, where $f(M)$ is roughly given by the sum of the fractional abundances of CO and O, $\sim 10^{-4}$, resulting in a fractionation ratio for H_2D^+ of about 0.1. This fractionation of H_2D^+ is passed on to other species through deuteron transfer:

$$H_2D^+ + N_2 \longrightarrow N_2D^+ + H_2 \qquad (2)$$

$$H_2D^+ + N_2 \longrightarrow N_2H^+ + HD, \qquad (3)$$

which happens in one-third of collisions. Thus one would expect the abundance ratio of N_2D^+ relative to N_2H^+ to be one-third that of H_2D^+ to H_3^+; that is, on the order of a few percent.

However, it is clear that if CO and other molecules are depleted from the gas phase so that $f(M)$ decreases, then the enhancement of deuterium in H_2D^+ can be larger than 10^5 and the H_2D^+/H_3^+ abundance ratio can exceed unity. This huge ratio is then reflected in very large ratios in other species. The most comprehensive observational test of this has been performed by Caselli et al. (2002a), who have used observations of DCO^+ and N_2D^+ to probe the ionization fraction within the starless core L1544. In the central dust core of this cloud, observations of $C^{17}O$ indicate that CO is severely depleted, by a factor of 100 or so. In this region, the lack of CO also results in weak HCO^+ and DCO^+ emission, with a DCO^+/HCO^+ ratio of 0.04. However, both N_2H^+ and N_2D^+ are detected in the core with a fractionation ratio of about 0.24, indicating that N_2 is more volatile than CO and that in such regions, estimates of the ionization fraction are better based on measurements of fractionation in N_2D^+ rather than DCO^+.

A number of authors have modelled such highly depleted regions, both analytically (Rodgers & Charnley 2001b) and numerically (Roberts & Millar 2000b; Lis et al. 2002), in an attempt to reproduce the extremely large fractionations detected in multiply deuterated species. In the protobinary source IRAS 16293, the ratio of doubly deuterated formaldehyde to the fully hydrogenated form is 0.40, an enhancement of about 10^9 over the cosmic ratio (Loinard et al. 2000), while ND_3 has an enhancement of about 10^{11} in Barnard 1 and NGC 1333 (Lis et al. 2002; Van der Tak et al. 2002), a truly remarkable result.

The origin of these huge fractionations is uncertain. Roberts & Millar (2000b) and Rodgers & Charnley (2001b) argued that they occur through gas phase chemistry in depleted gas. In this scenario, the molecules are heavily fractionated and observed before they are frozen onto dust grain surfaces. Rodgers & Charnley (2001b) argued that a factor of two depletion of CO can give rise to substantial fractionation and hence efficient multiple deuteration. Another interpretation comes from Loinard et al. (2000), who find that jumps in the absolute abundances of D_2CO and H_2CO towards IRAS 16293 are indicative of material released from the grains into the gas phase. Since the depletion model fractionates most efficiently when the absolute abundances of molecules are small, high gas phase fractionation occurs in molecules with low absolute abundances. Thus, when these molecules are eventually frozen out onto the dust, the high fractionation ratio is reduced as the accreted gas is diluted by the large abundance of the hydrogenated parent, which had frozen out with a lower fractionation. This is certainly true if depletions of an order of magnitude or more are required to produce high fractionation but this is not necessarily the case (Roberts & Millar 2000b; Rodgers & Charnley 2001b).

Furthermore, there are extremely cold clouds with little evidence of star formation which also contain doubly deuterated molecules. The cold dust cloud L134N is one such example. It contains NHD_2 (Roueff et al. 2000) at an abundance ratio of 0.005. Since L134N does not contain an embedded star and has a kinetic temperature of only about 10 K, there is no mechanism for returning grain mantle material to the gas. Hence the material seen in the gas has not yet interacted with the dust.

Figure 1 shows the calculated fractionation for selected species both with and without accretion of material onto the grains (Roberts & Millar 2000a). The left-hand panel shows the fractionation when accretion is ignored. The fractionation in H_2D^+ reaches about 10%, singly deuterated species, such as HDCO and NH_2D, reach a

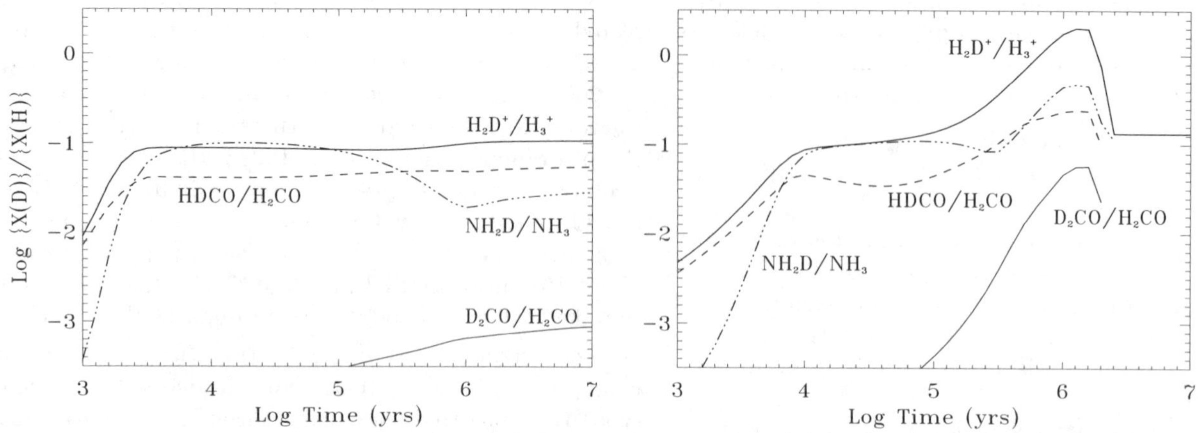

Figure 1. A comparison of selected molecular D/H ratios predicted by our gas phase (*left*) and depletion (*right*) models. $T_{kin} = 10$ K, $n(H_2) = 10^4$ cm^{-3} (Roberts & Millar 2000a).

level of a few percent, while doubly-deuterated species reach 0.1% at maximum. The right-hand panel shows the effect of including grain accretion using standard parameters, which deplete the gas entirely on a time scale of a few million years. Accretion begins to remove material to a significant extent at around a few times 10^5 yr and, as a result, the fractionation in H_2D^+ becomes larger than 1. This large enhancement is transferred to those species which still remain in the gas and causes large enhancements in both singly and multiply deuterated species.

Roberts & Millar (2000b) included a number of doubly deuterated molecules in a quantitative model of fractionation in cold clouds and Table 1 shows a comparison between the model results and the abundance ratios observed toward the cold, dark cloud L134N. We note that good agreement between the model and observations occurs at a time of about 10^5 yr in the accretion model. No such agreement can be found, at any time, in the model which neglects accretion. We note that, in L134N, the fractionations in N_2D^+ and DCO^+, which are both larger than 10%, are much greater than that of NHD_2, which is about 0.5%. This is the reverse of what is found in IRAS 16293, which contains an embedded protostar. Here, the abundance ratio of D_2CO to H_2CO is about 40%, much larger than seen in the ions. It is worth stressing that this difference could be an important indicator of whether or not grains can contribute actively to fractionation. Since grains cannot produce *fractionated ions*, these species must arise from gas phase chemistry. When the fractionation in N_2D^+ is seen to be large, then it implies that the fractionation of H_2D^+ is also large, since N_2D^+ can only be formed by deuteron transfer from the primary ion H_2D^+. Conversely, a low fractionation in ions with a large fractionation in neutrals, as seen in IRAS 16293, may indicate that grains are synthesising deuterated molecules. One should note that such a synthesis is likely to be a selective process. For example, Roberts & Millar (2000b) predicted the presence of D_2O in L134N at a relatively high fractionation but the species has not yet been detected in interstellar clouds, despite sensitive searches.

The most recent detection of a multiply deuterated species is CHD_2OH (Parise et al. 2002) with an abundance ratio of 0.2 in IRAS 16293 relative to the methanol abundance. Indeed, the total column density of deuterated methanol, in all forms, exceeds the column density of normal methanol in this source. Parise et al. (2002) discuss the formation of methanol via surface hydrogenation and deuteration of CO which will also produce H_2CO, HDCO and D_2CO as intermediate products. However, they find that reproduction of the large fractionation in doubly deuterated methanol produces too much CH_3OD and HDCO. In addition, the large fractionation requires a large atomic D to atomic H ratio, around 0.2–0.3, since it is reactions of D atoms on the surface which drive the synthesis of CHD_2OH. Such a large ratio is not predicted by the models (Roberts & Millar 2000a). It is possible that the inclusion of realistic activation energy barriers for some, but not all, of the surface reactions could reproduce the observed fractionations in both methanol and formaldehyde with a lower D/H abundance ratio. A preliminary calculation along these lines has been done (Caselli et al. 2002b), but full agreement with the observed ratios is still missing.

Table 1. A comparison of abundance ratios measured in L134N (Tiné et al. 2000; Roueff et al. 2000) with predictions from models at 10 K, $n(H_2) = 5\times10^4$ cm^{-3} (Roberts & Millar 2000b)

Ratio	Observed	Modelled		
		gas phase		accretion
		10^5 yr	10^7 yr	10^5 yr
DCO^+/HCO^+	1.8(-1)	1.6(-2)	4.7(-2)	1.6(-1)
DCO^+/CO	7.7(-6)	1.1(-7)	2.5(-7)	2.5(-5)
NH_2D/NH_3	1.0(-1)	7.5(-2)	2.0(-2)	1.6(-1)
NH_2D/CO	1.5(-4)	3.5(-5)	4.4(-6)	1.2(-4)
N_2D^+/N_2H^+	3.5(-1)	2.3(-2)	3.9(-2)	1.8(-1)
N_2D^+/CO	3.5(-6)	5.8(-9)	1.1(-8)	2.7(-6)
NHD_2/NH_3	5.0(-3)	1.3(-4)	8.4(-5)	4.1(-3)
NHD_2/CO	7.5(-6)	6.3(-8)	1.8(-8)	3.1(-6)

NOTE: $a(-b)$ signifies $a \times 10^{-b}$.

Surveys of D_2CO are now underway (Loinard et al. 2002). It appears that D_2CO is readily observed in many low-mass star-forming clouds but not in high-mass star-forming regions. Figures 2 and 3 show JCMT spectra at 231 GHz towards L1448mms and L1527, respectively (Roberts & Millar 2003, in preparation). The observed band includes the 4(0,4)–3(0,3) line of D_2CO and the 3–2 line of N_2D^+. Clearly the latter is missing in L1527 despite being very strong in L1448mms; D_2CO appears in both spectra. One explanation for the absence of N_2D^+ is that the N_2 molecule is completely depleted from the gas. However, it is then difficult to envisage why D_2CO, which has a higher binding energy than N_2, is not also missing from the gas.

A second possibility is that N_2D^+ is destroyed more rapidly than D_2CO in some clouds. Since ions are destroyed by dissociative electron recombination, a route not available to neutrals, then clouds in which the electron fraction is large could be depleted in N_2D^+. However, since the primary deuteration reactions all involve molecular ions, such as H_3^+ and H_2D^+, a high electron fraction will also destroy these ions and depress the level of fractionation which can be transferred to other species (Millar 2002). Hence, the D_2CO observed would need to be made on the grains and returned to the gas—difficult if the cloud does not contain a protostar.

Finally, a third mechanism to decrease fractionation occurs via the ion-neutral drift caused by Alfvén wave propagation in cold, dark clouds (Markwick et al. 2001). Here, small drift velocities can provide kinetic energy which drives slightly endothermic reactions, such as the reverse of reaction (1), and hence reduce fractionation. Similarly, it is possible that wave action also leads to some desorption of mantle material in cold, starless cores (Markwick et al. 2000). Such wave interactions reduce the gas phase fractionations and release deuterated molecules from the grains to the gas. Markwick et al. (2001) have shown that such waves can explain gradients in fractionation measured in several molecules along the ridge in the dark cloud TMC-1. Deuteration is particularly sensitive to nonthermal effects since the exothermicity of reaction (1) is only around 220 K; that is, modest ion-neutral kinetic effects can drive the reverse of reaction (1) and reduce fractionation. Because of the very low proton affinity of N_2, N_2D^+ can only be produced from H_2D^+, so that its fractionation is very sensitive to nonthermal effects. Formaldehyde, however, is made from CH_3, which derives its fractionation from the methyl ion, CH_3^+, another ion which is fractionated directly by reaction with HD. The endothermicity of the relevant back reaction is around 370 K, so that nonthermal, kinetic effects do not destroy CH_2D^+ as readily as they do H_2D^+. Hence, deuterated formaldehyde can be present in regions in which N_2D^+ is absent.

4. HIGH-MASS STAR FORMATION

Regions of high-mass star formation show levels of fractionation that are, in general, about an order of magnitude less than those in low-mass star-forming regions. This is not simply due to the higher kinetic temperatures of the former, since fractionation of about 0.1% or so is seen in gas with a temperature in excess of 100 K, a temperature at which fractionation is inefficient. Indeed, it is found that, with the exception of the nearby Orion Molecular Cloud, deuterated molecules are difficult to detect in such regions.

Hot molecular cores (HMCs)—small ($<$ 0.1 pc), hot (T $\sim$ 100–300 K), dense (n[H_2] $\sim 10^7 - 10^9$ cm^{-3}) clumps of gas—are often associated with high-mass star formation. They typically have very rich submillimeter line spectra, with tens of lines per GHz and contain large abundances of complex molecules such as methanol, ethanol and dimethyl ether. These also contain abundances of small hydrogenated species such as water, ammonia and hydrogen sulphide which are orders of magnitude larger than those in low-mass star-forming regions. The general consensus is that the gas phase composition of HMCs is determined by the evaporation of primary species from grain ice mantles together with secondary species formed in the hot gas from these primary molecules. Since the D/H abundance ratios observed in

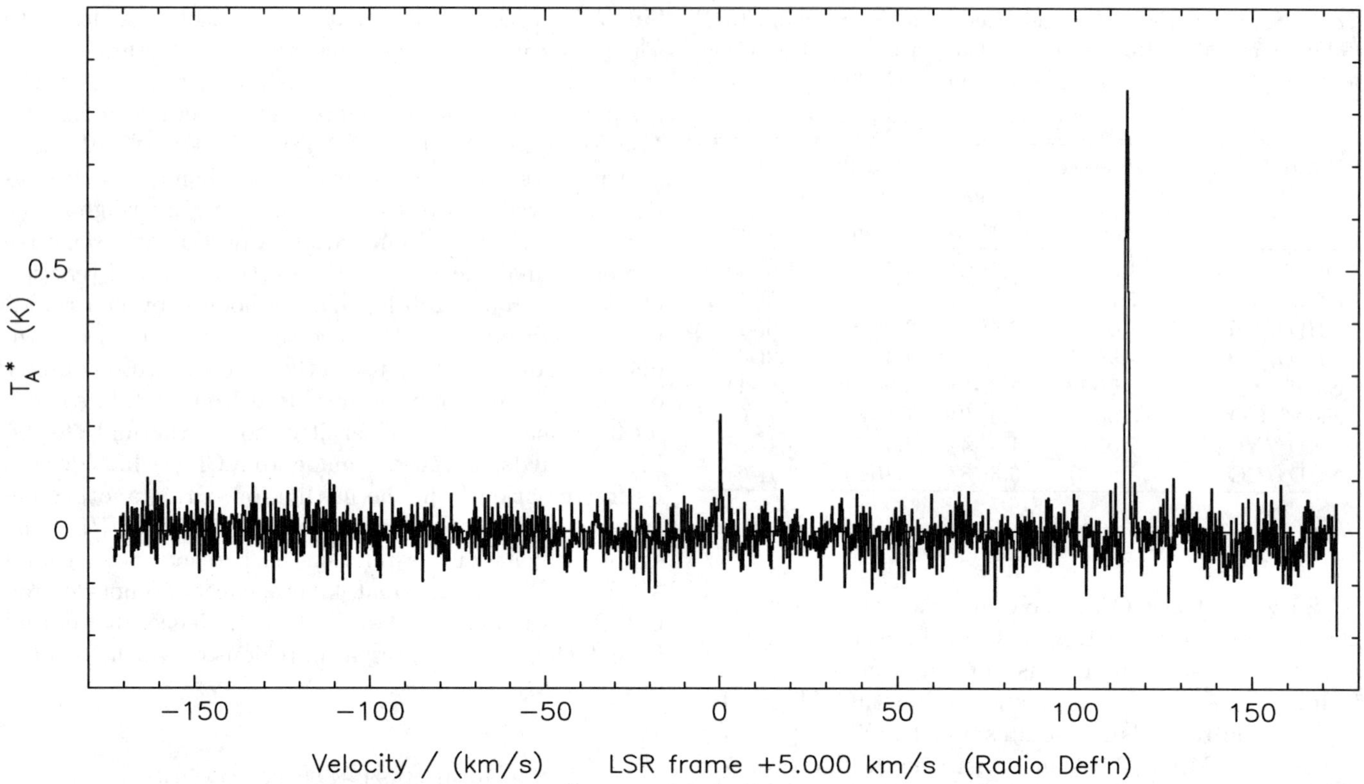

Figure 2. The 231 GHz spectrum of L1448mms taken with the JCMT. The two lines detected are D_2CO (at 231.41 GHz, v_{LSR} = 0 km s^{-1}) and N_2D^+ (at 231.32 GHz, $v_{LSR} \sim 115$ km s^{-1}).

molecules in HMCs cannot be produced at the high temperatures observed, it is believed that the ratios are set in a previous colder phase; that is, before the gas was warmed by the presence of a nearby massive star. These ratios, whether set primarily in gas phase or grain surface reactions, are then stored in grain mantles prior to evaporation.

It is therefore important to know which of the three synthetic pathways are contributing to molecule formation: gas phase, pre-accretion chemistry, in which case D/H ratios may give information on physical parameters; grain surface reactions, in which case composition and D/H ratios can inform us of the detailed surface chemistry; and post-evaporation chemistry, in which case observation may be interpreted as a clock to give an evolutionary probe of high-mass star formation.

Let us consider gas phase, pre-accretion chemistry. In low-mass star-forming regions, the low (10 K) gas temperature leads to very high fractionation and to a poor degree of hydrogenation in the observed molecules—gas phase reactions involving the abundant H_2 molecule are often endoergic, so that highly unsaturated carbon chains are readily observed. Thus, for example, in cold clouds the maximum fractional abundance of methanol is about 10^{-8}, ammonia about 3×10^{-8} and ethanol less than 10^{-11}, with associated D/H ratios of a few percent. Contrast these values to those in HMCs, which are about 10^{-6}, 10^{-5}, 10^{-7} and 0.1 percent, respectively. Thus, we conclude that the observed abundances in cold clouds are not preserved in HMCs. Indeed, although the observed D/H ratios could be matched if the gas temperature was about 50 K, the degree of hydrogenation observed implies that the major synthesis of molecules takes place on the grain surface.

Since a significant fraction of elemental C, N, O is observed in the form of methane, ammonia, and water, surface hydrogenation must be very efficient. Such reactions could contribute to the formation of larger, saturated molecules such as methanol, ethanol and dimethyl ether, but the details are unknown. Some simple models have been produced which relate the ease of hydrogenation to the availability and mobility of atomic hydrogen on the grain surface. One would then expect D atom re-

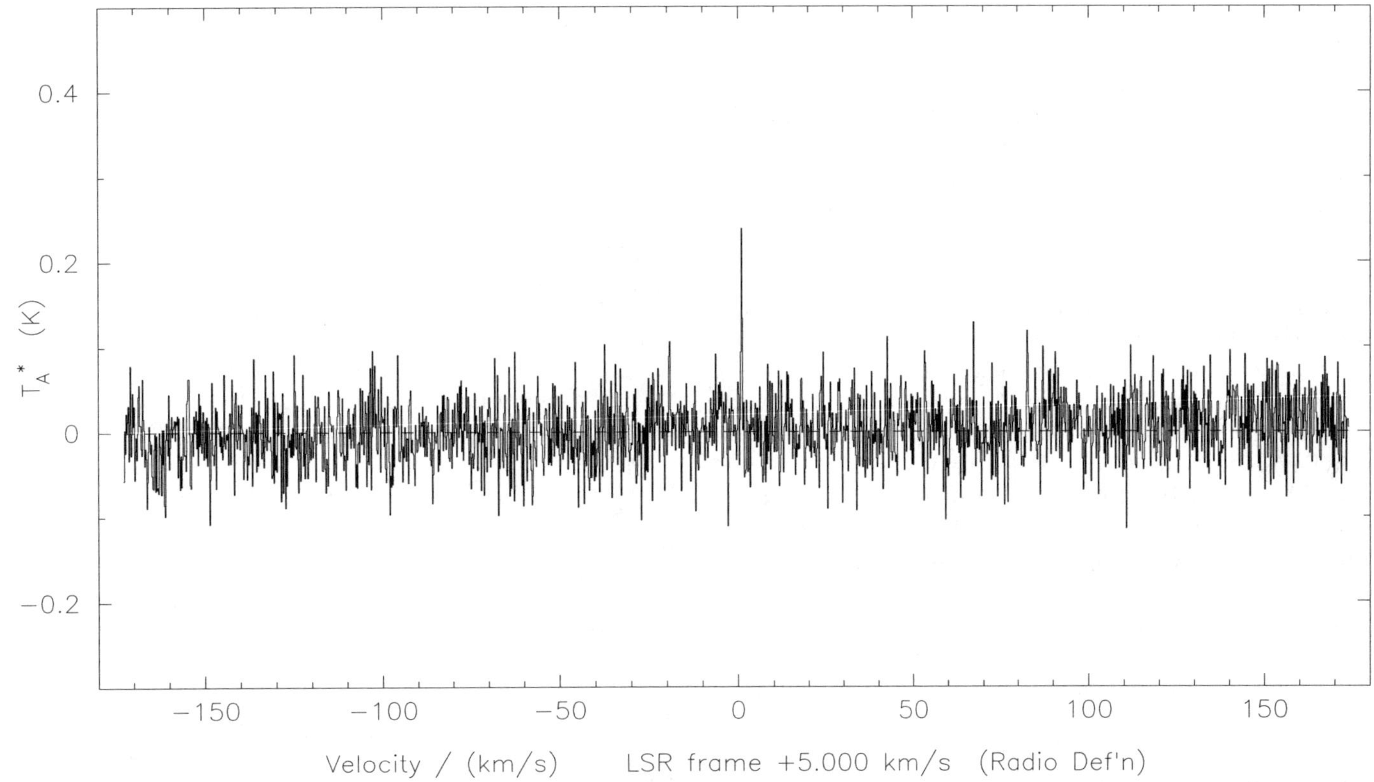

Figure 3. The 231 GHz spectrum of L1527 taken with the JCMT. The detected line is D_2CO (at 231.41 GHz).

actions to produce deuterated species. Brown and Millar (1989a,b) and Rodgers & Charnley (2001b) have used these simple surface chemistries to predict D/H ratios in NH_2D, NHD_2 and ND_3. To date, however, no one has determined these ratios in the same object. In broad terms, one expects that any molecular D/H ratios set by solid state chemistry would reflect the atomic D/H ratio in the ice—and therefore the atomic D/H ratio set in the gas—since the atoms accrete on the grains through collisions between the gas and dust. The lower fractionation observed in HMCs then implies that the D/H ratio is set at a temperature of about 40−60 K, and that the grain ices form at this temperature rather than at 10 K.

Finally, we discuss briefly the chemistry of the post-evaporation phase. The bulk of the molecules evaporated from the ices are stable, unreactive species such as H_2O, CO, CH_4 and so on. As a result, despite the relatively high temperature of the gas, HMCs are inert on a short time scale. Chemical reactions do eventually occur, on a time scale of 10^3−10^4 yr, driven by cosmic rays and cosmic ray-induced photons. The consequences of this hot chemistry, the production of secondary species, and their uses as clocks of the chemical evolution have been discussed by a number of authors (Charnley et al. 1992; Charnley et al. 1995; Charnley et al. 1997; Hatchell et al. 1998; Rodgers & Charnley 2001a).

5. Conclusions

The interaction between gas and dust is important in star-forming regions. This interaction removes the gas rapidly at high density and low temperature, thus making it difficult to observe clouds in the final stages of prestellar collapse. It is clear that even CO can be significantly depleted in such regions. Depleted regions do have an important observational tracer, namely high fractionation in deuterium. This fractionation is most clearly observed in N_2D^+ which can only be produced in gas phase reactions involving H_2D^+. Fractionation ratios in such cases can approach unity and may lead to large fractionation and multiple deuteration in other species. Whereas ions can only be fractionated in gas phase reactions, neutrals can be fractionated also on grain surfaces. However, difficulties in modelling such processes have meant that quantitative agreement with the observations is not yet possible.

In regions of massive star formation, and particularly in hot molecular cores, both the absolute abundances and the deuterium fractionations observed imply that an active surface chemistry has occurred. It appears that the gas which was incorporated into the grain mantles was frozen out at higher temperatures than the ices in low-mass star-forming regions. Many complex molecules detected in HMCs can be produced in the post-evaporative chemistry although the very large fractional abundances of some molecules, such as ethanol, likely imply that some complex species are produced directly by chemical reactions in the ices; see, e.g., Ikeda et al. (2001). These authors have surveyed the abundances of several complex molecules, including ethanol, dimethyl ether, ethyl cyanide (C_2H_5CN), ethylene oxide (c-C_2H_4O), and acetaldehyde (CH_3CHO) in several hot cores. Through comparing abundances and derived rotational temperatures with observed properties such as gas and dust temperatures and L_{IR}/M, a measure of star formation per unit mass, they have been able to show that the abundances of the former three increase with dust—but not gas—temperature, indicating that these species may be the products of grain mantle evaporation. While the excitation temperatures of ethylene oxide and acetaldehyde are low, their abundances correlate with that of ethanol, suggesting that they are the secondary products produced in reactions involving ethanol upon desorption from the grains. Observations such as these are key to identifying the parent mantle products from those which form in the gas phase and will help constrain theoretical models of grain surface chemistry. Progress in understanding surface chemistry is, in the author's view, more likely to come from observational studies of complex molecules in hot cores and deuterated species in regions of low-mass star formation than from laboratory experiments.

ACKNOWLEDGEMENTS
Astrophysics at UMIST is supported by a grant from PPARC.

REFERENCES

Brown, P.D., & Millar, T.J. 1989a, MNRAS, 237, 661

Brown, P.D., & Millar, T.J. 1989b, MNRAS, 240, 25P

Caselli, P., Walmsley, C. M., Zucconi, A., Tafalla, M., Dore, L., & Myers, P. C. 2002a, ApJ, 565, 344

Caselli, P., Stantcheva, T., Shalabiea, O., Shematovich, V. I., & Herbst, E. 2002b, Plan. Sp. Sci., 50, 1257

Charnley, S. B., Kress, M. E., Tielens, A. G. G. M., & Millar, T. J. 1995, ApJ, 448, 232

Charnley, S. B., Tielens, A. G. G. M., & Millar, T. J. 1992, ApJ, 399, L71

Charnley, S. B., Tielens, A. G. G. M., & Rodgers, S. D. 1997, ApJ, 482, L203

Hatchell, J., Thompson, M. A., Millar, T. J., & Macdonald, G. H. 1998, A&A, 338, 713

Ikeda, M., Ohishi, M., Nummelin, A., Dickens, J. E., Bergman, P., Hjalmarson, Å., & Irvine, W. M. 2001, ApJ, 560, 792

Lis, D. C., Roueff, E., Gerin, M., Phillips, T. G., Coudert, L. H., van der Tak, F. F. S., & Schilke, P. 2002, ApJ, 571, L55

Loinard, L., Castets, A., Ceccarelli, C., Tielens, A. G. G. M., Faure, A., Caux, E., & Duvert, G. 2000, A&A, 359, 1169

Loinard, L., Castets, A., Ceccarelli, C., Lefloch, B., Benayoun, J.-J., Caux, E., Vastel, Ch., Dartois, E., & Tielens, A. G. G. M. 2000, Plan. Sp. Sci., 50, 1205

Markwick, A. J., Millar, T. J., & Charnley, S. B. 2000, ApJ, 535, 256

Markwick, A. J., Charnley, S. B., & Millar, T. J. 2001, A&A, 376, 1054

Millar, T. J. 2002, Plan. Sp. Sci., 50, 1189

Parise, B., Ceccarelli, C., Tielens, A. G. G. M., Herbst, E., Lefloch, B., Caux, E., Castets, A., Mukhopadhyay, I., Pagani, L., & Loinard, L. 2002, A&A, 393, L49

Roberts, H., Herbst, E., & Millar, T. J. 2002, MNRAS, 336, 283

Roberts, H., & Millar, T. J. 2000a, A&A, 361, 388

Roberts, H., & Millar, T. J. 2000b, A&A, 364, 780

Rodgers, S. D., & Charnley, S. C. 2001a, ApJ, 546, 324

Rodgers, S. D., & Charnley, S. C. 2001b, ApJ, 553, 613

Roueff, E., Tiné, S., Coudert, L. H., Pineau des Fôrets, G., Falgarone, E., & Gerin, M. 2000, A&A, 354, L63

Tiné, S., Roueff, E., Falgarone, E., Gerin, M., & Pineau des Fôrets, G. 2000, A&A, 356, 1039

Van der Tak, F. F. S., Schilke, P., Müller, H. S. P., Lis, D. C., Phillips, T. G., Gerin, M., & Roueff, E. 2002, A&A, 388, L53

PAHS AND H_2: A CHEMICAL COUPLING? OBSERVATIONAL AND LABORATORY INSIGHTS

C. Joblin

CESR-CNRS, 9 Av. du Colonel Roche, 31028 Toulouse Cedex 04, France

ABSTRACT

Molecular hydrogen is the most abundant interstellar molecule. Still, its formation mechanism remains an open question, although it is clear that it must involve dust. Polycylic aromatic hydrocarbons (PAHs) are an abundant dust component and it is therefore of interest to evaluate their contribution to the formation of H_2. This paper discusses observational facts and laboratory experiments that are dedicated to these studies. The topic has gained importance recently due to the spatial correlation between PAH and H_2 vibrational infrared emissions that is observed in several photodissociation regions (PDRs), including massive star formation regions such as S106 IR (Joblin et al., in preparation), but excluding the prototype PDR of the Orion Bar. In a PDR model, these differences can be accounted for by differences in the physical conditions, the spatial correlation being best observed at $n/\chi \sim 10^2$ with n the gas density and χ the UV intensity scaling factor. These conditions may also correspond to those necessary for the catalytic production of H_2 by PAHs. Laboratory studies, which are in progress to address this question, are presented. They make use of the new experimental set-up PIRENEA, wherein the photophysics and chemistry of interstellar PAH candidates can be studied in physical conditions that approach those found in interstellar space. Recent results on the PAH cation $C_{24}H_{12}^+$ show that upon continuous UV-visible irradiation, the dehydrogenation proceeds mainly by loss of H atoms, all the way down to the bare carbonaceous cluster C_{24}^+ (Joblin et al., in prep.). A first attempt to measure the reactivity of the dehydrogenated species $C_{24}H_6^+$ with H_2 was found to be unproductive.

Key words: astrochemistry – molecular processes – ISM: molecules

Chemistry as a Diagnostic of Star Formation (2003). *Edited by Charles L. Curry and Michel Fich.* Published in 2003 by NRC Research Press, Ottawa ON, K1A 0R6, Canada. ISBN 0-660-19089-3.

H_3^+ IN DARK CLOUDS

T. R. Geballe[1], B. J. McCall[2], and T. Oka[3]

[1]Gemini Observatory, 670 N. A'ohoku Pl., Hilo, HI 96720, USA
[2]Department of Chemistry and Department of Astronomy, Univ. of California, Berkeley, CA 94720
[3]Department of Astronomy and Astrophysics, Department of Chemistry, and the Enrico Fermi Institute, The University of Chicago, Chicago, IL 60637, U.S.A.

Abstract

H_3^+, long thought to be the instigator of most interstellar gas phase chemistry, is now detected in about ten dark clouds, in all cases in absorption toward embedded young stellar objects. Analysis yields abundances consistent with predictions from known cloud parameters, thus providing a basic confirmation of the theory of cosmic ray-induced ion-molecule interstellar chemistry. The special characteristics of H_3^+ allow it to serve as a yardstick for measuring cloud dimensions and as a cosmic ray fluxmeter, in addition to being a simple thermometer in star forming regions. We present and discuss observational results in the light of recent models of dark clouds.

Key words: ISM: molecules – ISM: structure – ISM: clouds

1. Introduction

The existence of H_3^+ in the interstellar medium was predicted over forty years ago. About thirty years ago Herbst and Klemperer's (1973) and Watson's (1973) proposal that ion-molecule reaction networks result in production of many of the interstellar molecules that had then been discovered placed H_3^+ at the starting point of these networks.

Since then the importance of ion-molecule reactions in interstellar chemistry has been verified by countless observational and theoretical results. However, it was only after the detection of H_3^+ in two dark clouds (Geballe & Oka 1996), and subsequent detections in other dark clouds (McCall et al. 1999), that there was direct confirmation of the most fundamental premise of the hypothesis, that H_3^+ is the starting point of interstellar ion-molecule chemistry. A basic understanding is now developing of how observations of H_3^+ can be used in combination with observations of other molecules to learn more about the physical characteristics of dark clouds. In this paper we review the properties of H_3^+ and the unique ways in which it gives information about dark clouds, and then discuss recent observations of H_3^+ and their use.

2. The Significance of H_3^+ Measurements

There are two reasons why H_3^+ is fundamentally important in interstellar gas phase chemistry. First, it is generally true that in cold dark clouds ion-neutral chemistry is much faster than neutral chemistry. Second, the rate at which H_3^+ is produced from neutral material in dark clouds is far higher than the production rate of any other ionic species. This is because the origin of H_3^+ is the interaction between cosmic rays, which enter the cloud, and hydrogen molecules, which are by far the most abundant species in dark clouds and are the species with which cosmic rays interact most frequently. The cosmic rays readily ionize H_2 to H_2^+, which subsequently and quickly reacts with H_2 to form H_3^+. While the lifetime of a neutral H_2 molecule in a typical dark cloud is a billion years before it is ionized, in a typical molecular cloud the conversion to H_3^+ of that H_2^+ ion by interaction with H_2 requires only about a day. Hence, the production rate of H_3^+ may be written simply as $\zeta n(H_2)$, where ζ, the cosmic ray ionization rate, is generally thought to be $\sim 3\times10^{-17}$ sec^{-1}.

The destruction of H_3^+ in the cloud takes place whenever it interacts either with neutrals (other than H, H_2, He, N, and a few additional species), or with negatively charged species. The destruction rate can be written as:

$$R_d(H_3^+) = k_{CO}n(CO) + k_O n(O) + \ldots + k_e n(e) + \ldots .$$

This sum is dominated by the first term if CO is not heavily depleted. All of the other terms except that of atomic oxygen are 1–2 orders of magnitude smaller. Some have large reaction constants but small abundances, such as electrons; others, such as atomic oxygen, have more comparable abundances but lower reaction rates (see McCall 2001). When CO dominates, its constant abundance relative to H_2 over a wide range of conditions (Lee et al. 1996) results in a constant H_3^+ density of about 1×10^{-4} cm^{-3}, independent of the density of the cloud. Thus, the column

Chemistry as a Diagnostic of Star Formation (2003). *Edited by Charles L. Curry and Michel Fich.* Published in 2003 by NRC Research Press, Ottawa ON, K1A 0R6, Canada. ISBN 0-660-19089-3.

density of H_3^+ in a dark cloud directly yields the dimension of the cloud along the line of sight. That information cannot be obtained from other molecules so clearly; in combination with other molecular tracers it can yield accurate information about cloud densities.

In cool portions of dark clouds, where temperatures are below 20 K, gaseous CO will freeze onto grains. In those regions other predator molecules and atoms dominate the destruction, and the density of H_3^+ is expected to increase to well above 1×10^{-4} cm^{-3}, until the sum of their rates equalizes the production rate of H_3^+. Whether such regions dominate the H_3^+ column densities of some dark clouds depends on the relative column lengths of the warmer and cooler regions.

Finally, we note the ease, in principle, both of using H_3^+ to determine the cloud temperature and of measuring the column density of this molecular ion. The two lowest rotational levels of H_3^+, (J,K) = (1,1) and (1,0), are at much lower energy than all other levels (note that (0,0) is forbidden by the Pauli principle). That means that they are the only significantly populated levels in dark clouds, unless the cloud temperature is above 100 K. The only clouds observed to date where levels other than these two are significantly populated are in the galactic center (Goto et al. 2002). The lower lying (1,1) level corresponds to para-H_3^+, the other to ortho-H_3^+. Unlike molecules (such as NH_3) where ortho and para populations have independent existences, collisions of H_3^+ with H_2 result in proton hops or exchanges, both of which are opportunities for changes in nuclear spin modifications. Hence the two levels are in LTE, and measurements of the strengths of a single line from each level results in determinations of both the temperature and the total H_3^+ column density. The energy separation of these low-lying ortho and para levels is 33 K, comparable to typical cloud temperatures, making the relative strengths of lines from these levels an ideal thermometer. Fortuitously, a vibration-rotation doublet consisting of one line from each level is located near 3.67 μm and both lines have been observed with the echelle spectrographs built for UKIRT and for NOAO telescopes.

3. Basic Observational Results

H_3^+ was first detected in dark clouds by Geballe & Oka (1996) using the above doublet, toward the embedded sources W33A and AFGL 2136. Their spectra illustrate the difficulty of detecting H_3^+; the absorption lines are usually very weak, and careful attention must be given to correcting for nearby strong telluric absorption lines and instrumental effects. Recent detections of the doublet toward W51 IRS2 and W40 IRS1a are shown in Figure 1. Toward W51 IRS2 two velocity components are seen in each line. Table 1 lists all detections reported in the literature.

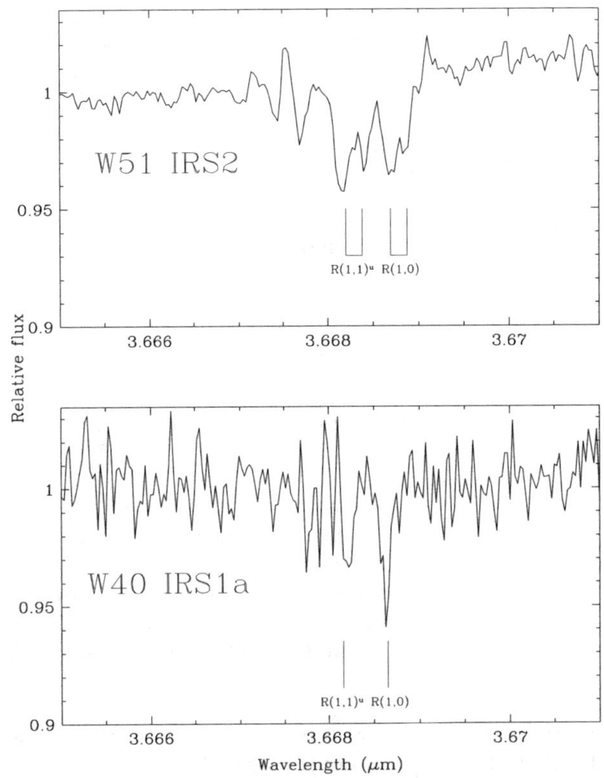

Figure 1. CGS4 (UKIRT) spectra of W51 IRS2 and W40 IRS1a at the wavelength of the H_3^+ ortho-para doublet, observed at UKIRT on 2001 May 25-26 UT.

The temperatures derived from the H_3^+ lines generally are well above the CO freezing point of about 20 K. Thus, although solid CO has been detected along the line of sight to some of these clouds, one can safely conclude that the bulk of the H_3^+ in each cloud is located in regions where CO is predominantly gaseous. The high temperatures indicate that if ζ is 3×10^{-17} s^{-1}, the densities of H_3^+ should be close to the canonical dark cloud value of 1×10^{-4} cm^{-3}. Using this value and the measured column densities, cloud dimensions of typically 1 parsec are derived (although there is considerable variation), roughly similar to the dimensions of the cloud cores on the plane of the sky. Moreover cloud densities, estimated from the above column lengths and the column densities of CO translated into H_2 column densities, are mostly in the

Table 1. Detections of H_3^+ in molecular clouds

Name	$N_{tot}(H_3^+)$ 10^{14}cm^{-2}	ζL^a cm s^{-1}	L^b pc	$N(H_2)^c$ 10^{22}cm^{-2}	$< n(H_2) >$ 10^4cm^{-3}	T^d K
AFGL 2136	3.8	110	1.3	22	6	47
W33A	5.2	150	1.7	27	5	36
MonR2 IRS3	1.4	41	0.5	1.3	9	31
AFGL 961E	1.7	49	0.6	3.2	2	25
AFGL 490	1.1	32	0.4	7.6	6	26
AFGL 2591	2.2	64	0.7	13	6	38
NGC 2024 IRS2[e]	0.4	12	0.13	3.5	7	32
W51 IRS2 (46 km s^{-1})	5.3	155	1.7	–	–	32
W51 IRS2 (62 km s^{-1})	2.4	70	0.8	–	–	39
W40 IRS1a	3.3	99	1.1	0.7	0.2	64

[a]Assumes $n(H_3^+) = 1 \times 10^{-4}$ cm^{-3}.
[b]Assumes $\zeta = 3 \times 10^{-17}$ s^{-1}.
[c]Derived from measured CO column densities (see McCall et al. 1999 for references for first seven sources and Shuping et al. 1999 for W40 IRS1a), using $n(CO)/n(H_2) = 1.5 \times 10^{-4}$ (Lee et al. 1996).
[d]From ratio of ortho and para H_3^+ line intensities.
[e]H_3^+ and H_2 data from Kulesa & Black (2000).

range 10^4–10^5 cm^{-3}, as expected based on other data. Both suggest that the abundance of H_3^+ is roughly as predicted for cosmic ray-induced ion–molecule chemistry, confirming the fundamental importance of H_3^+ in molecular cloud gas phase chemistry.

4. H_3^+ and Cloud Models

A measurement of N(H_3^+) gives the absorption path length times the cosmic ray ionization rate and, if the latter is assumed, then the former can be determined. In the hypothetical case of a cloud of constant density $n(H_2)$, the strength of an optically thin H_2 or CO absorption line is proportional to $n(H_2)L$, and one can determine the density accurately since L is determined from H_3^+. In reality of course, clouds are not of constant density, cloud chemistry is density-dependent, and the information provided by H_3^+ is only one piece of a complex puzzle.

Recently the Leiden University group has begun to develop thermal and chemical models of the envelopes of some of the same star forming regions that have been observed in H_3^+. In these models they constrain the density structure from measurements of the submillimeter continuum and a variety of lines. To date they have published one paper (Doty et al. 2002), modeling the envelope around AFGL 2591, and fitting a number of observations of the cloud using a cloud density decreasing as r^{-1}, with the cloud truncated at $n = 5 \times 10^4$ cm^{-3}. For H_3^+ they calculate about a factor of five less H_3^+ than is observed, using a cosmic ray ionization rate (i.e., an H_3^+ production rate) of almost twice the value adopted here. As the dark cloud would still be dark at much lower densities than 5×10^4 cm^{-3}, a lower truncation density would result in a considerably longer column and an H_3^+ column density closer to that observed. However, in their model of AFGL 2591 an extended lower density envelope would have a lower temperature, leading to a cooler average temperature than found for this cloud using the H_3^+ thermometer.

The discrepancy might be due to incomplete modeling of the outer envelope of the cloud, but it could also be due to significant additional contributions to the H_3^+ absorption from other material along the line of sight. It is possible that the observed H_3^+ absorption lines, which are fairly broad, arise in more than one discrete dark cloud. Indeed millimeter CO measurements (van der Tak et al. 1999) reveal more than one cloud along the line of sight. In addition, absorption may arise partly in the outflow, a considerable part of which lies along the line of sight to the young stellar object, judging from the relatively small displacement of the red- and blue-shifted wings of millimeter CO lines (e.g., Mitchell, Hasegawa & Schella 1992) and the extreme breadth of the infrared CO profiles (Geballe & Wade 1985; Mitchell et al. 1989). Correcting

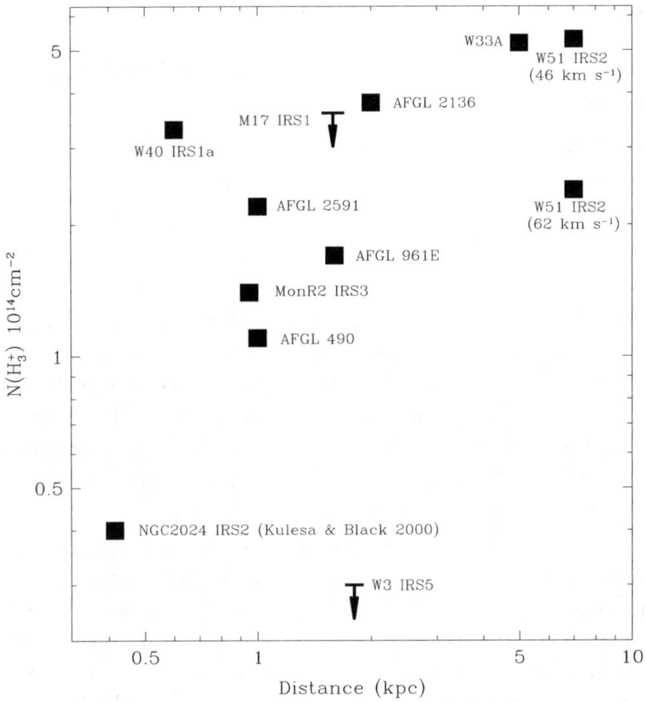

Figure 2. Column densities of H_3^+ in molecular clouds versus distances of sources. H_3^+ data are from McCall et al. (1999) or this paper, except where noted.

for these contributions could bring the observed H_3^+ column density and the model value into somewhat better agreement.

It also has been suggested by van der Tak & van Dishoeck (2000) that a diffuse or translucent cloud component could be responsible for a significant fraction of the H_3^+ absorption column density toward distant galactic sources. They noted a trend of increasing H_3^+ column density with increasing source distance. In Figure 2 we have updated their Fig. 3, adding the latest detections and removing their galactic center and Cyg OB2 No. 12 data points (as the former is a special case and the latter is not a dark cloud). A trend may still exist but it is much less apparent, particularly for the sources within 2 kpc (which include AFGL 2591). We also note that selection effects probably are important, as the number of sources at large distances is small and some of them were chosen because there were reasons to believe that they were particularly excellent possibilities for detecting H_3^+. That was not necessarily the case for the nearby sources, which often were selected because of their brightness as opposed to their being the very best H_3^+ candidates. The more distant sources tend to be more luminous objects, and could be more massive young stellar objects located in larger molecular clouds or cloud complexes, with longer path lengths within the cloud or complex, and thus stronger H_3^+ lines. In conclusion, we believe that it is premature to argue based on the present data for large foreground cloud components to the H_3^+ absorptions observed toward the more distant embedded sources.

Acknowledgements

The new observations reported here were obtained at the United Kingdom Infrared Telescope, which is operated by the Joint Astronomy Centre on behalf of the U.K. Particle Physics and Astronomy Research Council. TRG's research is supported by the Gemini Observatory, which is operated by the Association of Universities for Research in Astronomy, Inc., on behalf of the international Gemini partnership of Argentina, Australia, Brazil, Canada, Chile, the United Kingdom and the United States of America. BJM is supported by the Miller Institute for Basic Research in Science. TO is supported by NSF grant PHY-00-99442.

References

Doty, S.D., van Dishoeck, E.F., van der Tak, F.F.S. & Boonman, A.M.S. 2002, A&A, 389, 446

Geballe, T.R. & Oka, T. 1996, Nature, 384, 334

Geballe, T.R. & Wade, R. 1985, ApJ, 291, L55

Goto, M., McCall, B.J., Geballe, T.R., Usuda, T., Kobayashi, N., Terada, H., Oka, T. 2002, PASJ, 54, in press

Kulesa, C. & Black, J. 2000, NOAO Newsletter No. 62

Herbst, E. & Klemperer, W. 1973, ApJ, 185, 505

Lee, H.-H., Bettens, R.P.A. & Herbst, E. 1996, A&AS 119, 111

McCall, B.J. 2001, PhD Thesis, University of Chicago

McCall, B.J., Geballe, T.R., Hinkle, K.H. & Oka, T. 1999, ApJ, 522, 338

Mitchell, G.F., Curry, C., Maillard, J.-P. & Allen, M. 1989, ApJ, 341, 1020

Mitchell, G.F., Hasegawa, T.I. & Schella, J. 1992, ApJ, 386, 604

Shuping, R.Y., Snow, T.P., Crutcher, R. & Lutz, B.L. 1999 ApJ, 520, 149

van der Tak, F.F.S. & van Dishoeck, E.F. 2000, A&A, 358, L79

van der Tak, F.F.S., van Dishoeck, E.F., Evans, N.J., II, Bakker, E.J. & Blake, G. A. 1999, ApJ, 522, 991

Watson, W.D. 1973, ApJ, 183, L17

INTERSTELLAR TRIPLY DEUTERATED AMMONIA

Floris F.S. van der Tak[1], Dariusz C. Lis[2], Maryvonne Gerin[3,4], Evelyne Roueff[4], Peter Schilke[1], and Tom G. Phillips[2]

[1]Max-Planck-Institut für Radioastronomie, Bonn, Germany
[2]California Institute of Technology, Pasadena, USA
[3]Ecole Normale Supérieure, Paris, France
[4]Observatoire de Paris, Meudon, France

ABSTRACT

Research on interstellar ND_3 is reviewed and updated from the discovery papers. Results from observations of a dozen sources at centimeter and submillimeter wavelengths are presented. The two data sets are consistent, but do not constrain the excitation conditions in the ND_3-emitting gas. The column density ratios of NH_3/NH_2D, NH_2D/ND_2H and ND_3/ND_2H, observed in similar sized beams, are ≈10, and present a challenge for both the gas phase and the grain surface chemistry scenarios of deuterium fractionation. The role of shocks and of CO depletion in deuterium chemistry is discussed.

Key words: ISM: molecules – ISM: abundances

1. INTRODUCTION

Deuterium-bearing molecules are important as probes of the very cold phases of molecular clouds, prior to star formation. Moreover, the isotopic composition of molecules is a valuable clue to their formation mechanism. In the gas phase, deuterium fractionation occurs by reactions with H_2D^+, CH_2D^+ and C_2HD^+ (Millar, this volume). Alternatively, reactions on the surfaces of dust grains may enhance molecular D/H ratios. Both methods require low temperatures (< 20 K) and high densities (> 10^5 cm^{-3}), as are found in dark clouds and prestellar cores. The grain surface route also needs a mechanism to return the molecules to the gas phase where they can be observed by means of submillimeter spectroscopy. The gas phase route can account for the deuteration of HCO^+ and N_2H^+, while grain surface chemistry may be important for H_2CO and CH_3OH.

Although astronomical detections of both singly and doubly deuterated NH_3 exist (Roueff et al. 2000), the deuteration mechanism for NH_3 is not clear. Despite early claims, searches for NH_3 ice based on the 2.21 and 9.0 μm features have not been successful (Taban et al. 2002), but the 3.47 μm feature indicates abundances of up to 7 percent of H_2O ice (Dartois et al. 2002). However, existence in the solid state does not imply formation on grain surfaces, since NH_3 may just passively accrete after forming in the gas phase. Soon after Rodgers & Charnley (2001) proposed ND_3 as a test of theories of interstellar deuterium chemistry, searches were initiated.

The rotational energy levels of symmetric top molecules like ND_3 are labeled by the total angular momentum J and its projection K onto the molecular symmetry axis. Inversion splits these levels into states which are symmetric (s) and antisymmetric (a) upon reflection in the plane of the D atoms. Coupling of the ^{14}N nuclear spin I with the rotational angular momentum J causes additional hyperfine splitting ($F = I + J$). Current instrumentation cannot resolve the even finer splitting due to D, which, owing to its small quadrupole moment, is only ≈ 200 kHz (A complete line list is available from the CDMS catalog (Müller et al. 2001) at the URL `http://www.cdms.de`). Under the conditions favourable for deuterium enhancement, the strongest lines of ND_3 should be the ground state rotational line ($1_0^a \rightarrow 0_0^s$) at 309.9 GHz and the (1,1) inversion line ($1_1^a \rightarrow 1_1^s$) at 1589 MHz, which are both accessible from the ground.

Because the Pauli exclusion principle only holds for fermions, certain states are forbidden in NH_3 but allowed in ND_3. For example, the ND_3 ground state rotational line also has a $1_0^s \rightarrow 0_0^a$ component at 306.7 GHz, which does not exist for NH_3. However, spin statistics make this line 10 times weaker than the $1_0^a \rightarrow 0_0^s$ line, which is why it has escaped detection so far.

2. SUBMILLIMETER OBSERVATIONS

Emission in the $J_K = 1_0^a \rightarrow 0_0^s$ line has been detected in the dark cloud Barnard 1 (Lis et al. 2002b), two positions in the star-forming region NGC 1333 (van der Tak et al. 2002), and several positions in the molecular cloud L 1689N (Roueff et al, in prep.). Table 1 lists the results of a more extensive search, made with the Caltech Submil-

Chemistry as a Diagnostic of Star Formation (2003). *Edited by Charles L. Curry and Michel Fich.* Published in 2003 by NRC Research Press, Ottawa ON, K1A 0R6, Canada. ISBN 0-660-19089-3.

Table 1. Upper limits for ND_3 rotational line emission

Source	R.A. (1950) h m s	Dec. (1950) ° ′ ″	rms[a] mK
OMC-2	05 32 58.6	−05 11 42	24
HH1-C	05 33 51.5	−06 47 57	25
NGC 2264 C	06 38 26.3	+09 32 18	39
NGC 2264 G	06 38 25.7	+09 58 54	31
Serpens S68N	18 27 15.2	+01 14 47	30
Serpens FIRS1	18 27 17.3	+01 13 16	28

[a] Raw antenna temperature. Multiply by 1.5 to convert to $T_{\rm mb}$.

limeter Observatory. Sources were selected from Shah & Wootten (2001) to have high NH_2D column densities, but the positions used here are from Jijina et al. (1999) for Orion and Monoceros and from McMullin et al. (2000) for Serpens. System temperatures were ~500 K and integration times 30–50 minutes on-source, using beam switching with a throw of 240″. For an excitation temperature of 10 K, the noise levels correspond to column density upper limits of 1.4–5.0 $\times 10^{11}$ cm^{-2}.

Table 2. Upper limits for ND_3 inversion line emission and absorption

Source	R.A. (1950) h m s	Dec. (1950) ° ′ ″	rms[a] mK
Emission targets			
NGC 1333	03 26 03.6	+31 04 42	11-16
Barnard 1	03 30 15.0	+30 57 31	9-12
CB 17	04 00 35.0	+56 48 00	16-22
L 1400 K	04 26 51.0	+54 45 27	12-23
L 1544	05 01 15.0	+25 07 00	35-38
NGC 2264	06 38 24.9	+09 32 29	38-52
L 134N	15 51 32.0	−02 42 13	15-20
L 1689N	16 29 27.6	−24 22 08	30-46
S68 FIR	18 27 17.5	+01 13 23	9-12
Absorption targets			
W 33	18 11 19.5	−17 56 40	36-46
W 43	18 45 00.4	−01 59 16	101-120
37.763-0.216	18 58 33.1	+04 07 41	43-62
W 49A	19 07 52.1	+09 01 08	70-94

[a] Raw antenna temperature. Multiply by 1.7 to convert to $T_{\rm mb}$.

3. Centimeter observations

Measurements of the pure inversion spectrum of ND_3 go back to the early days of microwave spectroscopy (Nuckolls et al. 1953); many lines have been measured since (Fusina & Murzin 1994). However, only recently has the hyperfine structure of the (1,1) line near 1589 MHz been resolved (van Veldhoven et al. 2002).

In 2001–2002, we used the 100-m Effelsberg telescope to search for the (J, K) = (1,1) 1589.006 MHz, (2,2) 1591.695 MHz, (3,3) 1599.645 MHz, and (4,4) 1612.997 MHz metastable inversion lines. The front end was the 1.3-1.7 GHz primary focus receiver, while the back end was the 1024-channel autocorrelator. System temperatures were 18–33 K depending on elevation. Integration times are 4–10 hours for NGC 1333, CB 17, S68 FIR, Barnard 1, L 1400K and L 134N, and 1–2 hours for the other sources. Frequency switching with a throw of 1 MHz was used for all observations. The telescope has a beam size of 8′ at this wavelength. Search targets for ND_3 emission were selected to have $N(NH_2D) > 3 \times 10^{13}$ cm^{-2} in an 80″ beam (Shah & Wootten 2001). We excluded Orion, where at 19 cm, the H II region blocks our view of the molecular cloud.

The top section of Table 2 reports the range of rms noise levels obtained for the four lines. Only data unaffected by interference with terrestrial signals are listed. For the (2,2) and (3,3) lines, ≈10 percent of the data were affected by interference, but up to 50 percent were affected in the (4,4) line. In addition, OH 1612 MHz emission along the line of sight made some of the (4,4) data unusable. Fortunately, the (1,1) spectra are clear of interference, and a follow-up search with Arecibo may prove successful.

The inversion lines were also searched for in absorption towards bright Galactic H II regions. Targets for this search were selected to have $N(H_2CO) > 10^{14}$ cm^{-2} as measured in 6 cm absorption (Wadiak et al. 1988), and 20 cm flux densities, taken from the NVSS (45″ resolution) of >1 Jy, excluding Sgr B2 because of its low declination. The noise levels obtained for these sources (Table 2, bottom section) are higher than for emission sources, because the background H II regions contribute significantly to the system temperature.

The centimeter data limit $N(ND_3)$ to $\lesssim 0.3\text{–}1.5 \times 10^{12}$ cm^{-2}, consistent with the submillimeter data. However, the inversion lines not only provide independent estimates of the ND_3 column density, they can also constrain its excitation. Column densities derived from the $J_K = 1_0^a \rightarrow 0_0^s$ line are uncertain by factors of up to 2 through the excitation temperature. The lower frequency of the $J_K = 1_1^a \rightarrow 1_1^s$ line implies a much lower spontaneous decay rate and virtually ensures thermalization. However, being

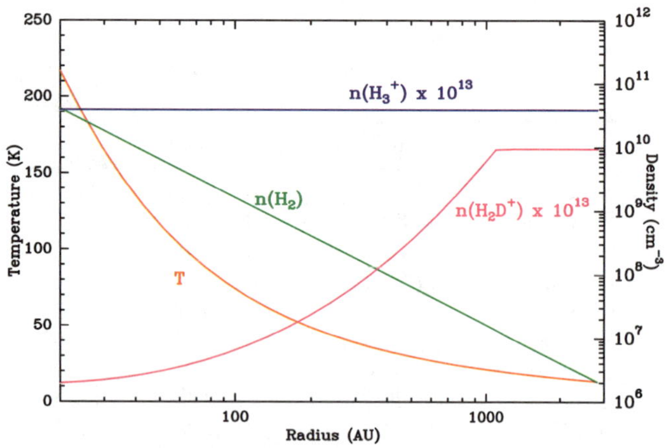

Figure 1. Model of NGC 1333 IRAS 4A used to determine the ND_3 abundance. Based on van der Tak et al. (2002).

beyond the capabilities of 100-m class telescopes such as Effelsberg and the VLA (an effective 130-m telescope), its detection requires 300-m class telescopes or larger.

4. Abundance of ND_3

If the excitation of ND_3 cannot be measured, it can be calculated through a model of the physical and chemical structure of the source. Figure 1 shows the model that van der Tak et al. (2002) used to determine the ND_3 abundance in NGC 1333 IRAS 4A. The density structure is obtained from interferometric dust continuum observations, while the dust temperature structure comes from a self-consistent calculation of the thermal balance as a function of radius. These quantities peak toward the center of the protostellar envelope. Cosmic-ray ionization produces H_3^+ at a rate that does not vary with radius. The reaction of H_3^+ with HD produces H_2D^+, and since it is faster at low temperatures, the H_2D^+ distribution peaks at large radii.

The distribution of ND_3 can either be assumed to follow that of H_2 or that of H_2D^+. These two situations may apply to the cases of grain surface and gas phase formation, respectively. The inferred abundance of ND_3 is 3×10^{-12} if ND_3/H_2 is constant, and 1×10^{-11} if ND_3/H_2D^+ is constant. Maps of the ND_3 emission will enable us to decide between these two possibilities. The Submillimeter Array (SMA) will soon provide the necessary resolution of $\lesssim 10''$ at 310 GHz.

5. Chemical implications

Table 3 lists measured column densities of NH_3 isotopomers. Only in the case of NGC 1333 IRAS 4A have abundances of NH_3 and ND_3 been derived by radiative transfer modeling, so we use column density ratios as proxies for abundance ratios. This approach is valid as long as the excitation conditions of all molecules are similar, which is likely true except perhaps for NH_3.

The column density ratios of NH_3/NH_2D, NH_2D/ND_2H and ND_3/ND_2H are all ≈ 10 (Table 3). This strong fractionation presents a challenge to grain surface chemistry, and requires an atomic D/H ratio in the gas phase of ≈ 0.15, which is more than 10 times higher than in the models of Roberts & Millar (2000). Unfortunately, observations do not well constrain the atomic D/H ratio in molecular clouds. While H I 21 cm data indicate $H/H_2 \sim 10^{-3}$ (Li & Goldsmith 2002), observing D I at 92 cm is very difficult. To determine the deuterium budget in dense interstellar clouds, measurements of the HD 119 μm line, with HIFI on *Herschel* and GREAT on SOFIA, will be crucial (see Phillips, this volume).

However, standard models of gas phase chemistry also have some difficulty explaining the results of Table 3. To reproduce the observed column density ratios, the models require deuteron transfer reactions to be much faster than proton transfers. Indeed, laboratory data indicate that in the dissociative recombination of partially deuterated ions, H is more easily ejected than D (Le Petit & Roueff 2002). On the other hand, recent experiments on the H_3^+ + HD reaction indicate a lower formation rate of H_2D^+ than the value used in most chemical models (Gerlich & Schlemmer 2002).

One potential problem with applying the surface chemistry route to prestellar cores is their temperatures, which are too low for significant evaporation of even the most volatile ices to occur. One alternative may be non-thermal desorption by cosmic rays (Najita et al. 2001), but quantitative predictions do not yet exist. However, Lis et al. (2002a) propose a connection between shocks and the chemistry of deuterium. Indeed, all ND_3 detections so far have been at or near bipolar outflow sources. Perhaps the shocks merely compress surrounding gas, which leads to the cold, dense conditions favourable for deuterium chemistry. However, the shocks may also desorb grain mantle material, while leaving the grain cores intact. Shock velocities of $\sim$10 km s^{-1} would be sufficient. Such shocks occur frequently in the interstellar medium. One possibility is that interstellar gas is periodically shocked by ongoing star formation activity.

Table 3. Observed column densities of NH_3 isotopomers

Source	$N(NH_3)$ 10^{14} cm^{-2}	$N(NH_2D)$ 10^{13} cm^{-2}	$N(ND_2H)$ 10^{12} cm^{-2}	$N(ND_3)$ 10^{11} cm^{-2}	T_{ex}[f] K
Barnard 1	21[c]	58[b]	48[b]	13[d]	5
NGC 1333 IRAS 4A	13.8[a]	39.0[a]	<4.6[b]	2.9[e]	10
NGC 1333 D-peak	19.2[a]	37.6[a]	<1.9[b]	5.9[e]	10
Beam size ($''$)	37	22	22	25	

[a] From Hatchell (2002).
[b] Preliminary values based on IRAM 30-m observations (Gerin et al., in prep.).
[c] From Effelsberg observations.
[d] Data from Lis et al. (2002b); non-LTE calculation at $n(H_2)=7\times10^4$ cm^{-3} and T=12 K.
[e] From van der Tak et al. (2002).
[f] Excitation temperature used to analyze the NH_2D and ND_2H data.

Another avenue worth exploring is the connection of deuterium chemistry with CO depletion. Freeze-out of CO, the major destroyer of H_2D^+, onto dust grains helps channel deuterium into heavy molecules (Roberts & Millar 2000) and correlations between enhanced deuteration and CO depletion have indeed been observed (Hatchell 2002; Bacmann et al. 2002). However, in the case of NH_3, such a trend does not provide direct evidence of formation in the gas phase. The formation of NH_3 and its isotopomers on grain surfaces requires a high abundance of N in the gas phase. While present in molecular clouds, most nitrogen is in the form of N_2 (Womack et al. 1992), depletion of CO suppresses its formation and enhances N instead (Charnley & Rodgers 2002). One key question is therefore that of the major nitrogen carrier in pre-stellar cores. Although neither N nor N_2 can be observed directly, N_2H^+ can serve as a tracer of N_2. A correlation between the fractionations of N_2H^+ and NH_3 would thus constitute an important test of the formation mechanism of interstellar NH_3 and ND_3.

Acknowledgements

This paper is based on observations with the CSO and Effelsberg telescopes. The CSO is supported by NSF grant AST 99-80846. The 100-m telescope at Effelsberg is operated by the Max-Planck-Institut für Radioastronomie. The authors thank Andrea Tarchi for carrying out some of the observations presented here.

References

Bacmann, A., et al. 2002, A&A, 389, L6
Charnley, S. B., & Rodgers, S. D. 2002, ApJ, 569, L133
Dartois, E., d'Hendecourt, L., Thi, W., Pontoppidan, K. M., & van Dishoeck, E. F. 2002, A&A, 394, 1057
Fusina, L., & Murzin, S. N. 1994, J. Mol. Sp., 167, 464
Gerlich, D., & Schlemmer, S. 2002, in *Deuterium in the Universe*, in press
Hatchell, J. 2002, A&A, submitted
Jijina, J., Myers, P. C., & Adams, F. C. 1999, ApJS, 125, 161
Le Petit, F., & Roueff, E. 2002, in *Deuterium in the Universe*, in press
Li, D., & Goldsmith, P. 2002, ApJ, in press
Lis, D. C., Gerin, M., Phillips, T. G., & Motte, F. 2002a, ApJ, 569, 322
Lis, D. C., et al. 2002b, ApJ, 571, L55
Müller, H. S. P., Thorwirth, S., Roth, D. A., & Winnewisser, G. 2001, A&A, 370, L49
McMullin, J. P., et al. 2000, ApJ, 536, 845
Najita, J., Bergin, E. A., & Ullom, J. N. 2001, ApJ, 561, 880
Nuckolls, R. G., Rueger, L. J., & Lyons, H. 1953, Phys. Rev. 89, 1101
Roberts, H., & Millar, T. J. 2000, A&A, 361, 388
Rodgers, S. D., & Charnley, S. B. 2001, ApJ, 553, 613
Roueff, E., et al. 2000, A&A, 354, L63
Shah, R. Y., & Wootten, A. 2001, ApJ, 554, 933
Taban, I. M., Schutte, W. A., Pontoppidan, K. M., & van Dishoeck, E. F. 2002, A&A, in press
van der Tak, F. F. S., et al. 2002, A&A, 388, L53
van Veldhoven, J., Jongma, R. T., Sartakov, B., Bongers, W. A., & Meijer, G. 2002, Phys. Rev. A, 66, 32501
Wadiak, E. J., Rood, R. T., & Wilson, T. L. 1988, ApJ, 324, 931
Womack, M., Ziurys, L. M., & Wyckoff, S. 1992, ApJ, 393, 188

CHEMISTRY AND TRANSPORT IN ACCRETION DISKS

Thomas Henning[1] and Martin Ilgner[2]

[1]Max-Planck-Institut für Astronomie, Königstuhl 17, 69117 Heidelberg, Germany
[2]Institute of Computational Physics, University of Tübingen, 72076 Tübingen, Germany

ABSTRACT

The influence of mass transport processes on the chemical evolution in a protoplanetary accretion disk is discussed. Chemistry may easily couple back to disk dynamics by determining ionization state and cooling rates via the abundance of electrons, metals, and key molecules. On the other hand, the coupling between the dynamics and the chemical evolution of accretion disks has a great significance for determining the initial chemical conditions for planet formation as well as for the chemical composition of primitive bodies in the solar system.

Two different approaches can be used to study the combined effects of dynamics and chemistry: Models based on a reduced chemical network and fully developed (magneto-) hydrodynamics, or models with extended chemical networks and simplified hydrodynamics. We used the latter approach to study the chemical evolution in accretion disks taking into account radial and vertical transport processes. The results demonstrate the strong influence transport processes have on disk chemistry.

Key words: ISM: molecules – ISM: abundances – stars: circumstellar matter – stars: planetary systems: protoplanetary disk – solar system: formation

1. DISK PHYSICS

The dynamics of accretion disks is closely connected to the angular momentum transport. Although there is still a fundamental discussion on the origin of this transport process (hydrodynamical vs. magneto-hydrodynamical origin), one has to take into account a very efficient transport mechanism such as turbulent motion. The most powerful ansatz to describe turbulent viscosity in disks was introduced firstly by Shakura & Sunyaev (1973). Disk models based on their α parameterization of the turbulent viscosity are often called α-disks. In addition, the angular momentum flux can be characterized as a viscous diffusion process. Both energy dissipation and angular momentum transport enable the mass transport (accretion flow) within the disk.

The evolution of protoplanetary accretion disks is governed by a set of six partial differential equations for density, momentum, thermal and radiation energy. Assuming axisymmetry, the equations written in cylindrical coordinates (r, φ, z) are

$$\begin{aligned}
\frac{\partial \varrho}{\partial t} + \nabla \cdot (\varrho \mathbf{u}) &= 0 \\
\frac{\partial(\varrho u_r)}{\partial t} + \nabla \cdot (\varrho u_r \mathbf{u}) &= -\frac{1}{\varrho}\frac{\partial p}{\partial r} - \frac{\gamma M_* r}{(r^2+z^2)^{\frac{3}{2}}} + r\Omega^2 \\
&\quad + \frac{1}{r}\frac{\partial}{\partial r}(r\tau_{rr}) + \frac{\partial \tau_{rz}}{\partial z} - \frac{\tau_{\varphi\varphi}}{r} \\
\frac{\partial(\varrho u_z)}{\partial t} + \nabla \cdot (\varrho u_z \mathbf{u}) &= -\frac{1}{\varrho}\frac{\partial p}{\partial z} - \frac{\gamma M_* z}{(r^2+z^2)^{\frac{3}{2}}} \\
&\quad + \frac{1}{r}\frac{\partial}{\partial r}(r\tau_{rz}) + \frac{\tau_{zz}}{z} \\
\frac{\partial(\varrho \Omega r^2)}{\partial t} + \nabla \cdot (\varrho \Omega r^2 \mathbf{u}) &= \frac{1}{r}\frac{\partial}{\partial r}(r^2 \tau_{\varphi r}) + r\frac{\partial \tau_{\varphi z}}{\partial z} \\
c_V \varrho \left[\frac{\partial T}{\partial t} + (\mathbf{u}\cdot\nabla)\, T\right] &= -p\nabla \cdot \mathbf{u} + \varrho\Phi - \Lambda \\
\frac{\partial E}{\partial t} + \nabla \cdot (E\mathbf{u}) &= \Lambda - \nabla \cdot \mathbf{F},
\end{aligned}$$

where $\mathbf{u}$ denotes the velocity field, Ω the angular velocity, ϱ the mass density, T the temperature, p the gas pressure, c_V the specific heat, γ the gravitational constant, and E, $\mathbf{F}$, and Λ the radiation energy, radiation flux, and heating and cooling rates per unit volume, respectively. Here Φ is the viscous dissipation rate per unit mass and τ_{ij} are the components of the viscous stress tensor.

In the context of the chemo-dynamical evolution of disks it is practically impossible to solve such a complex system numerically. For that reason one must simplify the description of the chemical network or of the disk physics. Previous studies have demonstrated that the most important processes in accretion disks can be reproduced in a

Chemistry as a Diagnostic of Star Formation (2003). *Edited by Charles L. Curry and Michel Fich.* Published in 2003 by NRC Research Press, Ottawa ON, K1A 0R6, Canada. ISBN 0-660-19089-3.

more qualitative manner by simpler models than the system above. For example, the equation of motion of the gas can be derived based on the equation of continuity and on the equation of conservation of angular momentum, assuming that velocities in the disk are independent of the height above the central plane. We shall further assume that transport is caused by shearing stresses because of the differential rotation of the disk. Since the paper by Lynden-Bell & Pringle (1974), the general nature of the evolution of such a disk is well understood and finally the picture characterizing the viscous evolution of the disk is very well known (see Fig. 1). In addition, the time scale

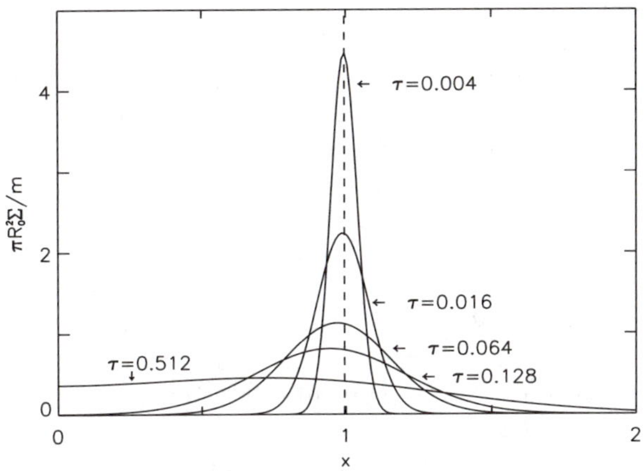

Figure 1. The viscous evolution of a ring of matter of mass m. The surface density Σ is shown as a function of $x = R/R_0$ and time $\tau = 12\nu t/R_0^2$, where R_0 and ν are the initial radius of the ring and the viscosity, respectively.

determining the viscous evolution is much longer than the time scales for the other relevant physical processes within the disk. With respect to the difference between the viscous time scale and the other time scales, one can switch from a 2D to a 1+1-D description of the complex system above. Finally, this simplified description corresponds with the assumption that a static atmosphere in the z-direction will be established instantaneously during the viscous evolution. Especially in the steady-state case, a number of 1+1-D α-disk models were developed; e.g., the models by Shakura & Sunyaev (1973), Meyer & Meyer-Hofmeister (1982), Bell et al. (1997), and D'Alessio et al. (1998). In this case, the dynamics is determined solely by the mass accretion rate $\dot{M}$ and the α value (i.e., the efficiency of the angular momentum transport).

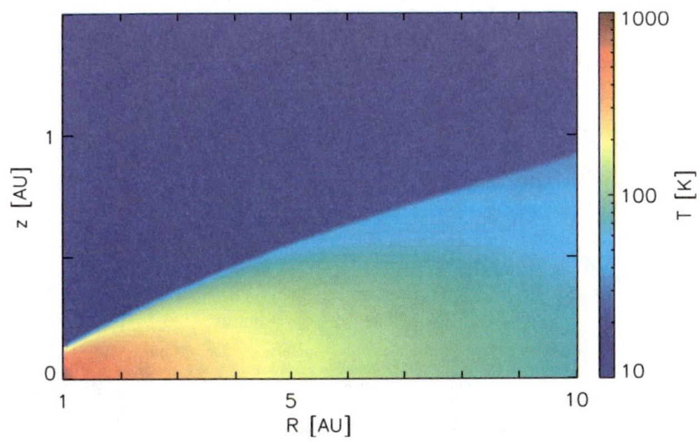

Figure 2. Typical temperature distribution of a steady 1+1-D α-disk model with a mass accretion rate $\dot{M} = 1 \times 10^{-7} M_\odot\, yr^{-1}$ and $\alpha = 0.01$.

2. CHEMISTRY

The kinetic model, which describes the time evolution of chemical species, is based on the assumption that chemical reactions in the gas phase, as well as processes on the dust surfaces, contribute to the formation/destruction of the chemical species. The corresponding differential equations can be written as

$$\frac{d}{dt} n_i^{\mathrm{g}} = n_{\mathrm{H}} G_i - \lambda_i n_i^{\mathrm{g}} + \xi_i n_i^{\mathrm{s}}$$
$$\frac{d}{dt} n_i^{\mathrm{s}} = \lambda_i n_i^{\mathrm{g}} - \xi_i n_i^{\mathrm{s}},$$

determining the gas and the grain-surface abundances of the chemical species i relative to H_2, n_i^{g}, and n_i^{s}, respectively. The quantity G_i denotes the net production/loss of the gas phase species due to chemical reactions in the gas phase, while λ_i and ξ_i, the accretion rate and the surface desorption rate, respectively, refer to processes on the dust surface. The basic processes occurring on dust grains that influence the kinetics are shown schematically in Figure 3. Here, the so-called "two-phase" approach was used, taking only the two reactive phases into account (gas phase and surface phase). Of course, one could also include the transition from the chemically reactive surface components to the chemically non-reactive mantle components, as proposed by Hasegawa & Herbst (1993). Nevertheless, the majority of astrochemical models containing grain surface processes are based on the simple two-phase description.

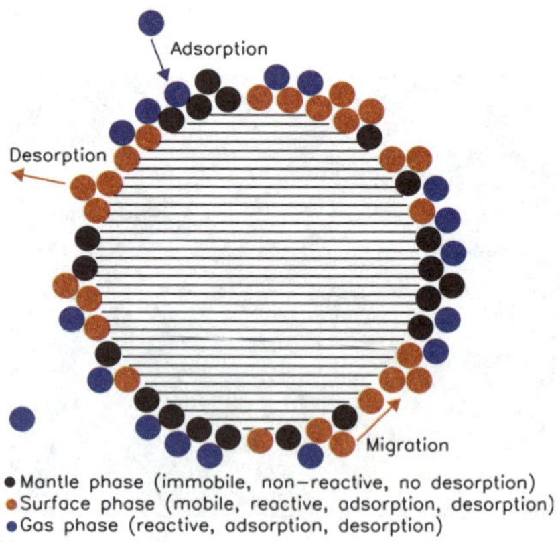

Figure 3. Basic classification of processes occurring on dust surfaces. Adapted from Hasegawa & Herbst (1993).

3. Coupling between kinetics and transport

To be able to study transport processes in chemically reacting flow systems, one needs information beyond thermochemistry and chemical kinetics already encountered in astrochemistry. The transport processes refer particularly to the transport of the mass of the chemical species, momentum, and energy. However, transport processes influence the local reaction rates as well as the ability of the reactants to be elevated to temperature regimes where the rates of important chemical reactions are appreciable.

The coupling between kinetics and transport is limited by the following two extreme situations: the transport-controlled situation, and the kinetically limited situation. In particular, the processes on the dust surfaces in the colder disk regions might be attributed to such a kinetically limited situation. Nevertheless, corrections for the mass fluxes are certainly required to interpret the chemical kinetic data. Different aspects of the chemical evolution in protoplanetary accretion disks can be described by different types of chemical reactors. For steady-flow reactors there are two important limiting cases, the so-called "plug-flow" reactor and the "well-stirred" reactor. The plug-flow reactor refers to the special case in which the reacting dust-gas mixture flows along a distinguished direction without any kind of recirculation. In contrast, recirculation, which is included in the well-stirred reactor, tends to establish no spatial gradients along the flow.

From the dynamical point of view, ongoing research on the chemical evolution of disks can be summarized in terms of these types of chemical reactors. Considering only non-equilibrium thermodynamics, we first mention models *without* any macroscopic transport. Aikawa et al. (1999) calculated the chemical evolution of a disk on the background of the minimum-mass solar nebula model, proposed by Hayashi et al. (1985). Another model of the disk—the Chiang-Goldreich model (Chiang & Goldreich 1997)—was used by Willacy & Langer (2000) to study the chemical evolution. The Chiang & Goldreich model does not have any constraints on dynamical processes. From this point of view, the model by Willacy & Langer provides a more coherent description than the model by Aikawa et al. It is remarkable that one still does not consider macroscopic transport processes although dynamical α-disk models are applied for disk chemistry (see, e.g., the latest model by Aikawa et al. [2002]). Various recent studies concerning the chemical disk evolution are based on a plug-flow reactor-type description; e.g., Bauer et al. (1997), Finocchi et al. (1997), Willacy et al. (1998), Aikawa et al. (1999), and Markwick et al. (2002). The differences among these models are determined much less by the disk model itself than by the chemical networks used. In all these publications a steady-state α-disk model was applied, resulting in a *steady* mass flow toward the central object. Because of this typical feature of a steady α-disk, it is no longer difficult to understand the link to the plug-flow reactor. The more advanced case of reactor types also takes diffusion processes into account. A first attempt to consider turbulent mixing processes was taken by Gail (2002), who calculated the stationary solution of a transport-diffusion-reaction equation in the case of a very restricted chemical network. Gail introduced radial mixing, assuming that the disk is vertically well-mixed. Vertical mixing operates on a much shorter time scale than that of radial mixing. The importance of vertical mixing on the chemical evolution will be clearly demonstrated below.

All models based on the plug-flow reactor calculated the chemical evolution along the streamline of the gas flow (Lagrangian description). From the computational point of view, the kinetics can be coupled to the radial transport very easily as long as radial, 1D disk models are used. This situation will change dramatically if the back-coupling of chemistry on the disk structure via the destruction of dust grains is included (Bauer et al. 1997). However, concerning the 1+1-D model, the situation becomes even worse because the disk structure itself is determined by a system of coupled differential equations. The kinetic equations and the equations of radiation hydrodynamics cannot be solved simultaneously and therefore some kind of decoupling is needed. The condition for a reasonable splitting can be specified very easily: The splitting between the kinetics and radial transport processes is only valid on length and time scales connected to a minor change in the physical conditions. Only in this case can the substantial time derivative d/dt be approximated by the local time derivative $\partial/\partial t$. Finally, the adaptive grid, shown in Figure 4, is based on this kind of splitting strategy.

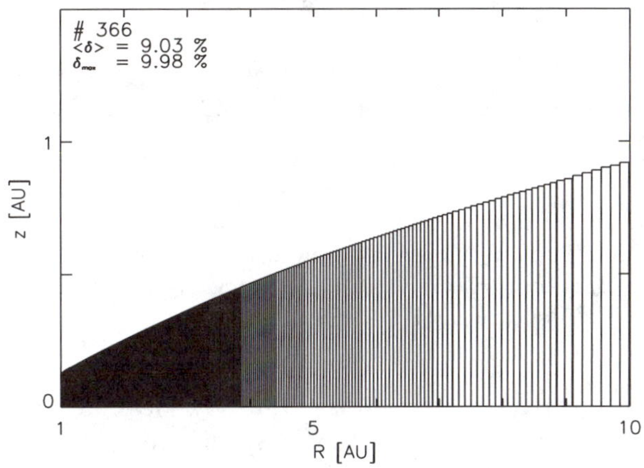

Figure 4. Typical decomposition of a trajectory within a steady 1+1-D α-disk model from 10 to 1 AU. The same parameters were chosen as in Fig. 2. More than 366 grid points are needed to construct the adaptive grid. The change in the physical quantities moving from one grid point to the neighboring grid point is 9.03%, on average.

4. Results

The results of our simulations are based on two different models. The first model determines the chemical evolution along a streamline of the gas on the background of a steady 1+1-D α-disk model. The second model differs from the first only in the time-dependent description of the disk structure itself. The chemical network is identical with the network used in Markwick et al. (2002), containing more than 240 chemical species and 2500 reaction pathways. Chemical reactions in the gas phase were considered as well as gas-grain interactions. The chemical model contains various types of reactions: adsorption, desorption, ion-neutral, neutral-neutral reactions, and photo-processes. More details can be found in Willacy et al. (1998) and Markwick et al. (2002).

The following set of parameters serves as a reference for all the simulations presented here: mass accretion rate $\dot{M}=1\times10^{-7}M_{\odot}\,yr^{-1}$, $\alpha=1\times10^{-2}$, $M_{\star}=1M_{\odot}$, $R_{\star}=3R_{\odot}$ and the opacity table after Henning & Stognienko (1996). The distribution of CS is a good example of the influence of transport processes on the chemical evolution. Because of its large dipole moment, CS is expected to trace high-density regions. Based on measurements of higher$-J$ CS transitions, one could get better constraints on the chemical and physical structure of disks.

Figure 5 shows the chemical evolution of CS along the trajectory from 10 AU to 1 AU. It is the only simulation where the turbulent vertical mixing is switched off (Simulation 1). Of course, the main formation and destruction processes of CS in the gas phase vary dramatically within the disk. Although it is beyond the purpose of this paper to explain the differences in more detail, we want to point out one remarkable feature: The increased relative abundance of CS (represented by the contour line of 10^{-10}; see Fig. 5) originates mainly from the desorption processes.

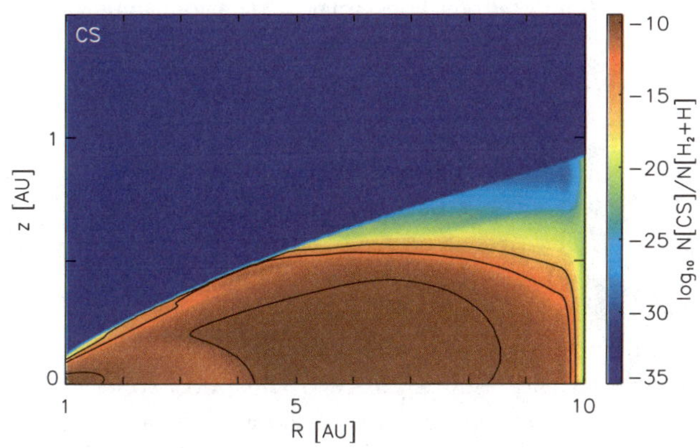

Figure 5. Chemical evolution of CS along the trajectory from 10 to 1 AU. The contour lines refer to the following abundance values relative to the total amount of hydrogen: 10^{-10}, 10^{-13}, and 10^{-16} (Simulation 1).

Including vertical mixing, one gets a different distribution of CS along the trajectory (see Fig. 6). One should keep in mind that, compared to the simulation, the values of the disk parameters are identical (Simulation 2). The differences between the two simulations were caused directly by the vertical mixing process. The differences are small at the beginning of the trajectories. The initially existing gradient in CS will be reduced by diffusion processes occurring in the $z-$direction. The efficiency of the diffusion process itself is determined by the spatial gradients in the molecular composition as well as by

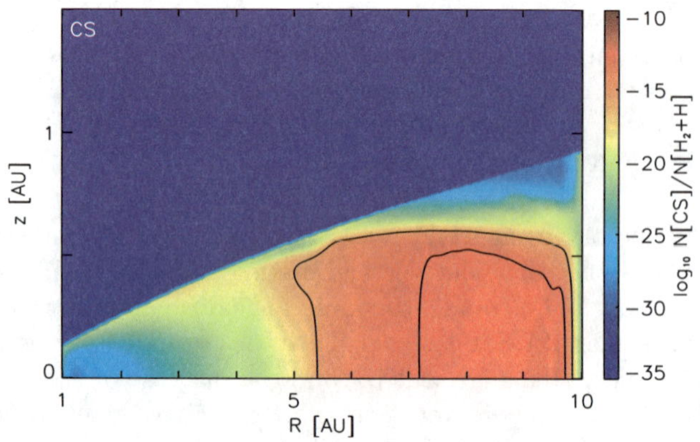

Figure 6. Chemical evolution of CS along the trajectory from 10 to 1 AU. The simulation is based on diffusion-reaction equations to model the vertical mixing processes of the chemical species. The contour lines refer to the following abundance values relative to the total amount of hydrogen: 10^{-13} and 10^{-16} (Simulation 2).

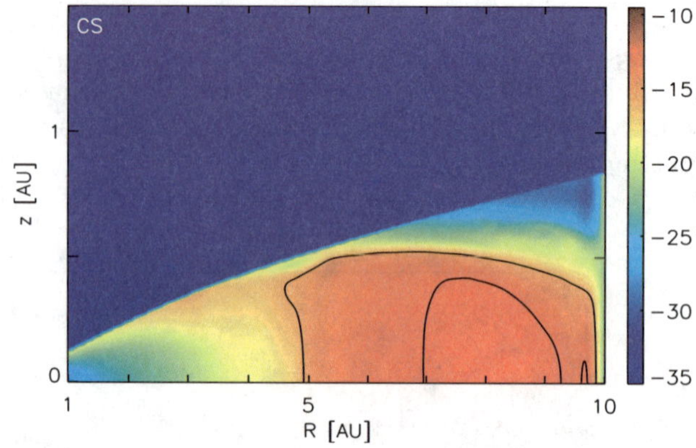

Figure 7. Same as Fig. 6, with the same abundance contour values. The only difference from the previous model (Simulation 2) is that another opacity table was used; i.e., the opacity table of Bell & Lin (1994) (Simulation 3).

the local value of the diffusion coefficient. In the case of CS, the mixing becomes important close to the midplane of the disk, leading to a reduction of the gradient of CS (see Fig. 6). There is a shift in the contour lines of the adsorbed counterpart of CS—i.e., CS onto grains—compared to the previous simulation. CS molecules will diffuse to the regions close to the photospheric surface. For these low-temperature regions, the adsorption process onto dust grains becomes very efficient indeed. In addition, the concentration of the adsorbed CS decreases close to the midplane. Now, it is obvious that one cannot find an increased abundance of CS in the gas phase around 8.5 AU (Fig. 5), because of the low concentration of adsorbed CS. The local change in molecular composition of the chemical species affects the global chemical evolution due to the radial advection process.

The above leads to a study of the chemical evolution of disks from the point of view of the dynamical parameters of the $\alpha-$model itself. Using different opacity tables, one changes the disk structure locally. However, in this case the time needed to transport the parcel of gas from 10 to 1 AU is quite similar. The differences between Simulation 2, using the opacity table of Henning & Stognienko (1996), and the current simulation (Simulation 3), using the opacity table of Bell & Lin (1994), is less than 10 percent of the total time ($\approx 8.5 \times 10^4$ yr). Thus, it is reasonable to assume that the differences in the chemical distribution of CS along the trajectory (Figs. 6 and 7, respectively). were caused only by local variations in the disk structure. It is easy to specify the regions where both opacity tables are different from each other. This is mainly the temperature region between $R = 4$ AU and $R = 5$ AU where the dust opacity is different in these two models. However, the local differences in the abundances do not influence the global chemical evolution of CS.

Changing the value of the mass accretion rate $\dot{M}$ or of α will more dramatically influence the radial transport process. For example, decreasing the mass accretion rate to a smaller value of $\dot{M} = 5 \times 10^{-8} M_{\odot}\,\mathrm{yr}^{-1}$ (Simulation 4), and α to 5×10^{-3} (Simulation 5) increases the time following the trajectory to $\approx 1.25 \times 10^5$ yr and $\approx 1.42 \times 10^5$ yr, respectively. Based on the disk structure, Simulations 2, 4, and 5 can be classified into two categories: firstly, the disk structure is quite similar (Simulations 2 and 5); and secondly, the radial transport operates on different time scales (Simulations 4 and 5). Based on this idealized classification scheme, one can explain the differences between the distributions of CS (see Figs. 6, 8, and 9).

The most advanced case represented here refers to a time-dependent description of a 1+1-D α-disk model. How does the chemical evolution change within a non-steady mass accretion flow? In the case of a steady disk model, one will always find the same chemical evolution, e.g., of CS. If the disk structure itself changes in time, then the chemical evolution depends on the evolutionary state of the disk. If one would follow the trajectory for two different times, one gets totally different temperature profiles along each trajectory. In Figures 10 and 11, the

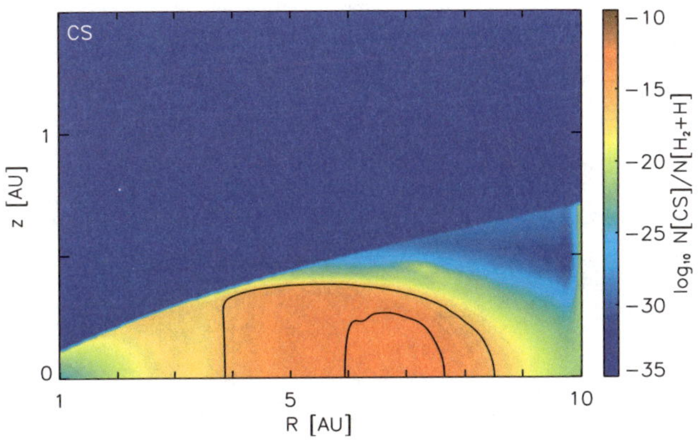

Figure 8. Same as Fig. 6, with the same abundance contour values. Compared to the reference values, only the mass accretion rate was changed, to $\dot{M} = 5\times10^{-8} M_\odot\ yr^{-1}$ (Simulation 4).

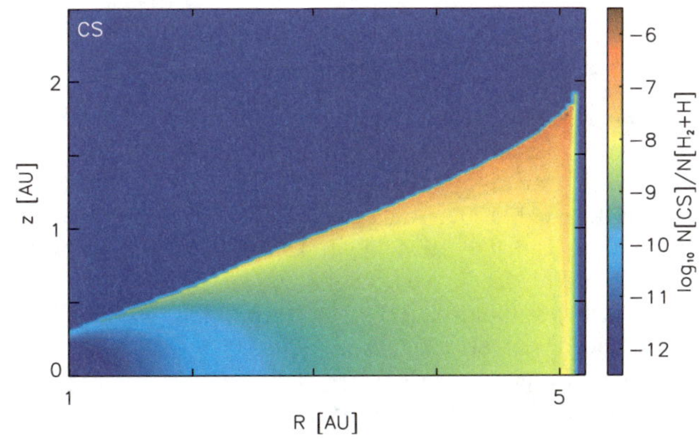

Figure 10. Chemical evolution of CS along the trajectory from 5.2 to 1 AU based on a time-dependent description of the 1+1-D α-disk model. Vertical mixing is included.

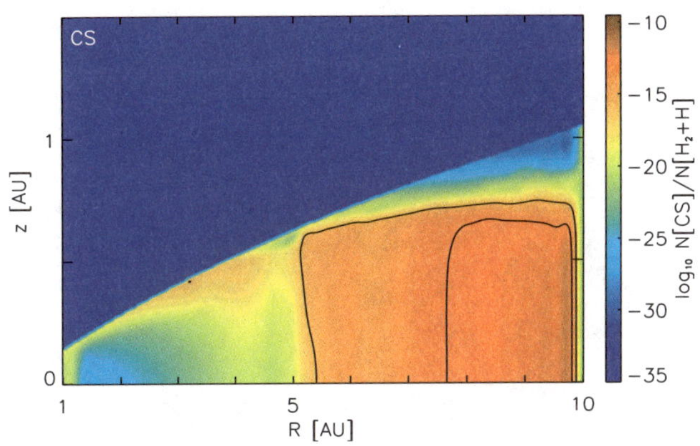

Figure 9. Same as Fig. 6, with the same abundance contour values. Compared to the reference values, only the α-value was changed, to $\alpha = 5\times10^{-3}$ (Simulation 5).

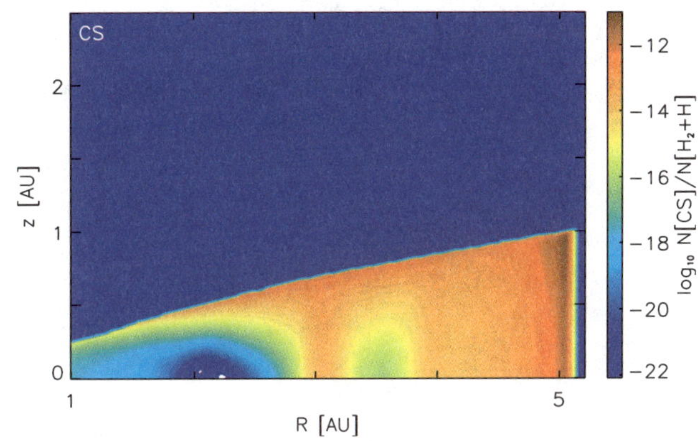

Figure 11. Same as Fig. 10, but using a different initial time.

initial time is separated by a period of 2.5×10^5 yr. In the earlier case, the temperatures along the trajectory were never below 100 K. Most of the regions passed by the later parcel of gas are, by contrast, dominated by lower temperatures. The huge differences, concerning the CS distribution, along the two trajectories indicate that the global chemical evolution can change with time (see Figs. 10 and 11). One can show that the main destruction and formation processes of CS along the two trajectories are quite different.

5. Outlook

The change from the Lagrangian description to the Eulerian description of disk dynamics is certainly one of the next steps. In the case of steady disk models, such simulations are currently under way. For several reasons, it seems not to be realistic to do the same step for time-dependent 1+1-D α-disk models. However, the transition to a fully 2D description of (magneto-) hydrodynamics might be the right next step. In this case, the coupling between the dynamics of the disk and the kinetics can be only realized on the basis of a reduced chemical network (see Wiebe et al. 2002).

ACKNOWLEDGEMENTS

This work was supported by DFG grant He 1935/17-1.

REFERENCES

Aikawa, Y., et al. 1996, ApJ, 467, 684
Aikawa, Y., et al. 1999, ApJ, 519, 705
Aikawa, Y., et al. 2002, A&A, 386, 622
Bauer, I., et al. 1997, A&A, 317, 273
Bell, R., & Lin, D. 1994, ApJ, 427, 987
Bell, R., et al. 1997, ApJ, 486, 372
Chiang, E., & Goldreich, P. 1997, ApJ, 490, 368
D'Alessio, P., et al. 1998, ApJ, 500, 411
Finocchi, F., et al. 1997, A&A, 325, 1264
Gail, H. 2002, A&A, 390, 253
Hasegawa, T., & Herbst, E. 1993, MNRAS, 263, 589
Hayashi et al. 1985, in *Protostars and Planets II*, ed. D. Black & M. Matthews (Tucson: Univ. of Arizona Press), 1100
Henning, Th., & Stognienko, R. 1996, A&A, 311, 291
Markwick, A., et al. 2002, A&A, 385, 632
Meyer, F., & Meyer-Hofmeister, E. 1997, A&A, 128, 34
Lynden-Bell, D., & Pringle, J. 1974, MNRAS, 168, 603
Shakura, N., & Sunyaev, R. 1973, A&A, 24, 337
Wiebe, D., et al. 2002, A&A, in press
Willacy, K., et al. 1978, A&A, 338, 995
Willacy, K., & Langer, W. 2000, ApJ, 544, 903

PART III:

CHEMISTRY, PHYSICAL CONDITIONS, AND STRUCTURE OF DARK CLOUDS

THE CHEMISTRY OF DARK CLOUDS: NEW ASTROCHEMICAL TOOLS FOR STAR FORMATION STUDIES

Edwin A. Bergin

Harvard Smithsonian Center for Astrophysics, 60 Garden St., Cambridge MA 02138

Abstract

The past decade has led to significant improvements in our understanding of the physical structure of the molecular cores of cold dark clouds. Observational efforts, in combination with improved knowledge of cloud structure, now provide clear evidence that the chemistry of dark clouds is dominated by the depletion of gaseous species onto grain surfaces. We outline the basis of these observational efforts and show how the abundance determinations have moved beyond single point analyses to the derivation of abundance profiles. We discuss the basic physics of the interaction between molecules and grain surfaces and show that when physics is coupled into a chemical model there is excellent agreement, for a limited set of species, between theory and observations. We discuss our improved understanding of cloud chemistry can be used as a new tool for studies of the formation of stars and planetary systems.

Key words: astrochemistry – ISM: molecules – ISM: globules – ISM: clouds – stars: formation

1. Introduction

The formation of dense molecular condensations, and eventually stars, involves large changes in the physical properties of the atomic and molecular gas. These changes also have very specific consequences on the chemical interactions of the gas and dust inside the forming cores/stars. In particular as the density increases molecules in the gas phase collide with dust grains with greater frequency and, if the molecules stick with any reasonable efficiency, they will deplete from the gas phase. Thus, the density gradient that results from core condensation is accompanied by chemical gradients, with the inner parts of the core representing high density chemistry (with freeze-out onto grains), and the outer parts representing the original low density molecular composition. Indeed, these molecular depletions have been suggested as the primary chemical indicator of the earliest stages of the star formation process (Mundy & McMullin 1997; Bergin & Langer 1997). Such chemical structure should be particularly evident in dark clouds because the low temperatures ($\leq$ 10 K) preclude thermal evaporation of the molecules frozen in the mantle.

The process of star formation is by nature a dynamic one and molecular emission is the primary method for obtaining information on kinematic motions inside molecular clouds. Extensive efforts have been placed towards using molecular tracers, typically CS and HCO^+, to search for the presence of star-forming infall due to gravitational collapse (see Evans 1999; Myers, Evans, & Ohashi 2000 and references therein). The freeze-out of molecules onto grain surfaces significantly reduces the effectiveness of using molecular emission as a tracer of motions. Indeed the inside-out collapse models of Shu (1977) predict that the highest infall speeds are found in the exact regions where molecules are expected to freeze-out.

Despite these arguments it is only recently that detections of gas-phase freeze-out have become commonplace. Until recently inherent difficulties in extracting molecular abundances from gas-phase emission prevented definitive detections of gas-phase freeze-out. Previous studies assumed local thermodynamic equilibrium (LTE) to estimate total column densities. However, molecular emission from tracers with high dipole moments is far from LTE. In addition, secondary tracers such as CO and its isotopic variants were previously used to estimate the total H_2 column density, since it cannot be observed directly. It is now known that depletion of CO in cold regions seriously hampers its utility as an estimator of the total hydrogen column density.

The recent advent of sensitive continuum and heterodyne arrays probing millimeter/submillimeter wavelengths, along with wide-field infrared imaging devices, has led to an explosion of clear detections of gas-phase depletion/freeze-out (Bacmann et al. 2002; Bergin et al. 2002; Tafalla et al. 2002; Hotzel et al. 2002; Caselli et

Chemistry as a Diagnostic of Star Formation (2003). *Edited by Charles L. Curry and Michel Fich.* Published in 2003 by NRC Research Press, Ottawa ON, K1A 0R6, Canada. ISBN 0-660-19089-3.

al. 2002; Jessop & Ward-Thompson 2001; Kramer et al. 1999; Alves et al. 1999). At this conference alone there are 8 contributions discussing gas-phase depletions (Carey et al., Di Francesco et al., Feldman et al., Savva et al., Kontinen et al., Lai et al., Lee et al., and Peng et al., this volume). Observations of dust in emission or absorption provide direct knowledge of the dust distribution. In addition, since the dust column density and mass is correlated with the H_2 column density and mass (Hildebrand 1983; Gordon 1995), these observations provide the clearest information to date on the spatial distribution of H_2 molecules. Furthermore, with some geometrical assumptions, the radial profiles of core density can be constructed; these methods have greatly increased our knowledge of the physical conditions throughout the star formation process (André et al. 2000; Alves, Lada, & Lada 2001). Knowledge of the density and column density distribution of H_2 also significantly aids the molecular observations in two ways: (1) by indirectly confirming the location of the H_2 density and column density peak and; (2) by providing the density profile which helps to unravel the similar effects of density and abundance on excitation. These advances have moved chemical analyses beyond surveys of objects that show evidence for complicated chemistry (e.g., TMC-1, L134N) towards studies of more centrally concentrated objects that are closer to collapse and star formation.

In this concise review we outline how combined studies of dust emission/absorption with molecular emission have improved our ability to estimate molecular abundances and, in consequence, how our picture of dark cloud chemistry has been changed to one dominated by the effects of freeze-out. In the following (Sec. 2 & 3) we briefly outline our current theoretical understanding of gas-grain interactions. Furthermore we discuss key observational efforts that demonstrate the systematic nature of gas-phase freeze-out in condensed cores and how these observations can be qualitatively understood by a coupling of gas-grain chemical models to sophisticated radiative transfer models. In Sec. 4 we discuss how our improved understanding opens the door towards using astrochemistry as a new and potent tool to study the process of star formation.

2. Gas-Phase Freeze-Out: Theory

The rate of deposition of a molecule in the gas phase onto a grain surface is

$$k_{\rm freeze-out} = \pi a^2 \overline{v} S n_{\rm gr} \quad {\rm sec}^{-1}, \tag{1}$$

where a is the grain radius, $\overline{v}$ is the mean thermal velocity, S is the sticking coefficient, and $n_{\rm gr}$ the space density of grains. If we use "classical" grains with $a = 1000$ Å and $n_{\rm gr} = 1.3 \times 10^{-12}\,{\rm n}(H_2)$ cm^{-3} (Spitzer 1978), then the timescale for a molecule to freeze-out onto a grain surface is:

$$\tau_{\rm freeze-out} \sim \frac{2 \times 10^9}{Smn({\rm H_2})} \quad {\rm yr}, \tag{2}$$

where m is the molecular weight and S is the sticking coefficient. CO is a typical tracer of molecular cores which have typical densities of $n({\rm H_2}) \sim 10^5$ cm^{-3} (Evans 1999). For a sticking coefficient of unity the freeze-out timescale is $< 10^4$ year, well above the estimated time for thermal evaporation ($\gg 10^8$ yr), but below estimated ages of $\sim 1 - 10$ Myr (Palla and Stahler 2000; Hartmann 2001; Lee and Myers 1999). Thus it is surprising that a gas phase is observed at all, instead the near total freeze-out of the gas phase would be expected (e.g., Iglesias 1977). This points to the existence of some non-thermal mechanism to desorb or remove molecules from the grains and allow for the observed active gas-phase chemistry.

Various mechanisms have been identified, including spot heating due to cosmic-ray or X-ray impacts, chemical desorption, and others. The reader is referred to Williams (1993) for more detailed discussion. The strength of the molecule bond to the grain surface is a key parameter that determines the efficiency of evaporation. Because low temperatures preclude the breaking of chemical bonds, molecules are not expected to be chemically bound to grain surfaces. Instead the approaching molecule has an induced dipole interaction with the grain surface or mantle and is bound through weaker Van der Waals-London interactions called physical adsorption (Kittel 1996). Van der Waals interactions are proportional to the product of the polarizabilities of the molecule and the nearby surface species ($E_{\rm b} \propto \alpha_{\rm mol}\alpha_{\rm surface}$). Because of this property some species are more tightly bound to the grain surface than others. For example CS is expected to be more tightly bound to the grain surface than CO or N_2. In addition, theoretical analysis has suggested that N_2 has a weaker bond to a water ice surface than CO (Sadlej et al. 1995).

Models which have achieved the best success in reproducing observations are those that incorporate these assumptions regarding bond strengths, resulting in a desorption process that preferentially removes more weakly bound species (Bergin & Langer 1997; Charnley 1997; Nejad & Wagenblast 1999; Aikawa et al. 2001; Li et al. 2002). As as sample in Figure 1 we provide chemical abundances as a function of time for a cloud with a centrally concentrated density profile. This result uses the model for B68

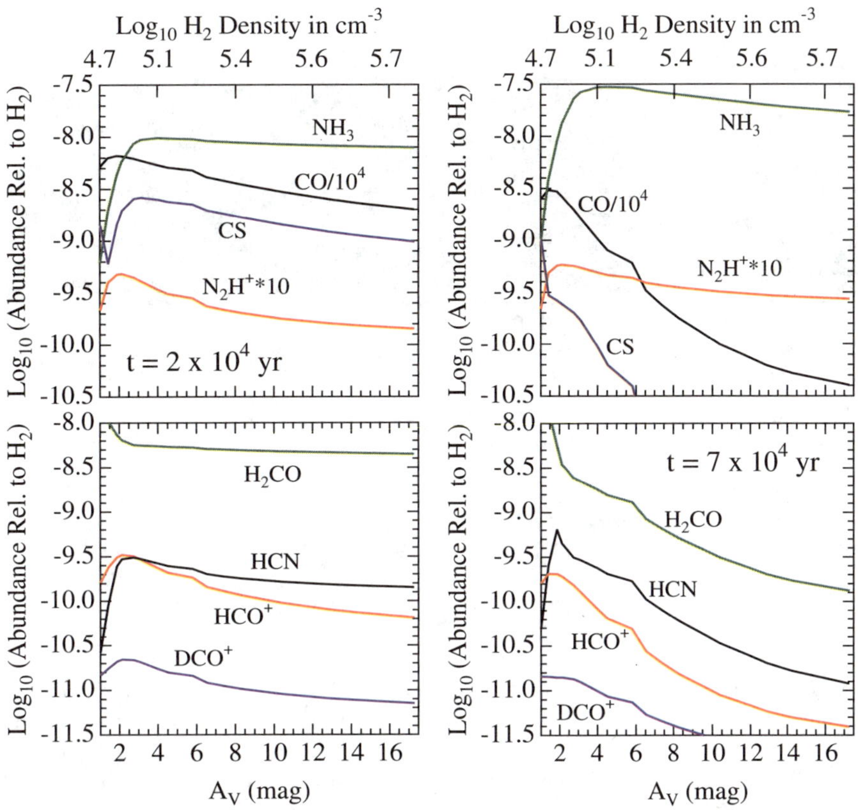

Figure 1. Left panels: Chemical abundances of select tracers as a function of cloud depth ($A_V = 1$ to 17 mag) from a time dependent model sampled at $t = 2 \times 10^4$ yr. *Right panels:* same, except for $t = 7 \times 10^4$ yr. The top axes show the density that corresponds to the extinction on the lower x-axis. The model shown is representative for the evolution of the chemistry in an object that is centrally concentrated such that the highest densities are present at the greatest extinctions. This gives rise to the observed chemical structure as the depletion rate increases with density.

discussed by Bergin et al. (2002) where the density profile is that for a Bonnor-Ebert sphere near-equilibrium that is kept constant with time as the chemistry evolves. This is a simplification as more realistic models couple the physical and chemical evolution (Bergin & Langer 1997; Aikawa et al. 2001; Li et al. 2002). Nonetheless, it is illustrative of the observed effects.

Figure 1 shows chemical abundances for species that are typically used to trace molecular clouds. At early times ($t = 2 \times 10^4$ yr) the effects of freeze-out are evident, but the overall effect is small. At later times ($t = 7 \times 10^4$ yr) strong gas-phase depletions are seen for nearly all species. CS has the largest abundance decrease followed by other carbon-bearing species. In contrast the nitrogen chemistry remains active at later times due to the presumed volatility of the N_2 molecule (see Bergin & Langer 1997). One by-product of CO depletion is an increase in the DCO^+/HCO^+ ratio (see Aikawa et al. 2001; Caselli et al 2002; Caselli 2002 for further discussion). In the following section we show that the pattern of differential depletions – early CS depletion, followed by CO, and finally N_2 – is observed in numerous sources. Similar effects would be observed in any model that incorporates these particular assumptions with regards to the gas-grain physics and evolves to a centrally concentrated state.

3. Gas-Phase Freeze-Out: Observations

Combined gas/dust investigations have resulted in the development of a new method to derive molecular abundances. This method derives abundance profiles as a function of radius, as opposed to a line of sight average. More specifically, the observed core density profile from the dust observations is used as an input to one-dimensional (1D) radiative transfer codes (Monte-Carlo is generally

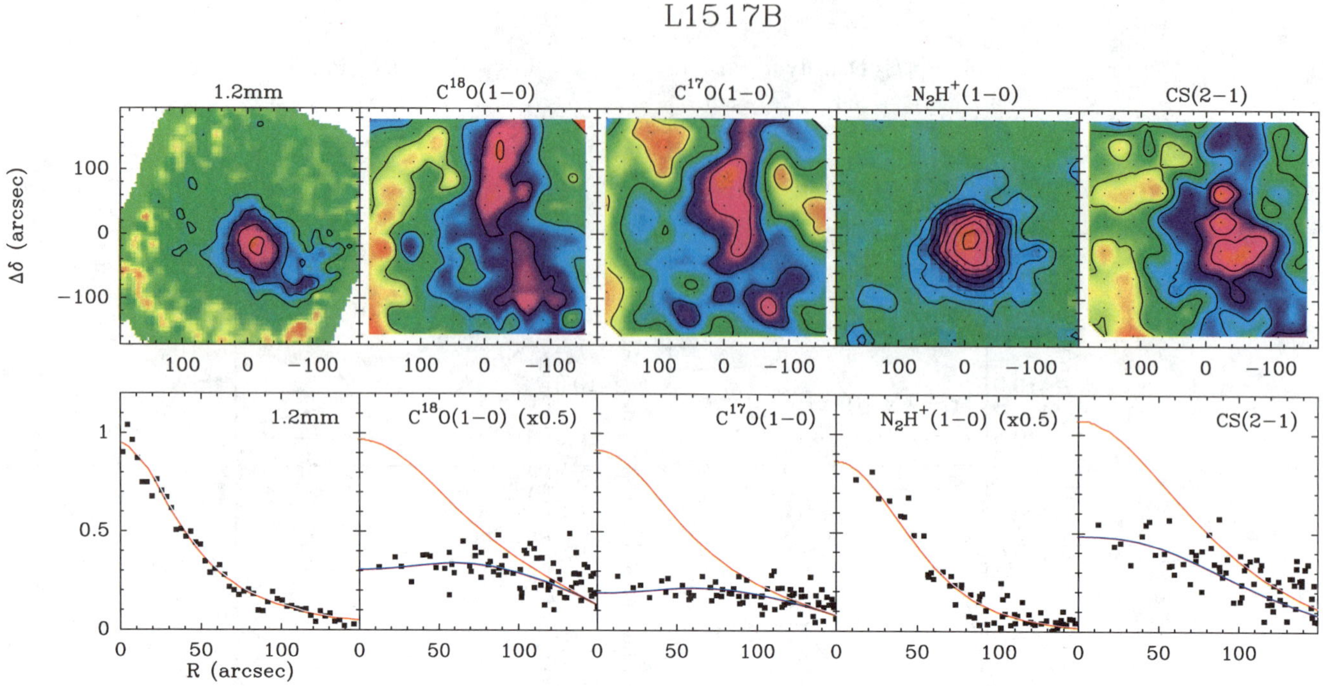

Figure 2. Top panels: Maps of 1.2 mm continuum, $C^{18}O$(1-0), $C^{17}O$(1-0), N_2H^+(1-0), and CS(2-1) emission in L1517B. *Bottom panels:* Radial profiles of continuum and integrated emissions. Observations are indicated by solid squares while the lines are results from models of the radiative transfer for two abundance profiles. The solid red lines are constant abundance models while the blue lines represent models with a rapid decrease in the central abundance. Taken from Tafalla et al. (2002).

used; e.g., Bernes 1979; Hogerheijde and van der Tak 2000). The temperature profile is another parameter that is constrained by dust thermal emission or by multi-transitional studies of molecular emission. Variables are the abundance profile and velocity structure (both bulk and turbulent). With these constraints, and iterating over the variables, the radiative transfer model predicts the radial profile of emission which can then be compared in detail to the observations. This technique has been outlined by van Dishoeck and Hogerheijde (1999) and applications can be found in van der Tak et al. (2000), Tafalla et al. (2002), Bergin et al. (2002), and Lee et al. (2002).

Below we discuss some recent investigations that highlight (1) the observed depletions are differential in nature and are systemic in the dense ISM (Tafalla et al. 2002); and (2) current gas-grain chemical models are capable of qualitatively matching observations of a limited set of key species (Bergin et al. 2002).

3.1. Tafalla et al. 2002

In the top panels of Figure 2 we show integrated molecular emission maps of the centrally concentrated molecular core L1517B for sample species with a map of the 1.2 mm continuum emission (Tafalla et al. 2002). The notable feature in this figure is the lack of correlation between the 1.2 mm continuum – a tracer of H_2 – with $C^{18}O$ and CS emission; where the continuum peaks the molecular emission exhibit local minima. In contrast, the N_2H^+ emission peak is coincident with the continuum emission maximum. An analysis of these data using the physical description of the source given by the dust continuum emission is shown in the lower panels (Tafalla et al. 2002). Here the radial profiles of emission (solid squares) are compared to model predictions. Models that assume constant abundance are shown as solid red lines; these models predict excess emission at the core center for the CO isotopes and CS. To reach agreement with observations the abundances of these species must be sharply reduced in the densest regions of the core (models shown as solid blue lines) – presumably through freeze-out. Constant abundance models provide a reasonable match for N_2H^+, while NH_3 requires a slight abundance increase towards the center.

Tafalla et al. also compare the observations directly to the chemical models of Bergin & Langer (1997), finding good agreement with models of fast-collapse. Most significantly similar results are found for all five cores exam-

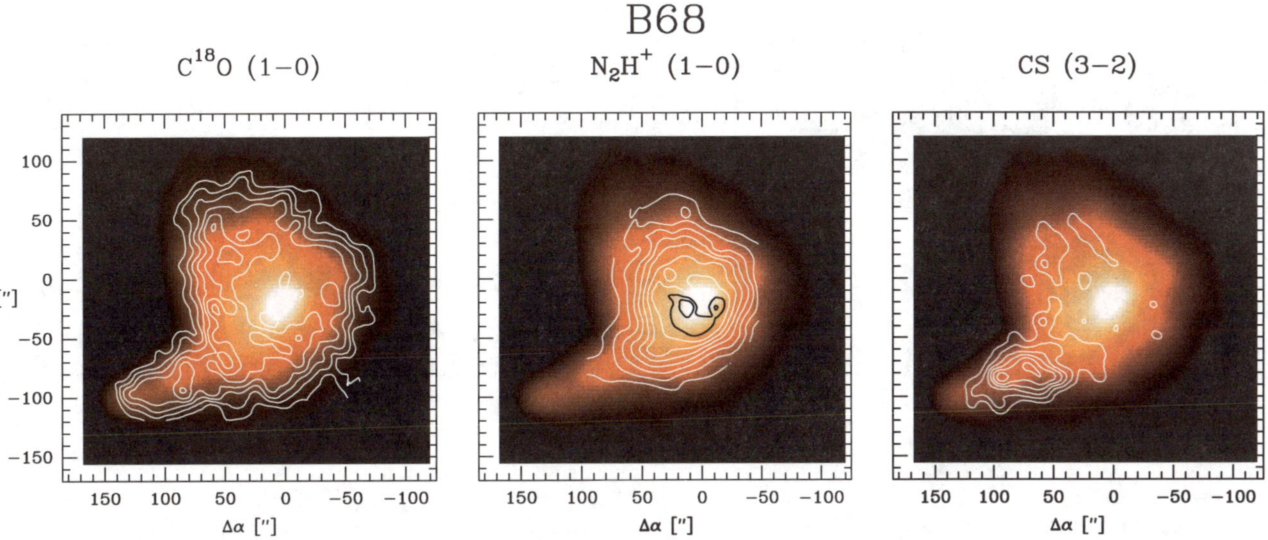

Figure 3. The observed emission distribution of visual extinction (*color*) and total column density (*gray scale*) in the B68 dark cloud. Contours are integrated emission maps of $C^{18}O$ (*left-hand panel*), N_2H^+ (*middle*), and CS (*right-hand panel*). These emission distributions show a sequence of increasing molecular depletion from CS, to $C^{18}O$, and even N_2H^+ (Bergin et al. 2002; Lada et al. 2002).

ined in this work, as well as in sources observed by other studies (Kuiper, Langer, & Velusamy 1996; Ohashi et al. 1999; Lee et al. 2002; Bergin et al. 2002). This demonstrates the systematic nature of depletions: they are observed in nearly all centrally concentrated low mass cores. *Thus, differential depletions in cold dark clouds must be a chemical signature of the earliest stages of star formation.*

3.2. Bergin et al. 2002: B68

The dark cloud B68 is another example of a centrally concentrated object that exhibits emission features which are attributed to gas-phase freeze-out. In this core the H_2 distribution is traced through a map of dust visual extinction (A_V) derived via near-infrared extinction mapping techniques (e.g., Alves, Lada, and Lada 2001). In Figure 3 we present the B68 A_V map along with a series of molecular emission maps. In this core we see that the N_2H^+ emission peaks in a shell partially surrounding the peak of dust extinction. Moreover, the N_2H^+ peaks inside the much larger $C^{18}O$ emission hole, which itself lies inside the CS emission depression. Analysis of these data differed from Tafalla et al. (2002) in that a gas-grain chemical model is directly linked to a radiative transfer model predicting the profile of integration emission. This allows for chemical considerations to determine how the abundance is structured with depth. For example, photodissociation limits the abundance at cloud edges and the competition between depletion and desorption determines how the abundance declines with density.

Figure 4 presents an analysis of the N_2H^+ and $C^{18}O$ emission; in this case the open squares show the integrated emission as function of visual extinction. This particular representation of the data is independent of cloud geometry, depending only on physical variables. Furthermore, if the emission is in LTE and optically thin the integrated emission from a source with constant abundance would appear as a straight line in this diagram (e.g., Frerking et al. 1982; Lada et al. 1994). Clearly the $C^{18}O$-A_V relation deviates from a straight line at high extinctions. Such structure in this relation could be due to freeze-out or high opacities. To discriminate between these possibilities, observations of a lesser abundant isotope are typically needed (and in this case confirm freeze-out). The dashed line is the "best fit" radial abundance profile predicted in the chemical model and the solid line is the resulting emission profile. For $C^{18}O$ the observations require significant freeze-out and also an abundance reduction at low extinction due to photodissociation. In the case of N_2H^+ an abundance reduction at low extinction is required, but only a small decrease in abundance at high A_V is needed to match observations.

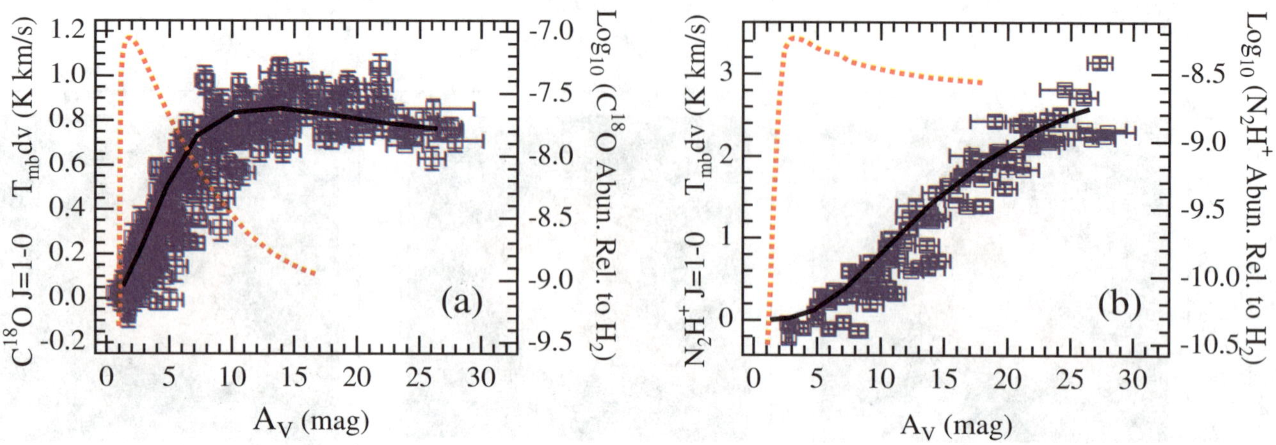

Figure 4. Point-by-point comparison of $C^{18}O$ J=1-0 and N_2H^+ integrated intensity with visual extinction for the entire B68 dark cloud. In all plots the data are presented as open squares with error bars, while solid curves represent the emission predicted by a model combining chemistry with a Monte-Carlo radiative transfer code. The dashed lines are the best-fit molecular abundance profiles with the axis labeled to the right (abundance of N_2H^+ is multiplied by 100). In a symmetric model, the abundance profiles represent the radial chemical structure within a pencil beam towards the extinction peak and thus are shown only for one-half of the total maximum extinction. Adapted from Bergin et al. (2002).

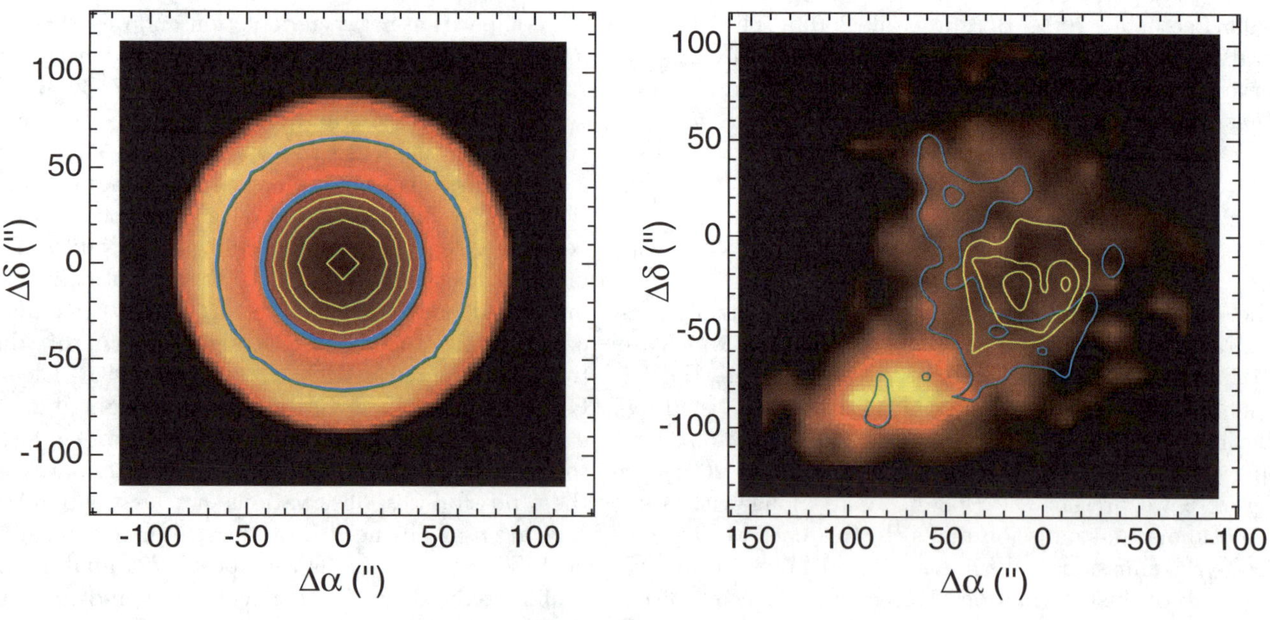

Figure 5. Comparison of model emission predictions with observational data. *Left:* Color scale map of CS ($J = 3-2$) emission predicted from the B68 chemical model. Also shown are the peak contours of $C^{18}O$ $J = 1-0$ (*blue*) and N_2H^+ $J = 1-0$ (*yellow*). *Right:* Similar plot of the observed emission in B68. Although the core clearly shows deviations from spherical symmetry there is a close correspondence between model predictions and observations.

In Figure 5 we present a more qualitative comparison of the structure predicted by the chemical models to the B68 data. Here the left-hand panel represents the peak contours predicted by a spherical radiative transfer model combined with the chemistry for transitions of CS, $C^{18}O$ and N_2H^+. The right-hand panel shows an identical representation of the observed emission in B68. Although clear divergences from spherical symmetry are evident there exists striking agreement between model and observation. The results presented here and elsewhere suggests that we are making progress in our understanding of the chemistry in these cold regions (e.g., Tafalla et al. 2002; Aikawa et al. 2001; Li et al. 2002; Lee et al. 2002). The decrease in the N_2H^+ abundance is also significant as this suggests that its parent molecule (N_2) is freezing onto grain surfaces. If this result is observed beyond B68 it could have important implications for our ability to use molecules as the tracers of star formation dynamics.

3.3. Summary of Trends

The above results are excellent examples of the trends that are becoming apparent in chemical studies of low mass dark clouds. These trends have suggested changes in the way molecular emission is used to probe star-forming cores.

1. The use of $C^{18}O$ as a probe of the H_2 column density in regions below the sublimation temperature of CO ($\sim$ 20 K) is no longer appropriate. Whether this extends to warmer cores in giant molecular clouds remains an open issue (see Mauersberger et al. 1992; Chandler & Carlstrom 1996; Gibb & Little 1998; Savva et al., this volume)

2. Due to freeze-out, the CS abundance can be sharply reduced when $n > 10^4$ cm^{-3} and is no longer a reliable tracer of motions in high density gas. However, this molecule is still a good probe of low density molecular gas prior to condensation and after star formation.

3. The emission from N_2H^+ and NH_3 most closely resembles the maps of dust emission/extinction. N_2D^+, DCO^+, and NH_2D are also found to be well correlated with the dust (Caselli et al. 2002; Shah and Wooten 2001). These tracers are emerging as the best probes of the innermost regions of dense cores prior to the formation of a star.

4. Once a star has been born the heating of the inner envelope returns some tracers to the gas phase. This has not specifically been addressed in this paper but the reader is referred to recent work in this area, including: Hogerheijde et al. (1999), Maret et al. (2002), Jørgensen et al. (2002), and references therein.

4. The Future of Dark Cloud Astrochemistry

For a few key species we are now finding a convergence between observation and theory. This opens the door towards using chemistry as a new tool to study the process of star formation. Below we discuss how this can be accomplished through both observational and theoretical efforts.

4.1. Observations

One new tool is the isolation of specific tracers that either do not freeze-out or freeze onto grain surfaces at late stages. Observations of these tracers (e.g., N_2H^+) over small or large scales can provide complementary kinematical information to continuum observations. For instance, in regions that are currently forming stellar clusters, we can investigated the core-core velocity dispersion and dense core gas motions relative to the overall structure (see Walsh; Hogerheijde, this volume). On similar scales combined studies of an "early" depleter (e.g., CS) and a high density tracer (e.g., N_2H^+) can also take advantage of the chemical evolution to detect cores prior to condensation (CS), during condensed stages (N_2H^+) and after star formation (see Lee et al. 2002; Lee, this volume).

The presence of differential depletions can certainly reduce the effectiveness of using molecular emission as a tracer of star formation kinematics, However, with careful choice of tracers one can use our expanded chemical knowledge to reconstruct the radial profile of kinematical motions. Indeed it is likely that *the key to fully unraveling star formation dynamics lies in our improved understanding of the chemistry.* One example is in B68 where the emission lines of $C^{18}O$ and N_2H^+ are characterized by a systematic and well defined velocity gradient across the face of the cloud. However, the magnitude and direction of the velocity gradient is different between the two tracers. The chemical analysis presented in Figure 4 shows that N_2H^+ emission traces the innermost regions of the B68 core, while emission from $C^{18}O$ predominantly traces the outer layers. Thus, if the gradient is due to rotation, then the combined chemical/dynamical analysis suggests that the core is differentially rotating (Lada et al. 2002).

Another example of the importance of chemistry to dynamics is shown in Figure 6. The left-hand panel shows

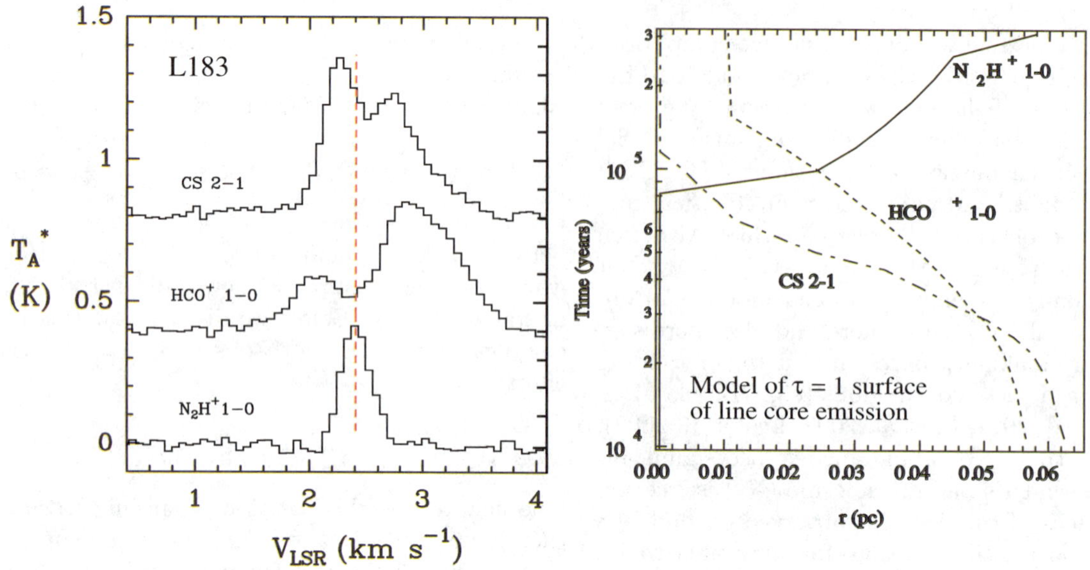

Figure 6. Left: Comparison of CS $(2-1)$, HCO^+ $(1-0)$, and N_2H^+ $(1-0)$ spectra taken towards the same position in the dark cloud L183. The vertical line shows the systemic velocity of the cloud as determined by the optically thin N_2H^+ emission. The CS and HCO^+ optically thick spectra show evidence for motions along the line of sight. However, for CS the spectrum is consistent with infall, while HCO^+ appears to be consistent with outflowing motions (Lee & Myers, in preparation). *Right:* Plot showing the emission surface for selected molecular transitions from a combined chemical and radiative transfer model of B68. The $x-$axis is the radius position in the core and the $y-$axis shows the chemical evolution. The curves delineate the $\tau = 1$ surface for the core of the spectral line of selected molecular transitions. All positions to the left of a given line trace greater opacities and are not probed by observations. When optically thick molecular line is used to trace motions via asymmetries in the profile, the emission preferentially probes the $\tau = 1$ surface and hence the motions at that depth. This figure demonstrates that as the chemistry evolves, each molecule is sensitive to motions in different regions along the line of sight.

spectra of 3 molecules taken towards the same position in the L183 molecular core. The double peaked structure observed in the optically thick CS and HCO^+ emission is now commonly used as a tracer of motions along the line of sight (e.g., Evans 1999). However, in L183 the CS profile is consistent with infalling motions, while HCO^+ is consistent with outflow. Since both of these transitions sample the same level of excitation this set of conflicting data must be due to chemistry. In particular, the answer must lie in the differences in the abundance structure of each molecule, and changes in the velocity field, as a function of depth.

A qualitative example of how these tracers could sample different motions along the line of sight is shown in the right-hand panel of Figure 8. This example uses the combined chemical/radiative transfer model constructed for B68. In the plot the x-axis is the radial position inside the cloud with a radius of 0.065 pc, and the y-axis shows the time evolution. The lines delineate the $\tau = 1$ surface of the line core emission. All positions to the left of a given line for a particular transition have higher opacities. For example, for $t < 8 \times 10^4$ yr, emission from N_2H^+ is optically thin and traces the entire line of sight through the cloud. At later times the abundance rises and the emission becomes thick and only probes a surface that is successively further from the core center.

When optically thick molecular transitions are used to probe motion via asymmetric self-reversals the emission preferentially probes the $\tau = 1$ surface and hence the magnitude and direction of motions at that depth. *This plot clearly demonstrates that as the chemistry evolves the various tracers are sensitive to different layers within the cloud.* For instance in this model CS $(J = 2-1)$ and HCO^+ $(J = 1-0)$ initially both probe the outer layers. As the cloud evolves, CS differentially depletes and the $\tau = 1$ surface retreats deeper into the cloud. Eventually (for $t > 3 \times 10^4$ yr), HCO^+ emission traces outer layers, while CS probes the inner layers. This plot is not supplied as a template for the actual evolution of any particular core, rather it is shown as an example of how evolutionary changes in the chemical composition can affect observed emission, and more importantly the interpretation. Therefore, with observation of several species, and knowledge of the chemical abundance and physical profiles, another powerful new tool is the ability to search for structure in the radial velocity field, a key to understanding star formation.

5. Theory

Our understanding of chemical theory in dark clouds also must evolve. Areas that clearly need further investigation include comparison of models to observations of similar cores (L1544, L1517B, B68) in species beyond the typical suite of tracers (e.g., CO, CS, N_2H^+). For example, Lee et al. (2002) find that the abundance of CCS is depleted in the the L1512 and L1544 cores, but find evidence for a small abundance rise in the center. This would be in agreement with the models of Ruffle et al. (1997) that suggest a rebound in the abundances of complex molecules due to CO freeze-out. Do other complex molecules provide similar results and can we further refine our understanding of gas-phase pathways? This moves the testing of chemical models beyond comparison to abundances derived in template sources, such as TMC-1 and L134N, to observational comparisons with more centrally condensed objects that have well defined physical properties. Abundances derived in these sources still have uncertainties due to calibration, collision rates, and geometrical assumptions, but can generally be considered more reliable than simple line of sight averages. Finally one challenge for future theory is to place the above observations in context of the non-detections of water vapor and molecular oxygen in cold cores (Snell et al. 2000; Bergin & Snell 2002). Some efforts have begun to examine this question (Viti et al 2001; Charnley et al 2001; Roberts & Herbst 2002).

Models that combine chemistry with dynamics have become increasingly more sophisticated. The ultimate goal of these efforts is to use chemical observations to refine our understanding of star formation theory. This goal is noteworthy and it remains to be seen whether the chemical unknowns regarding gas-grain interactions dominate (e.g., sticking coefficients, binding energies, grain sub-structure, surface chemistry, exact desorption process). Combined models have recently begun to attempt to reproduce not only the pattern of depletions, but also the column densities observed in L1544, a new template source (Aikawa et al. 2001; Li et al. 2002). This commendable step is needed to move theory beyond qualitative towards quantitative comparisons with observations. However, in Sec. 3.2 we demonstrated that we now have the capability to place model predictions into radiative transfer models and predict direct observables. This is important because column density estimates depend on assumptions regarding excitation that may not be reproduced in the model at the given time where model column densities best match observational estimates. Hence one could reproduce the column density and not the observed emission because of differences in the radial profile of physical conditions. This task is time consuming, but sophisticated radiative transfer codes are now readily available (see Hogerheijde & van der Tak 2002). The combination of chemical/dynamical models with radiative transfer truly offers the best opportunity to take chemical models and astrochemistry to the next level and potentially begin to test star formation theory.

6. Summary

The chemical processes that result in the formation of molecules play an important role in star and planetary formation through molecular contributions to the stability of the cloud against gravitational forces. Given the widespread nature of gas-phase freeze-out it is becoming increasingly evident that knowledge of the chemistry is also required in order to fully characterize the star formation process. The future of dark cloud chemical studies is promising as more complex and predictive models will be developed that can be compared to observations with increasing resolution and sensitivity. This will lead to crucial tests of chemical theory but also presents a completely new approach to star formation studies, bringing astrochemistry into a wider realm as a potent tool.

Acknowledgements

We are grateful to I. Bergin for proofing the manuscript and M. Tafalla for providing comprehensive figures for this presentation. We are also grateful for our collaboration with Charlie Lada, João Alves, and Tracy Huard which made the B68 analysis possible.

References

Aikawa, Y., Ohashi, N., Inutsuka, S., Herbst, E., & Takakuwa, S. 2001, ApJ, 552, 639

Alves, J., Lada, C. J., & Lada, E. A. 2001, Nature, 409, 159

Alves, J., Lada, C. J., & Lada, E. A. 1999, ApJ, 515, 265

André, P., Ward-Thompson, D., & Barsony, M. 2000, in *Protostars and Planets IV*, ed. V. Mannings, A.P. Boss, & S.S. Russell(Tucson: Univ. of Arizona Press), 59

Bacmann, A., Lefloch, B., Ceccarelli, C., Castets, A., Steinacker, J., & Loinard, L. 2002, A&A, 389, L6

Bergin, E.A. & Snell, R.L. 2002, ApJ, 581, L105

Bergin, E. A., Alves, J., Huard, T., & Lada, C. J. 2002, ApJ, 570, L101
Bergin, E. A. & Langer, W. D. 1997, ApJ, 486, 316
Bernes, C. 1979, A&A, 73, 67
Caselli, P. 2002, Plan. Sp. Sci., 50, 1133
Caselli, P., Walmsley, C. M., Zucconi, A., Tafalla, M., Dore, L., & Myers, P. C. 2002, ApJ, 565, 344
Caselli, P., Walmsley, C. M., Tafalla, M., Dore, L., & Myers, P. C. 1999, ApJ, 523, L165
Chandler, C. J. & Carlstrom, J. E. 1996, ApJ, 466, 338
Charnley, S. B., Rodgers, S. D., & Ehrenfreund, P. 2001, A&A, 378, 102
Charnley, S. B. 1997, MNRAS, 291, 455
Evans, N. J. 1999, ARA&A, 37, 311
Frerking, M. A., Langer, W. D., & Wilson, R. W. 1982, ApJ, 262, 590
Gibb, A. G. & Little, L. T. 1998, MNRAS, 295, 299
Gordon, M. A. 1995, A&A, 301, 853
Hartmann, L. 2001, AJ, 121, 1030
Hildebrand, R. H. 1983, QJRAS, 24, 267
Hogerheijde, M. R. & van der Tak, F. F. S. 2000, A&A, 362, 697
Hogerheijde, M. R., van Dishoeck, E. F., Salverda, J. M., & Blake, G. A. 1999, ApJ, 513, 350
Hotzel, S., Harju, J., Juvela, M., Mattila, K., & Haikala, L. K. 2002, A&A, 391, 275
Iglesias, E. 1977, ApJ, 218, 697
Jessop, N. E. & Ward-Thompson, D. 2001, MNRAS, 323, 1025
Jørgensen, J. K., Schöier, F. L., & van Dishoeck, E. F. 2002, A&A, 389, 908
Kittel, C. 1996, *Introduction to Solid State Physics* (New York: Wiley)
Kramer, C., Alves, J., Lada, C. J., Lada, E. A., Sievers, A., Ungerechts, H., & Walmsley, C. M. 1999, A&A, 342, 257
Kuiper, T. B. H., Langer, W. D., & Velusamy, T. 1996, ApJ, 468, 761
Lada, C. J., Bergin, E. A., Alves, J., & Huard, T. 2002, ApJ, 586, 286
Lada, C. J., Lada, E. A., Clemens, D. P., & Bally, J. 1994, ApJ, 429, 694
Lee, J.-E., Evans, N.J., Shirley, Y.L., & Tatematsu, K. 2003, ApJ, 583, 789
Lee, C. W. & Myers, P. C. 1999, ApJS, 123, 233
Li, Z., Shematovich, V. I., Wiebe, D. S., & Shustov, B. M. 2002, ApJ, 569, 792
Mauersberger, R., Wilson, T. L., Mezger, P. G., Gaume, R., & Johnston, K. J. 1992, A&A, 256, 640
Maret, S., Caux, E., & Ceccarelli, C. 2002, Ap&SS, 281, 139
Mundy, L. G. & McMullin, J. P. 1996, in *IAU Symp. 178: Molecules in Astrophysics: Probes & Processes*, ed. E.F. van Dishoeck, 183
Myers, P. C., Evans, N. J., & Ohashi, N. 2000, in *Protostars and Planets IV*, ed. V. Mannings, A.P. Boss, & S.S. Russell(Tucson: Univ. of Arizona Press), 217
Nejad, L. A. M. & Wagenblast, R. 1999, A&A, 350, 204
Ohashi, N., Lee, S. W., Wilner, D. J., & Hayashi, M. 1999, ApJ, 518, L41
Palla, F. & Stahler, S. W. 2000, ApJ, 540, 255
Roberts, H., & Herbst, E. 2002, A&A, 395, 233
Ruffle, D. P., Hartquist, T. W., Taylor, S. D., & Williams, D. A. 1997, MNRAS, 291, 235
Sadlej, J., Rowland, B., Devlin, J. P., Buch, V. 1995, J. Chem. Phys., 102, 4804
Shah, R. Y. & Wootten, A. 2001, ApJ, 554, 933
Shu, F. H. 1977, ApJ, 214, 488
Snell, R. L. et al. 2000, ApJ, 539, L101
Spitzer, L. Jr. 1978, *Physical Processes in the Interstellar Medium* (New York: Wiley)
Tafalla, M., Myers, P. C., Caselli, P., Walmsley, C. M., & Comito, C. 2002, ApJ, 569, 815
Williams, D. A. 1993, in *Dust and Chemistry in Astronomy*, ed. T.J. Millar & D.A. Williams (Bristol: Institute of Physics), 143
van der Tak, F. F. S., van Dishoeck, E. F., & Caselli, P. 2000, A&A, 361, 327
van Dishoeck, E. F. and Hogerheijde, M. R. 1999, in *The Origin of Stars and Planetary Systems*, ed. C.J. Lada & N.D. Kylafis (Dordrecht: Kluwer), 97
Viti, S., Roueff, E., Hartquist, T. W., Pineau des Forêts, G., & Williams, D. A. 2001, A&A, 370, 557

L134N (L183) REVISITED: DUST CONTENT AND GAS DEPLETION

Laurent Pagani[1], Juan-Ramon Pardo[2], Laurent Cambrésy[3], Frédérique Motte[4], Aldo J. Apponi[5], Bertrand Stepnik[6], and Michel Fich[7]

[1]LERMA & FRE 2460 du CNRS, Observatoire de Paris, 61, A. de l'Observatoire, 75140 Paris, France
[2]Instituto de Estructura de la Materia, Dpto de Física Molecular, CSIC, Madrid, Spain
[3]Observatoire Astronomique de Strasbourg, UMR 7550, Université Louis Pasteur, 67000 Strasbourg, France
[4]California Institute of Technology, Downs Laboratory of Physics, 320-47 Pasadena, CA 91125, USA
[5]University of Arizona, Steward Observatory, 933 N. Cherry Ave., Tucson, AZ 85721, USA
[6]IAS, Bât. 121, Université Paris-Sud, 91435 Orsay, France
[7]Physics Department, University of Waterloo, Waterloo, Ontario, Canada, N2L 3G1

Abstract

L134N is a cold, starless cloud, very high above the galactic plane, close to us and well-delineated in continuum dust emission maps. This cloud is considered to be representative of oxygen-rich dark clouds (with the presence of SO, SO_2, NO, ..., etc.). It is thus a good reference, together with TMC-1, to test astrochemical models. Thanks to *ISO*, SCUBA, MAMBO and near-IR wide-field cameras, the detailed study of the dust has become possible in such cold and dark clouds. In parallel, progress in radio telescope receiver sensitivity now permits the mapping of weak lines on large surfaces. We have thus started a project to study both dust and a few gaseous key species (CO, CS, SO and N_2H^+) to address several questions. As first steps we want to assess the quantity of dust and gas everywhere in the cloud, study possible $C^{18}O$ and/or $C^{17}O$ depletion towards dense cores, evaluate the structure of the gas and the abundance of CS and SO to possibly estimate the chemical age of the cloud (time-dependent models show that the CS/SO ratio diminishes with time), and evaluate the rare isotope abundances, especially ^{17}O and ^{34}S. To constrain the molecular abundances with the highest possible confidence, we have observed several transitions for each species and each isotopomer. Preliminary results have been reported in Pagani et al. (2002) and more results are presented here. A large core is seen in the SCUBA and MAMBO maps with a mass of about 1.5 $M_\odot$ and with a CO depletion factor of 35. This core is not seen in the 200 μm ISOPHOT map, making ISOPHOT a doubtful tool to search for really cold cores.

Key words: ISM: molecules – ISM: dust,extinction – ISM: abundances – ISM: individual: L183 – ISM: individual: L134N – Galaxy: abundances

1. Introduction

Studying starless clouds is a fundamental first step in understanding how stars form. Because these clouds possess no internal energy input, no large kinematic motions and no grain mantle evaporation, they are cold and unperturbed. As a result, the chemistry and radiative transfer in these clouds are much easier to understand and analyze. Still their structure remains complex, as shown by the mapping of different species; e.g., in L134N, where Swade (1989a) has shown that each species traces a different cloud shape.

L134N was recognized early on as a good candidate to test chemical models. Considered as an oxygen-rich cloud, it has often been compared to TMC-1 (a carbon-rich cloud) to test chemical model predictions. However, to test these models, species abundances need to be accurately quantified. Until recently, this has rarely been the case and astrochemistry relies on measurements which are sometimes erroneous by as much as an order of magnitude. Because these models were still far from maturity, the problem was of limited importance but recent progress on our understanding of the dust, depletion, and evaporation phenomena makes better quantification of at least the most abundant species a necessity. A better understanding of the physical structure of the source is also mandatory for both the chemistry (reaction speed depends on density, e.g.) and to help evaluate the excitation parameters of each species.

L134N is ideally suited for such studies because it is close (100 pc; Mattila 1979; Franco 1989), rather large, and high above the galactic plane ($b^{II}= 37°$). The reference position of the source in this study is : $\alpha_{2000}= 15^h54^m06.6^s$, $\delta_{2000}= -2°52'19.1''$ and the LSR velocity is 2.5 km s^{-1}. Among the studies that have focused on L134N, one finds several molecular studies (Swade 1989a; Swade 1989b; Swade & Schloerb 1992; Dickens et al. 2000) which have revealed the large diversity of spatial

Chemistry as a Diagnostic of Star Formation (2003). *Edited by Charles L. Curry and Michel Fich.* Published in 2003 by NRC Research Press, Ottawa ON, K1A 0R6, Canada. ISBN 0-660-19089-3.

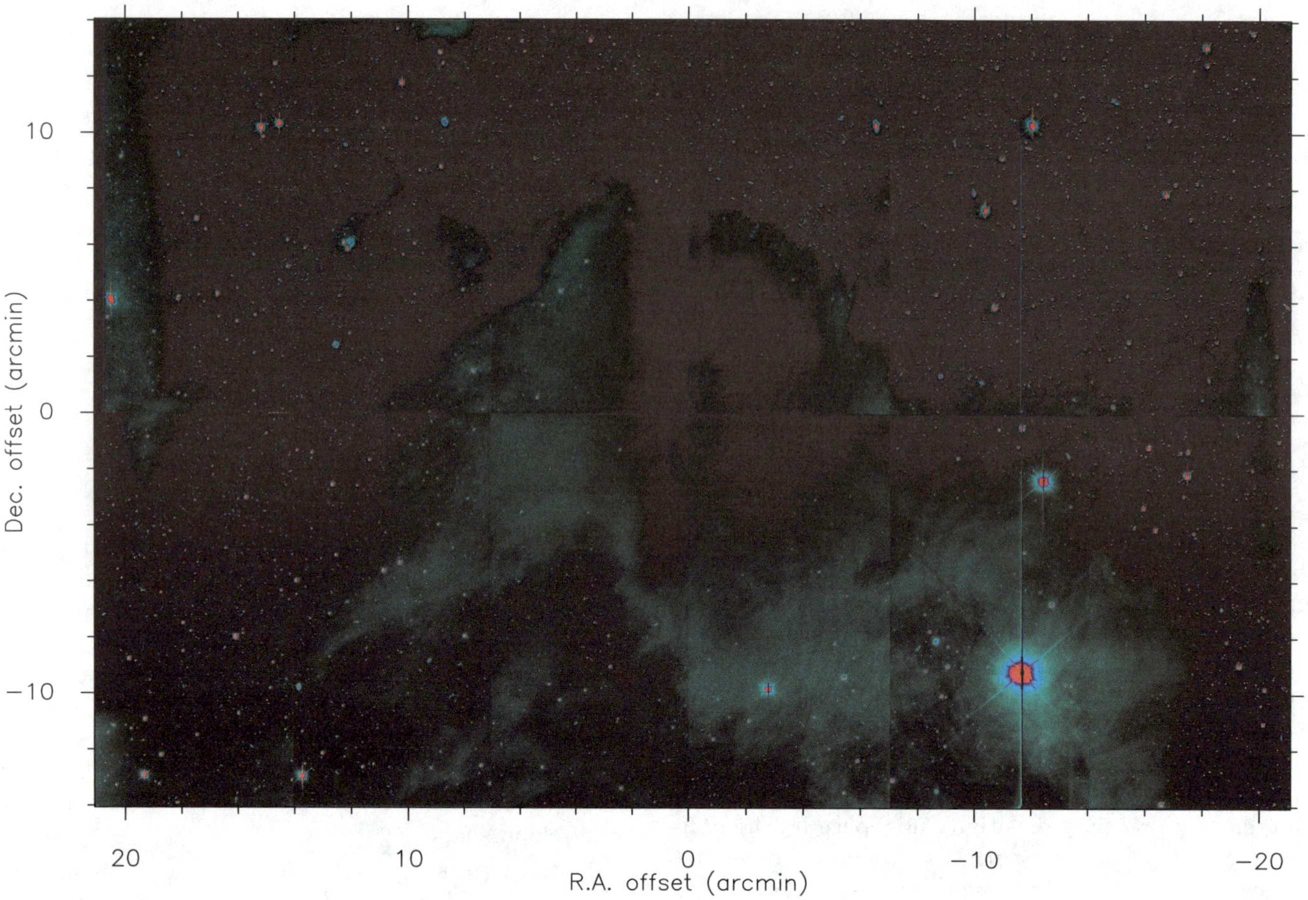

Figure 1. I-band image of L134N with the CFHT12K.

distribution for different molecules in L134N, while *IRAS* FIR (Laureijs et al. 1991; Laureijs et al. 1995), SCUBA and MAMBO work (Ward-Thompson et al. 1994; Ward-Thompson et al. 1999) have revealed a cold condensation coincident with the densest part of the molecular cloud. *ISO* has also observed this cloud with both ISOCAM (Pagani et al. 2003) and ISOPHOT (Pagani et al. 2002; Juvela et al. 2002; Ward-Thompson et al. 2002). With the low spatial resolution ISOPHOT data, Pagani et al. (2002) have shown that this source suffers from a CO depletion of at least a factor of 2 over a large area. This was subsequently confirmed by a re-analysis of the same ISOPHOT data by Juvela et al. (2002).

We have undertaken a large study of this cloud, covering the main part of the cloud core (15′×15′). For the gas component, we have concentrated on four species at the moment: namely, CO, CS, SO and N_2H^+. For each species, at least two isotopomers have been observed in at least two transitions (except N_2H^+). CO traces total column density including diffuse gas, while CS is a widespread tracer of dense gas. However, both suffer from depletion in the densest parts of the cloud and N_2H^+, which is less subject to depletion, is used to trace the very dense gas and dust (see Tafalla et al. 2002 for a similar study in other protostellar cores). To study the dust and its properties, we have analyzed the publicly available ISOPHOT data and have acquired large-scale maps with SCUBA (at 850 μm) and MAMBO (at 1250 μm) to measure the dust emission, and we have also performed near-infrared observations at CFHT (IHK bands) to measure the dust absorption. A series of papers in A&A will detail all this work. Preliminary results are presented here and in Pagani et al. (2002).

2. Preliminary results

2.1. Near-Infrared observations

Recent observations in the near-infrared have included a large scale (48′×36′) image in band I with the CFHT12K camera to measure the low extinction of the cloud envelope using star counts (Fig. 1). The results show that the method is valid up to $A_V \approx 15$ mag (by comparison with the H-K results). The inner core (the one mapped in $C^{18}O$) has been mapped with the CFHT infrared camera

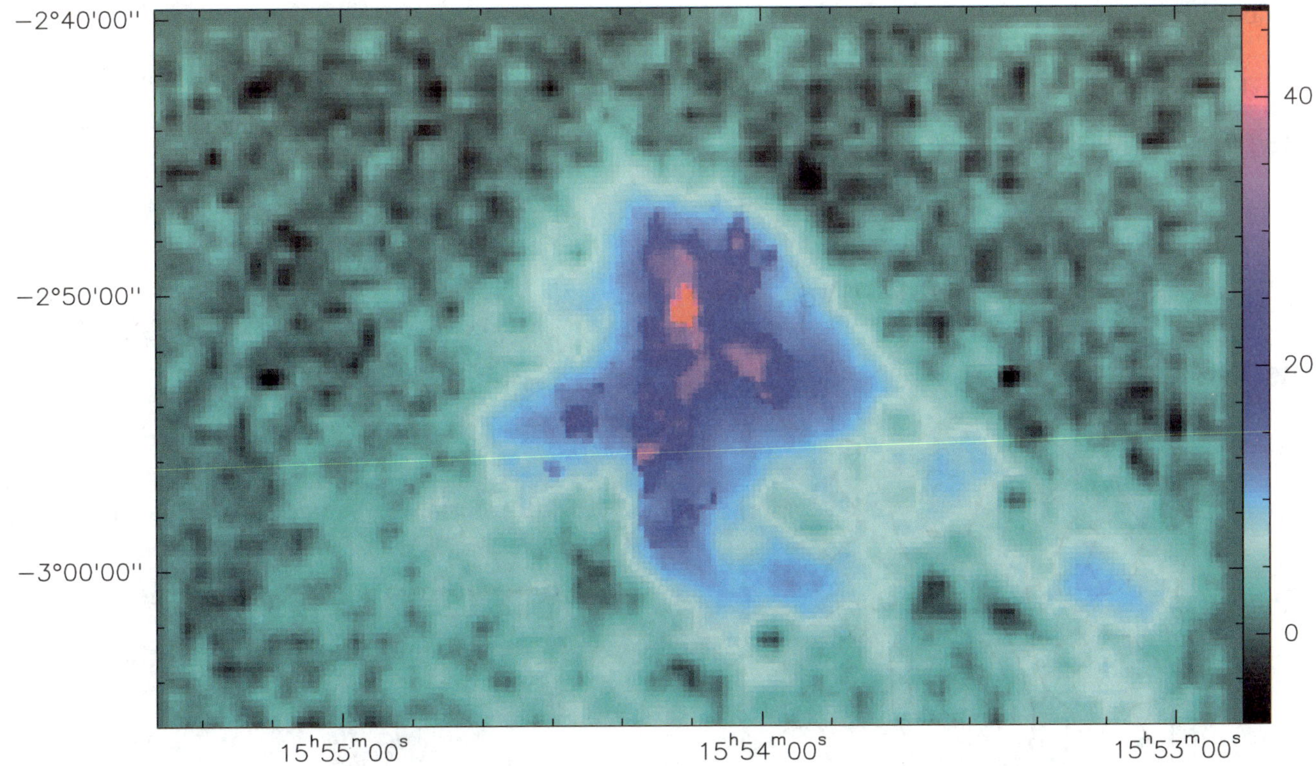

Figure 2. A_V image of L134N from the I- and HK-band image combination. Scale is in magnitudes.

in the H and K bands. The camera covers $3.6' \times 3.6'$ on the sky and we have covered the core with 25 different exposures, with exposure time adapted to the expected opacity (a few minutes at the borders, several hours in the center). I-band calibration has been done using the DENIS survey and HK calibration has been done with the 2MASS survey. The H-K reddening method described by Lada et al. (1994) has been used to measure the reddening. Because we are high above the galactic plane, the mean H-K value in the unreddened region is ≈ 0.5 mag, instead of the mean 0.13 mag which applies for stars in our local Galaxy. This is due to the fact that we are dominated by unresolved galaxies at high redshift. The global extinction map combining both I and H-K results is shown in Figure 2. Values above 40 mag in the inner core are not reliable because this part of the core shows no sources at all with a limiting extinction of $A_V \approx 40$ mag (Fig. 3). If we superimpose the MAMBO emission data on this core we find that the starless core corresponds to the 20 mJy per beam level of the MAMBO emission. Thus the MAMBO peak intensity of 70 mJy per beam is indicative of a dust opacity of ≈ 140 mag. This is an upper limit since large-scale emission is lost in the MAMBO chopper observing mode. If we lose as much as 20 mJy per beam, the contrast is then only a factor of 2.25 and the peak opacity would then be only 90 mag. This is much higher than the preliminary value ($A_V = 10$ mag, incorrectly underestimated by 40%) we reported in Pagani et al. (2002) and which was corrected (to $A_V = 17$ mag) by Juvela et al. (2002). As we warned at that time, the maximum A_V was probably underestimated due to both the low resolution and the short wavelength of the ISOPHOT measurements. The latter at 100 and 200 μm fall beyond a 12 K blackbody peak in the Wein regime and small temperature errors induce large opacity estimate errors. Another possibility is that the core may be much colder than 12 K and even the 200 μm emission would then fail to trace it.

The MAMBO core is $84''$ in diameter, with an average of 44 mJy per beam which converts to $A_V \approx 57$–71 mag, depending on the lost large-scale emission offset we reintroduce. This gives a mass of 1–1.5 $M_\odot$ for the core and, if the core is spherical, a volume density of 7–9 $\times 10^5$ cm^{-3}. This core was first seen by Ward-Thompson et al. (1994) at 850 μm with the JCMT but the coordinates are wrong by almost $2'$ in declination. With a size comparable to the ISOPHOT 200 μm resolution, and a contrast of around 2 with the surrounding material, there is thus no reason that ISOPHOT could miss this core unless it is optically thick and/or extremely cold. Finally, within a

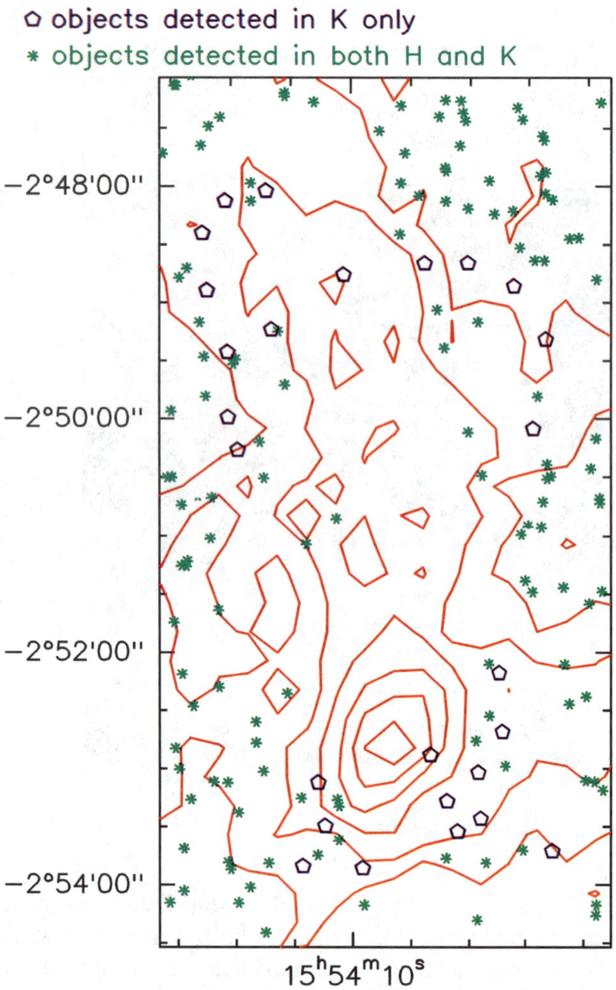

Figure 3. L134N core. Contours represent emission at 1250 μm (MAMBO). Levels run from 10 to 60 mJy per beam in steps of 10. Almost no source is seen inside the vertical 20 mJy per beam contour, which corresponds to $A_V \approx 40$ mag. On the right-hand side of the MAMBO peak, the increase of sources is indicative of lower extinction and thus of warmer dust.

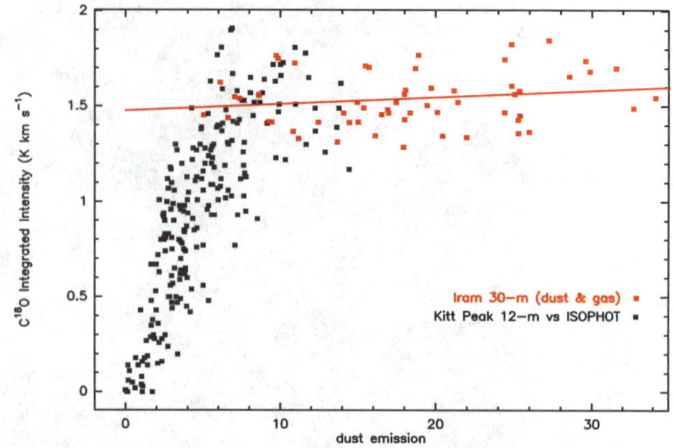

Figure 4. $C^{18}O$ depletion. Low (ISOPHOT vs. Kitt Peak 12-m) and high (IRAM 30-m) resolution data have been assembled in this plot. The dust emission axis is in arbitrary units.

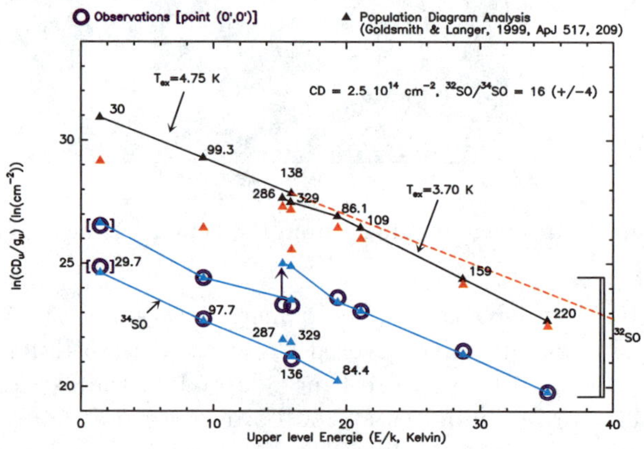

Figure 5. SO and ^{34}SO rotational population diagram. Black triangles represent the "optically thin" evaluation, red triangles the opacity correction, and blue triangles the 3 K background correction. In fact, the diagram only helps to visualize the fit. It cannot be used to find the solution (this would introduce a two order of magnitude error here). For ^{34}SO, only the solution (*cyan*) is shown

radius of 11′, which encompasses most of the cloud, the average extinction is $A_V = 7.1$ mag, which yields a total cloud mass of 36 $M_\odot$.

2.2. CO DEPLETION

CO depletion, which was predicted long ago but discovered only recently, was reported for the first time in L134N by Pagani et al. (2002) and confirmed by Juvela et al. (2002). However, because dust was not correctly traced by ISOPHOT measurements, the actual depletion factor was completely underestimated in both papers. With both high resolution $C^{18}O$ ($J = 1$–0) and dust measurements, both from the IRAM 30-m, we can now extend the original plot we published in Pagani et al. (2002) and show the dramatic effect of depletion in this core (Fig. 4). The horizontal scale is only qualitative (from preliminary estimates as discussed above, it represents roughly 1/4 of the actual A_V scale). Depletion appears at 7 in the horizontal scale, or $A_V \approx 28$ mag. The regression line fits the IRAM data. It indicates that between 10 and 70 mJy per beam, the $C^{18}O$ ($J = 1$–0) integrated intensity increases by 0.1 K km s^{-1}. For a 10 K core, this represents about a 9 percent increase in column density, while for a 7.5 K core, it represents 16 percent, the line being optically thick (with a simple LVG model). These are order of magnitude estimates and a detailed model should be developed to investigate this problem (Pagani et al. 2003). With only a 15 percent increase in intensity while the dust goes from 28 to 140 mag in extinction, we can deduce that CO is depleted by a factor of ≈ 35 in the core.

Another interesting result is that the $C^{18}O$ peak at the position (4.5′, -2.5′) from the center (see Fig. 1 of Pagani et al. 2002) is sitting on a dust peak ($A_V \approx 20$ mag from H-K measurements) of lukewarm (15 K) dust as shown by the ISOPHOT 100 μm peak. This $C^{18}O$ peak shows no sign of depletion (but there is strong absorption by foreground material in the $C^{18}O$ ($J = 2$–1) transition; see Fig. 2 of Pagani et al. 2002). A preliminary evaluation of density towards this peak gives about 10^4 cm^{-3}, while the $C^{18}O$ emitting core layer was evaluated at around 3×10^4 cm^{-3} at 10 K (Pagani et al. 2002). In fact, the density in the core layer is probably higher because the temperature is lower than 10 K in the core. The provisional conclusion is that in this cloud, depletion occurs only above $A_V \approx 25$ mag, in a very cold core only. Whether it is a density threshold or a temperature threshold or both is not yet clear. The opacity limit for depletion to occur is a factor of greater than 2 above usual determinations, but it looks from this result that determinations based on ISOPHOT data alone are probably much underestimated in many cases.

2.3. SO abundance in the core

SO is interesting because the CS/SO abundance ratio is supposed to trace the cloud chemical age. However, the collision coefficients published by Green (1994) are not suitable for low temperatures and lead to estimates which can be wrong by an order of magnitude (or more if one tries to fit the ground transition [1_0-0_1] at 30 GHz). In such a case and because SO has, like other nonlinear molecules, many lines in the millimeter regime sampling a full range of energy levels, it is relatively easy to study many of these lines and plot them on a rotational population diagram to retrieve the actual column density (Goldsmith & Langer 1999). We have thus mapped the 99, 109, 138 and 220 GHz lines of SO, and the 98 and 136 GHz lines of ^{34}SO. A few positions have been also obtained for the 159 GHz line of SO and the 106 GHz line of ^{34}SO (no detection for the latter). To test the population diagram analysis, we have included the 30 GHz SO and ^{34}SO data of Rydbeck et al. (1980) and the 86 GHz data of Swade (1989a) towards the reference position (or very close to it). The results are displayed in Fig. 5. In fact, because the 30, 99 and 138 GHz lines are optically thick and because the excitation temperature is close to the 3 K background, a direct fit of the values would lead to an underestimate of the true column density by two orders of magnitude. Instead we have to guess a possible column density and excitation temperature, estimate the resulting intensities and compare them with the observations. Note that the 3 K background is not subtracted in the standard rotational population diagram method. This becomes a problem at low temperature when the excitation temperature is close to the 3 K background. Thus even for optically thin lines, the rotational population diagram cannot be directly applied if the excitation temperature is low, which is the case for most molecules in dark clouds. The 3 K correction appears in Fig. 5 as the distance between the red and the cyan triangles. The column density we derive would be the one that is found by using the "optically thin estimates" and extrapolating them to zero energy. The value we derive in this diagram is 2.5×10^{14} cm^{-2} for SO and 16 times less for ^{34}SO. This is a factor of 5 above the result presented by Dickens et al. (2000) using extrapolated Green collision coefficients to low temperature (including a 30 percent over-correction they brought to their data by calibrating them in T_{MB} instead of T_r^*, which is more appropriate for extended sources). This shows how cautious one must be when using column density estimates from the literature as input to chemical models.

3. Conclusions

We have brought new insight to the L134N/L183 cloud through the work which we have described here and in Pagani et al. (2002). The new picture which emerges is the following: an opaque core ($A_V \approx 140$ mag), probably very cold (7 K?), with a mean density just below 10^6 cm^{-3} is depleted in CO (and CS) by a factor of 35. Its mass is in the range 1–1.5 $M_\odot$. It is also traced by N_2H^+, which does not coexist with CO in this cloud as predicted by chemical models. CO is apparent in both outer core layers at densities just below 3×10^4 cm^{-3} and diffuse gas at a density of around 10^3 cm^{-3}. The latter absorbs preferentially the $C^{18}O$ ($J = 2$–1) transition without signs of self-reversal and thus yields abnormally low $C^{18}O$ ($J = 2$–1)/($J = 1$–0) ratios. This can be solved with the use of the optically thin $C^{18}O$ ($J = 3$–2) line and a multi-layer Monte-Carlo radiative transfer model. This shows that the $C^{18}O$ ($J = 2$–1) transition is not a good tracer of gas in dark clouds. The mass in the envelope is an important fraction of the total mass of the cloud.

Another important result we can probably derive already is that the ISOPHOT 200 μm maps cannot reveal extremely cold, probably optically thick (!) dust cores such as the one described here. Caution must thus be used when deriving dust column density in cold clouds with only ISOPHOT data. We doubt that ISOPHOT has the capability to reveal very cold cores.

ACKNOWLEDGEMENTS

We thank A. Bacmann and M.-A. Miville-Deschênes for valuable discussions and help. We also thank the organizers for selecting this work for an oral presentation.

REFERENCES

Dickens, J.E., et al. 2000, ApJ, 542, 870
Franco, G. 1989, A&A, 223, 313
Frerking, M.A., Langer, W.D., & Wilson, R.W. 1982, ApJ, 262, 590
Goldsmith, P., & Langer, W.D. 1999, ApJ, 517, 219
Green, S. 1994, ApJ, 434, 188
Juvela, M., et al. 2002, A&A, 382, 583
Lada, C.J., Lada, E.A., Clemens, D.P., & Bally, J. 1994, ApJ, 429, 694
Laureijs, R.J., Clark, S.O., & Prusti, T. 1991, ApJ, 371, 602
Laureijs, R.J., et al. 1995, ApJS, 101, 87
Mattila, K. 1979, A&A, 78, 253
Pagani, L., et al. 2002, in *Infrared and Submillimeter Space Astronomy*, ed. M. Giard et al., EASP, Ser. 4, 145 (Les Ulis: EDP Science)
Pagani, L., et al. 2003, submitted
Rydbeck, O.E.H., et al. 1980, ApJL, 235, L171
Swade, D.A. 1989a, ApJS, 71, 219
Swade, D.A. 1989b, ApJ, 345, 828
Swade, D.A., & Schloerb, F.P. 1992, ApJ, 392, 543
Tafalla, M., et al. 2002, ApJ, 569, 815
Ward-Thompson, D., Scott, P.F., Hills, R.E., & André, P. 1994, MNRAS, 268, 276
Ward-Thompson, D., Motte, F., & André, P. 1999, MNRAS, 305, 143
Ward-Thompson, D., André, P., & Kirk, J.M. 2002, MNRAS, 329, 257

ODIN MAPPING OF THE H_2O DISTRIBUTION IN THE ORION KL OUTFLOW/CLOUD CORE

A. O. Henrik Olofsson[1] and the *Odin* Team[2]

[1]Onsala Space Observatory, SE-439 92, Onsala, Sweden
[2]http://www.snsb.se/Odin/Odin.html

Abstract

We briefly report on new water observations in the Orion KL region using the submm/mm–wave satellite *Odin* (beam size 2.1′ at 557 GHz). The ground state transitions ($J_{K_+K_-}=1_{1,0}\rightarrow 1_{0,1}$) of $H_2^{16}O$ (at 557 GHz) and $H_2^{18}O$ (at 548 GHz) were observed with an SSB receiver system (T_{sys} = 3200–3500 K), and the acquired spectra have channel separations down to 0.5 MHz and bandwidths up to 1.1 GHz. $H_2^{16}O$ was mapped over a $7'\times 7'$ region at a spacing of 1′, while $H_2^{18}O$ was observed in two positions (Orion KL and a position 2′ south of Orion KL). A few highlights of the scientific results are presented here (featuring, e.g., asymmetric line shapes in the molecular outflow in Orion KL, and a detection of $H_2^{18}O$ in the southern position). A more complete account for these results will appear in one of a number of forthcoming Letters (A&A, 2003) that present early *Odin* results.

Key words: ISM: molecules – ISM: structure – ISM: clouds

1. Introduction

Odin is a Swedish-led satellite project funded jointly by the Swedish National Space Board (SNSB), the Canadian Space Agency (CSA), the National Technology Agency of Finland (Tekes), and the Centre National d'Études Spatiales (CNES, France). The Swedish Space Corporation (SSC) has been the prime industrial contractor, and is also responsible for the satellite operation from its *Odin* Mission Control Centre at SSC in Solna and its *Odin* Control Centre at ESRANGE near Kiruna in northern Sweden.

Odin was successfully launched on 20 February 2001 from Svobodny in eastern Russia on a Start-1 rocket, and the observations decribed herein were performed in September–October 2001 and April 2002.

In the course of these observations, the ground state transitions of O_2 (at 119 GHz) and NH_3 (at 572 GHz) were also measured by complementary receivers on the spacecraft. A lower limit on the O_2 column density is presented by Pagani et al. (2003 and this volume), and detections of the NH_3 line in three positions (including the Orion Bar) are reported in Larsson et al. (2003).

2. Results

A subset of spectra from *Odin*'s water map is shown in Figure 1 where offsets are versus the map centre coordinates of α=05^{h}35^{m}14^s.4, δ=−05°22′30″(J2000). Towards the map centre at (0,0), the spectrum appears devoid of narrow line ambient cloud emission, in contrast to the one observed by *SWAS* (Melnick et al. 2000) with a larger beam size. In fact, when the *Odin* map is convolved with the *SWAS* beam, the centre spectra agree extremely well (Hjalmarson et al. 2003). Compared to an SiO spectrum ($v = 0$, $J = 2 \rightarrow 1$) scaled to match the high velocity wings of water (Fig. 2), it is evident that the H_2O line is very different at the line centre, notably by the absorption-like jump near the OMC-1 systemic velocity of +8 km s^{-1}. The broad part of this feature we tentatively attribute to self-absorption in the outflow itself, while the steep, narrow feature could be associated with absorption in low-density quiescent halo gas on the near side of the cloud complex. Deep, narrow self-absorptions at the respective systemic velocities are seen in H_2O lines towards most outflow sources observed by *Odin*.

In Figure 3, the likely $H_2^{18}O$ detection at 2′ south of Orion KL is shown. We have used the integrated intensity of this line to calculate the water abundance assuming LTE (T_{ex} = 25 K) and an $^{16}O/^{18}O$ ratio of 500. We find a very low value, $X(H_2O) \approx 10^{-8}$, in agreement with *SWAS* estimates. However, it is plausible that the $H_2^{18}O$ optical depth is not insignificant and a more sophisticated analysis then would be required.

Chemistry as a Diagnostic of Star Formation (2003). *Edited by Charles L. Curry and Michel Fich.* Published in 2003 by NRC Research Press, Ottawa ON, K1A 0R6, Canada. ISBN 0-660-19089-3.

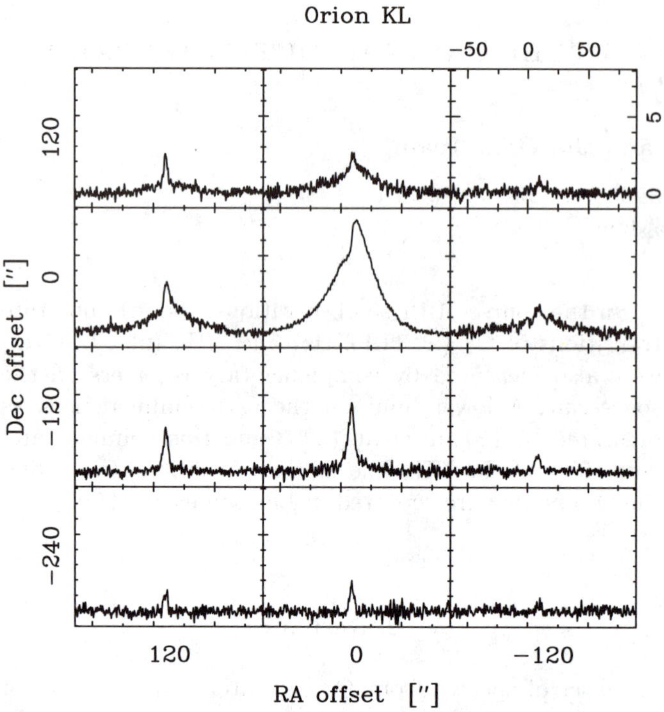

Figure 1. A subset of spectra from the *Odin* 557 GHz H_2O map. The intensity axis is uncorrected for beam efficiency. The increments of the velocity and intensity axes are given next to the spectrum at top right. The map origin corresponds to J2000 coordinates α=05^{h}35^{m}14^s.4, δ=−05°22′30″.

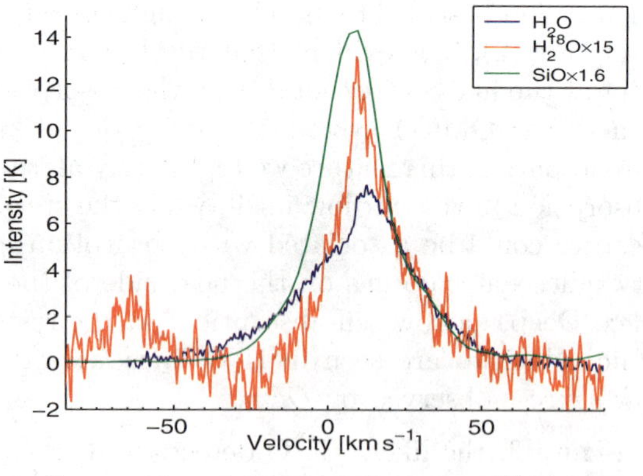

Figure 2. Comparison of the *Odin* H_2O and $H_2^{18}O$ profiles with an Onsala Space Observatory (OSO) SiO spectrum. The latter two have been scaled so that the red line wing will match that of H_2O. Note that the SiO profile was corrected for the OSO 20m beam efficiency before scaling, whereas the *Odin* water spectra are in units of antenna temperature.

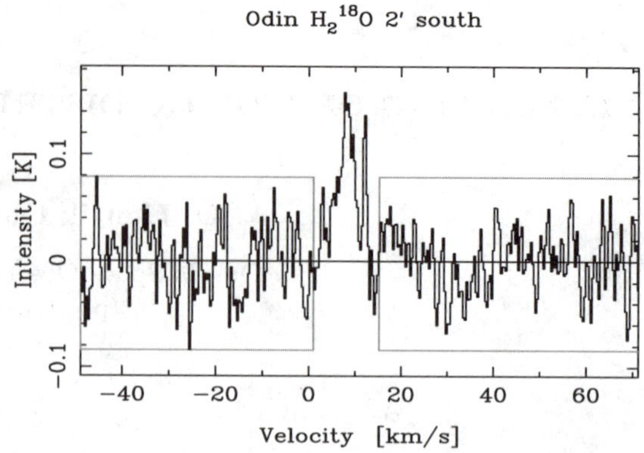

Figure 3. $H_2^{18}O$ towards 2′ south of Orion KL. The intensity axis is uncorrected for beam efficiency.

3. Outlook

A fuller report and a first analysis of these observations will appear in Olofsson et al. (2003). A more extended analysis is intended to be published later in 2003. At that time, we aim to include additional (more than five times more) *Odin* Orion water mapping data that have already been obtained. Work is also underway to make use of the MEM deconvolution algorithm to produce a full deconvolved spectral map that may provide new insights into the structure of the region.

Acknowledgements

Generous financial support from the Research Councils and Space Agencies in Canada, Finland, France and Sweden is gratefully acknowledged.

References

Hjalmarson, Å., et al., 2003, A&A, 402, L39
Melnick, G. J., Ashby, M. L. N., Plume, R., et al., 2000, ApJ 539, L87
Larsson, B., et al., 2003, A&A, 402, L69
Olofsson, A. O. H., et al., 2003, A&A, 402, L47
Pagani, L., et al., 2003, A&A, 402, L77

CARBON CHEMISTRY IN INTERSTELLAR CLOUDS

Maryvonne Gerin[1], David Fossé[1], and Evelyne Roueff[2]

[1]LERMA, Département de Physique de l'ENS et Observatoire de Paris, FRE2460 du CNRS, 24 Rue Lhomond, 75231 Paris, Cedex 05, France

[2]LUTH, Observatoire de Paris, FRE2462 du CNRS, Place J. Janssen, 92195 Meudon, Cedex, France

Abstract

We discuss new developments in interstellar chemistry, with particular emphasis on carbon chemistry. We confirm that carbon chains and cycles are ubiquitous in the ISM and closely chemically related to each other, and to carbon. Investigation of the carbon budget in shielded and UV-illuminated gas shows that the inventory of interstellar molecules is not complete and more complex molecules with four or more carbon atoms must be present. Finally, we discuss the consequences for the evolution of clouds and conclude that the ubiquitous presence of carbon chains and cycles is not a necessary consequence of a very young age for interstellar clouds.

Key words: ISM: molecules – ISM: clouds – ISM individual (IC 63, L 134N, Horsehead Nebula, TMC-1)

1. Introduction

Nearly 120 molecules have now been observed in various sources in the interstellar medium. Emission or absorption lines from these molecules carry important information on the physical conditions and the chemical properties of the various sources where they have been recognized. However, only a very limited number of molecules is used for that purpose, while much more information is available. The use of molecular lines detected with high resolution spectroscopy for studying the interstellar medium, and star formation processes, is still largely unexplored.

High resolution spectroscopy, either using heterodyne techniques in the radio domain, or with high dispersive power spectrometers, yields not only the line intensities but also information on the gas dynamics through the analysis of line profiles. Many lines can be analyzed together for studying physical conditions. Different approaches can be used to solve the radiative transfer equation, from the simplest LTE model to sophisticated multi-dimensional models. Also, other line properties can be exploited to get information on the magnetic field, either directly from the Zeeman effect on paramagnetic molecules, or indirectly with other techniques (e.g., Houde, this volume). And finally, the molecular abundances are the basic information for the chemistry. The question which arises then is how to select the best molecule and the best lines to get useful information.

All interstellar molecules cannot be detected in the various environments of the ISM. This evolution of the emerging spectrum, and hence of the chemical properties, is tightly connected to the physical conditions and environment of the source. With the advent of more sensitive receivers and better mapping techniques, it is now possible to study extensively the chemical properties of specific sources. For example, Turner, Herbst, & Terzevia (2000) have performed detailed observational studies and modeling of a large set of interstellar molecules in cloud cores. We present in this paper new studies of interstellar chemistry devoted to carbon chemistry and the condensation processes on dust grains, which make use of molecular line observations.

2. Hydrocarbons in interstellar chemistry

Carbon is the fourth most abundant element in the interstellar medium, and also the more versatile for building molecules. Carbon chemistry can therefore be viewed as the core of interstellar chemistry. Most interstellar molecules (74 percent, or 85 out of 114) have at least one carbon atom. The heaviest and most complex molecules are organic molecules with carbon. These statistics does not take into account the PAHs, nor the DIB carriers which most likely are large organic molecules (Herbig 1995). Carbon, neutral or ionized, is also one of the main reactants in interstellar chemistry networks, due to the large number of organic molecules, but also to the reactivity and versatility of carbon, which can participate in numerous chemical reactions at any temperature, from very cold dense cores to warm and hot gas. Therefore,

Chemistry as a Diagnostic of Star Formation (2003). *Edited by Charles L. Curry and Michel Fich.* Published in 2003 by NRC Research Press, Ottawa ON, K1A 0R6, Canada. ISBN 0-660-19089-3.

understanding carbon chemistry is of major importance in astrochemistry and for star formation. As a first step, we discuss diffuse clouds and photon-dominated regions, and determine the carbon budget. We then discuss dark cloud chemistry and compare the carbon budget in all environments. Finally, we compare observations and state of the art models.

2.1. Diffuse gas

While CH, CH^+ and CN are known in the ISM since the 1940s, the detection of other molecules in diffuse gas has paused for a while. But recent deep observations have revealed and/or confirmed the presence of diatomic, triatomic, and even more complex molecules in diffuse gas. Carbon clusters (C_2, C_3) are now almost routinely detected in the visible towards bright stars (Maier etal. 2001; Roueff et al. 2002). C_3 can be seen in the far infrared through its low-energy vibration mode (Giesen et al. 2001; Cernicharo, Goicoechea, & Caux 2000), and maybe also C_4 Cernicharo, Goicoechea, & Bénilan (2002). Detections in the far infrared are obtained towards the massive star forming region SgrB2 close to the Galactic Center, and sample more opaque clouds than typical diffuse clouds. Typical numbers for column densities and abundance ratios are given in Table 1.

Use of radio telescopes Lucas & Liszt (2000) has shown that CCH and c–C_3H_2 are ubiquitous in diffuse gas, confirming previous work by Cox, Walmsley, & Güsten (1989). These two species are spatially correlated and present a well-defined abundance ratio N(CCH)/N(c-C_3H_2) = 21±6 in diffuse clouds (Lucas & Liszt 2000). The previous work used strong background continuum sources such as quasars to search for molecular absorption lines. The supernova remnant, Cas A, one of the strongest radio sources in the sky, is a good target to search for absorption lines. Bell, Feldman, & Matthews (1983) and Bell & Matthews (1995) reported the detection of C_3N and C_4H along this line of sight. Though the physical conditions are not as accurately determined as in diffuse clouds, these observations confirm the presence of rather complex molecules in different environments exposed to UV radiation.

Table 1. Typical column densities and abundances in diffuse clouds and PDRs

Diffuse clouds (absorption lines)

Species	$N(X)$ cm^{-2}	$N(X)/N(H_2)$	$N(C_2)/N(X)$	Ref.
C	0.03–20 $\times 10^{15}$	$6.5 \pm 4 \times 10^{-6}$	0.008	1
CH	1–10 $\times 10^{13}$	$5 \pm 2 \times 10^{-8}$	1	1
C_2	1–10 $\times 10^{13}$	$5 \pm 2 \times 10^{-8}$	1	1
C_3	1.5–5 $\times 10^{12}$	3×10^{-9}	15	2
C_4 ?	1.2×10^{15}	–	–	3
CCH	0.17–3.5 $\times 10^{13}$	3×10^{-8}	4	–
c-C_3H_2	0.5–1.7 $\times 10^{12}$	1.5×10^{-9}	30	4
C_4H	0.9–2.5 $\times 10^{13}$	1.5×10^{-9}	30	5

PDRs

Species	$N(X)$ cm^{-2}	$N(X)/N(H_2)$	$N(C_2H)/N(X)$	Ref.
CCH	1.8×10^{14}	1.7×10^{-8}	1	4
c-C_3H_2	1.0×10^{13}	1.1×10^{-9}	18	4
C_4H	2.0×10^{13}	2.0×10^{-9}	9	6
c-C_3H	3.7×10^{12}	3.7×10^{-10}	50	6
l-C_3H	2.0×10^{12}	2.0×10^{-10}	90	6
l-C_3H_2	$\leq 3.2 \times 10^{11}$	–	≥ 500	6

References— (1) Federman et al. (1994), Rachford et al. (2002), Shuping et al. (1999); (2) Roueff et al. (2002); (3) Cernicharo, Goicoechea, & Bénilan (2002); (4) Lucas & Liszt (2000); (5)Bell, Feldman, & Matthews (1983); (6) This work.

2.2. Photodissociation regions

A natural question therefore arises: if carbon chains are present in the diffuse interstellar medium, what happens in photodissociation regions (PDRs)? As in diffuse clouds, chemical processes are dominated by the radiation in PDRs, but the gas is denser. The radiation field, also more intense than in diffuse clouds, is generally well measured. Also, with the new data obtained with *ISO*, and particularly the mid-IR emission due to PAHs, it is possible to estimate the gas density from the mid-IR data and compare with determinations from molecular lines (see, e.g., the case of the Horsehead Nebula (Abergel et al. 2003; Teyssier, this volume). We have detected simple carbon chains and cycles in four nearby PDRs: the interface near HD 147889 in Ophiuchus, NGC 7023, IC 63 and the Horsehead Nebula. We report here the results for the last two sources but similar conclusions can be drawn for all PDRs (Fossé 2003). The Horsehead Nebula showed the brightest lines and deserved a more extensive study (Teyssier, this volume).

IC 63 is a small cometary globule illuminated by the B0.5 star γCas (HD 5394), at a distance of 230 ±70 pc.

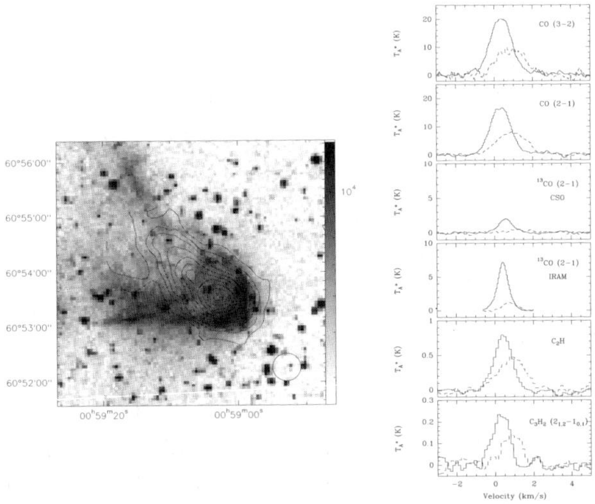

Figure 1. Left: Image of IC 63 from the Digital Sky Survey, with contours of the $^{12}CO(2-1)$ emission. *Right:* Comparison of spectra towards the CO peak (*full curve*) and the "tail" (*dashed curve*).

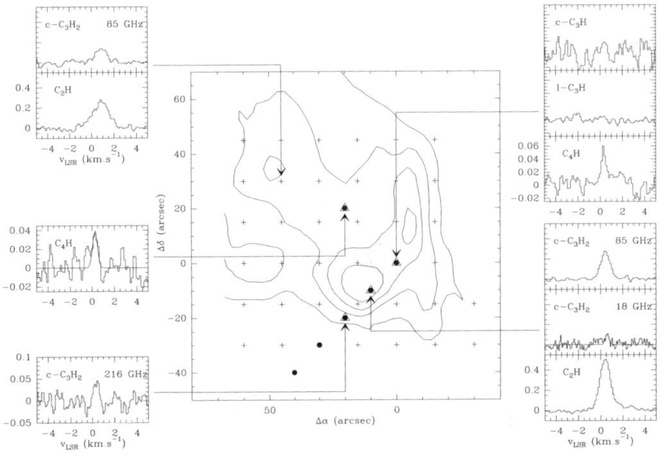

Figure 2. Examples of spectra of various molecules obtained towards IC 63, overlayed on a map of the 6.2 μm feature obtained with ISOCAM.

The PDR has been studied and characterized by Jansen, van Dishoeck, & Black (1994) and Jansen et al. (1996). The gas density is rather high in the "head" of the nebula, 5×10^4 cm^{-3}, but slightly lower in the "tail" (10^4 cm^{-3}). The gas is warm with $T_K = 50$ K due to the intense radiation field, $\chi = 1100$ in units of the Habing ISRF (Habing 1968). From the overall shape of the nebula, it is likely that the molecular gas is currently photodissociated and evaporated by the intense radiation. Indeed, the total column density of molecular hydrogen is rather moderate for a dense cloud, with a peak value of $N(H_2) = 5 \pm 2 \times 10^{21}$ cm^{-2} (Jansen, van Dishoeck, & Black 1994). Using the IRAM 30m telescope, we have detected CCH, c-C_3H_2 and C_4H in IC 63 and searched for c-C_3H and l-C_3H. Examples of spectra are shown in Figures 1 and 2.

We found that CCH and c-C_3H_2 are present everywhere in the nebula, both at the CO peak and in the tail where ^{13}CO disappears. Compared to the line profiles at the peak position, the lines are a factor of two weaker in the tail, and their centroid is redshifted. The same difference is also seen in ^{12}CO spectra while ^{13}CO is barely seen in the tail (Fig. 2). Both the large scale distribution and the similarity of profiles with ^{12}CO indicate that CCH and c-C_3H_2 are present in the external layers of the PDR, where the radiation field destroys ^{13}CO. C_4H looks similar to CCH and c-C_3H_2, though the S/N is rather low for this species. We derived column densities of N(CCH)$\sim 3 \times 10^{13}$ cm^{-2} and N(c-C_3H_2) $\sim 2.0 \times 10^{12}$ cm^{-2}, and $N(C_4H) = 3 \times 10^{12}$ cm^{-2} for this source. The upper limits are N(c-C_3H) $\leq 1.2 \times 10^{12}$ cm^{-2} and N(l-C_3H) $\leq 5.8 \times 10^{12}$ cm^{-2}.

Lines are stronger in the Horsehead Nebula, where both C_3H isomers, the cyclic and linear forms, are detected. As in IC 63, all hydrocarbon maps are very similar to each other, but rather different from those of CO and isotopes (Teyssier, this volume). A scatter plot between intensities of CCH, c-C_3H_2, C_4 and $C^{18}O$ is shown in Figure 3. While the correlation is perfect between CCH and both C_4H and c-C_3H_2, the scatter plot with $C^{18}O$ shows a larger dispersion. A closer inspection reveals that the signal from CCH remains constant near 1 K kms^{-1} at positions furthest away from the cloud edge, while the $C^{18}O$ lines get brighter there. From these correlations, and knowing the physical conditions in the Horsehead Nebula, we derived typical abundance ratios for the hydrocarbons, which can be compared with the results in diffuse clouds. They are listed in Table 1. C_4H and c-C_3H_2 have very similar abundances in PDRs, and are about one order of magnitude less abundant than CCH. Both C_3H isomers are less abundant than c-C_3H_2. The same is true for the linear isomer l-C_3H_2 compared to the cyclic isomer.

2.3. The carbon budget

Abundances and abundance ratios of carbon chains are very similar in PDRs and in diffuse gas. We present below

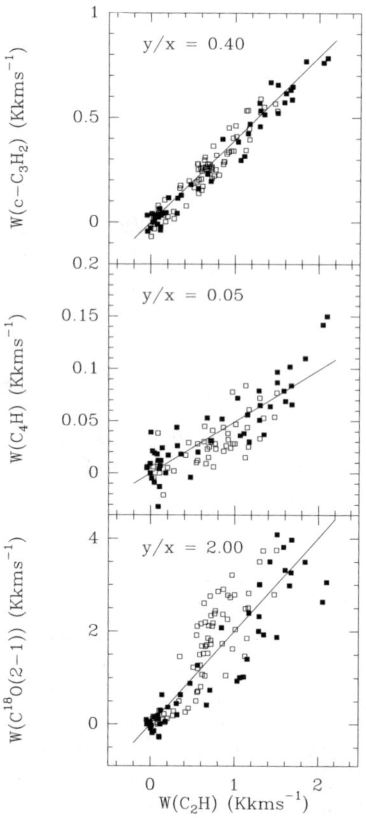

Figure 3. Scatter plot of the integrated emission of CCH (J=1−0) and c-C_3H_2 ($2_{12}-1_{01}$ (*top*), C_4H (9−8) (*middle*) and $C^{18}O$ (2−1) (*bottom*) in the Horsehead Nebula. The spatial resolution is 28 arcsec for all lines.

a tentative carbon budget, combining the information obtained at all wavelengths in diffuse clouds and in PDRs. Since all molecules are not observed in the same sources, this carbon budget has only a statistical meaning, and numbers cannot be very accurate. We have co-added typical abundances for molecules with a given number of carbon atoms, from one to four (Table 2). Nevertheless, the comparison of the fraction of carbon atoms locked into molecules of various sizes (column 3) shows a clear trend. It decreases steeply with the number of carbon atoms for small molecules, but then tends to flatten for molecules having three or more carbon atoms. If the trends continue for larger species, the detection of larger carbon clusters or carbon chains looks promising for the near future. The only work so far is the unsuccessful search for C_4 and C_5

Table 2. The carbon budget in PDRs

Number of Carbon Atoms		$N^a_{tot}/N(H_2)$	$N(C)^b/C_{tot}$
1	(C^+, C, CO, ...)	2.6×10^{-4}	1.0
1	(CH, CH^+, ...)	7.0×10^{-8}	2.7×10^{-4}
2	(C_2, C_2H, ...)	8.0×10^{-8}	6.0×10^{-4}
3	(C_3,c-C_3H_2, ...)	5.0×10^{-9}	6.0×10^{-5}
4	(C_4, C_4H, ...)	2.0×10^{-9}	3.0×10^{-5}

[a] Total abundance relative to H_2.
[b] Total number of carbon atoms locked in molecules this size compared to the available number of carbon atoms in the gas phase. We used a gas phase carbon abundance relative to H of 1.3×10^{-4}.

towards ζ Oph (Maier, Walker, & Bohlender 2002). Despite the brightness of this star, the limits achieved so far do not rule out the presence of these carbon clusters with an abundance ten times lower than C_3. Searches are necessary along more opaque lines of sight where larger column densities for all species are expected.

2.4. Models

Chemical models are now capable of dealing with very large reaction rate sets, including complex molecules. We used two different reaction rate sets, the UMIST-95 file (Le Teuff, Millar, & Markwick 2000) and the new standard model (NSM; Terzevia & Herbst 1998), coupled with the Meudon PDR code (Le Bourlot et al. (1993) and http://aristote.biophy.jussieu.fr/MIS/). Results are shown in Figure 4 for physical conditions adapted to the Horsehead Nebula—namely, a gas density of $n = 2 \times 10^4\,cm^{-3}$ and a radiation field 100 times more intense than the mean value, i.e., $G_0 = 100$. We show for each model the variations of selected molecular abundances as a function of the distance to the cloud edge, as well as the spatial variations of abundance ratio relative to CCH. A good spatial correlation with CCH (as c-C_3H_2 in Fig. 3) would produce a constant abundance ratio in these plots.

Predictions obtained from both the UMIST-95 and NSM sets follow the same general trend with a peak of carbon chains at $\sim$ 20 arcsec ($A_V = 2$ mag) and a second peak deeper into the cloud. Also the chemistry in the UV-dominated region is very simple, with pure carbon clusters (C_2, C_3, etc.) being more abundant than radicals (C_2H,C_3H, etc.), which are themselves more abundant than more hydrogenated species (C_2H_2, C_3H_2, etc.). Finally, abundance ratios of both C_3H_2 and C_4H with CCH do not vary much inside the cloud (for positions larger

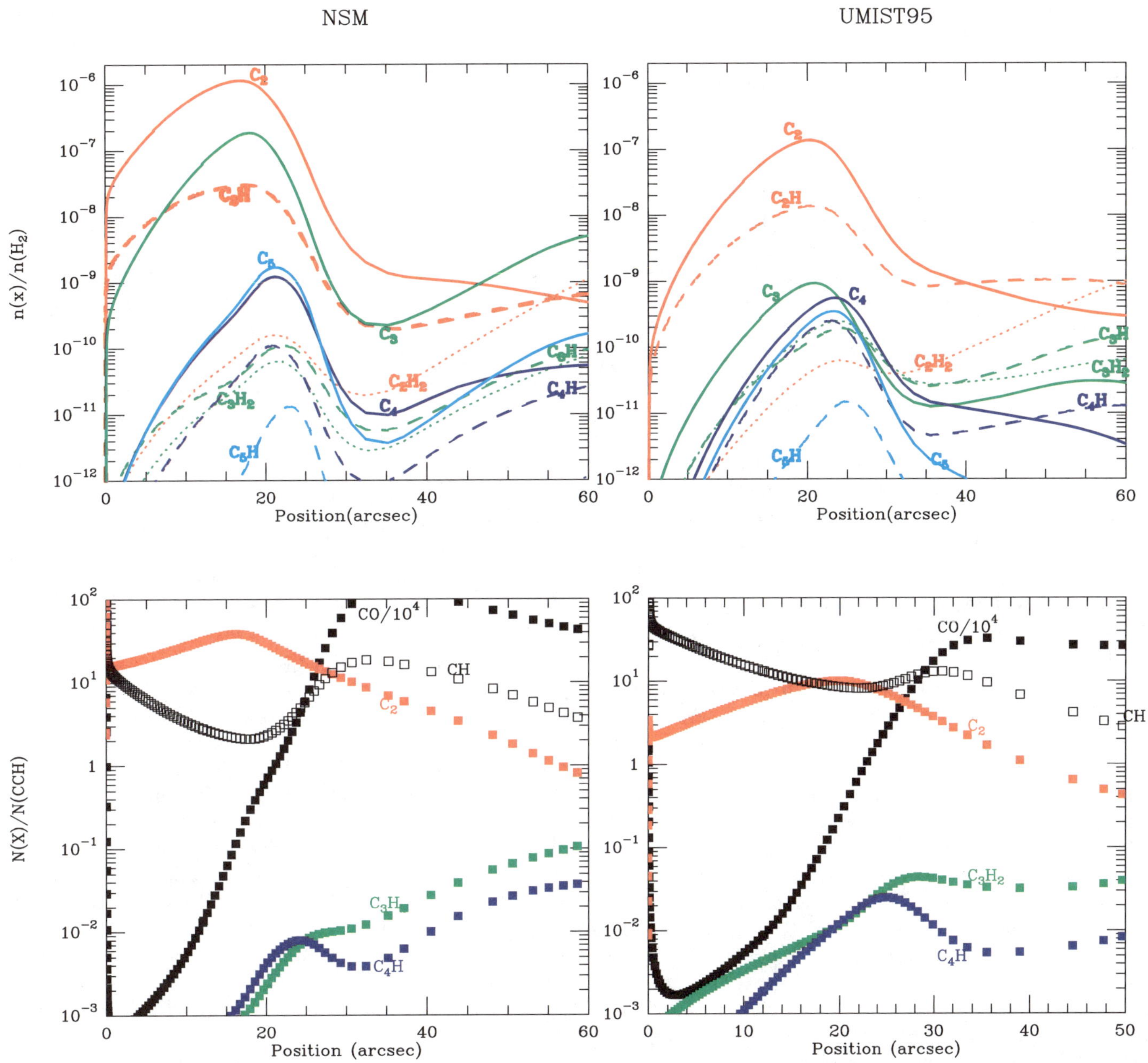

Figure 4. Predictions for the abundances of various hydrocarbons in a PDR with a gas density $n = 2 \times 10^4\,\mathrm{cm^{-3}}$ and a radiation field $G_0 = 100$. Predictions using the NSM reaction set are shown in the left panels, and using UMIST-95 in the right panels. Both models underpredict the abundances for c-C_3H_2 and C_4H, and also have spatial distributions differing from the observations. However, both models predict that CCH is present in the UV-illuminated gas, as is observed.

than 20 arcsec), but decrease sharply at the cloud edge. This simple behavior is in contrast to our measurements, viz.:

– C_3H_2 is always more abundant than C_3H. This is true for the cyclic isomer alone and when different isomers are combined.

– C_4H and c-C_3H_2 have almost the same abundance, $\sim 2 \times 10^{-9}$, at least a factor of ten larger than the predicted value.

– All hydrocarbons observed are detected in the UV-illuminated region where the PAH emission is observed, while the models predict a spatial offset.

While models are able to reproduce fairly well the formation of CO, and marginally the CCH abundance, severe problems remain for larger molecules. Since the main destruction process is photodissociation—an efficient mechanism in PDRs—new formation routes must be found for explaining the large abundances of C_4H and C_3H_2 observed in PDRs. It is likely that some reaction rates used in current models are inaccurate, and that important reactions are still missing. We need more observations of simple species to progress, especially of neutral and ionized carbon which are the main reactants in the chemical reaction networks.

3. Chemistry in dark clouds

3.1. Carbon chemistry

Since the same networks are used to probe dark cloud and PDR chemistry, we may learn of the formation processes by comparing abundances and abundance ratios in different environments. TMC-1 in the Taurus molecular cloud is known as the ISM "carbon factory," in which long carbon chains and cyanopolyyne have been detected. Neutral carbon itself is known to be abundant both in TMC-1 (Schilke et al. 1995; Maezawa et al. 1999) and in L134N (Tatematsu et al. 1999). According to these authors, the constancy of the antenna temperature of the ground state fine structure line of atomic carbon in dark clouds, $T_A^* \sim 2\,\mathrm{K}$, indicates that C is abundant in the sources and acts as a coolant. The corresponding kinetic temperature is about 9 K, a typical figure for dark clouds. Since the emission is very extended, it is likely that carbon is present in the molecular envelope, and helps in cooling it to 10 K.

Is the presence of long carbon chains related to neutral carbon? Are long carbon chains present in other dark clouds? We performed a restricted survey of carbon chains in L 134N, to have a comparison with TMC-1 and answer these questions. The results will be described in more detail elsewhere (Fossé 2003), and are only summarized below.

In deep integrations we obtained new detections of C_3H (cyclic and linear), C_4H and C_6H in L134N. A CCH map is shown in Dickens et al. (2000). Compared to molecules tracing the cloud core, such as NH_3 and N_2H^+ (Pagani; Bergin, this volume), CCH peaks are spatially separated from dense gas but coincide with those of C_4H and other hydrocarbons (Dickens et al. 2000; Fossé 2003). From the low line intensity and different spatial distribution, we conclude that carbon chains reside in cold but not very dense molecular gas in L134N: $n(H_2) = 3 \times 10^3\,\mathrm{cm^{-3}}$ and T = 10 K. The column densities are nevertheless rather large, and reach similar values as towards the cyanopolyyne peak in TMC-1, namely $N(\mathrm{CCH}) \sim 1.5 \times 10^{14}\,\mathrm{cm^{-2}}$, $N(\mathrm{C_4H}) \sim 1.3 \times 10^{14}\,\mathrm{cm^{-2}}$ and $N(\mathrm{C_6H}) \sim 1.2 \times 10^{12}\,\mathrm{cm^{-2}}$.

These conclusions are consistent with the presence of atomic carbon in dark cloud envelopes. Carbon chains, which are detected both in diffuse and dark clouds and also in PDRs, seem to be always seen in places with rather large abundances of carbon, either atomic or ionized. Though unexpected, since the formation of hydrocarbons is driven by chemical reactions involving C and C^+, the observed abundances and abundance ratios efficiently constrain current models.

3.2. Carbon budget in dark clouds

As for PDRs, we have computed the carbon budget for cold dark clouds. Due to the presence of ice mantles on dust grains, the absolute column densities of H_2 are uncertain in this case. Nevertheless, the numbers cited in Table 3 can be compared, at least relative to each other. We used TMC-1 as most molecules have been detected towards this source, and assume a total extinction of 30 magnitudes, a column density of $N(H_2) = 2 \times 10^{22}\,\mathrm{cm^{-2}}$ and a gas phase CO abundance relative to molecular hydrogen of 8.0×10^{-5}. Depletion on ices causes a deficit of gas phase carbon compared to the situation in diffuse clouds, as seen by the lower abundance of gas phase CO relative to H_2.

In addition to CO, gas phase carbon is locked in organic molecules with three or four carbon atoms in TMC-1 (Table 3). There is a clear decrease in the fraction of carbon tied into more complex molecules. It is likely that the available information on organic molecules with five or more carbon atoms is incomplete, since a lower number of molecules this size are detected, compared with smaller molecules. Other species might still remain to be identi-

Table 3. The carbon budget in dark clouds[a]

Number of Carbon Atoms	$\frac{N_{\rm tot}}{N(H_2)}$	$\frac{N(C)}{C_{\rm tot}}$
1 (C^+, C, CO, ...)	8×10^{-5}	1
1 (CH, HCN, CS ...)	$\leq 1.0 \times 10^{-7}$	$\leq 1.2 \times 10^{-3}$
2 (C_2H, C_2S...)	1.0×10^{-8}	2.5×10^{-4}
3 (c-C_3H_2, HC_3N, ...)	1.2×10^{-8}	4.5×10^{-4}
4 (C_4H, C_4H_2, ...)	1.7×10^{-8}	8.5×10^{-4}
5 (HC_5N, C_5H, ...)	1.2×10^{-9}	7.5×10^{-5}
6 (C_6H, C_6H_2...)	4.2×10^{-10}	3.1×10^{-5}
7 (HC_7N, ...)	5.5×10^{-10}	4.8×10^{-5}
8 (C_8H, ...)	1.1×10^{-11}	1.1×10^{-6}
9 (HC_9N, ...)	1.2×10^{-10}	1.3×10^{-5}
11 ($HC_{11}N$, ...)	1.4×10^{-11}	1.9×10^{-6}

[a] We assumed a gas phase CO abundance relative to H_2 of 8.0×10^{-5} and a total H_2 column density of 2×10^{22} cm^{-2}.
REFERENCES— Turner, Herbst, & Terzevia (2000); Pratap et al. (1997); Bell et al. (1998); Fossé et al. (2001); Dickens et al. (2001).

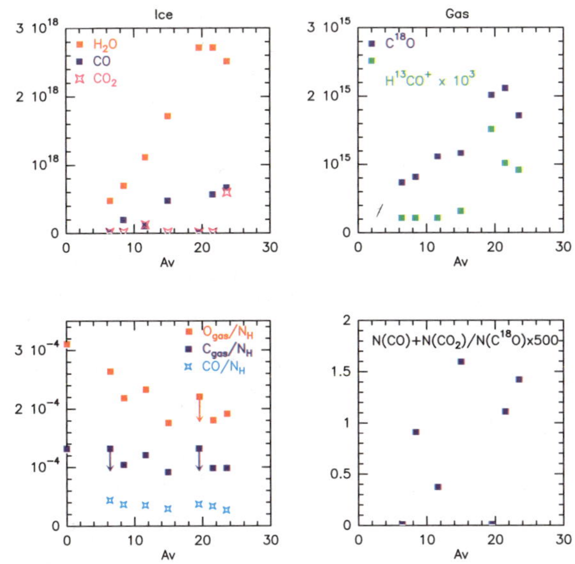

Figure 5. Column densities of molecular ices and gas phase species for background sources in the Taurus Molecular Cloud. We used $N_{\rm H} = 1.52 \times 10^{21}\, A_V\,{\rm cm}^{-2}$, $C_{\rm gas} = 1.3 \times 10^{-4}[N_{\rm H} - N_{\rm ice}({\rm CO}) - N_{\rm ice}({\rm CO_2})]/N_{\rm H}$ and $O_{\rm gas} = 3.09 \times 10^{-4}[N_{\rm H} - N_{\rm ice}({\rm H_2O}) - N_{\rm ice}({\rm CO}) - 2N_{\rm ice}({\rm CO_2})]/N_{\rm H}$, and assumed $N_{\rm ice}({\rm CO_2}) = N_{\rm ice}({\rm CO})$ for lines of sight with no CO_2 observations. Data from Teixeira & Emerson (1999), Nummelin et al. (2001), and Murakawa et al. (2000).

fied. For example, pure carbon clusters are likely present in dark clouds, but could not been taken into account in the above statistic due to the lack of definite detections.

3.3. Chemistry and depletion

The widespread presence of molecular ices on dust grains in shielded regions is now well established. These ices contribute to modifying the grain optical properties and favor grain coagulation. In the Taurus cloud, water ice is seen for lines of sight with extinction larger than a threshold of $A_V \sim 3$ mag, Teixeira & Emerson (1999), while CO ice appears for larger A_V, near 6 magnitudes. It is therefore certain than many gas phase molecules, including carbon monoxide and water, are condensed on grain mantles in cloud interiors, though a few of them only have a solid phase abundance high enough to be detectable in the infrared (the typical limit is a few percent of the water ice, or a few times 10^{-6} relative to hydrogen; Nummelin et al. 2001). The inventory of ices has been completed thanks to the *ISO* mission, toward both YSOs and background field stars. These lines of sight are particularly interesting for chemistry since they sample random lines of sight across clouds, which are likely representative of the bulk of shielded molecular gas. Both the total extinction and the chemical composition of the solid phase are fairly well known.

We selected a sample of seven lines of sight in the Taurus molecular clouds—two of them located in TMC-1—to investigate the relationship between solid and gas phase chemical composition. We used the IRAM 30m telescope to perform the observations in December, 2000. Properties of the background sources are summarized in Table 4. Figure 5 shows that, while CO isotopes are readily detected for all lines of sight with A_V between 6 and 21 magnitudes, molecular ion lines are weak or absent for $A_V \lesssim 15$ mag. At this point, the amount of carbon in the solid phase (condensed as CO or CO_2 ice) is larger than the amount of carbon in the gas phase (primarily as CO). Dense prestellar cores have similar $C^{18}O$ column densities as the lines of sight with the largest extinction in our sample, 2×10^{15} cm^{-2} for $A_V = 20$ mag but larger column densities of molecular ions. Because the total extinction through these dense cores is very large ($\gtrsim 50$ mag derived from star counts in L134N; Pagani, this volume), it is expected that the remaining fraction of molecules in the gas phase is low. Indeed, there is indirect evidence of large depletions of molecules onto grains, based on comparisons with the the continuum emission from dust grains, star counts, and the spatial distribution of different molecules (Pagani; Bergin, this volume), in agreement with the moderate depletions reported here for lower extinctions. Note that because CO gets converted into CO_2 and possibly other species on grains, the amount of carbon condensed on grains, and hence the impact of molecular ices on the chemistry, is larger than what can be derived from the concentration of CO ice only. An inventory of the ice composition, including CO_2 and possibly CH_3OH, is necessary for determining the carbon and oxygen budget of the solid phase.

It is interesting to analyze the chemical differences in TMC-1 itself. Two lines of sight through TMC-1 belong to our sample: Tamura 8 in the ammonia-rich northern region, and E 0484 in the carbon-rich southern region. Both have a similar total extinction of ~ 20 mag and water ice column densities $2.7 \times 10^{18}\,\mathrm{cm}^{-2}$, while the gas phase abundances we observed follow the general trends in TMC-1 of larger column densities of molecular ions in the ammonia-rich region compared to the carbon-rich region. Since the formation of molecular ices in clouds is a rather slow process (Ruffle & Herbst 2000; Ruffle & Herbst 2001), the presence of molecular ices everywhere in the Taurus cloud, with a well-defined dependence on total extinction but no dependence on the gas phase composition, is a strong argument against different ages for different regions. For most of the lines of sight, the abundance of water ice relative to molecular hydrogen (deduced from A_V) is comparable to or larger than 10^{-4}, which means that a significant fraction of the available oxygen is condensed. Other processes such as those discussed below are more likely to contribute in producing the marked gas phase chemical differences. A deeper investigation of the selected lines of sight in TMC-1 may bring clues to the origin of the chemical differentiation since both the total gas column density and the composition of dust grains are known.

Table 4. Column densities[a] of various species toward background stars in the Taurus molecular cloud

Name	A_V	H_2O	CO	CO_2	CO[b]
	mag	(ice)	(ice)	(ice)	(gas)
Elias 6	6.4	4.6	...	...	3.6
Elias 3	8.4	6.8	1.8	...	4.0
Elias 13	11.6	11	0.9	1.1	5.5
Elias 15	15.0	17	4.6	...	5.8
E 04384	19.5	27	...	...	10
Tamura 8	21.5	27	5.5	...	10
Elias 16	23.5	25	6.5	8.5	...

[a] In units of $10^{17}\,\mathrm{cm}^{-2}$.
[b] Column density of gas phase CO deduced from $C^{18}O$ data (see text).
REFERENCES— Teixeira & Emerson (1999); Nummelin et al. (2001); Murakawa et al. (2000).

4. Discussion and Perspectives

In all interstellar environments, dark and diffuse clouds, and PDRs, carbon chains and cycles are closely related to each other and to atomic carbon. They are abundant in the moderately dense molecular gas but avoid the darkest and densest cores. This conclusion is valid for the pure carbon chains and cycles. Other carbon species, like cyanopolyynes, are found in high density regions. Though their formation mechanism also involves carbon, the difference between nitrogen and carbon chemistry may contribute to the different spatial distribution. The difference of incident radiation between diffuse or UV-illuminated gas on one hand, and dark clouds on the other hand, affects somewhat the relative abundance of carbon molecules in permitting the growth of larger chains (compare the abundance of C_4H in TMC-1 and in the Horsehead Nebula), but the chemistry nevertheless looks very similar.

Another major difference between the two environments is the condensation of abundant gas phase molecules on dust grains. The main impact of condensation on the chemistry is the enhanced abundance of molecular ions in depleted regions. The reason for this is that the destruction channels for molecular ions are not only dissociative recombination reactions with electrons, but also reactions with abundant gas phase molecules. The depletion of CO on dust grains leads to lower destruction rates for molecular ions and therefore to larger abundances. While the composition of molecular ice depends mainly on the total extinction, marked chemical differences are found in lines of sight with similar column densities. Also, measured abundances of carbon chains and cycles are often larger than steady state predictions.

The usual explanation for understanding the large abundances is to invoke the so-called "early time chemistry," in which clouds are seen in an early evolutionary stage in which the available carbon, initially ionized, has not been completely transformed into CO. However, several arguments show that "early time chemistry" cannot be the sole explanation. Early time chemistry is efficient because the formation of molecular hydrogen and the conversion of carbon to CO takes a very long time

($\geq 10^6$ years) when starting from atomic gas. It is thus tempting to use molecular abundances in dark clouds as a chemical clock. However, the same arguments leading one to doubt "early time" chemistry, also argue against this "'chemical clock" idea for dense cores:

(1) Since the same molecules, with similar abundances, are seen in many different clouds at different evolutionary stages, such as PDRs, pre-stellar clouds, and clouds having already formed stars, how could we be viewing "early time" chemistry in all cases? For example, clouds having already formed stars and clouds associated with PDRs cannot be presented as "young," since they have already had time to collapse and form stars, and for PDRs the newborn stars have had time to disrupt their parent clouds. Even if the evolution is not uniform across a cloud (chemical time scales depend on the density, so that more diffuse regions evolve at a slower pace compared to dense cores), the requirement of "early time"—i.e., a cloud age lower than 10^5 yr—is very strong compared to the typical time scale for cloud formation and evolution towards star formation ($\geq 10^6$ years; Hartmann, Ballesteros-Paredes, & Bergin 2001).

(2) A second argument against "early time" chemistry is the heavy dependence of the predicted chemical abundances on the initial conditions (Lee et al. 1996), while observations show well-defined chemical properties.

(3) Finally, as stated above, molecular ices are commonly detected in dark clouds, indicating that the gas has had time to evolve, and to build gas phase and solid phase molecules (Roberts & Herbst 2002).

Other processes are known to strongly affect the chemistry, and may favor the presence of carbon and carbon chains in cloud envelopes. The propagation of UV radiation depends on the grain optical properties. Casu, Cecchi-Pestellini, & Aiello (2001) have shown that grain growth in dark clouds affects the grain optical properties and hence the propagation of UV radiation and the photodissociation rates in dense gas. This effect leads to higher abundances of neutral and ionized carbon compared to models using the grain properties valid for diffuse gas. It must be remembered that photons play a very important role in interstellar chemistry, even in dark clouds since the interaction of cosmic rays with hydrogen leads to the production of UV photons deep into the clouds. In addition, the chemical equilibrium for moderately dense gas can be carbon rich (e.g., the HIP solution; Le Bourlot et al. 1995) even without any incident UV photons.

The last two explanations rely on microscopic processes and the choice of elemental abundances. But the structure and dynamical state of interstellar clouds also affects the chemistry. Clumpy cloud models, such as those of Spaans & van Dishoeck (1997) or Störzer, Stutzki, & Sternberg (1996) can explain the widespread distribution of carbon by the accumulation of many clump surfaces, rich in carbon, in telescope beams. Since UV radiation can penetrate deeper into clumpy clouds than into homogeneous structures, each clump surface behaves as a mini-PDR in this context. Another idea is to include the effect of turbulence in homogeneous clouds. The main effect of turbulence is to increase the diffusion between the cloud surface and its interior. A small level of turbulent diffusion, compatible with the observations, can maintain higher abundances of atomic carbon and atomic hydrogen in cloud envelopes than in steady state models, in agreement with observations of atomic hydrogen in dark clouds (Willacy, Langer, & Allen 2002).

It is likely that all additional physical mechanisms described above, both microscopic (change of optical properties of dust grains, abundances) and macroscopic (turbulence, chemistry) and possibly others still to uncover, contribute significantly to the chemistry. There are so many unknowns in the physical properties and chemistry of dark clouds, that deriving the age of clouds from their chemical abundances is not reliable. Clues from the age of embedded stars (if any), geometry, environment, etc., must be used to retrace the history of interstellar structures. To progress, the whole evolution of interstellar clouds, from their formation in the warm neutral gas, to their concentration into molecular gas, and finally the birth of new stars, must be better understood than it is now.

Acknowledgements

The authors thank J. Cernicharo for letting them use his radiative transfer code, and D. Teyssier and A. Abergel for permission to use their work on the Horsehead Nebula. We also thank J. Pety for helping with the processing of IRAM Plateau de Bure data.

References

Abergel, A., et al. 2003, A&A, in press

Bell, M.B., Feldman, P.A., & Matthews, H.E. 1983, ApJ, 273, L35

Bell, M.B. & Matthews, H.E. 1995, ApJ, 438, 223

Bell, M.B., et al., 1998, ApJ, 518, 740

Casu, S., Cecchi-Pestellini, C., & Aiello, S., 2001, MNRAS, 325, 826

Cernicharo, J., Goicoechea, J.R., & Caux, E. 2000, ApJ, 534, L199

Cernicharo, J., Goicoechea, J.R., & Bénilan, Y. 2002, ApJ, 580, L157
Cox, P., Walmsley, C.M., & Güsten, R. 1989, A&A, 209, 382
Dickens, J.E., et al. 2000, ApJ, 542, 870
Dickens, J.E., Langer, W.D., & Velusamy, T. 2001, ApJ, 558, 693
Federman, S.R., et al. 1994, ApJ, 420, 756
Fossé, D., Cernicharo, J., Gerin, M., & Cox, P. 2001, ApJ, 552, 168
Fossé, D. 2003, PhD thesis
Hartmann, L., Ballesteros-Paredes, J., & Bergin, E.A. 2001, ApJ, 562, 852
Habing, H.J. 1968, BAN, 19, 421
Herbig, G.H. 1995, ARA&A, 33, 19
Giesen, T.F., Van Orden, A.O., & Cruzan, J.D. 2001, ApJ, 551, L181
Jansen, D.J., van Dishoeck, E.F., & Black, J.H. 1994, A&A, 282, 605
Jansen, D.J., et al. 1996, A&A, 309, 899 (erratum: 315, 327)
Le Bourlot, J., Pineau des Forêts, G., Roueff, E., & Flower, D.R. 1993, A&A, 267, 233
Le Bourlot, J., Pineau des Forêts, G., Roueff, E., & Flower, D.R. 1995, A&A, 302, 870
Lee, H.H., et al. 1996, A&A, 311, 690
Le Teuff, Y.H., Millar, T.J., & Markwick, A.J. 2000, A&AS, 146, 157
Lucas, R. & Liszt, H.S. 2000, A&A, 358, 1069
Maezawa, H., et al. 1999, ApJ, 524, L129.
Maier, J.P., Lakin, N.M., Walker, G.A.H., & Bohlender D.A. 2001, ApJ, 553, 267
Maier, J.P., Walker, G.A.H., & Bohlender D. 2002, ApJ, 566, 332
Murakawa, K., Tamura, M., & Nagata, T. 2000, ApJS, 128, 603
Nummelin, A., et al. 2001, ApJ, 558, 185
Pratap, P., et al. 1997, ApJ, 486, 862
Rachford, B.L., et al. 2002, ApJ, 577, 221
Roberts, H. & Herbst E. 2002, A&A, 395, 233
Roueff, E., Felenbok, P., Black, J.H., & Gry, C. 2002, A&A, 384, 629
Ruffle, D., & Herbst, E. 2001, MNRAS, 324, 1054
Ruffle D., & Herbst E., 2000, MNRAS, 319, 837
Schilke, P., et al. 1995, A&A, 294, L17
Shuping, R.Y., Snow, T.P., Crutcher, R., & Lutz, B.L. 1999, ApJ, 520, 149
Spaans, M., & van Dishoeck, E. 1997, A&A, 323, 953
Störzer, H., Stutzki, J., & Sternberg A. 1996, A&A, 310, 592
Tatematsu, K., et al. 1999, ApJ, 526, 295
Teixeira, T.C., & Emerson, J.P. 1999, A&A, 351, 292
Terzevia, R., & Herbst, E. 1998, ApJ, 501, 207
Turner, B.E., Herbst, E., & Terzevia, R. 2000, ApJS, 126, 427
Willacy, K., Langer, W.D., & Allen M. 2002, ApJ, 573, L119

THE MEASUREMENT OF THE MAGNETIC FIELD IN MOLECULAR CLOUDS

Martin Houde[1], Pierre Bastien[2], Jessie L. Dotson[3], C. Darren Dowell[4], Roger H. Hildebrand[5,6], Ruisheng Peng[1], Thomas G. Phillips[4], John E. Vaillancourt[7], and Hiroshige Yoshida[1]

[1]Caltech Submillimeter Observatory, 111 Nowelo Street, Hilo, HI 96720, U.S.A.
[2]Département de Physique, Université de Montréal, Montréal, QC, Canada H3C 3J7
[3]SETI Institute, NASA Ames Research Center, Moffett Field, CA 94035, U.S.A.
[4]California Institute of Technology, Pasadena, CA 91125, U.S.A.
[5]Department of Astronomy and Astrophysics and Enrico Fermi Institute, University of Chicago, IL 60637, U.S.A.
[6]Department of Physics, University of Chicago, Chicago, IL 60637, U.S.A.
[7]Physics Department, University of Wisconsin, 1150 University Avenue, Madison, WI 53706, U.S.A.

ABSTRACT

We discuss how the combination of Zeeman, polarimetry and ion-to-neutral molecular line width ratio measurements permits the determination of the magnitude and orientation of the magnetic field in the weakly ionized parts of molecular clouds. Zeeman measurements provide the strength of the magnetic field along the line of sight, polarimetry measurements give the field orientation in the plane of the sky and the ion-to-neutral molecular line width ratio determines the angle between the magnetic field and the line of sight. We show the first results obtained with this technique on the M17 star-forming region using a HERTZ 350 μm polarimetry map and HCO^+-to-HCN molecular line width ratios to provide the first view of the spatial orientation of the magnetic field in M17.

Key words: ISM: cloud — ISM: individual (M17) — ISM: magnetic field — ISM: polarization — ISM: radio lines

1. INTRODUCTION

As it turns out, the magnetic field can be specified with three parameters: its magnitude B, the viewing angle α defining its orientation relative to the line of sight, and the angle β made by its projection on the plane of the sky (as defined relative to a predetermined direction, east from north). Until now, of these three quantities only the third could be measured. At submillimeter wavelengths, this can be accomplished, for example, with polarization measurements of the continuum radiation emanating from elongated dust grains that are aligned by the local magnetic field (Davis & Greenstein 1952). The angle β is thus obtained from the angle of the polarization vector (with, however, an ambiguity of 180° in its determination) since the projection of the magnetic field vector in the plane of the sky is oriented at right angles to the polarization vector (Hildebrand 1988). The magnitude B cannot be directly measured, only the projection of the magnetic field vector to the line of sight $B_{\rm los}$ can be obtained with Zeeman measurements (e.g., Brogan & Troland 2001; Crutcher et al. 1993; Crutcher et al. 1999; Heiles 1997). For general cases, where the magnetic field lies out of the plane of the sky, a determination of the viewing angle α, in combination with the measurements for $B_{\rm los}$ and β, would provide a description of the magnetic field vector $\mathbf{B}$.

As shown by Houde et al. (2002), the determination of the viewing angle α can be accomplished through a comparison of the profiles of line spectra from coexistent ion and molecular species. The ion-to-neutral line width ratio, as defined by Houde et al. (2000a), is the fundamental parameter that, when combined with polarimetry data, holds the key to the determination of the viewing angle.

In this paper, we will summarize the technique and the first results of Houde et al. (2002) obtained for the M17 molecular cloud using HCO^+ and HCN spectroscopic data and an extensive 350 μm continuum polarimetry map obtained with the HERTZ polarimeter (Dowell et al. 1998) at the Caltech Submillimeter Observatory (CSO).

2. THE ION-TO-NEUTRAL LINE WIDTH RATIO

It was shown in Houde et al. (2000a,b) that a comparison of the line profiles of coexistent neutral and ion species can be used to detect the presence of the magnetic field in molecular clouds. For example, if we solve the equation of motion for an isolated ion immersed in a weakly ionized plasma and undergoing elastic collisions with the numerous neutral particles which take part in flows or turbulent motions in the region under study, we find that the ion velocity is in general smaller than the average velocity of the neutrals. The difference between the two velocities

Chemistry as a Diagnostic of Star Formation (2003). *Edited by Charles L. Curry and Michel Fich.* Published in 2003 by NRC Research Press, Ottawa ON, K1A 0R6, Canada. ISBN 0-660-19089-3.

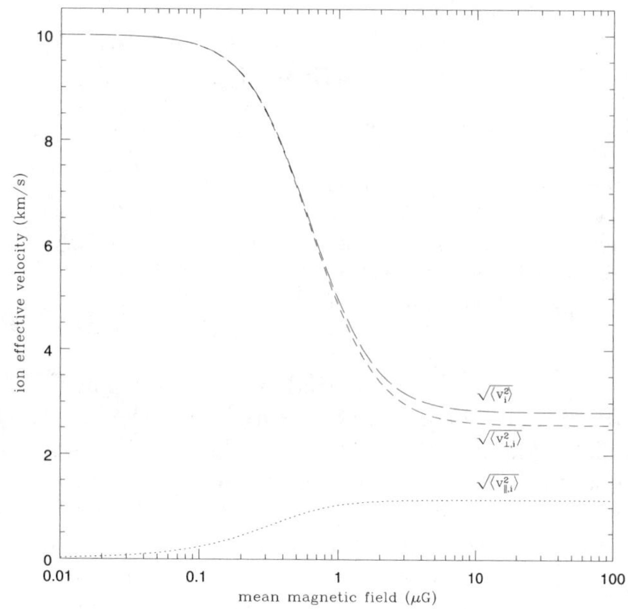

Figure 1. An ion's velocity in directions parallel ($\langle \mathbf{v}^2_{\parallel,i} \rangle^{\frac{1}{2}}$) and perpendicular ($\langle \mathbf{v}^2_{\perp,i} \rangle^{\frac{1}{2}}$) to the magnetic field, and the total velocity ($\langle \mathbf{v}^2_i \rangle^{\frac{1}{2}}$) as a function of the field strength when a neutral flow travels in a direction perpendicular to the field at 10 km s^{-1}. The density is $n = 10^6$ cm^{-3} and the masses are $A_i = 29$ and $A_n = 2.3$ for the ion and neutrals, respectively.

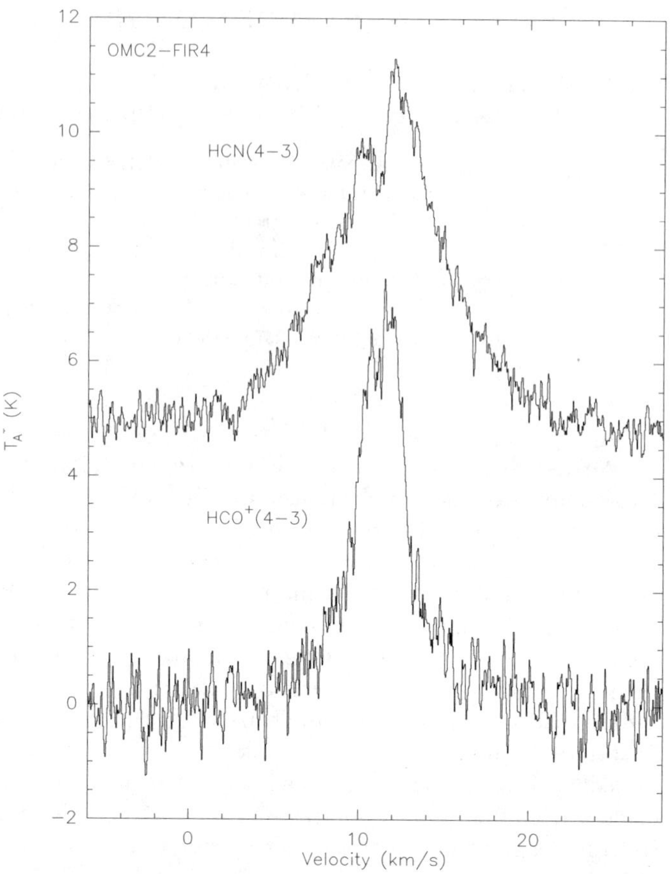

Figure 2. HCN *(top)*, and HCO$^+$ *(bottom)* observations at the position of peak intensity of OMC-2 FIR4 in the $J = 4 \rightarrow 3$ transition.

is maximum in cases where the neutral flows travel in a direction perpendicular to the direction of the magnetic field. An example of this is shown in Figure 1.

In generalizing the results of Figure 1 to cases where a large number of neutral flows are allowed to span any angles relative to the orientation of the magnetic field, irrespective of their velocity, it is possible to conclude that *in the core of molecular clouds, the width of line profiles of molecular ions should in general be less than that of coexistent neutral molecular species.* This fact has been verified in more than twenty molecular clouds so far (see Table 1 of Houde et al. [2002]). We show an example of a comparison of neutral and ion molecular species spectra (HCN and HCO$^+$ in the $J = 4 \rightarrow 3$ transition) for OMC-2 FIR4 in Figure 2.

It is also possible to give upper and lower limits for the ratio of the ion-to-neutral line width ratio. If $\sigma_{l,i}$ and $\sigma_{l,n}$ are the ion and neutral line widths, respectively, as defined by the standard deviation of the line profiles we find that the ion-to-neutral line width ratio is restricted to the following range:

$$0.26 \lesssim \frac{\sigma_{l,i}}{\sigma_{l,n}} \lesssim 1, \qquad (1)$$

for HCO$^+$.

Equation (1) can be verified by studying Figure 3 where we have plotted the ion line width against the corresponding neutral line width for every object and pair of molecular species studied so far (HCO$^+$ is plotted against HCN and H^{13}CO$^+$ against H^{13}CN, from Table 1 of Houde et al. [2002]). The two straight lines correspond to the upper and lower limits discussed above. The steeper of the two (with a slope of $\simeq 1$) arises when the magnetic field is oriented in a direction parallel to the line of sight while the other (with a slope of $\simeq 0.26$) when the field lies in the plane of the sky. As can be readily seen from this figure, the data obtained so far are in excellent agreement with the prediction made with equation (1). Even

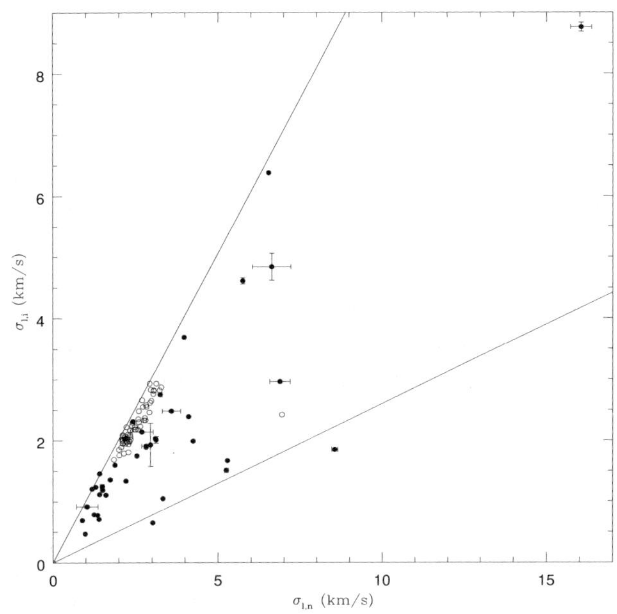

Figure 3. The ion line width vs the neutral line width for every pair of spectra obtained so far, using standard deviation measurements (from Table 1 of Houde et al. [2002]). HCO^+ is plotted against HCN and $H^{13}CO^+$ against $H^{13}CN$. The two straight lines correspond to the upper and lower limits discussed in the text where the line width ratio is $\simeq 1$ and $\simeq 0.26$, respectively. The cluster of open circles all pertain to spectra obtained on the same object (M17, which will be discussed later).

the lower limit of $\simeq 0.26$ for the line width ratio appears to be fairly accurate as only two of the more than ninety plotted points have a ratio which is lower than this value, and then only slightly.

These results have important implications for the study of the magnetic field in molecular clouds. Namely, it is possible to conclude that:

1. *In the weakly ionized regions of turbulent molecular clouds, the neutrals drive the ions.* If the opposite were true, we would expect the ion species to exhibit line profiles which would be at least as broad as those of coexistent neutral species and probably broader (Houde et al. 2001), contrary to observation.

2. The difference in the width of the line profiles of coexistent ion and neutral molecular species implies that *the coupling between ions and neutrals is poor in the core of molecular clouds*, at least at the scales probed by our observations (up to a few tenths of a parsec).

3. At the spatial resolution attained with our observations (we have a beam width of $\approx 20''$ in most cases), the diffusion between ions and neutrals can be studied through a comparison of the width of their line profiles. It then appears from our results that *the drift speed between ions and neutrals can often be significant in the core of molecular clouds* (on the order of a few km/s at the gas densities probed with the molecular species used here, i.e., $n \gtrsim 10^6$ cm^{-3}).

3. The Determination of the Viewing Angle α

Since the line width ratio is maximum at approximately unity when the field is aligned with the line of sight ($\alpha = 0$), and decreases to a minimum of $\simeq 0.26$ when the field lies in the plane of the sky ($\alpha = \pi/2$), we could be justified in hoping that it might be a well-behaved function which decreases monotonically with increasing α. It can in fact be shown that this is true, at least when considering reasonably general models of neutral flow configurations [see Fig. 2 of Houde et al. (2002) and the discussion that pertains to it]. When studying such models one obtains a family of curves of the ion-to-neutral line width ratio as a function of the inclination angle α. The curve that best fit the physical conditions of the region under study has to somehow be identified.

To do this, we use another parameter that is a function of the orientation of the magnetic field relative to the line of sight that can be readily obtained. This parameter is the polarization level that is measured, for example, from the continuum emission from dust at submillimeter wavelengths. Since the (elongated) dust grains are presumably aligned by the magnetic field, the polarization level P detected by an observer studying a given region where the field is oriented with a viewing angle α can be expressed as

$$P = P_{\max} \sin^2 \alpha \,, \tag{2}$$

where $P_{\max}$ is the maximum polarization level that can be detected, i.e. when the field lies in the plane of the sky ($\alpha = \pi/2$). Equation (2) can be used in conjunction with the relevant equation for the ion-to-neutral line width ratio in order to express the latter as a function of the normalized polarization level $P/P_{\max}$. By carefully taking into account the well-known depolarization effect observed in molecular clouds toward regions of higher flux (Weintraub, Goodman & Akeson 2000) and assigning a suitable value for $P_{\max}$, it then becomes possible to use polarization data to constrain the neutral flow model and, in effect, calibrate the ion-to-neutral line width ratio.

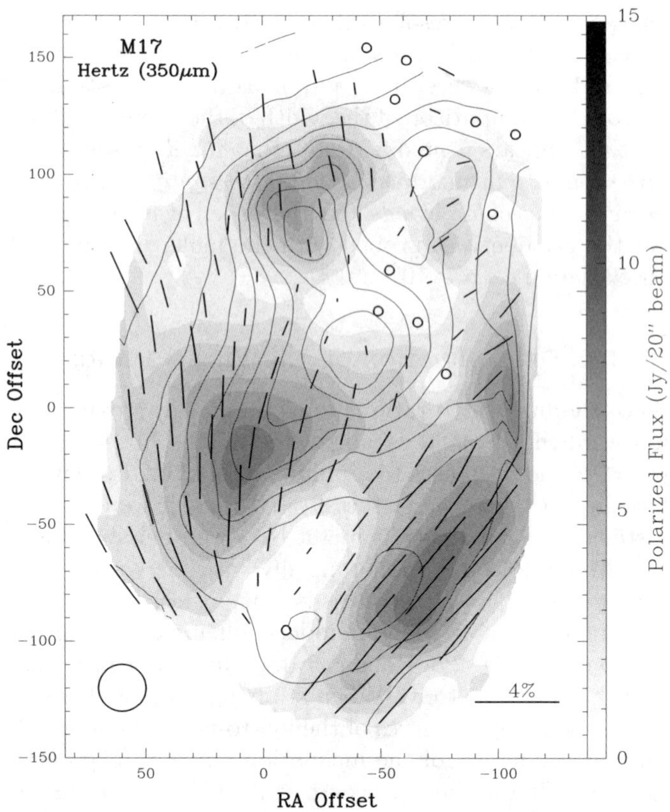

Figure 4. HERTZ polarization map of M17 at 350 μm. All the polarization vectors shown have a polarization level and error such that $P > 3\sigma_P$. Circles indicate cases where $P + 2\sigma_P < 1\%$. The contours delineate the total continuum flux (from 10% to 90% with a maximum flux of ≈ 700 Jy) whereas the underlying grey scale gives the polarized flux according to the scale on the right. The beam width ($\simeq 20''$) is shown in the lower left corner.

4. Observational Evidence

An extensive 350 μm polarimetry map of the M17 molecular cloud was obtained using the HERTZ polarimeter (Dowell et al. 1998) at the CSO on 1997 April 20 through 27 and 2001 July 19 and is presented in Figure 4. Besides the total flux (in contours) and polarized flux (in gray scale), this figure gives a detailed view (with a beam size of $\simeq 20''$) of the polarization vectors (or E vectors) across an area of more than $3'$ by $4'$. All the polarization vectors shown have a polarization level and error such that $P > 3\sigma_P$. Circles indicate cases where $P + 2\sigma_P < 1\%$.

As can be seen, both the magnitude and the orientation of the polarization vectors are "well-behaved" across the map in that, at this spatial resolution, the variations are smooth and happen on a relatively large scale. The amount of polarization is seen to vary from $\approx 0\%$ to a maximum of $\approx 4\%$ which is consistent with the bulk of observations made on other objects at this wavelength.

We have also made maps of the same region in the $J \rightarrow 4-3$ transition of both HCN and HCO^+ at the CSO, using the facility's 300-400 GHz receiver, during a large number of nights in the months of March, May, June and August 2001 (see Fig. 8 of Houde et al. [2002]). The two maps have a similar appearance and are also not unlike the 350 μm continuum map presented in Figure 4. The beam size for all sets of observations was about the same at $\approx 20''$. This is a nice feature as our analysis will rest on comparisons of polarimetry and spectroscopic data across the molecular cloud.

4.1. Calibration of the Line Width Ratio

With the polarization and spectroscopic (i.e., HCN and HCO^+) data at hand, it now becomes possible to constrain the neutral flow model and calibrate the ion-to-neutral line width ratio, as was discussed earlier. The line ratios were calculated after the ion and neutral line widths $\sigma_{l,i}$ and $\sigma_{l,n}$, respectively, were measured at every position of the corresponding maps. We plotted the line width ratios thus calculated against the normalized polarization levels in Figure 5.

In Figure 5, we have defined the normalized polarization level $P/P_{\rm max}$ with the maximum level of polarization set at $P_{\rm max} = 7\%$. This choice of $P_{\rm max}$ was not made arbitrarily nor was its value determined so as to provide a fit to the data. We have based its value on the extensive polarimetry data already obtained with HERTZ where we found that the highest levels of polarization detected so far at 350 μm were in the neighborhood of 7%. We assume that this applies well to M17 and that it corresponds to the hypothetical case where the magnetic field lies in the plane of the sky ($\alpha = \pi/2$) at a position unaffected by the depolarization effect. It should be noted that small variations in this parameter would not significantly change our results. The neutral flow model is constrained using the points that delineates the "outer" limit of the data in Figure 5. The rest of the data falling within the shaded area is seen to be significantly affected by the depolarization effect discussed earlier.

We show in Figure 6 a map of the orientation of the magnetic field in M17 at every observed position. The angle β, made by the projection of the magnetic field in the plane of the sky, was obtained from the polarimetry data by rotating the corresponding polarization angle PA by 90° and is represented on the map by the orientation

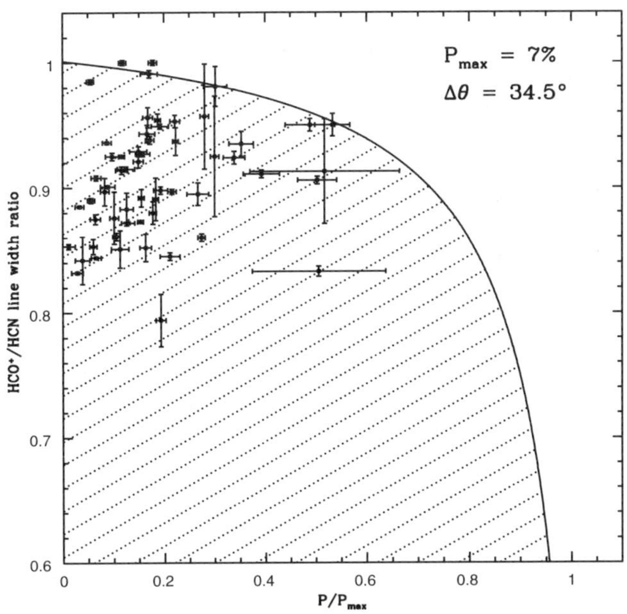

Figure 5. The HCO^+/HCN line width ratio versus the normalized polarization level $P/P_{\rm max}$ for M17. $P_{\rm max}$ is set at 7% and the data is shown against the model of neutral flow collimation that best fits the data. The polarization levels have $P > 3\sigma_P$ or $P + 2\sigma_P < 1\%$.

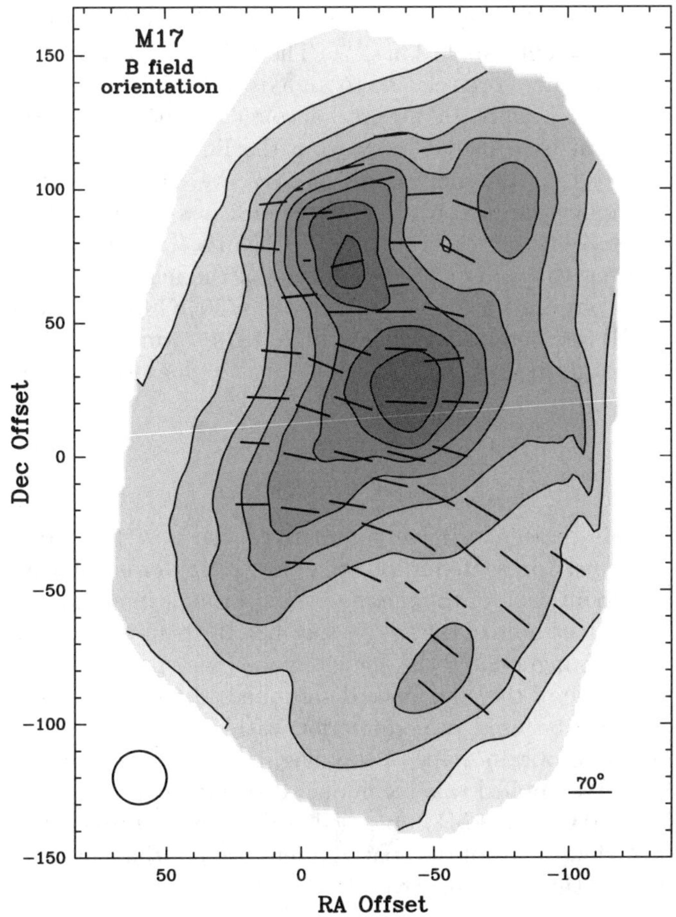

Figure 6. Orientation of the magnetic field in M17. The orientation of the projection of the magnetic field in the plane of the sky is shown by the vectors and the viewing angle is given by the length of the vectors (using the scale shown in the bottom right corner). The contours and the grey scale delineate the total continuum flux. The beam width ($\simeq 20''$) is shown in the lower left corner.

of the vectors. The viewing angle α, or the angle made by the magnetic field to the line of sight, was obtained by using the fit shown in Figure 5 along with equation (2) and can be read on the map by the length of the vectors (using the scale in the bottom right). Both angles are plotted on top of the 350 μm continuum flux obtained with HERTZ.

From Figure 6, we can observe some of the main features in the orientation of the magnetic field in M17. First, there is a gradual shift of some 40° in the orientation of the projection of the magnetic field on the plane of the sky from the south-west part of the map ($\beta \sim 50°$) to the north ($\beta \sim 90°$). On the other hand, the viewing angle is maximum at $\alpha \sim 65°$ in the neighborhood of the region of peak continuum emission and smoothly decreases south-westerly to a local minimum where the field is better aligned to the line of sight with $\alpha \simeq 30°$ at RA off $\simeq -50''$, Dec off $\simeq -50''$. The field gradually approaches the plane of the sky, once again, in the south of the map.

Most interestingly, there is an important and localized decrease in the viewing angle of roughly $30° - 40°$ close to the position of steepest change in continuum (or HCN and HCO^+) emission where α reaches a minimum at approximately 10°. This region, located at RA off $\simeq 0''$, Dec off $\simeq 80''$, is also nearly coincident with the locations of H_2O and OH masers and the ultracompact H II region of the Northern Condensation, where Brogan & Troland (2001) (see their Fig. 16) obtained a value of $B_{\rm los} \simeq -300\,\mu$G using OH Zeeman measurements at 20 km/s.

It is important to realize that α does not provide us with the information concerning the direction of the magnetic field relative to the plane of the sky (i.e., is it going into or coming out of the plane?): this will be provided by Zeeman measurements. We know that for M17 the magnetic field is actually coming out of the plane of the sky

(Brogan & Troland [2001], using their HI or OH Zeeman measurements at 20 $\mathrm{km\,s^{-1}}$). The values of α obtained here are therefore relative to an axis directed toward the observer. Finally, taking into account that the magnetic field can be directed away from the line of sight by as much as 60° in some parts of M17, we can see that a multiplicative factor of the order of 2 has to be applied to the Zeeman measurements of Brogan & Troland (2001) in order to evaluate the magnitude of the magnetic field. They obtained a maximum value of -750 μG for B_{los}, using HI Zeeman measurements at 20 $\mathrm{km\,s^{-1}}$, implying that the magnitude of the field could be as high as 1.5 mG for this object.

5. Conclusion

In this paper, we have summarized the new technique previously presented in Houde et al. (2002) for evaluating the orientation of the magnetic field in molecular clouds. The orientation of the field is specified by its inclination or viewing angle α and the angle β made by its projection on the plane of the sky. Once determined, these parameters can also be used in conjunction with Zeeman measurements to obtain maps of the magnitude of the magnetic field. We applied this technique to spectroscopic and polarimetry data of M17 obtained at the CSO to provide the first view of the of the spatial orientation of the magnetic field in this molecular cloud.

Acknowledgements

The Caltech Submillimeter Observatory is funded by the NSF through contract AST 9980846 and the observations made with HERTZ were supported by NSF Grants AST 9987441 and AST 0204886.

References

Brogan, C. L., Troland, T. H., 2001, ApJ, 560, 821

Crutcher, R. M., Troland, T. H., Goodman, A. A., Heiles, C., Kazès, I., & Myers, P. C. 1993, ApJ, 407, 175

Crutcher, R. M., Troland, T. H., Lazareff, B., Paubert, G., Kazès, I. 1999, ApJ, 514, L121

Davis, L. M. & Greenstein, J. L. 1952, ApJ, 114, 206

Dowell, C. D., Hildebrand, R. H., Schleuning, D. A., Vaillancourt, J. E., Dotson, J. L., Novak, G., Renbarger, T., Houde, M. 1998, ApJ, 504, 588

Heiles, C. 1997, ApJS, 111, 245

Hildebrand, R. H. 1988, QJRAS, 29, 327

Houde, M., Bastien, P., Peng, R., Phillips, T. G., and Yoshida, H.2000a, ApJ, 536, 857

Houde, M., Peng, R., Phillips, T. G., Bastien, P. and Yoshida, H. 2000b, ApJ, 537, 245

Houde, M., Phillips, T. G., Bastien, P., Peng, R., and Yoshida, H. 2001, ApJ, 547, 311

Houde, M., Bastien, P., Dotson, J. L., Dowell, C. D., Hildebrand, R. H., Peng, R., Phillips, T. G., Vaillancourt, J. E., and Yoshida, H. 2002, ApJ, 569, 803

Weintraub, D. A., Goodman, A. A., Akeson, R. L. 2000, in *Protostars and Planets IV*, ed. V. Mannings, A.P. Boss, & S.S. Russell (Tucson: University of Arizona Press), 247

CHEMICAL DIFFERENTIATION IN STRONGLY MAGNETIZED STARLESS CORES

Valery I. Shematovich[1], **Dmitri S. Wiebe**[1], **Boris M. Shustov**[1], **Yaroslav N. Pavlyuchenkov**[1] **and Zhi-Yun Li**[2]

[1]Institute of Astronomy of the RAS, 48 Pyatnitskaya str., Moscow, 119017, Russia
[2]Department of Astronomy, University of Virginia, Charlottesville, VA 22903, U.S.A.

Abstract

A detailed chemical model for the starless cores of strongly magnetized molecular clouds has been developed. In this model the ambipolar diffusion-driven evolution of the clouds is coupled to the chemistry through ion abundances. The model predictions of the peak values and spatial distributions of the column densities of CO, CCS, and NH_3 are compared with those inferred observationally for the well studied starless core L1544, which is thought to be on the verge of star formation. A reasonable overall fit to the available data on L1544 was found for a model cloud which is magnetically supported for several million years before collapsing dynamically. Several key molecules that seem to be useful for distinguishing between different dynamic and chemical scenarios were identified. The coupled dynamical and chemical model allows us to calculate simultaneously the spatial distribution of a given tracer molecule and the velocity field. The synthetic line profiles were calculated and used to constrain the chemical models of the starless core evolution leading to star formation.

Key words: ISM: molecules – ISM: structure – ISM: clouds

1. Introduction

Magnetic fields have long been considered an important factor in star formation. The ambipolar diffusion-driven condensation of cores in a strongly magnetized cloud is the first step in the widely discussed, by now "standard," scenario of isolated, low-mass star formation (Shu, Adams & Lizano 1987; Mouschovias & Ciolek 1999).

There have been many studies of the structure and evolution of magnetized, quiescent molecular cloud cores leading to star formation over the years (e.g., Lizano & Shu 1989; Basu & Mouschovias 1994; Li & Shu 1996). These studies elucidated the crucial role of a strong magnetic field in controlling the *dynamical* processes, such as cloud support and angular momentum removal, involved in *single* low-mass star formation. There is, however, little discussion of the abundances of molecules in these calculations, making it difficult for a direct comparison with observations of molecular lines.

Chemical inhomogeneities are observed in star-forming dark clouds on a range of scales (Langer et al. 2000). On the scale of individual dense cores, it is now becoming increasingly clear that many species are chemically differentiated (Tafalla et al. 2002). A notable example is the severe reduction in the abundances of the sulphur-bearing species CS and CCS toward the center of a number of starless cores, including L1498 (Kuiper, Langer & Velusamy 1996), L1544 (Tafalla et al. 2002), and L1512 (Lee et al. 2002). Such a reduction is generally attributed to the freeze-out of molecules onto dust grains (which occurs efficiently in the dense, cold interiors of starless cores, especially for sulphur-bearing species because of their strong bonds to the grain surface) and chemistry triggered by depletion (Bergin & Langer 1997).

Understanding the chemical differentiation in starless cores quantitatively is important for studies of low-mass star formation. The classical density tracers, such as CS, HCO^+, HCN, and H_2CO, have been widely used to probe the physical conditions in dense cores, the basic units of low-mass star formation (Evans 1999). They may, however, miss the high density, inner core region because of depletion. Perhaps more importantly, the observed pattern of chemical differentiation can put constraints on the timescale of cloud evolution leading to star formation (Bergin & Langer 1997; Aikawa et al. 2001; Li et al. 2002). In particular, it has the potential of constraining indirectly the dynamical role of magnetic fields in core formation envisioned in the standard scenario of isolated, low-mass star formation.

There is already some indication from molecular line studies of infall motion that dense, starless cores are magnetically supported to a large extent. Lee, Myers & Tafalla (2001) mapped 37 cores in both CS and N_2H^+, and estimated a typical one-dimensional infall speed in

Chemistry as a Diagnostic of Star Formation (2003). *Edited by Charles L. Curry and Michel Fich.* Published in 2003 by NRC Research Press, Ottawa ON, K1A 0R6, Canada. ISBN 0-660-19089-3.

the range 0.05–0.09 $km\,s^{-1}$, which is about half the isothermal sound speed (0.188 $km\,s^{-1}$ at 10 K) or less. The slow contraction is naturally achieved in a strongly magnetized core, whose evolution is driven by ambipolar diffusion over several dynamical times (Li 1999; Ciolek & Basu 2000). The relatively long evolution time should leave a strong imprint on the core chemistry, since the chemical and dynamical timescales are comparable. In a magnetic cloud, the interplay between dynamics and chemistry is highly nonlinear: the density evolution affects the chemistry, which determines the abundances of charged particles, which in turn controls the rate of ambipolar diffusion that regulates the density evolution. A coupled dynamical and chemical model is required for detailed data interpretation.

2. Model Development

Although there have been many studies of the evolution of magnetic clouds driven by ambipolar diffusion, most involve major simplifications: they are either two-dimensional (2D) but with the velocity field ignored (i.e., "quasi-static"; see Lizano & Shu 1989) or fully dynamic but spatially one-dimensional (e.g., Basu & Mouschovias 1994). There are a few 2D dynamic models assuming either axisymmetry (Desch & Mouschovias 2001) or a thin disk approximation (Li & Nakamura 2002). Such 2D models are computationally demanding by themselves, however, because to ensure numerical stability in the presence of ambipolar diffusion the time step must scale with the *square* of the grid size, unlike the usual Courant condition in which the time step scales *linearly* with grid size. This requirement is particularly stringent in the high density regions where chemical differentiation is expected to be most pronounced and where good spatial resolution is necessary. It is impractical at the present time to couple these 2D models to a chemical network that involves thousands of reactions (Millar, Farquhar & Willacy 1997). One needs to simplify either the dynamics or the chemistry or both to make progress.

We have made a start on developing a coupled dynamical and chemical model for magnetic cloud cores (Li et al. 2002; Shematovich et al. 2002), based on a simplified model for the dynamics, but with an extended chemical network incorporated. The simplification comes from an approximation introduced by Safier, McKee & Stahler (1997): by considering only the pressure component of the magnetic forces, it is possible to retain a spherical geometry even in the presence of a dynamically important magnetic field. Although the spherical models are approximate, they do capture the essential feature of ambipolar diffusion-driven cloud evolution: an initial slow contraction during the magnetically subcritical phase, followed by a more dynamic collapse leading to star formation (Li 1999). These models represent a significant improvement over the semi-analytic hydrodynamic models (decoupled from chemistry) that have been used in the chemical models of other research groups (Rawlings et al. 1992; Bergin & Langer 1997; Aikawa et al. 2001).

For chemistry, we included a network of about 2000 gas phase reactions between 59 neutral and 83 ion species, freeze-out of species onto dust grains, and a limited set of 53 grain surface reactions. The direct coupling between chemistry and dynamics is achieved through ions, the abundances of which are determined self-consistently from the chemical network. The ionic abundances control the time scale of cloud dynamic evolution via ambipolar diffusion.

In our first paper (Li et al. 2002), we concentrated on the role of magnetic fields on both the dynamics and the chemistry of dense cores, and compared model predictions with the dynamical and chemical data on L1544, arguably the best studied starless core (Caselli et al. 2002*a*, *b*). We found a good agreement between the available observations and a model in which the cloud initially had a central density of molecular hydrogen $n_{H_2} = 10^3\,cm^{-3}$, and was supported equally by magnetic fields and thermal motions. It takes about 6 million years for the cloud to develop a core that matches the density profile of L1544 with a central $n_{H_2} = 10^6\,cm^{-3}$ deduced from dust continuum observations (Ward-Thompson, Motte & André 1999; Evans et al. 2001). The predicted peak column densities of CO, CCS, N_2H^+ and HCO^+ are within a factor of two or so of the observationally inferred values, and the spatial distributions of these species are consistent with those observed. An alternative, non-magnetic model was also considered for comparison. With the same initial conditions as in the magnetic model, the non-magnetic cloud collapses promptly, reaching the observed state in merely 1.5 Myr. It produces a worse overall fit to the data on L1544. In particular, its predicted infall speed is too high, and the radial distributions of CS and CCS are centrally peaked, unlike the ring-like structure observed for these two species. Our initial study added some weight to the standard scenario of isolated low-mass star formation involving strongly magnetized clouds and ambipolar diffusion over several dynamic times. However, there are well known uncertainties in the initial conditions for core formation as well as the physical and chemical processes, particularly those involving dust grains (Herbst 2000).

In our second paper (Shematovich et al. 2002), we explored the effects of some of the uncertain parameters. These include the initial chemical composition, sticking probability of gas species onto grain surfaces, adsorption energies, cosmic ray ionization rate, cloud mass, and magnetic field strength. We found that for the purposes of modeling the data from L1544, the following combination is sufficient: an initial composition of "mixed" metal abundances with a heavy sulphur depletion (Bergin et al. 1995), a moderate sticking probability of 0.3, a "standard" cosmic ray ionization rate of 1.3×10^{-17} s^{-1}, and a 20 $M_\odot$ cloud supported equally by magnetic fields and thermal motions.

A major motivation for modeling chemical abundances of starless cores is to constrain their formation time scale and thus indirectly the dynamic importance of magnetic fields.

2.1. Dynamics

In this paper we emphasize the dependence of the developed model on the chemical exchange between the gas and dust phases. The so-called standard chemical and dynamical model of the magnetized starless core of Shematovich et al. (2002)—model StM+B+sp0.3 with a sticking probability of 0.3—is analyzed together with the following variants: (i) model StM+B+sp1.0—the same as StM+B+sp0.3, but with the sticking probabilities for both neutrals and ions set to 1.0—and (ii) model StM-B+sp1.0—the same as model StM+B+sp1.0, but without a magnetic field. Higher values of sticking probabilities cause both a stronger chemical coupling between the gas and dust and a larger depletion of gas-phase molecules. Model StM-B+sp1.0 is close to the non-magnetic models explored in Aikawa et al. (2001).

The main dynamical characteristics—the density distribution of hydrogen nuclei, infall speed, and ionization fraction—are plotted in Figure 1 as a function of radius for the above three models.

The non-magnetic core model StM-B+sp1.0 collapses promptly, reaching the observed state of central density $n_{H_2} = 10^6\,\mathrm{cm}^{-3}$ in 1.52 Myr, which is significantly shorter than the times of 4.98 Myr and 6.49 Myr for the StM+B+sp1.0 and StM+B+sp0.3 models, respectively. As expected, the non-magnetic model has a higher infall speed at this time, with a maximum value of $0.39\,\mathrm{km\,s^{-1}}$, or nearly twice the isothermal sound speed. This maximum speed is to be compared with those of StM+B+sp0.3 ($0.21\,\mathrm{km\,s^{-1}}$) and StM+B+sp1.0 ($0.23\,\mathrm{km\,s^{-1}}$) and that inferred observationally for L1544 ($\sim 0.1\,\mathrm{km\,s^{-1}}$). Therefore, the non-magnetic model provides a worse fit to the dynamical data on L1544 than the magnetic models, particularly StM+B.

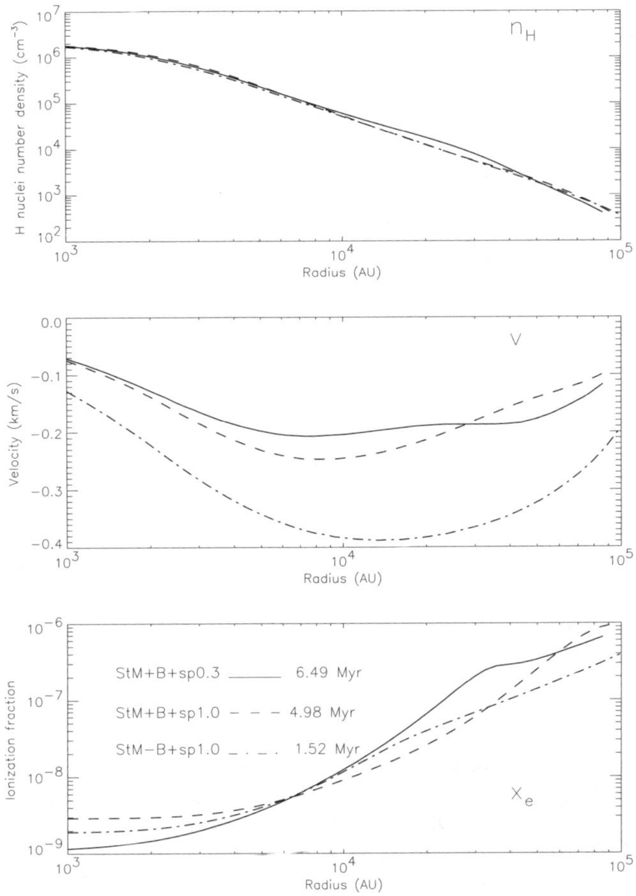

Figure 1. The radial distributions of hydrogen nuclei number density n_H (*top panel*), infall speed v (*middle panel*), and ionization fraction x_e (*bottom panel*) for the standard models with different sticking probabilities at a time when the H_2 central density reaches a value of 10^6 cm^{-3}. Here the solid, dashed, and dash-dotted lines correspond to models StM+B+sp0.3, StM+B+sp1.0, and StM-B+sp1.0, respectively.

The relatively low central values of ionization fraction x_e (bottom panel in Fig. 1) should not much affect the dynamics of the magnetically supercritical plateau region formed in the innermost part of the core (top panel in Fig. 1), which is controlled mostly by thermal, rather than by magnetic, pressure. All three models match the density distribution (see Tafalla et al. 2002, and Fig. 7 of Shema-

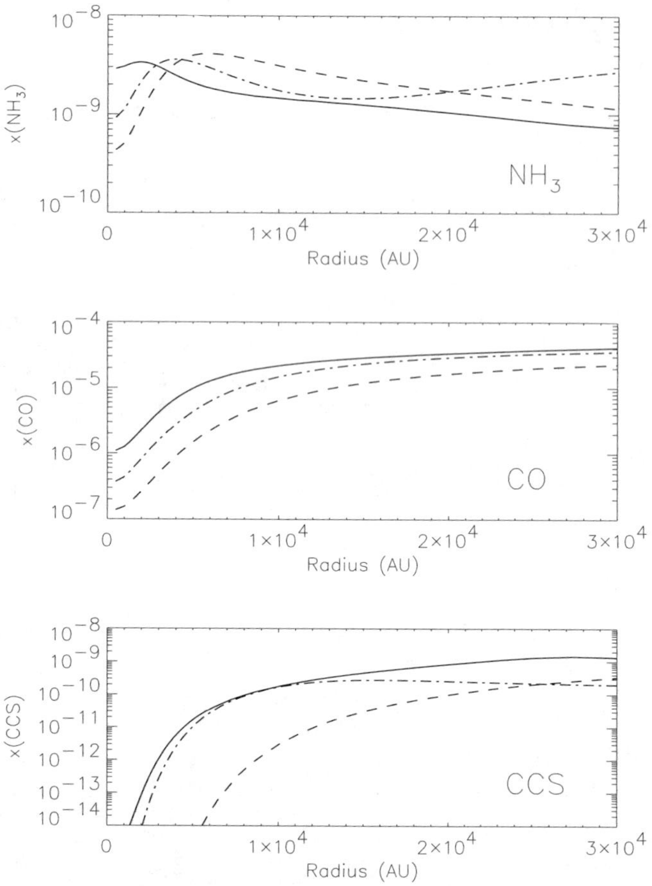

Figure 2. Radial profiles of the fractional abundances of the neutral species NH_3, CO, and CCS for the three models considered at a time when the H_2 central density reaches a value of 10^6 cm^{-3}. The line styles are the same as in Figure 1.

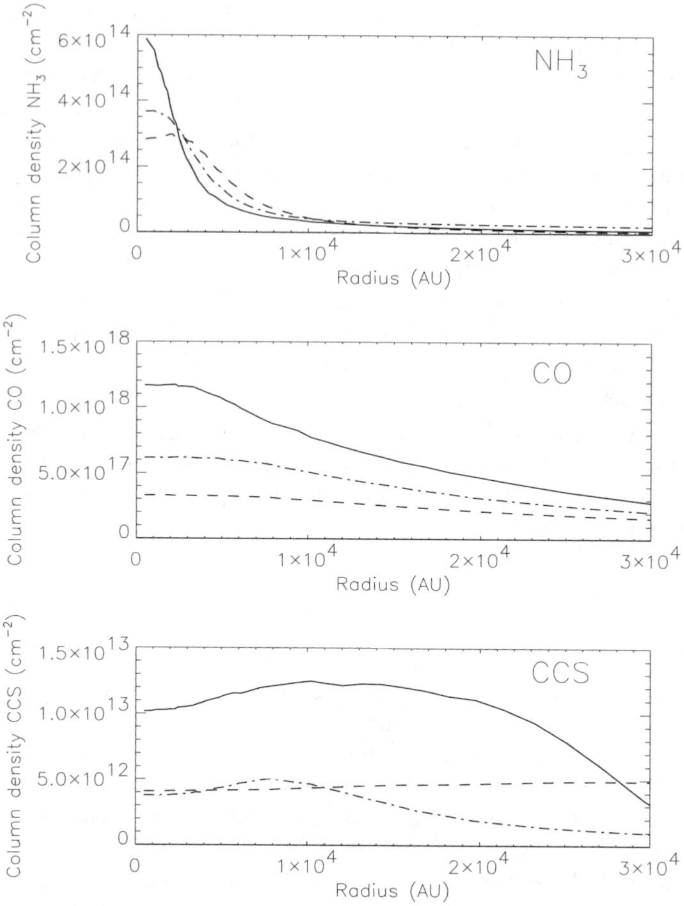

Figure 3. Same as Figure 2, but showing distributions of column density.

tovich et al. 2002) and ionization fraction near the center (Caselli et al. 2002b) inferred for L1544 reasonably well.

2.2. Chemical Differentiation

The chemical differentiation manifests itself observationally as a ring-like (or onion-like) space distribution of different neutral and ionized species in the starless core. The fractional abundances (relative to hydrogen nuclei) of the neutral species CO, CCS, and NH_3 are plotted as a function of radius in Figure 2. It is seen that for the sulphur-bearing molecule CCS the central depletion is particularly prominent. In contrast, nitrogen-bearing species, such as NH_3, show little depletion in the central region. CO remains abundant in the outer envelope, but decreases by more than an order of magnitude in the very central region. The calculated depletion pattern is in agreement with expectation and qualitatively consistent with that inferred for L1544 (Tafalla et al. 2002).

The column density distributions of NH_3, CO, and CCS for both the non-magnetic and magnetic models with different sticking probabilities are shown in Figure 3. These radial distributions of column densities clearly show the aforementioned chemical fractionation in the evolving starless cores: a centrally-peaked distribution of ammonia, a ring-like distribution of dicarbon sulphate, and the moderate depletion of carbon monoxide in the central plateau region of the core.

A comparison of fractional abundances predicted by the standard magnetic model StM+B+sp0.3 with those inferred directly from observations by Tafalla et al. (2002) is discussed in detail in Shematovich et al. (2002). There is reasonable agreement in both the mass distribution and the abundance distributions of various chemical species, particularly NH_3. The most notable discrepancy lies with $C^{18}O$ (with an abundance 1/490 of CO) at the smallest radii (inside 10^4 AU), although the distribution in the innermost region is not well constrained either observationally or in the numerical models because of finite spatial resolution. The predicted N_2H^+ abundance increases gradually inward, which is different from the constant distribution suggested by Tafalla et al. (2002). The difference is relatively moderate, however, within a factor of 2 or so.

The abundance of CCS is of special interest. This molecule is a well known "early-time" species. Its abundance should decrease rapidly after a few times 10^5 years, according to Suzuki et al. (1992), Bergin & Langer (1997) and Aikawa et al. (2001). In our magnetic model for L1544, it takes about 6 million years for the core to reach the observed state. Given such a long evolutionary time, one might expect very little CCS to be left in L1544, yet its observationally inferred column density is among the highest in the cores studied by Suzuki et al. (1992). This apparent puzzle is resolved by a "late-time" peak in the abundance of CCS, induced by CO depletion (Ruffle et al. 1997; Li et al. 2002; Lee et al. 2002). Using the standard UMIST RATE95 file (Millar et al. 1997), we obtained a peak CCS column density of $1.3 \times 10^{13}\,cm^{-2}$, which is a factor of 3 lower than that observed (Ohashi et al. 1999). Taking into account the newly recalculated, higher rate coefficient for the neutral-neutral reaction S + CCH $\rightarrow$ CCS + H (Millar et al. 2001) reduces the discrepancy to a factor of 2 (Shematovich et al. 2002). In view of the uncertainties involved in inferring the peak column density from observations, including the excitation temperature, non-spherical shape, and clumpiness in emission, we deem the agreement acceptable. A central hole is evident in the modeled CCS column density distribution, with size and depth in good agreement with those observed (Ohashi et al. 1999).

2.3. Synthetic Line Profiles

A unique strength of coupled dynamical and chemical models is that they allow for a simultaneous determination of the spatial distribution of a given chemical species *and* the velocity field, the two ingredients for line profile modeling. Line profiles of species such as CS and HCO^+ are routinely employed to deduce infall motions in dense cores. The fact that these species may be depleted poses a problem. Rawlings & Yates (2001) examined the line profiles in a collapsing core in some detail, and concluded that models assuming chemical uniformity or simple monotonic variations are unacceptable. Caselli et al. (1999, 2002*a*, *b*) noted that the optically thin lines of CCS and $HC^{18}O^+$ are doubly-peaked in L1544, which could result from a strong depletion near the center, leaving most of the emitting molecules in a (contracting) shell. A coupled dynamical and chemical model has the advantage of allowing one to determine the line profiles in a self-consistent way, with the effects of both dynamics and chemical differentiation taken properly into account.

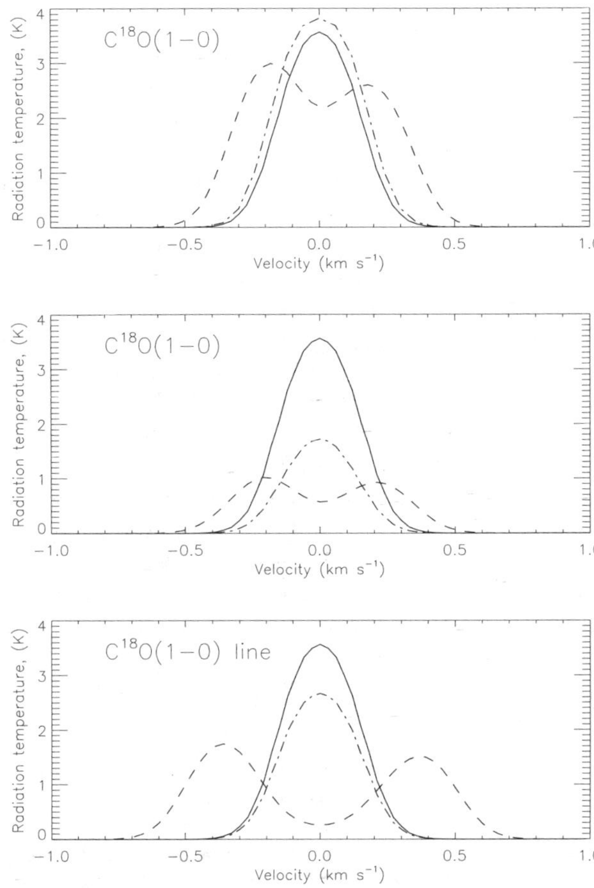

Figure 4. Observed (*solid line*) and calculated (*dashed line*) line profiles for $C^{18}O(J{=}1{-}0)$ in the L1544 core. *Top panel:* model StM+B+sp0.3; *middle panel:* model StM+B+sp1.0; *bottom panel:* model StM-B+sp1.0. The dash-dotted line in each panel corresponds to calculations with zero infall speed.

We have calculated the line profiles of $C^{18}O(J{=}1{-}0)$ transition with the radiative transfer code RATRAN (Hogerheijde & van der Tak 2000) for the models described above and compared these spectra with observed line profile for L1544 taken from Tafalla et al.(2002). The calculated spectra for the standard magnetic model are shown in the top panel of Figure 4. We see that the intensity of the calculated profile is close to the observed value but the line profile is double-peaked and more broad. These features are caused by the relatively high infall speed in the model core, and this could be confirmed with the line profile (dash-dotted line in Fig. 4) calculated for the same density distributions without infall. The synthetic line profiles obtained with and without infall for the magnetic model with a sticking probability of 1.0 are shown in the middle panel of Figure 4. The calculated profiles have a lower intensity because of the stronger molecular depletion. In the bottom panel the calculations for the non-magnetic model StM-B+sp1.0 are plotted. For this model the line intensity increases compared to the previous case. This is caused by the faster core contraction and shorter time for depletion. The calculated line profile is still too broad and double-peaked compared to the observed line. These differences between modeled and observed $C^{18}O(J{=}1{-}0)$ line profiles are mainly caused by the high infall speeds obtained in the models and are a shortcoming of the 1D model of the dynamical and chemical evolution of magnetized starless cores that will be overcome in the near future.

3. Conclusions

To summarize, we have begun constructing coupled dynamical and chemical models of magnetized dense cores. The initial results suggest that the chemical differentiation observed in L1544 can be explained quantitatively in the context of the standard scenario for isolated low-mass core formation involving ambipolar diffusion over several dynamical times. We plan to confirm this conclusion by refining our model, developing diagnostic tools, and applying it to more sources, for which detailed observations are becoming available.

Acknowledgements

This research is supported in part by RFBR Grants 01-02-16206 and 02-02-04008, and in part by NASA.

References

Aikawa, Y., et al. 2001, ApJ, 552, 639
Basu, S., & Mouschovias, T. Ch. 1994, ApJ, 432, 720
Bergin, E. A., & Langer, W. D. 1997, ApJ, 486, 316
Bergin, E. A., Langer, W. D., & Goldsmith, P. F. 1995, ApJ, 441, 222
Caselli, P., et al. 1999, ApJ, 523, L165
Caselli, P., et al. 2002a, ApJ, 565, 331
Caselli, P., et al. 2002b, ApJ, 565, 344
Ciolek, G. E., & Basu, S. 2000, ApJ, 529, 925
Desch, S. J., & Mouschovias, T. 2001, ApJ, 550, 314
Evans, N. J. II 1999, ARA&A, 37, 311
Evans, N. J. II, et al. 2001, ApJ, 557, 193
Herbst, E. 2000, in *Astrochemistry: From Molecular Clouds to Planetary Systems*, ed. C. Minh & A. F. van Dishoeck (Chelsea, Michigan: Sheridan Books), 147
Hogerheijde, M., & van der Tak, F. 2000, A&A, 362, 697
Kuiper, T. B. H., Langer, W. D., & Velusamy, T. 1996, ApJ, 468, 761
Langer, W. D., et al. 2000, in *Protostars and Planets IV*, ed. V. Mannings, A. P. Boss, & S. S. Russell (Tucson: Univ. of Arizona Press), 29
Lee, J.-E., et al. 2002, ApJ, in press
Lee, C.-W., Myers, P. C., & Tafalla, M. 2001, ApJS, 136, 703
Li, Z.-Y. 1999, ApJ, 526, 806
Li, Z.-Y., & Nakamura F. 2002, ApJ, 578, 256
Li, Z.-Y. & Shu, F. H. 1996, ApJ, 472, 211
Li, Z.-Y., et al. 2002, ApJ, 569, 792
Lizano, S., & Shu, F. H. 1989, ApJ, 342, 834
Millar, T. J., Farquhar, P. R. A., & Willacy, K. 1997, A&AS, 121, 139
Millar, T. J., Flores, J. R., & Markwick, A. J. 2001, MNRAS, 327, 1173
Mouschovias, T. Ch., & Ciolek, G. E. 1999, in *The Origin of Stars and Planetary Systems*, ed. C. J. Lada & N. D. Kylafis (Dordrecht: Kluwer Academic Publishers), 305
Ohashi, N., et al. 1999, ApJ, 518, L41
Safier, P. N., McKee, C. F., & Stahler, S. W. 1997, ApJ, 485, 660
Shematovich, V. I., et al. 2002, ApJ, submitted
Shu, F. H., Adams, F. C., & Lizano, S. 1987, ARAA, 25, 23
Suzuki, H., et al. 1992, ApJ, 392, 551
Rawlings, J. M. C., et al. 1992, MNRAS, 255, 471
Rawlings, J. M. C., & Yates, J. A. 2001, MNRAS, 326, 1423
Ruffle, D. P., et al. 1997, MNRAS, 291, 235
Tafalla, M., et al. 2002, ApJ, 569, 815
Ward-Thompson, D., Motte, F., & André, P. 1999, MNRAS, 305, 143

ABUNDANCE AND EXCITATION OF H_2, H_3^+ AND CO IN STAR-FORMING REGIONS

Craig A. Kulesa

Steward Observatory, University of Arizona, 933 N. Cherry Ave, Tucson, AZ 85721, USA

ABSTRACT

Although most of the 123 interstellar molecules reported to date have been detected through millimeter wave emission line spectroscopy, this technique is inapplicable to non-polar molecules like H_2 and H_3^+, which are central to our understanding of interstellar chemistry. Thus, high resolution infrared absorption line spectroscopy plays an important role in interstellar studies: chemically important non-polar molecules can be observed, and their abundances and excitation conditions reference the same "pencil beam" absorbing column of gas.

In this fashion, the abundance of cold H_2 in dark molecular clouds and star-forming regions can now be accurately measured and compared along the same line of sight with the abundance of its most commonly cited surrogate, CO. Also observed is the pivotal molecular ion H_3^+, whose abundance provides the most direct measurement of the cosmic ray ionization rate of H_2, a process that initiates the formation of many other molecules observed in dark clouds. For each of the distinct velocity components seen toward some embedded young stellar objects, we can determine the temperature, density, and abundance of CO, H_2, H_3^+, ^{13}CO and ^{12}CO thereby unraveling some of the internal structure of a star-forming cloud. These measurements represent significant tests of models of molecular clouds and star-forming regions.

Key words: ISM: molecules – ISM: clouds – Galaxy: abundances

1. INTRODUCTION

The hydrogen molecule, H_2, comprises the vast majority of the mass in star-forming molecular clouds and represents a large fraction of the interstellar mass of the Galaxy. Since H_2plays a significant role in the thermal balance and ionization of interstellar gas, it therefore partially *regulates* the dynamical evolution of the gas. All stars are born in these dense, dark molecular clouds, giving the hydrogen molecule special importance in the complicated cycle of the interstellar medium (ISM) and the stars that form from it. In a similar vein, H_3^+, or protonated H_2, is a cornerstone of interstellar chemistry; it is responsible for initiating the gas-phase chemical networks that form many of the observed molecules in dark clouds (Herbst & Klemperer 1973; Watson 1976).

The physical symmetry of H_2 and H_3^+ give them vanishingly small permanent dipole moments and therefore do not allow for a traditional rotational emission-line spectrum at radio wavelengths. Hopes for detecting the dominant H_2 molecule in cold clouds are frustrated further by its low moment of inertia, which yields large rotational energy level spacings that are energetically unreachable to cold hydrogen molecules in molecular clouds. Direct measurements of H_2 in its ortho- and para- ground states are made possible by a dipole permitted electronic absorption line spectrum in the vacuum ultraviolet between 912 and 1130 Å. Although such observations, made possible by space-borne telescopes like *Copernicus* and *FUSE*, are critical to assessing the chemistry and contents of diffuse lines of sight, they are inapplicable to studies of the dense, dark clouds where stars are formed. Thus, indirect measures of the abundance of H_2 have been necessary. These indirect observations have taken the form of millimeter-wave spectroscopy of CO as a surrogate for H_2, millimeter and submillimeter dust continuum emission, and infrared extinction mapping through multi-wavelength photometry of embedded and background stellar sources.

Direct measurement of the relationship between H_2 and CO, and the abundance and excitation of H_3^+, represent critical tests of our current theoretical understanding of gas-phase chemistry in the interstellar medium and the accounting of a significant component of the Galaxy. Although both hydrogenic species can be made to emit line radiation in energetic environments, the vast majority of

Chemistry as a Diagnostic of Star Formation (2003). *Edited by Charles L. Curry and Michel Fich.* Published in 2003 by NRC Research Press, Ottawa ON, K1A 0R6, Canada. ISBN 0-660-19089-3.

cold H_2 and H_3^+ molecules in dark interstellar clouds are in their respective ground states and hence are invisible to traditional emission line techniques. The case of H_3^+ is even more difficult: not only does it suffer from the lack of a permanent dipole moment, it also lacks a well-defined electronic spectrum, so UV observations are not possible as they are for H_2. Despite these setbacks, it is possible to observe H_3^+ in absorption through lines of its ν_2 vibration-rotation band in the infrared L-band at 3.5 – 4.0 μm, and the ground states of ortho- and para-H_2 can be probed by weak absorption line measurements of its rotational-vibrational spectrum at 2–2.5 μm toward heavily obscured lines of sight. The ro-vibrational $v = 1 \rightarrow 0$ and $v = 2 \rightarrow 0$ bands of both ^{12}CO and ^{13}CO lie at 4.6 and 2.3 μm, respectively, and unlike radio observations, entire ro-vibrational bands can be observed simultaneously. Since the chemistries of H_2, H_3^+ and CO are intricately related, and the abundances and physical conditions derived from such observations reference the *very same parcel of gas*, this technique represents a very powerful probe of molecular material in dense clouds and star-forming regions.

2. Observations and Analysis

Infrared observations of ^{12}CO, ^{13}CO, C^{18}O, H_2 and H_3^+ toward dark molecular clouds were taken with the Phoenix spectrometer at the $f/15$ Cassegrain focus of the 2.1-meter and 4-meter telescopes of the Kitt Peak National Observatory during the period of 1997 through 2001. Our data have an effective resolution of 4–7 km s^{-1} ($\lambda/\Delta\lambda = 45{,}000$–$75{,}000$). Spectra of CO, H_2 and H_3^+ are depicted in Figures 1–3.

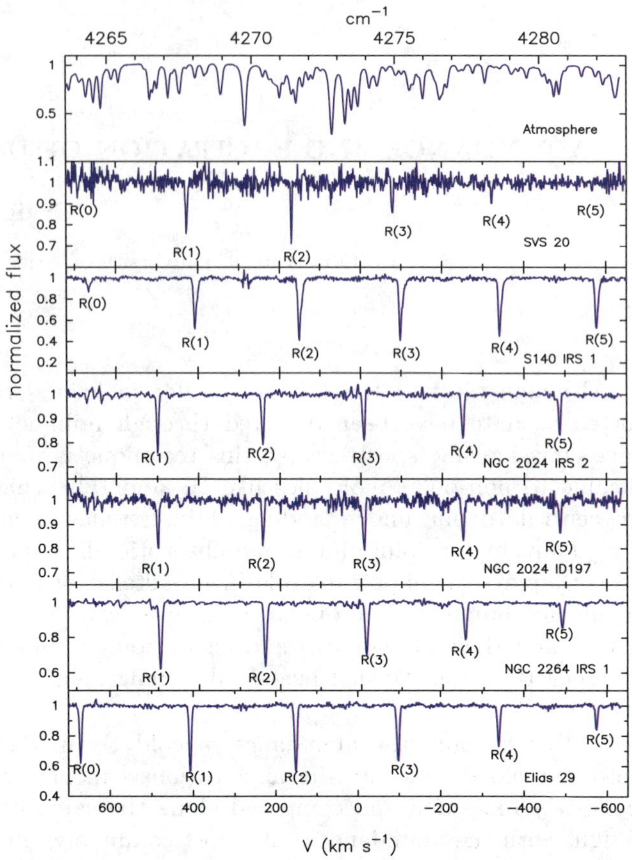

Figure 1. Sample spectra of ^{12}CO in the (2,0) R–band: $R(0)$ through $R(5)$. The uppermost spectrum is a typical comparison spectrum of the B2V star β_1 Sco that depicts the atmospheric transmission. The spectra are the ratioed (corrected) spectra of obscured young stellar objects that demonstrate prominent ^{12}CO absorption. The observed velocity shifts are due to the orbital motion of the Earth along the line of sight, and the velocity scale at the bottom is arbitrarily placed.

Lines of the ^{13}CO (1,0) and ^{12}CO (2,0) bands are often slightly saturated, so that column densities must be determined from curve of growth analyses (Spitzer 1978). The weak H_2 and H_3^+ lines detected in this study are unsaturated and can therefore be easily transformed into corresponding column densities. As H_2 comes in both ortho- and para-modifications due to Fermi-Dirac statistics of its nuclear spins, populations of both forms must be accounted for, and many of our detections of cold H_2 probe only the para- form of the molecule. However, the level populations between ortho- and para-H_2 are expected to be thermalized by reactive exchanges with hot protons, processes that are directly implied by the presence of H_3^+ in many of these sources.

This assertion can be tested directly, as ortho-H_2 is detected toward NGC 2024 IRS2 via observations of the (1,0) S(1) line. From the (1,0) S(0) equivalent width of 0.0023 cm^{-1}, a column density in the $J = 0$ level of $N_0 = 2.8 \times 10^{22}$ cm^{-2} is measured, and a total H_2 column density of $N(H_2) = 3.6 \times 10^{22}$ cm^{-2} is estimated at T=50 K from the best-fit CO temperature. From the (1,0) S(1) equivalent width of 0.00060 cm^{-1}, a column density in the $J = 1$ level of $N_1 = 1.2 \times 10^{22}$ cm^{-2} is measured, for a total of $N(H_2) = (4.0 \pm 0.2) \times 10^{22}$ cm^{-2}. This measurement of N(H_2) is only 11% higher than that derived under the assumption of H_2 thermalization. From the measured $J = 0$ and $J = 1$ column densities, a kinetic temperature of 57±3 K is derived, within 7 K of the CO temperature. A similar analysis confirms the thermalization of H_2 toward AFGL 2591.

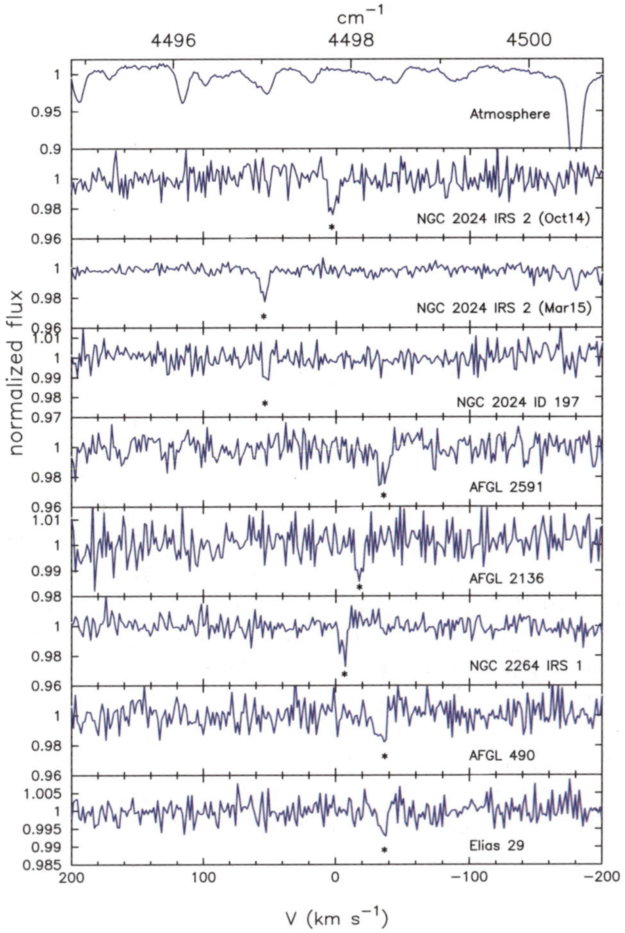

Figure 2. Detections of H_2 (1,0) S(0) absorption in dark molecular clouds. The top plot represents the atmospheric transmission; ratioed astronomical spectra follow. The quality of the spectra obtained toward NGC 2024 IRS2 at two different epochs can be compared to the landmark observations performed by Lacy et al. (1994). In this plot and those that follow, asterisks show the V_{LSR} of CO absorption in each source.

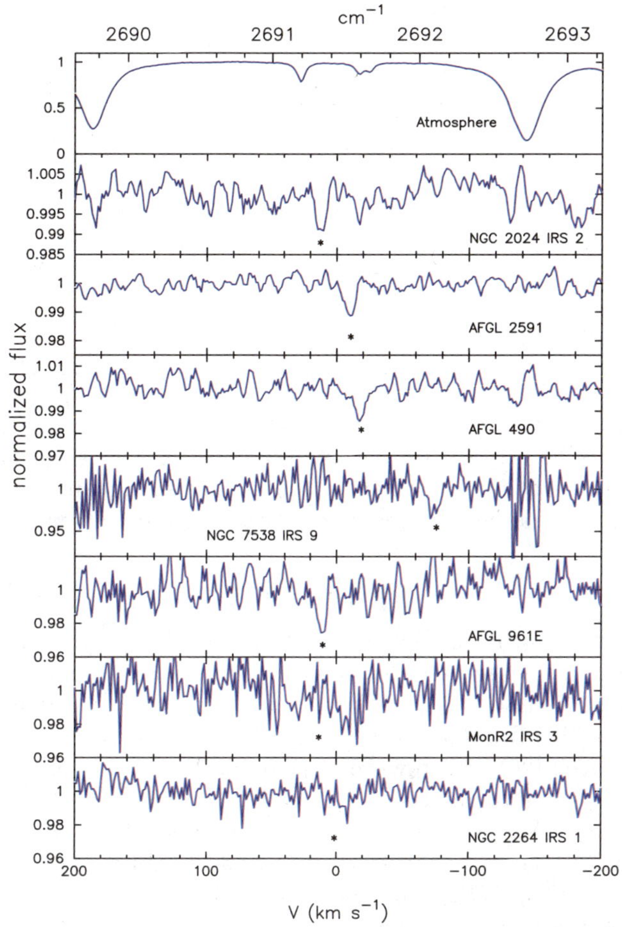

Figure 3. Sample observations of the $R(1,1)^l$ para- line of H_3^+ at 2691.45 cm^{-1} in dark molecular clouds. The top plot represents the atmospheric transmission; ratioed astronomical spectra follow. No detections are claimed for MonR2 IRS3 and NGC 2264 IRS1; the former was detected by McCall et al. (1999) with better sensitivity.

All other sources with (1,0) S(0) detections were either not observed at (1,0) S(1) or show small amounts of H_2 emission at that frequency. For the former, we assume that the ortho- and para- forms of H_2 are thermalized at the temperature indicated by CO, as this assumption seems to be valid for at least some lines of sight. For the latter, it is important to correct the (1,0) S(0) detections for contamination by emission.

As with H_2, the populations of ortho- and para-H_3^+ are expected to be in thermal equilibrium at the kinetic temperature of the gas. Thus the measurement of the ortho-to-para abundance ratio provides a sensitive interstellar thermometer which can be compared with thermalized low$-J$ level populations in CO and H_2. It is even more fortunate that two transitions, corresponding to the R(1,0) and $R(1,1)^u$ lines, probe the ground states of ortho- and para-H_3^+ and lie only 35 km s^{-1} apart from each other, though in a region rife with telluric absorption features of HDO and CH_4. We have achieved several detections of the ortho-para doublet, but observations presented in this brief overview are of the $R(1,1)^l$ singlet at 2691.44 cm^{-1}.

Many of our H_3^+ sources overlap the sample in McCall, Geballe, Hinkle, & Oka (1999); our derived column densities differ significantly only toward AFGL 2591 and AFGL 490, where our derived column densities are two-

Table 1. Abundances of ^{12}CO, ^{13}CO, H_2 and H_3^+

Source	T(^{12}CO) (K)	[^{12}CO/^{13}CO]	[$^{12}CO/H_2$] ($\times10^{-4}$)	[H_3^+/H_2] ($\times10^{-10}$)
NGC 2024 IRS2	50±0.8	56±17	1.5±0.2	8±2
NGC 2024 ID 197	46.9±2.1		1.7±0.4	
NGC 2264 IRS1	34.3±1.1	76±12	1.9±0.5	≤45 (3σ)
AFGL 490	(28±0.9)(108±0.7)	57,49±15	1.4±0.2	16±6
AFGL 2591: cloud	(32.5±0.8)(150±9)	>38,56±7	1.5±0.2	≤ 25±8
AFGL 2591: outflow	192±21	66±11		
AFGL 2136			1.6±0.6	≤ 82±20
S140 IRS1	41.6±0.9	54±8		
Elias 29	28±0.8		2.4±0.5	≤90 (3σ)

thirds as large. The uncertainties in the McCall et al. measurements toward those sources are significant, however, so the measurements themselves are in agreement.

3. Discussion

How do our measurements of ^{12}CO and H_2 compare with indirect measurements of the ^{12}CO and H_2 abundances in molecular clouds over the past two decades? Dickman (1978) produced the first quantitative comparison between ^{13}CO microwave emission at 2.6 mm wavelength and A_V, thought both then and now to be directly related to N_H over a wide range of cloud depths. The relation $N(H_2) = 5 \times 10^5$ N(^{13}CO) was established, which for [^{12}CO/^{13}CO]=60 leads to [$^{12}CO/H_2$]=1.8 $\times10^{-4}$. The study was expanded by Frerking, Langer, & Wilson (1982) by correlating rarer isotopes like $C^{18}O$ to A_V. They established [$^{12}CO/H_2$]=8 $\times10^{-5}$, which became widely recognized as the canonical abundance of ^{12}CO until recent years. Meanwhile, Watson, Genzel, Townes, & Storey (1985) directly measured [$^{12}CO/H_2$] = 1.2×10^{-4} using H_2 emission line techniques to probe small amounts of hot, shocked gas in Orion. However, it was recognized that the physical conditions in such shocked gas might not be representative of cold, quiescent, interstellar gas. Indeed, Wilson, Serabyn, Henkel, & Walmsley (1986) probed similar warm gas in Orion using $C^{18}O$ $J = 1 \rightarrow 0$ line emission, and estimated [$^{12}CO/H_2$]=5 $\times10^{-5}$. Finally, Lacy et al. (1994) performed the landmark measurements of [$^{12}CO/H_2$] toward the dense cloud NGC 2024 via infrared H_2 line absorption, estimating it to be 2.5×10^{-4}, albeit with large error bars. Most recently, Lee, Bettens, & Herbst (1996) performed a suite of chemical models that converge to [$^{12}CO/H_2$]=1.5 $\times10^{-4}$ for normal metallicities, densities of 10^3–10^5 cm^{-3} and kinetic temperatures between 10 and 50 K.

3.1. Extinction vs. hydrogen column density

How does the total column density of molecular hydrogen compare to dust extinction measurements? The principal motivations for exploring this question are the recently exploited techniques of near infrared multicolor imaging and extinction mapping as methods of deriving the high-resolution physical structure of molecular clouds (e.g., Alves, Lada, & Lada 1999).

Diffuse cloud measurements of atomic and molecular hydrogen, made possible for numerous sources by the *Copernicus* mission, established the H-to-extinction ratio for $A_V \leq 1$ (Bohlin, Savage, & Drake 1978):

$$\frac{N_H}{\mathrm{E(B-V)}} = 5.8 \times 10^{21}\ \mathrm{cm^{-2}\ mag^{-1}}$$
$$\frac{N_H}{A_V} = 1.9 \times 10^{21}\ \mathrm{cm^{-2}\ mag^{-1}}$$

Expectations are that this relation should extend deeply into the dark cloud regime without significant modification (Dickman 1978), but observational proof remains elusive. Recent results from FUSE, however, have extended the Bohlin, Savage, & Drake (1978) relation without modification to $A_V \leq 3$ (Rachford et al. 2002). With infrared absorption-line spectroscopy, we now have the capability of extending this relation to $A_V \sim 60$. Estimating A_V for complicated sources like AFGL 2591, AFGL 490, and even NGC 2024 IRS2 has proven very difficult, however.

We calculate A_V from the latest available infrared photometry for each of our objects, and assign maximal error bars that extend across all other known measures of A_V for each source in the literature. We adopt the Rieke & Lebofsky (1985) extinction law to derive A_V from the published observations independently, derive $N_H \approx 2H_2$from our observations, and graph the results in Figure 4 with the Bohlin et al. relation overlaid independently (i.e., unfitted).

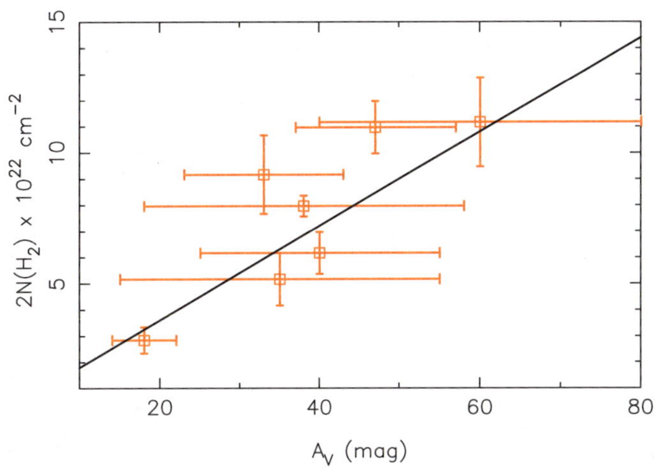

Figure 4. Photometrically-derived estimates of A_V for each source are plotted against the column of hydrogen nuclei estimated from the cold absorbing H_2 detected in this study. The straight line is the standard dust-to-extinction conversion of Bohlin et al. (1978), simply plotted (unfit) across the data. Error bars are maximal, $\gg 1\sigma$.

Although there is significant scatter and uncertainty in the estimates of A_V, the Bohlin et al. relation established for diffuse clouds appears to independently fit the data up to $A_V \sim 60$ (AFGL 2136). This is a significant and important empirical test that verifies the extension of this relation to high A_V (Dickman 1978) and its theoretical underpinnings, and validates the use of interstellar extinction mapping as a means of probing the H_2 structure of interstellar dark clouds. Nevertheless, the error bars are large; a joint collaboration of infrared H_2 absorption line work with large infrared imaging and extinction-mapping surveys would be an exceptionally productive way to calibrate both the method and the A_V-to-N_H relation with a minimum of systematic uncertainties.

3.2. The cosmic ray ionization rate of H_2

The formation of H_3^+ is a two-step process, stemming from the cosmic ray ionization of H_2 to H_2^+:

$$H_2 + \mathrm{cr} \rightarrow H_2^+ + e^-$$
$$H_2^+ + H_2 \rightarrow H_3^+ + H.$$

Once H_3^+ forms, it serves as the basis upon which the carbon, nitrogen, and oxygen ion-molecule networks initiate their chemistries, leading to the formation of many of the known interstellar molecules in dark clouds. Thus, the abundance of H_3^+ is a critical measure of the gas phase chemistry of molecular clouds.

Table 2. Measurements of ζ_0 in Dense Clouds

Source	Minimum ζ_0 ($\times 10^{-17}$ s^{-1})	Maximum ζ_0 ($\times 10^{-17}$ s^{-1})
NGC 2024 IRS2	1.1	3.5
AFGL 490	2	4
AFGL 2591	4	6
AFGL 2136	4.5	8

The rate of cosmic ray ionization of H_2 has been estimated indirectly to be around 10^{-17} s^{-1} (Herbst & Klemperer 1973), roughly 10^8 times slower than the second reaction from H_2^+ to H_3^+. The ionization rate of H_2 directly limits the rate of H_3^+ formation.

The destruction of H_3^+ is more complicated, however. Interactions with atomic or molecular hydrogen do not alter the number density of H_3^+; the next most abundant collision partner is CO. Interaction of H_3^+ and CO yields HCO^+, representing one of the few ways in which the robust CO molecule can be destroyed deep inside dark molecular clouds. That is,

$$H_3^+ + CO \rightarrow HCO^+ + H_2.$$

Interactions with atomic oxygen, nitrogen, free electrons, ...etc., all constitute common pathways by which H_3^+ may be destroyed.

These direct observations of H_2 and H_3^+, and the use of infrared and submillimeter CO, HCO^+ and H_2CO observations as densitometers, supply measurements of the thickness of the absorbing column L and the H_2/CO ratio. The theoretically predicted column density of H_3^+ is (Black, van Dishoeck, Willner, & Woods 1990)

$$\begin{aligned} N(H_3^+) &\leq 5.3 \times 10^8 \, \zeta_0 \, L \, n(H_2)/n(CO) \\ &\leq 5.3 \times 10^{13} \left(\frac{\zeta_0}{10^{-17}\ \mathrm{s}^{-1}}\right)\left(\frac{L}{10^{18}\ \mathrm{cm}}\right) \\ &\quad \times \left(\frac{[CO/H_2]}{10^{-4}}\right)^{-1} \mathrm{cm}^{-2}. \end{aligned}$$

Lower limits on the removal of H_3^+ come from assuming that only CO dissociatively recombines with H_3^+, and upper bounds can be derived by adopting the maximal equilibrium abundances of CO, O, and electrons for molecular clouds calculated in the chemical models of Lee, Bettens, & Herbst (1996). Our derived limits (Table 2) represent a remarkable validation of gas-phase chemical models in interstellar clouds.

4. Future Prospects

These observations exploit the power of a modern high-resolution infrared spectrometer on 2 to 4-meter class telescopes. Now that such instruments are being coupled with cross-dispersers, large-format infrared detector arrays, and 6 to 10-meter telescopes, the scientific throughput will be improved by a factor of 10^2 or more. This will allow practical direct measurement of the abundances of cold, quiescent H_2 and H_3^+ for a variety of fainter, true *background* sources within a single molecular cloud complex—sources which are less likely to have the additional complications of the current crop of YSOs with their accompanying circumstellar disks, winds and outflows. Though such phenomena are deeply interesting in their own right, they also contribute intervening hot H_2 which partially contaminates the H_2 absorption spectrum with line emission. We are at the dawn of establishing high resolution infrared spectroscopy as a technique of mapping the basic structure and chemistry of molecular clouds and star-forming regions, much as submillimeter telescopes and infrared imaging arrays are capable of doing today, but with far greater accuracy and with the added capability of observing pivotal species like H_2, H_3^+ and CO, which are unobservable by any other means in star-forming clouds.

References

Alves, J., Lada, C. J., & Lada, E. A. 1999, ApJ, 515, 265
Black, J. H., van Dishoeck, E. F., Willner, S. P., & Woods, R. C. 1990, ApJ, 358, 459
Bohlin, R. C., Savage, B. D., & Drake, J. F. 1978, ApJ, 224, 132
Dickman, R. L. 1978, ApJS, 37, 407
Frerking, M. A., Langer, W. D., & Wilson, R. W. 1982, ApJ, 262, 590
Herbst, E., & Klemperer, W. 1973, ApJ, 185, 505
Lacy, J. H., Knacke, R., Geballe, T. R., & Tokunaga, A. T. 1994, ApJL, 428, L69
Lee, H.-H., Bettens, R. P. A., & Herbst, E. 1996, A&AS, 119, 111
McCall, B. J., Geballe, T. R., Hinkle, K. H., & Oka, T. 1999, ApJ, 522, 338
Mitchell, G.F., Curry, C., Maillard, J.-P., & Allen, M. 1989, ApJ, 341, 1020
Rachford, B. L. et al. 2002, ApJ, 577, 221
Rieke, G. H., & Lebofsky, M. J. 1985, ApJ, 288, 618
Spitzer, L. 1978, Physical Processes in the Interstellar Medium (New York: Wiley-Interscience)
Watson, W. D. 1976, ApJ, 204, 47
Watson, D. M., Genzel, R., Townes, C. H., & Storey, J. W. V. 1985, ApJ, 298, 316
Wilson, T. L., Serabyn, E., Henkel, C., & Walmsley, C. M. 1986, A&A, 158, L1

CHEMISTRY AS A PROBE OF STRUCTURE IN DARK INTERSTELLAR CLOUDS

David A. Williams[1] **and Serena Viti**[2]

[1]Department of Physics and Astronomy, University College London, Gower Street, London, WC1E 6BT, UK
[2]CNR-istituto di Fisica dello Spazio Interplanetario, Area di Ricerca di Tor Vergata, via del fosso del Cavaliere 100, I-00133, Roma, Italy

Abstract

Molecular clouds are clumpy. Much of the clumpiness is unresolved by single-dish telescopes but is apparent in the data from array telescopes. Even in the unresolved observational data, the existence of clumpiness can be inferred from the distribution of molecular species. This evidence requires that "early-time" molecular species are widespread, while "late-time" species are more confined, and possibly associated with newly formed stars. The clumps are required to be transient.

Molecular clumps may also be observed close to Herbig-Haro (HH) objects. These clumps have a chemical composition different from that of clumps in a cloud interior, and the emissions from HCO^+ and NH_3 are strong. These HH clumps are dynamically quiescent; they have not been overtaken by the HH shock. We discuss here a detailed observational and theoretical study of one HH clump, and show that this clump must be transient and have a similar density and temperature to those clumps detected or inferred in the cloud interior. The anomalous chemistry is shown to be the result of HH illumination of regions that would otherwise be dark, so that in these regions a significant photochemistry develops. Thus, HH clumps are probably of the same family as the clumps in cloud interiors, even though they are chemically distinct.

It is a consequence of the ideas presented here that the clump mass distribution should be related to the formation of low-mass stars.

Key words: ISM: molecules – ISM: structure – ISM: clouds

1. Clumpiness within molecular clouds

Single-dish observations of molecular lines in dark clouds tend to reveal relatively smooth structures. However, there are some differences between the morphologies of different species. For example, emission from the CS molecule generally traces the full extent of the cloud and shows a fairly smooth structure, but ammonia emission is generally much more confined around regions where star formation is occurring (Zhou et al. 1989; Myers et al. 1991; Pastor et al. 1991; Morata et al. 1997). This behaviour is found in many dark clouds where low-mass star formation is taking place (see, e.g., Figs. 1–3, but it is the reverse of naive expectation. Since the critical density of the CS emission line is actually slightly larger than that of the ammonia line, CS and NH_3 should be tracing more or less the same material, or—if there is a slight difference—CS might be expected to favour the denser material that is, presumably, associated with star formation.

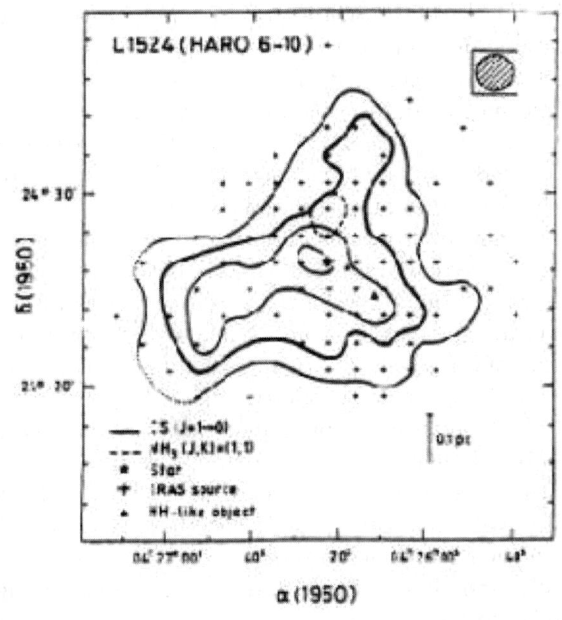

Figure 1. Figure 2 from Pastor et al. (1991): Contour map of the integrated line intensity of CS (1–0) from L1524. The half-power contour of NH_3 from Anglada et al. (1989) is also shown (*dashed line*).

Chemistry as a Diagnostic of Star Formation (2003). *Edited by Charles L. Curry and Michel Fich.* Published in 2003 by NRC Research Press, Ottawa ON, K1A 0R6, Canada. ISBN 0-660-19089-3.

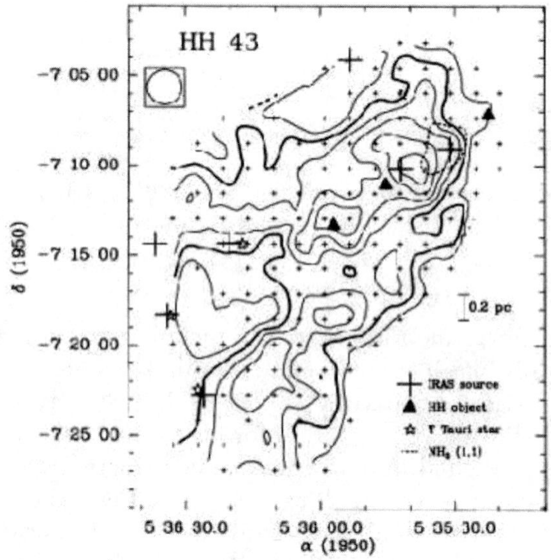

Figure 2. Figure 2 from Morata et al. (1997): as in Figure 1, but for HH 43.

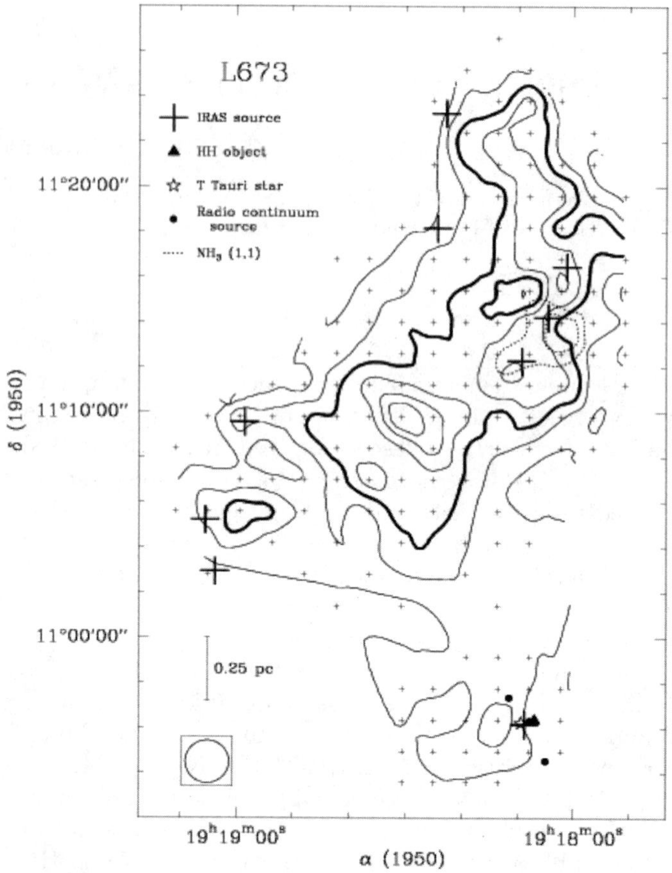

Figure 3. Figure 6 from Morata et al. (1997): as in Figure 1, but for L673.

One interpretation of this unexpected result was given by Taylor et al. (1996) in terms of the effects of time-dependent chemistry. Studies of low-temperature high density time-dependent chemistry, with initial atomic conditions (but with most of the hydrogen in H_2) show that some molecules rise towards their peak values distinctly earlier than others. The "early-time" species include CS, and the "late-time species" include NH_3. Consequently, Taylor et al. suggested that molecular clouds contain density fluctuations unresolved by the single-dish observations available at that time. These density fluctuations were required to have a duration of around one million years, so that CS would reach a high abundance while that of NH_3 was still rising. The fluctuations were assumed to decay to some undetermined background density in which molecular abundances would be expected to be lower than in the density enhancements. Not all the fluctuations would be expected to be identical. Occasionally a fluctuation might be of higher density and mass; such a fluctuation would certainly evolve faster chemically and dynamically, and may generate enough self-gravity to form a star.

In this picture, therefore, there would be a natural association of ammonia emission and star formation. The CS emission would appear wherever the unresolved density fluctuations appeared, and would therefore be a superposition of emission from many fluctuations. The CS line width would therefore be larger than that of the NH_3 which should arise in a few denser clumps. This is, in fact, observed.

Taylor et al. offered no explanation of the origin of such unresolved fluctuations, but very strong support for this model has come from the results of BIMA observations of the dark cloud L673 (Morata et al. 2003). These observations were of a small region of the cloud in lines of CS (Fig. 4), N_2H^+ (Fig. 5), and HCO^+ (Fig. 6), and showed that the cloud is indeed clumpy on small scales of the order of 10 arcsec and that molecular clouds are dynamically active on a relatively short timescale ($\sim$1 Myr). Such clumps are indeed unresolved by conventional single-dish observations. Smaller clumps than these may also exist, but were not resolved in the BIMA observations, nor could any background molecular emission be detected. The structure detected in the three molecular species is not identical, suggesting that the clumps are transient on a timescale corresponding to the epochs at which different species peak in abundance; i.e., on the order of a million years in the chemical evolution timescale. Thus, the BIMA observations in the three molecular lines may illustrate clumps in different stages of evolution.

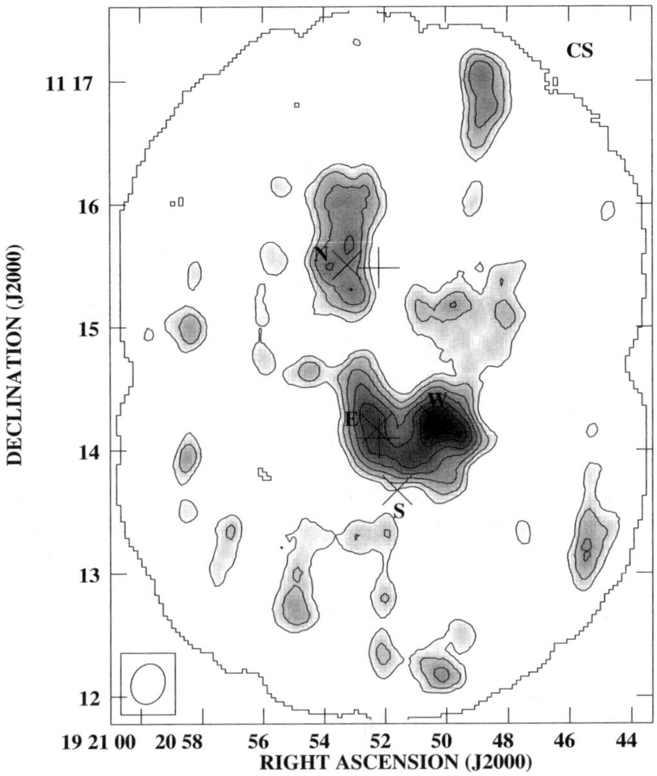

Figure 4. Contour map of CS in L673 from Morata et al. (2003).

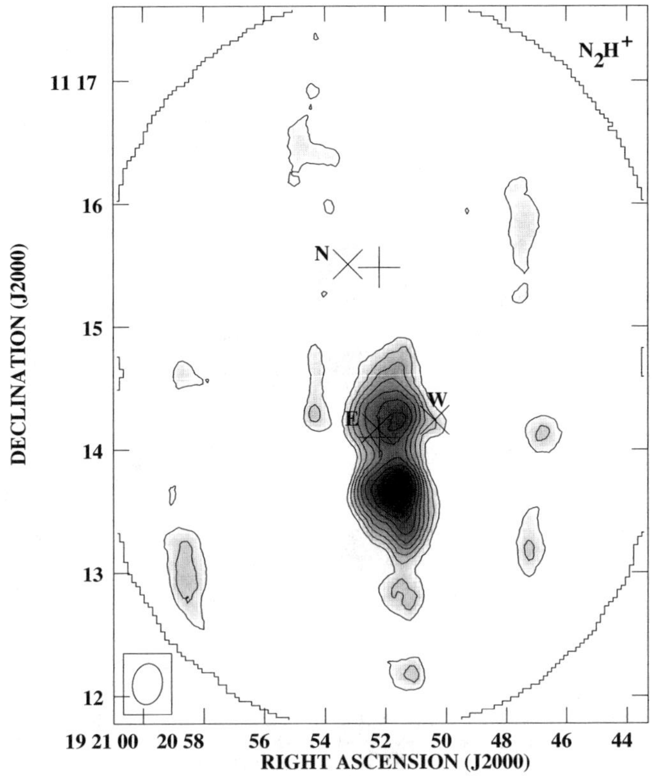

Figure 5. Same as Figure 4, but for N_2H^+.

Therefore, for at least one cloud, L673, the observational evidence supports the view that the cloud contains an assembly of transient clumps with lifetimes on the order of a million years. Such clumps should not show NH_3 or other late-time molecules, but only early time molecules like CS. At least some of these clumps are resolved in the BIMA observations. Thus, the picture revealed by the observations of Morata et al. is entirely consistent with the dynamical picture suggested by Taylor et al.

2. Origin of the density fluctuations

What causes the transient clumps? Evidently, they are some kind of representation of interstellar turbulence (cf. Franco & Carraminana 1999). However, it seems quite unlikely that they can arise in hydrodynamical processes (Hartquist et al. 2003) and that an MHD context is required. Falle & Hartquist (2002) have shown that under suitable conditions, slow-mode MHD waves can create large density enhancements, of the order of a factor of 30, and that the duration of the enhancement is on the order of 1 Myr (see Fig. 7). Thus, slow mode MHD waves are potential carriers of the transient density enhancements proposed by Taylor et al. (1996) and apparently observed by Morata et al. (2003). If so, then the implication from the work of Falle & Hartquist is that the ambient number density in the cloud is about 300 H nuclei cm^{-3} and that the maximum density in a fluctuation is about 10^4 cm^{-3}, as indicated by the CS observations in L673 and other clouds.

Small-scale density fluctuations have, in fact, been detected in the diffuse interstellar medium, and it is possible that these also arise in slow mode MHD waves, excited in this case by dissipation of Rayleigh-Taylor or other instabilities at diffuse cloud boundaries (Hartquist et al. 2003). These small-scale structures have been detected in the line of sight towards pulsars and double stars, and towards single stars where the structure has a transverse

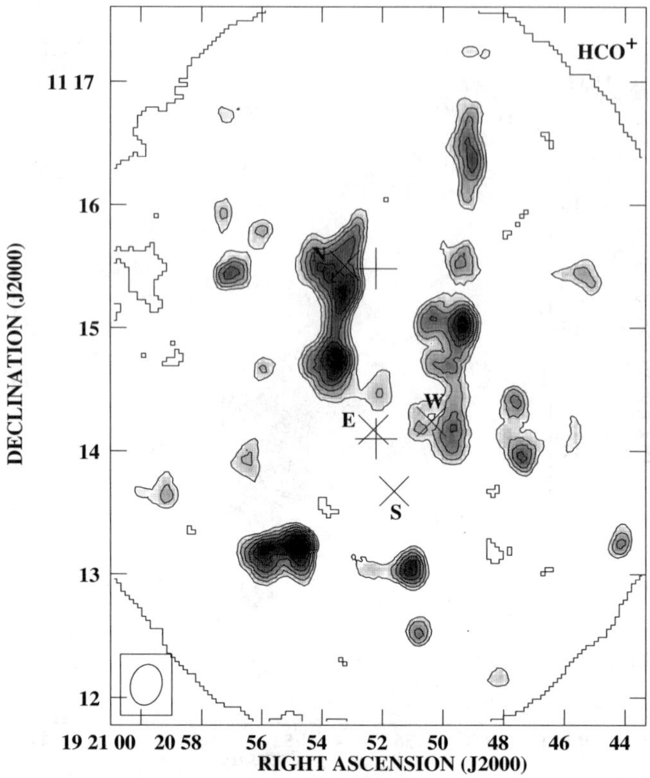

Figure 6. Same as Figure 4, but for HCO^+.

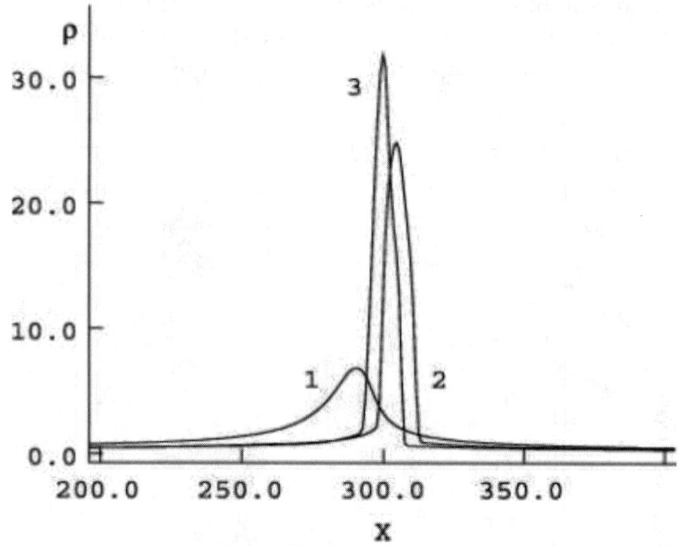

Figure 7. Time evolution (for three times) of the density enhancement created by the passage of a slow-mode MHD wave through a one-dimensional medium. The abscissa is normalized position, and the ordinate is density enhancement relative to the ambient density. The figure shows that for this case the peak density enhancement is around a factor of 30. See Falle & Hartquist (2002) for more details.

motion to the line of sight. For example, in one cloud component towards δ Orionis, the strength of the Na D1 line has been observed to vary by factors of two within a period of 6 years (Price et al. 2000, 2001). This is shown in Figure 8, where data obtained at three different times are overplotted with the model fit. The interpretation is that small-scale structure of size about 10 AU has passed through the line of sight with a transverse velocity of about 20 km s^{-1}. Such structures must have much larger densities than the mean density in the diffuse interstellar medium.

The mechanism of Falle & Hartquist is potentially of wide application. Its application to the small-scale structures arising in diffuse clouds remains to be confirmed, but it is highly plausible that these effects are playing a significant role. If so, then viable applications may be made to explain the clumps detected in the Rosette Molecular Cloud (Blitz & Stark 1986; Williams et al. 1995)—a giant molecular cloud—and also to the substructure which may be that observed by Morata et al. in L673. The chemistry that would arise in a cloud made up of an ensemble of such transient clumps is being evaluated by Garrod et al. (2003, in preparation).

The consequences of this model of molecular clouds are clear. Firstly, the chemistry will be mainly "young" (as defined by Taylor et al. 1996). Secondly, while freeze-out of molecules certainly occurs in these transient dense clumps, the freeze-out process will not continue indefinitely, and thus these regions will not become totally depleted of tracer molecules; this appears consistent with observations (Gibb & Little 1998). Finally, the low-mass star formation rate is likely determined by the mass distribution of these transient clumps. The more massive clumps are those that are associated with star formation.

3. A novel way to observe to observe clumpiness

It has been established over the last few years that a number of HH objects have associated with them distinct molecular clumps (Rudolph & Welch 1988; Dent et al. 1990; Rudolph & Welch 1992; Girart et al. 1994, 1998, 2000, 2001; Davis et al. 1990; Torrelles et al. 1992, 1994; Choi & Zhou 1997). Herbig-Haro (HH) objects are regions of hot, ionized post-shock gas where fast protostellar jets interact with a molecular cloud. An example is shown in

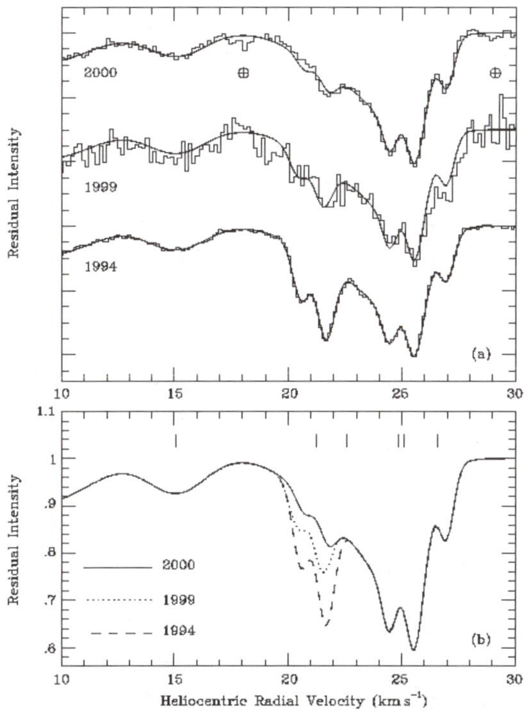

Figure 8. Top: Observations of Na D1 absorption towards Delta Ori on the indicated dates, each overplotted with the model fit. *Bottom:* Three model profiles overplotted to highlight the change that has occurred. See Price (2002) for more details.

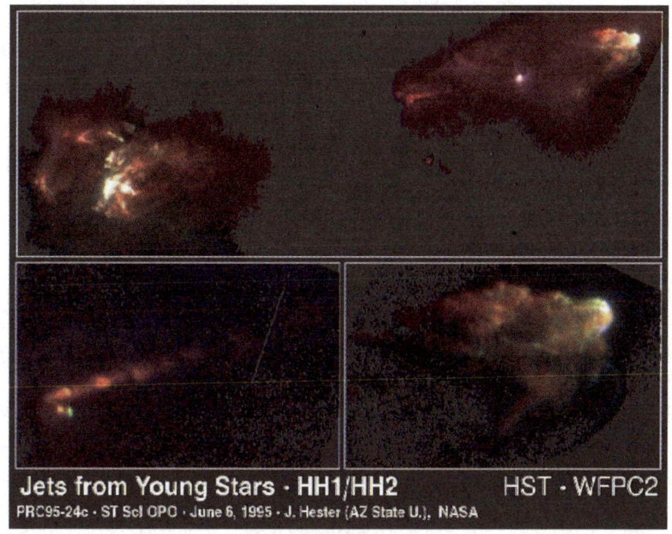

Figure 9. NASA *Hubble Space Telescope* image of the HH1/HH2 system, which lies 1500 light years away in the constellation Orion. The embedded young star lies more or less midway along the jet.

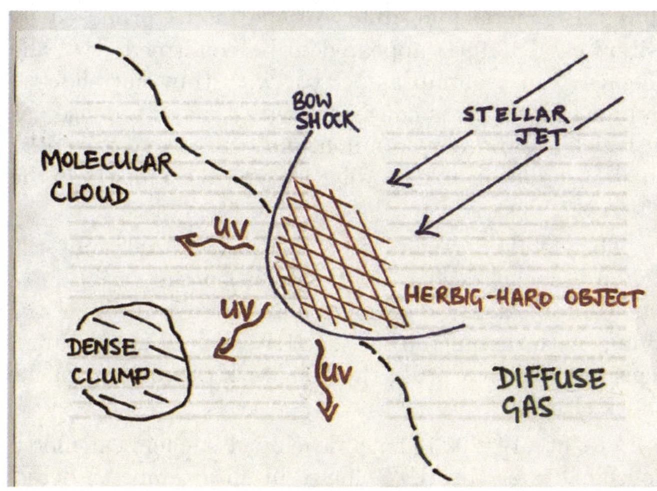

Figure 10. Simplified sketch of the proposed explanation for the enhanced molecular emission ahead of HH objects.

the *HST* images in Figure 9. These clumps are typically up to one tenth of a parsec from the HH objects, are dense and cool and show narrow linewidths. Evidently, they have not been shocked. They were first identified in lines of HCO^+ and NH_3; these molecules show line intensities anomalously strong compared to quiescent dark clouds. The clumps are found generally somewhat downstream from the HH jet, and are interpreted to be within the molecular cloud that the HH jet is penetrating.

To account for the anomalous chemistry, Girart et al. (1994) proposed that radiation generated in the HH shock triggers a photochemistry which helps to release the icy mantles on dust in the clump, and causes photoionization that generates an active chemistry (see Fig. 10). This is summarized in the following reactions:

$$(H_2O,\ CO,\ NH_3, \ldots)_{\rm ice} \rightarrow (H_2O,\ CO,\ NH_3, \ldots)_{\rm gas},$$
$$CO \rightarrow C \rightarrow C^+,$$
$$C^+ + H_2O \rightarrow HCO^+.$$

Taylor & Williams (1996) investigated numerically a simple model of such a process and confirmed that appreciable enhancements of HCO^+ and NH_3 should arise in the irradiated clump. Viti & Williams (1999) considered a more detailed model with a much more extended chemistry and predicted that, besides HCO^+ and NH_3, a wide range of molecular species would be expected to have enhanced abundances compared to those normally found in molecular clouds. These included CH_3OH, H_2CO, SO,

Table 1. HH2 clump column densities

Molecule	Observed (cm^{-2})	Predicted (cm^{-2})
CO	3.8 (17)	3.1 (17)
HCO^+	1.0 (13)–1.0 (14)	3.0 (13)
CH_3OH	6.4 (13)	7.8 (13)
H_2CO	5.7 (13)	3.3 (13)
SO	3.2 (13)	3.0 (13)
SO_2	1.7 (13)	1.8 (13)
HCN	4.1 (12)	4.0 (12)
CN	3.4 (12)	4.0 (12)
CS	2.8 (12)	8.0 (12)

NOTE: $a(b) \equiv a \times 10^b$.

SO_2, and CN. A very detailed observational study, using BIMA and CSO, of the clump associated with HH2 (Girart et al. 2002) measured the abundances of many molecular species in that source (see Fig. 12). In qualitative terms, the predictions of Viti & Williams (1999) agreed well with these measurements (see Table 1), and thus the general plausibility of the model proposed by Girart et al. (1994) appeared to be confirmed: viz., the chemistry in a clump associated with (but not shocked by) an HH object is consistent with that of a dense gas in which photodesorption and photochemistry driven by the HH radiation field modify the typical chemistry of the quiescent, dark gas.

The detailed results of Girart et al. (2002) for the clump associated with HH2 established the variation of the column density of molecular species through the clump with distance from the HH shock. These column densities rise to a peak, then fall with distance from the shock (Fig. 11).

Viti et al. (2003) have developed a chemical model specifically for the HH2 clump, in an attempt to determine the physical parameters of this region. This model incorporates as free parameters the clump density, the HH UV radiation field, the HH X-ray field, and the degree of freeze-out within the clump before the incidence of the HH radiation. Although this model is spatially simplistic (cf. Hester et al. 1998), it gives reasonable agreement with the observational results of Girart et al. (2002). They deduced that the X-rays did not significantly modify the chemistry; rather, the initial level of freeze-out onto dust and the UV radiation intensity were significant in determining the chemistry. Figure 13 shows the column density of selected species for the best-matching model taken from Viti et al. (2003). It is clear from this

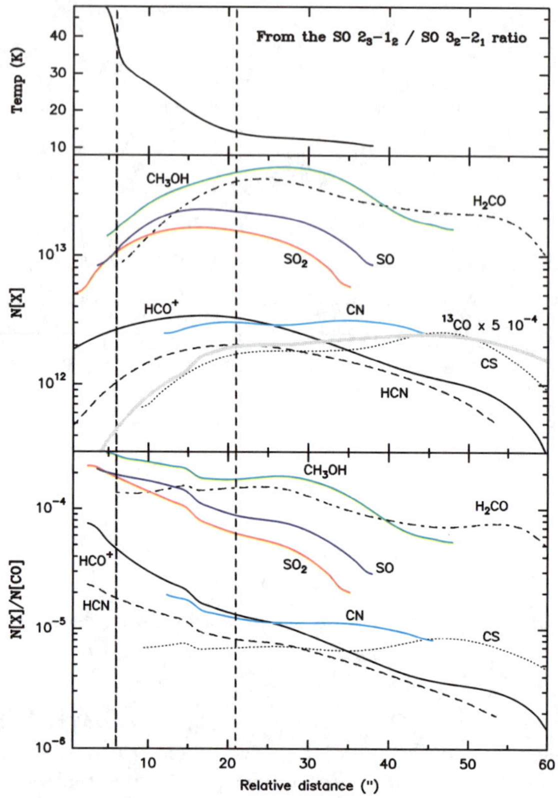

Figure 11. Slices of the rotational temperature (*top panel*), column density (*middle panel*) and relative abundance with respect to CO (*bottom panel*) as a function of distance to HH2. See Viti et al. (2003) for more details.

picture that the observed chemistry must be short-lived. In fact, the initial level of freeze-out is required to be an intermediate value, corresponding to there being comparable amounts of heavy molecules being in the gas and in the ices. The implication is that the clump must be transient; otherwise, at the inferred high clump density the freeze-out would be nearly total if the clump had a prolonged existence, i.e., longer than a few million years. Thus, the clump survival time must be on the order of one million years. Similarly, the radiation field must be large enough to promote a prompt photochemistry, but not so large as to destroy the products of that chemistry too rapidly.

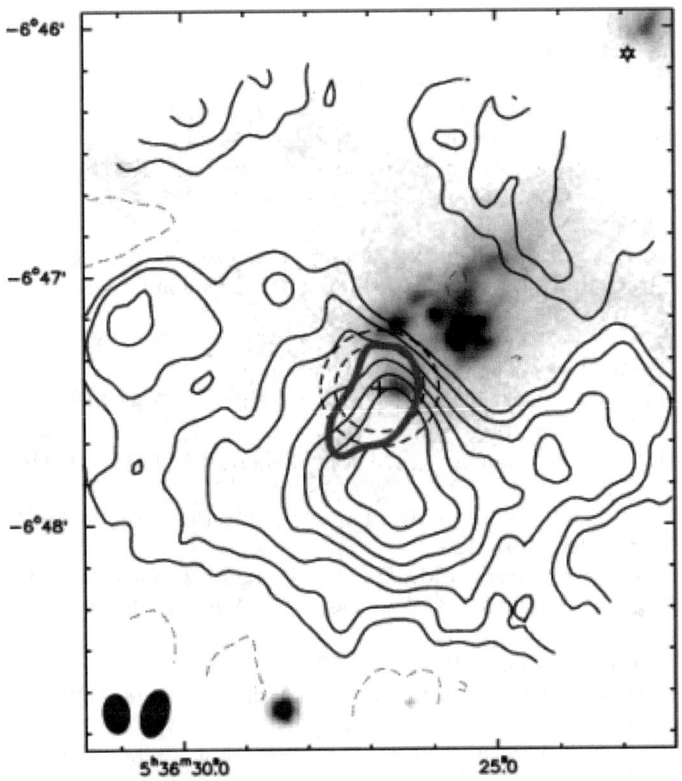

Figure 12. Figure 2 from Girart et al. (2002). The thick contour is the integrated emission of SO (3−2). See Girart et al. (2002) for more details.

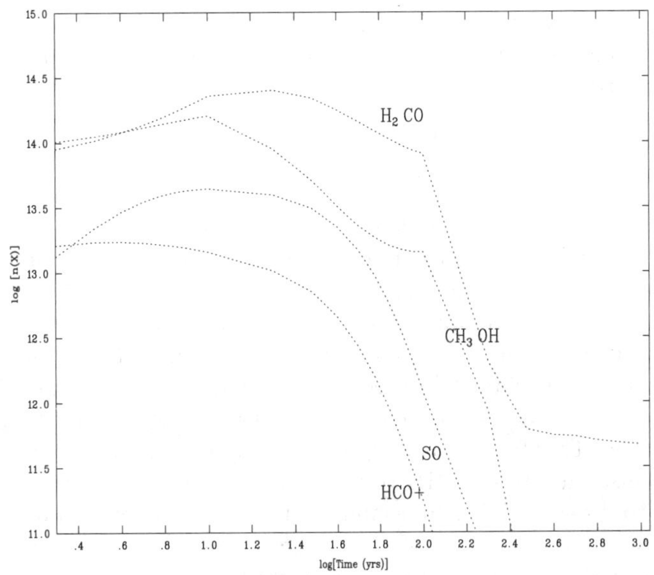

Figure 13. Fractional abundances (relative to hydrogen) of selected species as a function of log time for the best-matching model of the clump ahead of HH2. See Viti et al. (2003) for more details.

It was evident that although the Viti et al. model was too simple to represent the complexities of the spatial situation as indicated by Hester et al. (1998), some general conclusions could be drawn from this study. These are:

1. The clump has a density structure, with a peak number density of H nuclei of around 3×10^5 cm^{-3} and an average number density one order of magnitude smaller.

2. The lifetime of the clump after irradiation begins is on the order of 100 years; after that, molecular abundances in the clump decline.

3. The clump is probably overtaken by the HH shock front on a similar timescale.

4. The radiation field intensity incident on the clump is in the range of 100–1000 times the mean interstellar intensity;

5. Comparable amounts of 'heavy' molecules exist in the gas and in ices on the dust, before the clump is irradiated.

6. X-ray emission from HH2 does not appear to affect the chemistry in the clump significantly.

The question now arises: what is the relation between these clumps that have been illuminated by the HH radiation and the general clumpiness that may be present in molecular clouds? The clumps illuminated by HH objects appear, by density, size, and temperature considerations, to belong to the class of structures predicted by Taylor et al. and observed by Morata et al. Assuming that the conclusions of Viti et al. (2003) concerning the clump associated with HH2 are generally valid, then such clumps are necessarily transient on a timescale of about 1 My, their number density is about 10^4 H nuclei cm^{-3}, and they are cool and quiescent. Their chemistry is anomalous when compared to cool clouds simply because of the effects of the HH radiation in driving a photochemistry that enhances certain species, notably HCO^+. When a clump that is normally in a near-dark environment is subjected to the UV radiation from an HH shock, it responds in a characteristic way.

We conclude that clumps illuminated by HH objects are simply members of the same class of density fluctuations distributed throughout molecular clouds. If so, then the HH objects provide a means of studying individual clumps that happen to be close to the HH shock. Localized and anomalously large HCO^+ abundances are a clear signature of this physics. A search for density enhancements by this method is under way (Viti et al. 2003, in preparation).

4. Summary

There is strong evidence—both observational and theoretical—that molecular clouds are clumpy on a scale that is normally unresolved, and that most of these clumps simply dissipate into an interclump medium on a timescale on the order of 1 Myr. Those clumps that happen to be near an HH object are affected by its radiation field, and generate a characteristic variety of molecular species. This variety is a signature of photochemistry on a dense clump, and HH objects therefore provide a useful tool for studying molecular cloud density fluctuations through the response of these fluctuations to radiation. Although these clumps are generally too small to detect in single-dish studies, they may be readily detected in array telescope observations. The origin of these clumps is probably in slow-mode MHD waves; these waves may have other interstellar consequences, including the small-scale structure recently detected in the diffuse interstellar medium.

Clumpiness in molecular clouds is clearly of relevance to the initiation of low-mass star formation. In the picture presented here, most density fluctuations simply dissipate into the background gas, and the chemistry that they show is that of "early-time" species (Taylor et al. 1996). However, denser or more massive fluctuations may develop sufficient self-gravity to proceed to star formation. Regions associated with star formation should develop "late-time" species (Taylor et al. 1996). Thus the clump mass spectrum, a quantity that can be determined by the means discussed above, may be related to the stellar initial mass function.

Acknowledgements

The ideas presented here draw together contributions from many collaborators. We would particularly like to draw attention to the work of colleagues: at UCL, R. Garrod; at Barcelona, R. Estalella, J.M. Girart, and O. Morata; at Leeds, S.A.E.G. Falle and T.W. Hartquist; and at CfA, P.T.P. Ho. We are grateful to the Particle Physics and Astronomy Research Council for their support of our research.

References

Anglada, G., et al., 1989, ApJ, 341, 208
Blitz, L., & Stark, A.A. 1986, ApJ, 300, L89
Choi, M., & Zhou, S. 1997, ApJ, 477, 754
Davis, C.J., Dent, W.R.F., & Burnell, S.J.B 1990, MNRAS 244, 173
Dent, W.R.F., et al. 1993, MNRAS 262, L13
Falle, S.A.E.G., & Hartquist, T.W. 2002, MNRAS 329, 195
Franco, J., & Carraminana, A. 1999, eds. *Interstellar Turbulence* (Cambridge: Cambridge University Press)
Gibb, A. G., & Little, L.T. 1998, MNRAS, 295, 299
Girart, J.M., et al. 1994, ApJ, 435, L145
Girart, J.M., Estalella, R., & Ho, P.T.P. 1998, ApJ, 495, L59
Girart, J.M., Estalella, R., Ho, P.T.P., & Rudolph, A.L. 2000, ApJ, 539, 763
Girart, J.M., et al. 2001, ApJ, 562, L91
Girart, J.M., et al. 2002, A&A, 388, 1004
Hartquist, T.W., Falle, S.A.E.G., & Williams, D.A. 2003, Ap. Sp. Sci., in press
Hester, J.J., Stapelfeldt, K.R., & Scowen, P.A. 1998, AJ, 116, 372
Morata, O., Estalella, R., Lopez, R., & Planesas, P. 1997, MNRAS, 292, 120
Morata, O., Girart, J.M., & Estalella, R. 2003, A&A, 397, 181
Myers, P.C., Fuller, G.A., Goodman, A.A., & Benson, P.J. 1991, ApJ, 376, 561
Pastor, J., et al. 1991, A&A, 252, 320
Price, R.J., Crawford, I. A., & Howarth, I.D., 2001, MNRAS, 328, 555
Price, R.J., Crawford, I. A., & Barlow, M. J., 2000, MNRAS, 312, 43
Price, R.J., 2002, PhD thesis, University of London
Rudolph, A., & Welch, W.J. 1988, ApJ, 326, L31
Rudolph, A., & Welch, W.J. 1992, ApJ, 395, 488
Taylor, S.D. Morata, O., & Williams, D.A. 1996, A&A, 313, 269
Taylor, S.D., & Williams, D.A. 1996, MNRAS, 282, 1343
Torrelles, J.M., et al. 1992, ApJ, 396, L95
Torrelles, J.M., et al. 1994, ApJ, 435, 290
Viti, S., & Williams, D.A. 1999, MNRAS, 310, 517
Viti, S., et al. 2003, A&A, in press
Williams, J.P., Blitz, L., & Stark, A.A. 1995, ApJ, 451, 252
Zhou, S., et al. 1989, ApJ, 346, 168

OBSERVATIONAL APPROACH TO MOLECULAR CLOUD EVOLUTION WITH THE SUBMILLIMETER C I LINES

Tomoharu Oka, Satoshi Yamamoto, and the Mt. Fuji Submillimeter-wave Telescope Group

Research Center for the Early Universe and Department of Physics, The University of Tokyo, 7-3-1 Hongo, Bunkyo-ku, Tokyo 113-0033, Japan

Abstract

We have constructed a 1.2-m submillimeter-wave telescope at the summit of Mount Fuji for the exclusive purpose of surveying molecular clouds in two submillimeter C I lines, $^3P_1$–$^3P_0$ (492 GHz) and $^3P_2$–$^3P_1$ (809 GHz). The telescope has been operated successfully for four observing seasons since installation in 1998. We have revealed large-scale C I 492 GHz distributions of many molecular clouds, including Orion MC, Taurus MC, DR 15, ρ Oph, DR 21, NGC 2264, M17, W3, W44, W51, and the Rosette MC, covering more than 40 square degrees of the sky. The distribution of C I 492 GHz emission is found to be different from that of the ^{13}CO or C^{18}O emission in some clouds. We found C^0-rich areas (C/CO $\sim$ 1) in several dark clouds without strong UV sources. We also found the spatial ordering C$^+$/CO/C^0 with distance from FUV sources. This is a general property of clouds illuminated by intense FUV radiation, and it is apparently inconsistent with the standard photodissociation region picture. These results are discussed in relation to formation processes of molecular clouds and dense cloud cores.

Key words: ISM: clouds – ISM: atoms – ISM: molecules

1. Introduction

Diffuse pressure-bounded clouds in the Galactic disk gradually increase their density with decreasing temperature, and then evolve into self-gravitating molecular clouds. In the course of dynamical evolution, interstellar matter changes its form from ionized, to atomic, to molecular phase. Carbon, the fourth most abundant element in the Universe, also changes its major form from ionized (C$^+$) to atomic (C^0), to carbon monoxide (CO). Since the conversion timescale for C^0 into CO is comparable to the dynamical timescale of molecular clouds ($\sim$10^6 years), we might expect a C^0-rich era in the early stages of molecular cloud formation.

The electronic ground state of carbon is ^3P. Two magnetic dipole transitions, $^3P_1$–$^3P_0$ (hereafter C I 1–0) at 492 GHz and $^3P_2$–$^3P_1$ (hereafter C I 2–1) at 809 GHz, are allowed between the fine structure levels. Since the $^3P_1$ level is 23.6 K above ground, and the critical density is as low as 10^3 cm^{-3}, the 1–0 line is expected to be a good tracer of atomic carbon in molecular clouds or cloud-forming regions. The 2–1 line is rather difficult to detect because it requires high temperatures to excite the $^3P_2$ level, and because receiver performance drops steeply toward higher frequencies. If both C I lines are detected, we can estimate excitation temperature, line optical depth, and column density under the LTE assumption.

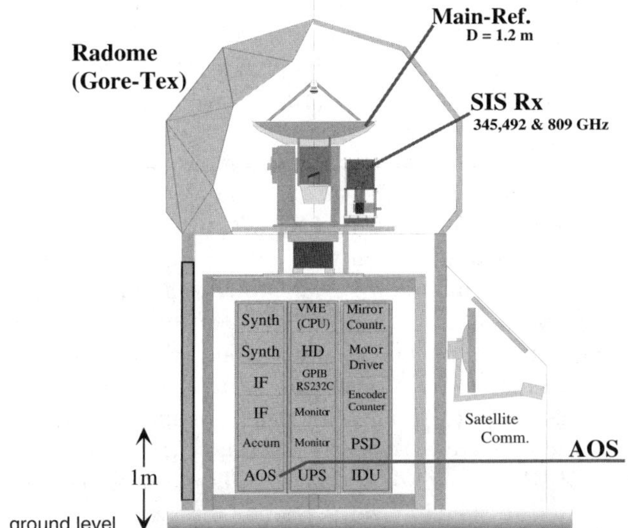

Figure 1. Schematic illustration of the Mount Fuji submillimeter-wave telescope.

Chemistry as a Diagnostic of Star Formation (2003). *Edited by Charles L. Curry and Michel Fich.* Published in 2003 by NRC Research Press, Ottawa ON, K1A 0R6, Canada. ISBN 0-660-19089-3.

2. The Mount Fuji Submillimeter-wave Telescope

With the exclusive intent of making C I survey observations, a submillimeter-wave telescope was constructed at the summit of Mount Fuji (Fig. 1; Sekimoto et al. 2000). Mt. Fuji is a dormant volcano with a 600 m diameter crater and is the highest mountain in Japan. In winter, we have temperatures as low as −20°C and a high percentage of fine days. The 220 GHz opacity at the summit is lower than 0.06 for 45 percent of the time from November to March; thus, the summit in winter is a site suitable for submillimeter-wave observations. The telescope was installed in August 1998, on Nishi-yasugawara (alt. 3725 m), about 200 m north of the highest peak, Kengamine (alt. 3776 m). The telescope was transported to the site by helicopter. The electric power is supplied from the weather station of the Meteorological Agency at Kengamine.

The main reflector of the telescope has a 1.2 m diameter and a 0.48 m focal length. The surface accuracy is better than 10 μm in rms. The half-power beamwidth of the telescope at 492 GHz is 2.2′. We developed a triple-band receiver, in which SIS mixers for 492, 809, and 345 GHz bands are installed. Nb-based, parallel-connected twin junction (PCTJ)-type SIS mixers, fabricated at Nobeyama Radio Observatory, are employed for all bands. All mixers are operated in double sideband mode. The receiver temperature is 120 K (DSB) at 492 GHz and 580 K (DSB) at 809 GHz (Noguchi et al. 2001). We use an acousto-optical radio spectrometer as a receiver backend. The total bandwidth of the spectrometer is 700 MHz, and the frequency resolution is 1.6 MHz.

In summer, the telescope site is accessible by a bulldozer which can transport a maximum load of 1 ton. It takes about 3 hours to the summit from the bulldozer base at Tarobo (alt. 1400 m). In winter, however, a bulldozer cannot reach the summit because of frozen snow. It is about 9 hours' walk to the summit (depending on the weather and physical conditions) with a professional mountain guide, and is sometimes highly dangerous. Thus, remote operation is indispensable to perform this project. All telescope instruments are controlled by a MicroSPARC-II-based CPU board with a VME bus via GP-IB and serial interfaces. The control system is connected to the Nobeyama Radio Observatory via commercial satellite communication system at a speed of 64 kbps. The telescope is operated remotely from the Hongo campus of the University of Tokyo.

Since the installation, we have successfully operated the telescope for 4 observing seasons. We have already surveyed more than 40 square degrees of the sky with the C I 1–0 line. We have also mapped a few representative sources with the C I 2–1 line. The excellent observing conditions at the summit of Mt. Fuji in winter made it possible to perform an efficient survey in the submillimeter C I lines.

3. C^0 in Dark Clouds

Dark clouds, which are free of intense far-ultraviolet (FUV) radiation, provide ideal sites for tracing chemical evolution of molecular clouds. A problem is, however, that C I emission from dark clouds is generally weak, about 2 K in antenna temperature. Thus we have taken only a small sample of C I distributions in dark clouds. To date, we have obtained C I distributions in Heiles' Cloud 2, Lynds 1495, and the Lynds 134 complex.

3.1. Heiles' Cloud 2

Heiles' Cloud 2 (HCL2) is a dark cloud in the Taurus molecular cloud complex that is well-studied because of its proximity to the Sun ($D = 140$ pc). Maezawa et al. (1999) observed the entire region of HCL2 in the C I 1–0 line. Figure 2 shows a velocity-integrated map of C I 1–0 emission in the HCL2 region, with the $C^{18}O$ integrated intensity map superimposed.

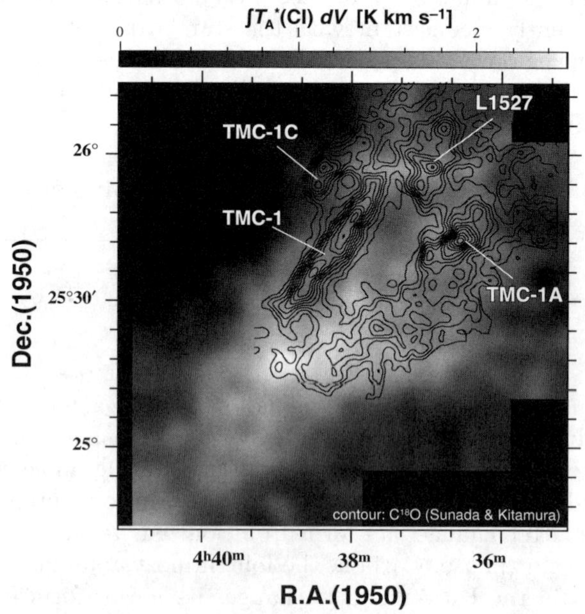

Figure 2. Maps of C I velocity-integrated intensity (*gray scale*) and $C^{18}O$ integrated intensity (*contours*) in HCL2.

The C I intensity is generally weak in the dense core region in the north, and is brightest in the south end of HCL2, where $C^{18}O$ emission is fairly weak. Toward the C I emission peak, the C^0/CO column density ratio is evaluated to be greater than 0.8. This ratio is comparable to those found in diffuse clouds (>1), although the visual extinction toward the C^0-rich area is higher than 6 magnitudes. The C^0-rich area lies between the dense cores and a diffuse cloud which extends to the southeast. The *IRAS* 60 μm/100 μm color map shows a lack of enhancement in warm dust fraction toward the C^0-rich area. These results indicate an origin of atomic carbon unrelated to photodissociation processes. It is most likely that the cloud is in an early stage of chemical evolution. If this is the case, the dense core formation is taking place from northwest to southeast in this region. This direction is consistent with the result inferred for the TMC-1 ridge from the molecular abundance gradient (Hirahara et al. 1992).

3.2. Lynds 1495

Lynds 1495 is a dark cloud in the Taurus molecular cloud complex, associated with a chain of small clouds (B211, B213, and B216). We observed the L1495 cloud in the C I 1–0 line. C I emission extends over the cloud, exhibiting a distribution similar to that of ^{13}CO emission. The C^0/CO column density ratio is $\sim$ 0.2 in the eastern periphery, and increasing to $\sim$ 0.5 on the western side of the L1495 core. Toward the L1495 core, where the visual extinction exceeds 10 mag, the ratio reaches $\sim$ 0.3. A diffuse cloud is associated with a western extension of the L1495 cloud, a situation which resembles HCL2. The spatial trend in C^0/CO column density ratio suggests that dense core formation is taking place from east to west in this region.

3.3. Lynds 134 Complex

Lynds 134 dark cloud complex consists of three dark clouds (L134, L169, and L183) with similar mass. This complex is spatially isolated in high Galactic latitude ($b \sim 35°$), and exhibits no signs of star formation. Ito (2003, in preparation) have delineated the C I distribution in the entire region of the L134 cloud complex. Two clouds in the complex (L169 and L183) show C I distributions similar to those of ^{13}CO emission, while L134 shows a slight difference. A relatively C^0-rich area ($C^0/CO \sim$ 0.2) was found at the southeast edge of the L134 cloud. This indicates that L134 differs chemically from the other two, possibly being in an earlier stage of chemical evolution.

4. C^0 in Giant Molecular Clouds

Giant molecular clouds (GMCs) are the formation sites of massive stars, which are sources of intense FUV photons. In the vicinity of intense FUV sources, C^0 is thought to be formed by photodissociation of CO in the cloud surface exposed to FUV radiation. Thus C I distributions in GMCs have been discussed in terms of the photodissociation region (PDR) models (Hollenbach & Tielens 1999, and references therein). When we think of a plane-parallel cloud illuminated by FUV radiation from one side, the major form of gas-phase carbon changes spatially from C^+ to C^0 then to CO with depth from the surface. According to this picture, atomic carbon is confined in a thin layer near the surface, and the spatial order of $C^+/C^0/CO$ is expected. Nevertheless, C I observations to date have not detected such a thin C^0 layer, nor have they revealed the predicted phase configuration.

4.1. Large-scale C^0 Distribution

We have obtained C I distributions in many GMCs including Orion A and B, DR 15, DR 21, NGC 2264, M17, W3, W44, W51, and the Rosette MC. Ikeda et al. (2002) observed the entire region of the Orion A and B GMCs in the C I 1–0 line. Orion A is the nearest GMC (D=450 pc) associated with a very active star-forming region (the Orion Nebula), and has been subject to extensive studies at various wavelengths. Figure 3 shows the C I velocity-integrated intensity maps of the Orion A and B molecular clouds. C I emission extends over the Orion molecular clouds irrespective of FUV field strength. The large-scale C I distributions in the Orion clouds are nearly identical to those of ^{13}CO obtained by Bally et al. (1987). C^0/CO column density ratios concentrate on a narrow range from 0.1 to 0.3. The similarity in C I /^{13}CO distributions is a general property of GMCs. These results suggest that C^0 in GMCs is not confined to a thin layer near the surface, possibly even coexisting with CO.

4.2. C^0 in the vicinity of FUV Sources

4.2.1. OMC-1 Region

In order to delineate distributions of C^0 column density and excitation temperature in a cloud adjacent to intense FUV sources, Kuboi et al. (2003, in preparation) mapped the OMC-1 region in the Orion A molecular cloud with

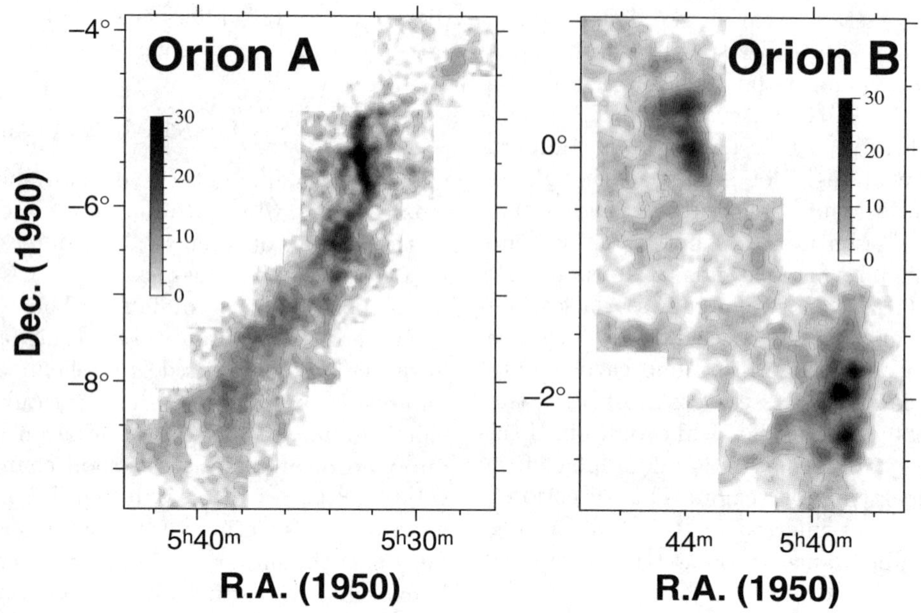

Figure 3. Maps of C I integrated intensity in the Orion A and B molecular clouds.

the C I 2–1 line. The distribution of the 2–1 emission is concentrated in the vicinity of the OB stars (θ^1 Ori and θ^2 Ori), while the 1–0 emission extends over the molecular cloud ridge running from north to south. From the two C I lines, Kuboi et al. determined the column density and excitation temperature of C^0 in the OMC-1 region under the LTE assumption. The column density distribution was found to be very similar to that of the C I 1–0 emission. Excitation temperature, on the other hand, is enhanced in the small region surrounding the OB stars, possibly because of the heating by the intense FUV field.

Figure 4 shows a comparison between C^0 and CO column density distributions in the OMC-1 region. C^+ 158 μm (C II) emission is known to be distributed in a small area around the OB stars. CO column density has a peak in the Orion KL region. C^0 column density, however, has peaks behind the CO column density peak with respect to the OB stars. Thus, the phase configuration is $C^+/CO/C^0$ from the OB stars. This is inconsistent with the standard PDR picture, which predicts the spatial order $C^+/C^0/CO$.

The phase configuration $C^+/CO/C^0$ is generally observed in various regions associated with OB stars, such as DR 15 (Oka et al. 2001a), ρ Oph (Kamegai et al. 2003, in preparation), M17, and NGC 1333 (Oka et al. 2003, in preparation). Thus, this configuration does not originate from a special situation of each source, but is rather a general property of clouds irradiated by intense FUV radiation.

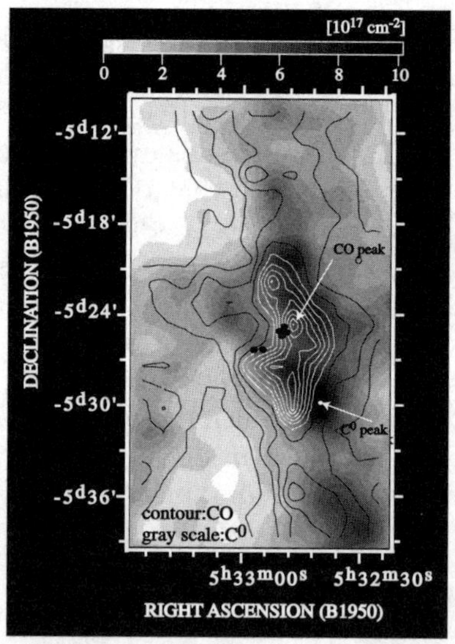

Figure 4. Distributions of C^0 (*gray scale*) and CO (*contours*) column density in OMC-1.

4.2.2. ρ Ophiuchus Molecular Cloud

The ρ Oph molecular cloud is a huge complex of dark clouds extending over more than 35 square degrees. Lynds

1688, in particular, is known to be an active star-forming region close to the Sun ($D = 160$ pc). The L1688 cloud is associated with a B2V star (HD 147889), and hence is a typical PDR. Kamegai et al. (2003, in preparation) conducted the C I 1–0 observations toward the entire region of L1688 (Fig. 5). C I emission extends over the L1688 cloud, with a peak at the ρ Oph A position where ^{13}CO emission has also a peak. Another C I peak is found in the southeast of ρ Oph A, where no ^{13}CO counterpart is seen. In Figure 5, the C I distribution is compared with the distributions of C II (Yui et al. 1993) and $C^{18}O$ emission (Umemoto et al. 2003, in preparation). The C II distribution, which traces an ionized layer of the PDR, is almost centered at HD 147889. Dense cores traced by $C^{18}O$ emission are located between the C II and C I distributions. Hence, we see the phase configuration $C^+/CO/C^0$ from the FUV source. The scale of the C^0/CO displacement is of the same order as that found in the OMC-1 region.

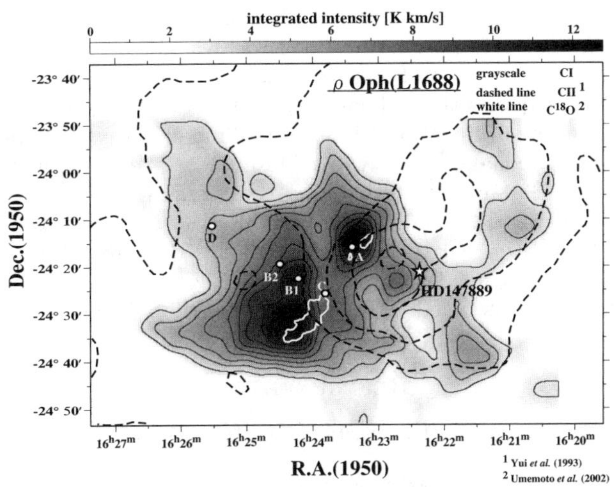

Figure 5. Distributions of C II (*dotted contours*), C I (*gray scale*) and $C^{18}O$ (*white contours*) emission in L1688.

5. Origin of the $C^+/CO/C^0$ Configuration

NGC 1333, a reflection nebula in the Perseus cloud complex, provides another example of the $C^+/CO/C^0$ configuration (Oka et al. 2003, in prep.). Why are C^0 and CO spatially inverted? To address this problem, we constructed a simple 3-dimensional, time-dependent PDR model for the NGC 1333 cloud (Fig. 6). After starting chemical reactions, C^+ inside the cloud rapidly recombined into C^0 within a thousand years, and then C^0 was slowly converted into CO. These reactions proceed more quickly in high-density cores than in low-density envelopes, thereby leaving a C^0 peak behind the CO cores at an age of 2×10^6 years. The C^0 peak behind the CO cores disappears after 10^7 years, to reach chemical equilibrium. Thus, the observed spatial inversion of C^0 and CO can be understood in terms of chemical evolution.

Some clouds with C^0-rich area or C^0/CO spatial inversions are associated with expanding H I shells. For example, both the ρ Oph cloud and the L134 complex are associated with an expanding H I shell with an age of $\sim 1.5 \times 10^6$ years, formed by Sco OB2 (de Geus 1992). The Perseus cloud complex, which includes the NGC 1333 cloud, is also associated with an expanding H I shell with a radius of ~ 25 pc and an age of 10^6 years (Sancisi et al. 1974). This age is consistent with that suggested by our model calculations. The association of the H I shell invokes a scenario of rapid molecular cloud formation by expanding shells. An expanding shell, which might be formed by a supernova explosion, sweeps up low density gas in diffuse clouds, where the major forms of carbon are C^+ and C^0. In the swept-up gas, molecular clouds and dense cores are formed. The dense cores evolve dynamically with increasing gravitational instability, to form massive stars in the densest core. The newly formed massive stars dissociate and ionize ambient molecular material to expose neighboring dense cores. Behind these exposed cores, there remains large amount of C^0 which has not been converted into CO: thus, the $C^+/CO/C^0$ configuration is created.

6. Correlation with Dynamical Evolution

Support for the chemical evolution scenario comes from a correlation with dynamical evolution. Using the C I 1–0 data sets, Ikeda et al. 2002 identified 38 cores in Orion A, and 42 cores in the Orion B molecular clouds. They derived the virial theorem mass ($M_{\rm VT}$) and LTE mass ($M_{\rm LTE}$), as well as C^0 and CO column densities for each core. Figure 7 is plots $M_{\rm VT}/M_{\rm LTE}$ versus the C^0/CO column density ratio. The ratios are well correlated. Since $M_{\rm VT}/M_{\rm LTE}$ is a measure of gravitational stability (e.g., Oka et al. 2001b), this relationship indicates that the C^0/CO ratio decreases with increasing gravitational instability. Cloud cores gradually increase their density by dissipating internal turbulent motion, then collapse to form stars. Therefore, this result demonstrates that the C^0/CO ratio could be a good tracer of molecular cloud evolution.

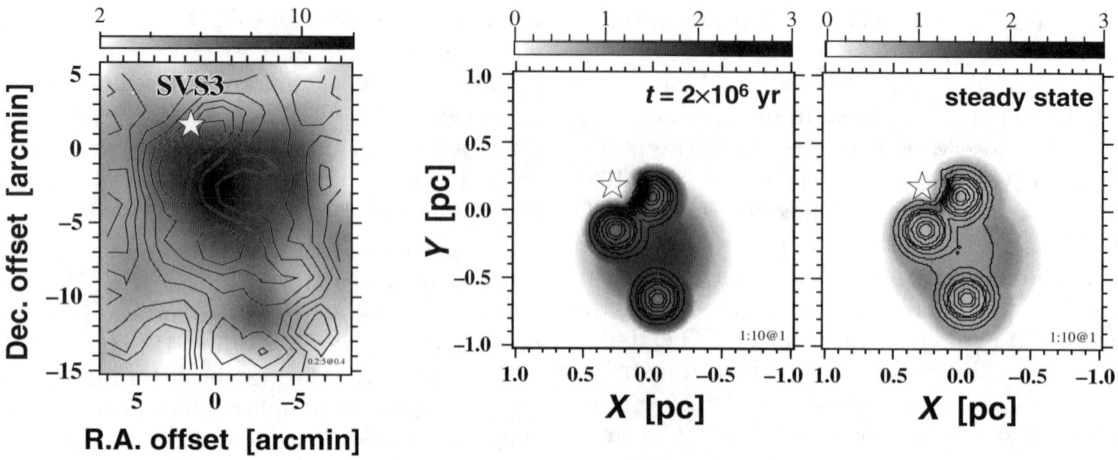

Figure 6. Left: Maps of C I (*gray scale*) and $C^{18}O$ (*contours*) integrated intensities toward NGC 1333. *Center and right*: Results of model calculations at $t = 2 \times 10^6$ yr and for a steady state. C I (*gray scale*) and CO (*contours*) column densities are shown in units of 10^{18} cm^{-2}.

7. Summary

We have performed large-scale C I surveys of nearby molecular clouds with the Mt. Fuji submillimeter-wave telescope. In these observations, we have discovered molecular clouds in the early stages of chemical evolution. The C^+/ CO/ C^0 configuration found in the vicinity of FUV sources can be understood in terms of a chemical evolution scenario. Virial analyses of cloud cores in Orion have demonstrated that the C^0/CO ratio could be a good tracer of molecular cloud evolution.

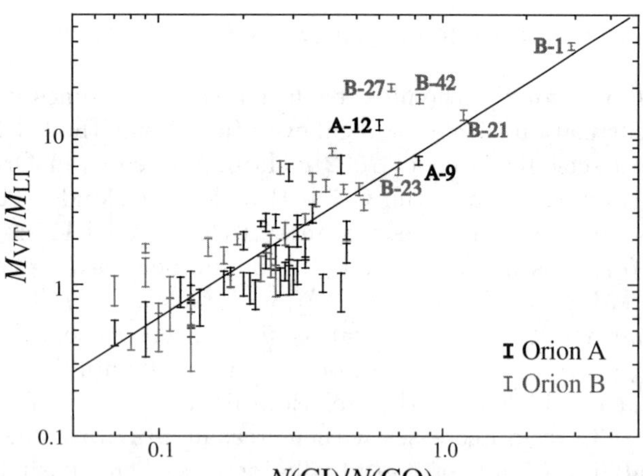

Figure 7. Mass ratio M_{VT}/M_{LTE} versus the C^0/CO column density ratio for cloud cores in the Orion A and B molecular clouds.

Acknowledgements

The authors are grateful to Professor Katsuhiko Sato and Professor Kazuo Makishima for their continuous encouragement and support throughout this work. This paper is based on the results of the Mount Fuji submillimeter-wave telescope project in collaboration with Masafumi Ikeda, Hiroyuki Maezawa, Tetsuya Ito, Kazuhisa Kamegai, Takeshi Sakai, Nobuyuki Kuboi, Kunihiko Tanaka, Ken Shimbo, Masaaki Hayashida, Ken'ichi Tatematsu, Yutaro Sekimoto, Takashi Noguchi, Keisuke Miyazawa, Hiroyuki Ozeki Junji Inatani, Masatoshi Ohishi, Hedeo Fujiwara, and Shuji Saito.

References

Bally, J., Langer, W. D., Stark, A. A., & Wilson, R. W. 1987, ApJ, 312, L45

de Geus, E. J. 1992, A&A, 262, 258

Hirahara, Y., et al. 1992, ApJ, 394, 539

Hollenbach, D. J., & Tielens, A. G. G. M. 1999, Rev. Mod. Phys., 71, 173

Ikeda, M., et al. 1999, ApJ, 527, L59

Ikeda, M., Oka, T., Tatematsu, K., Sekimoto, Y., & Yamamoto, S. 2002, ApJS, 139, 467

Maezawa, H., et al. 1999, ApJ, 524, L129

Noguchi, T., Maezawa, H., Shi, S. -C., & Yamamoto, S. 2001, IEEE Trans. Appl. Supercond., 71, 167

Oka, T., et al. 2001a, ApJ, 558, 176

Oka, T., Hasegawa, T., Sato, F., Tauboi, M., Miyazaki, A., & Sugimoto, M. 2001b, ApJ, 562, 348

Sancisi, R., Goss, W. M., Anderson, C., Johansson, B., & Winnberg, A. 1974, A&A, 35, 445

Sekimoto, Y., et al. 2000, Rev. Sci. Instr., 71, 2895

Yui, Y. Y., et al. 1993, ApJ, 419, L37

CHEMICAL PROBES OF THE EVOLUTIONARY STATUS OF YOUNG, LOW-MASS PROTOSTARS

Gary A. Fuller[1], Jane V. Buckle[1,2], Edwin F. Ladd[3], Helen Roberts[1,4], Tom J. Millar[1], and Jennifer Hatchell[5]

[1]Physics Dept., UMIST, Manchester, UK
[2]Joint Astronomy Centre, Hilo, Hawaii
[3]Physics Dept., Bucknell University, Lewisburg, PA, USA
[4]Dept. of Physics, Ohio State University, Columbus, OH, USA
[5]Max-Planck Institut für Radioastronomie, Bonn, Germany

Abstract

The results of a number of studies of Class 0 and I young stellar objects in a range of molecular tracers will be discussed. These surveys demonstrate how the chemistry of the circumstellar material around these sources can be used to study their evolution. Results from observations of rare isotopes of CO and methanol demonstrate that much of the molecular emission observed towards these objects is from material already being affected by the embedded sources, probably through the action of the outflows from the sources. Observations of sulphur- and deuterium-containing species will be presented which show how these species can provide potentially important clocks for the star formation process. For both sulphur and deuterium, key species are released into the gas phase from grain mantles; but whereas the sulphur species can provide an estimate of the time since the mantles have been released, the deuterium species provide an estimate of the conditions in the gas when the grain mantles were formed.

Key words: stars: formation – ISM: clouds, molecules

Chemistry as a Diagnostic of Star Formation (2003). *Edited by Charles L. Curry and Michel Fich.* Published in 2003 by NRC Research Press, Ottawa ON, K1A 0R6, Canada. ISBN 0-660-19089-3.

PART IV:
THE ROLE OF DUST CONTINUUM OBSERVATIONS

SUBMILLIMETER STUDIES OF PRESTELLAR CORES AND PROTOSTARS: PROBING THE INITIAL CONDITIONS FOR PROTOSTELLAR COLLAPSE

Philippe André[1], Jeroen Bouwman[1], Arnaud Belloche[1,2], and Patrick Hennebelle[2,3]

[1]CEA Saclay, Service d'Astrophysique, Orme des Merisiers, Bât. 709, F-91191 Gif-sur-Yvette Cedex, France
[2]Laboratoire de Radioastronomie, ENS, 24 rue Lhomond, F-75231 Paris Cedex 05, France
[3]Department of Physics & Astronomy, Cardiff University, Cardiff CF24 3YB, Wales, UK

Abstract

Improving our understanding of the initial conditions and earliest stages of protostellar collapse is crucial to gain insight into the origin of stellar masses, multiple systems, and protoplanetary disks. Observationally, there are two complementary approaches to this problem: (1) studying the structure and kinematics of prestellar cores observed prior to protostar formation; and (2) studying the structure of young (e.g., Class 0) accreting protostars observed soon after point mass formation. We discuss recent advances made in this area thanks to millimeter and submillimeter mapping observations with large single-dish telescopes and interferometers. In particular, we argue that the beginning of protostellar collapse is much more violent in cluster-forming clouds than in regions of distributed star formation. Major breakthroughs are expected in this field from future large submillimeter instruments such as Herschel and ALMA.

Key words: ISM: structure – ISM: gravitational collapse – Stars: protostellar accretion rates

1. Introduction

Although the formation of low-mass stars is now reasonably well understood in outline (see, e.g., Mannings, Boss, & Russell 2000 for a recent review), several important aspects remain poorly known, such as the initial stages of the collapse process, the mechanism(s) selecting stellar masses, or the formation of multiple systems. Some progress on the earliest stages of star formation was achieved over the last decade thanks to the use of sensitive receivers on large (sub)millimeter radio telescopes such as JCMT, CSO, and the IRAM 30m. Young protostars were identified at the beginning of the main accretion phase (Class 0 objects; André, Ward-Thompson, & Barsony 1993), and the starless dense cores extensively studied in NH_3 by Myers and collaborators (e.g., Benson & Myers 1989) were found to be characterized by flat inner density gradients (Ward-Thompson et al. 1994). Direct evidence for infall motions was observed toward a large number of Class 0 protostars and prestellar cores (Gregersen et al. 1997; see also Sec. 4 below and Evans, this volume). Class 0 objects were also found to drive more powerful outflows than more evolved (Class I) protostars, suggesting a marked decrease of the mass accretion/ejection rates in the course of protostellar evolution (Bontemps et al. 1996). As advocated by Henriksen, André, & Bontemps (1997), there may be a causal relationship between these results, in the sense that the accretion/ejection decline during the protostellar phase may be a direct consequence of the form of the density profile at the prestellar stage.

Further studying the detailed properties of prestellar cores and young protostars is of prime importance to distinguish between collapse models and shed light on the origin of stellar masses. Indeed, the effective reservoirs of mass available for the formation of individual stars may be largely determined at the prestellar stage, and it is during the main protostellar accretion phase that stars accrete some fraction of these reservoirs and build up the masses they will have on the zero-age main sequence.

After a brief introduction of the theoretical background in Sec. 2, we review recent observational advances concerning the density and velocity structure of cloud cores in Sec. 3 and 4, respectively. We conclude in Sec. 5 with a comparison between observations and theoretical models.

2. Collapse Initial Conditions: Theory

The inside-out collapse model of Shu (1977), starting from a singular isothermal sphere (SIS) or toroid (Li & Shu 1996, 1997), is well known and underlies the "standard" picture of isolated, low-mass star formation (e.g., Shu, Adams, & Lizano 1987). Other collapse models exist, however, which adopt different initial conditions. In particular, Whitworth & Summers (1985) have shown that

Chemistry as a Diagnostic of Star Formation (2003). *Edited by Charles L. Curry and Michel Fich.* Published in 2003 by NRC Research Press, Ottawa ON, K1A 0R6, Canada. ISBN 0-660-19089-3.

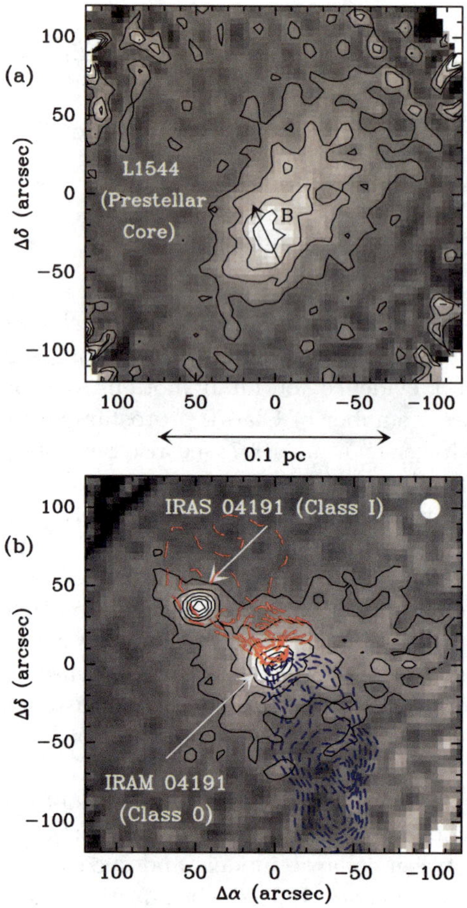

Figure 1. Dust continuum maps at 1.3 mm of: *(a)* L1544 (IRAM 30m telescope, Ward-Thompson et al. 1999); and *(b)* IRAM 04191 (MPIfR bolometer array, André et al. 1999). The effective resolution is 13″, and the base contour and contour steps are at 20 mJy/beam. The direction of the magnetic field measured in L1544 with the SCUBA polarimeter on JCMT (Ward-Thompson et al. 2000) and the collimated CO(2−1) bipolar flow emanating from IRAM 04191 are shown.

there is a two-parameter continuum of similarity solutions to the problem of isothermal spherical collapse. One of the parameters measures how close to hydrostatic equilibrium the system is initially, while the other parameter reflects how important external compression is in initiating the collapse. In this continuum, the Shu (1977) and Larson-Penston (Larson 1969; Penston 1969) solutions represent two extreme limits.

All of the similarity solutions share a universal evolutionary pattern. At early times ($t < 0$), a compression wave (initiated by, e.g., an external disturbance) propagates inward leaving behind it a $\rho(r) \propto r^{-2}$ density profile. At $t = 0$, the compression wave reaches the center and a point mass forms which subsequently grows by accretion. At later times ($t > 0$), this wave is reflected into a rarefaction or expansion wave, propagating outward through the infalling gas (at the isothermal sound speed a_s), and leaving behind it a free-fall $\rho(r) \propto r^{-1.5}$ density distribution. Several well-known features of the Shu model (such as the expansion wave) are thus in fact common to all solutions. The various solutions can be distinguished by the *absolute* values of the density and velocity at $t \sim 0$. In particular, the Shu (1977) solution has $\rho(r) = (a_s{}^2/2\pi G)\, r^{-2}$ and is static ($v = 0$) at $t = 0$, while the Larson-Penston (1969) solution is ~ 4.4 times denser and far from equilibrium ($v \approx -3.3\ a_s$). During the accretion phase ($t > 0$), the infall envelope is a factor of ~ 7 denser in the Larson-Penston solution. Accordingly, the mass infall rate is also much larger in the Larson-Penston case ($\sim 47\, a_s{}^3/G$) than in the Shu case ($\sim a_s{}^3/G$).

In practice, however, protostellar collapse is unlikely to be strictly self-similar, and the above similarity solutions can only be taken as plausible asymptotes. More realistic initial conditions than the SIS are provided by the so-called Bonnor-Ebert spheres (Bonnor 1956), which represent the equilibrium states for self-gravitating isothermal spheres and have a flat density profile in their central $\sim$ Jeans length region. Such spheres are stable for a center-to-edge density contrast of 14.0 and unstable for a density contrast > 14.0 (Bonnor 1956). Numerical hydrodynamic simulations of cloud collapse starting from such initial conditions (Foster & Chevalier 1993; Hennebelle et al. 2003) find that the Larson-Penston similarity solution is generally a good approximation near point-mass formation ($t = 0$) *at small radii*, but that the Shu solution is more adequate at intermediate $t \geq 0$ times, before the expansion wave reaches the edge of the initial, pre-collapse dense core. In general, the mass accretion rate is thus expected to be time-dependent.

Observationally, it is by comparing the (density and velocity) structure of prestellar cores such as L1544 (Fig. 1*a*) with the structure of the envelopes surrounding Class 0 protostars such as IRAM 04191 (Fig. 1*b*) that one may hope to constrain the initial conditions for collapse and to discriminate between the various existing models.

3. Density Structure

3.1. Prestellar cores

Two main approaches have been used to trace the density structure of cloud cores: (1) mapping the optically thin (sub)millimeter continuum *emission* from the cold dust contained in the cores; and (2) mapping the same cold core dust in *absorption* against the background infrared emission (originating from warm cloud dust or remote stars). Mapping the molecular gas component is generally less effective as most molecules tend to freeze out onto dust grains in the dense, cold inner parts of cloud cores (Walmsley et al. 2001; Bergin, this volume).

Ward-Thompson et al. (1994, 1999) and André et al. (1996) employed the first approach to probe the structure of prestellar cores (see also Shirley et al. 2000). Under the simplifying assumption of spatially uniform dust (temperature and emissivity) properties, they concluded that the radial density profiles of isolated prestellar cores were *not* consistent with the single $\rho(r) \propto r^{-2}$ power law of the SIS but were flatter than $\rho(r) \propto r^{-1}$ in their inner regions (for $r \leq R_{\rm flat}$), and approached $\rho(r) \propto r^{-2}$ only beyond a typical radius $R_{\rm flat} \sim$ 2500–5000 AU. The details of this conclusion have been challenged by Evans et al. (2001) and Zucconi et al. (2001) who made the important point that starless cores are not strictly isothermal. Accounting for a realistic temperature distribution inside the cores (see Sec. 3.3 below), these authors found smaller values for $R_{\rm flat}$ and even claimed that in some cases the observed profiles could not be distinguished from a SIS profile. We will return to this point below in Sec. 3.3.

More recently, it has been possible to use the *absorption* approach, both in the mid-IR from space (Bacmann et al. 2000; Siebenmorgen & Krügel 2000) and in the near-IR from the ground (Alves et al. 2001). In particular, using the ISOCAM infrared camera aboard the *ISO* satellite, Bacmann et al. (2000) carried out a 7 μm survey of 24 low-mass starless cores (all undetected by *IRAS*) and observed absorption features in 23 of them. As an example, Figure 2*a* shows the ISOCAM image obtained at 6.75 μm for the prestellar core L1689B in the Ophiuchus complex ($d \sim 150$ pc). The core is seen in absorption against the diffuse mid-IR background arising from the rear side of the parent molecular cloud (see, e.g., Bernard et al. 1993 and Sec. 3.3).

The column density structure of the absorbing cores can be derived from a simple model of the ISOCAM observations. The mid-IR intensity measured at projected radius $\bar{r}$ from core center may be expressed as:

$$I(\bar{r}) = I_{\rm back} e^{-\tau(\bar{r})} + I_{\rm fore}, \qquad (1)$$

where τ is the core optical depth (directly related to the column density N_{H_2} via the dust opacity) and $I_{\rm back}$ and $I_{\rm fore}$ are the background and foreground intensities, respectively. Both $I_{\rm back}$ and $I_{\rm fore}$ are uncertain but their values are well constrained when independent millimeter measurements of the H_2 column density in the centers and outer parts of the cores are available (see Bacmann et al. 2000 for details).

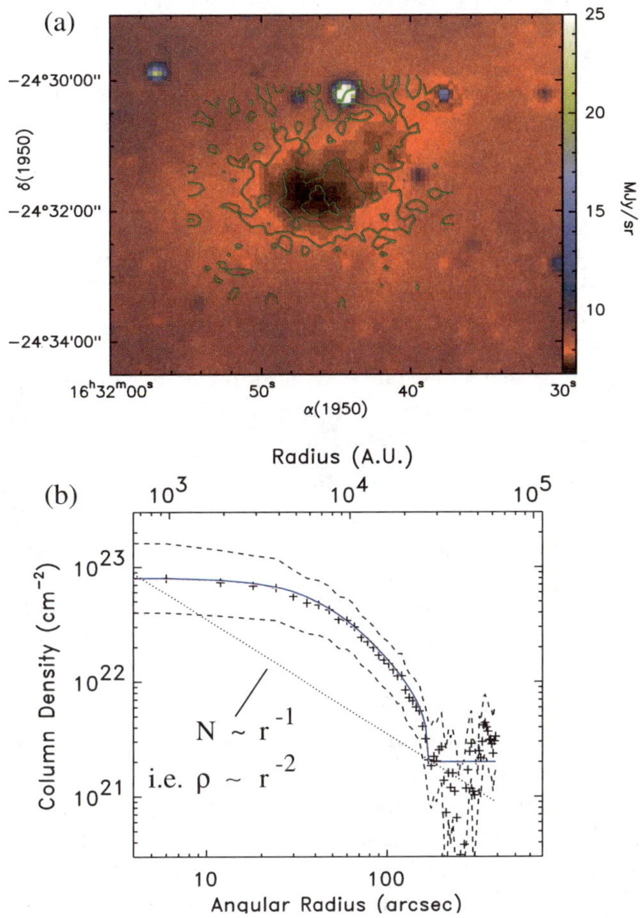

Figure 2. *(a)* ISOCAM 6.75 μm absorption image of the prestellar core L1689B (color scale at right in MJy/sr). The 1.3 mm continuum emission map obtained by André et al. (1996) at the IRAM 30m telescope is superposed as green contours (*levels*: 10, 30, and 50 mJy/13″beam). *(b)* Column density profile of L1689B (*crosses*) derived from the absorption map shown in *(a)* by averaging the intensity over elliptical annuli for a 40° sector in the southern part of the core. The dashed curves show the most extreme profiles compatible with the data given the uncertainties affecting the absorption analysis. The blue solid curve represents the best fit of a Bonnor-Ebert sphere model (embedded in a medium of uniform column density), obtained with the following parameters: $\rho_c/\rho_{\rm out} = 40 \pm 15$ (i.e., well into the unstable regime), $T_{\rm eff} = 50 \pm 20$ K, $P_{\rm ext}/k_B = 5 \pm 3 \times 10^5$ K cm^{-3}. For comparison, the dotted line shows the $N_{\rm H2} \propto \bar{r}^{-1}$ profile of a SIS at $T = 10$ K (adapted from Bacmann et al. 2000).

Using these constraints on $I_{\rm back}$ and $I_{\rm fore}$, a range of possible column density profiles can be derived for each core imaged in absorption with ISOCAM. This is shown in Figure 2*b* for L1689B, where the crosses display our 'best' estimate of the column density profile; i.e., derived using a revised estimate of $8 \times 10^{22}\,{\rm cm}^{-2}$ for the central column density from millimeter continuum data, in agreement with a central temperature of $\sim 8\,{\rm K}$ (Sec. 3.3) instead of 12.5 K as assumed by Bacmann et al. (2000). The dashed curves lying above and below the 'best' profile show the most extreme column density profiles consistent with the mid-IR absorption data, given the range of permitted values for $I_{\rm back}$ and $I_{\rm fore}$. In general, four different regimes can be distinguished on the prestellar column density profiles (see, e.g., Fig. 2*b*):

(1) a flat inner region (of radius $R_{\rm flat} = 5000 \pm 1000\,{\rm AU}$ for L1689B according to the fitting analysis shown in Fig. 2*b*); (2) a region roughly consistent with $N_{\rm H2} \propto \bar{r}^{-1}$ (corresponding to $\rho \propto r^{-2}$ for a spheroidal core); (3) an edge where the column density falls off with projected radius more rapidly than $N_{\rm H2} \propto \bar{r}^{-2}$ (suggesting a density gradient steeper than $\rho \propto r^{-3}$), until; (4) the end of the core is reached (at $R_{\rm out} = 28000 \pm 1000\,{\rm AU}$ for L1689B) and $N_{\rm H2}$ fluctuates about the mean value $N_{\rm out} \sim 1-2 \times 10^{21}\,{\rm cm}^{-2}$ characterizing the ambient molecular cloud.

A qualitatively similar column density profile was derived by Alves et al. (2001) for the Bok globule B68 (for which $R_{\rm flat} \approx 4000\,{\rm AU}$ and $R_{\rm out} \approx 12500\,{\rm AU}$), based on VLT and NTT measurements of the near-IR colors of background stars seen through, and suffering extinction from, this globule.

Thus, the *inner flattening of the density profiles of prestellar cores*, first observed in submillimeter emission (see above) is confirmed by infrared absorption/extinction measurements, which are essentially *independent of the temperature distribution.* Moreover, a new result emerges: at least in some cases, such as L1689B and B68, isolated prestellar cores appear to be characterized by *sharp edges* defining typical outer radii $R_{\rm out} \sim 20{,}000\,{\rm AU} \sim 0.1\,{\rm pc}$.

3.2. Comparison with Theoretical Models

These observed features—i.e., a flat inner region and sharp outer edge—set constraints on possible models of core structure. First, it is clear that self-similar, singular models such as the SIS or magnetized, non-spherical generalizations of it (Shu 1977; Li & Shu 1996) cannot account for the detailed density structure of prestellar cores. As pointed out by Bacmann et al. (2000), the logotropic models of McLaughlin & Pudritz (1996), which behave as $\rho \propto r^{-1}$ over a wide range of radii, are also inconsistent with the observations of (low-mass) starless cores. On the other hand, the circularly-averaged column density profiles can often be fitted remarkably well with models of pressure-bounded Bonnor-Ebert spheres, as first demonstrated by Alves et al. (2001) for B68. This is also the case in L1689B as illustrated in Fig. 2*b*. The quality of such fits shows that Bonnor-Ebert spheres provide a good first-order model for the structure of isolated prestellar cores.

In detail, however, there are several problems with these models. First, most starless cores exhibit elongated shapes (see, e.g., Fig. 1*a* and Fig. 2*a*) which are not consistent with spherical models beyond $\sim 0.03\,{\rm pc}$. Second, the inferred density contrasts (from center to edge) are generally larger (i.e., $\gtrsim$ 20–80—see Fig. 2*b*) than the maximum contrast of $\sim$14 for a stable Bonnor-Ebert sphere (Sec. 2.1). Third, the effective core temperature needed in these models for thermal pressure gradients to balance self-gravity is often significantly larger than both the average dust temperatures measured with ISOPHOT (Ward-Thompson et al. 2002) and the gas temperatures measured in NH_3 (Lai et al., this volume). In the case of L1689B, e.g., the effective temperature of the Bonnor-Ebert fit shown in Fig. 2*b* is $T_{\rm eff} \sim 50\,{\rm K}$, while the dust temperature observed with ISOPHOT is only $T_d \sim 11\,{\rm K}$ (Ward-Thompson et al. 2002). Finally, the physical process responsible for bounding the cores at some external pressure is unclear. These arguments suggest that prestellar cores cannot be described as simple isothermal hydrostatic structures and are either already contracting (see Sec. 4.1 below) or experiencing extra support from static or turbulent magnetic fields (Curry & McKee 2000).

One way of accounting for large density contrasts and high effective temperatures is to consider models of cores threaded by a static magnetic field and evolving through ambipolar diffusion (Ciolek & Mouschovias 1994; Basu & Mouschovias 1995). In these models, at any given time prior to protostar formation, the cores are expected to feature a uniform-density central region whose size corresponds to the instantaneous Jeans length. This agrees well with the characteristics of the flat inner regions seen in starless cores. Furthermore, the observed sharp edges (e.g., Fig. 2*b*) are consistent with the model predictions shortly after the formation of a magnetically supercritical core. Physically, this is because when a supercritical core forms, it collapses dynamically inward, while the outer, subcritical envelope is still efficiently supported by the magnetic field and remains essentially held in place.

As a result, a steep density profile develops at the outer boundary of the supercritical core (see Fig. 8 of Basu & Mouschovias 1995).

A problem, however, with ambipolar diffusion models involving only a static magnetic field is that they require fairly large field strengths ($\sim 30-100\ \mu$G; see Bacmann et al. 2000), which seem to exceed the few existing Zeeman measurements for low-mass dense cores (Crutcher 1999; Crutcher & Troland 2000). It is possible that more elaborate versions of the models, incorporating the effects of a non-static, turbulent magnetic field in the outer parts of the cores and in the ambient cloud, would be more satisfactory and could also account for the filamentary shapes often seen on large ($\gtrsim 0.25$ pc) scales (Curry 2000, 2002; Jones & Basu 2002).

3.3. Temperature distribution in starless cores

The models discussed in Sec. 3.2 assume that prestellar cores are isothermal, a good first approximation (Larson 1969; Tohline 1982). In actual fact, however, there are good reasons to believe that the central regions of starless cores are somewhat cooler than their outer regions. Indeed, the results of *ISO* imaging in the mid-IR (Bacmann et al. 2000) and far-IR (Ward-Thompson et al. 2002) are consistent with the idea that starless cores are heated only from outside by the local interstellar radiation field (ISRF): the amount of energy radiated by such cores in the far-IR is roughly equal to the fraction of ISRF energy absorbed at shorter (e.g., mid-IR) wavelengths, with no evidence for any central heating source (Ward-Thompson et al. 2002). In such a situation, recent dust radiative transfer calculations (Evans et al. 2001; Zucconi et al. 2001) predict that there should be a positive temperature gradient from the cores' centers (with T_d as low as $\sim 5-7$ K) to their edges (at $T_d \sim 15$ K).[1] The first calculations of Evans et al. (2001) and Zucconi et al. (2001) assumed rather a simplistic ISRF, based on current estimates of the average field in the solar neighborhood (Black 1994). We recently carried out similar calculations in which we use the levels of the diffuse mid-IR and far-IR backgrounds observed toward the cores to make more realistic estimates of the effective ISRF directly impinging on their surfaces.

Figure 3*a* gives an illustration for the prestellar core L1689B already discussed in Sec. 3.1 (see Fig. 2). In this case, the ISRF differs (in strength and spectrum) from the standard Black ISRF for several reasons: (1) due to the presence of early-type stars such as HD 147889

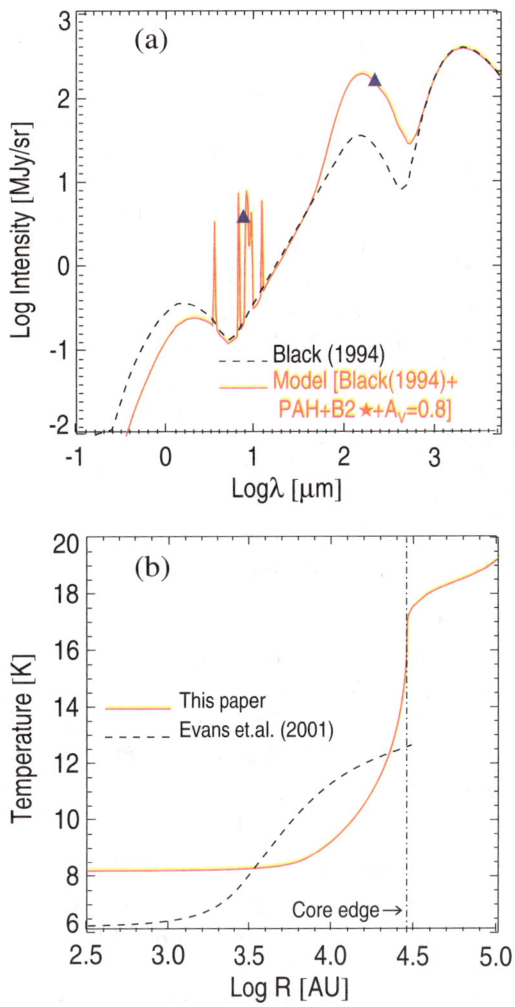

Figure 3. *(a)* Effective radiation field impinging on L1689B (*solid red curve*), as estimated from observed levels of the mid- and far-IR backgrounds (*blue triangles*). This is much stronger than the Black (1994) radiation field (*dashed black curve*). *(b)* Model dust temperature profiles calculated for L1689B with MODUST under two sets of assumptions. The solid red curve assumes the effective radiation field shown in *(a)*, the density profile derived by Bacmann et al. (2000) (with $R_{\rm flat} = 6600$ AU), and the dust opacity law of Preibisch et al. (1993) with thin ice mantles in the core and no ice mantles in the ambient cloud. The dashed black curve corresponds to the model favored by Evans et al. (2001) which has $R_{\rm flat} = 2000$ AU ($d = 160$ pc), no ambient cloud, the Black (1994) radiation field ×0.5, and the dust opacity law of Ossenkopf & Henning (1994) with coagulation (OH5).

[1] Note that, in the dense ($n_{H_2} > 3\times10^4$ cm^{-3}) inner regions, one expects the temperature of the gas to be well-coupled to that of the dust; e.g., Doty & Neufeld (1997).

(B2 V) only a few pc away from the Ophiuchus cloud (Liseau et al. 1999), the ambient far-UV field is significantly stronger ($G_0 \sim 12$) than the solar neighborhood average; (2) this strong FUV field in turn excites small PAH-like grains in the outer ($A_V \lesssim 1$) layers of the cloud which are responsible for a high level of diffuse mid-IR emission distributed in several spectral bands (Bernard et al. 1993 and Fig. 3*a*); (3) the L1689B core is itself embedded at some depth ($A_V \sim 1$) inside the cloud, so that a good fraction of the external FUV radiation is effectively reprocessed to longer wavelengths. Consequently, the effective ISRF of L1689B is about *one order of magnitude stronger* in both the mid- and far-IR than the Black ISRF (Fig. 3*a*), in agreement with the strong mid- and far-IR backgrounds measured by Bacmann et al. (2000) and Ward-Thompson et al. (2002) with ISOCAM and ISOPHOT (~ 4 MJy/sr at 7 μm and ~ 120 MJy/sr at 200 μm, as opposed to ~ 0.25 MJy/sr and ~ 30 MJy/sr, respectively, in the Black radiation field). Based on this estimate of the effective ISRF, we used the MODUST radiative transfer code (Bouwman 2001; Kemper et al. 2001) to compute the temperature distribution in L1689B for several plausible density distributions consistent with the 1.3 mm continuum emission map of André et al. (1996) and the 7 μm absorption map of Bacmann et al. (2000). The results are illustrated in Fig. 3*b*.

Our calculations confirm some of the conclusions of Evans et al. (2001) and Zucconi et al. (2001), namely that the temperature reaches a minimum < 10 K in the centers of prestellar cores and that the central temperature depends primarily on the central optical depth (directly related to the degree of shielding from the external ISRF). However, we find that the minimum temperature may not be as low as reported by Evans et al. (2001) and Zucconi et al. (2001) and that the shape of the temperature profile may be quite different: it can be seen in Fig. 3*b* that the major drop in temperature occurs in the outer parts of the core in our preferred model of L1689B (red curve), while it occurs closer to the center in the Evans et al. (2001) model (dashed curve). We also stress that all the models we have run have a relatively high mass-averaged dust temperature of ~ 11 K, comparable to the temperature estimated by fitting a greybody to the global SED (cf. Fig. 13 of Ward-Thompson et al. 2002). Furthermore, if our assumptions about the effective ISRF and the dust opacity law (see Fig. 3*b*) are correct, then the radius of the flat inner region in the density profile of L1689B must lie between $R_{\rm flat} = 3300$ AU (the smallest value consistent with the 1.3 mm constraints of André et al. [1996]) and $R_{\rm flat} = 6600$ AU (value found by Bacmann et al. 2000 assuming a central core temperature of 12.5 K—see Sec. 3.1). A value as small as $R_{\rm flat} = 2000$ AU

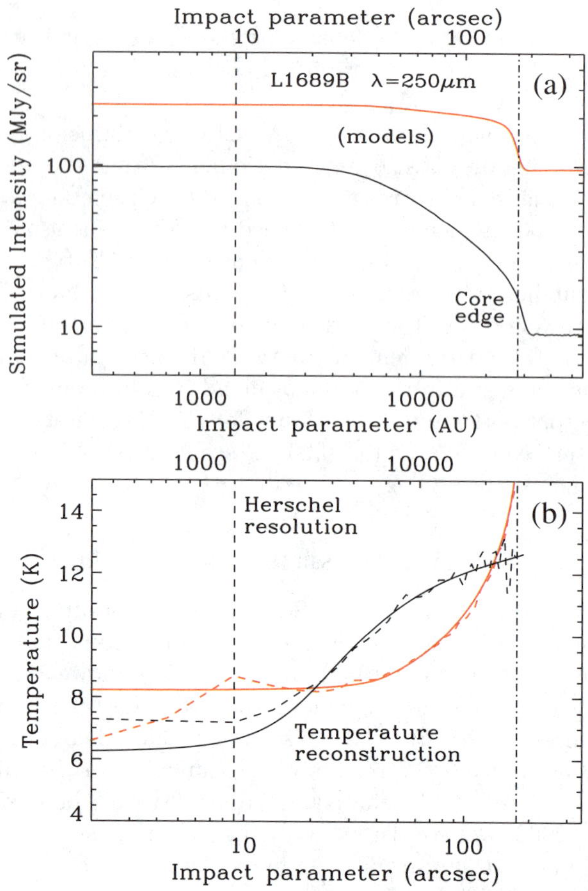

Figure 4. Simulated *Herschel* observations of the radial structure of L1689B. Panel *(a)* shows two model intensity profiles consistent with present millimeter data, as they could be observed by *Herschel* at 250 μm. The upper (*red*) curve corresponds to the model we favor based on our best estimate of the effective ISRF (Fig. 3), while the lower curve is for the model proposed by Evans et al. (2001) based on the Black (1994) ISRF $\times 0.5$. Note the large differences in shape and absolute scaling between the two models. Panel *(b)* shows the dust temperature distributions (*solid curves*) calculated with MODUST for these two models. The dashed curves (almost indistinguishable from the solid curves beyond 20″) show the reconstructed temperature profiles that could be derived from *Herschel* maps in 6 bands between 75 μm and 500 μm (plus ground-based data at 850 μm and 1.3 mm), assuming the 3D core geometry is known. The reconstruction becomes uncertain below the *Herschel* resolution limit (9″HP beam radius at 250 μm).

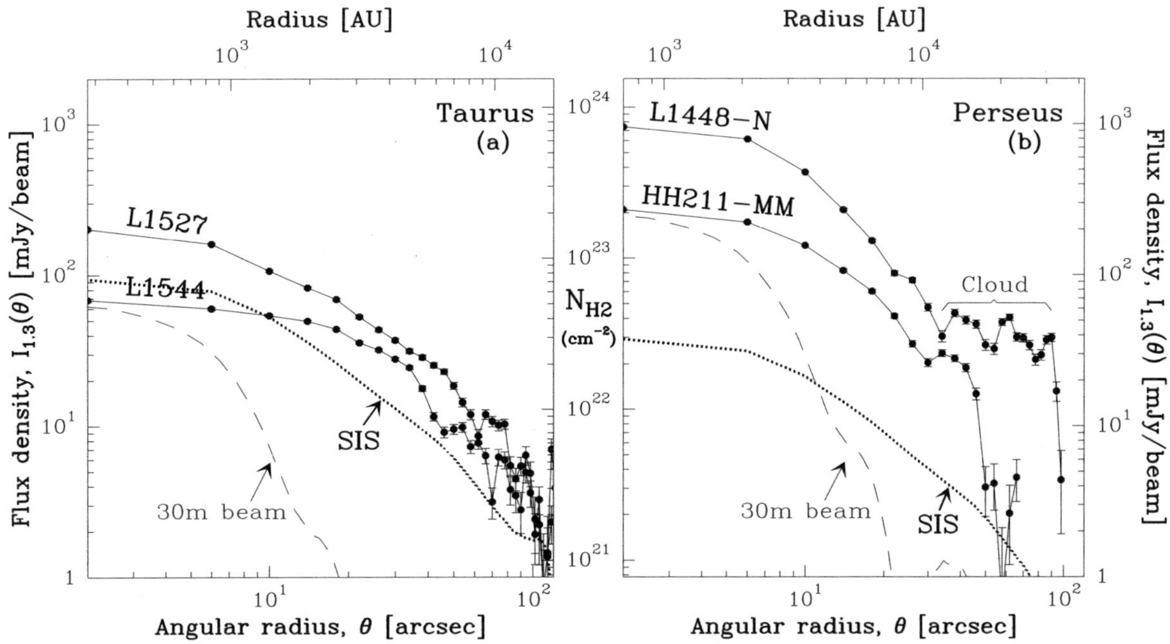

Figure 5. Averaged 1.3 mm radial intensity profiles of *(a)* L1544 (prestellar) and L1527 (Class 0) in Taurus ($d = 140$ pc), and *(b)* HH211–MM and L1448-N (Class 0s) in Perseus ($d = 350$ pc), compared to a simulated profile for a singular isothermal sphere at $T = 10$ K (see Appendix of Motte & André [2001] for details about the simulations). The 30-m beam profile is shown. An approximate column density scale, calculated assuming representative dust properties ($T_d = 15$ K, $\kappa_{1.3} = 0.0075$ cm^2 g^{-1}) is also indicated. (Adapted from Ward-Thompson et al. [1999] and Motte & André [2001]).

(Evans et al. 2001) is ruled out according to our analysis. Thus, conclusions derived from (sub)millimeter emission maps on the density structure of prestellar cores under the simplifying hypothesis that the dust temperature is uniform (Ward-Thompson et al. 1999) are probably not seriously flawed.

In our opinion, the final word on the structure of prestellar cores will come from high-resolution mapping at far-IR and submillimeter wavelengths with future space-borne telescopes such as *Herschel* to be launched by ESA in 2007 (Pilbratt et al. 2001). Using *Herschel* images at 75–500 μm (in combination with ground-based dust continuum mapping at longer submillimeter wavelengths) to construct SED *maps* for at least the nearest, spatially resolved prestellar cores it will be possible to reconstruct their *intrinsic temperature and density distributions simultaneously*, as illustrated in Figure 4. Such multi-band mapping observations will also set strong constraints on the *dust emissivity properties* (e.g., the opacity index β) and their spatial variation across dense cores.

3.4. Structure of protostellar envelopes

In the case of *protostellar* envelopes, the accretion luminosity released close to the central protostar provides an additional source of heating compared to starless cores. In practice, however, the thermal structure of the envelopes surrounding low-luminosity protostars ($L_{bol} \lesssim$ 1–10 $L_\odot$), including most Taurus embedded YSOs, is strongly affected by external heating from the ambient ISRF (Shirley et al. 2002), and the simple isothermal assumption (with T$\simeq$ 10 K) is often not too bad over a wide range of radii ($r \sim 3000 - 15{,}000$ AU; cf. Motte & André 2001).

The density structure of YSO envelopes has been studied primarily with the (sub)millimeter emission approach (although see Harvey et al. [2001] for a recent extinction study). In contrast to prestellar cores, protostellar envelopes are always found to be strongly centrally condensed (see Fig. 1*b*) and do *not* exhibit any marked inner flattening in their radial (column) density profiles. This is illustrated in Figure 5 which shows the circularly averaged radial intensity profiles measured at

1.3 mm for the circumstellar envelopes of the Class 0 objects L1527 in Taurus and HH211-MM and L1448-N in Perseus (Motte & André 2001). (Note that the inner disk contribution has been subtracted from the profiles based on high-resolution interferometric measurements with the IRAM Plateau de Bure interferometer). Apart from a small overdensity factor (~ 2), the profile of the L1527 envelope closely follows the *shape* of the SIS model profile (Fig. 5*a*), and is significantly steeper than the profile of the prestellar core L1544 at small radii.

More generally, several (sub)millimeter continuum studies indicate that protostellar envelopes in regions of isolated star formation such as Taurus have radial density gradients consistent with $\rho(r) \propto r^{-p}$ with $p \sim 1.5-2$ over more than $\sim 10,000-15,000$ AU in radius (Ladd et al. 1991; Chandler & Richer 2000; Hogerheijde & Sandell 2000; Shirley et al. 2000; Motte & André 2001; Jørgensen et al. 2002). The density gradient estimated for (Class 0 and Class I) protostars (i.e., $p \sim 1.5-2$) agrees with most collapse models which predict a value of p between 2 and 1.5 during the protostellar accretion phase (before and after the passage of the collapse expansion wave, respectively; Sec. 2.1). Furthermore, the median envelope mass measured by Motte & André (2001) for the bona-fide protostars of Taurus ($\overline{M_{\rm env}}[r < 4200\,AU] \sim 0.3\,M_\odot$) is roughly consistent with the predictions of the inside-out collapse model of Shu (1977) for $\sim 10^5$ year-old protostars.

The situation is markedly different in regions where stars form in tight groups or *clusters*, such as Serpens, Perseus, and the ρ Oph main cloud. In this case, the observed envelopes are clearly *not scale-free*: they merge with dense cores, other envelopes, and/or the diffuse ambient cloud at a *finite* radius $R_{\rm out} \lesssim 5000$ AU (Motte, André, & Neri 1998; Motte & André 2001; see also Fig. 5 of Mundy, Looney, & Welch 2000). Moreover, as can be seen in Fig. 5*b* for HH211-MM and L1448-N, the Class 0 envelopes mapped in cluster-forming regions are found to be *3 to 12 times denser* than the SIS at $T = 10$ K (the typical gas kinetic temperature expected in these clouds prior to massive star formation: Goldsmith & Langer 1978; Evans 1999; and Fig. 3*b*). Static magnetic fields can only account for moderate overdensity factors ($\lesssim 2$) compared to an unmagnetized SIS (Li & Shu 1997). Turbulence could in principle contribute to the support of the initial dense core (Myers & Fuller 1992; Mardones et al. 1997). However, the small-scale condensations corresponding to the precursors of protostars in star-forming clusters (Motte et al. 1998) appear to be largely devoid of turbulence (Sec. 4.1). A more likely explanation is that, in such regions, protostellar collapse is *induced by strong external disturbances* and starts from *non-singular initial conditions*, resulting in a *non-equilibrium density configuration* with a large overdensity factor ($\lesssim 10$) and significant inward velocities near $t \sim 0$ (see also Sec. 5 and Fig. 9).

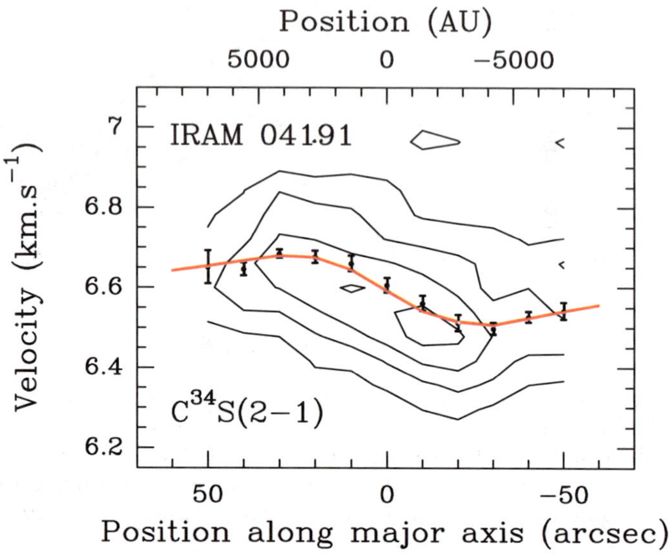

Figure 6. Position-velocity diagram along the major axis of the IRAM 04191 envelope (i.e., perpendicular to the flow) based on a $C^{34}S(2-1)$ map taken at the IRAM 30m telescope (Belloche et al. 2002). Contours: 0.2 to 0.8 by 0.2 K. Dots with error bars mark the observed velocity centroids. The solid curve shows the profile of a model with differential rotation ($V_{\rm rot}[r > 3500\,{\rm AU}] \propto r^{-1.5}$) also used in Fig. 7.

4. Velocity Structure

4.1. Internal motions in starless cores

On large ($\gtrsim$ pc) scales, the spectral line profiles/widths observed in molecular clouds are indicative of supersonic turbulent motions (Falgarone & Phillips 1990). The non-thermal component of the line width, σ_{NT}, roughly follows the scaling law $\sigma_{NT} \propto R^{0.5}$ (Fuller & Myers 1992), believed to arise from MHD turbulence (Mouschovias & Psaltis 1995). On intermediate ($\gtrsim 0.1$ pc) scales, there are strong differences between the "low-mass" cores observed in regions of distributed star formation (e.g., Taurus), whose line widths are dominated by thermal motions (Benson & Myers 1989), and the "massive", cluster-forming cores of, e.g., Ophiuchus and Orion, whose line

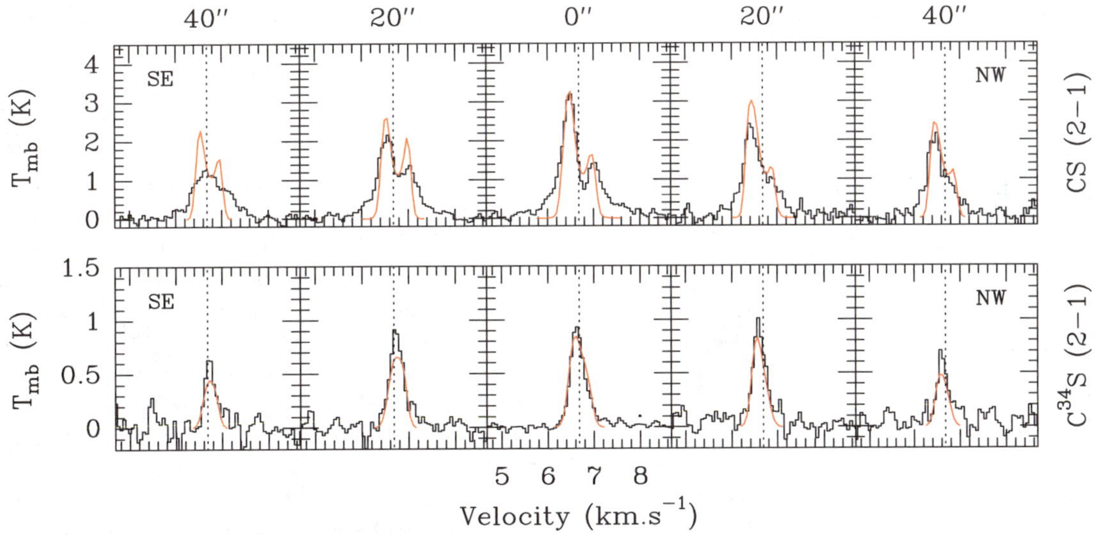

Figure 7. Spectra observed with the IRAM 30 m telescope toward IRAM 04191, along the axis perpendicular to the outflow, in the optically thick CS(2 − 1) and optically thin $C^{34}S(2-1)$ lines. Synthetic spectra corresponding to the 'best-fit' collapse model including rotation found by Belloche et al. (2002) are superposed (see text for parameters).

widths are still dominated by turbulent motions (Caselli & Myers 1995).

On small ($\lesssim 0.03$ pc) scales, however, the prestellar condensations of the Ophiuchus, Serpens, Perseus, and Orion cluster-forming regions are always characterized by very *narrow line widths* (Belloche et al. 2001; Myers 2001), reminiscent of the thermal cores of Taurus. In the ρ Oph protocluster, e.g., the nonthermal velocity dispersion is about half the thermal velocity dispersion of H_2 ($\sigma_{NT}/\sigma_T \sim 0.7$) toward the starless condensations of the dense cores Oph B1, C, E, and F (Belloche et al. 2001). This indicates that, *even in cluster-forming clouds*, the initial conditions for individual protostellar collapse are "coherent" (Goodman et al. 1998) and essentially free of turbulence. The dissipation, on small (< 0.1 pc) scales, of a significant fraction of the turbulent motions observed on large scales is thus a prerequesite for the formation of prestellar condensations (Nakano 1998; Myers 1998; Falgarone & Pety 2001). Apart from low levels of turbulence, low-mass prestellar cores are also characterized by small rotation rates in general ($\Omega \sim 10^{-14}$–10^{-13} rad s^{-1}; Goodman et al. 1993), and subsonic, extended infall motions (Lee, Myers & Tafalla 2001; Evans, this volume).

4.2. Rotation and infall in Class 0 objects

The velocity structure of only a few Class 0 objects has been studied in detail up to now. Here, we mostly discuss the example of the nearby ($d = 140$ pc) and relatively isolated Class 0 object IRAM 04191+1522 (IRAM 04191 for short) found by André, Motte, & Bacmann (1999) in the vicinity of the Taurus Class I source IRAS 04191+1523 (Tamura et al. 1991; see also Fig. 1*b*).

IRAM 04191 features a very high envelope mass-to-luminosity ratio ($M_{\rm env}^{<4200\,\rm AU}/L_{\rm bol} \sim 3\, M_\odot/L_\odot$) and a low bolometric temperature ($T_{\rm bol} \sim 18$ K), suggesting an age $t \lesssim 3 \times 10^4$ yr since the beginning of the accretion phase (André et al. 1999). It is also remarkable in that it has a very low bolometric luminosity ($L_{\rm bol} \sim 0.15\, L_\odot$), and no or only a tiny accretion disk ($R_{\rm disk} < 15$ AU, $M_{\rm disk} < 2 \times 10^{-3}\, M_\odot$). Accordingly, *IRAM 04191 is probably the youngest accreting protostar currently known in Taurus.*

IRAM 04191 is associated with a flattened circumstellar envelope, seen in dust continuum (Fig. 1*b*) and in dense gas tracers such as N_2H^+, C_3H_2, $H^{13}CO^+$, and DCO^+ (Belloche et al. 2002). All of the maps taken in optically thin lines show a clear velocity gradient across the envelope of ~ 7 km s^{-1} pc^{-1}, increasing from west to east along the long axis; i.e., roughly perpendicular to the outflow axis (see Fig. 1*b* and Fig. 6). Strongly suggestive of rotation, this velocity gradient corresponds to a mean angular velocity $\Omega \sim 3 \times 10^{-13}$ rad s^{-1} in the inner ~ 3000 AU region (after deprojection from a viewing angle $i = 50°$). Furthermore, this rotation does not occur in a rigid body, but rather differential, fashion: the inner $\sim$ 3500 AU-radius region seems to be rotating significantly faster than the outer parts of the core (see Fig. 6). The "S" shape of the position-velocity di-

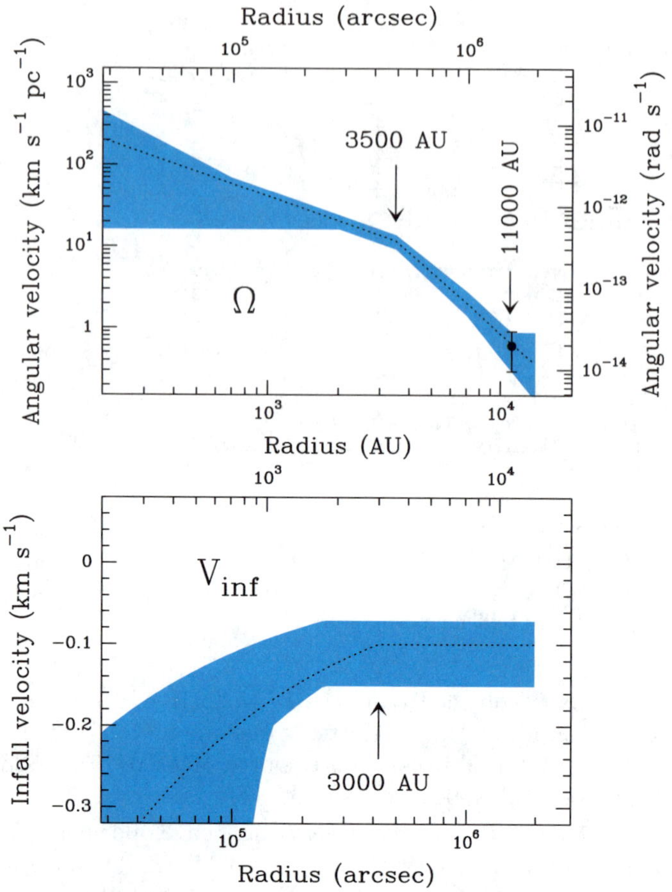

Figure 8. *(a)* Inferred rotational angular velocity, and *(b)* infall velocity in the IRAM 04191 envelope based on the radiative transfer modeling of Belloche et al. (2002). The shaded areas show the domains where the models match the CS and $C^{34}S$ observations reasonably well.

agram shown in Fig. 6 can be fitted with a rotation curve of the form $V_{\rm rot}(r) \propto r^{0.1\pm0.4}$ for $r < 3500$ AU and $V_{\rm rot}(r) \propto r^{-1.5\pm0.5}$ for $3500 < r < 11,000$ AU, corresponding to $\Omega(r) \propto r^{-0.9\pm0.4}$ and $\Omega(r) \propto r^{-2.5\pm0.5}$, respectively (see Fig. 8*a* and Belloche et al. 2002 for details).

Direct evidence for infall motions over a large portion of the IRAM 04191 envelope is also observed in optically thick lines such as CS(2–1), CS(3–2), $H_2CO(2_{12} - 1_{11})$, and $H_2CO(3_{12} - 2_{11})$. These lines are double-peaked and skewed to the blue up to $\gtrsim 40''$ from source center (see Fig. 7), which is indicative of infall motions up to a radius $R_{\rm inf} \gtrsim 5000$ AU from the center (Evans 1999; Evans, this volume). Radiative transfer modeling confirms this view, suggesting a relatively flat infall velocity profile ($V_{\rm inf} \sim 0.1\,{\rm km\,s^{-1}}$) for $3000 \lesssim r \lesssim 11,000$ AU and larger infall velocities scaling as $V_{\rm inf} \propto r^{-0.5}$ for $r \lesssim 3000$ AU (Fig. 8*b* and Belloche et al. 2002). The mass infall rate is estimated as $\dot{M}_{\rm inf} \sim 2 - 3 \times a_s^3/G \sim 3 \times 10^{-6}\ M_\odot\,{\rm yr^{-1}}$ (with $a_s \sim 0.15 - 0.2\,{\rm km\,s^{-1}}$ for $T \sim 6 - 10$ K), roughly independent of radius.

Another Class 0 object whose kinematics has been quantified in detail recently is the IRAS 4 system in the NGC 1333 protocluster (Di Francesco et al. 2001). IRAS 4 is a triple system (A/B/C) on scales of $\sim 10^4$ AU, whose primary component (A) itself breaks up into a ~ 600 AU binary (Looney et al. 2000). Using the IRAM Plateau de Bure interferometer, Di Francesco et al. observed inverse P-Cygni profiles in $H_2CO(3_{12} - 2_{11})$ toward IRAS 4A and IRAS 4B, from which they derived very large mass infall rates of $\sim 1.1 \times 10^{-4}\ M_\odot\ {\rm yr^{-1}}$ and $\sim 3.7 \times 10^{-5}\ M_\odot\,{\rm yr^{-1}}$ at $r \sim 2000$ AU for A and B, respectively. Even if a warmer initial gas temperature (~ 20 K) than in IRAM 04191 and some initial level of turbulence are accounted for (see Di Francesco et al. 2001), these values of $\dot{M}_{\rm inf}$ exceed the canonical $a_{\rm eff}^3/G$ value by greater than 15 and 6 times, respectively (where $a_{\rm eff} \lesssim 0.3\,{\rm km\,s^{-1}}$ is the effective sound speed). These very high infall rates result both from very dense envelopes (Motte & André 2001) and large, supersonic infall velocities ($\sim 0.68\,{\rm km\,s^{-1}}$ and $\sim 0.47\,{\rm km\,s^{-1}}$ for A and B; Di Francesco et al. 2001). Evidence of fast rotation, producing a velocity gradient as high as $\sim 40\,{\rm km\,s^{-1}\,pc^{-1}}$, was also found by Di Francesco et al. toward IRAS 4A.

5. Conclusion: Comparison with collapse models

In the case of low-mass, isolated dense cores, the SIS model of Shu (1977) describes global features of the collapse (e.g., the mass infall rate) reasonably well and thus remains a useful, approximate guide. In detail, however, the extended infall velocity profiles observed in prestellar cores (Lee et al. 2001) and in the very young Class 0 object IRAM 04191 (Sec. 4.2) are inconsistent with the pure inside-out collapse picture of Shu (1977). The *shapes* of the density profiles observed in prestellar cores are well-fit by purely thermal Bonnor-Ebert sphere models, but the *absolute* values of the densities are suggestive of some additional magnetic support (Sec. 3.2). The observed infall velocities are also marginally consistent with isothermal collapse models starting from Bonnor-Ebert spheres (Foster & Chevalier 1993; Hennebelle et al. 2003), as such models tend to produce somewhat higher velocities. This suggests that the collapse of "isolated" cores is essentially *spontaneous* and somehow moderated by magnetic effects in magnetized, probably not strictly isothermal Bonnor-

Ebert cloudlets. Furthermore, the contrast seen in Fig. 8 between a steeply declining rotation velocity profile and a flat infall velocity profile beyond ∼3500 AU suggests that angular momentum is *not* conserved in the outer envelope of IRAM 04191. Such behavior is very difficult to account for in the context of non-magnetic collapse models. In the presence of a relatively strong (∼60 μG) magnetic field, on the other hand, the outer envelope can be coupled to, and spun down by, the large moment of inertia of the ambient cloud (Basu & Mouschovias 1994). Based on a qualitative comparison with the ambipolar diffusion models of Basu & Mouschovias, Belloche et al. (2002) propose that the rapidly rotating inner envelope of IRAM 04191 corresponds to a magnetically supercritical core decoupling from an environment still supported by magnetic fields and strongly affected by magnetic braking (quantitatively, however, the published models are rotating too slowly to fit the observations). In this view, the inner ∼3500 AU radius envelope of IRAM 04191 would correspond to the effective mass reservoir (∼0.5 $M_\odot$) from which the central star is being built. Moreover, comparison of these results with the rotational characteristics of other prestellar and protostellar objects in Taurus (Ohashi et al. 1997) suggests that IRAM 04191 behaves in a typical manner and is simply observed particularly soon after point mass formation (i.e., at $t \gtrsim 0$). If this is correct, then *the masses of the stars forming in clouds such as Taurus may be largely determined by magnetic decoupling effects.*

In protoclusters, by contrast, the large overdensity factors measured for Class 0 envelopes compared to hydrostatic isothermal structures (Sec. 3.4), as well as the fast supersonic infall velocities and very large infall rates observed in some cases (Sec. 4.2), are inconsistent with self-initiated forms of collapse and require a *strong external influence.* This point is illustrated in Figure 9 with the results of recent SPH simulations by Hennebelle et al. (2003). These simulations follow the evolution of a Bonnor-Ebert sphere whose collapse has been induced by an increase in external pressure P_{ext}. Large overdensity factors (compared to a SIS), together with supersonic infall velocities and large infall rates ($\gtrsim 10\, a_s{}^3/G$), are found near $t = 0$ when (and only when) the increase in $P_{\rm ext}$ is strong and very rapid (Fig. 9*b*), resulting in a violent compression wave. Such a violent collapse may be conducive to the formation of both massive stars (through higher accretion rates) and multiple systems when realistic, non-isotropic compressions are considered. Future high-resolution studies with the next generation of (sub)millimeter instruments (e.g., ALMA) will greatly help test this view and shed further light on the physics of collapse in cluster-forming regions.

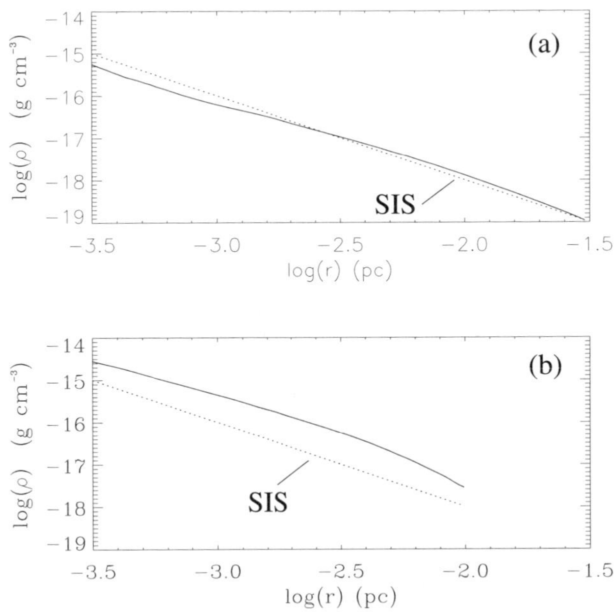

Figure 9. Density profiles (*solid curves*) obtained near point-mass formation ($t \approx 0$) in SPH numerical simulations of the collapse of an isothermal ($T = 10$ K) Bonnor-Ebert sphere induced by external compression (Hennebelle et al. 2003). For comparison, the dashed-dotted line shows the $\rho \propto r^{-2}$ profile of a SIS at 10 K. In *(a)*, the collapse was initiated quasi-statically by very slow compression and the density profile at $t = 0$ is very similar to that of the SIS. By contrast, in *(b)*, the collapse was induced by a very rapid increase in external pressure with $P_{\rm ext}/\dot{P}_{\rm ext} \sim$ a tenth of the initial sound crossing time, resulting in much larger densities at $t = 0$.

References

Alves, J.F., Lada, C.J., & Lada, E.A. 2001, Nature, 409, 159
André, P., Motte, F., & Bacmann, A. 1999, ApJ, 513, L57
André, P., Ward-Thompson, D., & Barsony, M. 1993, ApJ, 406, 122
André, P., Ward-Thompson, D., & Motte, F. 1996, A&A, 314, 625
Bacmann, A., André, P., Puget, J.-L., Abergel, A., Bontemps, S., & Ward-Thompson, D. 2000, A&A, 361, 555
Basu, S., & Mouschovias, T.C. 1994, ApJ, 432, 720
Basu, S., & Mouschovias, T.C. 1995, ApJ, 453, 271
Benson, P.J., & Myers, P.C. 1989, ApJSS, 71, 89
Belloche, A., André, P., & Motte, F. 2001, in *From Darkness to Light*, ed. T. Montmerle & P. André, ASP Conf. Ser., 243, 313

Belloche, A., André, P., Despois, D. & Blinder, S. 2002, A&A, 393, 927
Bernard, J.P., Boulanger, F., Puget, J.L. 1993, A&A, 277, 609
Black, J.H. 1994, in *The First Symposium on the Infrared Cirrus and Diffuse Interstellar Clouds*, ed. R.M. Cutri & W.B. Latter, ASP Conf. Ser., 58, 355
Bonnor, W.B. 1956, MNRAS, 116, 351
Bontemps, S., André, P., Terebey, S., Cabrit, S. 1996, A&A, 311, 858
Bouwman, J. 2001, PhD thesis, University of Amsterdam
Caselli, P., & Myers, P.C. 1995, ApJ, 446, 665
Chandler, C.J., & Richer, J.S. 2000, ApJ, 530, 851
Ciolek, G.E., & Mouschovias, T.C. 1994, ApJ, 425, 142
Crutcher, R.M. 1999, ApJ, 520, 706
Crutcher, R.M., & Troland, T.H. 2000, ApJ, 537, L139
Curry, C.L. 2000, ApJ, 541, 831
Curry, C.L. 2002, ApJ, 576, 849
Curry, C.L., & McKee, C.F. 2000, ApJ, 528, 734
Di Francesco, J., Myers, P.C., Wilner, D.J., Ohashi, N., & Mardones, D. 2001, ApJ, 562, 770
Doty, S.D., & Neufeld, D.A. 1997, ApJ, 489, 122
Evans, N.J. II 1999, ARA&A, 37, 311
Evans, N.J. II, Rawlings, J.M.C., Shirley, Y.L., & Mundy, L.G. 2001, ApJ, 557, 193
Falgarone, E., & Pety, J. 2001 in *From Darkness to Light*, ed. T. Montmerle & P. André, ASP Conf. Ser., 243, 53
Falgarone, E., & Phillips, T. 1990, ApJ, 359, 344
Foster, P.N., & Chevalier, R.A. 1993, ApJ, 416, 303
Goldsmith, P.F., Langer, W.D. 1978, ApJ, 222, 881
Goodman, A.A., Benson, P.J., Fuller, G.A., & Myers, P.C. 1993, ApJ, 406, 528
Goodman, A.A., Barranco, J.A., Wilner, D.J., Heyer, M.H. 1998, ApJ, 504, 223
Harvey, D.W.A., Wilner, D.J., Lada, C.J. et al. 2001, ApJ, 563, 903
Hennebelle, P., Whitworth, A.P., Gladwin, P.P., & André, P. 2003, MNRAS, 340, 870
Gregersen, E.M., Evans, N.J., Zhou, S., & Choi, M. 1997, ApJ, 484, 256
Henriksen, R.N., André, P., & Bontemps, S. 1997, A&A, 323, 549
Hogerheijde, M.R., & Sandell, G. 2000, ApJ, 534, 880
Jones, C.E., & Basu, S. 2002, ApJ, 569, 280
Jørgensen, J.K., Schöier, F.L., & van Dischoeck, E.F. 2002, A&A, 389, 908
Kemper, F., Waters, L.B.F.M., de Koter, A., & Tielens, A.G.G.M. 2001, A&A, 369, 132
Ladd, E.F., Adams, F.C., Casey, S. et al. 1991, ApJ, 382, 555
Larson, R.B., 1969, MNRAS, 145, 271
Lee, C.W., Myers, P.C., & Tafalla, M. 2001, ApJS, 136, 703
Li, Z.-Y., & Shu, F.H. 1996, ApJ, 472, 211
Li, Z.-Y., & Shu, F.H. 1997, ApJ, 475, 237
Liseau, R., White, G. J., Larsson, B. et al. 1999, A&A, 344, 342
Looney, L.W., Mundy, L.G., & Welch, W.J. 2000, ApJ, 529, 477
Mardones, D., Myers, P. C., Tafalla, M., Wilner, D. J., Bachiller, R., Garay, G. 1997, ApJ, 489, 719
McLaughlin, D.E., Pudritz, R.E. 1996, ApJ, 469, 194
Motte, F., & André, P. 2001, A&A, 365, 440
Motte, F., André, P., & Neri, R. 1998, A&A, 336, 150
Mouschovias, T.Ch., & Psaltis, D. 1995, ApJ, 444, L105
Mundy, L.G., Looney, L.W., & Welch, W.J. 2000, in *Protostars and Planets IV*, ed. V. Mannings, A.P. Boss, & S.S. Russell (Tucson: Univ. of Arizona Press), 355
Myers, P.C. 1998, ApJ, 496, L109
Myers, P.C. 2001, in *From Darkness to Light*, ed. T. Montmerle & P. André, ASP Conf. Ser., 243, 131
Myers, P.C., & Fuller, G.A. 1992, ApJ, 396, 631
Nakano, T. 1998, ApJ, 494, 587
Ohashi, N., Hayashi, M., Ho, P.T.P., Momose, M., Tamura, M., Hirano, N., & Sargent, A.I. 1997, ApJ, 488, 317
Ossenkopf, V., & Henning, Th. 1994, A&A, 291, 943
Penston, M.V., 1969, MNRAS, 144, 425
Pilbratt, G.L., Cernicharo, J., Heras, A.M., & Prusti, T. (eds.) 2001, *The Promise of FIRST*, ESA SP-460
Preibisch, Th., Ossenkopf, V., Yorke, H.W., & Henning, Th. 1993, A&A, 279, 577
Shirley, Y., Evans II, N.J., Rawlings, J.M.C., Gregersen, E.M. 2000, ApJS, 131, 249
Shirley, Y., Evans II, N.J., & Rawlings, J.M.C. 2002, ApJ, 575, 337
Shu, F. 1977, ApJ, 214, 488
Shu, F.H., Adams, F.C., & Lizano, S. 1987, ARA&A 25, 23
Siebenmorgen, R., & Krügel, E. 2000, A&A, 364, 625
Tamura, M., Gatley, I., Waller, W., & Werner, M.W. 1991, ApJ, 374, L25
Tohline, J. E. 1982, Fund. Cosm. Phys., 8, 1
Walmsley, M., Caselli, P., Zucconi, A., & Galli, D. 2001, in *The Origin of Stars and Planets: The VLT View*, ed. J. Alves & M. McCaughrean, ESO Astrophysics Symposia (Berlin: Springer-Verlag), 29
Ward-Thompson, D., André, P., & Kirk, J.M. 2002, MNRAS, 329, 257
Ward-Thompson, D., Kirk, J.M., Crutcher, R.M. et al. 2000, ApJL, 537, L135
Ward-Thompson, D., Motte, F., & André, P. 1999, MNRAS, 305, 143
Ward-Thompson, D., Scott, P.F., Hills, R.E., & André, P. 1994, MNRAS, 268, 276
Whitworth, A., & Summers, D. 1985, MNRAS, 214, 1
Zucconi, A., Walmsley, C.M., & Galli, D., 2001, A&A, 376, 650

ASTROCHEMISTRY ALONG THE ORION INTEGRAL-SHAPED FILAMENT

Doug Johnstone[1], **Annemieke M.S. Boonman**[2], **and Ewine F. van Dishoeck**[2]

[1]National Research Council Canada, Herzberg Institute of Astrophysics, Victoria, B.C., V9E 2E7, Canada
[2]Leiden Observatory, P.O. Box 9513, 2300 RA Leiden, The Netherlands

Abstract

The cornerstone molecules CO, H_2CO, CH_3OH, HCN, HNC, CN, CS, and SO were observed toward six submillimeter bright sources in the Orion molecular cloud. Four of the sources observed are protostellar, ranging in energy from $1-500\,L_\odot$, while the other two sources are located at a shock knot and within a photodissociation region. The abundance of each species is determined using statistical equilibrium calculations. A trend is found whereby the abundance of each molecule is lower in the colder sources.

Key words: ISM: molecules – ISM: structure – ISM: clouds

1. Introduction

We present a study of cornerstone molecules toward a variety of independent locations within the Integral-Shaped Filament (ISF) in Orion A, and also toward the strong infrared source NGC 2071IR in Orion B. A range of molecular lines were observed, in order to quantify the physical and chemical conditions at each position. This is primarily a morphological study, using simplified statistical equilibrium modeling, in order to search for obvious diagnostic features within the data. The list of sources observed includes both highly enshrouded, possibly protostellar, submillimeter bright sources, more evolved protostars, a bright photodissociation (PDR) knot, and a shock knot. Many of the young stellar objects are of intermediate mass and thus the observations bridge submillimeter studies of high mass (e.g., Hatchell et al. 1998; van der Tak et al. 2000) and low mass (e.g., Blake et al. 1994, 1995; van Dishoeck et al. 1995; Buckle & Fuller 2002; Jørgensen, Schöier, & van Dishoeck 2002) star-forming regions. The selection of molecules and molecular transitions is based on previous experience in detailed studies of individual objects.

2. The Individual Sources

The sources chosen for this study were primarily drawn from the 450 and 850 μm dust emission maps of the ISF (Fig. 1; Johnstone & Bally 1999). The six sources were chosen to cover a wide range of physical conditions and to test the hypothesis that molecular line observations provide sensitive diagnostics. A summary of the physical properties of the six sources is presented in Table 2. The sources are ordered from the coldest (least energetic) to the warmest (most energetic) with the PDR and shock knot noted separately.

The coldest submillimeter source in our sample is MMS6, as measured by the submillimeter flux from 350 to 1300 μm (Lis et al. 2001). MMS6 is also the brightest 850 μm source within the OMC-3 region of the ISF, with a peak flux of 7.4 Jy within a 15″beam (Johnstone & Bally 1999) and is surrounded by a massive envelope (Chini et al. 1997). There is no corresponding IRAS point source—suggesting that there is no strong internal heating source—although the background brightness of the region makes point source detection difficult. Despite the very cold appearance of MMS6 in the submillimeter, it has been classified as a Class 0 protostar (Chini et al. 1997). Yu et al. (1997) observed knots of H_2 directly north and south of MMS6 which are interpreted as bow shocks within a molecular outflow from the continuum source. VLA observations at 3.6 cm (Reipurth et al. 1999) located a radio source coincident with MMS6. Such coincidences are common among Class 0 sources. Thus, while MMS6 is possibly a pre-collapse clump, it is likely that the submillimeter source enshrouds a very young Class 0 protostar.

MMS9 appears to be the source of the most powerful outflow in the OMC2/3 region (Yu et al. 2000), despite being three times fainter than MMS6 at 850 μm. The spectral slope for MMS9 is less steep than observed in MMS6, suggesting a warmer temperature in the dust envelope (Chini et al. 1997). The total envelope mass is estimated to be three to four times lower than the envelope

Chemistry as a Diagnostic of Star Formation (2003). *Edited by Charles L. Curry and Michel Fich.* Published in 2003 by NRC Research Press, Ottawa ON, K1A 0R6, Canada. ISBN 0-660-19089-3.

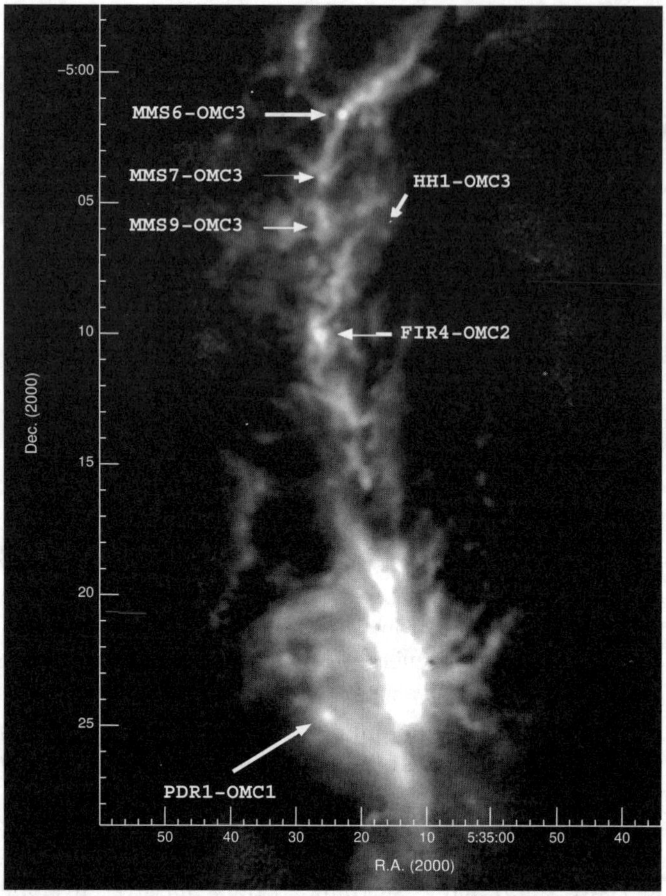

Figure 1. The ISF at the north end of the Orion A molecular cloud, observed in dust continuum emission at 850 microns. The greyscale ranges from 0 to 5 Jy with a logarithmic stretch. The positions of the six sources located in the ISF are indicated by arrows pointing at the peak continuum flux locations. Details of the sources are found in Table 1 (Johnstone et al. 2003, in preparation).

Table 1. Source list and submillimeter diagnostics

Name	T_d (K)	M ($M_\odot$)	L ($L_\odot$)	Object
MMS6-OMC3	15	36	1.2-60	Class 0
MMS9-OMC3	20	10	0.6-90	Class 0
FIR4-OMC2	15-40	34	400	Class 0/I
NGC 2071IR	20-50	30	500	Class 0/I
PDR1-OMC1	60			PDR
SK1-OMC3	10			Shock

mass of MMS6. The IRAS flux limits constrain the luminosity of the embedded source to be $< 94\,L_\odot$ (Chini et al. 1997). Despite having no coincident IRAS point source, MMS9 is expected to contain a well-advanced Class 0 protostar. Reipurth et al. (1999) observed 3.6 cm continuum emission at this location. The rising infrared background toward MMS9 may be responsible for the lack of an IRAS point source detection.

The far-infrared source FIR4 (Mezger, Zylka, & Wink 1990) lying further south along the ridge, near the center of the OMC2 molecular core, is also observed as an IRAS point source. The region around FIR4 harbors a cluster of near-infrared sources (Johnson et al. 1990) although none are explicitly coincident with the far-infrared source. FIR4 is the brightest submillimeter source in the OMC2 region and contains a VLA radio source (Reipurth et al. 1999), leading to its identification as a Class 0 object, although its proximity to the near-IR cluster suggests that it might be a more evolved envelope. The presence of far-infrared emission provides evidence for warm dust at $T_d \sim 40\,\mathrm{K}$ (Mezger et al. 1990), although the submillimeter emission traces the much colder outer envelope $T_d \sim 15\,\mathrm{K}$. The total envelope mass is $\sim 34\,M_\odot$ and the luminosity is $\sim 400\,L_\odot$ (Mezger et al. 1990).

NGC 2071IR in the Orion B North molecular cloud (Johnstone et al. 2001; Motte et al. 2001; Harvey et al. 1979) is the final protostellar source in our sample. Significant far-infrared emission (Butner et al. 1990) from the source provides evidence of an internal source and a warm dust component ($T_d > 50\,\mathrm{K}$) providing a reprocessed luminosity $\sim 500\,L_\odot$. The total envelope mass derived from the submillimeter dust emission is $12-30\,M_\odot$ (Johnstone et al. 2001). NGC 2071 contains one of the most powerful outflows known (Bally & Lada 1983).

Two regions which are not associated with individual stars were also observed. A submillimeter bright location, PDR1, along the PDR separating the HII region (M42, Orion Nebula) from the Orion A molecular cloud was included to monitor changes in the molecular line diagnostics in regions of enhanced UV fluxes and higher excitation temperatures (see also Hogerheijde, Jansen, & van Dishoeck 1995). Also, a strong Herbig-Haro shock knot, SK1, located via bright $2\,\mu\mathrm{m}$ H_2 emission (Yu et al. 1997), was included to test for shock diagnostics. The shocked region is also visible in the spectral index map of the ISF (Johnstone & Bally 1999) as a region with anomalously shallow slope, requiring either a *lower* temperature than the surrounding material, a shallower dust emissivity profile, or contamination of the $850\,\mu\mathrm{m}$ continuum measurement by molecular lines.

3. Molecular Line Survey

3.1. Representative Line Profiles

Many important differences between the sources are readily observed by simply looking at the spectra. The extremely strong, typically self-absorbed CO J=3–2 transition reveals information on the gas kinematics within each source (Fig. 2). Single-component Gaussian fits to the profiles are only possible for MMS6 and PDR1, with line widths of 4 and 3 km s^{-1}, respectively. MMS9 requires both a narrow component of 6 km s^{-1}, providing two-thirds of the integrated intensity, and a broad component of 15 km s^{-1}. FIR4 appears to be a scaled-up version of MMS9, with a narrow component of 9 km s^{-1} containing three-quarters of the integrated intensity and a broad component of 21 km s^{-1}. Both the shock region and NGC 2071 are poorly fitted by Gaussian components and are dominated by the extended wings.

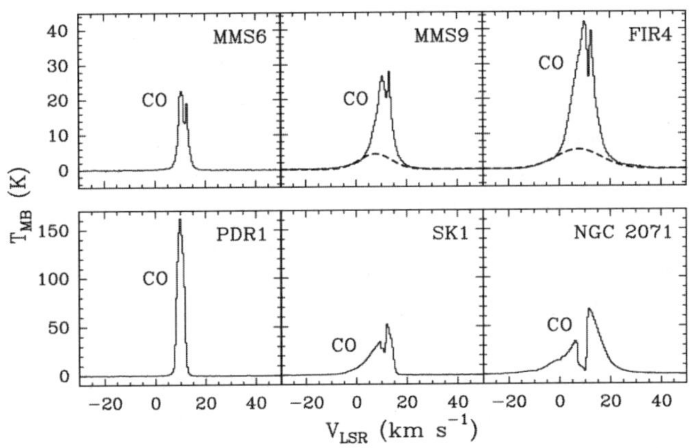

Figure 2. The ^{12}CO J=3–2 line profiles. Note the self-absorption in all sources except PDR1, and the extended line wings (e.g., dashed lines) in all sources except MMS6 and PDR1 (Johnstone et al. 2003, in prep.).

Eight of the methanol lines were observed in a single spectral setting and can be plotted for each source (see Fig. 3 for examples). Also observed in the same configuration was the sole SO transition. There is an obvious anti-correlation between the strength of the SO line and that of the methanol lines. As well, differences in the relative strengths of the methanol lines are observable between MMS6 and FIR4, indicating the more extreme temperature and density conditions in the energetic infrared source. Finally, the FIR4 methanol spectra show

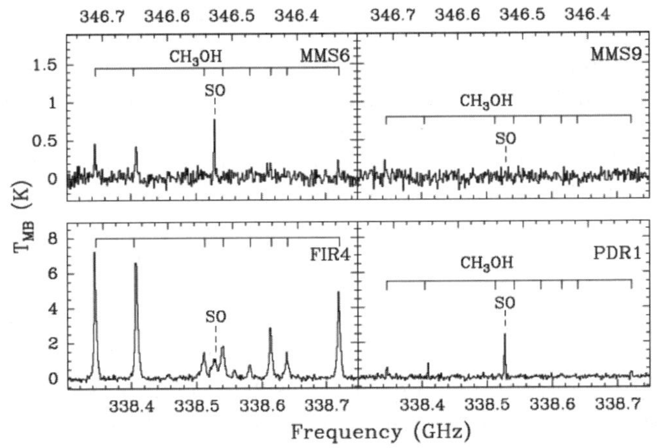

Figure 3. A selection of methanol lines in the $7_k - 6_k$ band observable with a single spectral setting, along with the SO $N_J = 8_9 \rightarrow 7_8$ line (Johnstone et al. 2003, in prep.).

evidence for two components, narrow and broad, similar to the CO spectra.

The HCN and HNC J=4–3 profiles were also observed in a single spectral setting (Fig. 4). A range of relative line strengths is seen in the spectra; however, optical depth considerations make simple abundance determinations more complicated. The substantial optical depth of the transitions in some sources allows for more detailed determination of source kinematics. With the exception of PDR1, the HCN line profile is broader than the HNC line profile or has evidence for more extended line wings.

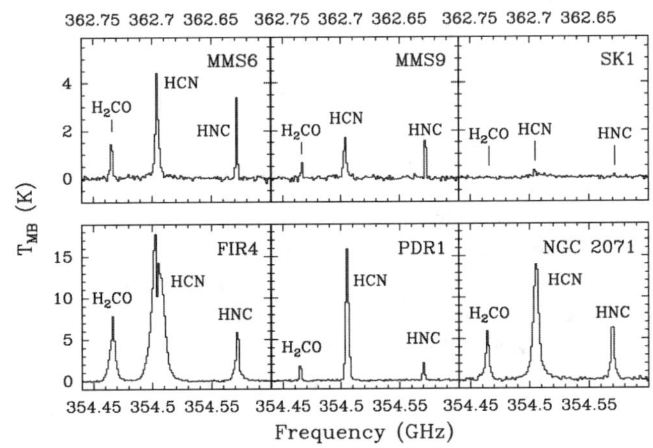

Figure 4. The HCN and HNC J=4–3 transitions, together with the H_2CO 5_{15}–4_{14} line (Johnstone et al. 2003, in prep.).

3.2. Statistical Equilibrium Calculations

The analysis of the data is carried out with statistical equilibrium calculations, using the numerical code RADEX[1]. The method is described in detail by Jansen et al. (1994), van Dishoeck et al. (1993), and Jansen et al. (1995). In this method the source is assumed to be a homogeneous spherical region. The decoupling of radiative transfer and level populations is done with the escape probability method. The optical depth τ at the line center is proportional to the ratio of the total column density and the line width of the molecule. The optical depth, and the line intensity as well, depends only on this ratio. All quantities can be calculated, using a geometrical assumption that they are constant throughout the region. The radiative excitation of rotational lines in the submillimeter region is usually dominated by the 2.73 K cosmic background radiation, which peaks at 1.871 mm. Therefore, this external radiation field is included in the calculations. No central radiation field is included, nor is infrared pumping by dust. Since the sources have rather low dust temperatures, the latter effect is expected to be negligible.

The kinetic temperature $T_{\rm kin}$ and density $n(H_2)$ for the homogeneous region can be estimated from line ratio calculations where available. The column density of a molecule X, $N(X)$, can then be estimated using the temperature and density as found from line ratio calculations, such that the line width, integrated line strength and main beam temperature of the emission lines can be fitted. The abundance of each molecule can be determined through comparison with the total gas column density (Table 2), derived from the 850μm flux measurements assuming an emissivity of $\kappa_{850} = 0.02\,{\rm cm^2\,gm^{-1}}$.

4. Discussion

4.1. Abundances

Despite the inherent danger in assuming a single temperature and density for the environments of each source, general trends were observed. It is worth noting that detailed modeling (van der Tak et al. 2000; Schöier et al. 2002a; Doty et al. 2002; Jørgensen et al. 2002) provides a more accurate determination of relative abundances, especially for lines which are excited only in part of the envelope. Also, abundance gradients or "jumps" can be determined if a sufficient number of lines of a given molecule covering a wide range of conditions has been observed. This detailed modeling requires a determination of the density and temperature profiles of the sources from the dust continuum emission, which has not yet been performed in this study. Schöier et al. (2002a) show that the inferred abundances using a constant temperature and density do not differ by more than a factor of a few from the detailed analysis as long as the adopted conditions are appropriate for the particular molecule or line.

Some strong trends were observed across the source list. The peak brightness and integrated line strength of both CO and ^{13}CO J=3–2 lines followed the energetics and warm dust temperatures of the envelopes. Despite the observed depletion of CO in the protostellar source envelopes (typically the abundance is a factor of 10 lower than the mean molecular cloud value), in all cases the objects are visible in all of the observed CO lines.

Our best tracer of the gas temperature is provided by the ratio of the H_2CO $(3_{03}–2_{02})/(3_{22}–2_{21})$ lines at 218 GHz. These relatively low-lying lines primarily trace the extended envelope conditions, where most of the mass resides (Schöier, private communication). The formaldehyde-derived temperature provides evidence for warm gas within each source ($T_k \sim 50-100$ K), higher than the submillimeter-derived dust temperatures. For sources with known internal heating (FIR4, NGC 2071), the formaldehyde temperature appears intermediate between the hot interior and cold exterior. Despite only a poor correlation between the dust and the formaldehyde temperatures, for the remainder of the discussion we adopt the formaldehyde conditions as representative of the bulk conditions within each source.

The abundances of both formaldehyde and methanol correlate well with the formaldehyde-derived temperature (Fig. 5). The changing abundance may reflect jump conditions in formaldehyde and methanol abundance with dust temperature in the inner warm part (e.g., van der Tak, van Dishoeck & Caselli 2000). However, ignoring the broad (outflow) methanol component in NGC 2071, the abundance of methanol may be plotted as a step function with cold sources ($T < 100\,{\rm K}$) having a low abundance and warm sources ($T > 100\,{\rm K}$) having a five times higher abundance. Although less observational data from the study are available, the CS and SO abundances also correlate with the temperature of the envelope. As well, the ratio of SO to CS increases with envelope temperature.

Unlike the other species, the abundance of HCN appears to be relatively insensitive to source energetics, with the only two exceptions being the enhanced abundance by a factor of three to four in the PDR region, and a

[1] This program was originally written by J.H. Black. Here a more recent version by D.J. Jansen is used.

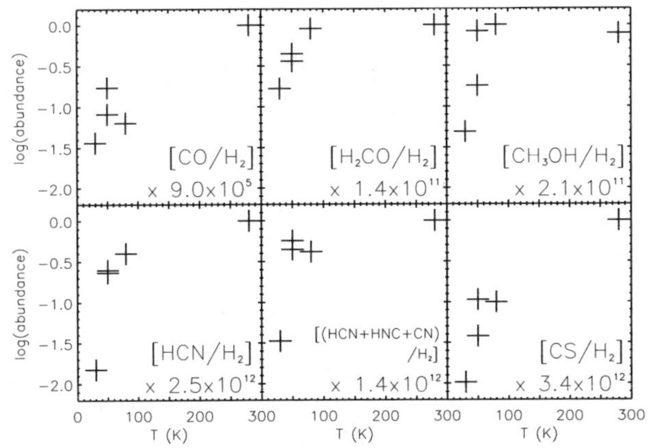

Figure 5. Abundances of the various molecular species versus the formaldehyde-derived source temperature. Note that in all cases the abundance increases significantly with temperature (Johnstone et al. 2003, in prep.).

significantly decreased abundance in the least energetic source MMS6. Considering the three CN-bearing species (HCN, HNC, and CN), the total abundance of these three molecules is remarkably constant across a wide range of conditions, varying by less than thirty percent between MMS9, FIR4, and NGC 2071 and by only a a factor of two when the PDR region is included. Only the cold, dense submillimeter source MMS6 appears to be severely depleted in CN-bearing molecules.

4.2. Protostellar Source Diagnostics

A large number of submillimeter continuum surveys have now been completed in star-forming molecular clouds, and hundreds of new submillimeter bright envelopes have been enumerated. However, an outstanding question is which fraction of these objects surround protostars and which are prestellar. Within the present study there are no clear prestellar objects although MMS6 appears somewhat ambiguously defined as a Class 0 source. Despite this lack of an obvious prestellar candidate, it is worth considering if any of the observed molecular signatures *requires* an embedded heating source.

For many of the continuum sources found within the mapping survey of Mitchell et al. (2001), additional ^{13}CO observations provided no indication of a coincident CO peak. This was taken as evidence for the youth, and possible prestellar condition, of the compact peaks. For the sample presented here, CO is clearly detected toward each source, although it is still depleted. The presence of CO may provide a clue to the protostellar nature of the sources, perhaps through a warming of the envelope and evaporation of CO back into the gas phase.

In addition, the appearance of molecular lines with relatively high excitation temperatures, and preferably also with high critical densities such that the warm region might be inferred to be deeper within the envelope, should provide a signature of Class 0 and later sources. Thus, the methanol lines are likely the strongest indicator of a warm, dense region within the envelope. For all sources observed in this study, except the known outflow source MMS9, the required internal temperature deduced from the methanol lines is > 50 K, a temperature unattainable without an energetic internal source. The CS J=7–6 line also provides evidence of a warm, dense interior region in all sources except MMS9. It is interesting to note that all the protostellar sources considered in this paper require high densities in their interior ($n > 10^7\,\mathrm{cm}^{-3}$). Standard models for star formation require a power-law density distribution toward the clump center during the collapse phase (Shu 1977; Henriksen, André, & Bontemps 1997); thus, the high densities measured here may be a result of the stage of evolution of the individual clumps.

Formaldehyde observations also provide important evidence for a large, warm envelope. While the low-lying lines observed in this study are predominantly excited in the massive outer envelope, the derived temperature places severe constraints on any possible protostar buried inside. In particular, envelope temperatures above 20 K are difficult to understand without a warm internal source. All of the sources observed in this study have two components associated with the formaldehyde measurements. However, this is not always the case for sources of strong submillimeter continuum (Mitchell et al. 2001; Tothill, private communication). It is possible that the broad component of the formaldehyde, for which the line ratios generally imply a warmer zone, is tracing an inner region within the envelope which is undergoing collapse or has become much more turbulent.

4.3. Line Contamination

One of the original motivations for this molecular line study was to determine the importance of line contamination within the broad SCUBA 850 μm passband. While theoretical calculations show that the influence of CO contamination can become exceedingly large in particular situations, it is also clear that typical conditions within molecular clouds are not so extreme. However, observational evidence presented in Table 2 does show that the

Table 2. Measured total line intensity within the 850 μm band and its contribution to the continuum flux

Name	CO Lines ($K\,km\,s^{-1}$)	Total Lines ($K\,km\,s^{-1}$)	Cont. %	$N(H_2)$ ($10^{22}\,cm^{-2}$)
MMS6	173	200	1.8	104
MMS9	287	301	8.2	14
FIR4	547	934	8.0	47
N2071	777	945	8.3	62
PDR1	765	853	11.1	10
SK1	483	487	63	6

contamination, while typically less than 10%, occasionally *dominates* the continuum flux. However, this only occurs in regions with warm molecular gas *and* large velocity gradients which allow for enhanced integrated line strengths in CO and its isotopes. Such conditions are most likely to exist around isolated shock fronts within the cloud, especially knots associated with jets at large distances from the embedded source. Contamination may also arise from lines of other molecules, especially around energetic sources. The best observed example for this is Orion KL, for which a forest of lines, primarily from SO and SO_2, produce between 28% and 50% line contamination at 850 μm (Serabyn & Weisstein 1995; Groesbeck 1994). Consideration of Table 2 shows that in our sample the line contamination is *never* dominated by lines other than CO. However, for the most energetic sources, line emission from HCN, HNC, CN, and methanol provides a significant fraction ($> 40\,\%$) of the total line contamination.

Acknowledgements

We wish to thank Floris van der Tak, James Di Francesco, and John Bally for helpful comments during the creation of this manuscript. DJ wishes to thank the Sterrewacht Leiden for its kind hospitality during the past three summers, and financial support from both an NSERC grant and NOVA. This work was supported by the Netherlands Organization for Scientific Research (NWO) through grant 614-041-003 and a Spinoza grant.

References

Bally, J., & Lada, C. J. 1983, ApJ, 265, 824

Blake, G.A., van Dishoeck, E.F., Jansen, D.J., Groesbeck, T.D., & Mundy, L.G. 1994, ApJ, 428, 680

Blake, G.A., Sandell, G., van Dishoeck, E.F., Groesbeck, T.D., Mundy, L.G., & Aspin, C. 1995, ApJ, 441, 689

Buckle, J.V., & Fuller, G.A. 2002, A&A, 381, 77

Butner, H.M., Evans, N.J.II, Harvey, P.M., Mundy, L.G., Natta, A., & Randich, M.S. 1990, ApJ, 364, 164

Chini, R., Reipurth, B., Ward-Thompson, D., Bally, J., Nyman, L.-Å., Sievers, A., & Billawala, Y. 1997, ApJ, 474, L135

Doty, S.D., van Dishoeck, E.F., van der Tak, F.F.S., & Boonman, A.M.S. 2002, A&A, 389, 908

Groesbeck, T.D. 1994, Ph.D. thesis, California Institute of Technology

Harvey, P. M., Campbell, M. F., Hoffmann, W. F., Thronson, H. A., Jr., & Gatley, I. 1979, ApJ, 229, 990

Hatchell, J., Thompson, M.A., Millar, T.J., MacDonald, & G.H. 1998, A&AS, 133, 29

Henriksen, R., Andre, P., & Bontemps, S. 1997, A&A, 323, 549

Hogerheijde, M.R., Jansen, D.J., & van Dishoeck, E.F. 1995, A&A, 294, 792

Jansen, D.J., van Dishoeck, E.F., & Black, J.H. 1994, A&A, 282, 605

Jansen, D.J., van Dishoeck, E.F., Black, J.H., Spaans, M., & Sosin, C. 1995, A&A, 302, 223

Johnson, J. J., Gehrz, R. D., Jones, T. J., Hackwell, J. A., & Grasdalen, G. L. 1990, AJ, 100, 518

Johnstone, D., & Bally, J. 1999, ApJ, 510, L49

Johnstone, D., Fich, M., Mitchell, G. F., & Moriarty-Schieven, G. 2001, ApJ, 559, 307

Jørgensen, J.K., Schöier, F.L., & van Dishoeck, E.F. 2002, A&A, 389, 908

Lis, D. C., Serabyn, E., Keene, J., Dowell, C. D., Benford, D. J., Phillips, T. G., Hunter, T. R., & Wang, N. 1998, ApJ, 509, L299

Mezger, P. G., Zylka, R., & Wink, J. E. 1990, A&A, 228, 95

Mitchell, G., Johnstone, D., Moriarty-Schieven, G., Fich, M., & Tothill, N. F. H. 2001, ApJ, 513, 139

Motte, F., André, P., Ward-Thompson, D., & Bontemps, S. 2001, A&A, 372, L41

Reipurth, B., & Aspin, C. 1997, AJ, 114, 2700

Reipurth, B., Bally, J., & Devine, D. 1997, AJ, 114, 2708

Reipurth, B., Rodriguez, L. F., & Chini, R. 1999, AJ, 118, 983

Schöier, F.L., Jørgensen, J.K., van Dishoeck, E.F., & Blake, G.A. 2002, A&A, 390, 1001

Serabyn, E., & Weisstein, E.W. 1995, ApJ, 451, 238

Shu, F. 1977, ApJ, 214, 488

van der Tak, F.F.S., van Dishoeck, E.F., & Caselli, P. 2000, A&A, 361, 327

van der Tak, F.F.S., van Dishoeck, E.F. Evans, N.J.II, & Blake, G.A. 2000, ApJ, 537, 283

van Dishoeck, E.F., Jansen, D.J., & Phillips, T.G. 1993, A&A, 279, 541

van Dishoeck, E.F., Blake, G.A., Jansen, D.J., & Groesbeck, T.D. 1995, ApJ, 447, 760

Wolstencroft, R. D., Scarrott, S. M., Warren-Smith, R. F., Walker, H. J., Reipurth, B., & Savage, A. 1986, MNRAS, 218, 1

Yu, K. C., Bally, J., & Devine, D. 1997, ApJ, 485, L45

Yu, K. C., Billawala, Y., Smith, M. D., Bally, J., & Butner, H. M. 2000, AJ, 120, 1974

NEW INSIGHTS INTO DUST GRAIN PHYSICS: THE POLARIZATION SPECTRUM OF THE OMC-1 CORE

Brenda Matthews[1], David Chuss[2], Jessie Dotson[3], C. Darren Dowell[4], Doug Johnstone[5], Roger Hildebrand[6] and John E. Vaillancourt[7]

[1]Department of Astronomy, University of California, Berkeley, CA 94720, U.S.A.
[2]NASA Goddard Space Flight Center, Greenbelt, MD, 20771, U.S.A.
[3]NASA Ames Research Center, CA, U.S.A.
[4]Department of Physics, California Institute of Technology, Pasadena, CA 91125, U.S.A.
[5]Herzberg Institute of Astrophysics, National Research Council, Victoria, BC, V9E 2E7, Canada
[6]Enrico Fermi Institute, University of Chicago, Chicago, IL 60637, U.S.A.
[7]Department of Physics, University of Wisconsin, Madison, WI, 53706, U.S.A.

Abstract

Although our understanding of the physics and chemistry of dust grains within molecular clouds impacts many areas of star formation (e.g., disk and envelope structure, magnetic field geometry, and the formation of planetary systems), the composition, shape and degree of alignment of grains is still poorly constrained. A primary tool in dust grain analysis is the polarization spectrum, or the variation of polarization percentage with wavelength. If two or more grain populations exist along a line of sight, the spectral index and polarization percentage of each can be separately determined from multi-wavelength flux and polarization data (Hildebrand et al. 1999; Greaves et al. 1999). However, investigations of the polarization spectrum have been limited by the paucity of data, particularly longward of 350 μm. This paper is a review of the importance of the polarization spectrum and what these data can reveal about the nature of grains in the envelope of the OMC-1 massive, star-forming core and molecular clouds in general. We also present a three wavelength spectrum for the NGC 2024 region of Orion B.

Key words: techniques: polarimetric – ISM: dust – ISM: structure – ISM: clouds

1. Introduction

The principal tool in the study of magnetic field geometries within molecular clouds is polarimetry of emission from rapidly spinning dust grains which are aligned by magnetic fields, provided their rate of rotation is rapid enough. Longward of 60 μm, dust emission is optically thin and absorption and scattering off the grains is negligible (Hildebrand 1988). Hence, interpretation of polarization data is relatively straightforward and the position angles of the polarization vectors trace magnetic field geometry (e.g., Schleuning 1998; Matthews, Wilson & Fiege 2001). However, no information is available about field strengths since the fractional polarization (the ratio of polarized to total emission from grains) is a function not only of magnetic field strength, but also of the grain composition and shape and the degree of alignment, all of which are unknowns that could differ through the cloud.

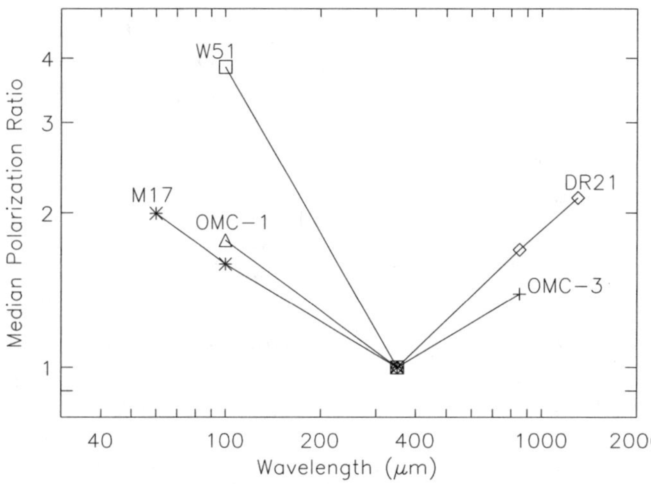

Figure 1. A composite polarization spectrum using all objects for which multiple wavelength data exist as of 2001 (Vaillancourt 2002).

The distribution of polarization percentage with wavelength is called the polarization spectrum. In Figure 1, we plot the median of the normalized fractional polarization, $P(\lambda)/P(350\ \mu\mathrm{m})$. The spectrum of each of the sources plotted exhibits similar variations.

Chemistry as a Diagnostic of Star Formation (2003). *Edited by Charles L. Curry and Michel Fich.* Published in 2003 by NRC Research Press, Ottawa ON, K1A 0R6, Canada. ISBN 0-660-19089-3.

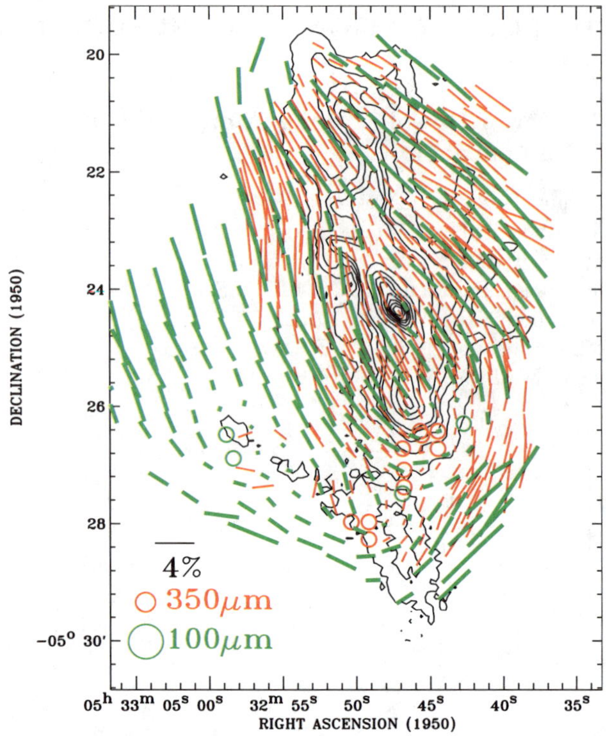

Figure 2. Comparison of the polarization data at 100 μm and 350 μm toward the OMC-1 core in Orion A. We are attempting to match this spatial coverage at 850 μm with SCUBA (Submillimeter Common User Bolometer Array) observations on the James Clerk Maxwell Telescope (JCMT) using scan mapping techniques. Data at 100 μm are from Schleuning (1998); data at 350 μm are from Hildebrand et al. (2000).

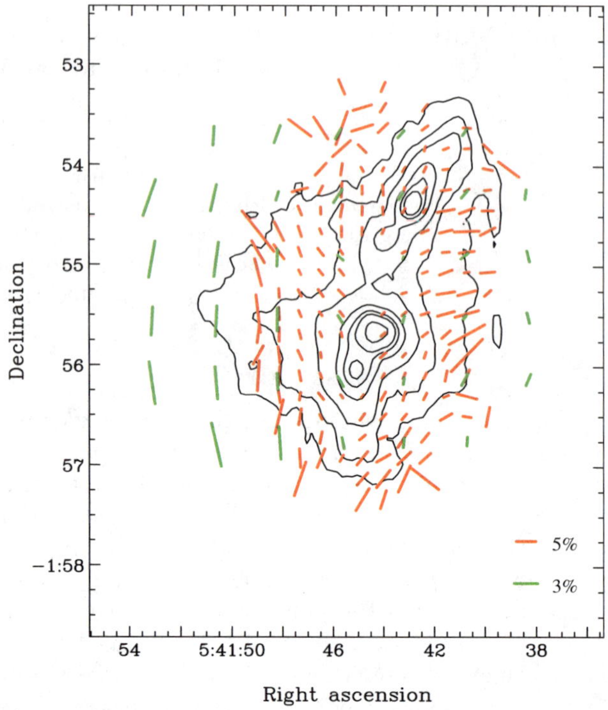

Figure 3. Polarized emission from the NGC 2024 ridge in Orion B has been mapped at 100 μm (green) with the Kuiper Airborne Observatory (Dotson et al. 2000) and at 850 μm (red) with SCUBA (Matthews, Fiege, & Moriarty-Schieven 2002). Strong agreement is evident between the position angles of the vectors at these two wavelengths, but the fractional polarization is clearly larger at 850 μm. It is critical to add the 350 μm data of this region, since the lowest polarization percentage should be measured between 100 and 850 μm, if the spectrum of Figure 1 is representative of envelopes of individual objects.

2. Increasing the Sampling in the Spectrum

To correctly characterize the polarization spectrum for a given molecular cloud, polarization data must exist at more than two wavelengths. However, as Figure 1 shows, it has proven challenging to measure polarized emission at a sufficient number of wavelengths to adequately sample the spectrum for individual sources. In order to meaningfully compare polarized emission data, Hildebrand et al. (1999) have defined a set of criteria that the data must meet. The vectors must of course be measured in the same regions of the clouds (spatial coincidence); the position angles at those positions must agree to within 10° (if differences are greater, then the vectors may sample different magnetic field geometries and hence different parts of the cloud); points within two beam diameters of bright peaks must be excluded (since optical depths and depolarization effects could alter results in these regions); and more than 5 pairs must remain for comparison.

These criteria are difficult to satisfy, but comparison of data at 350 and 100 μm toward the OMC-1 core in Orion A (see Fig. 2) and at 100 and 850 μm toward NGC 2024 in Orion B (see Fig. 3) shows that very similar polarization patterns can be produced at different wavelengths. The alignment of polarization position angles indicates that similar magnetic field geometries are being sampled at both wavelengths which removes the magnetic field strength as a variable from the problem, meaning that the changes in fractional polarization are due to grain physics alone.

2.1. OMC-1

OMC-1 is the best studied molecular cloud core. Due to its high brightness (its peak is 160 Jy at 850 μm) and large size, it also has the most extensive polarized emission data at large radii from its peak, as shown in Figure 2. We have begun using SCUBA's 850 μm capability at the JCMT to extend the spectrum of OMC-1 longward of 350 μm. SCUBA's limited chop throw (with a maximum of 3′) requires that we use scan mapping methods, which are challenging to reduce. An additional challenge was introduced when we learned (after this meeting) that the data presented there were taken without physical rotation of the waveplate due to a software problem. Since SCUBA is mounted at the Nasmyth focus, however, polarization can still be measured since the waveplate angle on the sky changes with sky rotation. We are currently re-reducing the data to compensate for this problem, and time has been allocated to allow us to expand the existing data set. Figure 2 shows a composite image of the 100 and 350 μm coverage of polarization data in OMC-1. This map shows how extensive the spatial coverage of the previously published polarimetry data is for OMC-1. Comparison of polarization data for several sources over four wavelengths in Figure 4 shows that the peak of the polarization percentage distribution for all vectors above 3σ decreases with increasing wavelength, at least to 350 μm.

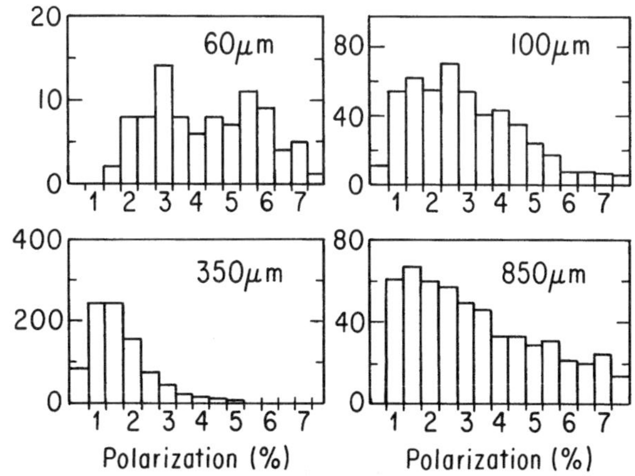

Figure 4. Histograms of all 3σ vectors at four wavelengths for an ensemble of sources show that the peak polarization percentage decreases with increasing wavelength to 350 μm. Data at 850 μm peak at similar polarization percentages as the 350 μm distribution, but exhibit many more vectors (as a fraction of the total) which have higher polarization percentages than the peak (Hildebrand 2001).

2.2. Other Sources

Improving our understanding of mixed grain populations within molecular clouds requires better sampling of both flux and polarization spectra and observations of more objects. These goals can be achieved with current and future detectors. Instruments on SOFIA (i.e., FORCAST, HAWC and the proposed Hale polarimeter [Dowell et al. 2002]) will cover the IR flux peak and new and existing ground-based instruments (e.g., SCUBA, SHARC-2, MAMBO and the Hertz polarimeter) will extend the spectrum out to millimeter wavelengths.

Comparison of data at 100 μm and 850 μm toward the NGC 2024 region of Orion B (see Fig. 3) shows strong correlation in polarization position angle. Comparison of these data with 350 μm data from the CSO's Hertz polarimeter produces the spectrum shown in Figure 5, which mirrors the behaviour of the assembled spectrum of Figure 1. We note that the five points of comparison between 100 and 350 μm, and 350 and 850 μm are not the same. In fact, only two points of comparison exist across all three wavelengths. The planned OMC-1 data set will provide many more points of overlap at three wavelengths.

Data at 850 μm also exist toward the NGC 2071 IR core in Orion B (Matthews, Fiege & Moriarty-Schieven 2002). This source is now being observed in polarized emission at 1.3 mm with the BIMA interferometer. However, due to the spatial filtering of the interferometer, comparisons between single dish and interferometric data must be made with additional care. Efforts are underway to calibrate the instrumental polarizations for the SCUBA short wavelength array at 450 μm, which will provide a short wavelength measurement for this source.

3. Why Does Polarization Vary?

Hildebrand et al. (1999) show that one grain species cannot produce the spectrum of Figure 1 for any source since idealized single grain population clouds uniformly produce flat spectra.

Multiple temperatures and/or spectral indices are required to reproduce measurements of flux spectra. The polarization spectra provide further evidence for multiple populations in that they require the different populations to also have different polarization properties. Given the current sparsely sampled state of both flux and polarization spectra, taking these two pieces of information

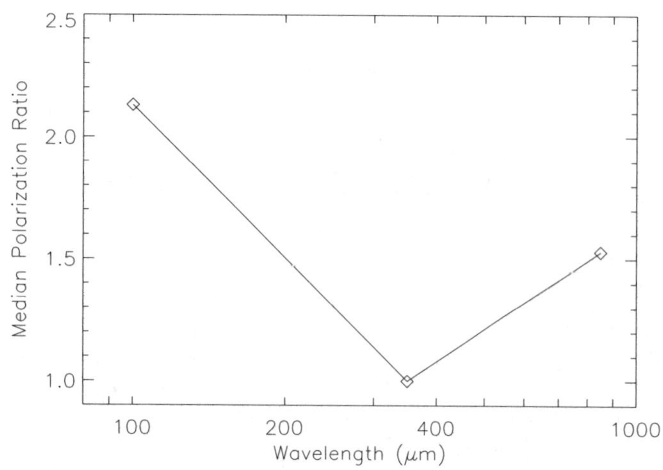

Figure 5. Polarization spectrum for NGC 2024 in Orion B at three wavelengths.

together can go a long way to simultaneously constraining all three properties (i.e., temperature, spectral index and polarization).

Polarization percentage can be reduced by the presence of poorly aligned grains which could reflect either the absence of aligning magnetic fields or the presence of a mechanism capable of destroying alignment, such as the growth of spherical grains. Populations could also differ in composition (graphite vs. silicates), geometry (compact vs. fluffy) and environment (warm vs. cold). It is important to remember that the radiation detected from the ensemble of grains will be mixed, even if the grain populations are not since the dust emission is optically thin. The combination of polarized emission from different grain species will follow the relation

$$P_{ABC...} = X_A P_A + X_B P_B + X_C P_C + \ldots \tag{1}$$

where X_i is the fraction of observed flux due to species i and P_i is the degree of polarization due to i alone (Hildebrand et al. 1999). As shown by Hildebrand et al. (1999), rising or falling spectra require two or more grain populations with differing properties. Equation (1) is linear, so the combination of two perfectly aligned species (each with a flat spectrum) could only produce polarization percentages equally invariant with wavelength.

Hildebrand et al. (1999) show that variations in spectra can only be produced if different populations of grains (in composition, temperature or alignment) exist within a molecular cloud. For example, a varying spectrum requires the P_i values to differ between the two populations and the X_i values of the two species to differ with wavelength. The latter condition can be met either by having different temperatures or different spectral indices.

OMC-1 has recently been analyzed for temperature variations using existing polarization data and a spectral energy distribution compiled from published flux data (Vaillancourt 2002). Information about temperature differences will help to interpret the polarization spectrum information.

4. Summary

The polarization spectrum reveals information about the dust grain populations within giant molecular and dark clouds. Knowledge of the temperature distribution is as important as the polarization data to determine the grain populations involved (Vaillancourt 2002). What is needed is multi-wavelength polarimetry of many molecular cloud envelopes (more sources and better wavelength coverage). The rise in the NGC 2024 spectrum beyond 350 μm could indicate a third grain population with low temperature and high polarization percentage. Once we have obtained our complete 850 μm data set, we will know whether or not a rising spectrum is detected longward of 350 μm in OMC-1.

References

Dotson, J., et al., 2000, ApJS, 128, 335

Dowell, C.D., et al., 2002, in *Polarimetry in Astronomy*, ed. S. Fineschi, Proc. SPIE, 4843

Greaves, J. et al., 1999, A&A, 344, 668

Hildebrand, R., 1988, QJRAS, 29, 327

Hildebrand, R. et al., 1999, ApJ, 516, 834

Hildebrand, R. et al., 2000, PASP, 112, 1215

Hildebrand, R., 2001, in *Astrophysical Spectropolarimetry*, ed. J. Trujillo (Cambridge: Cambridge University Press), 119

Matthews, B., Wilson, C., & Fiege, J. 2001, ApJ, 562, 400

Matthews, B., Fiege, J., & Moriarty-Schieven, G. 2002, ApJ, 569, 304

Schleuning, D. 1998, ApJ, 493, 811

Vaillancourt, J.E. 2002, ApJS, 142, 53

PRONAOS SUBMILLIMETER CONTINUUM OBSERVATIONS OF STAR-FORMING REGIONS: NEW INSIGHTS ON INTERSTELLAR DUST

I. Ristorcelli[1], B. Stepnik[1,2], X. Dupac[1], A. Abergel[2], J.P. Bernard[1], N. Boudet[1], M. Giard[1], A. Jones[2], J.M. Lamarre[3], C. Mény[1], F. Pajot[2], J.P. Torre[4], and G. Serra[1]

[1]CESR, 9 av. du Colonel Roche, BP 4346- 31028, Toulouse, Cedex 4, France
[2]IAS, UMR-CNRS, Université Paris Sud, 91405 Orsay, Cedex, France
[3]LERMA, 61 av. de l'Observatoire, 75014 Paris, France
[4]SA, UPR-CNRS, Verrieres-le-Buisson, France

Abstract

High sensitivity observations have been performed with the French balloon-borne experiment PRONAOS from 180 to 1100 μm in different sites of the interstellar medium (three flights in 1994, 1996, and 1999). These measurements have constrained both the dust temperature and the emissivity spectral index. The star-forming regions of ρ Ophiucus and Orion M42, M17 have been mapped over 1 square degree, and extended low-level emission was measured down to 2 MJy sr^{-1} (3σ) at 600 μm (i.e., $A_V \simeq 1$). Cold condensations have been discovered in the outskirts of the star-forming area, with the typical characteristics T=12 K, diam = 0.1–1 pc, M=5–30 $M_\odot$. Being invisible to *IRAS* at 100 μm , some of these clouds are likely to be the seeds for future star formation activity in the complex.

A number of questions have arisen about the nature and evolution of grain properties from diffuse to dense and cold media. Large variations of temperature (10 to 100 K) and emissivity spectral index (1 to 2.5) have been found in each complex, together with a significant anti-correlation between the two parameters. We suggest that this effect may be due to T−dependent quantum processes within the grains, arising at low temperature (low-energy structural transitions; Boudet et al., this volume). In addition, the modelling of dust emission observed with PRONAOS toward two thin and surprisingly cold clouds in the Taurus and Polaris regions shows an increase of the intrinsic emissivity of grains by a factor of 3.4 compared to standard dust models, which appears when $A_V \geq 2$. This result could be interpreted as the formation of fluffy dust aggregates.

Key words: dust, extinction — ISM: clouds — stars: formation

1. Introduction

The key role played by dust in the main galactic processes of star formation, energetic equilibrium, and the life cycle of interstellar matter makes a knowledge of its physical and chemical properties essential. An understanding of the very early phases of star formation requires better constraints on the physical and chemical properties of prestellar cold and dense cores. In particular, the opacity, which is dominated by dust grains, must play a dominant role in the phase preceding the collapse, and may influence the fragment sizes. It is therefore expected that global quantities such as the initial mass function and star formation efficiency should be strongly linked to the dust absorption and emission properties.

In the current dust models, three main components are considered, with a continuous size distribution, differing in composition, structure and emission mechanism (Desert et al. 1990). The smallest interstellar dust particles (large aromatic molecules [PAH] and very small grains [VSG], with sizes $\geq$ 15 nm) are transiently heated and emit essentially in the infrared, extensively observed with *ISO*. The largest grains ($\simeq$ 0.1 μm in size) are in thermal equilibrium and dominate the emission spectrum in the far-infrared and submillimeter regions. In this wavelength range, the emission can be well approximated by $I_\nu = \kappa\lambda^{-\beta}B_\nu(T)$, where T is the dust temperature, β is the grain emissivity spectral index, and κ is proportional to the column density. To constrain these parameters, it is necessary to perform multi-band observations covering the submillimeter from 200 μm to 1 mm, especially the critical range 200−300 μm , since it includes the emission peak for dust at $T \leq 30$ K. For this reason, and also because of the necessity of reaching high sensitivity and measuring low-gradient emission, space observations are needed. These measurements are particularly suited to characterize the interstellar cold component, to trace the morphology and structure of the clouds, and to study the early steps toward star formation, when the cloud opacity is a determining factor.

Chemistry as a Diagnostic of Star Formation (2003). *Edited by Charles L. Curry and Michel Fich.* Published in 2003 by NRC Research Press, Ottawa ON, K1A 0R6, Canada. ISBN 0-660-19089-3.

The objective of PRONAOS was to begin exploration of the submillimeter Galactic emission, so far poorly constrained by observations, except at the large angular scale of *COBE* (7°). Although the Desert et al. (1990) model gives a good fit to the IR-to-mm Galactic emission averaged over large angular scale, it is expected that significant changes happen at small scale, due to dust evolution. In particular, *IRAS* data show colour variations of the dust emission from place to place within the same molecular complex (Boulanger et al. 1990). *COBE* data extended this analysis, indicating that in the regions where the abundance of small particles is low, the temperature of large grains is $\simeq$15 K, significantly lower than the "standard" cirrus value of 17.5 K (e.g., Boulanger et al. 1996; Lagache et al. 1998). To study the physical properties of dust and its evolution, it was essential to trace and characterize the thermal equilibrium emission of the large grain component. By combining simultaneous measurements in four channels at 200, 260, 360, and 580 μm with *IRAS* 100 μm data, PRONAOS has been the first experiment to constrain both the dust temperature and its emissivity spectral index at an angular scale of a few arcminutes.

2. Observations

A description of the PRONAOS system and its focal instrument SPM is given in Lamarre et al. (1994) (see also Lamarre and Giard 2002; Serra et al. 2001). The main characteristics of the instrument are given in Table 1. Three launches were performed in Sept. 1994, 1996 and 1999, from the NASA balloon facility base at Fort-Sumner (USA). During its three flights, PRONAOS observed the various star-forming regions and molecular clouds of Orion M42, ρ Ophiuchus, M17, Cygnus, L134, L1689, IC 5146, L507, NCS (a very young massive star), diffuse filaments in Taurus, Serpens, and the high-latitude Polaris cirrus.

These observations were performed with a very high sensitivity to low brightness fluxes: about 30 minutes were needed for a 1° × 1° mapping, with a 3σ detection of I_ν=2 MJy sr^{-1} at 600 μm .

Table 1. Main instrument characteristics

Wavelength band	1	2	3	4
$\lambda_{\rm eff}$ (μm)^a	200	260	360	600
Field of view (arcmin)	2	2	2.5	3.5
NET^b (mK Hz$^{-1/2}$)	1.2	2.1	0.94	1.2
NEB^c (MJy sr^{-1} Hz$^{-1/2}$)	22	26	8.5	5.3
Primary mirror diameter	2045 mm, $f/10$			
Wobbling mirror	20 Hz, $\delta\alpha = 5.35'$			
Pointing accuracy	20″ absolute, 5″ relative			

^{a} $\Delta\lambda/\lambda \sim 30\%$.
^{b} Noise Equivalent Temperature.

3. Main results

3.1. star-forming regions

Several maps (typically of one square degree each) have been made in the nearby star-forming complexes M42 (Ristorcelli et al. 1998; Dupac et al. 2001), M17 (Dupac et al. 2002), ρ Ophiuchus (Ristorcelli et al. 2003, in preparation) and Cygnus (Mény et al. 2002). We were able to derive estimates of cloud masses in a way which is fully independent of the usual molecular tracers (e.g., CO). Moreover, the high sensitivity and wide angular coverage of PRONAOS allow one to properly recover the mass of the low density gas in such complexes, which is missing in ground-based observations due mainly to high-pass filtering in the data processing.

Almost all of these maps reveal the presence of cold extended condensations, with the following typical characteristics: T = 12 K, size $\sim$ 0.1–1 pc, M = 5–20 $M_\odot$. Some could correspond to a very early phase of prestellar object. The study of the physical properties of dust within those clumps could help provide a better understanding of the initial steps of the collapse and fragmentation processes toward star formation. Most of these cores are invisible in the 100 μm maps. Their emission at 100 μm is actually much fainter than the warmer surrounding halo contribution.

Figure 1 shows the emission mapped with PRONAOS in the ρ Ophiuchus main cloud. In the neighborhood of the rich star-forming core ρ Oph A, the emission of several condensations is visible, previously discovered with high density molecular tracers (e.g., $C^{18}O$, DCO^+). From PRONAOS measurements we have deduced a cold temperature and steep spectral emissivity in the ρ Oph B2 core: T = 11 $\pm$ 2 K, β = 2.1 $\pm$ 0.4; see Figure 1. The cut across the central part of B2 (Fig. 2) shows the emission profiles measured with PRONAOS, compared those at 100 μm from *IRAS* data, 1300 μm from IRAM (Motte 1998) and 6 μm from ISOCAM (Abergel et al. 1996). The condensation induces a deep absorption feature visible at 6 μm , against the near-infrared background of the illuminated background cloud (due to vibrational emission from large PAH molecules). This comparative cut also shows the complementarity between space and ground measure-

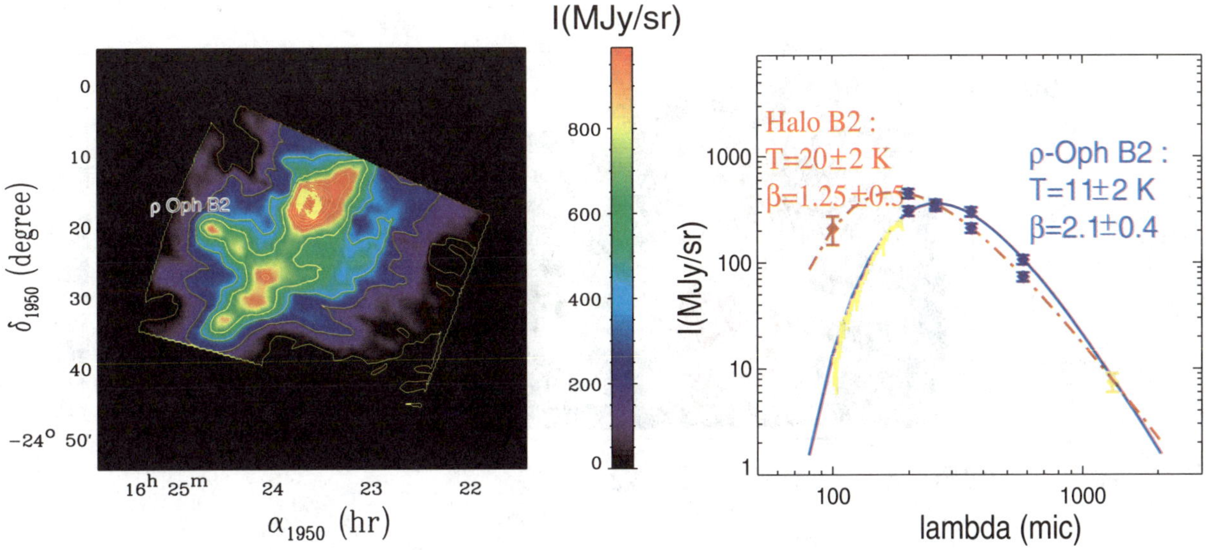

Figure 1. Left: PRONAOS map of the ρ Oph main cloud at $\lambda \simeq 260\,\mu$m (contour step $200\,\mathrm{MJy\,sr^{-1}}$). *Right:* Emission spectra deduced for the cold condensation B2, and its surrounding warmer halo. The fit to PRONAOS measurements is in good agreement with the differential spectrum obtained with *ISO*-LWS (*yellow curve*), and with the 1.3 mm IRAM data (*yellow point*; Motte et al. 1998).

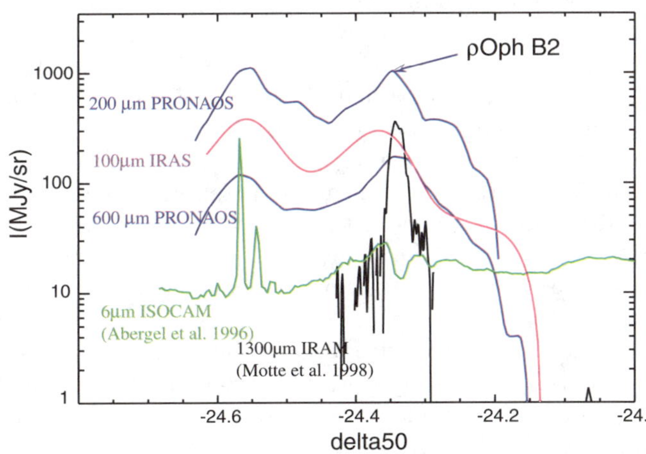

Figure 2. Emission profiles at $\delta = 16^{\mathrm{h}}\,24^{\mathrm{m}}\,25^{\mathrm{s}}$ (cut through B2 max) in two PRONAOS bands, compared to the $100\,\mu$m *IRAS* emission, the ISOCAM $5-8.5\,\mu$m band, and the IRAM 1.3 mm emission.

ments: the high resolution map performed at IRAM allows one to trace structure that could not be resolved with PRONAOS. On the other hand, the extended low gradient emission cannot be measured from the ground due to the subtraction of low spatial frequencies needed to remove the atmospheric noise.

Figure 3 presents an image of the M42 nebula in the $200\,\mu$m band (Dupac et al. 2001). The map enhances the very strong intensity contrast in the dust emission (more than five orders of magnitude in dynamic range) that exists in this kind of giant molecular complex. Three cold, extended condensations have been discovered in the outskirts of the star-forming area, 30′ to 1° northwest of the BN-KL region, that are good candidates for prestellar objects. A large variation exists in dust temperature within the complex, from 10 to 80 K, and also in the spectral index, which varies unexpectedly from 1 to 2.5. Such large variations have also been obtained from the M17 PRONAOS mapping (Dupac et al. 2002). Moreover, according to PRONAOS observations in different sites of the ISM, it seems that a significant anticorrelation exists between the dust temperature and the emissivity spectral index (Fig. 4). The correlation coefficients we obtain are −0.9 in Orion, −0.8 in M17, and −0.7 for the combination of the different observed clouds (Dupac et al. 2001, 2002). The investigation of the origin of this anticorrelation makes us support the idea of an intrinsic variation of the grain optical properties with temperature. We have shown that even taking into account a dust temperature

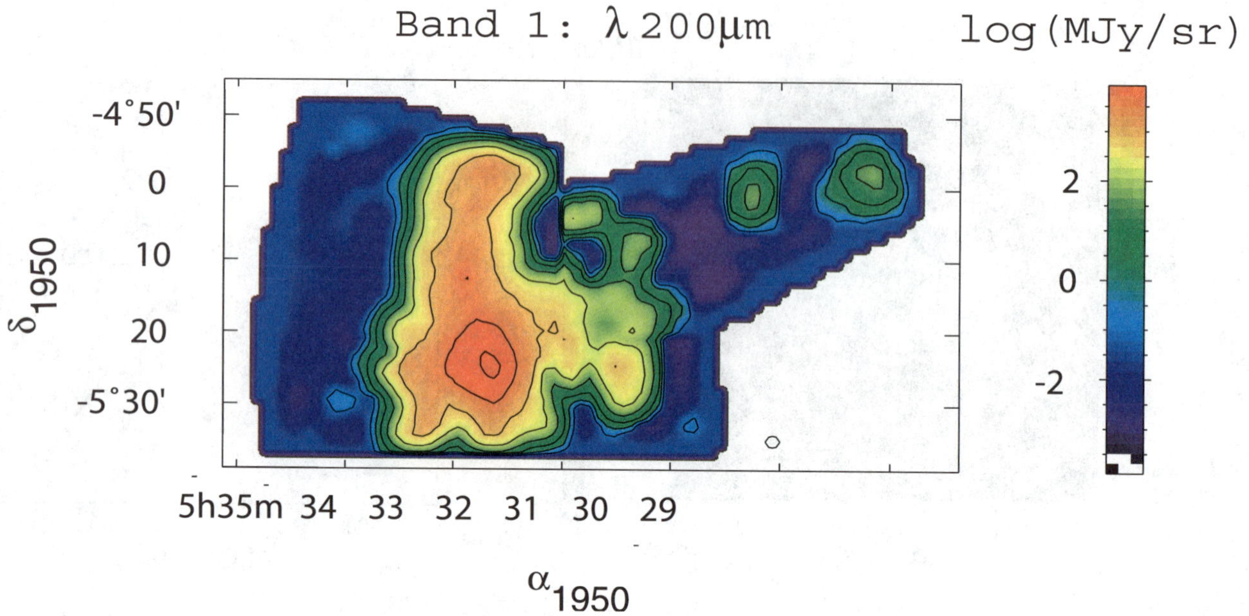

Figure 3. PRONAOS mapping of the M42 nebula in Band 1 at 200 μm (Dupac et al. 2001).

distribution along the line of sight, the anticorrelation cannot be reproduced: such a distribution clearly cannot explain high indices, and to get low values around 1, an unusually large amount of cold dust is needed. Recent laboratory measurements performed to characterize the optical properties of carbonaceous and silicate dust analogues at low temperature (10−20 K) show a similar anticorrelation (Agladze et al. 1996; Mennella 1998).

To progress in the understanding of the dust emission mechanisms at low temperature, it is necessary to study and characterize processes involving low-energy structural transitions in the solid state (Boudet et al., this volume).

3.2. Dust evolution

The low brightness dust emission of two diffuse clouds has been measured with PRONAOS in the Polaris flare and a filament in the Taurus cloud—both detected in molecular emission—with low measured extinction ($A_V = 0.8$ and 3.9 at maximum, respectively). Each cloud has a low 60-to-100 μm emission ratio. Surprisingly low temperatures were found in the inner parts: $T = 12.3 \pm 0.4$ K, $\beta = 1.9 \pm 0.2$ in the Taurus filament (Stepnik et al. 2003a), and $T = 13 \pm 0.7$ K, $\beta = 2.2 \pm 0.3$ in the Polaris cirrus (Bernard et al. 1999). Via radiative transfer modelling, we have shown that such temperatures cannot be explained with the standard grain optical properties used in Desert et al. (1990). This model only reproduces the emission from the surrounding warmer halo (T=15−17 K). The discrepancy in the inner part of the cloud coincides with very low values of the 60-to-100 μm emission ratio. This can be interpreted as a high deficiency of the VSG component, which dominates the emission in the 60 μm *IRAS* band. Stepnik et al. (2003a) have characterized the transition in the Taurus filament by a significant change in the dust properties from the halo toward the inner region: $n_H \geq 3 \times 10^3\ \mathrm{cm}^{-3}$, $A_V \geq 2$. This change leads to an increase of the grain submillimeter emissivity by a factor of 3.4 with respect to that of standard grains. This increase is correlated with a 80 to 100 percent reduction of the VSG abundance. Both variations can be explained by an efficient coagulation process between large grains and VSG, leading to porous aggregates (see Fig. 5 and Stepnik et al. 2003b). In order to reproduce the observations, we have shown that there is a need to coagulate at least $\simeq 20$ large grains per aggregate and to incorporate almost all of the available VSG into aggregates with a fractal dimension $\simeq 2$.

4. Open questions and perspectives

The submillimeter sky observed with PRONAOS has given rise to unexpected results, showing in particular that it cannot be simply extrapolated from shorter wave-

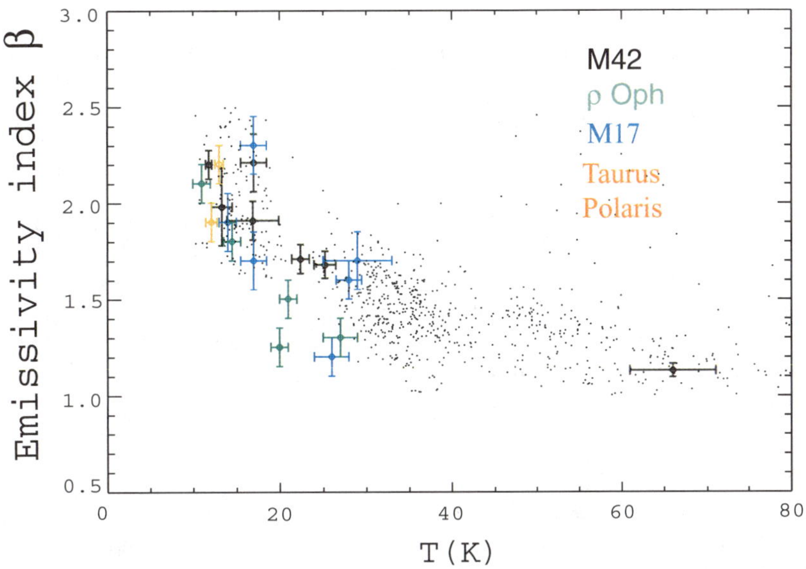

Figure 4. Anticorrelation between the dust temperature and emissivity spectral index in the regions observed with PRONAOS.

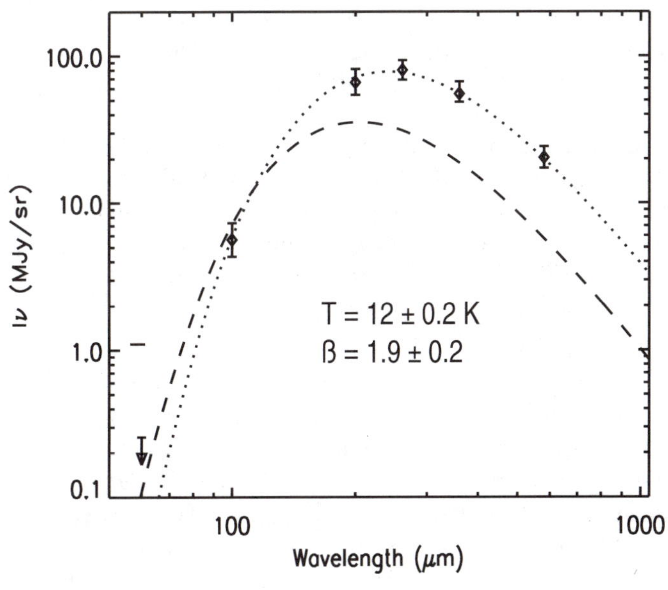

Figure 5. Left: Sketch of a mixed fractal aggregate composed of large grains (LG) and very small grains (VSG) [$N(\mathrm{LG}) = 20$, $D = 2$.]. *Right:* Dust emission in the Taurus filament observed with PRONAOS. The dashed curve represents the dust spectrum predicted by extinction using the standard dust properties. The dotted curve represents the dust spectrum modelled using the dust properties of coagulated mixed aggregates: $N(\mathrm{LG}) = 20$, $D = 1.8$, $N(\mathrm{VSG})/N(\mathrm{LG}) = 15$ (Stepnik et al. 2003a).

length measurements. A first point is that most of the cold condensations detected with PRONAOS were invisible on the *IRAS* maps. Many questions are addressed regarding the cold component of the interstellar medium in the Galaxy. In particular, surprisingly low temperatures ($\sim$ 12 K) have been measured with PRONAOS in some diffuse low density clouds: how representative is this at large scale in our Galaxy ? These "abnormally" cold dust was found in regions of low 60-to-100 μm *IRAS* ratio. Due to the limited number of observations, we cannot derive any mass estimate of such cold regions in the Galaxy—more statistics is clearly needed. Peculiar dust properties are needed to explain these low temperatures and the existence of a significant anticorrelation between the dust temperature and emissivity spectral index.

The observations in the Taurus filament and Polaris cirrus can be explained by models (Stepnik et al. 2003b) based on the existence of porous aggregates of grains which lead to enhanced submillimeter emissivity. Experimental work and more modelling are needed to characterize this new component. What is the physico-chemical efficiency for aggregate formation and what is the corresponding density transition? How is it related to the physical conditions (density, turbulence, etc.) and the other dust and gas components (e.g., gas-grain interaction, ice mantles)? In addition, a question has arisen regarding the origin of the temperature-spectral index anticorrelation seen with PRONAOS. Processes arising at low temperature could be responsible for the observed change of dust grain optical properties in terms of emissivity, $T-$dependent spectral index, and heat capacity. To progress, we need to study the low-energy structural transitions in the solid state, including lattice vibration-based mechanisms in disordered structures, and transitions in statistically distributed localized two-level tunnelling systems with induced quantum mechanical tunneling.

The dust evolution from diffuse to dense media and the new properties at cold temperature have to be taken into account in new dust models. These results will be particularly helpful in preparing and analysing future submillimeter Galactic observations scheduled with *Herschel* and the all-sky *Planck* survey. These will ultimately provide a full characterization of the far-infrared to millimeter dust emission properties.

In the meantime, the next step in our understanding of the far-infrared to millimeter emission properties of interstellar dust is expected to emerge from a large Galactic survey performed with the ELISA balloon-borne experiment (Bernard, this volume).

Acknowledgements

It is difficult to express how much we are indebted to Guy Serra who was the Principal Investigator of PRONAOS. Guy died in August 2000. He was one of the most active pioneers in the field of infrared and submillimeter space astronomy. He will remain in our minds as a great scientist, an extraordinarily generous humanist, and a highly esteemed friend.

We thank the French space agency Centre National d'Études Spatiales (CNES), which supported the PRONAOS project. We are very grateful to the PRONAOS technical teams at CNRS and CNES, and to the NASA-NSBF balloon-launching facilities group of Fort Sumner (New Mexico).

References

Abergel, A., et al., 1996, A&A, 315, L329
Agladze, N.I., et al. 1996, ApJ, 462, 1026
Bernard, J.P., et al. 1999, A&A, 347, 640
Boulanger, F., et al. 1990, ApJ, 364, 136
Boulanger, F., et al. 1996, A&A, 312, 256
Desert, F.X., Boulanger, F., & Puget, J.L. 1990, A&A, 237, 215
Dupac, X., et al. 2001, ApJ, 553, 604
Dupac, X., et al. 2002, A&A, 392, 691
Lagache, G., et al. 1998, A&A, 333, 709
Lamarre, J.M. et al. 1994, IR Phys. Technol., 35, 277
Lamarre, J.M., & Giard, M. 2002, EAS P. Series, Vol. 4
Mennella, V., et al. 1998, ApJ, 496, 1058
Mény, C., et al. 2002, EAS P. Series, Vol. 4
Motte, F., André, P., & Neri, R. 1998, A&A, 336, 15
Ristorcelli, I., et al. 1998, ApJ, 496, 267
Serra, G., et al. 2001, C.R. Acad. Sci. Ser. IV, 1, 1215
Stepnik, B., et al. 2003, A&A, in press
Stepnik, B., et al. 2003, A&A, submitted

PART V:
CHEMICAL DIAGNOSTICS OF INFALL, OUTFLOW AND DISKS

STUDYING INFALL

Neal J. Evans II

The University of Texas at Austin, Astronomy Department, 1 University Station C1400 Austin, Texas 78712–0259

Abstract

The study of protostellar collapse or infall has proven unusually difficult and controversial. Some historical context will be provided, against which recent advances can be measured. We now have a substantial number of objects with signatures that can be interpreted in terms of collapse, but a number of issues remain. One issue is the effect of chemical variations, especially depletion in the dense, cold interiors of cores that are likely to form low mass stars. Strategies for dealing with this issue depend on using dust emission to constrain the density and temperature distribution, leaving molecular line observations to constrain dynamics and abundance distributions. Recent progress in this area will be described and we will consider the next challenges to be overcome. Interferometric observations, especially with future instruments, will provide a powerful tool. In combination with chemical studies coupled with dynamical models, the observations made possible by interferometers should finally put this subject on a firm foundation.

Key words: ISM: molecules – Stars: formation

1. Introduction

While gravitational collapse is the accepted explanation for star formation, the observational evidence for it is very scarce. Establishing this picture *observationally* is a strong reason for studying infall. This subject has long been controversial (cf. Goldreich & Kwan 1974; Zuckerman & Evans 1974), and the burden of proof has rested heavily on those who interpret observations in terms of collapse. While it will always be difficult to rule out other explanations for observational signatures of collapse, the collapse interpretation is beginning to benefit from the principle of simplicity: other explanations are becoming increasingly contrived.

If we can establish the basic picture, we can move on to testing particular theories for how the collapse proceeds. There are many theoretical models that differ in their predictions of the density and velocity fields. In principle, these predictions may be subject to observational tests. As an example, collapse according to the Larson-Penston similarity solution (Larson 1969; Penston 1969) produces much wider linewidths than inside-out collapse from a singular isothermal sphere (Shu 1977). The line profiles expected for the Larson-Penston solution do not match data from regions forming low-mass stars (Zhou 1992) but may find application in regions forming massive stars.

Finally, observational constraints on the dynamics will provide an alternative way to assess the timescales for star formation. At present, these are determined only by statistical arguments based on the number of objects in different observational classes, assumed to be an evolutionary sequence. This method is subject to a number of uncertainties, including selection biases and the possible importance of factors other than age, such as environment, etc. For example, the usual view is that protostars spend 2×10^4 yr in the Class 0 phase and 2×10^5 yr in the Class I phase. The former number is based on the fact that the ρ Ophiuchus cloud has ten times as many Class I objects as Class 0 objects (André & Montmerle 1994). However, Visser, Richer, & Chandler (2001, 2002) have found as many Class 0 objects as Class I objects in a recent survey of other regions, implying a timescale for the Class 0 stage that is 10 times longer than usually assumed. Direct observation of infall speeds in various objects would provide an independent estimate of timescales.

The topic of infall has been reviewed by Zhou & Evans (1994), Myers (1997), Zhou (1999), Evans (1999), and Myers, Evans, & Ohashi (2000). In this paper, I intend mostly to update those reviews, but I will begin with some background.

Chemistry as a Diagnostic of Star Formation (2003). *Edited by Charles L. Curry and Michel Fich.* Published in 2003 by NRC Research Press, Ottawa ON, K1A 0R6, Canada. ISBN 0-660-19089-3.

2. Dark Ages and Renaissance

Partly because of the historical controversy referred to above, the study of collapse suffered through the "dark ages" from the mid-1970s to the mid-1990s. During this period, every claim to find evidence for collapse was followed by counter-claims with alternative explanations [cf. Goldreich & Kwan (1974), Liszt et al. (1974) vs. Zuckerman & Evans (1974); Snell & Loren (1977) vs. Leung & Brown (1977); Walker et al. (1986) vs. Mundy, Myers, & Wilking (1986) and Menten et al. (1987)]. For further historical discussion and references, see Evans (1991) and Myers et al. (2000). The net effect of all this contention was to make the study of infall seem disreputable.

In addition to sociological factors, physical factors have made the field difficult. Collapse motions tend to be small compared to other motions and hence easily obscured. A rough estimate of the infall speed is given by the following:

$$v_{\rm inf} \sim 1 \text{ km s}^{-1} \left[\frac{M_\star/\mathrm{M}_\odot}{r/1000\mathrm{AU}}\right]^{0.5}. \quad (1)$$

Thus, for low-mass stars, one needs to probe small radii to see substantial infall motion. Other motions are often substantially larger. For a sample of low-mass cores, Caselli & Myers (1995) found that the non-thermal (turbulent) contribution to the linewidth (FWHM) could be fitted by the following relation:

$$\Delta v_{\rm turb} = 1.3 \text{ km s}^{-1} r_{\rm pc}^{0.5}. \quad (2)$$

While this relation does not necessarily apply to the linewidth as a function of radius for a *given* region (see Goodman et al. 1998), there is some evidence that turbulent linewidths are less at small radii on average. Thus improved resolution is an obvious route to improve the detectability of infall. However, the decreases in linewidth will be limited as thermal broadening takes over. The minimum possible linewidth (aside from narrowing by maser emission) is

$$\Delta v_{\rm th} = \sqrt{\frac{8(\ln 2)kT_K}{m}} = 0.215 \text{ km s}^{-1}\sqrt{\frac{T_K}{m(\mathrm{amu})}}, \quad (3)$$

where T_K is the gas kinetic temperature and m is the mass of the species whose line is being observed. Based on this equation, one would gain by observing heavier molecules, but purely thermal broadening is not usually seen in heavier molecules. In fact, turbulence often persists at the subsonic level even on small scales. The sonic linewidth is

$$\Delta v_{\rm sonic} = \sqrt{\frac{8(\ln 2)kT_K}{m_{\rm mean}}} = 0.14 \text{ km s}^{-1}\sqrt{T_K}, \quad (4)$$

where $m_{\rm mean}$ is the mean mass per particle. In some very quiescent clouds, the turbulence is considerably less than this value.

In addition, rotational motions will increase as regions closer to the forming star are probed. If angular momentum is conserved,

$$v_{\rm rot} \propto r^{-1}. \quad (5)$$

Ohashi et al. (1997) found that angular momentum appears to be conserved inside about 6000 AU. On much smaller scales, where the circumstellar disk forms, angular momentum will be redistributed, but the rotational velocity does appear to increase with decreasing radius in the infall region.

Finally, and worst, outflows seem to begin very soon after a central source has formed, and they are very energetic at early times. These move matter at velocities ranging up to a few times 100 km s^{-1} in the wind and a few times 10 km s^{-1} in the molecular outflow. Most troubling for the study of infall is the *low* velocity motion of dense gas very near the base of the outflow. These motions can easily obscure or confuse infall signatures.

The renaissance in the study of infall can be traced to four developments. First, the discovery of objects in very early stages provided more fertile grounds for infall studies. Presaged by extensive surveys for dense gas in nearby dark clouds by Myers and colleagues (e.g., Myers, Linke & Benson 1983), the identification of Class 0 objects by André et al. (1993) and Class −1 objects (also called Pre-Protostellar Cores or PPCs) by Ward-Thompson et al. (1994) was a crucial development. Second, simple models for the density and velocity fields during collapse were presented by Shu (1977), but the idea of inside-out collapse from an initial, singular isothermal sphere really came to prominence after the seminal review article by Shu, Adams, &Lizano (1987). Third, Zhou (1992) converted the theoretical models of both the inside-out collapse and the Larson-Penston solution into simulated line profiles as a function of time, clarifying the characteristic line shapes expected for collapse. Finally, a credible example of collapse was found in the small globule B335 (Zhou et al. 1993).

The key signature of infall is a line profile that ideally is double-peaked with a central self-absorption dip, with the blue peak being stronger than the red peak (Leung & Brown 1977; Zhou 1992; Myers et al. 1996). Cartoons that demonstrate why this signature arises can be found in Figures 1 and 2. This signature will appear only if the line has a suitable optical depth and critical density. Observations of an optically thin line that peaks near the velocity of the self-absorption dip are necessary to rule out two separate velocity components. Some optically thick lines will not show a self-absorption dip, but will show a blue-skewed profile. The latter are referred to by Myers et al. (1996) as "red-shoulder" profiles, but they are the same as the "blue-skewed" profiles. For simplicity, I will refer to either of these as "blue profiles."

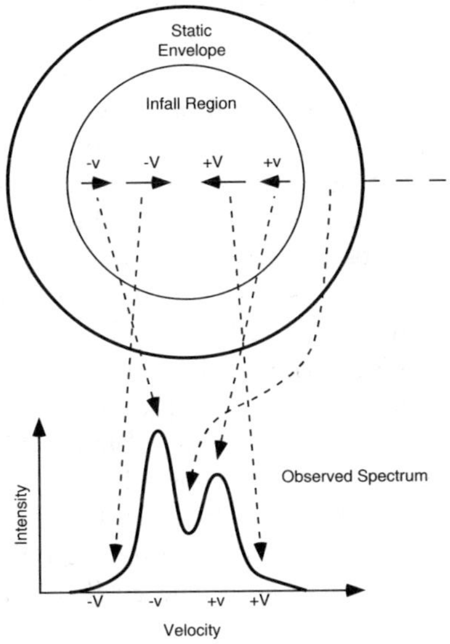

Figure 1. The origin of various parts of the line profile for a cloud undergoing inside-out collapse. The static envelope outside $r_{\rm inf}$ produces the central self-absorption dip, the blue peak comes from the back of the cloud, and the red peak from the front of the cloud. The faster collapse near the center produces line wings, but these are usually confused by outflow wings.

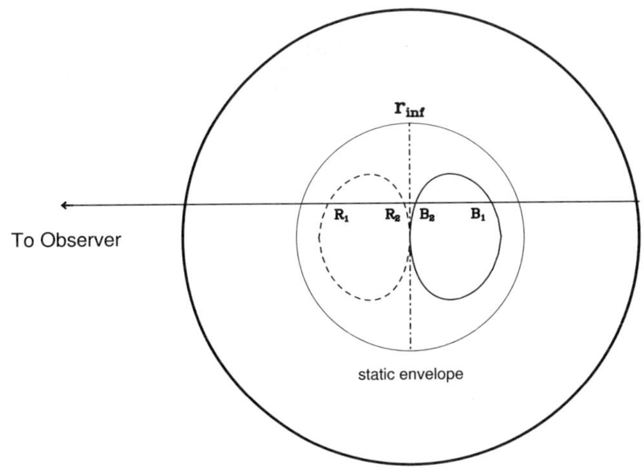

Figure 2. A schematic explanation of why line profiles of optically thick, high-excitation lines are skewed to the blue in a collapsing cloud. The ovals are loci of constant *line-of-sight* velocity, for $v(r) \propto r^{-0.5}$. Each line of sight intersects these loci at two points. The point closer to the center will have a higher $T_{\rm ex}$, especially in lines that are hard to excite, so that $T_{\rm ex}(R_2) > T_{\rm ex}(R_1)$ and $T_{\rm ex}(B_2) > T_{\rm ex}(B_1)$. If the line is sufficiently opaque, the point R_1 will obscure the brighter R_2, but B_2 lies in front of B_1. The result is a profile with the blue peak stronger than the red peak (Zhou & Evans 1994).

The full double-peaked signature was seen clearly in several lines toward B335 by Zhou et al. (1993) and modeled in more detail by Choi et al. (1995). While some issues have arisen later regarding B335 (Sec. 4), it was credible enough to return some respectability to the field, and it led to the next development.

3. Surveys for Infall Signatures

One example is insufficient to verify a theory for low-mass star formation, so a number of surveys began for blue profiles. These initially used the molecular tracers that showed blue profiles in B335, CS and H_2CO. Various lines were used for the optically thin tracer. In a survey of 12 globules, similar to B335, Wang et al. (1995) found 3 blue profiles using H_2CO as the optically thick tracer. Mardones et al. (1997) surveyed a large number of Class 0 and I sources, finding 14 out of 37 blue profiles in CS and 15 out of 47 in H_2CO. Most surprisingly, Lee et al. (1999) found blue profiles in 17/70 starless cores, using CS. In the inside-out picture, a central luminous object forms almost as soon as collapse begins, so a phase with collapse and no central luminosity would be very rare.

Meanwhile, it turned out that HCO^+ showed even deeper self reversals (Fig. 3) than CS and H_2CO toward B335, and searches commenced using HCO^+ $J = 3 \rightarrow 2$ as the optically thick line and $H^{13}CO^+$ as the optically thin line. Gregersen et al. (1997) found blue profiles toward 6 out of 18 Class 0 sources. Because the study by Mardones et al. (1997) using CS and H_2CO had found fewer blue profiles among the Class I sources than among Class 0 sources, it was important to put the Class 0/I searches on a common footing. A study of Class I sources by Gregersen et al. (2000) found blue profiles in 8 out of 16 Class I sources and Gregersen & Evans (2000, 2001) found blue profiles in 6 of 17 Class −1 cores (starless cores with submillimeter emission). Clearly the phenomenon of blue profiles was widespread, but far from universal. Worse yet, some lines showed **red** profiles!

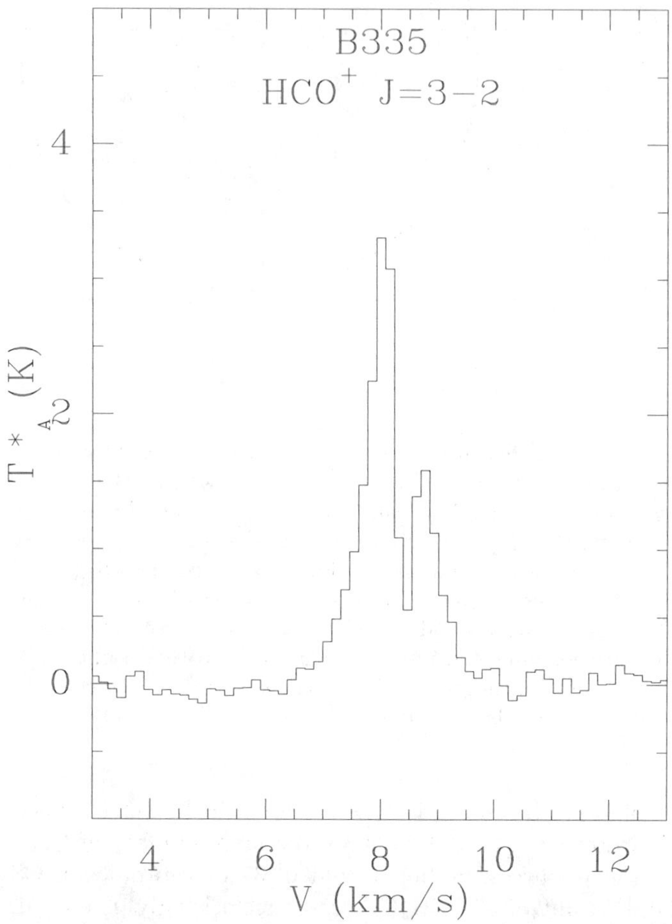

Figure 3. The line profile of the $J = 3 \rightarrow 2$ HCO^+ line toward B335 (Evans, Rawlings, & Choi 2002b, in preparation).

To quantify the results, Mardones et al. (1997) introduced the excess: $E = (N_{\rm blue} - N_{\rm red})/N_{\rm tot}$, where $N_{\rm blue}$ and $N_{\rm red}$ are the number of blue and red profiles in the total sample of $N_{\rm tot}$ objects. A line was counted as blue if the peak of the optically thick line was shifted blueward of the peak of the optically thin line by a quarter of the linewidth of the optically thin line: $\delta v = (v_{\rm thick} - v_{\rm thin})/\Delta v_{\rm thin} < -0.25$. Similarly, a red line would have $\delta v > 0.25$. The excess should be zero on average if the skewing of line profiles were caused by random outflow blobs or other effects without some overall inward motion.

Table 1. Excess versus Class

Class	−1	0	I
Excess	0.30	0.31	0.31

The surveys did reveal positive excesses, suggesting a systematic tendency toward inward motions. One question that arose was whether there was any tendency for the excess to vary with time. In some pictures, the main infall occurs during the Class 0 stage, with much less infall occurring during the Class I stage (e.g., André et al. 2000). Studies with CS and H_2CO did find a lower value of E among Class I sources (Mardones et al. 1997), but studies with HCO^+ found the same $E = 0.31$ in Class I sources as in Class 0 sources (Gregersen et al. 2000). Because HCO^+ is more opaque than CS or H_2CO, it may be that it was better able to reveal infall at later stages when less material remains in the envelope. Using HCO^+, a surprisingly large excess was found for Class −1 cores by Gregersen & Evans (2000), but the sample was very small. The value in Table 1 is based on counting only δv that are less than −0.25 by at least one σ; if all those with $\delta v < -0.25$ are counted, $E = 0.50$.

Another approach taken with surveys is to ask if different tracers agree. Gregersen et al. (2000) assembled a table of 40 sources, most with data in CS, H_2CO, HCO^+, and HCN. There were some cases in which the tracers disagreed, but considerably more in which they agreed. In quite a few cases, the HCO^+ showed blue profiles, while the others showed no asymmetry. Gregersen et al. (2000) suggested that the HCO^+ remained optically thick enough to develop a blue profile at later stages, when the CS and H_2CO were no longer opaque enough. They presented some simple simulations of line profiles to support this suggestion. A better understanding of chemical effects in these regions is needed to put this suggestion on firmer footing.

4. Complications

While the surveys indicated a substantial excess of blue profiles over red profiles, the presence of red profiles raises a red flag. Without understanding those profiles, how can we be sure what the blue profiles mean? The red profiles are likely to be caused by dense clumps in outflowing gas. Clearly some of the blue profiles could be caused by the same thing. While the statistics suggest that the predominance of blue profiles indicate collapse, a blue profile in an individual core may not be a reliable signature of collapse.

In view of the troubled history, modern students of infall have generally been quite cautious, subjecting a blue profile to a number of tests before describing it as a credible example of collapse. This process reminds me of the classic work of Puritan allegory (Bunyan 1688), so I offer a modern allegory for the "Profile's Progress" in Figure 4. A blue profile could just represent the presence of two velocity components, which must be ruled out by observations of an optically thin line that peaks near the dip in the optically thick line, rather than showing two peaks. If there is an embedded source (Class 0 or I), a map of the source should reveal a peak on or near the source; the peak could be in the "blueness," measured in various ways, or in the depth of the dip. The peak of the blue component need not be on the source if rotation is also present, but a pattern consistent with expectations for rotation should be seen. Now, the hardest test appears; a blue profile must resist the temptation to just be an outflow blob! The profile that passes this test may be considered a candidate for collapse. Agreement with someone's detailed model is required to achieve the status of a credible example. Opinions differ as to whether any regions have achieved this status.

If no embedded source is known (Class −1), the path is slightly different (no outflow to worry about) and is not shown in Figure 4. First, without a source, one has to check instead that some indicator is at least centered on the peak of the submillimeter continuum emission. In these cases, one may be seeing core formation, rather than star formation, and the pattern of blue profiles may be quite extended (e.g., Tafalla et al. 1998). The blue profiles cannot be tracing infall onto a central object, so they must be compared to models of core formation by ambipolar diffusion or turbulent dissipation (e.g., Myers & Lazarian 1998) or to models of collapse before a central object forms (Lee et al. 2003).

Because infall velocities should increase at smaller radii, it seemed that interferometers might provide clearer evidence of infall. Indeed, some studies did look promising (Chandler & Sargent 1993, Zhou et al. 1996, Wilner et al. 1997). However, Choi et al. (1999) found some discrepancies. In particular, the red peak was very weak or even absent in some cases. This was an extreme example of a trend already seen in single-dish data toward higher blue/red peak ratios in the data than predicted by the models (Gregersen et al. 1997). In addition, interferometers can artificially produce what appear to be

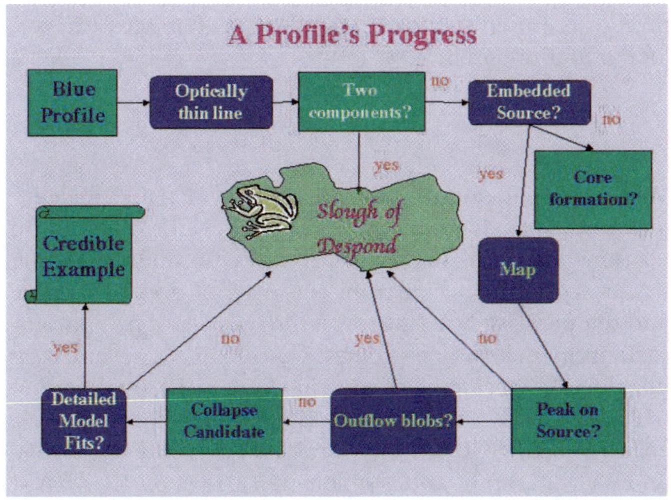

Figure 4. The progress of a blue profile through the many pitfalls on the path toward "salvation," as a credible example of collapse (with apologies to John Bunyan).

self-absorption dips by resolving out the extended emission from a static or slowly moving envelope (e.g., Gueth & Guilloteau 1999).

When the CS $J = 5 \rightarrow 4$ line was imaged in B335 by Wilner et al. (2000), they found that the line profile, while blue, was narrower than predicted from the best fitting model to the single dish data (Choi et al. 1995). At a minimum, this result suggested lower velocities at small radii than predicted from the inside-out collapse model. Even worse, the map showed that the blue emission was displaced from the infrared source and thus more likely an outflow blob.

Around the same time, it was becoming clear that many molecules can be heavily depleted in the dense, cold regions of Class −1 sources (e.g., Caselli et al. 1999), and that these depletions might affect the line profiles strongly (e.g., Rawlings and Yates 2001) if they persist into the infalling stage.

The detection of blue profiles in many Class −1 cores also called into question the basic model of inside-out collapse, suggesting instead that inward motions of some kind are commonly seen before formation of a central luminosity source. Maps of the lines indicated extended regions of blue profiles, inconsistent with very early stages of an inside-out collapse (Tafalla et al. 1998; Lee, Myers, & Tafalla 1999, 2001).

All these developments suggested the need for a more robust tracer of the density structure than provided by molecular lines because detailed simulation of the line profiles required models of the density, temperature, velocity, and abundance structure.

5. Using Dust and Gas

While dust emission or absorption provides no direct information on velocities, studies of dust can be quite useful for the infall problem. Studies of dust continuum emission or absorption can constrain the distribution of density and temperature, cutting down the number of free parameters in models of line profiles. Conveniently, several new instruments became available in the late 1990s that revolutionized observations of dust emission. For low-mass, cold regions, SCUBA on the JCMT, operating at 850 and 450 μm, and various bolometer arrays on the IRAM 30-m dish have provided many maps of submillimeter or millimeter emission (e.g., Shirley et al. 2000; Ward-Thompson et al. 1999; Visser et al. 2001, 2002).

Studies of cores through absorption (or scattering) by dust are complementary to emission studies. These have a long heritage, but one has to use absorption in the infrared to probe the denser regions. Studies in the near-infrared by Lada et al. (1994) and in the mid-infrared using ISO (Bacmann et al. 2000) are examples of this technique. The use of the mid-infrared will be extended by observations with SIRTF of dense cores (Evans et al. 2002a, in preparation).

Proper interpretation of the emission data requires knowledge of the dust temperature distribution [$T_d(r)$] because at the low temperatures around low luminosity sources, the dust emission is quite sensitive to temperature. Unlike the case of luminous sources, the emission is not in the Rayleigh-Jeans limit and $T_d(r)$ is not a power law (Shirley et al. 2000); consequently, convenient analytical approaches are inappropriate. Fortunately, there are publicly available codes to compute $T_d(r)$ self-consistently for a given density distribution, including both internal and external heating (Egan, Leung, & Spagna 1988; Nenkova, Ivezić, & Elitzur 2000).

For line profile modeling, one needs to know $T_K(r)$, the distribution of gas kinetic temperature. For much of the relevant range of radii, the density is high enough that collisions of molecules with dust grains makes $T_K \sim T_d$, but in the lower density envelope, a separate energetics calculation is needed to obtain the kinetic temperature. While one might worry that heavy depletion of molecular coolants deep in the interiors might affect T_K, Goldsmith (2001) has shown that the effects are small. Once the density is high enough to affect the coolants significantly, it is high enough to enforce $T_K \sim T_d$ to reasonable accuracy.

The results of models with self-consistent calculations of $T_d(r)$ (Leung 1975; Evans et al. 2001) show that Class -1 cores are very cold in the center ($T_d \sim 7$ K) and warmer on the outside ($T_d \sim 14$ K if exposed to a typical interstellar radiation field). Density distributions that are constant in the center but approach power laws at large radii, such as Bonnor-Ebert spheres fit the radial intensity profile well (Ward-Thompson, et al. 1999; Evans et al. 2001). The low central T_d decreases the emission from the center, even at millimeter and submillimeter wavelengths. When the lower temperatures in the center are accounted for, the central densities of Class -1 cores with reasonably strong submillimeter emission range from 3×10^5 to 10^6 cm^{-3} (Evans et al. 2001).

At the very low temperatures and high densities deep in Class -1 cores, substantial depletion of the molecules onto dust grains is expected on relatively short timescales. The observational evidence clearly confirms this expectation (e.g., Rawlings 2000; Caselli 2002 for reviews). Many Class -1 cores are nearly invisible in the usual molecular tracers of column density, such as $C^{18}O$. Other species that form from less condensible precursors or that profit from the depletion of their primary destroyers follow the dust emission more closely (e.g., Lee et al. 2003). Especially favored are molecules derived from N_2, ions, and deuterated species. On these grounds, N_2D^+ may be the last gaseous molecular tracer still working after the others have chilled out (Caselli et al. 2002).

This depletion makes it very hard to trace the velocity field in the interior, but studies at larger radii show widespread regions of blue profiles, as described earlier. These profiles can be successfully modeled in some cases with a Plummer-like model (Whitworth & Ward-Thompson 2001); in particular, Lee et al. (2003) have fit the HCO^+ line profiles toward L1544 at several positions with such a model. The density and temperature distributions of this model are nearly identical to the best fitting Bonnor-Ebert sphere and thus predict dust emission that matches observations, but the Plummer-like model also provides a velocity distribution needed to compare to line profiles. In the case that fits L1544, the inward velocity peaks around 0.015 pc. An alternative approach (Bourke et al. 2002, in preparation) to the same core treats each molecular transition with a two-layer model; when applied to different molecular transitions, one infers a velocity field larger at smaller radii, qualitatively consistent with the modeling by Lee et al. (2003). The exciting thing about these results is that we are beginning to probe velocity fields in very early stages. These will set timescales and initial conditions for the formation of central objects.

However, these are only initial steps and much work remains, even to show that these motions are necessarily related to collapse.

Once a central object forms, the core is heated from the inside as well as the outside, and T_d rises in the core. The submillimeter emission from dust becomes stronger and easier to map. Analysis of a substantial sample of Class 0 and I objects (Shirley et al. 2002; Young et al. 2003) finds that the radial intensity profile of the submillimeter emission is fitted well by power laws in density $[n(r) = n_f(r/r_f)^{-p}]$. An interesting dichotomy appears: for sources that are nearly spherical, $\langle p \rangle = 1.8$; for the small number of sources that are quite elongated, typical $p \sim 1$. While the mean value of p is similar to that of the singular isothermal sphere, these objects should have evolved into the inside-out collapse stage, which predicts $p \leq 1.5$ inside $r_{\rm inf}$. No sign of a shift to a smaller value of p at small radii is apparent in the data from single dishes, which probe $r \geq 2000$ AU, in Class 0 sources (Shirley et al. 2002). More surprisingly, most Class I sources also show no sign of a change in p outside 2000 AU (Young et al. 2003), even though $r_{\rm inf}$ should be larger in these sources.

For B335, the inside-out collapse model that fits the molecular line data (Choi et al. 1995) does not fit the dust continuum data, which prefer a power law density. Studies of extinction in the near-infrared using *HST* observations (Harvey et al. 2001) find a power law with $p = 2$ in the outer parts of B335, with some evidence for a turnover near the $r_{\rm inf}$ found by Choi et al. (1995). The extinction method runs out of stars inside $r \sim 3500$ AU, but recent studies of dust emission with the IRAM interferometer probing scales from 500 to 5000 AU were fitted well by a power law with $p = 1.65 \pm 0.05$ (Harvey et al. 2002). These results suggest that the inside-out model does reproduce the density distribution, returning the issue to the models of molecular lines.

Combining the constraints from dust continuum emission with those of molecular lines in a single, coherent model is clearly necessary. Hogerheijde & Sandell (2000) demonstrated the power of this technique; they found good fits to models of inside-out collapse for three of four sources. The fourth source, L1489 IRS, was later modeled as a more evolved object with a rotating, contracting disk (Hogerheijde 2001). Such studies are clearly necessary to reduce the large number of free parameters once molecular depletion is considered.

6. The Role of Interferometers

While data from interferometers must be interpreted with care (see discussion in Sec. 4), they provide the only probe of the velocity fields on small scales. Choi (2002) has explored some of the systematic effects and simulated line profiles from infall as seen by interferometers such as the SMA. As interferometers acquire more antennae, their imaging fidelity will improve, but it will remain important to account for extended emission.

In addition to looking for infall signatures in optically thick lines, interferometers have been used to probe velocity fields in optically thinner tracers (e.g., Ohashi et al. 1996; Ohashi 1997; Momose et al. 1998). The signature of infall is a velocity gradient along the projected *minor* axis of a flattened structure that is larger than the disk. Rotation produces a velocity gradient along the projected *major* axis, allowing one to separate these two motions. Typically, both infall and rotation are inferred from the data. As always, confusion with the outflow is a problem, but this technique is likely to prove powerful as we begin to understand which species do *not* trace the outflow.

In addition to studies of emission, interferometers provide a new possibility. Because they are sensitive to emission from disks, the continuum emission from the disk can serve as a background lamp and, with even higher spatial resolution and sensitivity, an occulting disk for the back half of the cloud (Evans 2001). For disk emission that is sufficiently bright, the front of the cloud will appear in absorption while the rest of the cloud appears in emission, producing an inverse P-Cygni profile. With instruments like ALMA, one may be able to resolve the opaque part of the disk, occulting the back of the cloud altogether, leaving only the absorption line from the front of the cloud. A redshift in this absorption line is the "smoking gun" for collapse. Hints of this absorption were seen toward NGC 1333 IRAS 4A by Choi et al. (1999), but the IRAM observations of Di Francesco et al. (2001) show a much more dramatic inverse P-Cygni profile. The remaining question in this source is whether the absorption arises from the infalling envelope or from a foreground cloud that just happens to be redshifted. Whatever the outcome in this particular case, this technique will be routine with instruments like ALMA and will allow detailed study of the velocity field (Evans 2001).

7. Studies of the Velocity Field

For the moment, most studies of the velocity field are still done with single dishes. By matching line profiles to models over a map of a source or by modeling the line profiles toward the center for lines with different excitation and opacity, it is possible to decode the velocity field, albeit with ample uncertainties, in part due to unknown depletion profiles. There are two general approaches to this problem. One is to take a well-defined theoretical model and try to constrain it with many observations to determine whether it fits the data with any reasonable set of input parameters (e.g., Zhou et al. 1993; Choi et al. 1995; Lee et al. 2003). The other approach is to decode the velocity field empirically by treating each molecular line in a fairly simple way, but using the ensemble to constrain $v_{\rm inf}(r)$.

The latter approach forms the basis for a study of IRAM 04191 by Belloche et al. (2002); this paper represents the current state of the art. This source appears to be in a very early stage of infall. The line profiles of optically thick, hard-to-excite lines toward the center show blue profiles, with the self-absorption dip moving farther redward as the excitation requirements of the line increase. By excitation requirements, I mean some combination of temperature and density. Belloche et al. interpret these line profiles in terms of a constant $v_{\rm inf} = 0.1$ km s^{-1} for $r > 3000$ AU, and $v_{\rm inf}$ increasing inward for $r < 3000$ AU. From line profiles away from the center, they find evidence of rotation as well, with $v_{\rm rot}$ increasing with decreasing radius in to $r = 3500$ AU, but constant (with large uncertainties) inside 3500 AU. This source is destined to become one of the standard test cases for infall studies, along with B335 and L1544.

8. Future Prospects

While many questions remain and skepticism is justified, the field seems to be moving from arguing about infall to studying it. The combined constraints of dust and molecular line data will be important, as will a deeper theoretical understanding of the depletion of molecules and other chemical effects. Future study of velocity fields, just getting under way, will require self-consistent modeling tools, especially 2D or 3D radiative transfer. Observations will need to probe a wide range of spatial scales, and theory will have to provide testable models that can be generated by observers from publicly available codes to compare against their data. If we acquire a better understanding of infall in regions forming low-mass stars, we may be in a position to address the question in the much more complex regions forming massive stars.

Acknowledgements

I am grateful to M. Choi, T. Bourke, P. Myers, and J. Di Francesco for providing early versions of work in progress and for comments. L. Loinard provided helpful discussions. I thank the State of Texas, NASA (Grant NAG5-7203), and the NSF (AST 9988230) for support.

References

André, P. & Montmerle, T. 1994, ApJ, 420, 837
André, P., Ward-Thompson, D., & Barsony, M. 1993, ApJ, 406, 122
André, P., Ward-Thompson, D., & Barsony, M. 2000, in *Protostars and Planets IV*, ed. V. Mannings, A. Boss, & S. Russell (Tucson: Univ. of Arizona Press), 59
Bacmann, A., André, P., Puget, J.-L., Abergel, A., Bontemps, S., & Ward-Thompson, D. 2000, A&A, 361, 555
Belloche, A., André, P., Despois, D., & Blinder, S. 2002, A&A, 393, 927
Bunyan, J. 1688, *Pilgrim's Progress, pt. 1*, as reprinted in 1907, Cambridge University Press
Caselli, P. 2002, Plan. Sp. Sci., 50, 1133
Caselli, P. & Myers, P. C. 1995, ApJ, 446, 665
Caselli, P., Walmsley, C. M., Tafalla, M., Dore, L., & Myers, P. C. 1999, ApJL, 523, L165
Caselli, P., Walmsley, C. M., Zucconi, A., Tafalla, M., Dore, L., & Myers, P. C. 2002, ApJ, 565, 344
Chandler, C. J. & Sargent, A. I. 1993, ApJL, 414, L29
Choi, M. 2002, ApJ, 575, 900
Choi, M., Evans, N. J., II, Gregersen, E. M., & Wang, Y. 1995, ApJ, 448, 742
Choi, M., Panis, J. ;., & Evans, N. J., II 1999, ApJS, 122, 519
Di Francesco, J., Myers, P. C., Wilner, D. J., Ohashi, N., & Mardones, D. 2001, ApJ, 562, 770
Egan, M. P., Leung, C. M., & Spagna, G. R. 1988, Comput. Phys. Commun., 48, 271
Evans, N. J. 1991, in *Frontiers of Stellar Evolution*, ASP Conf. Ser. 20 (San Francisco: Astronomical Society of the Pacific), 45
Evans, N. J., II 1999, ARAA, 37, 311
Evans, N. J., II 2001, in *Science with the Atacama Large Millimeter Array*, ASP Conf. Ser. 235 (San Francisco: Astronomical Society of the Pacific), 25
Evans, N. J., II, Rawlings, J. M. C., Shirley, Y. L., & Mundy, L. G. 2001, ApJ, 557, 193
Goldreich, P. & Kwan, J. 1974, ApJ, 189, 441
Goldsmith, P. F. 2001, ApJ, 557, 736
Goodman, A. A., Barranco, J. A., Wilner, D. J., & Heyer, M. H. 1998, ApJ, 504, 223
Gregersen, E. M. & Evans, N. J., II 2000, ApJ, 538, 260
Gregersen. Erik M. & Evans, N. J.,II 2001, ApJ, 553, 1042
Gregersen, E. M., Evans, N. J., II, Mardones, D., & Myers, P. C. 2000, ApJ, 533, 440
Gregersen, E. M., Evans, N. J., II, Zhou, S., & Choi, M. 1997, ApJ, 484, 256

Gueth, F. & Guilloteau, S. 1999, A&A, 343, 571
Harvey, D. W. A., Wilner, D. J., Lada, C. J., Myers, P. C., Alves, J., & Chen, H. 2001, ApJ, 563, 903
Harvey, D. W. A., Wilner, D. J., Myers, P. C., Tafalla, M., & Mardones, D. 2002, ApJ, 583, 809
Hogerheijde, M. R. 2001, ApJ, 553, 618
Hogerheijde, M. R. & Sandell, G. 2000, ApJ, 534, 880
Larson, R. B. 1969, MNRAS, 145, 271
Lada, C. J., Lada, E. A., Clemens, D. P., & Bally, J. 1994, ApJ, 429, 694
Lee, C. W., Myers, P. C., & Tafalla, M. 1999, ApJ, 526, 788
Lee, C. W., Myers, P. C., & Tafalla, M. 2001, ApJS, 136, 703
Lee, J.-E., Evans, N. J., II, Shirley, Y. L., & Tatematsu, K. 2003, ApJ, 583, 789
Leung, C. M. 1975, ApJ, 199, 340
Leung, C. M. & Brown, R. L. 1977, ApJL, 214, L73
Liszt, H. S., Wilson, R. W., Penzias, A. A., Jefferts, K. B., Wannier, P. G., & Solomon, P. M. 1974, ApJ, 190, 557
Mardones, D., Myers, P. C., Tafalla, M., Wilner, D. J., Bachiller, R., & Garay, G. 1997, ApJ, 489, 719
Menten, K. M., Serabyn, E., Guesten, R., & Wilson, T. L. 1987, A&A, 177, L57
Momose, M., Ohashi, N., Kawabe, R., Nakano, T., & Hayashi, M. 1998, ApJ, 504, 314
Mundy, L. G., Myers, S. T., & Wilking, B. A. 1986, ApJL, 311, L75
Myers, P. C. 1997, in *Star Formation Near and Far*, ed. S.S. Holt and L.G. Mundy (Woodbury N.Y.: AIP Press), 41
Myers, P. C., Evans, N. J., II, & Ohashi, N. 2000, in *Protostars and Planets IV*, ed. V. Mannings, A. Boss, & S. Russell (Tucson: Univ. of Arizona Press), 217
Myers, P. C. & Lazarian, A. 1998, ApJL, 507, L157
Myers, P. C., Linke, R. A., & Benson, P. J. 1983, ApJ, 264, 517
Myers, P. C., Mardones, D., Tafalla, M., Williams, J. P., & Wilner, D. J. 1996, ApJL, 465, L133
Nenkova, M., Ivezić, Ž., & Elitzur, M. 2000, in *Thermal Emission Spectroscopy and Analysis of Dust, Disks, and Regoliths*, ASP Conf. Ser. 196 (San Francisco: Astronomical Society of the Pacific), 77
Ohashi, N. 1997, in *Star Formation Near and Far*, ed. S.S. Holt and L.G. Mundy (Woodbury N.Y.: AIP Press), 93
Ohashi, N., Hayashi, M., Ho, P. T. P., Momose, M., & Hirano, N. 1996, ApJ, 466, 957
Ohashi, N., Hayashi, M., Ho, P. T. P., Momose, M., Tamura, M., Hirano, N., & Sargent, A. I. 1997, ApJ, 488, 317
Penston, M. V. 1969, MNRAS, 144, 425
Rawlings, J. M. C. 2000, in *Astrochemistry: From Molecular Clouds to Planetary Systems*, IAU Symposium 197, ed. Y. C. Minh & E.F. van Dishoeck, 15
Rawlings, J. M. C. & Yates, J. A. 2001, MNRAS, 326, 1423
Shirley, Y. L., Evans, N. J., II, Rawlings, J. M. C., & Gregersen, E. M. 2000, ApJS, 131, 249
Shirley, Y. L., Evans, N. J., II, & Rawlings, J. M. C. 2002, ApJ, 575, 337
Shu, F. H. 1977, ApJ, 214, 488
Shu, F. H., Adams, F. C., & Lizano, S. 1987, ARAA, 25, 23
Snell, R. L. & Loren, R. B. 1977, ApJ, 211, 122
Tafalla, M., Mardones, D., Myers, P. C., Caselli, P., Bachiller, R., & Benson, P. J. 1998, ApJ, 504, 900
Visser, A. E., Richer, J. S., & Chandler, C. J. 2001, MNRAS, 323, 257
Visser, A. E., Richer, J. S., & Chandler, C. J. 2002, AJ, 124, 2756
Walker, C. K., Lada, C. J., Young, E. T., Maloney, P. R., & Wilking, B. A. 1986, ApJL, 309, L47
Wang, Y., Evans, N. J., II, Zhou, S., & Clemens, D. P. 1995, ApJ, 454, 217
Ward-Thompson, D., Scott, P. F., Hills, R. E., & André, P. 1994, MNRAS, 268, 276
Ward-Thompson, D., Motte, F., & André, P. 1999, MNRAS, 305, 143
Whitworth, A. P. & Ward-Thompson, D. 2001, ApJ, 547, 317
Wilner, D. J., Mardones, D., & Myers, P. C. 1997, in *Star Formation Near and Far*, ed. S.S. Holt and L.G. Mundy (Woodbury N.Y.: AIP Press), 109
Wilner, D. J., Myers, P. C., Mardones, D., & Tafalla, M. 2000, ApJL, 544, L69
Young, C. H., Shirley, Y. L., Evans, N. J., II, & Rawlings, J. M. C. 2003, ApJS, 145, 111
Zhou, S. 1992, ApJ, 394, 204
Zhou, S. 1999, in *Millimeter-Wave Astronomy: Molecular Chemistry*, ASSL Vol. 241, *Physics in Space* (Kluwer Academic Publishers), 199
Zhou S., Evans N. J., II 1994, in *Clouds, Cores, and Low Mass Stars*, ed. D.P. Clemens and R. Barvainis, ASP Conf. Ser. 65 (San Francisco: Astronomical Society of the Pacific), 183
Zhou, S., Evans, N. J., II, Koempe, C., & Walmsley, C. M. 1993, ApJ, 404, 232
Zhou, S., Evans, N. J., II, & Wang, Y. 1996, ApJ, 466, 296
Zuckerman, B. & Evans, N. J., II 1974, ApJL, 192, L149

CHEMICAL MODELS OF PROTOPLANETARY DISKS

Yuri Aikawa[1], **Gerd-Jan van Zadelhoff**[2], **and Ewine F. van Dishoeck**[2]

[1]Department of Earth and Planetary Sciences, Kobe University, Kobe 657-8501, Japan
[2]Leiden Observatory, P.O. Box 9513, 2300 RA Leiden, The Netherlands

Abstract

We discuss recent theoretical work on two-dimensional molecular distributions in the outer regions of protoplanetary disks, which are currently observed via radio telescopes. Gaseous molecular abundances are sensitive to the vertical structure of the disk; i.e., the distribution of temperature and ultraviolet radiation flux. From a chemical point of view, disks are schematically divided into three layers: a surface PDR layer, a gaseous molecular layer at intermediate height, and a cold midplane layer in which most species are frozen onto grains. Detailed calculation of radiation transfer is required in order to accurately determine the boundary between the surface and the intermediate layer, and thus the molecular column densities. The latest theoretical model is in reasonable agreement with the observation of DM Tau, but underestimates molecular column densities and line intensities in LkCa 15. Molecular column densities depend on ultraviolet and X-ray fluxes of the central star, but not so much on disk mass and age. Dust coagulation and/or sedimentation will also alter molecular abundances.

Key words: ISM: molecules – stars: T Tauri – circumstellar matter – protoplanetary disks

1. Introduction

Stars are formed by gravitational collapse of molecular cloud cores. If the cores have angular momentum, circumstellar disks are formed around newborn stars. It is established by observations that most young T Tauri stars are surrounded by a disk (e.g., Beckwith & Sargent 1996). These disks are considered to be birth sites of planetary systems and hence are called protoplanetary disks. Their masses are estimated from dust continuum emission to be in the range of $10^{-3} - 10^{-1}$ $M_{\odot}$, which is consistent with the solar nebula model proposed for the origin of our own solar system.

While dust continuum emission has been used since 1980 to derive the size, temperature and density distribution in disks, emission lines of CO were first detected by Skrutskie et al. (1993) and Kawabe et al. (1993) in the disk around GG Tau. So far, several molecular lines have been observed via single-dish telescopes and interferometers (Blake, this volume). Line observation is useful in order to investigate both the chemical and physical properties of disks. Firstly, it reveals the chemical composition of disk matter, which is the raw material of planetary systems. So far, observers have found that gaseous molecular abundances in disks differ from those in molecular clouds: while relative abundances of organic molecules to hydrogen are lower in disks, the CN to HCN ratio is higher. Secondly, lines are used to estimate the temperature and density distribution, because line ratios depend on temperature, and because each molecular line has a critical density for emission (Dutrey, Guilloteau, & Guelin 1997; van Zadelhoff et al. 2001). Thirdly, line observations will reveal how and when the gaseous component dissipates (Thi et al. 2001). Being formed from cloud cores, disks initially consist of gas and dust with a mass ratio of 100. In order to make a planetary system like the solar system, the gaseous component should segregate from dust grains and dissipate at a certain evolutionary stage. The mechanism and timing of the gas dissipation are not yet understood.

Derivation of these chemical and physical properties from line observations is not straightforward. Comparison with theoretical models is therefore important. In this paper, we discuss recent theoretical models of disk chemistry. Since the spatial resolution of present molecular line observations is larger than $\sim 1''$, we consider the regions outside $\sim$100 AU, while chemistry in inner regions is discussed by Henning (this volume).

2. Physical Model of Disks

The physical structure of protoplanetary disks is a very extensive research area (e.g., McCaughrean et al. 2000;

Chemistry as a Diagnostic of Star Formation (2003). *Edited by Charles L. Curry and Michel Fich.* Published in 2003 by NRC Research Press, Ottawa ON, K1A 0R6, Canada. ISBN 0-660-19089-3.

Stone et al. 2000). Here we restrict ourselves to classical T Tauri stars, and briefly discuss the temperature distribution, which is one of the most important parameters for disk chemistry. Since disks are geometrically thin, many models have assumed that they are isothermal in the vertical direction. For example, the temperature is given by $T = 28\ (R/100\mathrm{AU})^{-1/2}$ K in the Kyoto model (Hayashi, Nakazawa, & Nakagawa 1985), one of the classical models of the solar nebula. Kenyon & Hartmann (1987) noted that radiation from the central star is more efficiently absorbed if the disk is more flared. The temperature distribution depends on the flaring of the disk surface, which in turn depends on temperature. In Kenyon & Hartmann's simple iterative model, the temperature is $T \sim 20\ (R/100\mathrm{AU})^{-1/2}$ K.

In more recent studies, the temperature variation in the vertical direction is considered. Chiang & Goldreich (1997) noted that dust particles at the disk surface are "super-heated", because their absorptivity in the visible exceeds their emissivity in the infrared. They constructed an analytical two-layer disk model, in which the surface temperature is as high as $87(R/100\mathrm{AU})^{-2/5}$ K, while the lower layer is heated by the radiation from the surface layer and reaches a temperature of $\sim$21 K at $R = 100$ AU. D'Alessio et al. (1998) solved for the vertical distribution of density and temperature at each radius via numerical calculation. Although they included mass accretion, it is effective as a heat source only at radii inside several AU as long as a typical accretion rate of $10^{-8} M_{\odot}$/yr is assumed. Irradiation from the central star is the dominant heat source at the outer radii. Temperatures in the uppermost layer and in the midplane are similar to those in the super-heated layer and the lower layer of Chiang & Goldreich (1997). Similar results are also obtained via independent or more sophisticated radiation transfer calculations (Nomura et al. 2002; Dullemond et al. 2002).

Density is determined by hydrostatic equilibrium. Assuming a typical disk mass of several times $10^{-2} M_{\odot}$, the hydrogen density is as high as $\sim 10^9$ cm^{-3} in the midplane at $R = 100$ AU.

3. Molecular Distribution in Disks

3.1. vertical distribution

Theoretical studies show that molecular abundances are highly inhomogeneous in accordance with the distribution of temperature and flux of ultraviolet (UV) radiation. T Tauri stars are strong UV emitters; the UV flux from a T Tauri star is about 10^4 times larger than that of the interstellar UV field at a radius of 100 AU (Herbig & Goodrich 1986). Aikawa & Herbst (1999) calculated the 2D (radial and vertical) distribution of molecular abundances in the Kyoto model. Since the temperature at $R \gtrsim 200$ AU is lower than 20 K, the sublimation temperature of the dominant carbon reservoir CO, abundances of organic molecules are low in the gas phase. Aikawa & Herbst had to assume an artificially low sticking probability ($S \sim 0.03$) when molecules hit grain surfaces in order to reproduce the CO line intensity and other molecular abundances observed in the disk around DM Tau.

Willacy & Langer (2000) performed a similar calculation for the two-layer model of Chiang & Goldreich (1997). Although the temperature in the super-heated layer was high enough ($\gtrsim$ 20K) to sublimate CO, molecules were not abundant there because of photodissociation by UV. A high efficiency of photo-desorption in the lower layer was required to reproduce the observed molecular abundances.

Aikawa et al. (2002) calculated molecular distributions in the model of D'Alessio et al. (1999). Since the vertical temperature variation is more gradual than that in the two-layer model of Chiang & Goldreich (1997), the model contains a layer in which the gas is warm but where the UV is sufficiently attenuated for molecules to be abundant. The model reproduces column densities of HCN, CN, CS, H_2CO, HNC and HCO^+ in DM Tau without assuming any nonthermal desorption, while CO is overestimated and C_2H is underestimated. Figure 1 shows their 2D molecular distribution schematically. The disk surface is a PDR layer; molecules and atoms are subject to photodissociation and photoionization by UV from the central star and interstellar field. The temperature in the surface layer is $\gtrsim$ 50 K, and the density is $\sim 10^4 - 10^5$ cm^{-3}. Gaseous molecules are most abundant in the intermediate layer, where UV radiation is not so harsh, and the temperature is high enough to keep molecules in the gas phase. The typical gas density in the layer is $\sim 10^6 - 10^7$ cm^{-3}. In the midplane, molecules are mostly depleted onto grains because of the low temperature ($\lesssim$ 20 K) and high density ($n_{\mathrm{H}} \gtrsim 10^7$ cm^{-3}). The model gives a good explanation for the observed characteristics of disk chemistry: the high CN/HCN ratio is caused by photo-reactions in the surface regions, while the total abundances of gaseous organic molecules are low because of depletion in the midplane, which contains most of the disk mass.

Figure 1 indicates that UV attenuation is important in determining the boundary between the surface PDR layer and the intermediate molecular layer, and thus influences the total molecular column densities. In previous

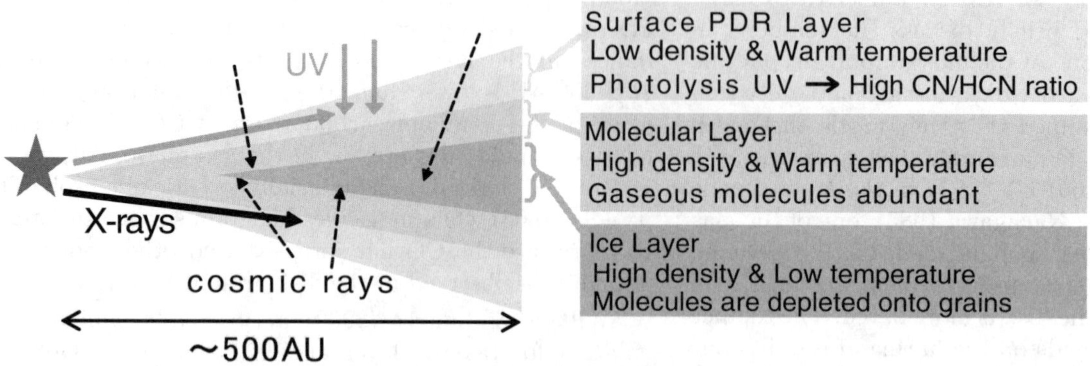

Figure 1. A disk can be schematically divided into three layers from chemical point of view.

work, UV attenuation was calculated via the following 1D ray-tracing method. At each point in the disk, hydrogen column densities from the central star and disk surface are calculated. They are converted to attenuation magnitude A_V, assuming that the optical properties and abundance of dust are the same as those of interstellar dust. Photodissociation and photoionization rates of species i are obtained using the depth-dependent function $A_i \exp(-\beta_i A_V)$, which was originally obtained for the 1D interstellar PDR model (van Dishoeck & Black 1988). This 1D ray-tracing method, however, cannot take scattering into account, which can enhance the penetration of UV photons by redirecting them in the vertical direction. In order to obtain a more accurate radiation field and molecular distribution, van Zadelhoff et al. (2003) performed 2D transfer calculations of UV radiation in the disk model of D'Alessio et al. (1999). At each point in the disk, the UV spectrum, photo-reaction rates, and molecular abundances are calculated. UV penetration is indeed enhanced, and the boundary between the surface layer and the intermediate layer is located at a lower height compared with that in Aikawa et al. (2002). The vertical distribution of molecules at $R = 100$ AU is shown in Figure 2.

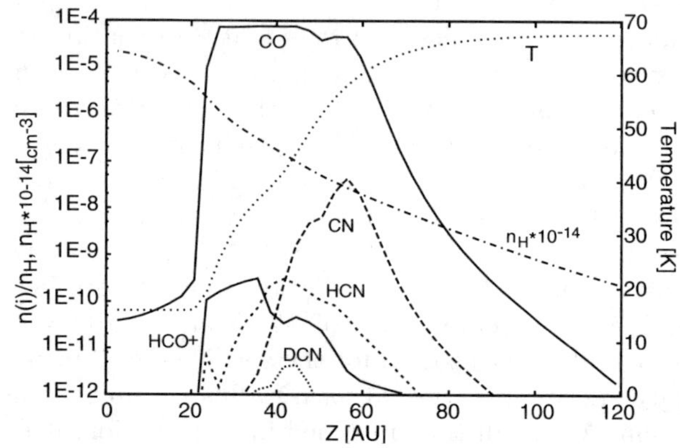

Figure 2. Vertical distribution of density, temperature, and relative abundance of gaseous molecules to hydrogen nuclei at $R = 105$ AU.

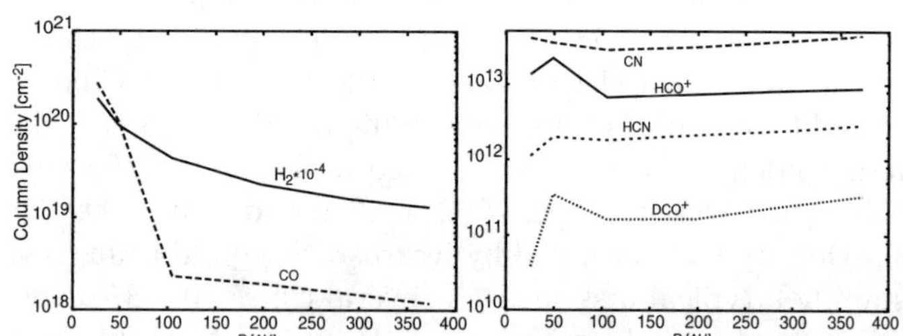

Figure 3. Radial distribution of molecular column densities.

3.2. RADIAL DISTRIBUTION

The radial distributions of molecular column densities obtained by van Zadelhoff et al. (2003) are presented in Figure 3. While the total hydrogen column density varies by an order of magnitude between $R = 26$ and 373 AU, the column densities of gaseous organic species are almost constant with radius, except for CO and HCO^+ at an inner radius $R \lesssim 100$ AU. Each molecular species is abundant in a layer with certain physical conditions. For example, CN and HCN have an abundance peak at $A_V \sim$ 0.2–0.4 mag and $A_V \sim$ 0.4–2 mag from the disk surface, respectively. The masses in these layers do not vary much with radius. The column densities of CO and HCO^+ jump at $R \sim 100$ AU, because the midplane temperature is higher than 20 K at inner radii.

The column density of deuterated species (DCO^+) is also shown in Figure 3. The molecular D/H ratios (e.g., DCO^+/HCO^+) are higher than the elemental abundance ratio of D/H $= 1.5 \times 10^{-5}$, because of the exothermic exchange reactions at low temperatures (e.g., H_3^+ + HD $\rightarrow$ H_2D^+ + H_2 + $\triangle E$). Molecular D/H ratios decrease at inner radii because of the higher temperatures.

3.3. DEPENDENCE ON PHYSICAL PARAMETERS

3.3.1. Disk mass

As described in the previous subsection, at the outer radii with low midplane temperature $T \lesssim 20$ K, gaseous molecules are abundant in certain layers at intermediate height. The mass contained in such layers, and hence the molecular column density, does not vary much with the total disk column density as long as the disk is thick enough; i.e., $A_V \gtrsim 20$ mag.

3.3.2. Age

A time-dependent molecular evolution code is adopted in calculating 2D models of disk chemistry. The molecular distributions shown in the previous subsection are obtained at $t = 1 \times 10^6$ yr, a typical age for classical T Tauri stars. Their dependence on time is not significant as long as we consider $t \gtrsim 10^5$ yr. In the midplane, depletion proceeds in the very short time scale 10 $(10^9 \mathrm{cm}^{-3}/n_H)$ yr, while chemistry reaches a steady state within $\sim 10^4$ yr in the surface PDR layer. Although the time to reach steady state is longer in layers at lower height with lower photo-reaction rates, the column densities of abundant organic species at $t = 1 \times 10^5$ and 1×10^6 yr differ by only a factor of $\lesssim 2$ in the model of van Zadelhoff et al. (2003).

3.3.3. Stellar UV and X-rays

T Tauri stars are strong emitters of UV and X-rays, and their UV and X-ray luminosities vary with time. Stellar UV significantly affects molecular abundances. It dissociates molecules, but does not necessarily decrease the molecular abundance, since dissociation of CO supplies more carbon to the chemical network. Whether the abundance is enhanced or decreased by UV is dependent on the species and its position in the disk. van Zadelhoff et al. (2003) investigated disk chemistry with and without stellar UV. The disk column densities are higher without stellar UV, except for CN at inner radii, which is significantly lower due to a reduced carbon supply.

X-rays also cause dissociation and ionization by directly hitting atoms, via energetic secondary electrons, and via induced UV photons. Since the attenuation length of X-rays is larger than that of UV, X-rays affect chemistry in deeper layers of the disk. Irradiation by X-rays enhances the abundances of gaseous organic species other than CO (Aikawa & Herbst 1999).

4. Comparison with Observation

Table 1 lists molecular column densities obtained via model calculations by van Zadelhoff et al. (2003) at 1 Myr and at 373 AU, and those in DM Tau and LkCa 15 estimated from observations. These authors adopted the disk models of D'Alessio et al. (1999) with a mass accretion rate of $\dot{M} = 10^{-9} - 10^{-7} M_\odot$/yr and an α viscosity parameter of 0.01. The chemical model is calculated, however, without considering hydrodynamic flows, so the three models are presented only to show the dependence on disk mass. The total disk mass is 0.06 $M_\odot$ for the disk with $\dot{M} = 10^{-8} M_\odot$/yr and is larger (smaller) in the model with larger (smaller) accretion rate by about an order of magnitude. Molecular column densities in DM Tau are obtained from Dutrey et al. (1997), who performed a molecular line survey toward the object using the IRAM 30 m telescope, and constructed a simple disk model to account for the line intensities. The model with $\dot{M} = 10^{-9} M_\odot$/yr agrees fairly well with DM Tau, except in the case of CO. Carbon monoxide is chemically stable, and its column density depends mainly on desorption rate. Since thermal desorption is the dominant desorption mechanism in van Zadelhoff et al. (2003), the temperature of DM Tau's disk should be lower than that of the model in order to account for its CO column density.

For LkCa 15, molecular column densities are estimated from interferometer observations using OVRO and NMA (Qi 2001; Blake, this volume; Aikawa et al. 2003), and from single dish observations by Thi (2002) using the IRAM 30 m and the JCMT. It should be noted that the disk cannot be resolved by single dish observations; Thi (2002) estimated molecular column densities N_{100} assuming a disk radius of 100 AU, so they are scaled by $N_{100}(100 \text{ AU}/R_{\text{disk}})^2$. The radius of LkCa 15's disk is estimated to be 435 – 600 AU from CO observations (Duvert et al. 2000; Qi 2001). If other molecules are as extended as CO, their column densities are smaller than those listed by a factor of 20–40. In the analysis of the interferometer observations, on the other hand, column density is estimated by considering the size of the observed emission area for each species. Even if we take into account the uncertainties in the radial distributions, the molecular column densities in LkCa 15 seem to be higher than those in DM Tau and in the models. The molecu-

Table 1. Molecular column densities in theoretical models and observed disks

Species	$\dot{M}$ [$M_\odot$ yr^{-1}]			DM Tau[a]	LkCa 15	
	10^{-7}	10^{-8}	10^{-9}		interferometer[b]	single dish[c]
H_2	1.6(24)	1.3(23)	4.5(21)	3.8(21)		
CO	1.2(18)	1.2(18)	6.0(17)	5.7(16)	1.6(18)	8.8(17)
HCN	2.8(12)	2.9(12)	1.8(12)	2.1(12)	0.02-1.2(15)	7.8(13)
CN	4.9(13)	4.4(13)	4.2(13)	9.5-12(12)	9.1-30(13)	6.3(14)
CS	1.3(12)	1.7(12)	1.7(12)	6.5-13(11)	1.9-2.1(13)	2.2(14)
H_2CO	3.3(12)	3.6(12)	2.7(12)	7.6-19(11)	7.2-19(12)	3.0-22(13)
HCO^+	7.3(12)	8.9(12)	4.6(12)	4.6-28(11)	1.5(13)	1.4(13)
C_2H	2.1(13)	1.9(13)	1.3(13)	4.2(13)		
HNC	2.4(12)	2.5(12)	1.6(12)	9.1(11)	<5.4(12)	
OCS	2.3(10)	6.3(10)	2.0(10)		<2.9(13)	
CH_3OH	4.2(8)	4.8(8)	5.7(8)		7.3-18(14)	< 9.4(14)
DCN	5.6(10)	6.1(10)	3.3(10)		1.0(13)	
HDO	6.5(11)	8.8(11)	4.1(11)		2.3-6.8(14)	
N_2H^+	2.4(12)	2.1(12)	3.5(9)	< 7.6(11)	<5.7(12)	< 5.9(13)

REFERENCES—[a] Dutrey et al. (1997); [b] Qi (2001), except for H_2CO, which is from Aikawa et al. (2003); [c] Thi (2002).

lar line emission from model disks was calculated by van Zadelhoff et al. (2003) for a more direct comparison. Intensities of CN and HCN lines from the model disk are indeed lower than those observed in LkCa 15. One possible explanation for this disagreement and the difference between DM Tau and LkCa 15 would be a higher X-ray flux in LkCa 15, which needs to be checked by Chandra.

5. Future Work

Chemistry in protoplanetary disks is a relatively new research area in astrochemistry, and further developments are expected in the near future. On the observational side, new instruments with high sensitivity and spatial resolution such as SMA, CARMA and ALMA will enable us to observe fainter lines and disks, and investigate the inner regions ($R \lesssim 100$ AU), where planet formation takes place (Blake, this volume). On the theoretical side, the following topics, among others, will be of special interest.

5.1. Coagulation and Sedimentation of Dust

Since planets and the cores of planets are made of solid material, coagulation of dust particles and segregation of dust from gas are essential in the formation of planetary systems. So far, 2D disk models have assumed that optical properties and abundance of dust in disks are the same as those of interstellar dust. However, spectral energy distributions and images of edge-on disks indicate that dust coagulation is taking place in classical T Tauri stars (D'Alessio et al. 2001; Chiang et al. 2001). Temperature and UV flux—the two major parameters controlling molecular abundances in disks—are significantly dependent on the size and distribution of dust particles. Depletion of small grains enhances penetration of UV and lowers the disk temperature by decreasing the angle between the disk surface and rays from the central star. Although dust growth and sedimentation can be investigated by continuum observations, the effects of these two processes are degenerate and hard to distinguish (D'Alessio et al. 2001). Molecular line observations and chemical studies could distinguish dust growth from sedimentation. In order to investigate disk chemistry with dust growth and sedimentation, we need to combine detailed radiation transfer calculations and chemistry codes, since the gas and dust temperatures could be different in such disks.

5.2. Diffusion and Dynamics

So far, hydrodynamic effects such as accretion and diffusion have been ignored for simplicity in the 2D models of disk chemistry used to reproduce the molecular abundances observed by radio telescopes. On the other hand, these dynamical effects are already included in disk chemistry models involving comets and meteorites in our solar system, although the models are reduced to 1D (i.e., considering only the midplane) and/or chemical reaction networks are significantly reduced (Drouart et al. 1999; Finocchi& Gail 1997; Willacy et al. 1998; Aikawa et al. 1999). It is very important and interesting to combine detailed chemical network calculations with 2D hydrodynamic disk models. At this conference, the first result of such attempt was reported by Henning & Ilgner (this volume), who included a detailed chemical network at $R \lesssim 10$ AU in accretion disks with $\alpha-$viscosity. Calculations for the outer radius are desirable for comparison with current radio observations.

Disk dynamics, chemistry, and dust coagulation and sedimentation will ultimately be coupled in future theoretical work. Magneto-rotational instability (MRI) is now considered to be the most promising mechanism controlling the turbulence in protoplanetary disks (Stone et al. 2000). Turbulence controls the distribution and coagulation of dust, which in turn determines chemistry, ionization degree, and hence the stability of disks against MRI (Sano et al. 2000). In order to reach this ultimate goal, further study of each physical and chemical process and their coupling are required.

Acknowledgements

The authors are grateful to the Organizing Committee for giving us a chance to present this work at the conference. Y.A. is supported by the Grant-in-Aid for Scientific Research of the Ministry of Education, Culture, Sports, Science and Technology (13011203 and 14740130). Astrochemistry in Leiden is supported by a SPINOZA grant from the Netherlands Organization for Scientific Research (NWO).

References

Aikawa, Y., & Herbst, E. 1999, A&A, 351, 233

Aikawa, Y., Umebayashi, T., Nakano, T., & Miyama, S. M. 1999, ApJ, 519, 705

Aikawa, Y., van Zadelhoff, G. J., van Dishoeck, E. F. & Herbst, E. 2002, A&A, 386, 622

Aikawa, Y., Momose, M., Thi, W. F., van Zadelhoff, G. J., Qi, C., Blake, G. A. & van Dishoeck, E. F. 2003, submitted to PASJ

Beckwith, S. V. W., & Sargent, A. I. 1996, Nature, 383, 139

Chiang, E. I., & Goldreich, P. 1997, ApJ, 490, 368

Chiang, E.I., Joung, M.K., Creech-Eakman, M.J., Qi, C., Kessler, J.E., Blake, G.A., & van Dishoeck, E.F. 2001, ApJ, 547, 1077

D'Alessio, P., Cantó J., Calvet, N.,& Lizano, S. 1998, ApJ, 500, 411

D'Alessio, P., Calvet, N., Hartmann, L., Lizano, S., & Cantó, J. 1999, ApJ, 527, 893

D'Alessio, P., Calvet, N., & Hartmann, L. 2001, ApJ, 553, 321

Drouart, A., Dubrulle, B., Gautier, D., & Robert, F. 1999, Icarus, 140, 129

Dullemond, C. P., van Zadelhoff, G. J., & Natta, A. 2002, A&A, 389, 464

Dutrey, A., Guilloteau, S., & Guélin. M. 1997, A&A, 317, 55

Duvert, G., Guilloteau, S., Ménard, F., Simon, F., & Dutrey, A. 2000, A&A, 355, 165

Finocchi, F. & Gail, H. -P. 1997, A&A, 327, 825

Hayashi, C., Nakazawa, K., & Nakagawa, Y. 1985, in *Protostars and Planets II*, ed. D.C. Black & M.S. Matthews (Tucson: Univ. of Arizona Press), 1100

Herbig, G. H., & Goodrich, R. W. 1986, ApJ, 309, 294

Kawabe, R., Ishiguro, M., Omodaka, T., Kitamura, Y., & Miyama, S. M. 1993, ApJ, 404, L63

Kenyon, S. J. & Hartmann, L. 1987, ApJ, 323, 714

McCaughrean, M. J., Stapelfeldt, K. R. & Close, L. M. 2000, in *Protostars and Planets IV*, ed. V. Mannings, A.P. Boss, & S.S. Russell (Tucson: Univ. of Arizona Press), 485

Nomura, H. 2002, ApJ, 567, 587

Qi, C., 2001, PhD thesis, California Institute of Technology

Sano, T., Miyama, S. M., Umebayashi, T., & Nakano,T. 2000, ApJ, 543, 486

Skrutskie, M. F., Snell, R. L., Strom, K. M., Strom, S. E., Edwards, S., Fukui, Y., Mizuno, A., Hayashi, M., & Ohashi, N. 1993, ApJ, 409, 422

Stone, J. M., Gammie, C. F., Balbus, S. A. & Hawley, J. F. 2000, in *Protostars and Planets IV*, ed. V. Mannings, A.P. Boss, & S.S. Russell (Tucson: Univ. of Arizona Press), 485

Thi, W. F., van Dishoeck, E. F.,Blake, G. A., van Zadelhoff, G. J., Horn, J., Becklin, E. E., Minnings, V., Sargent, A. I., van den Ancker, M. E., Natta, A., Kessler, J. 2001, ApJ, 561, 1074

Thi, W. F. 2002, PhD thesis, University of Leiden

van Dishoeck, E. F., & Black, J. H. 1988, ApJ, 334, 771

van Zadelhoff, G. J., van Dishoeck, E. F., Thi, W. F., & Blake, G. A. 2001, A&A, 377, 566

van Zadelhoff, G. J., Aikawa, Y., Hogerheijde, M. R. & van Dishoeck, E. F. 2003, A&A, in press

Willacy, K., Klahr, H. H., Millar, T. J. & Henning, Th. 1998, A&A, 338, 995

Willacy, K. & Langer, W. D. 2000, ApJ, 544, 903

TRACING PROTOSTELLAR EVOLUTION BY OBSERVATIONS OF ICES

A. C. A. Boogert[1], G. A. Blake[2], and M. R. Hogerheijde[3]

[1]Department of Astronomy 105-24, California Institute of Technology, Pasadena, CA 91125, USA
[2]Division of Geological and Planetary Sciences 150-21, California Institute of Technology, Pasadena, CA 91125, USA
[3]Steward Observatory, University of Arizona, 933 North Cherry Avenue, Tucson, AZ 85721, USA

Abstract

The energetic phenomena associated with protostars alter the physical and chemical state of the material in circumstellar envelopes and disks. In particular, the structure and composition of icy mantles on dust grains can potentially be strongly influenced by processes such as thermal heating, ultraviolet irradiation, cosmic rays, and shocks. In this paper we discuss the evidence for these processes by analyzing observations of ice bands in the infrared spectra of high and low-mass protostars. *ISO* observations have shown that thermal processing, i.e. sublimation and crystallization, of ices plays an important role in the envelopes of high-mass protostars. In fact, the ice bands may be used as a tracer of the evolution of the protostar. Toward low-mass protostars, the unprecedented sensitivity and spectral resolution of the NIRSPEC spectrometer at the Keck II telescope allow a detailed analysis of ice band profiles. The first observations of ices in circumstellar disks around low-mass protostars are discussed. We show that the apolar, volatile, CO ices in the upper layers of the slightly inclined edge-on circumstellar disk of L1489 IRS have evaporated. The energetic formation of new molecules, such as CO_2 and 'XCN', does not occur, however. New insights into the nature of apolar ices are evident from the newly detected band of ^{13}CO. In this paper we also present complementary observations obtained with (sub-)millimeter single dish telescopes and interferometers along the same 'icy' lines of sight. These are essential to characterize the line of sight conditions and determine the location of the ices observed in a pencil absorption beam.

Key words: astrochemistry – infrared: ISM – ISM: molecules – planetary systems: protoplanetary disks – stars: formation

1. Introduction

The condensation of atoms and molecules on grains in cold dense clouds leads to the formation of icy mantles. This is a chemically active process in which new molecules are formed. The type of molecules and the composition of the ice mantles being created depends critically on the composition of the accreting gas (Tielens & Hagen 1982). At high atomic hydrogen abundances, H_2 and H_2O are easily formed. While H_2 is released due to its low binding energy on the grains, H_2O creates an icy mantle, likely mixed with hydrogenated forms of C (e.g. CH_4) and CO (e.g., H_2CO and CH_3OH). At later stages, when the accreting gas is primarily molecular, a mantle of different composition is formed, likely CO– and perhaps N_2– or O_2–rich. Thus, it is not unlikely that during the evolution of the molecular cloud, layers of different composition are being formed on the same grain. Also, the ice composition may depend on the location of the grain within the cloud, as the cloud becomes increasingly more atomic from the inner regions to the outskirts. Movement of grains through such a cloud may also lead to the formation of ice layers of different composition. In general, the different types of ices, CO–rich and H_2O–rich, are referred to as "apolar" and "polar" ices due to the different molecular dipole moments. These ices have very different sublimation temperatures: typically 18 K for apolar ices, and 90 K for polar ices on interstellar time scales (Nakagawa 1980).

Once such an icy grain has been formed, it can be subject to a range of processes that destroy or alter the composition and structure of the ice. Heating of the grains by optical or near infrared light can sublimate the ices, and of course the apolar ices evaporate before the polar ices. Heating can also lead to a rearrangement of the molecules in the ice; i.e., crystallize the deposited amorphous ice and segregate the different molecules within (Blake et al. 1991). This is generally referred to as *thermal* processing of the ices. Higher energy photons (ultraviolet, X–ray) and cosmic rays can actually break chemical bonds and

Chemistry as a Diagnostic of Star Formation (2003). *Edited by Charles L. Curry and Michel Fich.* Published in 2003 by NRC Research Press, Ottawa ON, K1A 0R6, Canada. ISBN 0-660-19089-3.

create radicals which will react and form simple (CO_2, CH_3OH) or more complex species (e.g., polycyclic aromatic hydrocarbons or PAHs; hexamethylene tetramine or HMT; OCN^-; Schutte 1999). This is referred to as *energetic* processing of the ices.

It is expected that as a grain with an icy mantle migrates from a dense molecular cloud to a protostellar envelope and circumstellar disk and to a comet or a planetesimal, it is affected by at least some and perhaps all of these processes. Which processes are most important is still largely unknown and thus the question as to whether the pristine ices of dense molecular clouds are included mostly unaltered or heavily processed in present-day solar system ices is as yet unanswered.

We will discuss the observational signatures of ice processing that have been found toward well studied high-mass protostars in Sec. 2. In Sec. 3 we will present observations of ices in low-mass protostars, and in particular address the location of ices along the line of sight. In Sec. 4 we focus on observations of ices in disks of low-mass protostars and we address the question as to whether the ices in disks show signs of processing. Finally, in Sec. 5 we present the discovery of a new molecule in the ices, ^{13}CO, and show that its absorption band significantly constrains the composition of apolar CO ices. Directions for future work are discussed in Sec. 6.

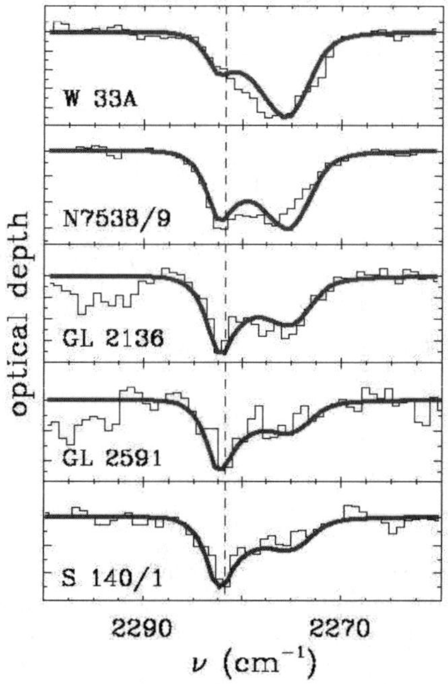

Figure 1. The absorption band of solid $^{13}CO_2$ as a tracer of heating of the envelopes of massive protostars. The thick, smooth lines represent laboratory fits of H_2O, CH_3OH and CO_2 mixed in about equal quantities at temperatures increasing from top to bottom, showing the crystallization effect.

2. Ice-Processing Massive Protostars

There is a fair amount of observational evidence that massive protostars process the ices in their envelopes. Sharp features were observed in the 3.0 μm H_2O ice band toward several massive protostars, indicating ice crystallization (Smith, Sellgren & Tokunaga 1989). Crystallization was also inferred from the $^{12}CO_2$ bending mode at 15 μm (Gerakines et al. 1999), and the $^{13}CO_2$ stretching mode at 4.38 μm (Boogert et al. 2000a). Based on this, an evolutionary sequence has been proposed in which massive protostars progressively thermally process the ices in their envelopes (Fig. 1). This may well occur on a time scale of a few 10^4 years, the typical hot core life time (Boogert et al. 2000a). Such a heating trend is also seen in several other tracers, such as in an increased gas/solid abundance ratio (van Dishoeck et al. 1998), in an increased dust temperature (Boogert et al. 2000a), and in the temperature and density structure of these envelopes (van der Tak et al. 2000). We thus conclude that thermal processing of ices is readily observed toward high-mass protostars and in fact the observations of ice bands seems to be a good *tracer* of the evolution of the envelopes of massive protostars.

3. Low-Mass Protostars: Where are the Ices?

Are the effects of ice processing observed toward high-mass protostars important toward low-mass protostars as well? Like high-mass protostars, the infrared spectra of low-mass protostars are often dominated by ice absorption features, as is shown in the *ISO*/SWS spectrum of Elias 29, a protostar in the ρ Ophiuchi cloud (Fig. 2). The ice bands do not show signs of crystallization and although the solid CO abundance is low, most of the CO is in the apolar phase (Boogert et al. 2000b). Interestingly, abundant warm molecular gas is present in the immediate environment of Elias 29 itself (Boogert et al. 2000b; Ceccarelli et al. 2002). The lack of ice processing can only be properly interpreted if the location of the ices along the pencil absorption line of sight is known. A technique for determining this is illustrated in Figure 3. Bright $C^{18}O$ 2–1 emission is detected in this line of sight in a 32″ single dish beam, at a range of velocities. No emission peak is seen toward Elias 29, however. In contrast, weak CO emis-

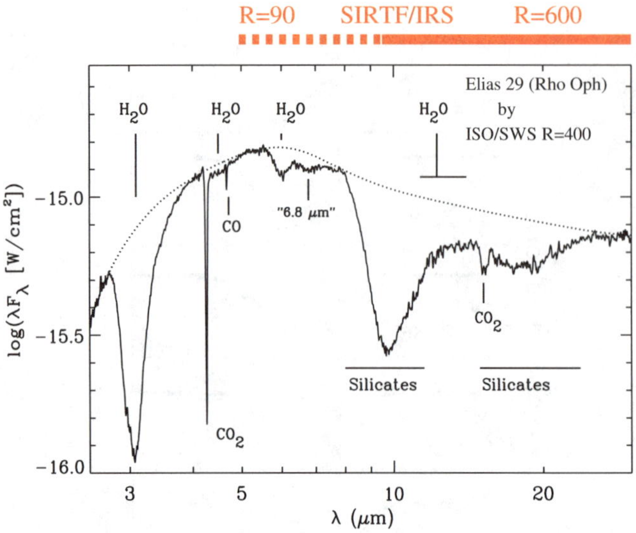

Figure 2. *ISO*/SWS infrared spectrum of Elias 29, a low-mass protostar in the ρ Oph cloud, showing the dominance of ice and dust absorption features. At the top we indicate the wavelength coverage and corresponding resolving power of the IRS spectrometer on board the SIRTF satellite, to be launched in 2003.

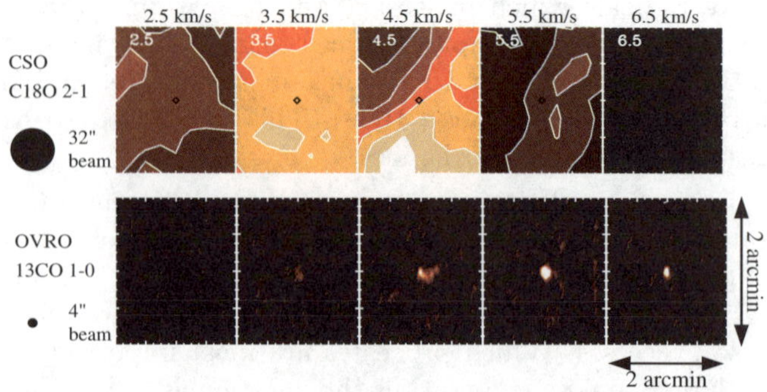

Figure 3. Single dish channel maps of $C^{18}O$ 2−1 emission (top) compared to interferometer ^{13}CO 1 − 0 emission (bottom) toward Elias 29 in the ρ Oph cloud. Note the larger v_{LSR} velocity of Elias 29 in the interferometer map. Also, corrected for the isotope ratio and beam dilution, the expected signal of Elias 29 in the single dish map is a factor of 60 less than the detected emission of extended foreground clouds.

sion is detected on the near infrared position at higher velocities in a 4″ interferometer beam. It is concluded that in this particular line of sight, the foreground clouds have a column corresponding to $A_V = 20$ and are responsible for at least 75% of the detected ices (Boogert et al. 2002c). Hence, this line of sight is less suited to study ices in the direct environment of low-mass protostars. The situation is however quite different in other lines of sight, as is shown in the next section.

4. Low-Mass Protostars: Ices in Disks

A low-mass protostar with a well characterized line of sight is L1489 IRS in the Taurus molecular cloud. Millimeter interferometer maps show a large 2000 AU radius flattened structure around this object (Hogerheijde 2001). Detailed models of the velocity structure in combination with continuum maps indicate that this structure is in fact an edge-on disk in (almost) Keplerian rotation, but with a noticeable infall component. This is thus a highly interesting target for infrared absorption studies of circumstellar gas and ices. We have obtained high spectral resolution ($R = 25,000$; $\Delta v = 12$ km s^{-1}) observations in the atmospheric M band of L1489 IRS with the NIRSPEC spectrometer at the Keck II telescope (McLean et al. 1998). The 4.7 μm spectrum shows deep ^{12}CO $v = 1{-}0$ ro-vibrational absorption lines with extensive red-shifted wings, up to 100 km s^{-1}, confirming the infalling nature of this edge-on disk (Boogert et al. 2002b). The same 4.7 μm Keck/NIRSPEC spectrum also shows a prominent absorption band of solid CO (Fig. 4a). This is a rare case in which ices in a protostellar disk can be directly seen[1]. Other claims for the detection of ices in edge-on disks have been made for Elias 18 (Shuping et al. 2001), RNO 91 (Weintraub et al. 1994), and CRBR 2422.8-3423 (Thi et al. 2002).

4.1. Ice Processing in Disks

The detection of solid CO in the edge-on disk of L1489 IRS allows a direct investigation of the composition and processing history of ices in a protostellar disk. Our approach is a phenomenological one (Boogert et al. 2002a). We compare the CO band profile of L1489 IRS with that observed in a line of sight which is most likely pristine: the massive protostar NGC 7538:IRS9. This protostar has a cold envelope with unprocessed ices and a modest hot core (Mitchell et al. 1990; van der Tak et al. 2000; Boogert et al. 2000a). In fact, its ice spectrum very much resembles that of background stars tracing quiescent molecular cloud material. Figure 4c shows a comparison of the CO band profiles of L1489 IRS and NGC 7538:IRS9. Both spectra have a narrow central peak surrounded by blue and red wings. The red wing is the polar ice component (CO mixed with H_2O), the central peak is the apolar component (likely pure CO; see Sec. 5), and the newly dis-

[1] Foreground clouds toward L1489 IRS have an extinction of $A_V = 2$ (Hogerheijde 2001) and do not harbor ices.

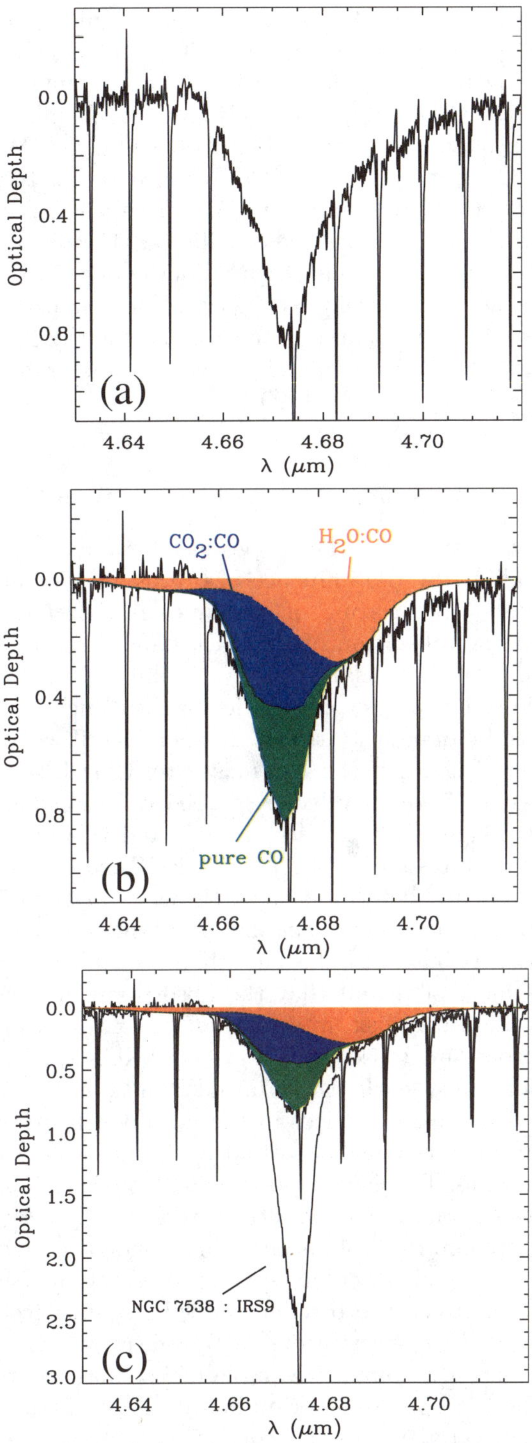

Figure 4. The M band $R = 25,000$ Keck/NIRSPEC spectrum of the low-mass protostar L1489 IRS, showing narrow ^{12}CO absorption lines and a broad multi-component band of solid CO **(a)**. Both the gas and ices reside in the edge-on disk of L1489 IRS. The colored surfaces indicate the laboratory fits discussed in the text **(b)**. Panel **(c)** shows a comparison with the massive protostar NGC 7538:IRS9.

covered blue wing may originate in a CO:CO_2 ice. When both spectra are scaled such that the wings match, it is obvious that the central, apolar peak of L1489 IRS is much less prominent (Fig. 4c). This empirical approach thus leads to the straightforward conclusion that the ices in the L1489 IRS disk are processed in the sense that the most volatile ices have evaporated. At first blush, this is consistent with Hubble Space Telescope images, which indicate a non-negligible disk inclination (0-30° with 0° being edge-on). The pencil absorption beam therefore traces warm upper disk layers where the most volatile ices have evaporated.

Are there, besides evaporation, other processing mechanisms at work in the upper disk layers? If the ices are hit by energetic photons or cosmic rays, it is expected that new molecules such as CO_2 are efficiently formed (e.g., Gerakines, Schutte & Ehrenfreund 1996). This would give rise to a significant broadening of the central absorption band profile. Such broadening is not observed. At high CO_2/CO mixing ratios, a shift to shorter wavelengths is expected. However, the column density of the CO responsible for the blue shoulder, normalized to the H_2O ice column, is the same in both sources (~3%). Hence, in the upper disk layers traced in the L1489 IRS line of sight no energetic formation of CO_2 has occurred. Similarly, the 'XCN' band at 4.62 μm, often thought to be an indicator of energetic processing, is not detected in this line of sight (Chiar et al. 1998). To conclude, in the upper layers of the disk around L1489 IRS thermal processing (sublimation) of the most volatile, apolar, CO–rich ices has taken place, but energetic processing has not. Perhaps the line of sight traces the upper layers of the *outer* flaring disk and the radiation field is soft.

5. Detection of Solid ^{13}CO

The evaporation of apolar ices will alter the gas phase composition and chemistry and thus it is important to accurately know the composition of the apolar ices. However, determining the origin of the ^{12}CO absorption band is complicated by the interaction of light with the interstellar grains. Polarization of the molecular dipoles induces a grain surface charge, which resonates with the dipoles themselves and can lead to large shifts of the peak absorption wavelength and to a broader profile. This effect is largest at large dipole concentrations, i.e. for pure ices. The central feature of the ^{12}CO band may well result from a CO–rich ice (Fig. 4), and thus grain shape effects lead to an ambiguity in detailed fits to this band. However, ^{13}CO molecules are always diluted in a ^{12}CO

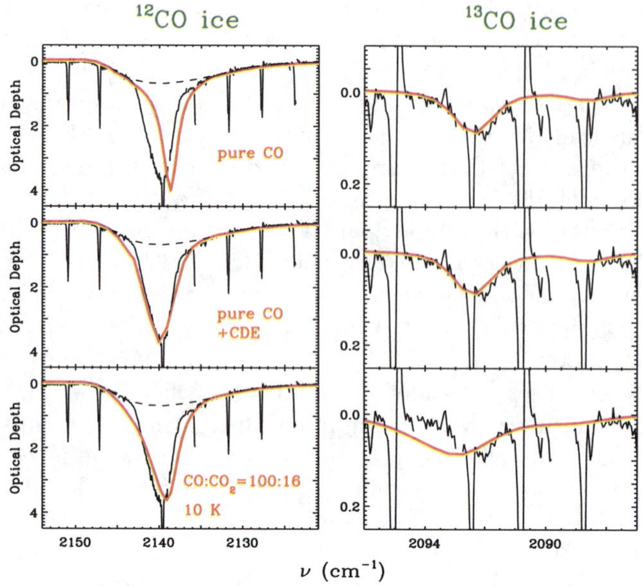

Figure 5. Detection of solid ^{13}CO toward NGC 7538:IRS9 (*right panels*). The left panels show the ^{12}CO band. From top to bottom different laboratory fits are shown: pure CO, pure CO on ellipsoidal grains, and a $CO:CO_2$ mixture

mixture and thus the absorption band of ^{13}CO is not influenced by the particle shape. Therefore, in principle, it is a better tracer of the composition of interstellar CO ices.

We accordingly observed the ^{13}CO stretching mode at 4.78 μm in the massive protostar NGC 7538:IRS9 with Keck/NIRSPEC (Boogert et al. 2002a). The discovery of this band is shown in Figure 5. As can be seen, the high spectral resolution of $R = 25,000$ is crucial in order to separate the ^{13}CO ice band from narrow ^{12}CO and ^{13}CO emission and absorption lines along the line of sight as well as telluric absorption lines.

5.1. New Clues to the Nature of Apolar Ices

We use laboratory simulations to confine the composition of interstellar ices (Fig. 5). The interstellar ^{13}CO band is remarkably narrow (1.50 cm^{-1}). In fact, it is narrower than the ^{13}CO band in most available laboratory mixtures. Best fits are obtained by pure CO ices. Slightly worse, but still reasonable fits are also obtained by mixtures of CO with N_2. $CO:O_2$ mixtures only give good fits when they are heated, which may not be reasonable given the large apolar ice abundance in this line of sight. Interestingly, small contaminations with CO_2 broaden the ^{13}CO band so much that $CO:CO_2$ mixtures can be entirely excluded as the carrier of the prominent narrow central peak of the interstellar CO band (Fig. 5, bottom panels). Finally, as mentioned above, simultaneous fits of the ^{12}CO and ^{13}CO bands indicate the importance of grain shape corrections for the ^{12}CO band. While a pure CO ice matches the ^{13}CO band, it completely mismatches the ^{12}CO band (Fig. 5, top panels), unless we assume the ice mantles are on ellipsoidal grains (Fig. 5, middle panels). A model with a continuous distribution of ellipsoidal shapes, simulating irregularly shaped grains, fits the ^{12}CO band best. Interestingly, the $^{12}CO_2$ bending mode at 15 μm is matched by the same grain shape model (Gerakines et al. 1999)!

6. Conclusions and Future Work

We have shown that the shapes of interstellar ice bands are modified by the influence of protostars. For massive protostars, the heating of envelopes is reflected in increased crystallization of the CO_2 bands. For low-mass protostars it is crucial to study the line of sight conditions in order to interpret the ice bands. Much work still needs to be done in this area. In some detail we analyze the line of sight of the low-mass protostar L1489 IRS. This source is surrounded by a 2000 AU radius edge-on disk, through which we see the gas and ices in absorption in high resolution 4.7 μm Keck/NIRSPEC spectra. We empirically determine that the most volatile CO ices (apolar component) in this disk are largely evaporated. This is attributed to the fact that the disk is inclined (by at most 30°) and that the absorption line of sight traces the warm disk upper layers. In contrast, no energetic processing is seen; i.e., no CO_2 or XCN species have been formed. Clearly, these initial results on the importance of processing of ices in the neighbourhood of low-mass protostars must be tested against a larger sample of sight lines. This is now possible with the availability of high resolution M–band spectrometers on large telescopes (Pontoppidan, this volume). An independent view on the ice processing history of low-mass protostars (and thus on the link between protostellar and solar system ices) will be provided by observations of the sensitive CO_2 bending mode for a complete sample of low-mass protostars with the IRS spectrometer on board the SIRTF satellite. Finally, we report the first detection of the absorption band of solid ^{13}CO toward the massive protostar NGC 7538:IRS9. This offers a new view on the nature of apolar ices. It is concluded that apolar ices are likely purely CO, although $CO:N_2$ mixtures cannot be excluded, on top of ellipsoidally or irregularly shaped grains. Mixtures of CO with CO_2 can be excluded with a high degree of confidence. As is true above, more lines of sight need to be studied in the sensitive band of ^{13}CO. High spectral resolution spectrometers at large telescopes will allow us to do so.

ACKNOWLEDGEMENTS

The research of A. C. A. B. and G. A. B. is supported by the SIRTF Legacy Science program and by the Owens Valley Radio Observatory through NSF grant AST 99-81546.

REFERENCES

Blake, D., Allamandola, L., Sandford, S., Hudgins, D., & Freund, F. 1991, Sci 254, 548

Boogert, A. C. A., et al. 2000a, A&A 353, 349

Boogert, A. C. A., Tielens, A. G. G. M., Ceccarelli, C., Boonman, A. M. S., van Dishoeck, E. F., Keane, J. V., Whittet, D. C. B., & de Graauw, Th. 2000b, A&A 360, 683

Boogert, A. C. A., Blake, G.A., & Tielens, A. G. G. M. 2002a, ApJ 577, 271

Boogert, A. C. A., Hogerheijde, M. R., & Blake, G. A. 2002b, ApJ 568, 761

Boogert, A. C. A., Hogerheijde, M. R., Ceccarelli, C., Tielens, A. G. G. M., van Dishoeck, E. F., Blake, G. A., Latter, W. B., & Motte, F. 2002c, ApJ 570, 708

Ceccarelli, C., Boogert, A. C. A., Tielens, A. G. G. M., Caux, E., Hogerheijde, M. R., & Parise, B. 2002, A&A 395, 863

Chiar, J. E., Gerakines, P. A., Whittet, D. C. B., Pendleton, Y. J., Tielens, A. G. G. M., Adamson, A. J., & Boogert, A. C. A. 1998, ApJ 498, 716

Gerakines, P. A., Schutte, W. A., & Ehrenfreund, P. 1996, A&A 312, 289

Gerakines, P. A., et al. 1999, ApJ 522, 357

Hogerheijde, M. R. 2001, ApJ 553, 618

McLean, I. S., et al. 1998, Proc. SPIE 3354, 566

Mitchell, G. F., Maillard, J.-P., Allen, M., Beer, R., & Belcourt, K. 1990, ApJ 363, 554

Nakagawa, N. 1980, in *IAU Symp. 87, Interstellar Molecules* (Dordrecht: Reidel), 365

Schutte, W. A. 1999, in *Laboratory Astrophysics and Space Research*, ed. P. Ehrenfreund et al. (Dordrecht: Kluwer), 69

Shuping, R. Y., Chiar, J. E., Snow, T. P., & Kerr, T. 2001, ApJ 547, L161

Smith, R.G., Sellgren, K., & Tokunaga, A.T. 1989, ApJ 344, 413

Thi, W. F., Pontoppidan, K. M., van Dishoeck, E. F., Dartois, E., & d'Hendecourt, L. 2002, A&A 394, L27

Tielens, A. G. G. M., & Hagen, W. 1982, A&A 114, 245

van der Tak, F. F. S., van Dishoeck, E. F., Evans, N. J., & Blake, G. A. 2000, ApJ 537, 283

van Dishoeck, E. F., et al. 1998, in *Star Formation with the Infrared Space Observatory*, ASP Conf. Ser. 132, ed. J.L. Yun & R. Liseau (San Francisco: ASP), 54

Weintraub, D. A., Tegler, S. C., Kastner, J. H., & Rettig, T. 1994, ApJ 423, 674

CHEMISTRY IN CIRCUMSTELLAR DISKS AS PROBED BY HIGH RESOLUTION MILLIMETER-WAVE TO INFRARED SPECTROSCOPY

Geoffrey A. Blake

Division of Geological and Planetary Sciences and Division of Chemistry and Chemical Engineering, MS150-21, California Institute of Technology, Pasadena, CA 91125, USA

Abstract

The role of high spectral and spatial resolution spectroscopy in understanding the evolution of the gaseous component of the circumstellar accretion disks encircling low mass stars is described. Millimeter-wave emission lines from trace constituents such as CO, CN, HCO^+, and HCN can be used to probe the kinematic and physicochemical properties in the near-surface regions of disks beyond 100 AU, but, thanks to extensive molecular depletion in the midplane, they are not a reliable proxy for the disk mass. Future observations of the pivotal deuterated species HD and the molecular ion H_2D^+ may provide more robust access to the outer disk midplane.

In the critical planet-forming region from 1 to 30 AU, mid-infrared observations of the pure rotational transitions of molecular hydrogen would alleviate many of the concerns raised by the (sub)millimeter spectral line observations, and results from the *ISO*–SWS instrument on several "transitional" disks are presented. The measurements of these weak lines used beams substantially larger than the disk angular diameter, and so must be verified or refuted by high angular resolution spectroscopy from the ground. Finally, the high-resolution M-band (5 μm) spectroscopy of CO in disks is outlined. Emission lines that are likely optically pumped by hot dust in the inner disk ($R \lesssim 1$ AU) are seen toward inclined systems, while the *absorption* spectra of edge-on disks clearly reveal the molecular depletion inferred at millimeter wavelengths. A wide variety of molecules can be studied with high-resolution spectroscopy in the near- to mid-infrared, and such observations would permit the first direct comparison of cometary and circumstellar ices.

Key words: astrochemistry – infrared: ISM – ISM: molecules – planetary systems: protoplanetary disks – stars: formation

1. Introduction

As the means by which matter and angular momentum are exchanged between molecular clouds and young stars and as the birth sites of planets, circumstellar disks provide the pivotal observational and theoretical link between star formation and (exo-)planetary science. Though disks must be established very early in the star formation process, this overview is concerned primarily with those systems in which the vast majority of accretion onto the central star has been completed. As the accretion process wanes, reprocessing of the central stellar radiation becomes the dominant source of energy release, and in these so-called passive circumstellar disks an examination of the re-radiated longer wavelength photons can be used to learn a great deal about the physical and chemical properties of the disk.

The basic geometry is outlined in Figure 1. Material near the star and especially at the disk surface radiates strongly in the infrared, while the more distant and interior portions of the disk can only be probed at longer millimeter wavelengths where the bulk of the cold dust is optically thin. Near- and mid-infrared imaging can efficiently search for the presence or absence of such excess emission, and has been used to infer the presence of disks around hundreds of young stars. Infrared surveys also yield time scale estimates of order a few million years for the disappearance of small dust grains in the inner regions of circumstellar disks (Skrutskie et al. 1990; Robberto et al. 1999)—a span similar to that measured for the creation of differentiated planetesimals in the early solar nebula from the absolute ages of chondritic meteorites (Amelin et al. 2002). The time scale and associated mechanism(s) by which individual optically thick disks dissipate to become the optically thin, or debris, disks is not yet quantitatively known, but is exceedingly important to nebular chemistry and to the formation of planetary systems. The more tenuous debris disks themselves are estimated to survive for several tens, perhaps hundreds, of millions of years in at least 15 percent of young

Chemistry as a Diagnostic of Star Formation (2003). *Edited by Charles L. Curry and Michel Fich.* Published in 2003 by NRC Research Press, Ottawa ON, K1A 0R6, Canada. ISBN 0-660-19089-3.

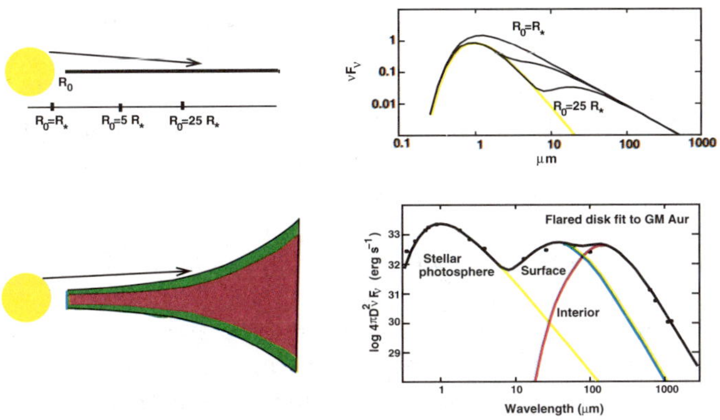

Figure 1. Schematic disk structure cross sections for passive flat versus flaring (hydrostatic equilibrium) disks, and the spectral energy distributions (SEDs) that result. Conceptual models of the latter predict that the infrared portion of the SED is dominated by optically thin radiation from the disk surface layers exposed to the stellar radiation field, while the millimeter-wave emission is dominated by the cooler disk interior (Chiang & Goldreich 1997). From van Zadelhoff (2002).

main sequence stars, as judged from mid- and far-infrared continuum excesses (Habing et al. 1999).

The spatio-temporal properties of disks as inferred from SEDs— namely, masses of 0.001-0.1 $M_\odot$ and sizes of 10's to 100's of AU—are similar to those inferred for the solar nebula, and have been verified for a *handful* of objects by detailed imaging (Mundy et al. 2000; McCaughrean et al. 2000). Direct optical/IR imaging and, especially, interferometry (Monnier & Millan-Gabet 2002) has begun to probe closer to the central stars, but cannot directly constrain critical disk properties such as the surface mass density or accretion flows, while millimeter-wave studies can only probe the outer regions of disks, as described below. Similarly, disk dissipation studies based on infrared excesses tell us little about the gas content of such objects. The discovery of extrasolar 'hot Jupiters' some 0.05 AU from their parent stars (Marcy, Cochran & Mayor 2000) proceedings) has highlighted the important role of star-disk-protoplanet interactions and time scales, and demands new tools that can investigate the critical planet-forming zone of disks, but the advent of the next generation of mid-infrared and millimeter-wave arrays lies several years in the future.

High-resolution, gas phase molecular spectroscopy with existing telescopes and arrays offers a means of addressing many of these issues, and forms the focus of this overview. The present capabilities will be examined as a function of wavelength; Sec. 2 deals with millimeter-wave observations, Sec. 3 presents mid-infrared spectroscopy such as that pioneered by *ISO* and soon to be available with *SIRTF*, and Sec. 4 examines the possibilities of high resolution 2−5 μm spectroscopy with large-format echelle spectrographs on 8−10 m-class, ground-based telescopes.

2. Mm-wave Spectroscopy of Disks

At long wavelengths, where the dust emission is largely optically thin, rotational line emission from disks forms a powerful probe of their physics and chemistry. Indeed, disk material at a fiducial radius of 100 AU, with a temperature of 20−30 K, will radiate preferentially at (sub)millimeter wavelengths. The morphology and kinematics of the circumstellar gas, along with an independent estimate of disk mass, can therefore be derived from spectral lines at millimeter wavelengths. The observational challenges of studying isolated accretion and transitional disks in this region are highlighted in Figure 2, in which it can be seen that even the strongest lines from the most optically thick species are strongly beam-diluted by the resolution achievable with single telescopes such as the IRAM 30 m, the CSO, and the JCMT. These telescopes operate over a wide frequency range, however, and so the combination of millimeter-wave imaging with submillimeter-wave spectroscopy forms a powerful means of constraining the temperature, density, and molecular abundances in disks (van Zadelhoff et al. 2001).

The double-peaked nature of the emission lines from LkCa 15 is consistent with Keplerian rotation in an inclined disk, and the separation of the peaks can be used to estimate a disk radius of $R \sim 300$ AU for an inclination of 60°. TW Hydra is thought to be viewed face-on, hence the narrow lines. These estimates can be checked directly with interferometer maps of the millimeter-wave continuum emission from circumstellar particles to constrain the dust morphology and mass, and with high-resolution molecular line maps to probe the disk kinematics, temperature and density structure, and chemistry. The age, large size, and mass of the LkCa 15 disk make it an important system for further study since it may represent an important transitional phase in which viscous disk spreading and dispersal competes with planetary formation processes (TW Hya, DM Tau and GM Aur likely present additional examples). Further, only the disk gas beyond $R \sim$ 100 AU can be studied with present (sub)millimeter-wave arrays and single dish telescopes due to sensitivity limitations and beam dilution (Dutrey, Guilloteau, & Guélin 2000), and so such large systems provide the only op-

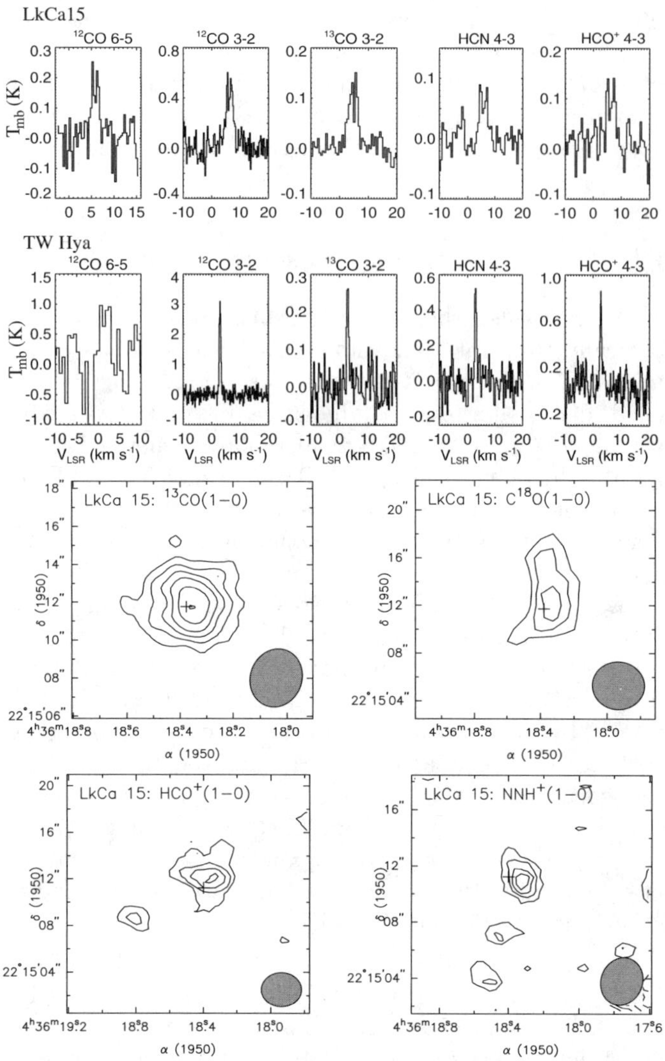

Figure 2. Top: Submillimeter-wave emission lines from the LkCa 15 and TW Hydra circumstellar disks, observed with the JCMT and CSO telescopes atop Mauna Kea (van Zadelhoff et al. 2001). *Bottom:* OVRO Millimeter Array interferometric images of the ^{13}CO, $C^{18}O$, HCO^+, and N_2H^+ $J = 1$–0 emission from the disk encircling LkCa 15 (Qi et al. 2003). The hatched ellipses at lower right present the synthesized beams obtained with natural weighting.

portunity at present to shed light on the chemical and physical gradients in disks.

As Fig. 2 shows, it is now possible to image the emission from several isotopologues of CO and ions such as HCO^+/N_2H^+ in optically thick circumstellar disks. The combined millimeter-wave images and higher$-J$ spectra have been examined with detailed 2D Monte Carlo radiative transfer statistical equilibrium models which reveal that the molecular emission arises from near-surface disk layers with $n_{H_2} \approx 10^6$–$10^8\,\mathrm{cm}^{-3}$, $T \gtrsim 30\,\mathrm{K}$ that are strongly influenced by the reprocessing of (inter)stellar optical, UV, and X-ray photons (van Zadelhoff et al. 2001; Aikawa et al. 2002). The strength of the isotopically substituted CO lines indicates $\tau(C^{18}O) < 1$. Still, the estimated disk mass using "standard" dark cloud $C^{18}O$ abundances is nearly two orders of magnitude less than that deduced from millimeter-wave emission from dust (Qi et al. 2003). This discrepancy is most often attributed to the depletion of molecules onto the icy mantles of dust grains near the dense, cold disk midplane (a topic to which we shall return in Sec. 4) and shows that while millimeter-wave rotational line emission is an excellent tracer of the outer disk velocity field, it is not a robust tracer of the disk mass. Despite the potentially large uncertainties associated with processes such as grain coagulation and growth, dust emission at long wavelengths remains the best tracer of disk mass presently available.

That further surprises await the new observational capabilities at millimeter-wavelengths to be deployed over the next decade (especially arrays such as the SMA and, ultimately, ALMA) is reinforced by the OVRO LkCa 15 observations of CO, CN, and HCN presented in Figure 3. While the CO is peaked at the stellar position, as expected, the CN and HCN emission show maxima offset from the star along the major axis of the disk, as measured by the CO velocity distribution (emission along the minor axis is heavily beam-diluted). This offset demands that the HCN and CN be formed *in situ*. As described in further detail by Kessler et al. in these proceedings, 2D Monte Carlo spectral line models with a radial CN/HCN annulus reproduce the observations quite well, provided $R_{\mathrm{CN,HCN}} \gtrsim 250\,\mathrm{AU}$.

The location of the maxima may be related to the sublimation behavior of various molecules, for the release of CO from grain mantles can drive a carbon-rich gas phase and grain mantle chemistry in the near-surface regions of certain disk models (Aikawa & Herbst 1999). The sharpness of the inner boundary of the observed morphology should reflect both the temperature profile and radial transport in the disk. Further, CN and HCN appear quite similar at the present resolution. CN is made photochemically from HCN, and for the distributions of these species to be similar there must be sufficient vertical mixing occurring within the disk. At higher resolution and at appropriate disk orientations, it should be possible to measure the vertical mixing rate directly from the images and the estimated chemical lifetimes of CN and HCN. In more face-on disks, the ratio of CN/HCN itself

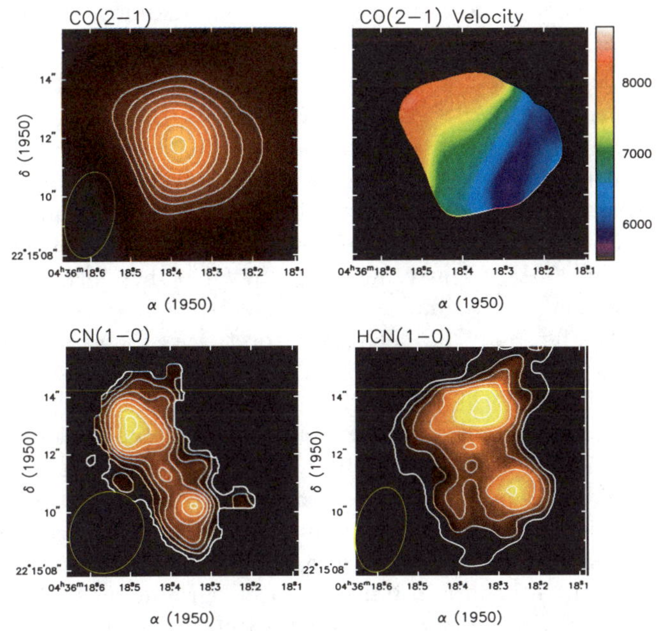

Figure 3. OVRO Millimeter Array maps of the CO $J = 2$–1 and CN/HCN $J = 1$–0 emission from the LkCa 15 circumstellar disk. The ellipses in each panel depict the synthesized beams obtained with robust weighting. The disk velocity field as traced by CO is plotted at upper right, red velocities lie to the NE and blue to the SW (data kindly provided by D. Koerner and A. Sargent; see also Simon, Dutrey, & Guilloteau 2001). Despite the similar single-dish line shapes, the radial distributions of these species are clearly quite distinct.

is a sensitive function of the radiation field and vertical mixing time scale. Such depletion and photochemical processes should also operate in older transitional and debris disks, and may well mean that CO is again not a good tracer of the disk mass (Zuckerman, Forveille, & Kastner 1995; van Zadelhoff 2002). Depletion will tend to increase the fractional ionization near the disk midplane since the H_3^+ abundance is now controlled by the dissociative recombination with electrons, and so future observations of the H_2D^+ ion at submillimeter wavelengths may provide better access to the dense, cold gas in the outer disk midplane (Kessler et al., these proceedings).

3. H_2 Spectroscopy of Circumstellar Disks

High sensitivity mid-infrared observations of the pure rotational lines of H_2, such as those made possible by the Infrared Space Observatory (*ISO*), have a number of advantages over more commonly used indirect tracers for assessing the molecular gas content of optically thick and debris disks. For a gas of solar composition, molecular hydrogen is the overwhelmingly most abundant gaseous species and does not deplete significantly onto grain mantles, obviating the uncertainties associated with the gas phase conversion factors needed for other molecules. Further, the self-shielding nature of molecular hydrogen with respect to photodissociation renders it sufficiently stable at low visual extinctions that it should persist in disks with gas masses as small as several $M_\oplus$.

Because H_2 is a homonuclear, diatomic molecule, the intrinsic Δv=0, ΔJ=2 transition strengths are quite small and the optical depths the mid-infrared lines are therefore low. While this can greatly simplify the radiative transfer analysis, it becomes critical to understand the dust and gas thermal structure in the near-surface regions of disks. Recent models of the disk temperature field predict that the gas temperature will exceed that of the dust, and so the lines are expected to appear in emission and to be optically thin. Except for edge-on disks (Sec. 4), nearly any orientation can be studied. Substantially higher temperatures are required for excitation of the H_2 vibration-rotation manifold near 2 μm, while special geometries are needed to conduct either near-infrared or vacuum ultraviolet (electronic) absorption line measurements. Thus, observations of the H_2 mid-infrared lines are the most direct means to directly probe the bulk of the warm molecular gas in the inner ~100 AU of disks.

We have therefore utilized the *ISO*–SWS instrument, which provided the first opportunity to conduct measurements of the H_2 J=2→0 S(0) 28 μm and J=3→1 S(1) 17 μm pure rotational transitions above the Earth's atmosphere. While these are the lowest-energy transitions of the H_2 molecule and so the most appropriate with which to study disks, they are difficult (17 μm) to impossible (28 μm) to observe with ground-based spectrographs due to telluric absorption. While a number of classical T Tauri stars, Herbig Ae stars, and young (pre-)main sequence stars with confirmed circumstellar disks have been observed (Thi et al. 2001a), here the emphasis is on three sources previously classified as debris disks; that is, low-mass disks with a dust population that is presumed to be dominated by second generation grains, but for which there is also evidence of gas within the disk. These may then be better thought of as "transitional" disks, and the detection of molecular hydrogen in such objects illustrates its potentially unique role in assessing the gas content of protoplanetary accretion disks versus age.

It is important to stress that the aperture of the *ISO*–SWS observations is $20'' \times 27''$ at the S(0) line and $14'' \times 27''$ at the S(1) transition, corresponding to 400×540 AU and 280×540 AU, respectively, at the dis-

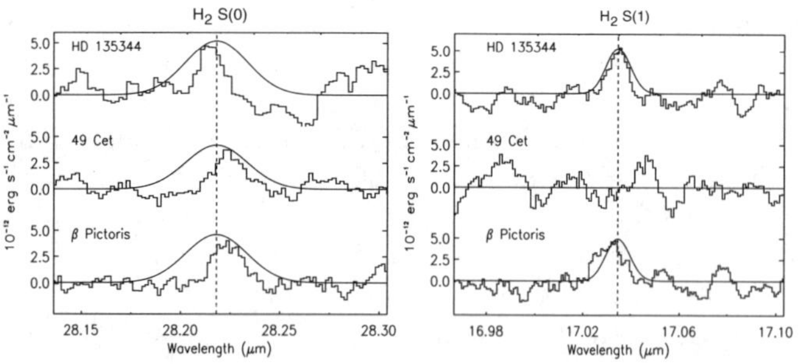

Figure 4. *ISO*–SWS mode 2 spectra of the H_2 S(0) and S(1) pure rotational lines toward the three transitional disk candidates HD 135344, 49 Ceti, and β Pictoris (Thi et al. 2001b). Similar spectra are observed toward the more massive, optically thick disks encircling classical T Tauri and Herbig Ae stars (Thi et al. 2001a).

tance of β Pictoris. The dashed vertical line in Figure 4 depicts the rest wavelengths of the H_2 transitions, while the solid curve illustrates the range of potential wavelength shifts induced by grating repositioning errors or pointing offsets of the spacecraft. All line shapes are consistent with the instrumental spectral resolution of $\lambda/\Delta\lambda \sim 2000/2400$ at $28.218/17.035\,\mu$m. Typical integration times were 600−1000 sec per line, in which the 12 SWS detectors were scanned several times over the 28.05−28.40 and $16.96-17.11\,\mu$m intervals around the H_2 lines. Individual scans are examined for cosmic ray glitches, dark current drifts, and readout nonlinearities before being summed to produce a final spectrum (Valentin & Thi 2000). Off-source integrations toward 49 Ceti and HD 135344 reveal no H_2 S(1) emission to a limiting flux of $8 \times 10^{-15}\,\mathrm{erg\,s^{-1}cm^{-2}}$ (root mean square).

As Fig. 4 shows, the measured H_2 intensities lie close to the *ISO* sensitivity; indeed, their detection required the development of special software to supplement the standard SWS Interactive Analysis Package (see Valentin and Thi 2000 for a thorough discussion of this software). The S(1)/S(0) ratios yield excitation temperature estimates of $T \approx 100\,$K assuming thermal equilibrium, and so the total gas mass estimates may have large error bars since the transition upper states lie at 510 and 1015 K above ground, respectively. Values range from a minimum of $0.17\,M_J$ (Jupiter mass) for β Pictoris to just over $6.6\,M_J$ for HD 135344. Only an upper limit to the S(1) flux is available for 49 Ceti, which constrains the temperature to be less than $\sim 100\,$K, corresponding to gas masses of at least $0.35\,M_J$. The uncertainties in the inferred masses are estimated to be a factor of three, and stem mainly from the weakness of the lines and the unknown molecular hydrogen ortho/para ratio in such disks.

The gas masses of these disks actually enhance the need for the *in situ* generation of $1-100\,\mu$m dust grains because such particles can be efficiently ejected from the system in the presence of radial gas gradients (Klahr & Lin 2001). Clearly, massive Jovian planet formation is no longer possible in such systems, and so as noted above they may be better interpreted as tenuous remnants of substantially more massive accretion disks. If so, HD 135344 and 49 Ceti could be extremely interesting transition objects in which Jovian planet formation may no longer be possible but for which the residual gas provides the opportunity to image gravitational perturbations in the disks due to existing planets formed at earlier stages. Such observations were, of course, impossible with *ISO*, but echelle mid-infrared spectrographs on 8−10 m telescopes provide the necessary spectral and spatial resolution. The software used to extract the lines presented in Fig. 4 has been carefully checked with observations of low-luminosity PDRs, and the agreement with theory is found to be excellent (Thi 2001). Still, the large-beam nature of the measurements must be stressed along with the compelling need to verify or refute the *ISO* observations. The *ISO* spectrometers were exquisitely sensitive to low-level H_2 emission from extended objects that will be nearly impossible to measure from the ground due to the bright sky at $17/28\,\mu$m, and so the detected lines need not arise from the disks themselves if they are surrounded by tenuous envelopes.

The initial steps in the ground-based search for pure rotational line emission have been taked with the TEXES high-resolution echelle spectrograph. While a number of the sources detected with *ISO* have not been detected with high spectral and spatial resolution observations (see Richter et al. 2002), as Figure 5 shows, narrow H_2 emission lines have been detected from the Class I protostars T Tau and HL Tau and from the Herbig Ae star AB Aur (which was detected in the *ISO*–SWS observations; Richter et al., in preparation). These ground-based observations on 4 m-class telescopes are difficult and require exceptional conditions both in terms of seeing and atmospheric transparency, but future observations with spectrographs such as TEXES and MICHELLE on 8−10 m-class telescopes will enable a wide variety of sources to be studied in these pivotal mid-infrared transitions.

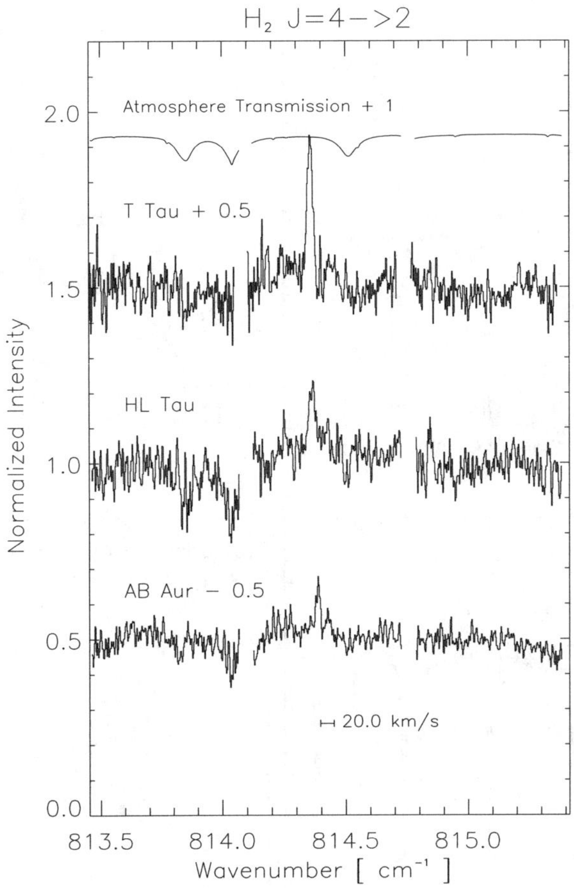

Figure 5. TEXES echelle mode spectra of the S(2) transition of molecular hydrogen toward T Tauri, HL Tauri and AB Aur, acquired at the IRTF (Richter et al., in preparation). The lines are very narrow, and at least in the case of AB Aur likely arise from the surface regions of the circumstellar disk.

4. M-Band Spectroscopy of Circumstellar Disks

The pure rotational spectroscopy outlined above principally traces the near-surface disk regions, and with current capabilities only the disk beyond 100 AU is detectable in pure rotational molecular transitions at millimeter wavelengths. CARMA (the Combined Array for Research in Millimeter Astronomy, the joining of the Caltech Owens Valley and Berkeley-Illinois-Maryland Millimeter Arrays) will improve this limit by factors of two to three, and ALMA will be able to examine disks in the nearest star-forming clouds with resolution as fine as 1 AU in the continuum and at several AU in molecular lines. Similarly, K-band interferometry has now resolved the brightest circumstellar disks (Monnier & Millan-Gabet 2002), and the VLTI and KI systems are targeting the regions of disks with $R < 5$–10 AU via 10 μm interferometry, for this is where the bulk of planet formation occurs.

Indeed, recent theoretical studies (Armitage et al. 2002) suggest that a relatively constant rate of planet formation near 5 AU followed by migration is consistent with the presently known distribution of orbits in extraterrestrial planetary systems (Marcy, Cochran & Mayor 2000). The models further suggest that giant planets can migrate inward from their formation zones after they have opened gaps in the disk (Lin et al. 2000; Ward & Hahn 2000), and so the observational characterization of gaps and of the fraction of disks containing Jovian protoplanets therefore forms an pivotal counterpoint to the highly successful radial velocity extrasolar planet searches.

Only high-resolution spectroscopy with 8–10 m-class telescopes permits robust access to the disk physical conditions and velocity fields on these spatial scales at present (see Najita et al. 2000). Indeed, the potential kinematic signatures are very large. Jupiter induces a stellar velocity wobble of only 13 m s^{-1}, for example, while alterations to the disk kinematics can be a large fraction of the orbital velocity of ~13 km s^{-1} at 5 AU. Gas in and near the $\lesssim$ AU gaps opened up by protoplanets can be heated by the star and, if the planet is massive enough, by shocks (Bryden et al. 1999; Kley 1999). The total amount of hot, dense gas is small, but it radiates strongly in the IR since it is adjacent to the low-density gap. Abundant species such as H_2, H_2O, and CH_4 are excellent candidate tracers, but they are very difficult to observe from the ground.

CO is widespread through the disk (and any surrounding envelope) thanks to its stability, and while the pure rotational lines of CO at millimeter wavelengths trace the outer disk, the $\Delta v = 2$ overtone emission near 2.3 μm can only arise from the several thousand degree gas immediately adjacent to the young star (Scoville et al. 1983; Najita et al. 1996)—but this can only be observed in a handful of stars. CO absorption at 2.3 and 4.6 μm traces cold gas and ice with moderate optical depths, leaving *emission* from the the CO vibrational fundamental near 4.6 μm as a potential gap/protoplanet tracer. In a pioneering study of 8 spectroscopic binary systems, Najita et al. (2000) show that 5 have measurable CO $v = 1 \rightarrow 0$ emission that is proposed to arise from a gap in the *circumbinary* disk, an interpretation consistent with the CO line shapes. In DQ Tau (Carr, Mathieu, & Najita 2001), for example, the 4.6 μm integrated intensities are ~10^{-16} W m^{-2} with a FWHM of ~50–55 km s^{-1}. The total amount of radiating material is only 10^{-5} earth

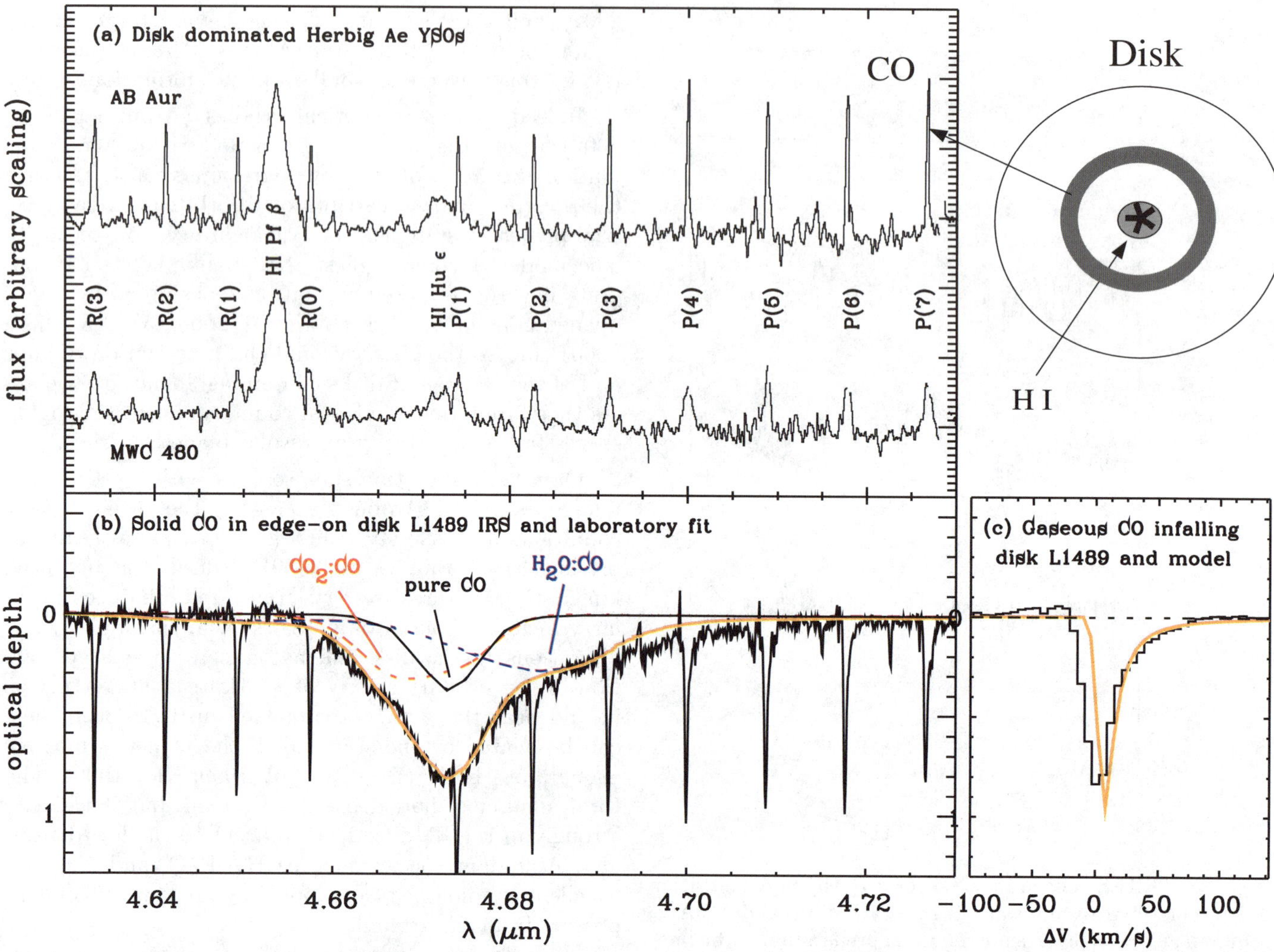

Figure 6. Keck/NIRSPEC R=25,000 spectra of the CO v= 1–0 band toward Herbig Ae stars (AB Aur, MWC 480) along with the nearly edge-on, collapsing disk surrounding L1489 IRS. The two H I lines, Pfund β and Humphries ϵ, arise from the accretion flow onto the star in AB Aur and MWC 480, while the red-shifted wings in the L1489 gas phase lines trace infall to within $\sim$0.1 AU of the central star (Boogert, Hogerheijde, & Blake 2002). Interestingly, no H I emission is seen in L1489 despite the measured radial inflow from the disk. The fits to the solid state CO feature show that the grain mantles possess both polar and apolar components, which can be used to examine the thermal state and history of the disk material.

masses, with $T \gtrsim 1100$ K. For $T \sim 300$ K, the same line strength would correspond to a mass of only $\sim 0.01\,M_\oplus$ at the distance of Taurus!

We have sought to expand upon this work by carrying out an extensive CO M-band spectroscopic survey of T Tauri and Herbig Ae stars in the nearest molecular clouds, which complements the ongoing work of Rettig and co-workers (see, for example, Brittain & Retting 2002). As is shown below, the disk inclination is critical in establishing the nature of the CO spectrum, and here we note that infrared spectroscopy has the pivotal advantage over millimeter-wave studies that it can examine *both* the gas phase and solid state.

Figure 6 shows the power of high-resolution, high dynamic range IR spectroscopy of isolated disks, where CO *emission* is commonly observed (Brittain & Rettig 2002; Blake & Boogert 2003). The velocity widths yield characteristic radii of several × 0.5–3 AU that are well correlated with, but larger than, the 2 μm sizes derived from interferometry (Monnier & Millan-Gabet 2002). The measured linewidths also follow the disk inclination, with face-on systems having the narrowest velocity widths,

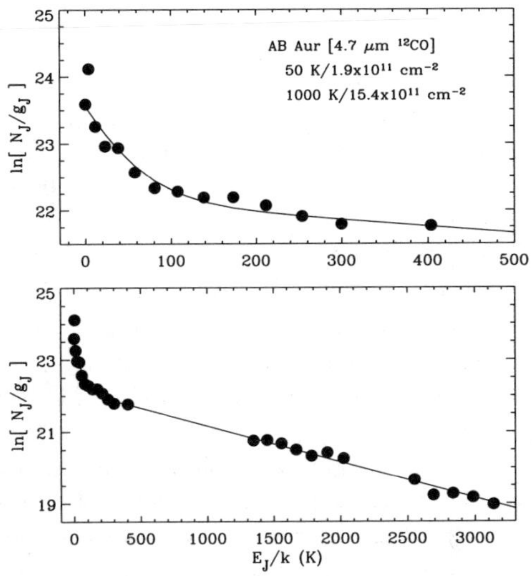

Figure 7. Rotation diagram of the ^{12}CO M-band emission from the AB Aur circumstellar disk. A two-temperature fit with $T_{\rm rot}$= 50 and 1000 K can formally fit the data, but it is likely that a wide range of excitation conditions are present within the NIRSPEC slit width of 0.42″.

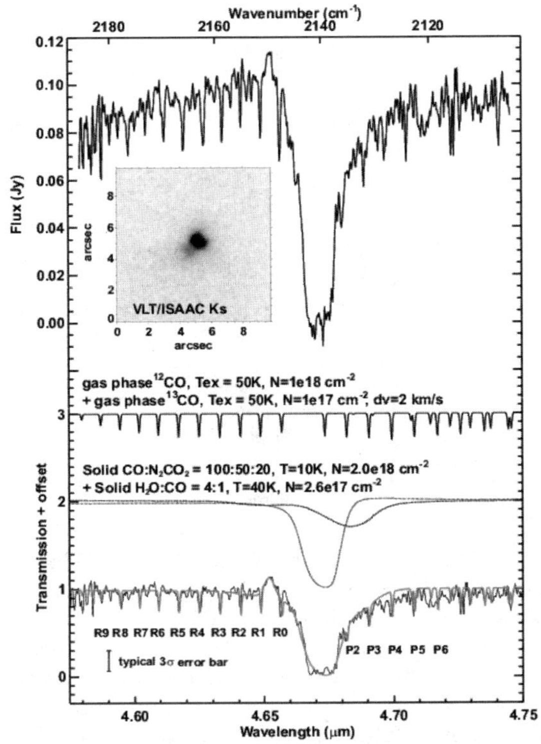

Figure 8. VLT ISSAC R=10,000 M-band spectra of the edge-on disk CRBR2422.8-3423, from Thi et al. (2002). The VLT K-band image of this source is shown in the upper inset (Brandner et al. 2000), while the lower panel displays the solid state and gas phase fits that yield an ice-to-gas CO ratio $\sim$1. Fits to the solid state profile yield the fractions of CO present in apolar and polar matrices (cf. Fig. 6, Boogert et al., these proceedings).

so it is unlikely that these lines originate in an outflow or disk wind. A range of excitation temperatures, $\sim$50−1500 K, are found even within single disks, but the bulk of the gas lies at temperatures of 500−1000 K (see Fig. 7). The total line emission is typically less than 10 percent of the 5 μm continuum flux. Neither the continuum nor line emission is resolved at the typical 0.4″−0.5″ seeing conditions in M-band.

These properties are consistent with the irradiation of a shadowed region of the disk *surface* by a "puffed up" ring of hot dust responsible for the near-IR continuum in these sources (Dullemond, Dominik, & Natta 2001) whose extent has been measured by K-band interferometry. The total amount of radiating material is difficult to estimate since the source is unresolved, but using the slit width ($\sim$0.45″) results in very small masses—typically $\ll$ M$_\oplus$, since IR radiative pumping is very efficient at low optical depths. This same process should occur for any molecule with active vibrations, and the examination of several species would provide exceptional probes of the disk physical conditions at radii near 1 AU. In this regard, the recent detection of H_3^+ emission in HD 141569 by Brittain & Rettig (2002) heralds the advent of an exciting era of discovery in which specific tracers of disk/planetary system evolution (the assembly of proto-Jovian sub-nebulae, gap formation, planet migration, etc.) can be investigated in large samples of objects with spectroscopy. Interferometry could then be used to follow up these surveys with high spatial resolution images.

As the inclination angle of the disk with respect to the observer increases, eventually it will begin to occult the central star and the hot dust near it. Such edge-on orientations, while statistically rare, provide a unique opportunity for several reasons: (1) As Figures 8 and 9 show, the disk can now be probed via *absorption* spectroscopy, which means that both the gas phase and solid state components of the disks can be traced. The depletion of molecules onto grains has long been suspected and is vital to planetesimal growth, but can only be directly tested via observations of edge-on disks. (2) Due to the large filling factor of the inner disk and the high column density at large radii, infrared spectroscopy of edge-on disks is sensitive to all disk regions beyond the hot dust zone. (3) Thanks to the observational geometry,

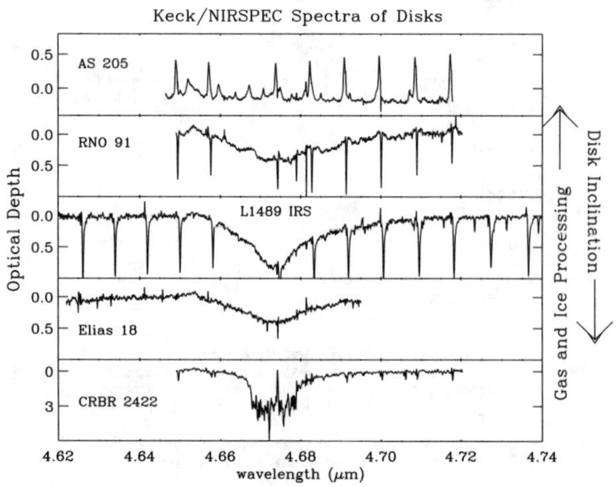

Figure 9. NIRSPEC R=25,000 CO spectra of the edge-on disks RNO91, L1489, Elias 18, and CRBR2422 on an optical depth scale, showing both gas phase and solid state features (the ice-to-gas ratio increases downward as successively denser regions near the disk midplane are probed). RNO91 shows broad emission (inner disk) and narrow absorption (outer disk), while the CO ice band in CRBR2422 is highly saturated (note the large change in the optical depth scale). Depletion estimates for this source are therefore lower limits. Finally, the rich emission line spectrum from the isolated T Tauri star AS205 is plotted at the top for comparison with the spectra observed from disk surfaces.

the disk *radial* velocity field can be examined over a wide range of distances, as we have shown for L1489 (Boogert, Hogerheijde, & Blake 2002). Radial mass transport is obviously a critical function of accretion disks, but only absorption spectroscopy of edge-on disks provides access to $R > 0.1$ AU at present. (4) With sufficient sensitivity, regions of the disk much closer to the midplane can be probed than are sampled by near-/mid-infrared emission spectroscopy.

Observations of edge-on disks at slightly different inclination angles make it possible to sample different heights within the disk, with slightly tilted disks offering access to near-surface layers and nearly perfectly edge-on disks sampling closer to the midplane. Molecular depletion should vary systematically with inclination angle, and as molecules deposit themselves onto grains there should be substantial changes in the gas fractional ionization. This has implications for mass transport in disks via magnetic instabilities (Balbus & Hawley 1991), and so it will be essential to probe both neutral and ionic species in future observations. Of the various edge-on disks discovered over the past decade, only a handful of these can be studied with even the excellent sensitivity of Keck/Gemini/VLT, thanks to the large extinction provided by the disk itself.

Fortunately, of the edge-on disks amenable to study, a variety of disk heights (inclinations) appear to be probed. The superb resolution available with echelle instruments such as NIRSPEC (and ISAACS on the VLT) not only permits the detection of gas phase lines, but also enables the shapes of ice bands to be examined in detail. The shape contains important information about the nature of the ices containing CO, and both Keck and VLT studies have shown that the CO ice-band profile is remarkably similar in various lines of sight, and can be interpreted in terms of the fundamental molecular interactions within the ice (Boogert, Blake, & Tielens 2002; Pontoppidan et al. 2003). With more extensive spectral line surveys of edge-on disks, such as will be possible via the combination of ground-based KLM-band surveys with results from the SIRTF IRS instrument, a detailed comparison of disk and cometary ices will become possible for the first time. Disks such as CRBR2242 (Brander et al. 2000) will be especially interesting in this regard, since the very deep CO ice band (Thi et al. 2002, Fig. 8) promises the detection of ice features from a wide suite of minor ice constituents observable in cometary comae.

Acknowledgements

The results outlined here form the thesis and postdoctoral research of several young scientists at Caltech and in Prof. Ewine van Dishoeck's astrochemistry group at Leiden University with whom it has been my great privilege to learn these past several years. Their dedication and camaraderie are gratefully acknowledged. The OVRO Millimeter Array and Caltech Submillimeter Observatory are operated with support from the National Science Foundation grants AST 99-81546 and AST 99-80846. The NIRSPEC data were obtained at the Keck II telescope of the W.M. Keck Observatory, which is operated as a scientific partnership among the California Institute of Technology, the University of California and the National Aeronautics and Space Administration. The Observatory was made possible by the generous financial support of the W.M. Keck Foundation.

References

Aikawa, Y., & Herbst, E. 1999, A&A, 351, 233

Aikawa, Y., van Zadelhoff, G.J., van Dishoeck, E.F., & Herbst, E. 2002, A&A, 386, 622

Amelin, Y., Krot, A.N., Hutcheon, I.D., & Ulyanov, A.A. 2002, Science, 297, 1678

Armitage, P.J., Livio, M., Lubow, S.H., & Pringle, J.E. 2002, MNRAS, 334, 248

Balbus, S.A. & Hawley, J.F. 1991, ApJ, 376, 214

Blake, G.A. & Boogert, A.C.A. 2003, ApJ, submitted

Boogert, A.C.A., Hogerheijde, M.R., & Blake, G.A. 2002a, ApJ, 568, 761

Boogert, A.C.A., Blake, G.A., & Tielens, A.G.G.M. 2002b, ApJ, 577, 271

Brandner, W. et al. 2000, A&A, 364, L13

Brittain, S.D., & Rettig, T.W. 2002, Nature, 418, 57

Bryden, G., Chen, X., Lin, D.N.C, Nelson, R.P., & Papaloizou, J. 1999, ApJ, 514, 344

Carr, J.S., Mathieu, R.D., & Najita, J.R. 2001, ApJ, 551, 454

Chiang, E.I. & Goldreich, P. 1997, ApJ, 490, 368

Dullemond, C.P., Dominik, C., & Natta, A. 2001, ApJ, 560, 957

Dutrey, A., Guilloteau, S., & Gúelin, M. 2000, in *Astrochemistry*, IAU Symp. 197, ed. Y.C. Minh & E.F. van Dishoeck (San Francisco: ASP), 415

Habing, H.J. et al. 1999, Nature, 401, 456

Klahr, H.H., & Lin, D.N.C. 2001, AJ, 554, 1095

Kley, W. 1999, MNRAS, 303, 696

Lin, D.N.C., Papaloizou, J., Terquem, C., Bryden, G., & Ida, S. 2000, in *Protostars & Planets IV*, ed. V. Mannings, A.P. Boss, & S.S. Russell (Tucson: Univ. Arizona Press), 1111 (*PPIV*)

Marcy, G.W., Cochran, W.D., & Mayor, M. 2000, *PPIV*, 457

McCaughrean, M.J., Stapelfeldt, K., & Close, L. 2000, *PPIV*, 485

Monnier, J.D. & Millan-Gabet 2002, ApJ, 579, 694

Mundy, L.G., Looney, L. & Welch, J. 2000, *PPIV*, 355

Najita, J., Carr, J.S., Glassgold, A.E. Shu, F.H., & Tokunaga, A.T. 1996, ApJ, 462, 919

Najita, J.R., Edwards, S., Basri, G., & Carr, J. 2000, *PPIV*, 457

Pontoppidan, K.M. et al. 2003, A&A, submitted

Qi, C., Kessler, J.E., Koerner, D.W., Sargent, A.I., & Blake, G.A. 2003, ApJ, submitted

Richter, M.J., Jaffe, D.T., Blake, G.A., & Lacy, J.H., 2002, ApJ, 572, L161

Robberto, M., Meyer, M.R., Natta, A., & Beckwith, S.V.W. 1999, in *The Universe as Seen by ISO*, ESA SP-427, 195

Scoville, N.Z., Kleinmann, S.G., Hall, D.N.B., & Ridgway, S.T. 1983, ApJ, 275, 201

Simon, M., Dutrey, A., & Guilloteau, S. 2001, ApJ, 545, 1034

Skrutskie, M.F., Dutkevitch, D., Strom, S.E., Edwards, S., Strom, K.M., & Shure, M.A. 1990, AJ, 99, 1187

Thi, W.F. 2001, Ph.D. thesis, University of Leiden

Thi, W.F. et al. 2001a, Nature, 409, 60

Thi, W.F. et al. 2001b, ApJ, 561, 1074

Thi, W.F. et al. 2002, A&A, 394, L27

Valentin, J.A. & Thi, W.F. 2000, Exp. Astron., 10, 215

van Zadelhoff, G.J., van Dishoeck, E.F., Thi, W.F., & Blake, G.A. 2001, A&A, 377, 566

van Zadelhoff, G.J. 2002, Ph.D. thesis, University of Leiden

Ward, W.R., & Hahn, J.M. 2000, *PPIV*, 1135

Zuckerman, B., Forveille, T., & Kastner, J.H. 1995, Nature, 373, 494

DEUTERIUM IN PROTOPLANETARY DISKS

Jacqueline E. Kessler[1], Geoffrey A. Blake[1,2], and Chunhua Qi[3]

[1]Division of Chemistry and Chemical Engineering, California Institute of Technology, Pasadena, CA, 91125, USA
[2]Division of Geological and Planetary Sciences, California Institute of Technology, Pasadena, CA, 91125, USA
[3]Harvard-Smithsonian Center for Astrophysics, Boston, MA, 02138, USA

ABSTRACT

The deuterium to hydrogen ratio is a sensitive tracer of the conditions of the star formation process and observations of deuterium in disks can be used to constrain the location and timescales of comet formation. In this study, we searched for the deuterated species DCN and HDO toward the T Tauri protostar LkCa 15 and the Herbig Ae stars HD 163296 and MWC 480 using the Owens Valley Radio Observatory Millimeter Array. HDO was detected toward LkCa 15 and upper limits for the HDO in MWC 480 and DCN in LkCa 15 and HD 163296 were obtained. The DCN to HCN fraction in LkCa 15 was calculated from the resulting upper limit for the DCN abundance and previously observed HCN and $H^{13}CN$ abundances toward the disk. HDO/H_2O was estimated through comparison with predicted H_2O abundances from chemical models, scaled by the HCN(modeled)/HCN(observed) abundances. The resulting deuterium fractionations in LkCa 15 were found to be similar to those found toward the T Tauri star disk TW Hya and those observed for molecular clouds and hot cores. The structure of emission toward LkCa 15 was modeled and is discussed.

Key words: Planetary systems: protoplanetary disks – ISM: molecules – stars: individual(LkCa 15)

1. INTRODUCTION

Deuterium abundances have been measured in several objects ranging from planets and meteorites to comets to hot cores and molecular clouds (Fig. 1). The D/H ratio varies by up to three orders of magnitude between molecular clouds and the planets in our solar system, indicating that deuterium can be a sensitive tracer of evolution of protoplanetary material. In particular, the D/H ratio indicates the temperature structure of the observed source, as deuterium atoms replace hydrogen most efficiently at low temperatures ($T \leq 200$ K). In the case of comets and meteorites, the deuterium abundance traces the temperature of the formation site when these planetesimals were assembled. The origin of comets has thus been described as being largely "interstellar" as opposed to "nebular" through comparison of cometary D/H values ($\sim 10^{-3} - 10^{-4}$) with those of molecular clouds ($\sim 10^{-2}$) and that estimated for the protosolar nebula (10^{-5}).

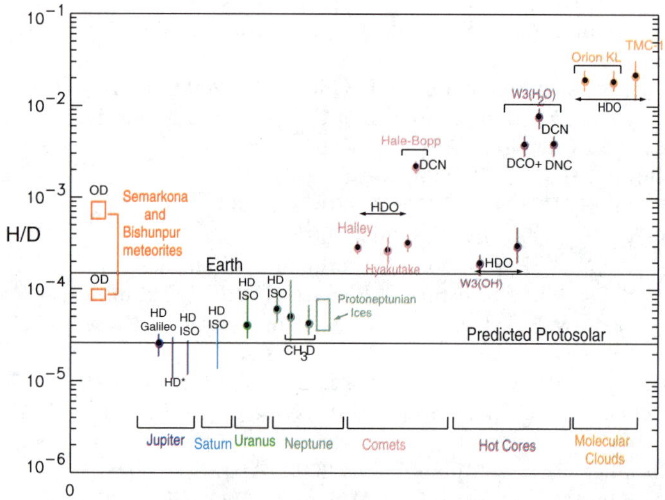

Figure 1. This graph is modeled after that presented in Bockelée et al. (1998) and Drouart et al. (1999) for solar system objects. The deuterium abundances for hot cores (Helmich, van Dishoeck & Jansen 1996) and molecular clouds (Mangum, Plambeck & Wooten 1991; Wooten 1987) were added for comparison with the accretion disks observed in this study.

As analogs to the solar nebula, circumstellar disks offer a unique opportunity to study the conditions during the planet formation process. Observations of deuterated species in circumstellar disks are particularly important in the investigation of the origin of primitive solar system bodies in that they should directly constrain the D to H fractionation in the outer (colder) regions of these disks, where cometary ices are likely formed. Spatially resolved

Chemistry as a Diagnostic of Star Formation (2003). *Edited by Charles L. Curry and Michel Fich.* Published in 2003 by NRC Research Press, Ottawa ON, K1A 0R6, Canada. ISBN 0-660-19089-3.

observations of the D/H ratio in disks would be even more informative, enabling the comparison of the fractionation measured in comets such as Hale-Bopp (Blake et al. 1999) with specific conditions at each disk radius.

Single-dish and millimeter-wave interferometric imaging of disks of gas and dust surrounding several young T Tauri and Herbig Ae/Be stars have dramatically improved our understanding of disk physical and chemical structure. As discussed by Blake (this volume), disk structure is complex, and molecular lines can be used to probe the effects of UV fields, ionization and grain surface reactions. They can also help us get a better handle on disk ages and evolution. In order to interpret observational results, we have been working on a model which combines the two-dimensional radiative transfer code of Hogerheijde & van der Tak (2001), the chemical model of Willacy & Langer (2000), and the physical disk model of D'Alessio et al. (2001). In Sec. 2 we present the results of our search for deuterium and then in Sec. 3 describe how the model can help in the interpretation of the images that we obtained. Finally, in Sec. 4 we will discuss ongoing observations and possible future studies.

2. Observations

Here we present the results of a search for HDO and DCN toward protoplanetary disks surrounding one T Tauri star, LkCa 15, and two Herbig Ae stars, HD 163296 and MWC 480. Observations of the HDO 1_{11}-2_{12} and 3_{12}-2_{11} and DCN J=3−2 lines were performed at the Owens Valley Radio Observatory (OVRO) millimeter array. Only HDO 1_{11}-2_{12} emission was marginally detected (2σ) toward LkCa 15 (Fig. 2), and upper limits for the HDO in MWC 480 and HD 163296 and DCN in HD 163296 and LkCa 15 were obtained. The upper limits for DCN abundances in LkCa 15 are combined with previously observed $H^{13}CN$ toward this source to obtain a DCN/HCN ratio of $< 2\times10^{-3}$. Water has not been detected toward this source, so we use model predictions for the H_2O column density and scale by the observed-to-modeled HCN column density,

$$\frac{N(\mathrm{HDO})}{N(\mathrm{H_2O})} = \frac{N_{\mathrm{obs}}(\mathrm{H_2O})}{N_{\mathrm{mod}}(\mathrm{H_2O})} \times \frac{N_{\mathrm{obs}}(\mathrm{HCN})}{N_{\mathrm{mod}}(\mathrm{HCN})},$$

to obtain an HDO/H_2O estimate for this disk of 0.064.

The D/H ratio calculated above for LkCa 15 is quite similar to that observed for DCO^+ toward the T Tauri star disk TW Hya (0.035 ± 0.003) using the JCMT (Thi et al. 2002). The D/H ratios in the two disks can be compared to those of molecular clouds, hot cores and our solar system. The ratios vary slightly from disk to disk and lie within the range of D/H ratios found in molecular clouds and hot cores. All three of these D/H ratios are much larger than that of the early solar nebula (10^{-5}) and are consistent with deuterium abundances in comets. Therefore, conclusions from previous observations, which stated that comets were assembled from interstellar material, can be adapted to state that comets are more likely to have formed in the outer regions of disks, where the conditions are more similar to interstellar clouds, than in nebular regions near proto-Jovian planets.

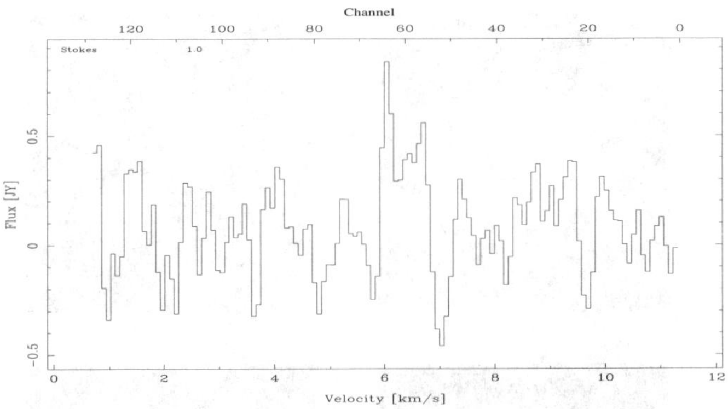

Figure 2. OVRO observation of the HDO 1_{11}-2_{12} emission toward LkCa 15. The spectrum shows the double-peaked profile of Keplerian rotation of an inclined disk centered at V_{LSR} = 6.3 km s^{-1}.

3. Image Modeling

Molecular distributions are essential to understanding the chemistry taking place in disks. The HDO emission observed here, as well as previous observations toward LkCa 15 (Qi et al., in preparation) indicate large differences in the morphology of the emission in integrated intensity maps (Fig. 3). Both the CO and HCO^+ emission peak at the source position, while the HCN, CN and HDO emission peak approximately 2″away along the major axis of the disk. This is especially interesting because the spectra for CO J=2−1, HCO^+, HDO and HCN do not differ significantly. Lack of emission toward the star results in only small changes to the line wings, which are buried in the noise in the HDO spectrum, but which may be slightly more visible in the higher signal-to-noise HCN spectrum. Clearly, imaging is necessary to study this type of disk structure and there is a need for detailed models which can be used to reproduce and interpret these images.

For this reason, we have created a model to simulate our observed channel maps and images of protoplanetary

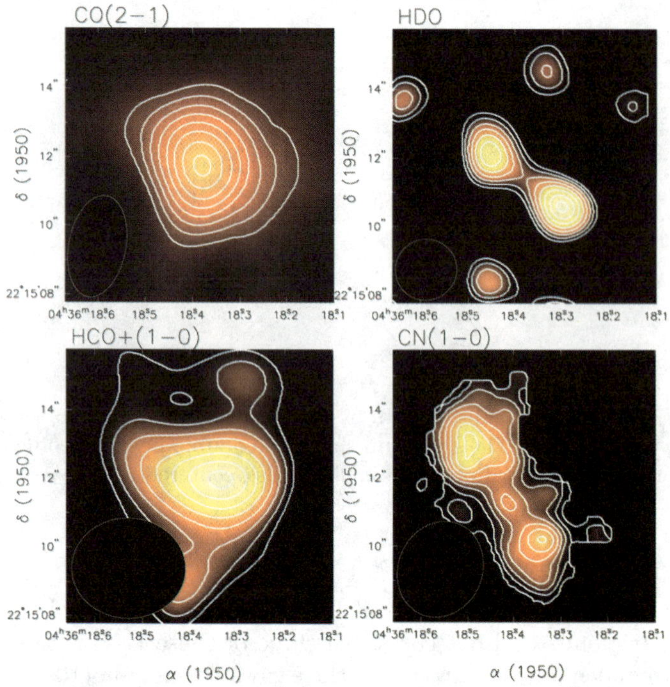

Figure 3. OVRO observations of CO (kindly provided by Koerner and Sargent), HDO, HCO^+ and HCN toward the T Tauri star LkCa 15. While CO and HCO^+ are centered at the stellar position, CN, HDO and HCN (not shown) peak at $\sim 2''$ away.

disks. The temperature and hydrogen density distributions are determined using the physically self-consistent disk models of D'Alessio et al. (2001), assuming a power law distribution of dust sizes, $n(a) = a^{-3.5}$, an accretion rate of 10^{-8} $M_\odot$/yr, and equal dust and gas temperatures. The stellar parameters are specific to LkCa 15: M_*=1.0 $M_\odot$, R_*=1.64 $R_\odot$, T_*=4395 K. The resulting physical disk model is then combined with the distribution of the molecule of interest, and the radiation transfer in the disk is calculated using a non-LTE Monte Carlo code (Hogerheijde & van der Tak 2001). The input parameters for this model are the mass and inclination of the disk, determined from previous observations of the continuum emission toward this source (Beckwith et al. 1990; Duvert et al. 2000). The inner and outer radius of the disk (R_{out} = 430 AU) and turbulent velocity width (0.1 km s^{-1}) are parameters which are adjusted to fit our observed CO J=2−1 emission spectrum (Qi et al., in prep.).

The initial results of our model are equivalent to the observations of a telescope with infinite resolution and complete coverage in the *uv* plane (Fig. 4). The map is then convolved with the *uv* sampling obtained, in order to simulate the OVRO observations more explicitly. The initial results clearly show the standard velocity pattern of a disk in Keplerian rotation. Although the details of this shape are not as clear in the model with the observed *uv* sampling, the characteristic channel-to-channel shift in the position of the peak emission with velocity can still be seen. We will now use these models to interpret our HCN and HDO observations.

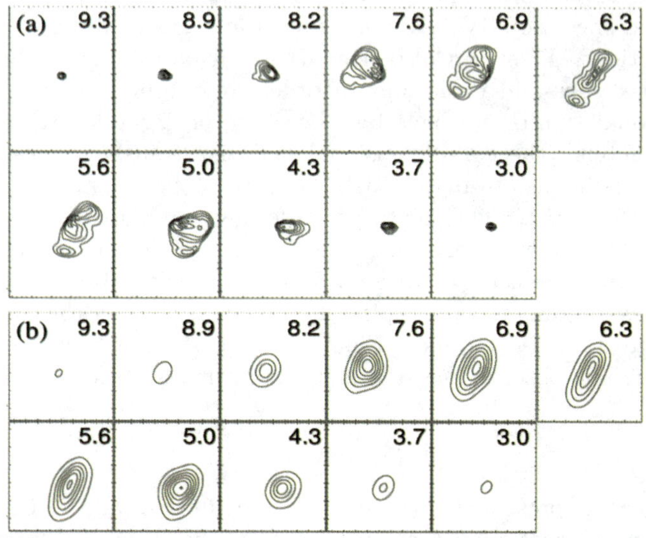

Figure 4. Modeled channel maps of CO J=2−1 emission in LkCa 15. Each channel is integrated over 0.16 km s^{-1} (V_{LSR}=6.3 km s^{-1}). The upper panel, *a*, is a simulation of the emission as detected from a telescope with infinite resolution and complete sampling. The lower panel, *b*, displays how the image changes due to the resolution and sampling of the OVRO observations.

The chemical models of Aikawa & Herbst (2001) and Willacy & Langer (2001) suggest that the radial distribution of HCN and CN in disks is determined by the processes of photodissociation by interstellar and stellar UV combined with desorption from grain surfaces. The newest models (Aikawa et al. 2002) suggest that the competition between interstellar and stellar UV in particular determines the variation in the amount of HCN as a function of radius. We simulate the emission morphology discussed above by concentrating the HCN in a ring around the star. We use a simple model, with an outer region in which the column density is constant and the volume density is proportional to that of hydrogen, and an inner region in which the column density is zero. We then set the outer radius of the annulus to the disk radius as de-

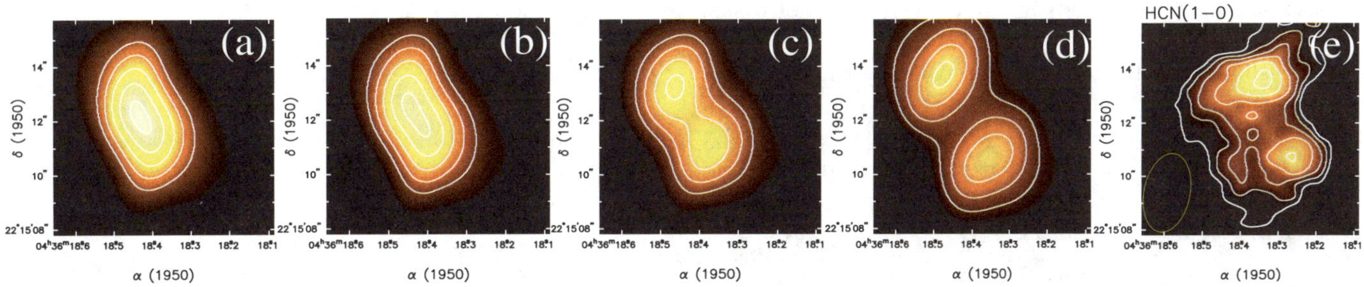

Figure 5. Model fits to the HCN J=1−0 emission from LkCa 15. The observed HCN emission in *(e)* is compared to a model of an annulus of HCN with an outer radius of 400 AU and an inner radius of: *(a)* 50 AU; *(b)* 100 AU; *(c)* 200 AU; and *(d)* 300 AU. In each case, the total flux of the model image is set to match the observed HCN map. The observed emission is consistent with an inner radius of 200−300 AU [i.e., between *(c)* and *(d)*].

termined from the CO J=2−1 observations (~ 430 AU) and vary the inner radius to match the observed integrated intensity maps. In each case, the column density of HCN in the ring is set by matching the integrated flux of the modeled map to that observed. As we vary the inner radius, we can see that these models reproduce the structure of the HCN emission only with large depletion zones in the disk center (R_i ~200−300 AU; see Fig. 5). More realistic models will be used in future to simulate the combined effects of desorption from grains and photodissociation of HCN from interstellar and stellar UV.

Although we have not yet modeled the HDO emission, the observed double-peaked emission map may also result from a ring-like distribution. Here models suggest that the annulus could be formed due to reduced production of HDO near the central star as H_2D^+, the primary precursor to HDO, is created less efficiently due to the increasing temperature gradient in the inner disk. This causes the HDO abundance to peak at $R \approx 250$ AU from the star (Aikawa & Herbst 2001). As the previous HCN models show, this morphology is consistent with the observed emission.

4. Future Observations

It is clear from this study that observations of deuterated molecules in disks are quite difficult with the sensitivity of existing millimeter arrays. One way to observe deuterated molecules with the currently available sensitivity would be to target ions, which remain in the gas phase at colder temperatures than neutral species and may probe a larger fraction of the disk. In addition, because reaction with CO is the main destruction mechanism for deuterated ions, chemical models predict that D/H should increase in disk midplanes as CO is depleted onto grain surfaces. Among the deuterated ions, H_2D^+, DCO^+ and N_2D^+ are the most abundant and are also precursors to many other deuterated species, including HDO and DCN. H_2D^+ resides in its ground para state at disk temperatures and should be observable with CASIMIR on SOFIA. N_2D^+ emits in the 1.3 mm receiver window at OVRO. Based on observations of N_2H^+ toward LkCa 15, N_2D^+ should be observable toward this source with an antenna temperature of 0.6 K. Additionally, with the detection of HDO toward LkCa 15, observations of $H_2^{18}O$ would be very valuable for the determination of an accurate D/H ratio. Therefore, we are currently pursuing OVRO observations of emission from $H_2^{18}O$ at 203 GHz toward LkCa 15.

With future arrays such as CARMA and, of course, ALMA, we will be able to observe many deuterated species at higher resolution. With its improved sensitivity in the 1.3 mm window, for example, CARMA will allow us to detect deuterated species such as DCN and HDO in 5 hours, as compared to the nearly 40 hours of integration time represented by the HDO observations presented here. Thus, we can dramatically increase the number of species and disks observed. ALMA, with its much more complete *uv* coverage, will allow us to recover almost all of the disk structure which is unresolved by current arrays.

5. Summary

We present the first observations of DCN and HDO toward circumstellar disks. Images of HDO in the T Tauri star LkCa 15 show emission which is consistent with an annulus of inner radius 200 AU and outer radius ~400 AU. This morphology is consistent with our current understanding of disk chemistry. D/H ratios in disks, including those reported here, and observations of DCO^+ toward the T Tauri star TW Hya, are similar to those of hot cores and of molecular clouds and comets, and are consistent with comet formation in the outer regions of disks. Future observations of deuterated ions (e.g., H_2D^+ and N_2D^+) should probe much closer to the disk midplane and constrain the deuteration in these regions. With more sensitive high-resolution arrays such as CARMA and ALMA, we will be able to expand our studies to include more molecules and probe the inner, planet-forming regions of circumstellar disks.

ACKNOWLEDGEMENTS

This work was supported by the NASA Graduate Student Researchers Program, NGT5-50231. OVRO observations were supported by the National Science Foundation, AST 9981546. The authors are also grateful to Michiel Hogerheijde, Paola D'Alessio and Yuri Aikawa for very useful discussions on this topic.

REFERENCES

Aikawa, Y., van Zadelhoff, G.-J., van Dishoeck, E. F. & Herbst, E. 2002, A&A, 386, 622

Aikawa, Y. & Herbst, E. 2001, A&A, 371, 1107

Blake, G. A., Qi, C., Hogerheijde, M. R., Gurwell, M. A. & Muhleman, D. O. 1999, Nature, 398, 213

Beckwith, S. V. W., Sargent, A. I., Chini, R. S. & Guesten, R. 1990, AJ, 99, 924

Bockelée-Morvan, D., et al. 1998, Icarus, 133, 147

D'Alessio, P., Calvet, N. & Hartmann, L. 2001, ApJ, 553, 321

Drouart, A., Dubrulle, B., Gautier, D. & Robert, F. 1999, Icarus, 140, 129

Duvert, G., et al., 2000, A&A, 355, 165

Helmich, F. P., van Dishoeck, E. F. & Jansen, D. J. 1996, A&A, 313, 657

Hogerheijde, M. R. & van der Tak, F. F. S. 2000, A&A, 362, 697

Mangum, J. G., Plambeck, R. L. & Wooten, A. 1991, ApJ, 369, 169

Thi, W.-F. et al. 2002, submitted to ApSS

Willacy, K. & Langer, W. D. 2000, ApJ, 544, 903

Wooten, A. 1987, IAUS, 120, 311

MODELS OF H_2 FLUORESCENCE IN THE UV SPECTRUM OF THE CTTS TW HYA

Gregory J. Herczeg[1], Brian E. Wood[1], Jeffrey L. Linsky[1], Jeff A. Valenti[2], and Chris M. Johns-Krull[3]

[1]JILA/University of Colorado
[2]Space Telescope Science Institute
[3]Rice University

Abstract

We model the Lyman-band H_2 fluorescence observed in *HST*/STIS E140M and *FUSE*spectra of the classical T Tauri star TW Hya. The H_2 fluorescence occurs because Ly α emission, produced in an accretion shock, irradiates a warm disk surface layer within 2 AU of the central star. We simulate a 1D plane-parallel atmosphere to reproduce fluxes for the 146 observed H_2 emission lines and reconstruct the Ly α emission profile incident upon the warm H_2. The H_2 lines with lower levels $E < 1$ eV are optically thick and, consequently, weaker than predicted by branching ratios. Dust extinction to TW Hya is negligible, based on the hydrogen column density towards the star, and does not affect H_2 line fluxes. Our best-fit parameters for the molecular layer are a temperature $T = 3000 \pm 400$ K, a column density of $\log N(H_2) = 17.9 \pm 0.5$, and a filling factor of H_2 as seen by the accretion shock of 0.3 ± 0.1, although geometrical effects complicate the interpretation of the column density and filling factor. We compare the inner disk of a classical T Tauri star to a photodissociation region. Formation of H_2 in highly excited states, by the dissociative attachment of H_3^+, may be important for those emission lines that are not well explained by the model of Ly α fluorescence of thermal H_2.

Key words: stars: pre-main-sequence — stars: TW Hydrae — ultraviolet: stars — accretion, accretion disks

1. Introduction

Identifying the nature of H_2 emission from classical T Tauri star (CTTS) has been complicated. Thi et al. (2001) used a large aperture on ISO to detect H_2 emission in pure rotation lines from CTTSs and, based on CO maps, claimed that this emission was produced in protoplanetary disks. Richter et al. (2002) could not reproduce these results using a long slit with the TEXES grating on IRTF, indicating that the H_2 emission detected by Thi et al. (2001) may have been produced in ambient molecular material surrounding these young stars. Bary et al. (2002) detected H_2 emission in the 1-0 S(1) line from a sample of CTTSs and inferred from the lack of surrounding molecular material for their sample that the H_2 emission originates in circumstellar disks.

Fluorescent H_2 emission was detected around T Tau with *IUE* by Brown et al. (1981). Those H_2 lines are spatially extended beyond the circumstellar disks and are interpreted as shock-heated emission from the interaction of the star/disk wind/jet with the surrounding cloud.

We detect Lyman-band H_2 fluorescence, photoexcited by Ly α, in the UV spectra of the CTTS TW Hya. The high spatial (0.″05) and spectral ($R = 45,000$) resolution available with STIS allowed Herczeg et al. (2002; hereafter Paper I) to determine that the H_2 emission from TW Hya is most likely produced in the protoplanetary disk. In this paper we present models of the H_2 fluorescence, which we use to reconstruct the intrinsic Ly α profile that irradiates the H_2, and we deduce a temperature of 3000 ± 400 K for the H_2 gas.

2. Observations of H_2 Fluorescence

In Paper I, we presented *HST*/STIS and *FUSE* observations of TW Hya, spanning 912 Å–2900 Å, that show 146 H_2 lines between 1100 Å–1640 Å, some of which are shown in Figure 1. A strong, broad Ly α emission line, produced at the accretion shock, photoexcites many upper levels of H_2 (Fig. 2), from which the H_2 molecule can decay via a variety of transitions (Fig. 3). We presented several lines of evidence that suggest a disk origin for the H_2 emission: (1) the emission is not spatially extended beyond a point source in the cross-dispersion direction; (2) the H_2 lines are emitted interior to TW Hya's wind because emission in one H_2 line is suppressed by CII wind absorption (see Fig. 1); (3) the H_2 line centroids are equal to the radial velocity of TW Hya; (4) no H_2 absorption is detected against the Ly α emission; and (5) the TW

Chemistry as a Diagnostic of Star Formation (2003). *Edited by Charles L. Curry and Michel Fich.* Published in 2003 by NRC Research Press, Ottawa ON, K1A 0R6, Canada. ISBN 0-660-19089-3.

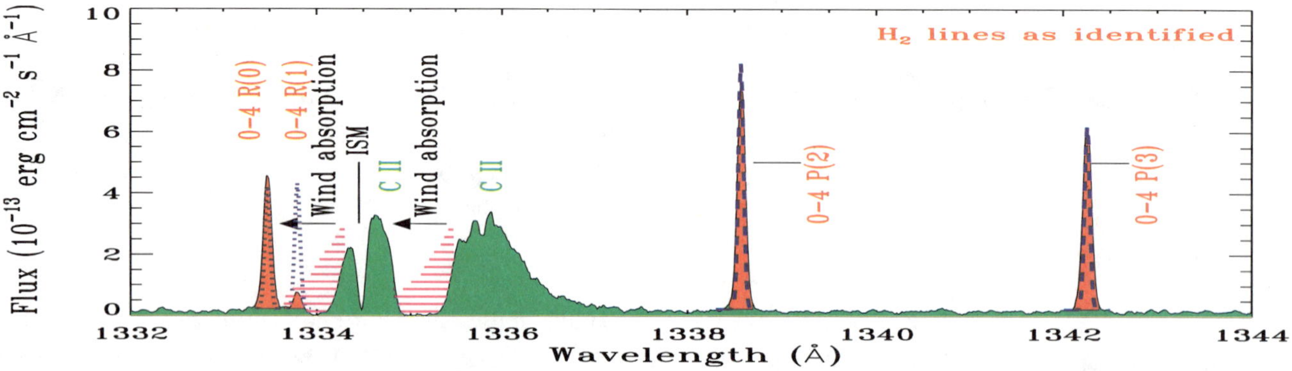

Figure 1. A segment of the *HST*/STIS E140M spectrum of TW Hya. H_2 emission is shaded red while model line fluxes, described in Sec. 3, are shown in blue. The H_2 line at 1333.8 Å is weak due to absorption by C II in the wind of TW Hya.

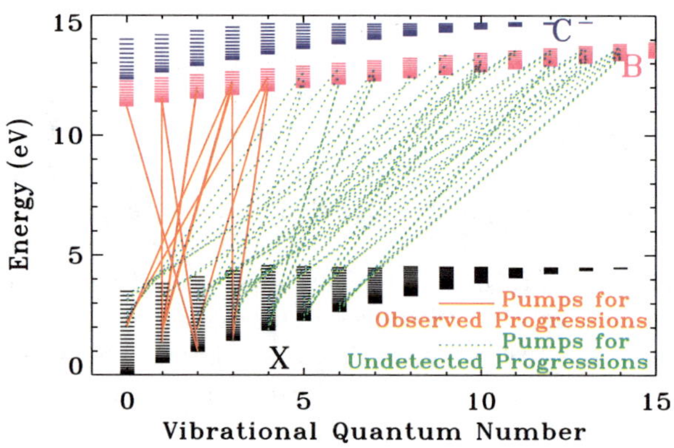

Figure 2. The H_2 molecule has many Lyman-band transitions coincident with Ly α. We detect emission from the upper levels that are photoexcited from lower levels with energies of $1-2$ eV (*solid red*), but do not detect photoexcited emission from highly-excited levels (*dotted green*), implying a thermalized gas.

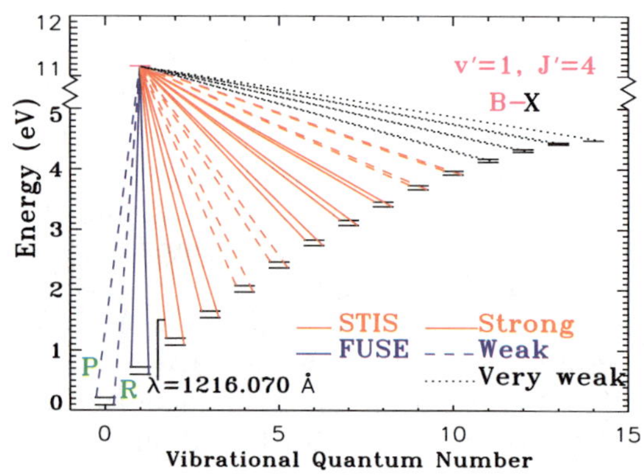

Figure 3. Ly α photoexcites the H_2 into the B electronic state, from which it cascades via many different routes, generating emission in a series of H_2 lines.

Hya association is isolated from large reservoirs of interstellar molecular material, making a circumstellar origin for the H_2 emission unlikely. The small spatial extent of H_2 emission in the cross-dispersion direction indicates the emission is not extended beyond 2 AU from the star.

3. Models of H_2 line fluxes

Figure 4 demonstrates that the line ratios from a single upper level are not consistent with the optically thin branching ratios calculated by Abgrall et al. (1993). Fluxes for lines at wavelengths $\lambda > 1350$ Å show some scatter but are generally consistent with the theoretical branching ratios because the warm H_2 layer is optically thin at those transitions. Fluxes of lines with low-energy lower levels, which occur at wavelengths $\lambda < 1350$ Å, are considerably weaker than predicted by their branching ratios because these lower levels can become optically thick for 1000–4000 K gas.

Dust extinction in our line of sight towards TW Hya could in principle produce a similar effect. The observed Ly α profile presented in Figure 5 is characterized by a strong, broad emission profile produced at the accretion shock and a dark absorption feature. The red side of the Ly α absorption profile is created by interstellar and circumstellar material, while the blue side of the absorption is created by TW Hya's wind. We find by fitting absorption profiles to the red wing that $\log N(\mathrm{H\,I}) = 19.05 \pm 0.15$ in our line of sight to TW Hya. We use the standard interstellar relationship between extinction and N(H I) (Savage & Mathis 1979) to determine that dust extinction towards TW Hya is negligible.

We follow the procedure of Wood et al. (2002) to model H_2 fluorescence in a 1D, isothermal, plane-parallel atmosphere of H_2 irradiated by normally incident Ly α photons using a Monte Carlo method. A Ly α photon enters the slab to photoexcite an H_2 molecule, which then de-excites to the ground electronic level through one of the many available routes, or dissociates with a probability calculated by Abgrall, Roueff, & Drira (2000). The resulting photon may then scatter off another H_2 molecule or may emerge from the layer, depending on the optical depth of the H_2 in the level that can absorb this photon. We tabulate photons that emerge on the same side that the Ly α photon entered the slab to roughly approximate a disk geometry. The free parameters of this model are the H_2 excitation temperature and effective column density of the slab, and a normalization factor for each of the 19 upper levels that scales with the total flux from each upper level. A best-fit global solution of $N(H_2)$ and T to all 19 upper levels is then determined by comparing the model fluxes to the observed fluxes from our STIS E140M data. Contours for $1-5\sigma$ fits are shown by the solid dark curves in Figure 7. In Sec. 5.1 we will compare the results from this method with results presented in Sec. 4 to constrain the physical parameters of the warm H_2.

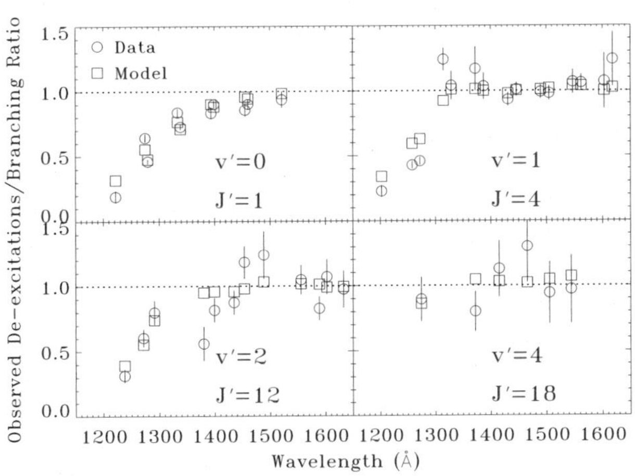

Figure 4. The observed number of decays in a line versus the branching ratios of an optically thin medium, as calculated by Abgrall et al. (1993), for four upper levels. The short wavelength lines are systematically weaker due to optical depth effects described in Sec. 3. Lines from v'=4, J'=18 do not show this effect because the lower levels, with high energies, are not significantly populated.

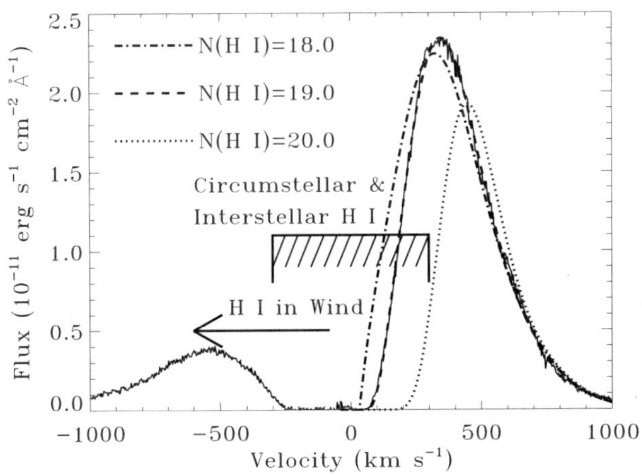

Figure 5. We fit the red wing of the Ly α absorption profile to determine an H I column density of 19.05 ± 0.15 in our line of sight to TW Hya. On the basis of interstellar extinction laws, we conclude that dust extinction is negligible.

4. Reconstruction of the Ly α profile

The emission from an upper level depends on the Ly α flux at the pumping wavelength of the fluorescent transition, the filling factor of H_2 around the star, and the amount of H_2 in each level that is available to absorb the Ly α photons. We can therefore reconstruct the Ly α profile at 18 wavelengths where the fluoresced H_2 is photoexcited from the amount of emission in each H_2 progression and the temperature and column density of the H_2 gas.

We first calculate the flux of Ly α that must be reprocessed into H_2 emission to produce the observed H_2 flux within each progression, correcting for the dissociation of H_2, unseen H_2 lines, and effects due to geometry. We then compute the equivalent width for the lower level of H_2 that is photoexcited by Ly α, given the T and $N(H_2)$ of the slab. We calculate the filling factor η of H_2 as seen from the accretion shock by comparing the calculated and observed flux at each wavelength.

We reconstruct the Ly α profile for the same grid of assumed T and $N(H_2)$ values used in Sec. 3 for comparison with the observed profile. Figure 6 shows three examples of reconstructed Ly α profiles (*circles*), with upper limits to the Ly α flux at certain wavelengths based on the undetected emission lines (*arrows*) that could in principle be photoexcited by Ly α. The three reconstructed Ly α profiles shown are fairly similar to each other and to most other Ly α profiles that we have reconstructed. We evaluate reconstructed profiles based on the scatter in the red wing of Ly α(compare the top and middle panels of Fig. 6), and on the consistency of upper limits to the Ly α flux for upper levels that are not detected. Each profile shows a clear depression near line center of Ly α, although not nearly as dark or as wide as the observed H I absorption (bottom panel of Fig. 6). This dip could be due to self-reversal of the intrinsic Ly α profile or H I between the accretion shock and the warm H_2. The colors in Figure 7 show 1–5σ $\Delta\chi^2_\nu$ contours for a fit with 6 degrees of freedom for $\chi_\nu = \chi/6$. The filling factors, with contours shown by dashed lines in Figure 7, are typically about 0.3 for acceptable models. This factor represents the solid angle of the sky subtended by H_2, as seen from the accretion column on TW Hya, although other geometrical effects may influence the filling factor.

5. Discussion

5.1. Properties of the H_2 emission region

Figure 7 compares results from models that were constructed to fit individual H_2 line fluxes with models constructed to fit the Ly α profile. Based on the overlap region of acceptable models for both the individual H_2 line fluxes and the reconstructed Ly α profile, we conclude that a column density of $\log N(H_2) = 17.9 \pm 0.5$ is heated to a temperature of 3000 ± 400 K, with a filling factor of $\eta = 0.3 \pm 0.1$ around TW Hya. The temperature $T \approx 3000$ K describes the excitation of the H_2 molecule. This temperature also describes the kinetic temperature because we expect the gas density in the disk surface is $n_H = 10^6$–10^8 cm^{-3} (Gomez & D'Alessio 2000), far larger than the critical density of 10^4 cm^{-3} at which collisions typically dominate over quadrupole radiative de-excitation in highly excited levels, and we would have found evidence for a gas density below the critical density. The dissociation rate of the H_2 in the region studied here is $(6.8 \pm 0.5) \times 10^{-10} M_\odot$ yr^{-1}, although this does not lead to any disk dissipation. We find that the filling factor is 0.3 ± 0.1, which leads to a geometrical disk height of $H = 0.6$ $(R/1$ AU) AU, where R is the distance of the H_2 emission region from the accretion column on TW Hya. However, we caution that significant uncertainties in the geometry, including projection effects due to an optically thick Ly α source and the amount of Ly α emission obscured from us by TW Hya but not to the H_2, lead to large uncertainties in the application of the filling factor to the star–disk system.

If we assume a surface temperature dependence of $r^{-0.4}$, as in the models of Chiang & Goldreich (1997), with $T = 3000$ K at 1 AU, we calculate a flux of 5×10^{-16} ergs cm^{-2} s^{-1}, roughly consistent with that observed by Weintraub et al. (2000). Some of the emission in this line may be produced deeper in the disk or farther from the star, as a result of X-ray ionization and subsequent collisional excitation of H_2 by nonthermal electrons. While we can explain the emission in the 1–0 S(1) line, the many uncertainties in the model prevent us from determining whether or not most of this emission arises in the warm molecular layer close to the star.

We find that $v \sin i < 7.3$ km s^{-1} by fitting the H_2 lines with a Gaussian, convolved with instrumental, thermal and rotational broadening for a stellar mass of $M = 0.7 M_\odot$ Webb et al. (1999). This broadening corresponds to a rotational velocity of $v_{\rm rot} < 42^{+42}_{-13}$ km s^{-1}, for an inclination angle of $10° \pm 5°$. If the H_2 is in Keplerian rotation, we constrain its location to be a distance greater than $0.4^{+0.4}_{-0.3}$ AU from the star. The study of H_2 emission from CTTSs with edge-on disks will place more stringent limits on this distance.

5.2. A line-driven photodissociation region?

In Paper I, we concluded that the H_2 fluorescence observed from TW Hya is most likely produced in the surface layers of the disk, within 2 AU of the star. We have determined here that this H_2 is heated to roughly 3000 K, although significant temperature gradients may exist. Based on photodissociation region (PDR) models, Hollenbach et al. (2000) predict that a warm atmosphere-like structure, heated by UV radiation, should surround the inner regions of the disk, with a temperature between 500 and 3000 K.

We measure a total flux of 6.4×10^{-11} ergs cm^{-2} s^{-1} between 1160 and 1700 Å, which corresponds to 6×10^{-3} $L_\odot$ assuming that TW Hya is located at a distance of 56 pc. At least 65% and as much as 90% of this emission occurs in Ly α, depending on the correction for H I Ly α absorption. Between 6–14% of the UV flux occurs in continuum emission, and the remaining emission occurs in emission lines other than Ly α. The UV radiation field at 1 AU from TW Hya is equivalent to the UV radiation in PDRs. Ly α radiation can ionize Si I from its ground state and C I from a low-energy metastable level. The properties of this PDR-like disk atmosphere will be dominated by emission lines, as evidenced by the fluorescent H_2 emission. In contrast, a continuum-driven PDR would exhibit nearly continuous H_2 emission because hundreds of H_2 levels would be electronically excited.

Other differences between the protoplanetary disk atmosphere and a PDR may result in different thermal

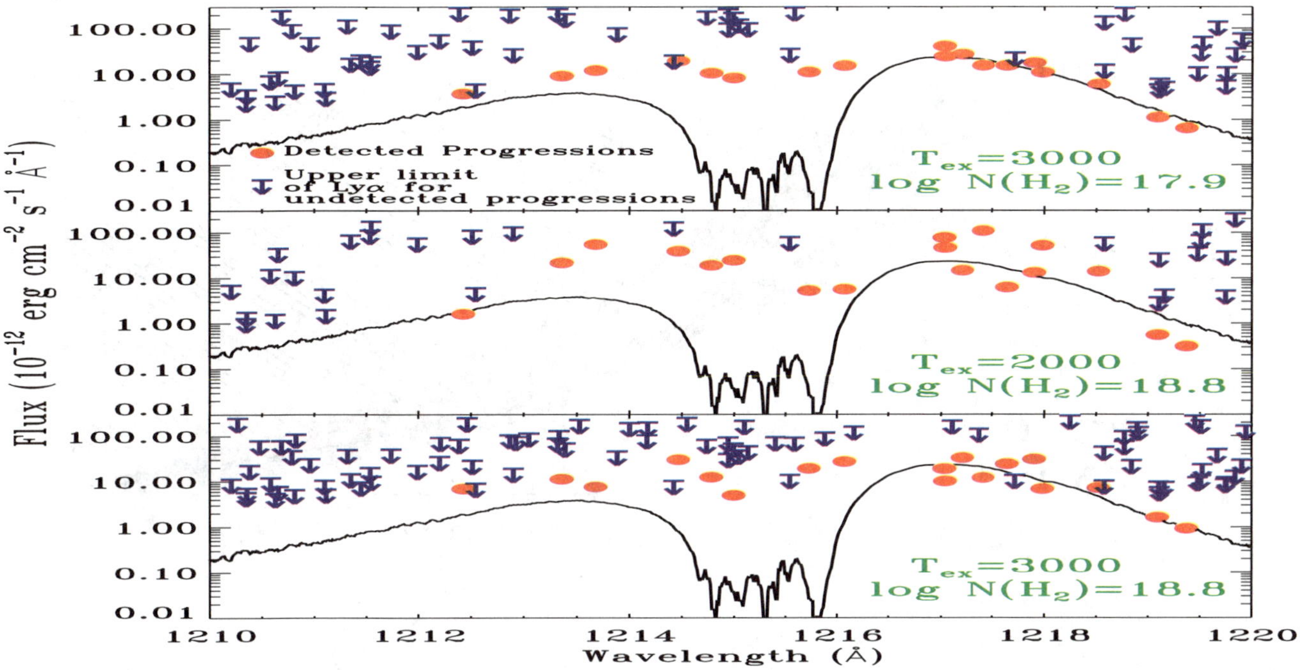

Figure 6. Three reconstructed Ly α profiles (*red dots*) compared with the observed Ly α profile (*solid curve*), for three different models of T and $N(H_2)$. The blue arrows indicate upper limits to the Ly α flux determined by the non-detection of emission lines. The top panel ($\log N[H_2] = 17.9$, $T = 3000$ K, and $\eta = 0.31$) represents our best-fit model. The middle panel ($\log N[H_2] = 18.8$, $T = 2000$ K, and $\eta = 0.13$) is inconsistent with the data because of too much scatter in the red wing of the Ly α profile. The bottom panel ($\log N[H_2] = 18.8$, $T = 3000$ K, and $\eta = 0.37$) is inconsistent with the data because at 1214.421 Å the Ly α emission required to photoexcite the level $v' = 7, J' = 4$ to the limit of detectability is significantly less than the Ly α flux required to photoexcite $v' = 3, J' = 16$ at 1214.465 Å.

structures. Calvet et al. (2002) found that the dust grains in the inner 5 AU of TW Hya's disk have evolved into larger grains and settled to the midplane. In a typical PDR, only 0.1–1% of the UV radiation heats the gas, because PAHs and small grains absorb most of this emission and quickly re-radiate the energy in the IR. If grains are not available to absorb the UV radiation, the UV radiation may penetrate the gas to greater depths and deposit more of its energy into the gas, which cools less efficiently than dust. The large X-ray flux of $L_X = 3.6 \times 10^{-4}\ L_\odot$ (Kastner et al. 2002) may also heat the disk surfaces by an additional 500 K (Igea & Glassgold 1999) by ionizing atoms to produce a large reservoir of energetic nonthermal electrons.

5.3. Formation of H_2 in highly excited states

We have presented a model that explains most of the H_2 emission as originating in a warm, thermalized layer, presumably at the surface of the inner disk. However, we also noted in Paper I that certain progressions are pumped from lower levels with very high energies, such as $v'' = 3, J'' = 25$ ($E'' = 4.2$ eV) and $v'' = 5, J'' = 18$ ($E'' = 3.8$ eV), which could only be thermally populated at temperatures that would dissociate H_2. The fluorescence process and multiple pumping (Ly α pumping of a highly-excited level that is populated by the Ly α pumping of a lower level) also cannot explain the populations in these levels because the required efficiency for the H_2 to populate those levels is too high to explain the observations. In a warm, dense environment H_2 probably forms by dissociative recombination of H_3^+, rather than on grains. Bertoldi & Draine (1996) found that if the ionization and dissociative regions of a PDR are mixed, due to either velocity gradients or a large Lyman continuum flux ($E > 13.6$ eV), then H_2^+ is present in large quantities and can react with H_2 to produce H_3^+. Also, Maloney et al. (1996) suggested that X-ray irradiated molecular

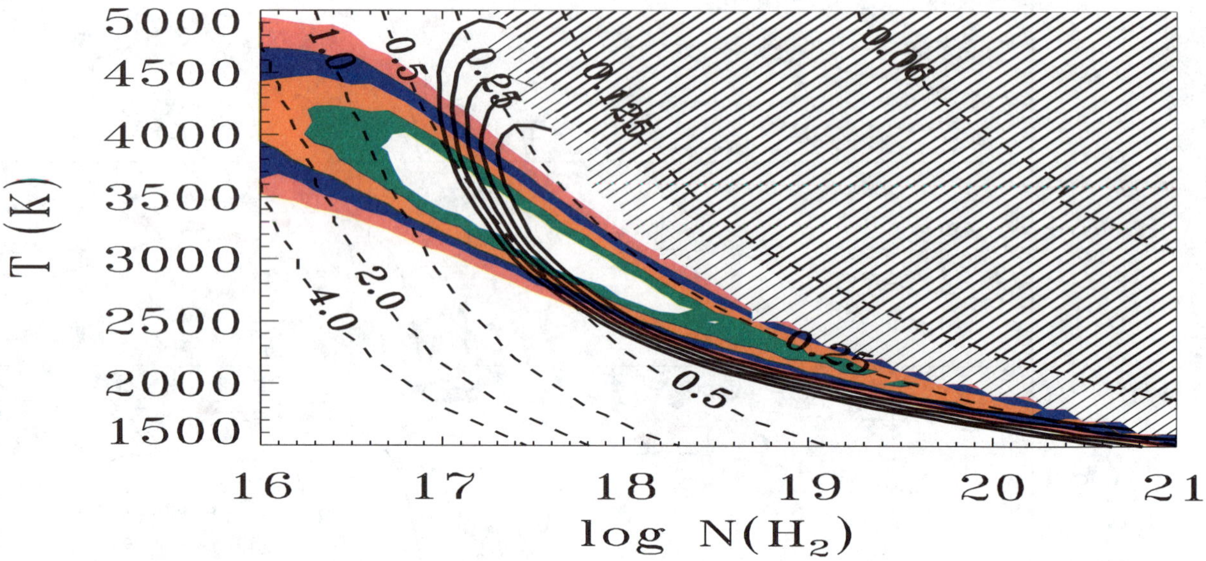

Figure 7. Here we compare results from the different methods and determine that a column density of $\log N(H_2) = 17.9 \pm 0.5$ is heated to a temperature of 3000 ± 400 K, with filling factors of $\eta = 0.3 \pm 0.1$ around TW Hya. The solid curves show $1-5\sigma$ fits to individual H_2 line fluxes, described in Sec. 3. The colors show $1-5\sigma$ fits of the reconstructed Ly α profile to the observed Ly α profile, described in Sec. 4. The dashed curves show filling factor contours, as labeled. The diagonal lines at upper right show models that we rule out based on the non-detection of other H_2 lines, as described in Sec. 4.

regions will also generate the H_3^+ ion. Calculations by Kokoouline et al. (2001) and experiments by Strasser et al. (2001) indicate that the vibrational distribution of H_2 after forming via dissociative recombination of H_3^+ peaks at $v = 5$ and $v = 6$. These formation reactions may indeed produce a population of H_2 in highly excited states and could in principle explain the population of H_2 in the $v'' = 5, J'' = 18$ and $v'' = 3, J'' = 25$ levels. Although further calculations of the H_2 formation reactions are required for a quantitative model, these initial results are promising.

Acknowledgements

This research is supported by NASA grant S-56500-D to the University of Colorado and NIST, and by a grant to the University of Colorado of NASA funds through the Johns Hopkins University.

References

Abgrall, H., et al. 1993, A&AS, 101, 273
Abgrall, H., Roueff, E. & Drira, I. 2000, A&AS, 141, 297
Bary, J.S., Weintraub, D.A., & Kastner, J.H. 2002, ApJ, 576, L73
Bertoldi, F., & Draine, B.T. 1996, ApJ, 458, 222
Brown, A., et al. 1981, Nature, 290, 34
Calvet, N., et al. 2002, ApJ, 568, 1008
Chiang, E.I., & Goldreich, P. 1997, ApJ, 490, 368
Gomez, J.F., & D'Alessio, P. 2000, ApJ, 535, 943
Herczeg, G. J., Linsky, J. L., Valenti, J.A., Johns-Krull, C.M. 2002, ApJ, 572, 310 (Paper I)
Hollenbach, D. J., Yorke, H. W., & Johnstone, D. 2000, in *Protostars & Planets IV*, ed. V. Mannings, A.P. Boss, & S.S. Russell (Univ. of Arizona Press: Tucson), 401
Igea, J., & Glassgold, A.E. 1999, ApJ, 518, 848
Kastner, J.H., et al. 2002, ApJ, 567, 434
Kokoouline, V., Greene, C.H., & Esry, B.D. 2001, Nature, 412, 891
Maloney, P.R., Hollenbach, D.J., & Tielens, A.G.G.M. 1996, ApJ, 466, 561
Savage, B.D., & Mathis, J.S. 1979, ARA&A, 17, 73
Strasser, D., et al. 2001, Phys. Rev. Lett., 86, 779
Thi, W. F., et al. 2001, ApJ, 561, 1074
Webb, R. A., et al. 1999, ApJ, 512, L63
Weintraub, D.A., Kastner, J.H., & Bary, J.S. 2000, ApJ, 541, 767
Wood, B.E., Karovska, M., & Raymond, J.C. 2002, ApJ, 575, 1057

PART VI:
HIGH-MASS STAR FORMATION

CHEMICAL CHANGES DURING STAR FORMATION: HIGH vs. LOW-MASS YSOs

Ewine F. van Dishoeck

Leiden Observatory, P.O. Box 9513, 2300 RA Leiden, The Netherlands

Abstract

Recent observational studies of high- and low-mass YSOs at (sub)millimeter and infrared wavelengths are reviewed, and chemical diagnostics of the different physical components are summarized. Methods for determining the temperature, density and abundance profiles in the envelopes are outlined, and are illustrated for one high-mass and one low-mass YSO. The combination of (sub)millimeter and infrared data gives a nearly complete chemical inventory of the gas and solid state material. In both high- and low-mass YSOs, the chemical characteristics are dominated by freeze-out in the cold outer part of the envelope and evaporation of ices in the warm inner part. Abundance jumps of factors of ~100 in selected molecules are found in the warm gas for both types of objects. Potential differences include: (i) the complex hot core chemistry, which has been observed so far only for high-mass YSOs; (ii) the high level of deuterium fractionation seen only in low-mass YSOs; (iii) the effects of internal or external UV and X-rays; (iv) the relative importance of shocks versus thermal heating of the envelope; and (v) the importance of geometrical effects. Prospects for future observational facilities are mentioned.

Key words: ISM: molecules – ISM: chemistry – star formation: high-mass – star formation: low-mass

1. Introduction

The formation of low- and high-mass stars is accompanied by large changes in the temperature and density structure in their surrounding envelopes. The chemistry responds to these changes by opening up different chemical routes (e.g., van Dishoeck & Blake 1998; Hartquist et al. 1998; Langer et al. 2000). For example, in the coldest ($\leq$10 K), densest regions the time scales for freeze-out are so short that most molecules stick onto the grains. Close to the young protostars, the grains are heated to at least 100 K, resulting in the evaporation of ices which can subsequently drive a rich gas-phase chemistry. At even higher temperatures ($\geq$230 K), most of the oxygen is driven into water and a different chemistry ensues. In this paper, a brief overview is given of recent systematic studies of the chemistry in the envelopes around young stellar objects (YSOs), with special attention to the similarities and differences between low- and high-mass sources.

Progress in this field has been driven by several new developments in the last decade. First, submillimeter observations routinely sample smaller beams ($\sim 15 - 30''$) and higher critical densities ($\geq 10^5\,\mathrm{cm}^{-3}$) than earlier millimeter studies, assuring less confusion with the lower-density cloud material. Bolometer arrays now readily provide maps of the continuum emission from cold dust, constraining the mass and the density structure of the envelope. High spatial resolutions down to $\sim 1''$ can be obtained with interferometers, although chemical data on YSOs are still limited. For comparison, most low-mass objects observed with these facilities are located at ~150 pc, where $20''$ corresponds to ~3000 AU —the typical size of an envelope—and $1''$ to 150 AU—the typical size of a circumstellar disk. The observed high-mass YSOs have distances of 2–4 kpc, so that much larger scales are probed. Nevertheless, it is possible to derive information on smaller scales by careful modeling of the line emission (see Sec. 2).

Second, infrared observations have matured and now provide essential complementary information to the millimeter data. Not only gas phase molecules, but also ices, PAHs and atomic fine-structure lines can be observed at mid-infrared wavelengths—unique probes of different evolutionary stages (e.g., van den Ancker et al. 2000a,b; Nisini et al. 2002). The single-dish submillimeter emission is more sensitive to the colder outer envelope, whereas the infrared absorption is weighted toward the warmer inner region. Together, they allow a nearly complete inventory of the gas phase and solid state material and can firmly establish abundance gradients across the envelope.

Chemistry as a Diagnostic of Star Formation (2003). *Edited by Charles L. Curry and Michel Fich.* Published in 2003 by NRC Research Press, Ottawa ON, K1A 0R6, Canada. ISBN 0-660-19089-3.

Third, the tools for the analysis of the observational data have improved substantially. Continuum and line radiative transfer codes are being extended from one dimension (1D) to three dimensions (3D) (e.g., Juvela 1997; Wolf et al. 1999). More importantly, the 1D codes have been tested critically against each other, establishing their accuracy (van Zadelhoff et al. 2002).

Several systematic studies of the chemistry in samples of low- and high-mass YSOs have been carried out in the last five years or are in progress (e.g., Hatchell et al. 1998a,b; Hogerheijde et al. 1999; van der Tak et al. 2000a,b; Lahuis & van Dishoeck 2000; Shah & Wootten 2001; Buckle & Fuller 2002; Jørgensen et al. 2002; Boonman et al. 2003b,c; Johnstone et al. 2003; Maret et al., in preparation). Questions to be addressed with these surveys are: (i) What is the physical structure of the envelope $[T(r),\ n(r)]$? Do continuum and line data give the same results? (ii) Which molecule/line traces which component? Can one uniquely distinguish envelope, outflow, disk and surrounding low-density cloud? (iii) How does the chemical composition change between different sources? Is this related to evolution or do other parameters play a role? (iv) What are the differences between high- and low-mass YSOs? (v) What do the data tell us about basic chemical processes, in particular gas-grain interactions, grain surface chemistry and high temperature gas phase chemistry?

In this paper, a brief review and progress report of recent results is given, focusing on items (i), (iii) and (iv). In Sec. 2, the analysis methods are outlined in more detail. In Sec. 3, the results on high-mass YSOs are summarized and illustrated by discussing one source in detail (AFGL 2591). In Sec. 4, the same recipe is followed for low-mass YSOs, using IRAS 16293-2422 as the example. In Sec. 5, hot core and envelope models are summarized. Similarities and differences between high- and low-mass YSOs are discussed in Sec. 6, followed by a summary in Sec. 7. For more extensive reviews of older work, see van Dishoeck & Blake (1998), van Dishoeck & Hogerheijde (1999) and van Dishoeck & van der Tak (2000).

2. Analysis methods

Traditionally, chemical abundances have been derived using simple tools, such as rotation diagrams assuming a constant excitation temperature T_{ex} or statistical equilibrium calculations using an escape probability for a single temperature T and density n. In both cases, comparison with a "standard" molecule such as CO is needed to derive the abundance with respect to H_2. It is well known that this method can lead to abundances that are in error by more than an order of magnitude. For example, the use of $n = 10^5$ versus 10^6 cm^{-3} for the interpretation of the HCN J=4–3 line with $n_{cr} \approx 10^7\,\mathrm{cm}^{-3}$ gives a factor of 10 difference in derived HCN abundance. Densities are known to range from 10^4 to $> 10^7$ cm^{-3} across protostellar envelopes, so which density is the appropriate one for this line? Neglect of beam dilution and optical depth effects can result in similar or even larger errors. For example, the emission from the complex organic molecules may originate in the inner 1″, much smaller than the observational beam of $\sim 15''$.

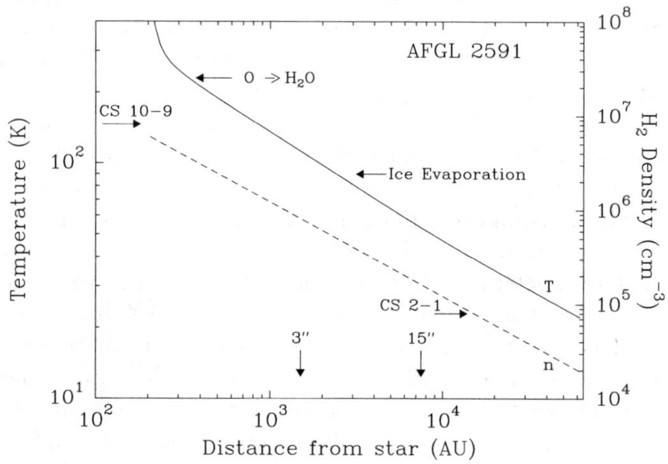

Figure 1. Derived temperature and density structure for the envelope around the massive YSO AFGL 2591 (L=$2\times10^4\,L_\odot$, d=1 kpc). The critical densities of the CS 2–1 and 10–9 lines are indicated, as are typical single-dish beam sizes ($\sim 15''$) and interferometer resolutions ($\sim 3''$). Note that the different chemical regimes are not resolved in single-dish data (based on van der Tak et al. 1999).

The modern analysis method starts with a physical model of the source, constrained by observations. The excitation and radiative transfer are subsequently calculated through this model and the resulting sky brightness distribution is convolved with the beam profile, or, in the case of interferometer data, analyzed with the same spatial filtering. The abundance of the molecule is then adjusted until agreement with the observational data is reached (see Fig. 8 of van Dishoeck & Hogerheijde [1999] for outline). To constrain the physical models, two approaches are taken. Both assume a 1D spherically symmetric model with a power-law density profile—the simplest

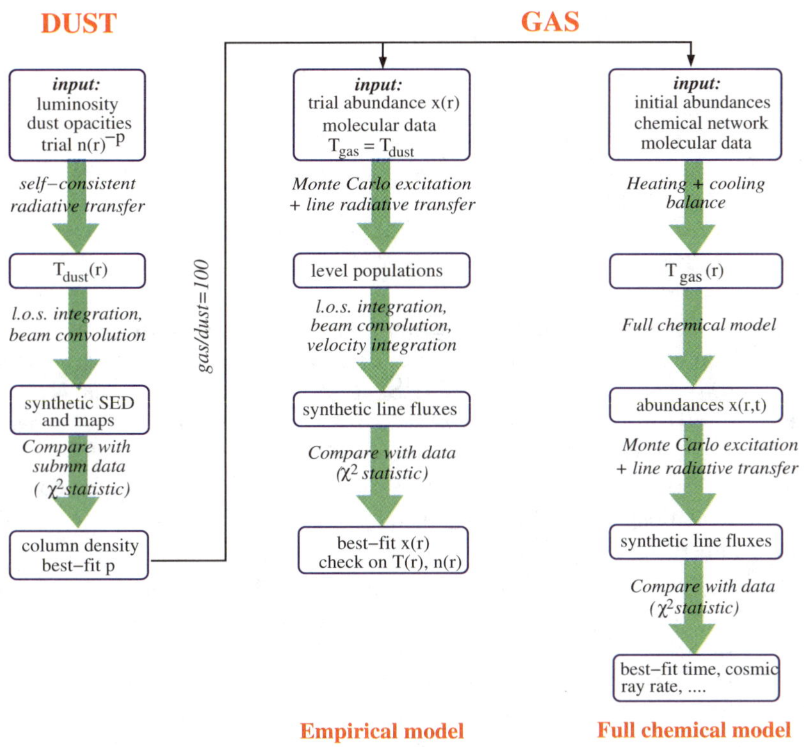

Figure 2. Overview of methods used for constraining the physical and chemical structure of YSO envelopes. The chemical analysis can proceed in two ways. In the "empirical" analysis, $T_{\rm gas} = T_{\rm dust}$ is assumed and the abundance of each species is derived by a χ^2-fit to the observed fluxes. In the "full chemical model" analysis, the physical structure is coupled with a detailed chemical network and $T_{\rm gas}$ and the abundances of each species are calculated explicitly at each radius as a function of time. The emerging fluxes are then calculated for each molecule and compared with observations for consistency. The basic model parameters (e.g., cosmic ray ionization rate, time, initial abundances) are adjusted to obtain the best fit (based on van Dishoeck & van der Tak 2000, Schöier et al. 2002, and Doty et al. 2002).

realistic source model—and calculate the dust temperature profile through the envelope explicitly. In the first method, lines of the same molecule (e.g., CS 2–1, 3–2, 5–4, 7–6, 10–9) are used to constrain the power-law index of the density structure (Zhou et al. 1994; van der Tak et al. 2000a) (see Fig. 1). The critical assumption is that the abundance of the molecule does not vary with position in the envelope. The second, now more robust approach uses the observed dust continuum maps and spectral energy distribution to constrain the density profile (see Fig. 2) (Shirley et al. 2002; Jørgensen et al. 2002). For each trial density structure, the dust temperature is calculated and the preferred power-law index is determined by a χ^2-fit to the observational data. This method assumes that the dust emissivity does not vary across the envelope.

The accuracy in derived power-law index is typically ± 0.2 and the density gradient is generally smooth, with little evidence for clumping. Within the inner envelope out to a few thousand AU, various analyses of the same sources usually agree well. The major differences are in the outer part, especially in the extent of the envelope and its merging with the surrounding cloud. Also, external UV radiation can increase the outer dust temperature. Deviations from spherical geometry due to, e.g., outflow cavities may affect the optical depths at infrared wavelengths by up to a factor of three.

Once the physical structure is fixed, the line emission of different molecules can be calculated to determine the abundances, assuming a gas-to-dust ratio of 100 and a gas temperature equal to the dust temperature. The latter assumption is known to fail in the outer, lower density part of the envelope, where the gas temperature can either fall below the dust temperature in the absence of external heating (Doty & Neufeld 1997) or be above the dust temperature in the case of external ultraviolet radiation (Kaufman et al. 1998). Test models which calculate the gas temperature explicitly generally give very similar results for the inferred abundances, so that this difference can be neglected to first order. For small beams and high-frequency lines, the differences in extent of the model envelopes do not affect the abundance analyses.

In the baseline models, the abundance of the molecule is assumed to be constant with envelope radius and is the single parameter to be fitted to the data. If lines originating from a large range of energies are available, abundance *profiles* can be constrained. The simplest assumption here is a "jump" in the abundance at a temperature where a chemical change is expected; e.g., around 100 K where evaporation of ices occurs. H_2CO, CH_3OH and SO_2 are three well-known examples of molecules for which lines with upper level energies ranging from 10 to 300 K are commonly observed in the 150–370 GHz atmospheric windows and for which such abundance jumps have been inferred (Ceccarelli et al. 2000; van der Tak et al. 2000b; Schöier et al. 2002). The free parameters are now the inner and outer abundances and the location of the jump. The latter parameter is usually taken to be fixed at the radius where $T = 90$ K. Accurate collisional rate coefficients for the various molecules are an essential ingredient for these analyses (Schöier et al., this volume). Such data are often lacking for more complex molecules, where LTE excitation must be assumed.

The scheme discussed above describes the "empirical" modeling to derive chemical abundances directly from observations. Alternatively, one can couple chemical models with the physical structure derived from observations to calculate the abundances explicitly with radius and time in a "full chemical model." The radiative transfer for the different molecules then provides line intensities which can be compared with observations for consistency. The parameters to adjust in these models are those that enter the chemical models, such as the cosmic-ray ionization rate, the initial elemental and molecular abundances, and the time since the abundances were last reset by some physical event (Doty et al. 2002).

3. High-mass YSOs

3.1. Recent observations

A number of recent submillimeter line surveys have highlighted the large variations in chemical composition toward different YSOs. Toward Orion-KL, the recent 650 and 850 GHz surveys illustrate the chemical richness of this source, even at high frequencies (Schilke et al. 2001). Observations towards different positions have shown large chemical variations, for example in oxygen- and nitrogen-bearing molecules between the "hot core" and "compact ridge" (e.g., Sutton et al. 1995) and in chemical complexity with the line-poor Orion-S position. Such differences are also reflected in infrared spectra toward the Orion-KL region, e.g., by selective absorption or emission of species like HCN, C_2H_2 and CO_2 (Evans et al. 1991; Boonman et al. 2003a). For Sgr B2, the spectra toward the N, M and NW positions are dominated by organic molecules, sulfur-bearing species and simple species including carbon chains, respectively (Nummelin et al. 1998, 2000a,b). Similar characteristics are seen towards the W3(H_2O), W3 IRS5 and W3 IRS4 positions (Helmich & van Dishoeck 1997). These variations have been interpreted as being due to differences in time scales since evaporation of the ices and variations in initial ice compositions (Charnley et al. 1992; Rodgers & Charnley 2001; see Sec. 5).

The chemical variations are known to occur on small scales down to 2000 AU or less. For example, Wyrowski et al. (1999) observed significant differences between the W3(OH) and W3(H_2O) positions, only 7″ (1500 AU) apart. Watson et al. (2002) find strong CH_3CN emission toward only one of the two continuum sources separated by less than 4″ (8000 AU) in W75N, a phenomenon also observed for other sources. Liu & Blake (in preparation) mapped various molecules in a chain of 8 ultracompact (UC) H II regions along G9.62+0.19, with the complex and sulfur-bearing molecules appearing only in one or two "hot cores". Thus, care has to be taken in the interpretation of unresolved single-dish data.

The above observations have led to an operational definition of a "hot core" as a compact (< 0.1 pc), warm ($T > 100$ K) and dense ($n > 10^6$ cm^{-3}) parcel of gas characterized by rich molecular line emission (e.g, Walmsley 1992). It is currently not clear whether the "hot core" is a physically distinct component or whether it is simply the warm part of the inner envelope at a certain stage of chemical evolution. For massive YSOs, a compressed, dense shell may be formed at the edge of the expanding ultra- or hyper-compact H II region due to the pressure from the ionized gas. Chemically, there is also a difference between simple evaporation of ices (as observed for W3 IRS5) and the subsequent "hot core chemistry" between evaporated species leading to more complex organic molecules and highly crowded submillimeter line spectra (as observed for W3[H_2O]). In this paper, only the latter case will be termed a "hot core".

Comprehensive surveys of a larger sample of objects in selected molecules are only just appearing. At submillimeter wavelengths, Hatchell et al. (1998a,b) observed a set of 14 UC H II regions in selected settings with the

Table 1. Chemical characteristics of young stellar objects

Component	Chemical characteristics	Submillimeter diagnostics	Infrared diagnostics	Examples High-mass	Examples Low-mass
Surrounding cloud	Low-T chemistry	Ions, long-chains (HC_5N, ...)	Simple ices (H_2O, CO)	Sgr B2 (NW)	TMC-1
Cold envelope	Low-T chemistry, Heavy depletions	Simple species (CS, N_2H^+)	Ices (H_2O, CO_2, CH_3OH)	N7538 IRS9, W 33A	N1333 IRAS 4B, IRAS 16293 (out)
Inner warm envelope	Evaporation	High T_{ex} (CH_3OH, H_2CO)	High gas/solid, High T_{ex}, Heated ices (C_2H_2, H_2O, CO_2)	GL 2591, GL 2136	N1333 IRAS4A (in), IRAS 16293 (in)
Hot core	High-T chemistry	Complex organics (CH_3OCH_3, CH_3CN, vib. exc. mol.), high HCN	Hydrides (OH, H_2O)	Orion hot core, Sgr B2(N), G34.3, W 3(H_2O)	?
Outflow: Direct impact	Shock chemistry, Sputtering	Si- and S-species (SiO, SO_2)	Atomic lines, Hydrides ([S I], H_2O, H_2 high-J)	W3 IRS5, Sgr B2(M)	L1448-mm, L1157
Outflow: Turbulent entrainment	Evaporation	Morphology (CH_3OH, HCO^+, HCN)	Hydrides (H_2 low$-J$)	?	L1527, B5 IRS1 Serpens SMM1
PDR, Compact H II regions	Photodissociation, Photoionization	Ions, radicals (CN/HCN, CO^+)	Ionic lines, PAHs, H_2 ([Ne II], [C II])	S 140, W3 IRS4	?

JCMT, with about half the sources showing the characteristic crowded hot core spectra. Ikeda et al. (2001) surveyed a few characteristic complex organic molecules toward a number of massive YSOs and found trends with increasing dust temperature for some, but not all, species, allowing molecules which originate in the inner envelope from high temperature chemistry or evaporation of grain mantles to be separated from those which occur mostly in the colder outer envelope.

Van der Tak et al. (2000a) selected a sample of 14 massive young stars for JCMT observations, most of which are bright at mid-infrared wavelengths. The infrared bright sources do not show strong emission from complex organic species, suggesting that they have not yet reached the hot core stage. The data have been analyzed systematically along the lines illustrated in Figure 2, so that reliable empirical abundances are available for this sample. For some sources, jumps in the abundances of two orders of magnitude of selected molecules (CH_3OH, SO_2) have been found in the inner envelope (van der Tak et al. 2000b, 2003).

The submillimeter data of the infrared-bright sample have been complemented by mid-infrared surveys with the *ISO*-SWS. The abundances of species like HCN and C_2H_2 (Lahuis & van Dishoeck 2000), CO_2 and H_2O (Boonman et al. 2003b,c) show systematic increases, which correlate well with other temperature tracers such as the far-infrared color, the gas/solid ratios and the heating signatures in the ice profiles (e.g., Gerakines et al. 1999; Boogert et al. 2000). Sulfur-bearing species present a more confusing picture, both at infrared (Keane et al. 2001) and submillimeter wavelengths (van der Tak et al. 2003), with both hot core chemistry and shocks likely playing a role.

If higher temperatures are indeed due to increased dispersion of the envelope (i.e., decreasing $M_{\rm env}/L_*$), this "global warming" may be interpreted as an evolutionary effect and the above features can then be used as powerful tracers of the earliest stages of massive star formation, even before the hot core and UC H II phases. In this respect, the analysis of massive YSOs appears ahead of that of low-mass YSOs, where systematic trends *within* the equivalent Class 0 or I stages are still difficult to discern, partly due to a lack of infrared spectroscopy (see Sec. 4).

3.2. AFGL 2591 AS AN EXAMPLE

The case of AFGL 2591 presents a good example of the power of the analysis techniques and the information that can be derived on the chemical structure, both from the "empirical" and "full chemical modeling" approaches. This source has a luminosity of $2 \times 10^4\,L_\odot$, and is relatively nearby ($d \approx 1$ kpc) and isolated. Figure 1 shows the best-fitting temperature and density structure, derived from the CS line ratios (van der Tak et al. 1999). The different chemical regimes are indicated. In the cold outer envelope (<90 K), significant freeze-out occurs and the abundances are expected to be low. In the inner envelope, ice evaporation dominates the chemistry. The temperatures at which ices evaporate differ from species to species and range from as low as 20 K (pure CO ice) to

80–100 K (CH_3OH and H_2O) (e.g., Sandford & Allamandola 1993; Fraser et al. 2001). At even higher temperatures, most of the oxygen is driven into H_2O and high abundances of species like HCN occur.

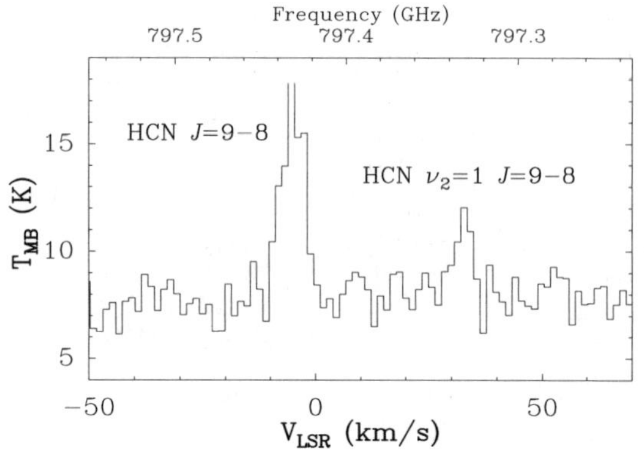

Figure 3. Detection of the HCN ν_2=0 and ν_2=1, J=9–8 line toward AFGL 2591 with the MPIfR/SRON 800 GHz receiver on the JCMT. The vibrationally excited emission indicates a jump of a factor of ~100 in the HCN abundance in the inner warm envelope, consistent with the high abundance derived from infrared absorption lines (from Boonman et al. 2001).

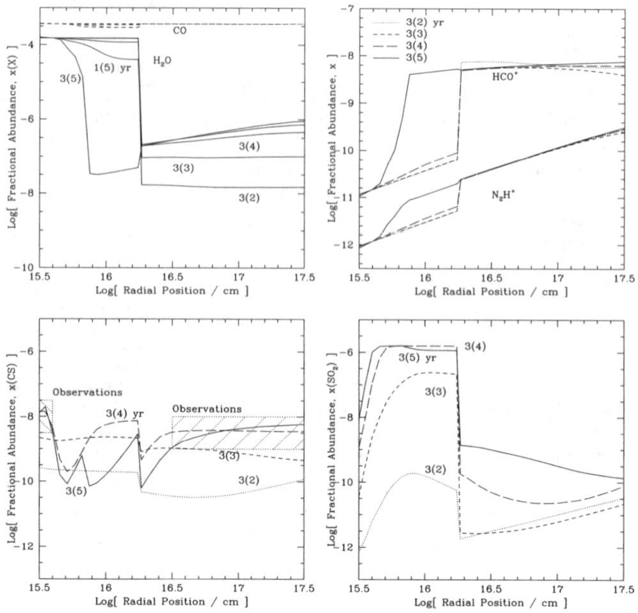

Figure 4. Examples of abundance profiles as functions of time for the AFGL 2591 model. *Top left:* CO and H_2O; *top right:* HCO^+ and N_2H^+; *bottom left:* CS; *bottom right:* SO_2. The physical structure is taken to be static. For some species, a constant (e.g., CS), jump (e.g., H_2O) or anti-jump (e.g., N_2H^+) abundance model is seen to be a good representation, whereas other molecules vary more strongly with radius and time (from Doty et al. 2002).

Figure 3 shows the HCN J=9–8 line in the ν_2=0 and 1 states detected with the MPIfR/SRON 800 GHz receiver on the JCMT (Boonman et al. 2001). The low-J HCN J=1–0 through 4–3 lines give an abundance with respect to H_2 of only 10^{-8} in the outer envelope. The HCN ν_2=1–0 infrared absorption lines indicate a value of $\sim 10^{-6}$ (Lahuis & van Dishoeck 2000). A similarly high abundance is needed to fit the $\nu_2 = 1$ J=9–8 data. Together, the data give irrefutable evidence for an abundance jump of two orders of magnitude between the outer and inner regions.

A second example of the "empirical" modeling is provided by the analysis of the H_2O data. Strong infrared absorption within the ν_2=1–0 band at 6 μm has been detected with the *ISO*-SWS by Helmich et al. (1996). H_2O emission in the $1_{10} - 1_{01}$ line at 557 GHz with *SWAS* is, however, weak (Snell et al. 1999). Boonman et al. (2003d) analyzed the combined *ISO*-SWS, *ISO*-LWS and *SWAS* data and tested different H_2O abundance profiles. The data are only consistent with a low abundance of $\sim 10^{-8}$ in the outer region caused by freeze out, and a high abundance up to 10^{-4} in the inner region due to evaporation.

For AFGL 2591, the "full chemical model" analysis has also been performed. Doty et al. (2002) have taken the density structure of van der Tak et al. (1999) and solved self-consistently for the gas temperature. They then ran a time-dependent chemistry as a function of position r in the envelope, keeping the physical structure fixed. The initial abundances are assumed to be low in the outer envelope (< 90 K), but high for selected species in the inner envelope due to evaporation of ices. The resulting abundances $x(r,t)$ are compared with those derived from observations in the inner and outer regions, respectively, and a best-fitting cosmic-ray ionization rate and time is determined. For AFGL 2591, the inferred young chemical "age" of 3×10^4 yr is consistent with the fact that this source does not yet show strong radio continuum emission.

Examples of abundance profiles as functions of time are shown in Figure 4. For some species, the "jump" or "anti-jump" model used in the empirical model is seen to be a good approximation, but in other cases a more complex chemical profile is more appropriate. Such detailed chemical models can be taken as guidance for the trial abundance profiles to be used in the empirical modeling.

4. Low-mass YSOs

4.1. Recent observations

The observed low-mass YSOs are closer than their high-mass counterparts, so that higher spatial resolution can be obtained allowing different chemical components to be resolved. However, their envelope masses are 1–2 orders of magnitude lower and their lines generally weaker, especially at higher frequencies. Many observations have therefore been performed at lower frequencies in large beams, but care has to be taken that such data do not apply mostly to extended, lower-density cloud material rather than the YSO envelope. Examples of recent chemical surveys of selected molecules toward low-mass YSOs include Hogerheijde et al. (1999), Shah & Wootten (2001), Buckle & Fuller (2002), Jørgensen et al. (2002; see also this volume), and Maret et al. (in preparation). While simple molecules such as HCN, HCO^+, H_2CO and even CH_3OH are readily detected, none of these sources have the crowded line spectra characteristic of high-mass hot cores. They also do not show strong lines due to unsaturated long carbon chains, such as seen toward some pre-stellar cores like TMC-1S. Extremely high deuterium fractionation is observed for most sources (van Dishoeck et al. 1995), including the detection of doubly deuterated (Loinard et al. 2001) and even triply deuterated species (van der Tak et al. 2002).

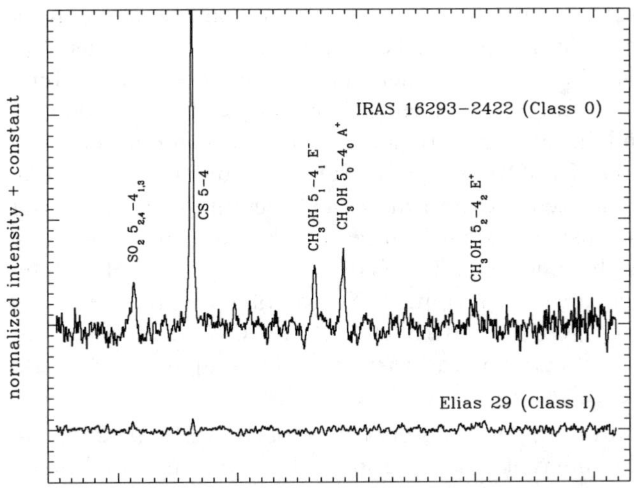

Figure 5. Comparison of the emission around 241.8 GHz from the Class 0 low-mass YSO IRAS 16293–2422 with that from the Class I object Elias 29. Strong lines from CH_3OH, CS and SO_2 are seen in the Class 0 object (from Boogert et al. 2002).

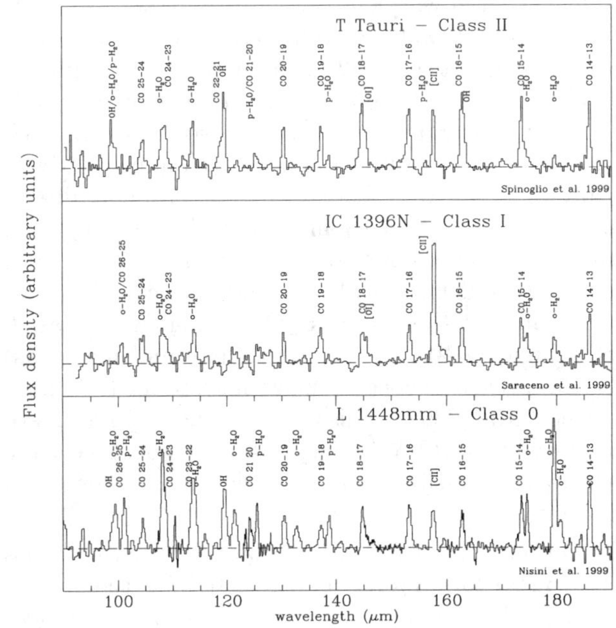

Figure 6. *ISO*-LWS spectra of low-mass YSOs at different stages of evolution. The continuum has been subtracted (based on Saraceno et al. 1999).

The sample of 16 YSOs by Jørgensen et al. (2002) has been analyzed systematically using the techniques outlined in Figure 2 and covers both deeply embedded Class 0 objects and more evolved Class I sources. The figures of Jørgensen et al. (this volume) show some of the trends in the derived abundances. For the Class 0 objects, significant freeze-out is inferred, even for species like CO which start to evaporate at relatively low temperatures. Class I objects generally show very weak lines (see Fig. 5), but the abundances are closer to general molecular cloud values. Surprisingly large variations in the HCN, HNC and CN abundances and abundance ratios are found between sources (see also Sec. 6). Johnstone et al. (2003, this volume) have observed a sample of 7 intermediate-mass YSOs and find clear trends in the abundances with temperature of the source.

For the Class 0 objects, sufficient lines of H_2CO and CH_3OH can be observed to test the presence of abundance jumps. Indeed, jumps of factors of 100 or more have been inferred for several species and sources (Ceccarelli et al. 2000; Schöier et al. 2002; Maret et al., in prep.). Such large enhancements are not only found in the innermost envelope, but can also occur at positions in outflow lobes, which can be spatially separated for some low-mass YSOs (Bachiller & Pérez-Gutiérrez 1997; Garay 2000).

At far-infrared wavelengths, 45–197 μm spectra of 28 low-luminosity YSOs have been measured with the *ISO*-LWS (see Nisini et al. 2002 for a summary). In spite of the low spectral resolution ($R \approx 200$), lines of H_2O, OH and CO are clearly detected in several sources with H_2O strong toward Class 0 objects and OH becoming relatively more prominent toward Class I objects, probably due to enhanced UV radiation (Spinoglio et al. 2000; Fig. 6). Because these data refer to large beams ($\sim$90″), both the envelope and large-scale outflows can contribute to the emission, but the overall H_2O abundance in Class 0 YSOs is high, $\sim 10^{-5}$. For intermediate and high-mass YSOs, the *ISO*-LWS spectra show much fewer H_2O lines (Lorenzetti et al. 2002). This is partly due to photodissociation by enhanced UV radiation, but also because these objects are more distant so that the data suffer more from beam dilution.

Bergin et al. (2002) have mapped the low-mass NGC 1333 star-forming region with *SWAS* in the ground state o-H_2O line at 557 GHz and find H_2O abundances of $> 10^{-6}$ in outflows and up to 10^{-7} in quiescent gas close to luminous external heating sources. A detailed analysis of the *ISO*-LWS H_2O data for one object, NGC 1333 IRAS 4A, shows low H_2O abundances in the outer envelope, increasing to high values of at least 5×10^{-6} in the inner envelope (Maret et al. 2002). Thus, similar H_2O abundance changes are seen in the envelopes of low- and high-mass YSOs (compare the case of AFGL 2591).

4.2. IRAS 16293–2422 AS AN EXAMPLE

The best studied low-mass YSO in terms of chemistry is IRAS 16293–2422, a 27 $L_\odot$ protobinary object located in the Ophiuchus molecular cloud at a distance of $\sim$160 pc. It has been coined "the low-mass equivalent of Orion-KL," although full spectral surveys are still lacking and many fewer molecules have been detected with the sensitivities of the observations published to date. Blake et al. (1994) and van Dishoeck et al. (1995) performed partial line surveys in the 230 and 345 GHz windows using matched beams with the JCMT and CSO, respectively. Deeper integrations on selected settings have been carried out, e.g., by Ceccarelli et al. (2000).

Schöier et al. (2002) have recently reanalyzed the available JCMT and CSO data using the methods outlined in Figure 2. The best-fitting temperature and power-law density profiles are very similar to those of high-mass YSOs as shown in Figure 1, except for a scale factor. For IRAS 16293–2422, the radius at which the

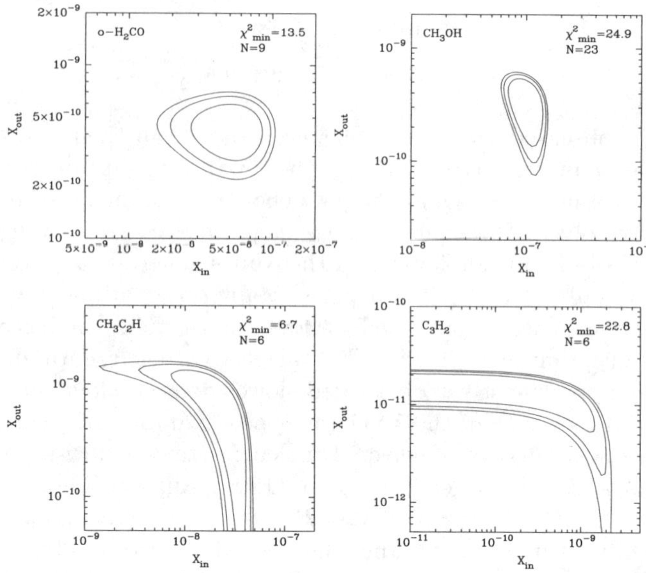

Figure 7. Examples of χ^2 plots for determining the abundances of species in the envelope of the low-mass source IRAS 16293-2422 according to the empirical model. For H_2CO and CH_3OH, many lines covering a wide range of excitation conditions have been observed, so that both the inner ($X_{\rm in}$) and outer ($X_{\rm out}$) abundances are well constrained, with a jump around 90 K. For other species, only the inner or outer abundance is well determined (from Schöier et al. 2002).

dust temperature reaches 90 K is 150 AU, compared with 2500 AU for AFGL 2591. For many species, sufficient lines covering a large range in excitation conditions are available to fit jump models, and χ^2-fits have been used to constrain both the inner and the outer envelope abundances. Large enhancements of factors of $\sim$100–1000 are found in the inner region above $\sim$90 K for species like H_2CO, CH_3OH, CH_3CN, SO, SO_2 and OCS (see Fig. 7). The temperature at which the jump occurs is not well constrained but is at least 50 K. For other molecules, only the inner (e.g., CH_3C_2H) or outer (e.g., C_3H_2) abundance is well determined. No complex organic molecules like CH_3OCH_3 and $HCOOCH_3$ have been reported but the limits are not yet very stringent compared with high-mass YSOs.

Interferometer observations of IRAS 16293-2422 reveal different distributions for various molecules. Molecules like CS and HCO^+ are good tracers of the envelope, whereas N_2H^+ and HNC clearly avoid the inner warm region. H_2CO, on the other hand, traces the inner envelope, especially in the higher excitation lines (Schöier et al., this volume). Such spatially-resolved data are important to test whether the enhanced abundances in low-mass YSOs are primarily due to thermal evaporation in the hot inner envelope or whether grain-grain collisions in turbulent regions caused by the interaction of the outflow with the inner envelope can also release grain mantles.

Doty et al. (this conference and in preparation) have performed full chemical modeling for this source. Starting from the (static) temperature and density profile derived by Schöier et al. (2002) and using the same chemical model as applied to AFGL 2591, the abundances of many species have been computed as functions of position and time in the envelope. Subsequently, the emerging line fluxes have been obtained by calculating the excitation and radiative transfer through the model envelope and comparing them directly with the observational data after convolution with the appropriate beams. For a best-fitting time of a few $\times 10^4$ yr, the model fluxes generally agree within 50% with the data, much better than the factor of 3–10 agreement generally cited for dark clouds (e.g., Terzieva & Herbst 1998).

5. Models of hot core and envelope chemistry

5.1. Static models

In the static envelope or "cocoon" models described in Sec. 3.2 and 4.2, the temperature and density structures are taken to be constant in time and the molecules do not move from one position to another within the envelope, e.g., due to infall. These 1D models can contain a "hot core" in the center, where evaporation of ices followed by high-temperature chemistry occurs. Static hot core models have been extensively described by Charnley et al. (1992, 1995), Charnley (1997), Caselli et al. (1993) and Rodgers & Charnley (2001) (see Millar 1997 for review). The choice of molecules to be evaporated and their abundances are inspired by the observed chemical composition of ices toward high-mass YSOs, which consists largely of H_2O, CO_2, CH_3OH and H_2CO (e.g., Gibb et al. 2000). NH_3 and H_2S are often assumed to represent the major nitrogen and sulfur reservoirs in the ices, although this has not yet been confirmed by observations (see Taban et al. [2003] for the case of NH_3). The hot core models are particularly sensitive to the initial CH_3OH/NH_3 ratio.

In most hot core models, all ices are assumed to evaporate instantaneously at t=0, but in some cases a more gradual evaporation is used (e.g., Viti & Williams 1999). The evaporated ices subsequently drive a rapid high temperature chemistry in the warm gas, leading to more complex organic molecules such as CH_3OCH_3 and $HCOOCH_3$ on a time scale of 10^4 yr. Nitrogen-bearing molecules require longer timescales up to 10^5 yr and favor higher temperatures. Similarly, the H_2S/SO_2 ratio can serve as a chemical clock, if it is assumed that most of the gas phase sulfur is initially in H_2S. Indeed, one of the main limitations in the use of these molecules to constrain the source age is that it is not known whether they are "first generation" molecules present in the ices or whether they represent "second generation" species produced in the hot gas.

The combination of a hot core with a more extended envelope was first performed in a 3-layer model for G34.3 by Millar et al. (1997) and subsequently within a single power-law model for AFGL 2591 by Doty et al. (2002). In the latter case, the density and dust temperature profiles are taken directly from observations. The gas temperature is calculated explicitly taking all heating and cooling mechanisms of the gas into account. The chemistry is then solved as a function of time at each position using an extensive chemical network containing both low- and high-temperature reactions and separating the envelope into a warm inner part where the ices are evaporated and a cold outer part where molecules are still heavily depleted.

5.2. Dynamical models

In the dynamical models, the temperature $T(r,t)$ and density $n(r,t)$ of the source change according to some physical prescription and the gas moves inward or outward according to a velocity profile $v(r,t)$. The best studied case is that of the popular Shu (1977) collapse model appropriate for low-mass YSOs. Rawlings et al. (1992) were the first to follow the chemistry $x(r,t)$ in a parcel of gas as it is falling into the center in an effort to explain the difference in chemistry and line profiles for species such as NH_3 and CS. Ceccarelli et al. (1996) subsequently computed the full thermal balance of the dust and gas as a function of time, coupled with a simple chemistry aimed at predicting the [O I] and H_2O emission from low-mass YSOs. More recently, Rodgers & Charnley (in preparation) coupled a full chemical network with a Shu-type dynamical model and compared the results with those of a static model. Interestingly, they concluded that small-scale differentiation and high abundances of complex species can only occur if the gas is not falling rapidly into the protostar; i.e., observations of such species require a more or less static model. In a dynamical model, the chemical timescales become longer than the dynamical timescales so that the abundances are frozen at their initial values.

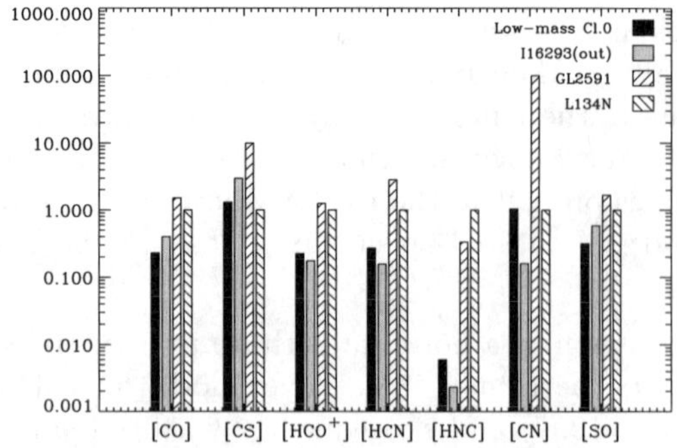

Figure 8. Comparison of the abundances found in the cold outer envelope of the low-mass YSO IRAS 16293-2422 (Schöier et al. 2002) with those derived for a larger set of Class 0 low-mass YSOs (Jørgensen et al., in prep.) and the high-mass AFGL 2591 envelope (van der Tak et al. 2000a, 2003). Note the strong variations in the HCN/HNC/CN abundances and ratios (from Jørgensen et al., in prep.).

6. High versus low-mass YSOs

It is clear from the preceding sections that there are several similarities but also some potential differences between the chemistry in low- and high-mass YSOs. Since statistically significant data sets analyzed in a homogeneous way are still lacking, some of the apparent differences may well become similarities and vice versa in future studies.

6.1. Similarities ?

Physical structure: The density structure of the envelopes of both low- and high-mass YSOs can be well described by a smooth power law whose index can be constrained observationally from dust continuum and/or line observations. Some analyses suggest a slightly shallower index for high-mass YSOs compared with low-mass YSOs (van der Tak et al. 2000a), whereas other studies find similar indices within the observational errors (Mueller et al. 2002; Beuther et al. 2002), but this potential difference does not have significant chemical consequences to first order. In both sets of sources, the dust temperature follows an optically thin $\propto r^{-0.4}$ law in the outer envelope, increasing more steeply in the inner part to temperatures above 100 K. The only difference is a scale factor, resulting in a larger physical size of the warm region for high-mass YSOs. The good fits to the data imply that outflows and shocks give only a limited contribution to the heating in both types of objects. Because of the higher temperatures throughout, the abundances in the outer envelopes of high-mass objects may be somewhat higher than those in low-mass Class 0 YSOs (see Fig. 8).

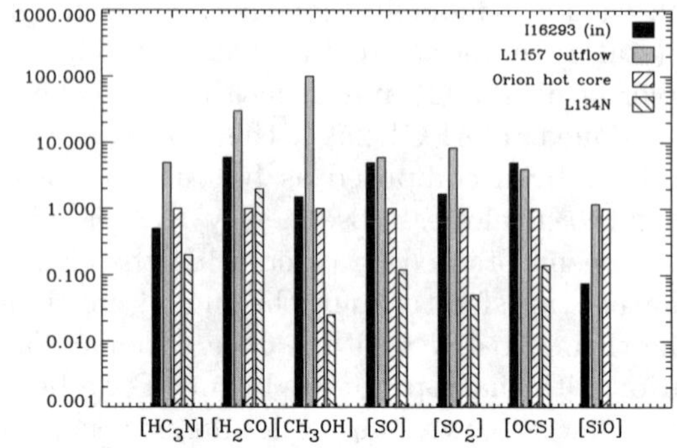

Figure 9. Comparison of the abundances found in the inner warm region of the low-mass YSO IRAS 16293-2422 with those derived for the high-mass Orion hot core and for the outflow position offset from the low-mass YSO L1157. The abundances are normalized to the Orion hot core (based on Schöier et al. 2002).

Abundance jumps: For both low- and high-mass YSOs, jumps in the abundances of selected molecules are found (e.g., CH_3OH, H_2O). The jumps are of comparable magnitude (about a factor of 100) and their location is consistent with the temperature of ~90 K at which most ices are thought to evaporate. Figure 9 compares the abundances of a few characteristic species in the Orion-KL hot core with those found in the inner warm region of IRAS 16293–2422 (see Table 7 of Schöier et al. 2002 for details). The number of objects for which such detailed abundance analyses have been done is still small however, and statistics on larger samples are needed.

6.2. Differences

Hot core chemistry: Both low- and high-mass YSOs are known to have an inner warm and dense region where ices have evaporated. But do they have a similar hot core chemistry? Complex organic molecules such as CH_3OCH_3 and $HCOOCH_3$, which are thought to be secondary species produced by high-temperature reactions, are found toward high-mass YSOs only. These species have not yet been detected toward low-mass YSOs in published spectra, although the abundance limits are comparable to those in high-mass YSOs if the complex molecules are located only in the inner warm region.

Deeper searches are needed for low-mass YSOs. Potential differences in hot-core chemistry could be due to several factors. First, the dynamical infall timescales in the warm gas in low-mass YSOs are much shorter ($< 10^3$ yr) compared with high-mass YSOs, so that the complex chemistry may not have enough time to develop unless the collapse is slowed down. Second, the chemical composition of the ices may be different due to different H/H_2 ratios in the collapsing envelope and/or different grain temperatures in high- versus low-mass star-forming regions. Third, the size of the hot core and the temperature region above 230 K may be so small in low-mass YSOs (<100 AU; i.e., comparable to the size of the disk) that it is negligible in practice. On the other hand, expansion of the ultracompact H II region may create a large compressed shell of warm molecular gas for high-mass YSOs.

Deuterium fractionation: The observed ratios of deuterated molecules with respect to their normal counterparts are at least an order of magnitude higher on average for low-mass YSOs (Roberts et al. 2002). This can be due to two factors. First, the temperature of the pre-stellar cores may be lower for low-mass YSOs, enhancing the deuterium fractionation especially if CO is frozen out. In some high-mass star-forming regions, the temperatures may never have been cold enough for significant CO depletion. Second, the time scales of the pre-stellar phase may have been longer for low-mass YSOs, leading to more complete CO freeze-out across the core and further favoring the production of H_2D^+.

Effects of UV or X-rays: For both low- and high-mass YSOs no clear diagnostics of UV radiation or X-rays have yet been found in the deeply embedded phase. The UV radiation can be either externally or internally produced. In the former case, the dust and gas temperatures in the outer envelope are raised and molecules are photodissociated. For example, HCN is photodissociated to CN, a well-known photodissociation product in PDRs (e.g., Fuente et al. 1998). The internal UV radiation or X-rays can originate either from the young star itself or from a hot accretion boundary layer between the inner disk and the star. Characteristic products are CO^+, CN and OH. Moreover, the radiation can affect the composition of the ices, e.g., by producing a feature of OCN^- at 4.62 μm (but see van Broekhuijzen et al., this volume). UV radiation also excites PAH features. Deep searches for photodissociation and X-ray products and PAH features in both low- and high-mass YSOs are needed. It is clear that UV radiation plays an important role in the chemistry of high-mass YSOs in later stages during the evolution from the ultracompact to the classical H II region stage, where PAHs and photodissociation products are widely observed.

Effect of shocks: Outflows are known to be associated with both low- and high-mass YSOs and their effects on the chemistry are well documented at *off-source* positions in the outflow lobes, where enhancements of species like SiO, CH_3OH and H_2O have been seen (see also Fig. 9). However, their importance in the inner dense envelope relative to thermal evaporation remains to be determined. The observed line widths toward both low- and high-mass YSOs are typically a few km s^{-1} and only the strongest lines of a few species show clear evidence for line wings (see also Buckle & Fuller 2002). It has been argued that shocks may be relatively more important in low-mass YSOs, not only for the dispersion of the envelope but also for releasing grain mantles in the turbulent shear zones between the outflow and the quiescent envelope. This can be tested by mapping the distribution of grain surface products such as H_2CO. High angular resolution observations of a few low-mass sources provide evidence for both scenarios (e.g., Schöier et al., in prep.). Schreyer et al. (2002) show that the presence of low-mass outflows within a high-mass envelope does not have noticeable effects on the chemistry, indicating that only a small fraction of the gas is affected. On the other hand, the absence of abundant gas phase CO_2 in the inner envelope has been cited as evidence for shock chemistry throughout high-mass YSO envelopes (Charnley & Kaufman 2000). Hatchell et al. (2001) argued based on SiO data that shocks do not affect the hot core of the massive YSO G34.3, although they are present in the more extended envelope. Further observations of shock diagnostics, such as the NS/CS ratio proposed by Viti et al. (2001), are needed (see also Hatchell & Viti 2002).

Geometry: The effects of geometry and foreground clouds are thought to be more important for the analysis of low-mass YSO data than for high-mass sources, where the envelope often overwhelms all other emission or absorption. For example, a significant fraction of the ices along the line of sight toward low-mass YSOs in Ophiuchus may originate in unrelated foreground material (e.g., Boogert et al. 2002). Low-excitation millimeter lines often have their origin primarily in the surrounding clouds, where they may or may not be distinguished by their velocity structure. Finally, disks are known to exist in the embedded Class I phase for low-mass YSOs, whereas their presence for high-mass YSOs remains to be firmly established. Depending on the viewing angle, disks can change the observed mid-infrared characteristics of the source. They also provide a natural cut-off for (part of) the inner warm envelope and may shield some of the envelope from heating. Their chemical structure is expected to be significantly different from that of the inner envelope due to the high densities and low midplane temperatures.

7. Conclusions

The combination of submillimeter and infrared observations allows a nearly complete inventory of gas- and solid state species of high- and low-mass YSOs. Combined with detailed modeling using a new generation of flexible radiative transfer codes, quantitative analyses of both their physical and chemical structure are now possible. Clear variations in the gas phase abundances and the gas/solid ratios are found with increasing temperature in the envelope, indicating that the chemistry is controlled by freeze-out and ice evaporation. Systematic studies of statistically significant samples of low- and high-mass YSOs are only just appearing, and some similarities and differences are emerging. Due to the limited samples and poor sensitivity and angular resolution of the existing data, much more work is needed to draw firm conclusions.

Progress in this field will be greatly aided by new observational facilities. *SIRTF* and eventually JWST will allow mid-infrared imaging and spectroscopy of a much larger sample of embedded low-mass YSOs. *Herschel* will provide unique data on H_2O, OH and other hydrides in both low- and high-mass YSOs, which, among many other applications, will help to disentangle the importance of shocks and thermal evaporation. Finally, ALMA will give maps of many species at unprecedented angular resolution and sensitivity, essential to constrain the small-scale structure of the envelope and outflows and probe the circumstellar disks. Together, they will allow the full chemical history of material from molecular clouds to planetary systems to be traced.

Acknowledgements

The author is grateful to Geoff Blake, Adwin Boogert, Annemieke Boonman, Rogier Braakman, Steven Doty, Helen Fraser, Michiel Hogerheijde, Doug Johnstone, Jes Jørgensen, Fred Lahuis, Klaus Pontoppidan, Fredrik Schöier, Willem Schutte, Fleur van Broekhuizen and Floris van der Tak for many enjoyable discussions and for providing input and figures for this review. The reader is strongly encouraged to read their original papers on these topics, some of which are summarized in this volume. Astrochemistry in Leiden is supported by grants no. 614.41.003 and 614.041.004 as well as a Spinoza grant from NWO and by the Netherlands Research School of Astronomy (NOVA).

References

Bachiller, R. & Pérez-Gutiérrez, M. 1997, ApJ, 487, L93
Bergin, E.A., et al. 2002, ApJ, 582, 830
Beuther, H., et al. 2002, ApJ, 566, 945
Blake, G.A., et al. 1994, ApJ, 428, 680
Blake, G.A., Sutton, E.C., Masson, C.R., & Phillips, T.G. 1987, ApJ, 315, 621
Boogert, A.C.A., et al. 2002, ApJ, 570, 708
Boogert, A.C.A., et al. 2000, A&A, 353, 349
Boonman, A.M.S., et al. 2001, ApJ, 553, L63
Boonman, A.M.S., et al. 2003a, A&A, in press
Boonman, A.M.S., van Dishoeck, E.F., Lahuis, F., & Doty, S.D. 2003b, A&A, in press
Boonman, A.M.S., & van Dishoeck, E. F. 2003c, A&A, submitted
Boonman, A.M.S., et al. 2003d, A&A, submitted
Buckle, J.V., & Fuller, G.A. 2002, A&A, 381, 77
Caselli, P., Hasegawa, T.I., & Herbst, E. 1993, ApJ, 408, 548
Ceccarelli, C. , Hollenbach, D. J. & Tielens, A. G. G. M. 1996, ApJ, 471, 400
Ceccarelli, C., et al. 2000, A&A, 357, L9
Charnley, S.B. 1997, ApJ, 481, 396
Charnley, S.B., & Kaufman, M.J. 2000, ApJ, 529, L111
Charnley, S.B., Tielens, A.G.G.M., & Millar, T.J. 1992, ApJ, 399, L71
Charnley, S.B., Kress, M.E., Tielens, A.G.G.M., & Millar, T.J. 1995, ApJ, 448, 232
Doty, S.D. & Neufeld, D.A. 1997, ApJ, 489, 122
Doty, S.D., van Dishoeck, E.F., van der Tak, F.F.S., & Boonman, A.M.S. 2002, A&A, 389, 446
Evans, N.J., Lacy, J.H., & Carr, J.S. 1991, ApJ, 383, 674
Fraser, H.J., Collings, M.P., McCoustra, M.R.S., & Williams, D.A. 2001, MNRAS, 327, 1165
Fuente, A., et al. 1998, A&A, 334, 253
Garay, G. 2000, in *Astrochemistry: From Molecular Clouds to Planetary Systems*, IAU Symposium 197, ed. Y.C. Minh & E.F. van Dishoeck (San Francisco: ASP), 203
Gerakines, P.A., et al. 1999, ApJ, 522, 357
Gibb, E.L., et al. 2000, ApJ, 536, 347

Hartquist, T. W., et al. 1998, in *The Molecular Astrophysics of Stars and Galaxies*, ed. T.W. Hartquist & D.A. Williams (Oxford: Oxford University Press), 101

Hatchell, J., & Viti, S. 2002, A&A, 381, L33

Hatchell, J., Fuller, G.A., & Millar, T.J. 2001, A&A, 372, 281

Hatchell, J., Thompson, M.A., Millar, T.J., & Macdonald, G.H. 1998a, A&AS, 133, 29

Hatchell, J., Thompson, M.A., Millar, T.J., & Macdonald, G.H. 1998b, A&A, 338, 713

Helmich, F.P., & van Dishoeck, E.F. 1997, A&AS 124, 205

Helmich, F.P., et al. 1996, A&A, 315, L173

Hogerheijde, M. R., van Dishoeck, E. F., Salverda, J. M., & Blake, G. A. 1999, ApJ, 513, 350

Ikeda, M., et al. 2001, ApJ, 560, 792

Johnstone, D., Boonman, A.M.S., & van Dishoeck, E.F. 2003, ApJ, submitted

Jørgensen, J.K., Schöier, F.L., & van Dishoeck, E.F. 2002, A&A, 389, 908

Juvela, M. 1997, A&A, 322, 943

Kaufman, M.J., Hollenbach, D.J., & Tielens, A.G.G.M. 1998, ApJ, 497, 276

Keane, J.V., Boonman, A.M.S., Tielens, A.G.G.M., & van Dishoeck, E.F. 2001, A&A, 376, L5

Lahuis, F., & van Dishoeck, E.F., 2000, A&A 355, 699

Langer, W.D., et al. 2000, in *Protostars & Planets IV*, ed. V. Mannings, A.P. Boss, & S.S. Russell (Tucson: University of Arizona Press), 29

Loinard, L., et al. 2001, ApJ, 552, L163

Lorenzetti, D., et al. 2002, A&A, 395, 637

Maret, S., et al. 2002, A&A, 395, 573

Millar, T.J. 1997, in *Molecules in Astrophysics: Probes and Processes*, IAU Symposium 178, ed. E.F. van Dishoeck (Dordrecht: Kluwer), 75

Millar, T.J., Macdonald, G.H., & Gibb, A.G. 1997, A&A, 325, 1163

Mueller, K.E., Shirley, Y.L., Evans, N.J., & Jakobson, H.R. 2002, ApJS, 143, 469

Nisini, B., Giannini, T., & Lorenzetti, D. 2002, ApJ, 574, 246

Nummelin, A., et al. 1998, ApJS, 117, 427

Nummelin, A., et al. 2000a, ApJS, 128, 213

Nummelin, A., et al. 2000b, ApJS, 132, 127

Rawlings, J.M.C., Hartquist, T.W., Menten, K.M., & Williams, D.A. 1992, MNRAS, 255, 471

Roberts, H., et al. 2002, A&A, 381, 1026

Rodgers, S.D., & Charnley, S.B. 2001, ApJ, 546, 324

Sandford, S., & Allamandola, L.J. 1993, ApJ, 417, 815

Schilke, P., et al. 2001, ApJS, 132, 281

Schöier, F.L., Jørgensen, J.K., van Dishoeck, E.F., & Blake, G.A. 2002, A&A, 390, 1001

Schreyer, K., et al. 2002, A&A, 394, 561

Shah, R.Y., & Wootten, A. 2001, ApJ, 554, 933

Shirley, Y.L., Evans, N.J., & Rawlings, J.M.C. 2002, ApJ, 575, 337

Shu, F. 1977, ApJ, 214, 488

Snell, R.L., et al. 2000, ApJ, 539, L101

Spinoglio, L., et al. 2000, A&A, 353, 1055

Sutton, E.C., et al. 1995, ApJS 97, 455

Taban, I.M., Schutte, W.A., Pontoppidan, K.M., & van Dishoeck, E.F. 2003, A&A, in press

Terzieva, R., & Herbst, E. 1998, ApJ, 501, 207

van den Ancker, M., Tielens, A.G.G.M., & Wesselius, P.R. 2000a, A&A, 358, 1035

van den Ancker, M., Wesselius, P.R., & Tielens, A.G.G.M. 2000b, A&A, 355, 194

van der Tak, F.F.S., Boonman, A.M.S., Braakman, R., & van Dishoeck, E.F. 2003, A&A, submitted

van der Tak, F.F.S., et al. 1999, ApJ, 522, 991

van der Tak, F.F.S., van Dishoeck, E.F., Evans, N.J., & Blake, G.A. 2000a, ApJ, 537, 283

van der Tak, F.F.S., van Dishoeck, E.F., & Caselli, P. 2000b, A&A, 361, 327

van der Tak, F.F.S., et al. 2002, A&A, 388, L53

van Dishoeck, E.F., & Blake, G.A. 1998, ARAA 36, 317

van Dishoeck, E. F., Blake, G. A., Jansen, D. J., & Groesbeck, T. D. 1995, ApJ, 447, 760

van Dishoeck, E.F. & Hogerheijde, M.R. 1999, in *Origin of Stars and Planetary Systems*, ed. C.J. Lada & N. Kylafis (Dordrecht: Kluwer), 97

van Dishoeck, E.F. & van der Tak, F.F.S. 2000, in *Astrochemistry: From Molecular Clouds to Planetary Systems*, IAU Symposium 197, ed. Y.C. Minh & E.F. van Dishoeck, (San Francisco: ASP), 97

van Zadelhoff, G.–J., et al. 2002, A&A, 395, 373

Viti, S., & Williams, D.A. 1999, MNRAS, 305, 755

Viti, S., Caselli, P., Hartquist, T.W., & Williams, D.A. 2001, A&A, 370, 1017

Walmsley, C.M. 1992, in *Chemistry and Spectroscopy of Interstellar Molecules*, ed. D.K Bohme et al., Tokyo Univ., 267

Watson, C., Churchwell, E., Pankonin, V., & Bieging, J.H. 2002, ApJ, 577, 260

Wolf, S., Henning, Th., & Stecklum, B. 1999, A&A, 349, 839

Wyrowski, F., Schilke, P., Walmsley, C.M., Menten, K.M. 1999, ApJ, 514, L43

Zhou, S., et al. 1994, ApJ, 428, 219

SUBARCSECOND IMAGING OF HOT CORES WITH BIMA

Andy G. Gibb[1], Friedrich Wyrowski[2], and Lee G. Mundy[1]

[1]Department of Astronomy, University of Maryland, College Park, MD 20742, USA
[2]Max-Planck-Institut für Radioastronomie, Auf dem Hügel 69, 53121 Bonn, Germany

Abstract

We present 1.4-mm BIMA observations with subarcsecond resolution of the bright dust and molecular line emission from hot cores associated with a sample of four ultracompact H II regions: G9.62+0.19, G10.47–0.03, G29.96–0.02, G31.41+0.31. Density power laws can reproduce the observed continuum emission but break down on scales smaller than 2000 AU. A total of 38 transitions from 18 species are detected, with G10.47 and G31.41 showing the greatest number of lines. In particular, these sources display emission from two collisionally-excited transitions of methanol lying more than 950 K above the ground state. Outflows traced by H_2S emission provide evidence for embedded exciting sources and the observed morphology of molecular lines is consistent with internal heating of the cores.

Key words: ISM: molecules – ISM: structure – ISM: clouds

1. Introduction: hot cores

Hot cores are dense condensations of gas found in the vicinity of ultracompact H II (UCH II) regions. Their hot compact nature was first revealed by ammonia observations (e.g. Pauls et al. 1983), which also provided the first clues to the different chemistry occurring in hot cores (Walmsley et al. 1987). Compared with dark clouds, the abundances of several species showed orders of magnitude enhancements which led to the proposition that evaporation of grain mantles was important (Menten et al. 1986; Henkel et al. 1987). The inclusion of freeze-out onto dust grains in chemical models yielded good agreement and gave the first clues to the formation process of massive stars, namely that they must go through a cold collapse phase of evolution (Brown, Charnley & Millar 1988).

Most studies of hot cores to date have been limited either spatially or spectrally. Single-dish studies of the rich chemistry of hot cores can yield valuable information on the species present and allow column densities to be estimated (e.g. Turner 1991; Macdonald et al. 1996). However, in order to derive spatial information, high resolution observations must be made. The VLA has led the field here with subarcsecond imaging of ammonia (Cesaroni et al. 1998) which permits temperature and density information to be derived on scales of a few thousand AU. However, it is millimetre/submillimetre interferometry which holds the key to simultaneously probing the structure and chemistry of hot cores. Furthermore, it is in the 1.3-mm window that current instruments can make substantial gains with the wide variety of molecular lines that are accessible and the fact that the continuum emission will be dominated by emission from dust rather than free-free emission from ionized gas in an H II region. To date 1.3-mm observations have been made towards the Orion hot core (Blake et al. 1996) and W3(OH) (Wyrowski et al. 1997, 1999), as well as two of our targets (G29.96 and G31.41), albeit with lower resolution (Maxia et al. 2001).

The proximity of hot cores to H II regions led early studies to suspect that hot cores may be heated by radiation from the newly-formed OB stars within the H II region. However, there is now mounting evidence that at least some hot cores mark the sites of massive star formation (e.g. Hofner et al. 2001, Testi et al. 2000). In the next section, results are presented of a new 1.4-mm BIMA study toward UCH II/hot core regions, with the aim of separating H II regions and hot dust from the molecular cores. On the whole, our data support the notion that our target hot cores contain embedded sources.

2. BIMA observations

The observations were made with the 9 elements of the BIMA array at Hat Creek, California, equipped with 1.3-mm receivers. A sample of four hot cores near ultracompact H II regions (G9.62+0.19, G10.47–0.03, G29.96–0.02, G31.41+0.31) was observed with angular resolution vary-

Chemistry as a Diagnostic of Star Formation (2003). *Edited by Charles L. Curry and Michel Fich.* Published in 2003 by NRC Research Press, Ottawa ON, K1A 0R6, Canada. ISBN 0-660-19089-3.

ing from 0.4 (G10.47) to 2.8 arcsec (G9.62). The correlator windows were set to cover a continuous bandwidth of 800 MHz centred at 216.4 and 219.9 GHz with spectral resolution of 4.3 $km\,s^{-1}$. In addition, $C^{18}O$ and H_2S were observed with higher spectral resolution (1.1 $km\,s^{-1}$). The 1.4-mm continuum images were made by summing the line-free channels in each band. All four regions were classified as 'line-rich' by Hatchell et al. (1998).

3. Results I: 1.4-mm continuum

In each case continuum emission was detected towards the hot core positions at 1.4 mm, clearly resolved from the respective UCH II regions. Comparing our 1.4-mm images with VLA and our own BIMA 3-mm data shows that the 1.4-mm emission is due to warm dust grains. Only in G10.47 is the peak 1.4-mm continuum coincident with a UCH II region. In G9.62 and G29.96 continuum emission is also detected towards the UCH II region(s), which appears to be free-free emission from the ionized gas, although in G29.96 there does appear to be some warm dust within the shell of the UCH II region. The water masers (Hofner & Churchwell 1996) in each region are associated with the dust continuum emission. Images comparing the dust continuum with the free-free emission are given in Wyrowski et al. 2002 (2002).

The peak continuum flux ranges from 0.2 to 0.7 Jy per beam, equivalent to brightness temperatures of up to 98 K (in a beam of 0.5 arcsec towards G10.47). Thus, the dust emission is warm. Masses of a few hundred to a thousand solar masses are inferred from these data, which translate to molecular hydrogen column densities of $\sim 10^{25}\,cm^{-2}$ towards the cores. Volume densities as high as $10^8\,cm^{-3}$ are derived for G10.47 and G31.41.

The high angular resolution data enables us for the first time to study the structure of the dust cores on scales of 0.1 pc and below. Previous, single dish studies were limited in the modelling to larger scales, hence probing only the outer envelope (see Hatchell et al. 2000). We have used the infrared radiative transfer models of Wolfire & Cassinelli (1986) to model the inner structure of the cores. Results of spherical dust models containing embedded heating sources with luminosity of $10^5\,L_\odot$ were 'reobserved' with the spatial-frequency coverage of the BIMA observations and then compared to the observations directly in Fourier space. Density power laws (with indices ranging from –1.5 to –2.0) can be fit to the data, but they break down on scales of approximately 2000 AU. It is not yet clear what happens within this radius. Currently, we cannot distinguish between, e.g., optically thick embedded disks (cf Cesaroni et al. 1998) or, on the other extreme, a dust free inner bubble.

4. Results II: Spectral line data

Emission was detected from 38 transitions from 18 species, although not all lines and species were detected in each source. G9.62 showed the fewest lines (~12 for E, 15 for the hot core source G9.62F) and the least number of high excitation transitions. In contrast, G10.47 displayed 38 lines within the 1.6 GHz passband, a line density of 25 lines per GHz. By way of comparison, single-dish studies of hot cores have detected only half this quantity (Macdonald et al. 1996), demonstrating not only the richness the spectrum of hot cores but also how much potential information millimetre interferometry is capable of yielding.

The lowest energy line detected was the 2→1 line of $C^{18}O$; the highest energy is the 25(3)→24(4) E transition of CH_3OH which lies 986 K above the ground state. Note too that this methanol line is a high-J line and thus is probably collisionally excited in hot gas, rather than pumped by infrared radiation from the warm dust.

In Figure 1, we present images of H_2S 2(2,0)→2(1,1), CH_3OH 7(1)→8(0) E and HNCO 10(0,10)→9(0,9) superimposed on the 1.4-mm continuum emission.

4.1. G9.62+0.19

G9.62 contains up to 9 H II regions (Testi et al. 2000; Cesaroni et al. 1994), A to I. Figure 1 marks the location of H II regions D to G. The hot core is G9.62F, while G9.62E also appears to be a deeply embedded massive young stellar object. All the lines detected towards G9.62 peak at the position of F although most are also detected towards E. SO and H_2S are the brightest lines towards both E and F. The linewidths are consistently higher towards the hot core source (typically 8 $km\,s^{-1}$ for F compared with 5 $km\,s^{-1}$ for E), possibly due to the presence of the outflow from F (Hofner et al. 2001).

4.2. G10.47–0.03

G10.47 may be the most evolved of the sources here since there does appear to be two UCH II regions embedded within the hot core. The UCH II region marked B in Figure 1 splits into two (B1 and B2) at higher resolution (Cesaroni et al. 1998). G10.47 has the brightest and broadest emission lines in our sample. Since the continuum brightness temperature is of order 100 K (see above) the presence of 40–50 K lines indicates gas temperatures in ex-

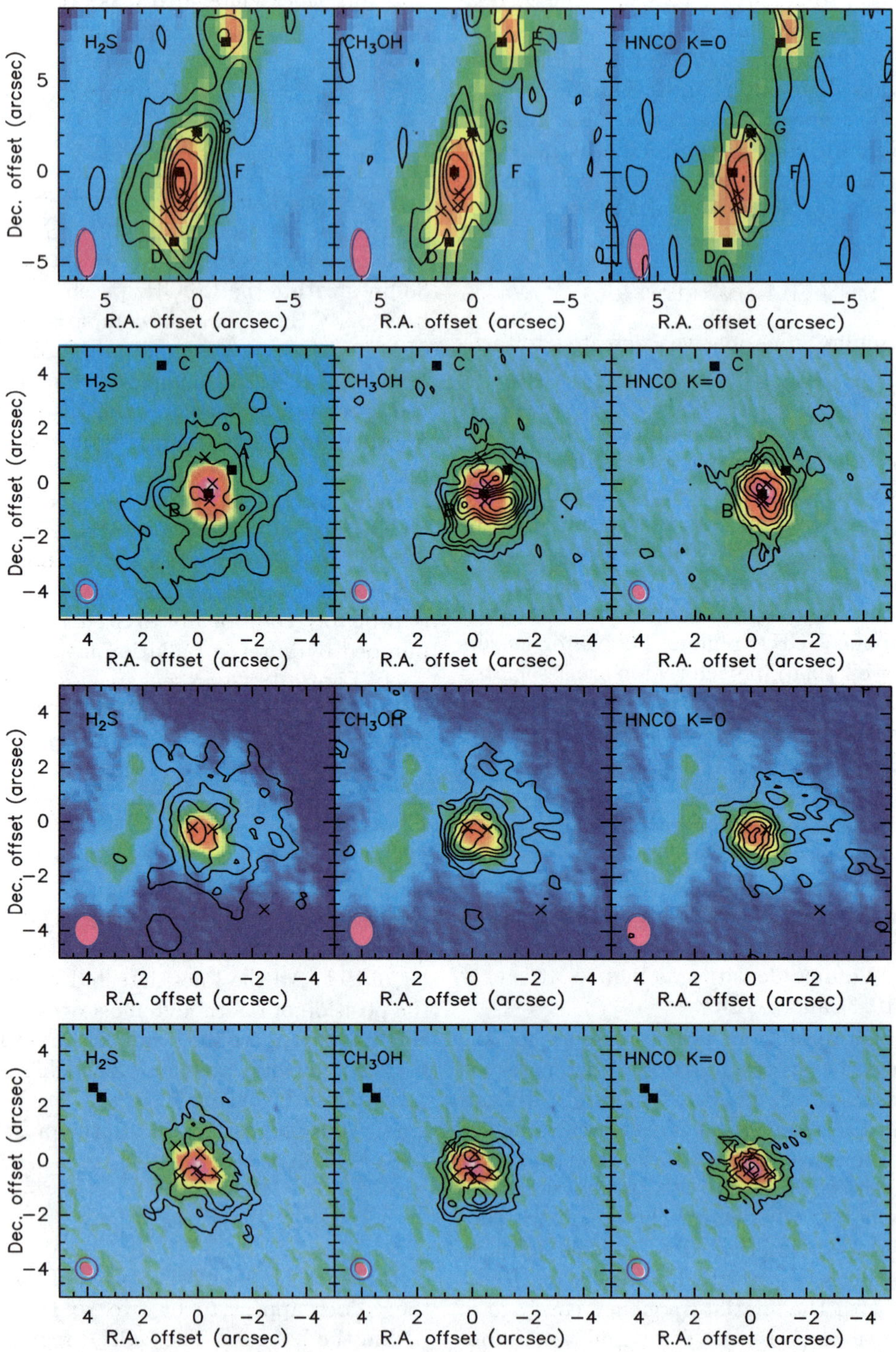

Figure 1. Top to bottom: G9.62, G10.47, G29.96 and G31.41. Contours are of molecular line emission superimposed on a linear colourscale representation of the 1.4-mm continuum. Filled squares mark the position of UCH II regions and crosses mark the positions of water masers. The filled ellipses represent the continuum beam dimensions; the outline ellipses are the spectral-line beams.

cess of 150 K. The typical linewidth is $\sim$13 km s^{-1}. The brightest line in the passband is the 9(1,8)$\rightarrow$9(1,9) transition of H_2CO (59 K). Surprisingly, the second brightest is the 24(4,20)$\rightarrow$23(4,19) transition of ethyl cyanide (52 K). Two very high excitation lines of methanol (with upper energy levels exceeding 950 K above ground) were detected towards G10.47.

As shown in Figure 1, the emission from H_2S and methanol is extended. Although no clear outflow is seen, the extension of the emission is asymmetric and could be part of a bipolar outflow. The ethyl cyanide (not shown) distribution resembles that of a shell, supporting the outflow interpretation. On the whole, the nitrogen-bearing species have a more compact distribution than oxygen-bearing species. In particular, the HNCO lines show a more compact appearance with increasing energy.

4.3. G29.96–0.02

G29.96 is a spectacular cometary H II region (Wood & Churchwell 1991). Like G34.26+0.15 (Macdonald et al. 1996; Watt & Mundy 1999), the hot core lies close to the tip of the cometary bow shape and is thus a prime candidate for testing whether hot cores are internally heated by an embedded source or externally heated by radiation from OB stars in the nearby H II region. In our observations all line emission originates from the hot core. No line emission is detected towards the UCH II region. Furthermore, all the line emission peaks on the hot core itself with no evidence for stratification of species by type or excitation. Most notably, the HNCO lines (believed to be pumped by mid- and far-infrared radiation) all peak on the dust core, as do optically thinner species such as $H_2^{13}CO$. Thus, in G29.96 all evidence points towards the hot core being internally heated, presumably by a luminous embedded protostar. The lack of line emission towards the UCH II region and the presence of line emission to the west of the hot core further supports the hypothesis that the cometary H II region is a champagne flow.

4.4. G31.41+0.31

G31.41 is the second richest source in our sample with 37 detected lines. The two high-energy lines of methanol were detected towards G31.41 as well. The lines are slightly narrower than those of G10.47 ($\sim$ 10 km s^{-1}), although our spectral resolution is not sufficient to distinguish such differences to high accuracy. The continuum emission is slightly elongated in the same direction as the ammonia (4,4) core detected by Cesaroni et al. (1998) suggesting they are tracing the same structure.

Intriguingly, and more clearly than in G10.47, the methanol and H_2S shown in Figure 1 appear to trace a shell of emission perhaps indicative of an outflow. Alternatively, the emission south of the hot core (seen in the methanol image) may be a separate source since it is evident in several other lines. The HNCO peaks on the hot core, presumably at the position of the strongest infrared radiation field.

4.5. Outflows from hot cores

The higher-spectral resolution observations of H_2S revealed non-Gaussian line wings towards all four targets. It is known that G9.62F houses an embedded source which drives an outflow seen in HCO^+ (Hofner et al. 2001), although our H_2S does not show a clear bipolar distribution. The most striking result is for G29.96 (shown in Fig. 2) in which spatially-bipolar red- and blue-shifted line wings are seen centred on the hot core. The SO $6_5\rightarrow5_4$ emission is also extended in the direction of the H_2S lobes. Furthermore, the methanol masers of Minier, Booth & Conway (2000) are aligned with this flow. This result provides further evidence for the presence of luminous protostellar sources embedded within hot cores.

5. Discussion: embedded sources in hot cores

In G10.47 and G31.41, three lines of methanol and four transitions of HNCO were detected. Three transitions of HNCO were detected in G29.96. We have constructed rotation diagrams (e.g. Macdonald et al. 1996) which are shown in Figure 2. These confirm the expectation from the above results that G10.47 and G31.41 contain hotter cores than the other sources in our sample with G10.47 being the hottest. Clearly more data points are desirable, especially of methanol, in order to constrain these fits further. The temperature of 505 K in G10.47 is the highest yet determined from molecular line observations of hot cores. However, it should be noted that the lowest energy line of methanol used in this derivation may be optically thick, which would give rise to an artificially high rotation temperature.

As pointed out above, the high energy lines of methanol are probably excited by collisions as they come from levels with high J (=24 and 25). Thus, they reflect the actual gas kinetic temperature on radii of 1500 to 2000 AU from the central source. It is not certain whether these temperatures represent those of the ambient gas or merely a shocked component.

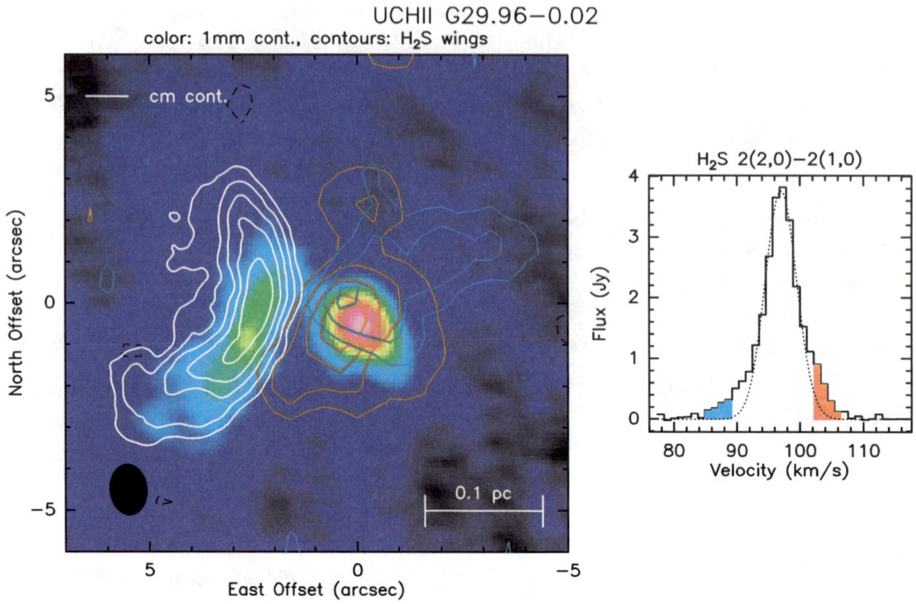

Figure 2. H_2S outflow towards G29.96 *(red and blue contours)* compared with the 1.4 mm continuum *(colourscale)* and 1.3 cm free-free emission from Cesaroni et al. (1998) *(white contours)*. The right panel shows the spectrum towards the hot core along with a Gaussian fit. The excess high-velocity wing emission (shaded for clarity) marks the velocity range for each lobe.

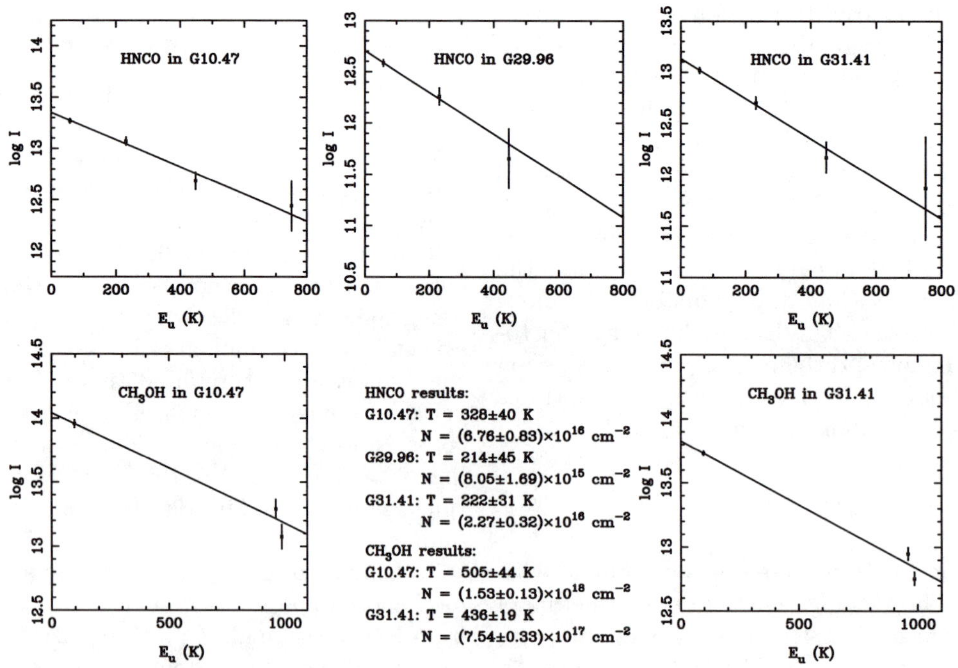

Figure 3. Rotation diagrams for HNCO and methanol in G10.47, G29.96 (HNCO only) and G31.41. The error bars represent the 1-σ uncertainties. The results of linear least-squares fits to these data are shown at centre in the bottom row.

For the HNCO lines, it is likely that their excitation is via a strong mid-infrared radiation field (Churchwell et al. 1986). All the HNCO lines we have detected are from the same (upper) J−state, but lie in different K_{-1} ladders (Zinchenko, Henkel & Mao 2000). Transitions between different K_{-1}-levels are very rapid and require either very high densities (exceeding 10^9 cm^{-3}) or a mid- to far-infrared radiation field. Lower values of K_{-1} are excited at longer wavelengths. The detection of the $K_{-1} = 4$ transition at 750 K above ground indicates a significant 20-μm radiation field. De Buizer et al. (2002) have detected the hot core in G29.96 at 18 μm supporting the HNCO detections.

6. Conclusions

We have used the BIMA millimetre array to observe a sample of four UCH II regions at a wavelength of 1.4 mm with subarcsecond resolution. In each case we detect compact dust cores spatially distinct from the UCH II regions, showing that the hot cores mark the location of a hot dense condensation of gas and dust. Dust emission is also detected within the shell of the cometary UCH II region G29.96. Modelling of the dust emission reveals density power laws with indices of –1.5 to –2 beyond a radius of ~2000 AU. It is tempting to speculate that within this radius either no dust is present or the geometry changes from spherical to axisymmetric, although further observations with higher sensitivity on long baselines are required to validate either of these options.

Line emission from 38 transitions of 18 molecular species has been detected. G10.47 shows the greatest number of lines, with G31.41 coming a close second. In each of these sources two high-energy lines of CH_3OH are observed with upper-level energies in excess of 950 K above the ground state. The high J−number for these transitions suggests that they are collisionally excited and are thus tracing highly excited gas. The spectral line emission peaks towards the dust cores with no strong evidence of excitation and/or chemical gradients, which indicates that the hot cores are probably not externally heated. Further evidence for internal heating comes from the observation of spatially-resolved bipolar wing emission from H_2S in G29.96. Our observations point towards these hot cores as housing massive, luminous protostellar objects which have yet to develop a UCH II region.

Acknowledgements

Research with BIMA is funded by NSF grant AST-9981289.

References

Blake, G.A., Mundy, L.G., Carlstrom, J.E., Padin, S., Scott, S.L., Scoville, N.Z., & Woody, D.P., ApJ, 472, L49

Brown, P.D., Charnley, S.B., & Millar, T.J., 1988, MNRAS, 231, 409

Cesaroni, R., Churchwell, E., Hofner, P., Walmsley, C.M., & Kurtz, S., 1994, A&A 288, 903

Cesaroni, R., Hofner, P., Walmsley, C.M., & Churchwell, E. 1998, A&A 331, 709

Churchwell, E., Wood, D., Myers, P.C., & Myers, R.V. 1986, ApJ 305, 405

De Buizer, J.M., Watson, A.M., Radomski, J.T., Piña, R.K., & Telesco, C.M., 2002, ApJ, 564, L101

Hatchell, J., Thompson, M.A., Millar, T.J., & Macdonald, G.H., 1998, A&AS, 133, 29

Hatchell, J., Fuller, G.A., Millar, T.J., Thompson, M.A., & Macdonald, G.H., 2000, A&A, 357, 637

Henkel, C., Mauersberger, R., Wilson, T.L., Snyder, L.E., Menten, K.M., & Wouterloot, J.G.A., 1987, A&A, 182, 299

Hofner, P., & Churchwell, E., 1996, A&AS, 120, 283

Hofner, P., Wiesemeyer, H., & Henning, T., 2001, ApJ, 549, 425

Macdonald, G.H., Gibb, A.G., Habing, R.J., & Millar, T.J., 1996, A&AS, 119, 333

Maxia, C., Testi, L., Cesaroni, R., & Walmsley, C.M., 2001, A&A, 371, 287

Menten, K.M., Walmsley, C.M., Henkel, C., Wilson, T.L., Snyder, L.E., & Hollis, J.M., Lovas, F.J., 1986, A&A, 169, 271

Minier, V., Booth, R.S., & Conway, J.E., 2000, A&A, 362, 1093

Pauls, T.A., Wilson, T.L., Bieging, J.H., & Martin, R.N., 1983, A&A, 124, 23

Testi L., Hofner P., Kurtz S., & Rupen M., 2000, A&A, 359, L5

Turner, B.E., 1991, ApJS, 76, 617

Walmsley, C.M., Hermsen, W., Henkel, C., Mauersberger, R., & Wilson, T.L., 1987, A&A, 172, 311

Watt, S., & Mundy, L.G. 1999, ApJS 125, 143

Wood, D.O.S., & Churchwell, E., 1989, ApJS, 69, 831

Wood, D.O.S., & Churchwell, E., 1991, ApJ 372, 199

Wolfire, M.G., & Cassinelli, J.P., 1986, ApJ, 310, 207

Wyrowski, F., Gibb, A.G., Mundy, L.G., 2002, in *The Earliest Stages of Massive Star Birth*, ed. P.A. Crowther (Astronomical Society of the Pacific: San Francisco), 43

Wyrowski, F., Hofner, P., Schilke, P., Walmsley, C.M., Wilner, D.J., & Wink, J.E. 1997 A&A 320, L17

Wyrowski, F., Schilke, P., Walmsley, C.M., & Menten, K., 1999, ApJ 514, L43

Zinchenko, I., Henkel, C., & Mao, R.Q., 2000, A&A, 361, 1079

INFRARED SPECTROSCOPY OF INTERSTELLAR MOLECULAR GAS

J. Lacy[1], C. Knez[1], N. Evans[1], and M. Richter[2]

[1]University of Texas, Austin
[2]University of Texas, Austin and University of California, Davis

Abstract

Infrared vibration-rotation spectroscopy provides unique information about interstellar molecules in star formation regions, which complements information obtained at longer wavelengths. Several important molecules, including H_2, CH_4, C_2H_2, and CH_3, which have no permanent dipole moments, can only be observed in the infrared. These and other molecules, including CO, SiO, HCN, and H_2O, have closely spaced lines from many rotational levels, which provide information about abundances and physical conditions from a few spectra measured with consistent spectral and spatial resolution. The principal drawback to infrared observations is that they usually must be made in absorption toward bright infrared continuum sources, restricting observations to lines of sight toward luminous embedded stars. Fortunately, many of the molecules observable in the infrared are abundant in hot-core gas near such objects.

Recent observations with TEXES, the Texas Echelon-Cross-Echelle Spectrograph, provide new spectra of interstellar molecules in high-mass star formation regions. TEXES obtains 4 km s^{-1} and $1''$ resolution spectra with sufficient sensitivity to observe the best-known embedded young stars. Lacy et al. have studied H_2, SiO, HCN, and C_2H_2 toward the Kleinmann-Low region in Orion. Knez et al. have observed HCN, C_2H_2, and possibly CH_3 toward NGC 7538 IRS 1. Finally, Boonman et al. have studied HCN and C_2H_2 absorption toward AFGL 2591. These observations and the potential for future similar observations will be discussed.

Key words: stars: formation – ISM: molecules

Chemistry as a Diagnostic of Star Formation (2003). *Edited by Charles L. Curry and Michel Fich.* Published in 2003 by NRC Research Press, Ottawa ON, K1A 0R6, Canada. ISBN 0-660-19089-3.

PROBES OF HIGH-MASS STAR FORMATION

Malcolm Walmsley

Osservatorio Astrofisico di Arcetri, Largo E.Fermi 5, I-50125 Firenze, Italy

Abstract

This review considers current ideas about how and where high-mass stars are formed and the observations which might test these ideas. In particular, I summarize recent work aimed at determining the physical characteristics of the high-density clumps where O-B star formation is either going on currently or there is reason to believe it will in the future. I briefly consider recent ideas linking the "hot cores" found in regions of high-mass star formation to stellar mergers. I also discuss the attempts which have been made to use outflows as probes of high-mass star formation and comment on the interpretation of enhanced abundances for sulfur- and silicon-containing species.

Key words: ISM: molecules — ISM: structure — ISM: clouds

1. Introduction

Progress in the field of high-mass star formation or indeed progress in the associated chemical models requires an evolutionary scheme. One needs to know where the high-mass stars come from and how their surroundings may evolve. In the case of the chemistry, one has to confront the fact that chemical and dynamical time scales are of the same order in the regions of interest. This has the advantage that one can sometimes use presently observed chemical abundances to infer something about the history of the clouds in which star formation is taking place. It has the disadvantage that there are many free parameters in the problem leading to considerable uncertainty in any conclusions which may be drawn. It is necessary therefore to make working hypotheses about the evolution of the gas which eventually gives rise to O-B stars even if such hypotheses are presently extremely arbitrary.

Thus in the case of high-mass star formation, one assumes that the first act is the formation (probably in a spiral arm) of a giant molecular cloud or GMC with masses upward of 10^5 solar masses and extent of order 50 parsecs or more. Together with the GMC or perhaps shortly afterward forms what has been called a "protocluster core" (Evans et al. 2002), with mass roughly a few thousand solar masses and size of order a parsec. It is from such dense (10^5 to 10^6 cm^{-3}) clumps (the nomenclature in this field is confused and I will use the term "clump" for parsec–scale structure and "core" for entities on smaller scales) that dense clusters of stars such as the Orion Nebula Cluster or ONC (Hillenbrand & Hartmann 1998) appear to form. All of this must happen in a relatively short timescale of perhaps 10^7 years or so for the formation of a GMC, and an order of magnitude less for protocluster clumps since there seem to be fewer of them (roughly 500 in the galaxy according to Plume et al. [1997] compared with maybe 10^4 GMCs of 10^5 solar masses).

Subsequently and again on timescales of 10^6 years or several free–fall times form stellar clusters and associations including a few luminous O-B stars. It is these few massive objects which dominate the action in the further development but it is useful to remember that they are a small minority among the newly formed stars and indeed that most of the stellar mass in clusters like the ONC appears to be in the form of low-mass stars (see, e.g., Testi et al. 1999). However, most of the luminosity is provided by the massive stars and they also ionize their surroundings producing the H II regions which have traditionally been the tracers of star formation both in our galaxy and in external galaxies. This ionization together with the destruction wrought by O-B star winds presumably causes the destruction of the core on timescales which again are likely to be of order a few million years.

Verifying the picture sketched above requires studies of many different types. Clearly, one should try to distinguish between "young" clumps and "old" clumps. One approach to this has been to search for ultra–compact (UC) H II regions associated with the dense gas. Since the presence of ionized gas implies the presence of O-B stars

Chemistry as a Diagnostic of Star Formation (2003). *Edited by Charles L. Curry and Michel Fich.* Published in 2003 by NRC Research Press, Ottawa ON, K1A 0R6, Canada. ISBN 0-660-19089-3.

ionizing their surroundings, it is assumed that clumps associated with UC H II regions are relatively evolved. Conversely, clumps lacking UC H II regions are thought to be young, perhaps suggesting that one should look for differences in chemical composition between the two classes. For instance, one might naively expect carbon-rich molecules to be abundant in the early stages of the evolution of a core (because if carbon was atomic in the phase prior to core formation, there will remain for relatively large timescales free C capable of forming C-rich species). Last but not least, there is evidence that maser emission (usually water or methanol) is present in the early phases of the evolution of a high-mass protostar and its associated outflow (Furuya et al. 1999; Moscadelli et al. 2000).

Consequently, some of the variety of surveys which have been carried out with the aim of studying the early phases of a protocluster have used maser emission as a tracer. Another approach has been to use *IRAS* colors; then, of course, one can use various combinations of the above criteria. Such surveys (Plume et al. 1997; Churchwell et al. 1990; Cesaroni et al. 1999; Molinari et al. 1996; Brand et al. 2001; Sridharan et al. 2002) have aimed at determining physical parameters of the cores as well as parameters of the outflows and other phenomena caused by the young protostars. There is beginning to exist also a corpus of near infrared data which allows some insight into the number and type of young stellar objects YSOs in the vicinity of the clumps. This is beginning to allow for example photometric distance estimates and spectral type determinations.

More detailed reviews on many of the topics discussed or mentioned here are those by Churchwell (1999, 2002), Garay & Lizano (1999), Evans (1999), Kurtz et al.(2000), and Stahler et al. (2000). For more chemical aspects, some of the articles in the recent IAU Symposium 197 (Minh & van Dishoeck 2000) may be useful. Section 2 of this review describes recent observational studies of the high-density cores associated with high-mass star formation and summarizes some recent theoretical models. The last part of this gives a brief description of what outflows tell us about star formation. In Section 3, I review some recent observations of outflows with particular emphasis on the abundances of molecules containing Si and S. Finally, in Section 4, I briefly consider possible future avenues of research.

2. High-Mass Star Formation and Outflows

High-mass star formation as noted above is, essentially, low-mass star formation, and this paradox is likely to be the key to understanding the complex processes which give rise to regions as diverse as Orion, W49, and NGC 3603. By this I mean that the process seems to involve the sudden formation of a cluster of stars with an IMF not greatly dissimilar from that of field stars. The simultaneity involved in this suggests a concerted effort perhaps stimulated by external events. Such external events are likely in many cases to be related to neighbouring groups of OB stars (Elmegreen & Lada 1977; Blaauw 1964), whose relative positions and ages are suggestive of sequential star formation. All of this is very different from our picture of low-mass star formation (Shu et al. 1987) and so one must probably search for a completely different "paradigm" from that applied to the formation of solar mass objects.

It is useful in this case to consider what we know about the cores or clumps from which dense star clusters and O-B associations form and this will be discussed in the next section.

2.1. Cold high-mass cores

The first question one might ask about high-mass cores is whether they have a "life" prior to star formation. That is, do such entities as cold massive cores with no star formation going on exist? A variation on the question is whether they have a phase where low-mass protostars are present but high-mass precursors are not. The most obvious candidates presently for such early phases are the infrared dark clouds (IRDCs) found by the *ISO* and *MSX* satellites due to their extinction of background small particle emission in the mid-IR. These have been shown in many cases to have counterparts in millimeter dust emission and masses ranging up to 1000 $M_\odot$ (Carey et al. 2000; Perault et al. 1996). Their sizes and densities are in some cases at least of the same order as in the cores associated with massive star formation. There are some indications of phenomena such as outflows associated with star formation but temperatures are low and high-mass stars do not appear to be present.

Studies of molecular abundances in these objects would be useful since if they are really young, one expects that chemical equilibrium will not have been reached. There is beginning to be evidence that depletion of CO and other species onto dust grain surfaces plays a big role in determining the observed chemical abundances in high-mass cores, just as has been shown to be the case in low-mass cores in the solar vicinity (Mitchell et al. 2001). For instance, Wyrowski et al. (2002) find good agreement between maps of dust emission from IRDCs (Carey et al. 2000) and emission in N_2H^+(1-0). This is similar to the pattern seen in several Taurus and Ophiuchus cores and there is good reason to believe that it indicates a high abundance of molecular nitrogen in the gas phase while CO has substantially depleted out onto grain surfaces (Walmsley et al. 2001; Caselli et al. 1999; Bergin & Langer 1997). This process requires a low dust temperature of order 15 K or less and high density. It is thus not compatible with the presence of nearby luminous stars but is indicative of future star formation. Wyrowski et al. (2002) find, for example, a cold clump several parsecs in projected distance away from the UC H II region G29.96, where N_2H^+ emission coincides with a source at 850 μm. It seems possible that the G29.96 core and the associated cold core are coeval but have developed at different rates. In any event, chemical studies aimed at comparing the abundance distributions in the two clumps seem warranted.

2.2. Density and Temperature Structure

Also of relevance when considering the evolution of the dense clumps which give rise to dense star clusters are the density and temperature structure, velocity field, and magnetic field in the dense gas. The temperature dependence with position has often been inferred from models which fit the spectral energy distribution of the observed dust emission (e.g., van der Tak et al. 1999; Doty et al. 2002). Here, one as a rule assumes spherical symmetry and a single luminosity source at the center. It is useful therefore to check these models using molecular line probes such as CH_3CCH (Fontani et al. 2002). The chemistry is also sensitive to the temperature structure, and van der Tak et al. (2000b) present evidence that some sources show a rapid increase in methanol abundance in regions with temperatures above 100 K.

Once one knows the temperature, one can consider the density structure of cores. This has recently been studied by a variety of authors using observations of both the millimeter dust emission and molecular tracers such as CS (Hatchell et al. 2000; van der Tak et al. 2000a; Evans et al. 2002; Fontani et al. 2002; Beuther et al. 2002a) These authors find density distributions behaving with radius r roughly as $r^{-1.5}$ (although with a large spread) for different samples of high-mass star-forming regions. This is reminiscent of the distribution expected for free-fall onto a point mass. However, the Beuther et al. (2002a) maps in particular show that spherical symmetry is the exception rather than the rule. Moreover, a sizable minority of the objects studied have flatter distributions consistent with most of the mass in these systems being at large "radius" (rather than concentrated at the center). This evidence suggests that we are dealing with young systems and it is possible that the power law index of the density distribution is an evolutionary indicator.

In this case, one might expect to see evidence for infall (onto the clump rather than the "protostar") but there is little direct evidence for this in the line profiles. However, one does occasionally find foreground redshifted absorption (Rudolph et al. 1990; Welch et al. 1987) in large clumps such as W51 and W49, and it is probable that there are more cases of this type. In the case of W51, Rudolph et al. have modelled their observations in terms of the collapse of a $4 \times 10^4\,M_\odot$ cloud on a timescale of 10^6 years. It would be interesting to investigate whether there is a correlation between the presence of infall and the density distribution. In this context, it is interesting that Fontani et al. (2002) find that the CH_3C_2H linewidth increases in clump centers, which again is suggestive of infall.

2.3. Accretion Rates and Outflows

The natural time scale for star formation in a dense clump is the free-fall time scale $\tau_{ff} \sim 10^5 n_5^{-1/2}$ years, where n_5 is the molecular hydrogen density in units of $10^5\,cm^{-3}$. Present evidence suggests that massive and low-mass stars in young clusters form together. We have not found, for instance, evidence for young clusters of Class 0 and I protostars without accompanying massive objects; although this could be because searches have not been made with this in mind. This then suggests that accretion rates $\dot{M}_a$ are proportional to final stellar masses M_f and thus $\dot{M}_a = M_f/\tau_{ff}$ to within factors of a few. Detailed models based on the idea that accretion is determined by the external pressure (McKee and Tan 2002; Tan and McKee 2002) arrive at a similar rate. The net result is that producing, say, a 30 $M_\odot$ star requires an accretion rate of $3 \times 10^{-4}\,M_\odot yr^{-1}$ (or somewhat less) for a density of $10^5\,cm^{-3}$. The Tan–McKee model forms a 100 $M_\odot$ object in 2×10^5 years with a final accretion rate of $10^{-3}\,M_\odot yr^{-1}$.

Accretion is likely linked to outflow and outflow is the most obvious tracer of young embedded protostars in the early phases of the evolution of a cluster. Crudely speaking, one expects outflow rates proportional to inflow rates and hence the arguments sketched above would suggest outflow rates proportional to the final stellar mass. The Tan–McKee model in fact finds that outflow momentum p_w, which observationally is likely to be the best-determined quantity, should be approximately $68\,M_f^{1.4}$ $M_\odot$km s^{-1} (with M_f in $M_\odot$). Thus, for a Salpeter-type IMF ($\propto M^{-2.35}$), different mass intervals make roughly equal contributions to outflow momentum. The implication of this is that if a protocluster is unresolved, one will observe an appreciable contribution to CO outflow wings from relatively low-mass objects. On the other hand, if one can resolve individual outflows and in particular if one sees evidence for bipolarity, one can attribute this to an individual (presumably high-mass) protostar. This underscores the importance of high angular resolution in general and millimeter interferometry in particular when attempting to deduce the characteristics of high-mass protostars from outflow observations.

2.4. Hot Cores

One curious phenomenon whose nature is still not completely clear is that of the "hot cores" (see Kurtz et al. [2000] for a more detailed discussion). These objects, as their name suggests, are compact (<0.1 pc), hot ($>$ 100 K), and massive (up to several thousand solar masses, though ranging down to 10). The problem of how so much gas can be maintained at such high temperatures is very puzzling, and in fact gave rise to a "luminosity paradox" (Cesaroni et al. 1994, 1998) since the inferred temperatures from molecular line measurements appeared to imply luminosities higher than observed. It seems likely that the explanation of this is partly geometrical (edge–on oblate structures do not radiate isotropically and hence may appear less luminous than they are) but it would be useful to have higher resolution images both in molecular lines and in the mid–IR. The latter technique is beginning to give useful results (De Buizer et al. 2002; Stecklum et al. 2002) and should eventually throw light on the situation.

It has usually been assumed that hot cores such as the Orion hot core are heated by a single early O–star (Kaufman et al. 1998; Kurtz et al. 2000) and that they are precursors of UC H II regions. In the case of Orion, however, it has been tantalizingly difficult to find this heating source (Blake et al. 1996). One solution might be that the source of the energy is gravitational due to stellar mergers in a dense cluster (Bonnell 2002; Bally 2002) rather than nuclear burning. The predictions of this type of model are rather unclear but a highly variable luminosity seems likely. For instance, Bally predicts that a collision between a 20 $M_\odot$ and a 1 $M_\odot$ star would give rise to a luminosity of 4×10^5 $L_\odot$ over a timescale of 100 years. Although a large fraction of this energy might be radiated at low effective temperature, a considerable fraction also is likely to emerge at UV and X–ray wavelengths. This type of event if repeated could perhaps account for the high temperatures observed in Orion and other hot cores. Might it also mean that Orion is unusual in having had an "important merger" in the last 100 years? Although this may at first sight seem improbable, Orion has some unusual features including its SiO maser emission (Genzel et al. 1980).

It is usually assumed that hot cores are the result of chemical evolution subsequent to the "switch–on" of a young O-star. The sudden increase in temperature of the dust grains causes ice mantle evaporation and subsequent evolution of the gas phase chemistry. The merger scenario, however, with a more distributed luminosity source and with rapid variability would have similar observational consequences since the chemistry would not "follow" the variability. One difference might be the occasionally high X–ray luminosity which stellar mergers might produce. The X-rays would leave their mark in terms of high ion concentrations and while these might disappear fairly rapidly after a flare, they should leave chemical traces. It might be useful to test whether energy input of this type affects chemical models such as those of Charnley (1997). It would also then become important to estimate the frequency of hot cores in regions of high-mass star formation and indeed a start on this has been made by Wilner et al. (2001) towards W49.

3. Outflow Observations

Various groups have attempted to isolate young massive protostars prior to the point at which they ionize their surroundings and emerge from the clump where they are born. One expects naively that such objects should be accompanied by disks both from the point of view of angular momentum conservation and because the problem of accreting material in spite of radiation pressure is substantially alleviated (Yorke 2002). As discussed above, one can also expect in these circumstances a large jet-outflow system with outflow rates of order 10^{-4} to 10^{-3} $M_\odot$yr^{-1}. Such jets may be observed in molecular hydrogen emission, in the radio continuum, and in outflow tracers such as SiO and CO.

An example of a source with these characteristics is the high-mass protostar associated with IRAS 20126+4104 (Cesaroni et al. 1999; Zhang et al.

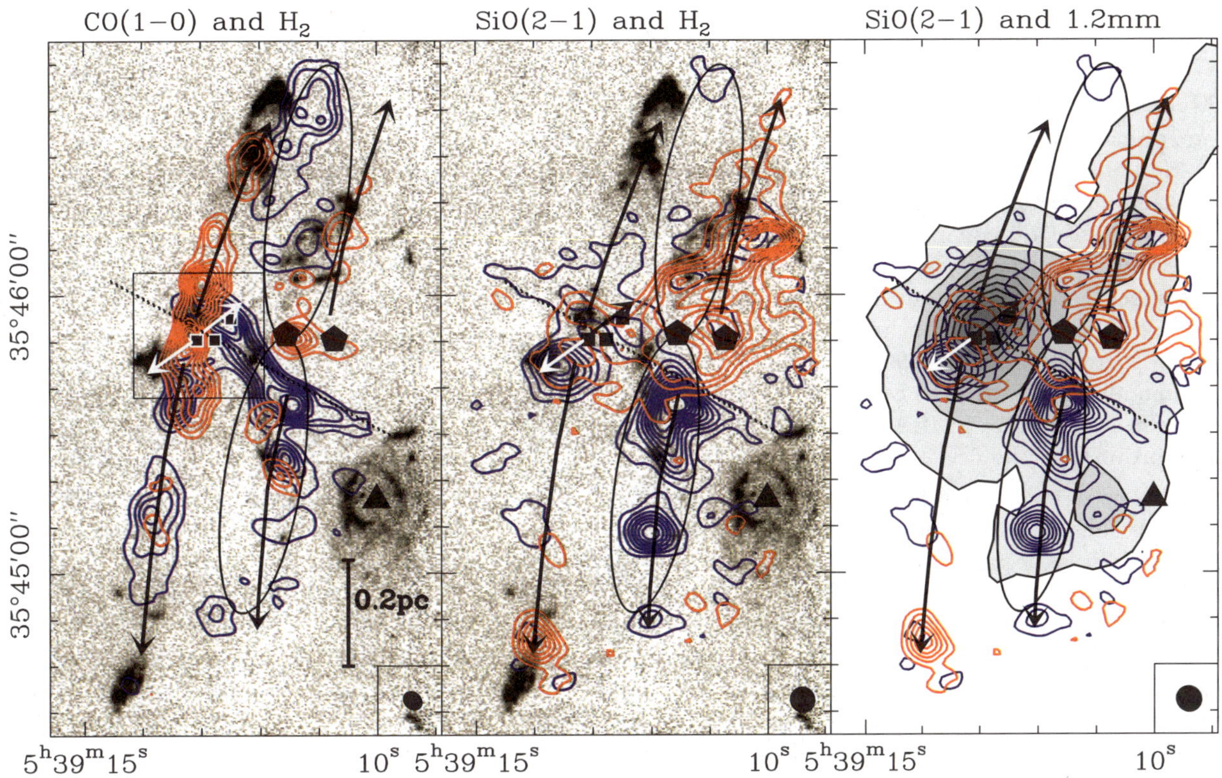

Figure 1. Maps of the outflows in IRAS 05358+3543 from Beuther et al.(2002b). The left-most panel shows the CO(1−0) map of the line wings taken with the IRAM Plateau de Bure interferometer (*red contours* [−14, −7] km s^{-1}, *blue contours* [−30, −21] km s^{-1}) superposed on an image of the emission in the 2 μm $H_2S(1)$ 1-0 line. The triangle shows the *IRAS* 12 μm position, the two pentagons compact emission in $H^{13}CO^+(1-0)$, and the small, filled black squares are millimeter continuum sources. There is evidence for four flows with orientations marked by the arrows. The center panel shows the corresponding map in SiO(2−1). In the right-most panel, the black contours are of the 1.3 mm dust continuum emission observed with the IRAM 30-m telescope.

1999; Shepherd et al. 2000; Zhang 2002). An elongated massive disk (10 $M_\odot$) seen in ammonia emission with the VLA is oriented roughly perpendicularly to a "jet" seen in SiO, in H_2 emission, and in the radio continuum. The size scale of this disk is roughly 10^4 AU and the inferred rotation period 10^5 years. It therefore seems unlikely that the disk is a stable structure and it also seems probable that the driving jet is precessing (Shepherd et al. 2000).

IRAS 20126+4104 is unlikely to be a "unique" object. It is thus not too surprising that a recent survey of outflows from high-mass young protoclusters (Beuther et al. 2002c) finds that collimated outflows are relatively common. The "collimation factor" which one finds is highly correlated with the angular resolution applied, and objects observed using interferometry typically show multiple relatively collimated bipolar flows. Figure 1 illustrates this by showing maps of the 5000 $L_\odot$ source IRAS 05358+3543 (or S 233; IRAS 05358 in what follows) mapped by Beuther et al. (2002b).

IRAS 05358 is notable because it consists of two young clusters: one which has emerged from its natal clump centred on the filled triangle in Fig. 1 and a younger embedded cluster centred on the 1.3 mm peak. Emanating from the embedded cluster, one can trace 3 (or perhaps 4) flows seen either in CO, SiO, or H_2 2 μm emission, all of which appear to be associated with the embedded cluster (age < 2 Myr). Could these flows be caused by solar-mass protostars, or can one definitively attribute them to massive ($> 10\ M_\odot$) objects? The associated momentum based on the interferometer data (with missing spacing information from IRAM 30-m maps) is roughly 50 $M_\odot km\,s^{-1}$, which would be consistent with an embedded solar-mass protostar according to Tan and McKee (2002). On the other hand, the evidence from nearby low-mass Class I outflows (Bachiller & Tafalla 1999) suggests values an order of magnitude lower. One concludes that while it is plausible that the flows in IRAS 05358 are produced by massive protostars, the case has not been proven.

In other cases such as HH 80-81 (Heathcote et al. 1998; Reipurth & Raga 1999), where one observes a collimated jet of dimensions ~ 2 parsec with [O III] emission lines as broad as 700 $km\,s^{-1}$, the case for a massive protostar producing the jet is strong. Yet another case is W75N (Shepherd et al. 2003) where the observed flow momentum is much larger (2000 $M_\odot km\,s^{-1}$) and the flow is clearly elongated. Thus in W75N, high-mass protostars likely are the cause of the observed outflow but the nature of the flow and whether objects of this type are a "scaled–up version" of lower mass flows is controversial.

Part of this controversy has revolved around the plot of mass loss rate versus bolometric luminosity discussed by Churchwell (1999, Fig. 13). This covers both high- and low-mass star-forming regions and appears to show a correlation between mass loss rate (here defined as flow mass from CO divided by age; i.e., size divided by flow velocity) and bolometric luminosity (Churchwell 2002 and references therein). The plot shows a dependence of mass loss rate going roughly as $L^{0.75}$ below luminosities L of $10^5\ L_\odot$, with some flattening at higher L. The relation has been used as input to theoretical models (Behrend & Maeder 2001) and as a consequence, one needs to understand whether or not the proposed correlation is significant. Larger statistical samples (Beuther et al. 2002c) suggest that the Churchwell (2002) mass loss rate relation is better seen as an upper envelope to that observed; certainly, what one finds seems to depend heavily upon the choice of source sample. Ridge and Moore (2001) have carried out an "unbiased survey" but with a sample of only 11 objects and it seems clear that more work in this direction is needed. It will be important for such studies to have sufficient angular resolution to distinguish "collimated flows," where present.

3.1. Abundance Enhancements in Outflows

It is also notable in Fig. 1 that the identified "outflows" have differing characteristics. For example, the western N-S oriented flow (denoted by ellipses in Fig. 1) is more clearly defined in SiO than in CO. What is the significance of this? In low-mass flows such as that associated with L1157, SiO is highly enhanced in features within the outflow apparently associated with shocks in swept–up material (Bachiller & Tafalla 1999; Bachiller & Pérez–Gutierrez 1997; Nisini et al. 2002). In that case, the enhancements in SiO abundance coincide with smaller increases in the abundances of a number of species such as methanol, ammonia, and water, all thought to be important constituents of dust grain ice mantles. One explanation of these data is that the ices (as well as, perhaps, some moderately refractory material) are ejected into the gas phase by sputtering in C-shocks (Schilke et al. 1997). It seems plausible that a similar mechanism is operating in IRAS 05358. This could explain the fact that the SiO in Fig. 1 seems to be coincident with the 1.3 mm dust emission (see also Hatchell et al. 2001 for the case of G34.26), suggesting that the conditions needed for the C-shock are only present in the dense gas.

A related question is that of abundance enhancements in S-bearing molecules. There is evidence for correlation between SO and SiO (Chernin et al. 1994; Codella et al. 2002) and hence reason to believe that the observed sulfur has its origin in dust grain mantles. Sulfur has the advantage from the perspective of chemical studies that one can directly observe most S-bearing molecules whose abundance is thought to be relevant at millimeter wavelengths. One might therefore expect that constraints on parameters such as outflow age could be obtained by comparing model predictions with observed abundances (see Charnley 1997 for a model of hot core chemistry which may however differ from that expected in shocks).

In practice, despite much useful work, the results have not been conclusive. It is clear that both in the Orion outflow as well as in flows from lower mass protostars (Blake et al. 1987; Bachiller & Tafalla 1999; Beltrán et al. 2002), S-bearing species are enhanced in abundance by one to two orders of magnitude (more in Orion where SO and SO_2 have abundances in the plateau feature of 5×10^{-7} relative to H_2). It has been suggested that this is due to destruction of H_2S ice, but there is no direct evidence for this. In fact, *ISO* data give no clear evidence of solid state sulfur except for traces of OCS (Palumbo et al. 1995; Gibb et al. 2000). In particular, solid H_2S is not detected and thus is presumably below 1 percent of the water abundance.

Thus, less than ten percent of the cosmic S abundance in high-density cores is in the form of solid H_2S. However, the other 90 percent does not appear to be in the gas phase, since the summed gas phase abundance of observed species is typically much less than one might expect on this basis. So where is the missing sulfur? One possibility is that it is present in chemisorbed sites with binding energy larger than water ice but less than those occupied by highly depleted elements such as iron. This would explain a behavior qualitatively similar to that of silicon and the depletion mechanism could be onto small grains as discussed by Weingartner and Draine (1999). However, this is controversial and one can only say that it seems likely that the explanation of this puzzle and that of the enhanced abundance of S-bearing species in outflows like the Orion "plateau" are related.

Finally, it is worth noting that one area where much recent progress has been made has been the study of iron line emission associated with outflows. There is good reason to believe that the near-IR Fe II lines which have now been imaged in several high-mass star-forming regions (Hodapp & Davis 2002; Shepherd et al. 2003) are associated with dissociative shocks of higher velocity than the shocks discussed above responsible for SiO and related emissions. The $1.6\,\mu$m Fe II line observed by Hodapp & Davis towards W51 has a width to zero power of $300\,\mathrm{km\,s^{-1}}$. This is much larger than seen in H_2 and clearly the two probe completely different conditions. Understanding the physical conditions probed by Fe II is an important future observational objective.

4. Future Prospects

Future studies of high-mass star-forming regions clearly need to assess the role of low-mass protostars. Indeed, as mentioned earlier, the "merger model" for forming massive stars assumes the coalescence of lower mass objects in an extremely high density cluster. This model makes a series of predictions (Bonnell 2002; Bally 2002) which it is worth attempting to verify and which I mentioned in the context of hot cores. This seems a promising avenue for the future and another observable might be high proper motions due to low-mass objects ejected from such a high-density cluster (a velocity of $20\,\mathrm{km\,s^{-1}}$, or roughly twice the velocity dispersion expected in such a high density cluster, corresponds to a proper motion of roughly $8\,\mathrm{mas\,yr^{-1}}$ at the distance of Orion. Plambeck et al. [1995] measured a $50\,\mathrm{km\,s^{-1}}$ relative motion of the BN object relative to source I which perhaps is consistent with this picture).

It is clear, in any case, that what one needs to study high-density clusters is high angular resolution. The densest clusters we know of have roughly 10^5 stars per cubic parsec and observing this at distances of a few kiloparsecs requires angular resolutions of order 0.1 arcsecond or better. The new generation of X-ray satellites have demonstrated that both Class I and Class II protostars radiate hard X-rays with luminosities occasionally as high as 10^{31} erg $\mathrm{s^{-1}}$ and these are being detected in sources as distant as W3 (Hofner et al. 2002). At least in nearby regions like Orion, such observations over time should constrain theorists' options. Another potential tracer of protostars embedded within cores is the millimeter emission from their associated disks. However, these may be tidally stripped in dense clusters leaving the observers once again frustrated. Perhaps the best bet is still to trace narrow jets such as those of Fig. 1 back to their origin, But doing this with precision will require highly sub-arcsecond angular resolution at millimeter/submm wavelengths, as well as high sensitivity. This is just as well since it would be sad to solve the problems of high-mass star formation prior to the arrival of ALMA (the projected Atacama Large Millimeter Array).

Acknowledgements

I would like to thank Antonella Natta and Riccardo Cesaroni for their comments on this manuscript. Thanks are also due to Henrik Beuther for supplying Figure 1 and for many discussions about outflows. I have also pleasure in acknowledging a travel grant from the Italian Ministry of Education (MIUR) "Dust and Molecules in the Astrophysical Environment," as well as travel support from the Max Planck Institut für Radioastronomie.

References

Bachiller, R., & Tafalla, M. 1999, in *The Origins of Stars and Planetary Systems*, ed. C.J. Lada & N.D. Kylafis (Dordrecht: Kluwer), 540
Bachiller, R., Pérez–Gutiérrez, M. 1997, ApJ, 487, L93
Bally, J. 2002, in *The Earliest Stages of Massive Star Birth*, ed. P.A. Crowther (San Francisco: ASP), 267
Behrend, R., & Maeder, A. 2001, A&A, 373, 190
Beltrán, M.T., et al. 2002, ApJ, 573, 246
Bergin, E.A., & Langer, W.D. 1997, ApJ, 486, 316
Beuther, H., et al. 2002a, ApJ, 566, 945
Beuther, H., et al. 2002b, A&A, 387, 931
Beuther, H., et al. 2002c, A&A, 383, 892

Blaauw, A. 1964, ARA&A, 2, 213

Blake, G.A., et al. 1987, ApJ, 315, 621

Blake, G.A., et al. 1996, ApJ, 472, L49

Bonnell, I. 2002, in *The Earliest Stages of Massive Star Birth*, ed. P.A. Crowther, ASP Series 267 (San Francisco: ASP)

Brand, J., Cesaroni, R., Palla, F., & Molinari, S. 2001 A&A, 370, 230

Carey, S.J., et al. 2000, ApJ, 543, L157

Caselli, P., et al. 1999, ApJ, 523, L165

Cesaroni R., Churchwell, E., Hofner, P., Walmsley, C.M., & Kurtz, S. 1994, A&A, 288, 903

Cesaroni, R., Hofner, P., Walmsley, C.M., & Churchwell, E. 1998, A&A, 331, 709

Cesaroni, R. et al. 1999a, A&A, 345, 949

Cesaroni, R., Felli, M., Walmsley, C.M. 1999b, A&AS, 136, 333

Charnley, S.B. 1997, ApJ, 481, 396

Chernin, L.M., Masson, C.R., Fuller, G.A. 1994, ApJ, 436, 741

Churchwell, E., Walmsley, C.M., Cesaroni, R. 1990, A&AS, 83, 119

Churchwell, E. 1999 in *The Origin of Stars and Planetary Systems*, ed. C.J. Lada & N.D. Kylafis (Dordrecht: Kluwer)

Churchwell, E. 2002 ARA&A, 40, 27

Codella, C., Scappini, F., Bachiller, R., & Benedettini, M. 2002, MNRAS, 331, 893

De Buizer, J.M., et al. 2002, ApJ, 564, 327

Doty, S.D., van Dishoeck, E.F., van der Tak, F.F.S., & Boonman, A.M.S. 2002, A&A, 389, 446

Elmegreen, B.G., & Lada, C.J. 1977, ApJ, 214, 725

Evans, N.J. II 2000, ARA&A, 1999, 37, 311

Evans, N.J II, Shirley, Y., Mueller, K.E., & Knez, C. 2002 in *The Earliest Stages of Massive Star Birth*, ed. P.A. Crowther ASP Series 267 (San Francisco: ASP)

Fontani, F., Cesaroni, R., Caselli, P., & Olmi, L. 2002, A&A, 389, 603

Furuya, R.S., et al. 1999, ApJ, 525, 821

Garay, G., & Lizano, S. 1999, PASP, 111, 1049

Genzel, R., et al. 1980, ApJ, 239, 519

Gibb, E.L., et al. 2000, ApJ, 536, 347

Hatchell, J., et al. 2000, A&A, 357, 637

Hatchell, J., Fuller, G.A., & Millar, T.J. 2001, A&A, 372, 281

Heathcote, S., Reipurth B., & Raga, A.C. 1998, AJ, 116, 1940

Hillenbrand, L.A., & Hartmann, L. 1998, ApJ, 492, 540

Hodapp, K.W., & Davis, C.J. 2002, ApJ, 575, 291

Hofner, P., et al. 2003, ApJ, in press

Kaufman, M.J., Hollenbach, D.J., & Tielens, A.G.G.M. 1998, ApJ, 497, 276

Kurtz, S., et al. 2000, in *Protostars and Planets IV*, ed. V. Mannings, A.P. Boss, & S.S. Russell (Tucson: Univ. of Arizona Press)

McKee, C.F., & Tan, J.C. 2002, Nature, 416, 59

Minh, Y.C., & van Dishoeck, E.F. eds. 2000, *Astrochemistry: From Molecular Clouds to Planetary Systems*, IAU Symposium 197, (San Francisco: ASP)

Mitchell, G., et al. 2001, ApJ, 556, 215

Molinari, S., Brand, J., Cesaroni, R., & Palla, F. 1996, A&A, 308, 573

Moscadelli, L., Cesaroni, R., & Rioja, M.J. 2000, A&A, 360, 663

Nisini, B., Giannini, T., & Lorenzetti, D. 2002, ApJ, 574, 246

Palumbo, M.E., Tielens, A.G.G.M., & Tokunaga, A. 1995, ApJ, 449, 674

Perault, M., et al. 1996, A&A, 315, L169

Plambeck, R.L., Wright, M.C.H., Mundy, L.G., & Looney, L.W. 1995, ApJ, 455, L189

Plume, R., et al. 1997, ApJ, 476, 730

Reipurth, B., & Raga, A. 1999, in *The Origins of Stars and Planetary Systems*, ed. C.J. Lada & N.D. Kylafis (Dordrecht: Kluwer)

Ridge, N.A., & Moore, T.J.T. 2001, A&A, 378, 495

Rudolph, A., Welch, W.J., Palmer, P., & Dubrulle, B. 1990, ApJ, 363, 528

Schilke, P., Walmsley, C.M., Pineau des Forêts, G., & Flower, D.R. 1997, A&A, 321, 293

Shepherd, D.S., Yu, K.C., Bally, J., & Testi, L. 2000, ApJ, 535, 833

Shepherd, D.S., Testi L., & Stark D.P. 2003, ApJ, in press

Shu, F.H., Adams, F.C., & Lizano, S. 1987, ARA&A, 25, 23

Sridharan, T.K., et al. 2002, ApJ, 566, 931

Stahler, S., Palla, F., & Ho, P.T.P 2000 in *Protostars and Planets IV*, ed. V. Mannings, A.P. Boss, & S.S. Russell (Tucson: Univ. of Arizona Press)

Stecklum, B., et al. 2002, A&A, 392, 1025

Tan, J., & McKee, C.F. 2002, in *The Earliest Stages of Massive Star Birth*, ed. P.A. Crowther, ASP Series 267 (San Francisco: ASP)

Testi, L., Palla, F., & Natta, A. 1999, A&A, 342, 515

van der Tak, F.F.S., et al. 1999, ApJ, 522, 911

van der Tak, F.S., van Dishoeck, E.F., Evans, N.J II, & Blake, G.A. 2000a, ApJ, 537, 283

van der Tak, F.F.S., van Dishoeck, E.F., & Caselli, P. 2000b, A&A, 361, 327

Walmsley, C.M., Caselli, P., Zucconi, A., & Galli, D. 2001, in *The Origins of Stars and Planets : The VLT View*, ed. J. Alves & M. McCaughrean (Berlin: Springer)

Weingartner, J.C., & Draine, B.T. 1999, ApJ, 517, 292

Welch, W.J., et al. 1987, Science, 238, 1550

Wilner, D.J., De Pree C.G., Welch, W.J., Goss, W.M. 2001, ApJ, 550, L81

Wyrowski, F., Gibb, A.G., & Mundy, L.G. 2002, in *The Earliest Stages of Massive Star Birth*, ed. P.A. Crowther, ASP Conf. Series 267 (San Francisco: ASP)

Yorke, H., in *The Earliest Stages of Massive Star Birth*, ed. P.A. Crowther, ASP Conf. Series 267 (San Francisco: ASP)

Zhang, Q., Hunter, T.R., Sridharan, T.K., & Cesaroni, R. 1999, ApJ, 527, L117

ICE IN HIGH-MASS STAR-FORMING REGIONS: SIGNATURES OF GRAIN SURFACE CHEMISTRY

Jacqueline V. Keane[1] **and Alexander G.G.M. Tielens**[2]

[1]NASA-Ames Research Center, Mail Stop 245-3, Moffett Field, CA 94035, USA
[2]Kapteyn Institute, P.O. Box 800, 9700 AV Groningen, The Netherlands

ABSTRACT

We have developed a model for grain surface reactions among accreting atoms, radicals, and molecules. This model is centered around key hydrogenation and oxidation reactions: $H + CO$, $H + H_2CO$, and a number of CO_2-forming reactions. Here, we examine the distribution of the carbonaceous species CO, CO_2, H_2CO, CH_3OH. According to our results the reaction $O + CO$ can only accommodate the observed CO/CO_2 ratio if a significant fraction of atomic O in accreting gas is made unavailable (i.e., locked up in O_2). The abundance ratios of CO, H_2CO, and CH_3OH reflect the ease with which H_2CO is hydrogenated as compared to CO. Our models are in good agreement with the observations if there is an equal probability for these two reactions (on a per species basis). The models best approximate the observed ice mantle compositions for densities $10^4 - 10^5 \mathrm{cm}^{-3}$, which is the density at which accretion becomes comparable to the free-fall timescale.

Key words: ISM: chemistry – ISM: clouds – ISM: molecules

1. INTRODUCTION

Interstellar grains were often believed to only play a passive role in interstellar chemistry by acting simply as reservoirs of molecules that previously condensed out from the gas phase. A well known exception was H_2, which had been shown to be formed exclusively on grains (Hollenbach & Salpeter 1971). However, it is now generally accepted that a substantial fraction of the cataloged gas phase species either cannot in fact be exclusively formed through gas phase chemistry, or have abundances that are considerably greater than what a gas phase origin can account for. Notably, CH_3OH can not be produced efficiently in gas phase models and a surface chemistry origin through the successive hydrogenation of CO is now proposed as the major route to CH_3OH (Tielens & Allamandola 1987; Hiraoka et al. 1994; Charnley et al. 1997). Moreover, as the process of D-fractionation is very temperature sensitive (Millar et al. 1989), high abundances of deuterated species in warm gas (e.g., HDO, HDCO, D_2CO, CH_2DOH, CH_3OD) betray the credible role of grains as cold storage "cells" in the preservation of these species from a previous cold dark phase. The much higher abundance of the hydrogenated parents of these species confirms that grain surface chemistry is a likely formation route.

Grain surface chemistry models differ greatly in their treatment of the diffusive surface chemistry. In the first approach, the surface chemistry is governed by the rate at which species are transported to the grain surface (i.e., "accretion-limited": Hollenbach & Salpeter 1971; Allen & Robinson 1977; Tielens & Hagen 1982; Tielens & Allamandola 1987). An accreted gas phase species "selects" among possible coreactants in a Monte Carlo framework weighted according to the activation barrier and the surface concentration of the coreactants involved. This approach is fundamentally different from the semi-empirical "reaction limited" models which, in an attempt to develop time-dependent models, treat the grain surface chemistry in the same manner as the gas phase chemistry (i.e., by rate equations: Pickles & Williams 1977; d'Hendecourt et al. 1985; Hasegawa et al. 1992; Caselli et al. 1998; Ruffle & Herbst 2000). In this case the chemistry occurring on the grain surface is controlled by reaction rates that are primarily dependent upon the collisional frequency (i.e., migration time) of the reactants. Recently, a new method based upon a stochastic treatment of the master equation governing the reaction kinetics has been developed (Charnley 1998, 2001). This type of approach may be more suited to coupling grain surface and gas phase chemistries.

In this work we present a simple, accretion-limited model for the chemistry occurring on grain surfaces. Here we examine the distribution of carbon over the carbonaceous species CO, CO_2, H_2CO, and CH_3OH.

Chemistry as a Diagnostic of Star Formation (2003). *Edited by Charles L. Curry and Michel Fich.* Published in 2003 by NRC Research Press, Ottawa ON, K1A 0R6, Canada. ISBN 0-660-19089-3.

Table 1. Abundances of interstellar ices for different lines of sight and typical abundances seen in comets and hot cores[a]

Species	NGC 7538:IRS9 high mass	W 33A high mass	Elias 29 low mass	Elias 16 dark cloud	Comets	Orion hot core
H_2O	100	100	100	100	100	100
CO_2 (total)	24	14	22	18	3-20	2-10
(polar)	16	13	—	18	—	—
(apolar)	8	1	—	—	—	—
CO (total)	16	9	5.6	25	5-7	1000
(polar)	2	6	0.8	3	—	—
(apolar)	14	2	4.8	22	—	—
CH_3OH	4	22	< 4	<3	0.3-5	2
CH_4	2	2	<1.6	—	0.2-1.2	—
H_2CO	2-4	2-6	—	—	0-5	0.1-1
HCOOH	< 3	< 3-7	—	—	0.05-0.1	0.008
OCS	0.05	0.3	<0.08	—	—	0.5
OCN^-	1	10	<0.24	<2	0.02-0.1	—
References	1,2,3,4,5,6	1,3,4,5,6	7	1,8	9,10,11	12,13

[a] Relative to H_2O ($H_2O = 100$), except in the case of the cometary abundances which are normalized to CO.

REFERENCES—(1) Whittet (1997) and references therein; (2) Schutte et al. (1996); (3) Gerakines et al. (2000); (4) Gibb et al. (2000); (5) Keane et al. (2001); (6) Boogert et al. (1996, 1998); (7) Boogert et al. (2000); (8) Whittet et al. (1998); (9) Mumma et al. (1993, 1996); (10) Crovisier et al. (1996); (11)Irvine et al. (2000); (12) van Dishoeck & Blake (1998); (13) Irvine et al. (1999).

2. Observational constraints on Ices

Current abundances of interstellar ices in contrasting environments, along with typical cometary abundances, are summarized in Table 1. One of the most notable accomplishments of *ISO*-SWS was the discovery of the ubiquitous presence of CO_2 ice along all lines of sight (de Graauw et al. 1996; Gerakines et al. 1999; Boogert 1999; Boogert et al. 2000a). In addition, carbon-bearing species such as CO, CH_3OH are detected, along with trace amounts of CH_4, H_2CO, and HCOOH (Boogert et al. 1996,1998; Dartois et al. 1999; Ehrenfreund et al. 1998; Gerakines et al. 1999; Gibb et al. 2000; Keane et al. 2001). NH_3 is also observed, though there is considerable uncertainty about its abundance (Gibb et al. 2000; Dartois & d'Hendecourt 2001). Table 1 also compares interstellar ice abundances with cometary abundances. Although there are some discrepancies in the details between the composition of cometary and interstellar ices (e.g., OCS), in general there is remarkable agreement in the species present as well as their global abundances indicating that cometary ices are a mixture of original interstellar material and material processed in the solar nebula (Mumma et al. 1993; Irvine et al. 2000).

Careful analysis and modeling of the spectral signatures of interstellar ices have revealed many aspects of their molecular environments. In particular, it has been shown that there are a number of independent ice components along most lines of sight which are generally classified as distinct polar (H_2O-rich) and apolar (CO-rich or H-poor) phases (Tielens et al. 1991; de Graauw et al. 1996b; Boogert et al. 1999; Gerakines et al. 1999). Notably, CO_2 is located almost exclusively in the polar phase implying that it formed simultaneously with the H_2O. This indicates that the grain mantle had oxidizing conditions while being formed in a reducing environment (e.g., H_2O, CH_3OH). CO, on the other hand, is found predominantly in an apolar state, though a small fraction is also found residing in the polar phase. This would suggest that both species (i.e., CO and H_2O) condensed simultaneously at low temperatures = 10 K (Sandford & Allamandola 1988). The existence of different ice phases may reflect variations in the chemical and/or physical conditions during the accretion phase (Tielens et al. 1991). It is equally likely that the different ice types noted above (i.e., polar and apolar) might be attributed to thermal processing of the ice by an embedded protostar.

3. The chemical model

We have adopted a "scoundrel approach" (Charnley et al. 1997) and consider that chemical reactions occurring on the grain surface are solely responsible for the formation of the molecular species investigated here.

The model follows the Monte Carlo, accretion-limited method of Tielens & Hagen (1982), which is based upon two essential simplifying assumptions:
(1) The accretion timescale is much greater than the timescales for reaction of the accreted atoms (H, N, C, or O) with other species.
(2) There is no desorption of grain mantle species and thus the surface chemistry is not coupled to the gas phase.

We refer the reader to Keane & Tielens (2003) for a more detailed description of the chemical model, gas phase composition, surface reaction network, and parameters governing the chemistry, along with a thorough analysis of the results.

4. Results

4.1. Oxidation versus Hydrogenation

Figure 1 shows the calculated abundance ratios of CO_2 and CH_3OH relative to CO for different gas phase O_2 abundances. Along each curve, the gas phase density increases from bottom to top and the oxidizing conditions parallel this. The reducing conditions increase from left to right. The parameter $f_{O_2} \equiv O_2/O$ controls the atomic O/H ratio and hence the oxidizing and reducing conditions.

Superimposed are the interstellar abundance ratios observed toward a number of high mass embedded protostars. The solid diamonds indicate the abundance ratios derived using the observed solid state CO abundances. Because CO is a very volatile molecule, a small increase in the dust temperature may cause some of the solid state CO to evaporate into the gas phase (Sandford & Allamandola 1988). There is evidence of thermal processing in the observed profiles of the CO_2 band toward massive protostars. Hence, some of the gaseous CO along the line of sight may have cycled through the grain mantles. We therefore also show the data observed toward massive protostars normalized to the gaseous CO column density. Comparison of the observations with the data raises two points. First, the high abundance of solid H_2O attests to the importance of hydrogenation reactions. However, the observed CO_2/CH_3OH ratios show that oxidation reactions have also been of great importance. Depending on whether outgassing has been important or not, the derived densities at which ice mantle formation occurred were 10^4 and 10^5 cm^{-3}, respectively. Second, the observed mantle composition requires a rather high O_2 abundance. This essentially reflects the adopted high efficiency of the $O + CO \rightarrow CO_2$ reaction on the grain surface.

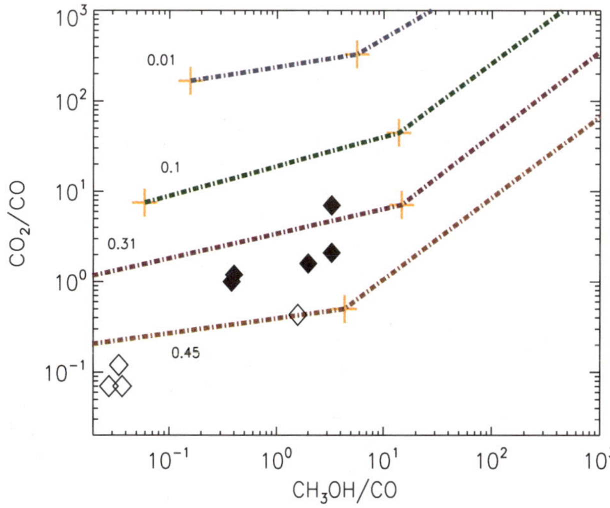

Figure 1. Abundance ratios of CO_2 and CH_3OH relative to CO for different O_2 abundances ($f_{O_2} \equiv O_2/O$). Along each curve, the total density varies from low (*upper right*) to high (*lower left*). f_{O_2} controls the atomic O/H ratio and hence the oxidizing and reducing conditions. The diamonds denote the observed abundance ratios derived using the solid state CO (*filled*) and gas phase CO abundances (*open*).

4.2. H_2CO and CH_3OH

The formation of CH_3OH on grain surfaces involves the successive hydrogenation of CO, viz.

$$CO P_1 1 \longrightarrow HCO 2 \longrightarrow H_2CO P_3 3 \longrightarrow CH_3O 4 \longrightarrow CH_3OH \quad (1)$$

Steps 2 and 4 in equation (1) are barrierless and atomic H reacts readily, ensuring that the radicals HCO and CH_3O do not accumulate on the surface. Steps 1 and 3, on the other hand, are believed to have appreciable reaction barriers. The accumulation of CO and H_2CO therefore depends on the probability of H reacting with either species. In Figure 2, model results are presented varying the H_2CO activation barrier from 230 K to 1850 K. Essentially all the H_2CO in Figure 2a is reduced to CH_3OH as the barrier against reaction is very low. As the H_2CO activation barrier is increased to 1000 K, the amount of H_2CO on the surface also increases. In this model, as the conditions change from reducing to oxidizing, the reaction terminates more often at H_2CO. Of course, for high densities all hydrogenation reactions decrease in importance. These effects are even more pronounced when the activation barrier for hydrogenation of H_2CO is 1850 K (Fig. 2c). Then H_2CO can accumulate at high levels ($\sim$20 percent) on the grain surface. Strikingly, the overall chem-

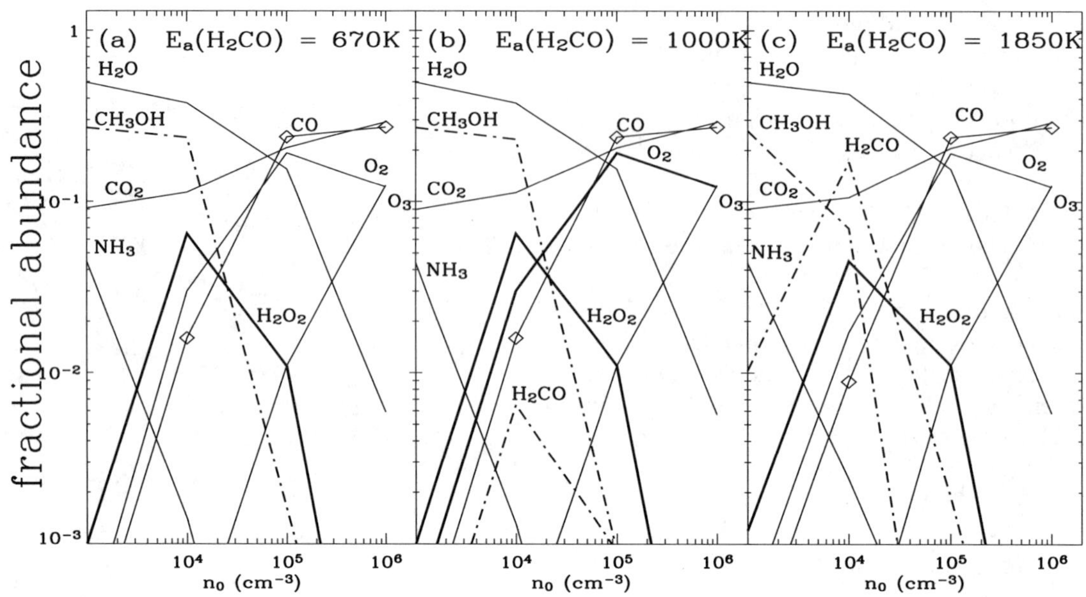

Figure 2. Fractional abundances of relevant grain surface species versus density for different H_2CO activation barriers. The dot-dashed lines indicate species that are significantly affected by varying the H_2CO activation barrier.

istry is very similar in all cases. We note that when the activation barrier for the H_2CO reaction increases, the H that now does not react with H_2CO reacts instead with H_2O_2. Thus, the abundance of the latter decreases slightly from panels a through c of Figure 2.

The CH_3OH chemistry can be expressed in terms of the ratio of the probability for H to react with CO relative to H_2CO; i.e., $P = P_1/P_3$ in equation (1) (Charnley et al. 1997). Figure 3 shows the abundance ratios of H_2CO/CH_3OH and CO/CH_3OH for different reaction probability ratios. The diamonds in the figure denote the H_2CO/CH_3OH and CO/CH_3OH interstellar ice abundance ratios observed toward a number of high mass embedded protostars. The observed abundance ratios of CO/CH_3OH is based upon the total solid CO column density, including the apolar component. However, in our model the CH_3OH and H_2CO are largely limited to the water-rich (polar) phase. If we had normalized to the polar CO abundance only, the filled diamonds would, in general, shift only slightly to the left, with only one object shifting significantly. Since the solid state CO abundance is sensitive to outgassing, the observational values may not accurately reflect the true CO abundance. Alternatively, gas-phase CO abundances can be used (open diamonds).

Also plotted in Figure 3 are a number of abundance ratios derived for the Orion Compact Ridge (Sutton et al. 1995; Wright et al. 1996). The variations seen in the observations for this particular source might well reflect the fact that the Orion molecular cloud OMC-1 is a complex region and observations might suffer from confusion (i.e., observations of the Compact Ridge might be slightly contaminated by nearby regions). Within the context of our models, the conditions that best account for the composition of ices toward the embedded protostars were accretion during an intermediate density ($n \sim 10^4$ cm^{-3}) gas-phase and $P \sim 1$; i.e., it is equally probable that an H atom can react with CO as with H_2CO on a per-species basis (i.e., assuming equal θ_i). Note also that the Orion Compact Ridge observations are outliers with regard to all other sources and indicate a value of $P \sim 10^{-1}$.

In contrast to ice studies, observations of gaseous CO, H_2CO, and CH_3OH in hot cores (van der Tak et al. 2000a,b), reveal high density, warm environments that have little atomic H and are thus not conducive to hydrogenation (Fig. 4). The model can reproduce the observed ratios in the gas phase of hot cores if we extend the curves to high densities ($\geq 10^6$cm^{-3}). However, the agreement is largely artificial because abundances of all hydrogenated species are very low at high densities. These gaseous ratios are difficult to reconcile with observed interstellar ice abundances unless only a minor fraction of the CO observed in hot cores has been cycled through the ice phase in massive star forming regions. Under the conditions in-

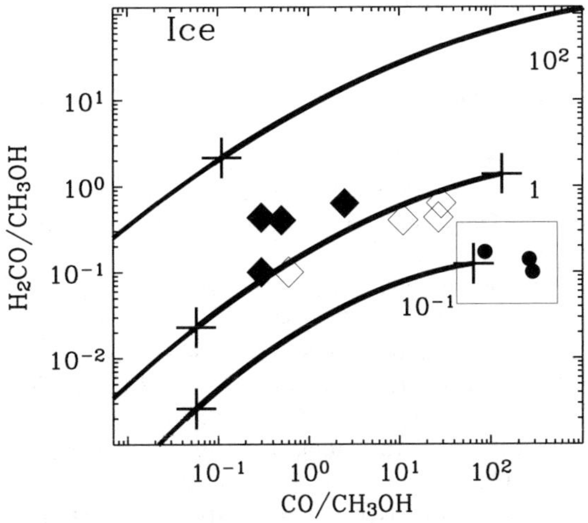

Figure 3. The abundance ratios of H_2CO and CO relative to CH_3OH. The curves show the results for the ratio of the probability of H to react with CO relative to H_2CO. Along each curve, the CO/H accretion probability varies from small to large (*left to right*). The crosses denote number densities of 10^4 and 10^5 cm^{-3}. The diamonds indicate observed abundances toward a number of embedded young stellar objects, using solid state (*filled diamonds*) and gas phase CO (*open diamonds*). The filled circles denote the gas phase composition of the Orion Compact Ridge and the surrounding box shows the range of published values (see text for details).

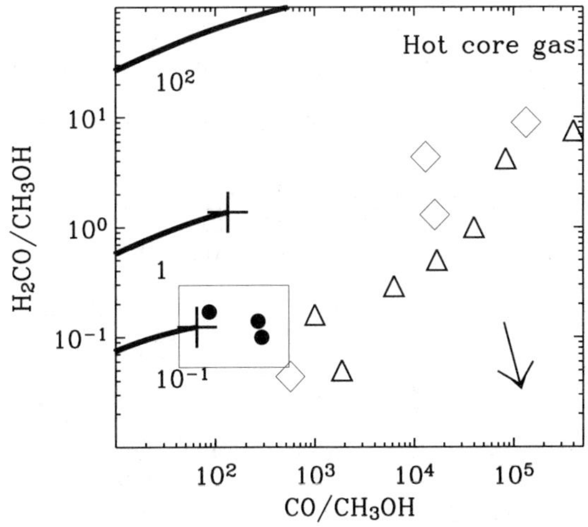

Figure 4. As in Figure 3, but in this case the abundance ratios are determined from gas phase data (van der Tak et al. 2000a,b). The diamonds indicate the observed gas phase abundances toward the embedded young stellar objects shown in Figure 3. The triangles are additional objects from the study of van der Tak et al.

dicated by the CO, H_2CO, and CH_3OH abundances, the accretion is mainly of inert species and, in addition, there is little atomic O present. As a result, little H_2O will be formed through standard surface chemistry and any H_2O (or its derivative, H_3O^+) in the hot core must be formed locally (e.g., through the reactions $O + H_2$ and $OH + H_2$) in the warm gas.

The observed ice composition and hot core compositions can possibly be reconciled if, unlike the hot core material, ices sample quiescent material, much further away from the collapsing core, which has never experienced the extreme conditions associated with the protostellar core-forming process. Alternatively, the gas phase abundance ratios might have fallen, at one stage, close to the $P \sim 1$ curve favored by the solid state abundance ratios and simply evolved with time to their current positions. In this respect, the gas phase abundance ratios might reflect a later stage of the star-forming process whereby these species are converted into other molecules. Indeed, models of the gas phase chemistry of warm molecular material around protostars show that after the mantles have been injected into the gas phase, the abundances of CH_3OH and H_2CO decrease, with H_2CO decreasing at a more rapid rate (Charnley et al. 1992). As a result, the data points in Figure 4 will shift to the lower right corner, i.e., in the direction indicated by the arrow. In this way, the bulk of the CO can still cycle through the ice phase in massive star-forming regions.

5. Comparison with observations

In Table 2, we compare the abundances of selected species for two models (low and high f_{O_2}) with abundances of observed interstellar ices. Model abundances of distinct polar and nonpolar phases are given to compare more directly with the observations. Generally the observations, especially the distribution of carbon over the (polar) CO, CO_2 and CH_3OH species, are in reasonable agreement with the models for densities $n \sim 10^4$–10^5 cm^{-3}. We note that at these densities, the accretion timescale becomes comparable to the dynamical timescale of protostar formation (10^4–10^5 yr). Furthermore, comparison of Figure 2 with the observed solid state H_2CO abundances suggests that the activation barrier of H_2CO is $\sim$1500 K. However, there is a considerable discrepancy between the abundance of CH_3OH produced by the models and what is actually observed. To a large extent, this results from the high adopted gaseous C/O ratio, which in the models is ~ 0.45 as opposed to an observed ratio of, in the case of Elias 16, ~ 0.25 in the ices. Models run with lower

Table 2. Comparison of model abundances with observed ice abundances[a]

		Low f_{O_2}		High f_{O_2}	
Species	Observed[b]	10^4	10^5	10^4	10^5
H_2O	100	100	100	100	100
CO_2 (total)	14–24	64	106	30	133
(polar)	13–16	64	69	30	—
(apolar)	1–8	—	106	—	133
CO (total)	5–25	1	14	6	153
(polar)	1–6	1	—	6	—
(apolar)	2–14	—	14	—	153
CH_3OH	4–22	20	1	61	0.9
NH_3	<10	1	—	0.5	—
CH_4	2	2	0.4	4.4	0.5
H_2CO	2–6	0.8	0.2	0.5	
HCOOH	0–7	0.4	3	—	—
NH_2CHO	—	8	7	19	30
H_2O_2	—	—	—	21	7

[a] Relative to H_2O ($H_2O = 100$).
[b] Based on the objects listed in Table 1.

C/O ratios result in CH_3OH abundances which are quite comparable to the observations. The observed apolar CO reflects high densities ($> 10^5 cm^{-3}$) in the models when the gas phase is H-poor. NH_3 and CH_4 abundances are set by the adopted atomic N and C abundances in the gas phase. Their good match with the observations is therefore not surprising. Although HCOOH is observed toward protostars, in our models it is not produced. Oxidation of the HCO radical does not appear to be efficient as atomic O generally prefers routes that terminate in H_2O. On the other hand, H_2O_2 and NH_2CHO have not yet been observed, but both species are present at substantial levels for $n \sim 10^4$–10^5 cm^{-3}. At even higher densities ($10^6 cm^{-3}$; not shown in Table 2) and in the case of the high f_{O_2} models, HNCO and OCN are present at just below the 1 percent level. A concerted effort should be made to search for signatures of these trace molecules which can better constrain the chemical reactions that occur on grain surfaces.

References

Allen, M., & Robinson, G.W. 1977, ApJ, 212, 396
Boogert, A.C.A., et al. 1996, A&A, 315, L377
Boogert, A.C.A., et al. 1998, A&A, 336, 352
Boogert, A.C.A. 1999, Ph. D. Thesis, University of Groningen
Boogert, A.C.A., et al. 2000, A&A, 353, 339
Caselli, P., Hasegawa, T.I., & Herbst, E. 1998, ApJ, 495, 309
Charnley, S.B. 1998, ApJ, 509, L121
Charnley, S.B. 2001, ApJ, 562, L99
Charnley, S.B., Tielens, A.G.G.M., & Rodgers, S.D. 1997, ApJ, 482, L203
Crovisier, J., & Bockelée-Morvan, D. 1999, Sp. Sci. Rev., 90, 19
Dartois, E., & d'Hendecourt, L. 2001, A&A, 365, 144
Dartois, E., et al. 1999, A&A, 342, L32
Ehrenfreund, P., Dartois, E., Demyk, K., & d'Hendecourt, L. 1998, A&A, 339, L17
Gerakines, P.A., et al. 1999, ApJ, 522, 357
Gibb, E.L., et al. 2000, ApJ, 536, 347
de Graauw, T., et al. 1996a, A&A, 315, L345
Hasegawa, T.I., Herbst, E., & Leung, C.M. 1992, ApJS, 82, 167
d'Hendecourt, L.B., Allamandola, L.J., & Greenberg, J.M. 1985, A&A, 152, 130
Hollenbach, D., & Salpeter, E.E. 1971, ApJ, 163, 155
Hiraoka, K., et al. 1994, Chem. Phys. Lett., 229, 408
Irvine, W.M., et al. 1999, Earth, Moon, Planets, 78, 29
Irvine, W.M., et al. 2000, in *Protostars and Planets IV*, ed. V. Mannings, A.P. Boss, & S.S. Russell (Tucson: University of Arizona Press), 1159
Keane, J.V., & Tielens, A.G.G.M. 2003, A&A, submitted
Keane, J.V., Tielens, A.G.G.M., Boogert, A.C.A., Schutte, W.A., & Whittet, D.C.B. 2001, A&A, 376, 254
Millar, T.J., Bennett, A., & Herbst, E. 1989, ApJ, 340, 906
Mumma, M.J., Weissmann, P.R., & Stern, S.A. 1993, in *Protostars and Planets III*, ed. E.H. Levy & J.I. Lunine (Tucson: University of Arizona Press), 1177
Mumma, M.J., et al. 1996, Science, 272, 1310
Pickles, J.B., & Williams, D.A. 1977, ApSS, 52, 443
Ruffle, D.P., & Herbst, E. 2000, MNRAS, 319, 837
Sandford, S.A., & Allamandola, L.J. 1988, Icarus, 76, 201
Schutte, W.A., et al. 1996, A&A, 315, L333
Sutton, E.C., et al. 1995, ApJS, 97, 455
Tielens, A.G.G.M., & Hagen, W. 1982, A&A, 114, 245
Tielens, A.G.G.M., & Allamandola L.J. 1987, in *Physical Processes in Interstellar Clouds*, ed. G.E. Morfill & M. Scholer (Kluwer: Dordrecht), 333
Tielens, A.G.G.M., Tokunaga, A.T., Geballe, T.R., & Bass, F. 1991, ApJ, 381, 181
van Dishoeck, E.F., & Blake, G.A. 1998, ARA&A, 36, 317
van der Tak, F.F.S., van Dishoeck, E.F., Evans, N.J. II, & Blake, G.A. 2000a, ApJ, 537, 283
van der Tak, F.F.S., van Dishoeck, E.F., & Caselli, P., 2000b, A&A, 361, 327
Whittet, D.C.B. 1997, in *Planetary and Interstellar Processes Relevant to the Origins of Life*, ed. D.C.B. Whittet (Kluwer: Dordrecht), 101
Whittet, D.C.B., et al. 1998, ApJ, 498, L159
Wright, M.C.H., Plambeck, R.L., & Wilner, D.J 1996, ApJ, 469, 216

PART VII:
POSTER PAPERS

CHEMISTRY IN PRESTELLAR CORES: CAN WE DISTINGUISH BETWEEN CORES WITH AND WITHOUT MAGNETIC SUPPORT VIA CHEMISTRY?

Yuri Aikawa

Department of Earth and Planetary Sciences, Kobe University, Kobe 657-8501, Japan

Abstract

We have investigated the evolution and distribution of molecules in collapsing prestellar cores via numerical calculation. Molecular abundances and distributions in a collapsing core are determined by the balance among the dynamical, chemical and adsorption time scales. Therefore, molecular distributions could be a probe to investigate how the collapse proceeds. We adopt the Larson-Penston solution and its delayed analogues as a physical model of cores. Chemical model includes gas-phase reactions, deuterium fractionation, adsorption onto grains, thermal and nonthermal desorption, and modified rate grain surface reactions. Molecular column densities obtained with the Larson-Penston solution show reasonable agreement with the observed values in L1544. Column density ratios of DNC/HNC observed in the Taurus molecular cloud are also reproduced by the Larson-Penston collapse. Our model indicates that variation of molecular abundances among cores can be explained by core evolution and different timescales of core collapse.

Key words: ISM: molecules – stars: formation

1. Introduction

Stars are formed by gravitational collapse of molecular cloud cores. Molecular lines have long been observed in molecular clouds as a probe of kinetics, density and temperature. Since the dominant component of molecular clouds, H_2, does not emit radiation at the temperature of prestellar cores ($\sim 10\,K$), heavy-element molecules such as CO are observed. Hence their abundances relative to hydrogen are essential in interpreting line observations.

Recently, chemical fractionations are found in several prestellar cores. The distribution of N_2H^+ is centrally peaked, while CO, CS, and CCS are depleted at inner radii (Bergin, this volume). Such fractionations are predicted and explained by theoretical models. Bergin & Langer (1997) (hereafter BL97) calculated molecular evolution in a fluid element, in which density increases as time evolves. Because of the low temperature and increasing density, species such as CO deplete onto grains. The molecular ion N_2H^+ survives for a relatively long time because of the lower binding energies of its parent species N_2 onto grain surfaces compared with other heavy-element molecules. These fractionations make it difficult for us to estimate mass distribution from line observations, but could be a useful probe of the evolutionary stage of cores.

Aikawa et al. (2001) (hereafter A01) ran similar chemical models to BL97 but with multiple fluid elements in order to obtain the spatial distribution of molecules in collapsing cores. They showed molecular abundances and their spatial distributions depend on how the gravitational collapse proceeds, since the abundances are determined by a balance between the chemical timescale of molecules and the dynamical timescale of cores. If cores are supported by magnetic fields, they collapse more slowly and hence their molecular abundances are different from those of non-magnetized cores (see also Shematovich, this volume).

The present work is an update of A01. We solve for the molecular evolution in the Larson-Penston core and its delayed analogues. We also extend the chemical model of A01 to include grain surface reactions.

2. Model

As a physical model of the collapsing core, we adopt a semi-analytical solution called Larson-Penston collapse (Larson 1969; Penston 1969). The distribution of density is almost flat at inner radii, and follows a power law $\rho \propto r^{-2}$ at outer radii. The initial central density of the model core is $n_H = 3 \times 10^4$ cm^{-3} (i.e., $n[H_2] = 1.5 \times 10^4$ cm^{-3}). The core collapses in almost a free-fall time; namely, the central density increases to $n_H = 3 \times 10^7$ cm^{-3} in 2.00×10^5 yr. Hydrodynamic simulations show that the spatial density distribution and time variation of the central density of the Larson-Penston core are good

Chemistry as a Diagnostic of Star Formation (2003). *Edited by Charles L. Curry and Michel Fich.* Published in 2003 by NRC Research Press, Ottawa ON, K1A 0R6, Canada. ISBN 0-660-19089-3.

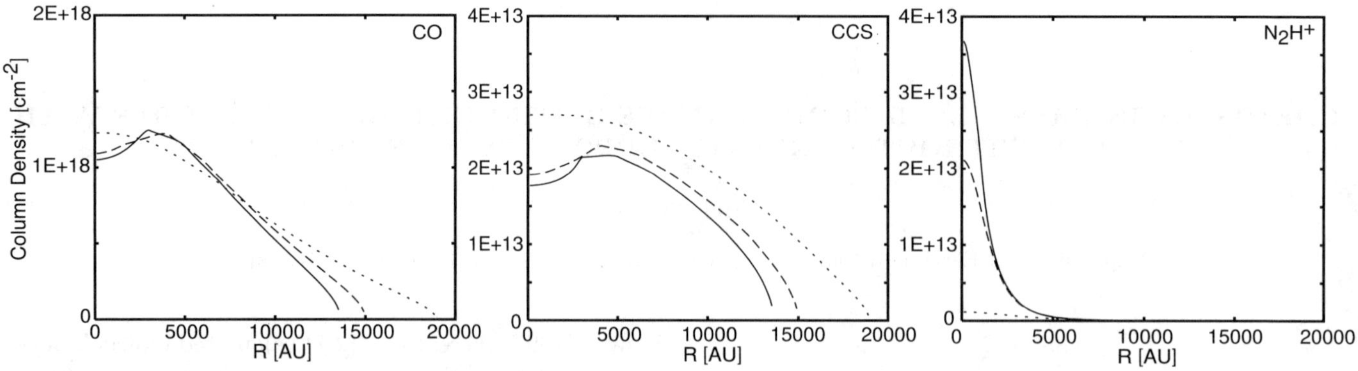

Figure 1. Molecular column densities as a function of impact parameter from the core center, when the central density of the core ($n_{\rm H}[R=0]$) is 3×10^5 cm^{-3} (*dotted curves*), 3×10^6 cm^{-3} (*dashed curves*), and 3×10^7 cm^{-3} (*solid curves*).

approximations to those in non-magnetized cores (Masunaga, Miyama, & Inutsuka 1998). If a core is supported by magnetic fields and/or turbulence, collapse will be delayed. In order to investigate their effects on molecular abundances, we adopt delayed analogues of the Larson-Penston core; the infall velocity is artificially decreased by a constant factor f.

Our chemical reaction network includes a set of gas-phase reactions known as the New Standard Model (e.g., Terzieva & Herbst 1998), deuterium fractionation, gas-dust interaction (i.e., adsorption and desorption), and grain surface reactions. All neutral species are assumed to be adsorbed when they hit grains with a sticking probability S of unity, unless stated otherwise. Species are desorbed by the thermal desorption and impulsive heating of grains by cosmic rays (Léger, Jura, & Omont 1985; Hasegawa & Herbst 1993). We adopt the set of grain surface reactions from Ruffle & Herbst (2000), and calculate the modified rate coefficients following Stancheva, Caselli, & Herbst (2001). Adopting the semi-analytical collapse model, it is easy to solve a large chemical network. In total our chemical network includes 878 species and 11,774 reactions.

3. Results

3.1. Distribution of Molecular Column Densities

Figure 1 shows molecular column densities integrated along the line of sight as a function of impact parameter from the core center at the three representative evolutionary stages. While the central density of the core increases from 3×10^5 cm^{-3} to 3×10^7 cm^{-3}, the total hydrogen column density is enhanced by about an order of magnitude. However, column densities of CO and CCS in the central regions remain almost constant or decrease. Their column densities are initially centrally peaked, but have a central hole in later stages. On the other hand, N_2H^+ is centrally peaked even in the latest stage, and its column density increases significantly during the collapse. While these qualitative features are the same as those in A01, the absolute value of N_2H^+ column density is larger than that in A01 by two orders of magnitude because of the grain surface formation of its parent species, N_2.

3.2. Dependence on Collapse Time Scale

The three leftmost panels in Figure 2 show molecular column density as a function of central density of the model cores with various collapse timescales and sticking probabilities. The column densities are averaged within a radius of 2000 AU at the core center. The filled circles with error bars represent molecular column densities in L1544 estimated from observations (Suzuki et al. 1992; Benson et al. 1998; Ohashi et al. 1999; Caselli et al. 2002). The Larson-Penston model (*solid curves*) is in reasonable agreement with L1544.

Line observations have been performed towards several prestellar cores other than L1544, and have revealed that molecular column densities vary among objects. Figure 2 indicates that these variations could be caused by core evolution and/or different collapse timescales. Column density of N_2H^+ increases with central density, i.e. as the core evolves, regardless of the model parameters. Hence, it could be a probe of core evolution. On the contrary, column densities of CO and CCS are more dependent on the collapse timescale and sticking probability rather than evolutionary stage.

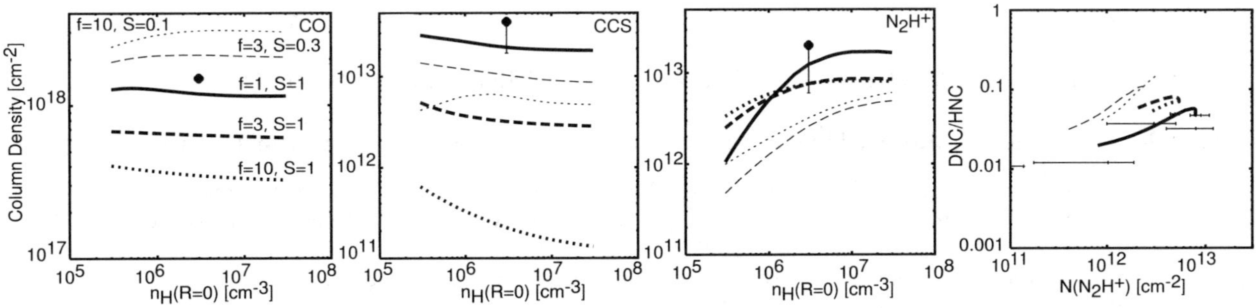

Figure 2. The first three panels show molecular column densities as a function of the central density of the cores. The rightmost panel shows column density ratio of DNC to HNC versus N_2H^+ column density. Solid curves represent the Larson-Penston core. Dashed and dotted curves represent more slowly collapsing cores by a factor f of 3 and 10, respectively. The sticking probability S is 1.0 for the thick curves, and $1/f$ for the thin curves. Filled circles and crosses with error bars represent observed values in prestellar cores.

3.3. D/H RATIOS

The rightmost panel in Fig. 2 shows the column density ratio of DNC to HNC versus N_2H^+ column density, following the above argument that N_2H^+ could be an indicator of core evolution. The crosses with error bars represent observations of prestellar cores in Taurus molecular cloud by Hirota et al. (2001), Hirota et al. (in preparation), Caselli et al. (2002) and Ohashi et al. (in preparation). The curves represent the trajectories of our model cores when their central densities increase from 3×10^5 cm^{-3} to 3×10^7 cm^{-3}. The molecular column densities are averaged over a radius of 1500 AU and 3500 AU at the core center for HNC (DNC) and N_2H^+, respectively, considering the beam sizes of the observations.

We emphasize that such a plot of D/H versus N_2H^+ column density is useful in comparisons with observations, because we do not have to estimate the central densities of the cores. The column density of N_2H^+ does not depend much on the total mass of the core, because of its centrally peaked distribution. The D/H column density ratio would not significantly depend on the total mass either, or its dependence is at least smaller than that of the molecular column densities.

The observed ratios are best reproduced by the model with Larson-Penston collapse. However, this result should be taken with caution. In our model, rate coefficients of deuterium exchange reactions are taken from Millar, Bennet & Herbst (1989). Recently, Gerlich, Herbst & Roeff (2002) measured these rate coefficients, which suggests deuterium fractionation is less efficient than previously thought. Uncertainties in these rate coefficients should be taken into account.

ACKNOWLEDGEMENTS

The author is grateful to Drs. N. Ohashi and T. Hirota for useful discussions. This work is supported by the Grant-in-Aid for Scientific Research of the Ministry of Education, Culture, Sports, Science and Technology (13011203 and 14740130). Numerical calculations were carried out at the Astronomical Data Analysis Center of National Astronomical Observatory in Japan.

REFERENCES

Aikawa, Y., Ohashi, N., Inutsuka, S., Herbst, E., & Takakuwa, S. 2001, ApJ 552, 639

Benson, P. J., Caselli, P., & Myers, P. C. 1998, ApJ, 506, 743

Bergin, E. A., & Langer, W. D. 1997, ApJ, 486, 316

Caselli, P., Benson, P. J., Myers, P., & Tafalla, M. 2002, ApJ, 572, 238

Caselli, P., Walmsley, C. M., Zucconi, A., Tafalla, M., Dore, L., & Myers, P. C. 2002, ApJ, 565, 344

Gerlich, D., Herbst, E.,& Roeff, E. 2002, Plan. Space. Sci, in press

Hasegawa, T. I., & Herbst, E. 1993, MNRAS, 261, 83

Hirota, T., Ikeda, M., & Yamamoto, S. 2001, ApJ, 547, 814

Larson, R. B. 1969, MNRAS, 145, 271

Léger, A., Jura, M., & Omont, A. 1985, A&A, 144, 147

Masunaga, H., Miyama, S. M., Inutsuka, S. 1998, ApJ., 495, 346

Millar, T. J., Bennet, A., & Herbst, E. 1989, ApJ, 340, 906

Ohashi, N., Lee, S. W., Wilner, D. J., & Hayashi, M. 1999, ApJ, 518, L41

Penston, M. V. 1969, MNRAS, 144, 425

Ruffle, D. P. & Herbst, E. 2000, MNRAS, 319, 837

Stancheva, T., Caselli, P., & Herbst, E. 2001, A&A, 375, 673

Suzuki, H., Yamamoto, S., Ohishi, M., Kaifu, N., Ishikawa, S., Hirahara, Y., & Takana, S. 1992, ApJ, 392, 551

Terzieva, R. & Herbst, E. 1998, ApJ, 501, 207

LARGE-SCALE MULTI-TRANSITION CO OBSERVATIONS OF GMCs

Katelyn N. Allers[1], Daniel T. Jaffe[1], Neal J. Evans II[1], René Plume[2], and Ewine F. van Dishoeck[3]

[1]Department of Astronomy, The University of Texas at Austin, Austin, Texas 78712–1083
[2]Department of Physics and Astronomy, University of Calgary, 2500 University Drive NW, Calgary, Alberta, Canada T2N 1N4
[3]Sterrewacht Leiden, P.O. Box 9513, 2300 RA Leiden, the Netherlands

Abstract

Multi-transition molecular line observations have given us a good indication of the range of physical conditions in dense cloud cores around massive protostars. There is little information, however, about conditions in the surrounding GMCs, on scales of 5–25 pc, where the bulk of the molecular material resides. We present large-scale observations of ^{12}CO 6 $\rightarrow$ 5, 4 $\rightarrow$ 3, 3 $\rightarrow$ 2, and 2 $\rightarrow$ 1; ^{13}CO $J = 2 \rightarrow 1$; C^{18}O $J = 2 \rightarrow 1$; and [C I] $^3P_1 \rightarrow {}^3P_0$ in 6 GMCs. Using current PDR models and radiative transfer codes, we derive estimates of density and the interstellar radiation field for positions both close to the dense cores and in more diffuse outer regions of the clouds. We find that the ^{12}CO $J = 6 \rightarrow 5$ line is much more sensitive to changes in physical conditions (even with larger calibration uncertainties) than lower$-J$ lines commonly used to obtain large-scale temperatures and densities in the outer regions of GMCs.

Key words: ISM: clouds – H II regions

Chemistry as a Diagnostic of Star Formation (2003). *Edited by Charles L. Curry and Michel Fich.* Published in 2003 by NRC Research Press, Ottawa ON, K1A 0R6, Canada. ISBN 0-660-19089-3.

CO DEPLETION AND DEUTERIUM FRACTIONATION IN PRESTELLAR CORES

A. Bacmann[1,2], C. Ceccarelli[3,4], B. Lefloch[4], J. Steinacker[2], A. Castets[3] and L. Loinard[5]

[1]European Southern Observatory, Karl-Schwarzschild-Str 2, D-85748 Garching bei München, Germany
[2]Astrophysical Institute and University Observatory, Schillergächen 2-3, D-07745 Jena, Germany
[3]Observatoire de Bordeaux, 2, rue de l'Observatoire, BP 89, F-33270 Floirac, France
[4]Laboratoire d'Astrophysique de l'Observatoire de Grenoble, BP 53, F-38041 Grenoble Cédex 9, France
[5]Instituto de Astronomía, Universidad Nacional Autónoma de México, Apartado Postal 72-3 (Xangari), 58089 Morelia, Michoacán, Mexico

Abstract

We present the results of a search for the doubly deuterated form of formaldehyde, D_2CO, in a sample of prestellar cores. D_2CO was detected in most cores and the $[D_2CO]/[H_2CO]$ ratio was found to lie between 1 and 10%, levels similar to those found in low-mass protostars, confirming the hypothesis that D_2CO forms during the cold and dense pre-collapse stage and is not a gas-phase product of the warm protostellar environment. According to available chemical models, high levels of deuteration can only be obtained if the gas phase is depleted of CO. A study of CO depletion in our sample cores showed that the cores were indeed affected by the freezing out of CO onto grains, and that the CO depletion factor increases with increasing core density. Plots of the $[D_2CO]/[H_2CO]$ fraction against the depletion factor suggest an increasing trend. Although more data are needed to definitely answer that question, these new results already set strong constraints on chemical models of molecule formation, which cannot fully account for such high levels of deuteration in this molecule.

Key words: ISM: molecules – ISM: structure – ISM: clouds

1. Introduction

Doubly deuterated molecules have drawn much attention in the past few years, with the detections of D_2CO (Turner 1990) and ND_2H (Roueff et al. 2000). D_2CO was found to be particularly abundant in low-mass protostellar environments (e.g., Loinard et al., 2002). The mechanisms leading to such high degrees of deuteration remain debated, however. According to Tielens (1983), e.g., high deuterium fractionation can be obtained via active grain chemistry: the deuteration of the molecules, stuck to the grains is regulated by the *atomic* D/H ratio of the gas accreting onto the mantles. The other major way to form deuterated molecules is through pure gas-phase reactions (Roberts & Millar 2000a): the deuterium is propagated to other molecules mostly by H_2D^+, and the $[H_2D^+]/[H_3^+]$ ratio determines the molecular deuterium ratio. In both approaches, a low temperature ($T \leq 20\,K$) is necessary to obtain enhanced D/H or $[H_2D^+]/[H_3^+]$ ratios and therefore enhanced levels of molecular deuteration.

The gas-grain interaction mechanism was favoured to interpret the observations towards IRAS 16293 (e.g., Ceccarelli et al. 2001). In this scheme, D atoms react with the CO molecules on the grain surface *during the colder, pre-collapse stage*, and the molecules formed are later released into the gas phase through heating from the newly-formed protostars. Pure gas-phase models (Roberts & Millar 2000b) could, however, account well for the observed $[ND_2H]/[NH_3]$ ratios found by Roueff et al. (2000). Recently, D_2CO was detected in L1689N, the cloud in which IRAS 16293 is embedded, with an abundance similar to that found in the envelope of the protostar (Ceccarelli et al. 2002). This detection posed new problems, since none of the aforementioned models can account for the abundance of D_2CO in that cloud. In order to have a high molecular D/H ratio, both models require strong CO and O depletion, which is not consistent with the high atomic O abundance observed (e.g., Caux et al. 1999).

Prestellar cores, as (starless) progenitors of low-mass protostars, offer the opportunity to study the cold, dense evolutionary stage in which the deuteration of molecules supposedly occurred. In order to gain insight into the depletion and the deuteration processes, and the links between the two, we have studied the CO depletion and searched for D_2CO, the doubly deuterated form of formaldehyde, in a sample of seven prestellar cores. This project is aimed at further constraining the aforementioned chemical models.

Chemistry as a Diagnostic of Star Formation (2003). *Edited by Charles L. Curry and Michel Fich.* Published in 2003 by NRC Research Press, Ottawa ON, K1A 0R6, Canada. ISBN 0-660-19089-3.

2. CO depletion

2.1. Evidence for CO depletion

The distribution of several molecular tracers ($C^{18}O$, $C^{17}O$, and N_2H^+) and of the 1.3 mm continuum (which traces the emission of cold dust and is used as an H_2 column density tracer) is presented in Figure 1. Whereas the millimeter continuum is centrally peaked, the $C^{17}O$ and $C^{18}O$ distributions are more or less flattened depending on the cores, or the peak completely disappears as in Fig. 1. The N_2H^+ distribution follows that of the millimeter continuum for all cores. This suggests that CO is underabundant at the centre of the cores (note that the $C^{17}O$ emission is optically thin).

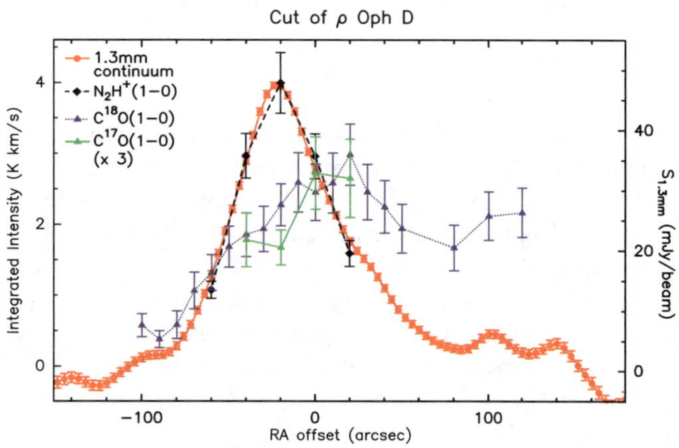

Figure 1. Cut through ρ Oph D in the millimeter continuum (*dots*) and various molecular tracers. The millimeter continuum and N_2H^+ integrated intensity (*diamonds*) are centrally peaked, whereas the CO isotopes (*triangles*) show a flat distribution.

The CO depletion factor was defined as $f = X_{can}/X$, where X is the CO abundance and X_{can} the "canonical" CO abundance as measured towards dark molecular clouds (Frerking et al. 1982). This ratio was found to lie between 4 and 15 (within a factor of 2) in the various cores at the central position. The depletion factor decreases towards the external parts of the cores, where the density becomes smaller.

2.2. Depletion Relation: Density

According to a simple model, the CO abundance should decrease by a factor $\exp(t/\tau)$ after a time t, where τ is the characteristic depletion timescale. Since $\tau \propto 1/n(H_2)$, the CO depletion factor f should increase as $\exp[n(H_2)]$. Figure 2 shows the depletion factor versus the central density for each core of the sample (the parallelograms represent the error boxes). The model mentioned above is plotted as dotted and dash-dotted curves for different times.

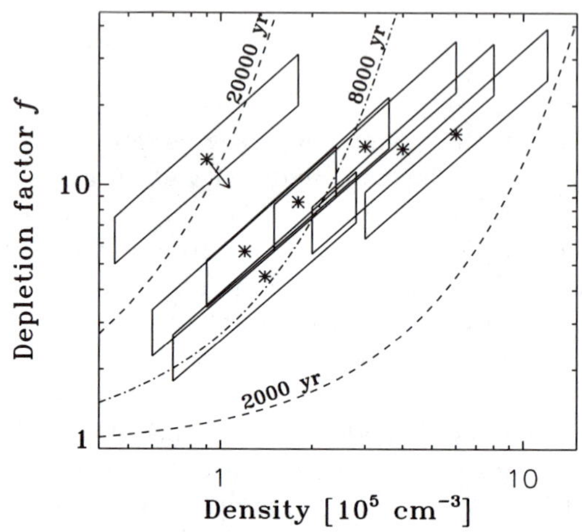

Figure 2. Depletion factor as a function of density for each core (errors contained within parallelograms). The dashed lines represent the simple model $f \propto \exp[n(H_2)\,t]$ for different ages t.

The data hint at an increasing trend between depletion and core density. The ages inferred from the comparison with the dotted curves are around 10^4 years, 1 to 2 orders of magnitude smaller than the ages estimated from core statistics (Beichman et al. 1986) or ambipolar diffusion models (Ciolek & Basu 2001). This suggests that desorption (not taken into account in the simple model described above) takes place even in such cold ($\sim 10\,K$) environments (Bacmann et al. 2002a).

3. Deuterium fractionation and implications

We searched for the doubly deuterated form of formaldehyde, D_2CO, in the same seven prestellar cores for which we had studied CO depletion. D_2CO was clearly detected in five of the cores.

The $[D_2CO]/[H_2CO]$ ratio is found to lie between 1 and 10%, after correcting for opacity effects in H_2CO. This extreme fractionation is similar to the values found in low-mass protostars by Loinard et al. (2002). This confirms the hypothesis that the D_2CO seen in protostars (where T is too high to account for such high fractionations) originates in the cold, prestellar stage (Bacmann et al., 2002b).

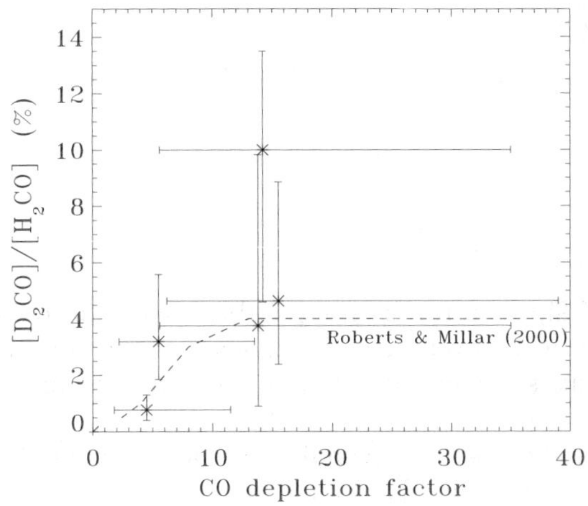

Figure 3. Relative deuterium fractionation as a function of CO depletion. The data are consistent with an increase in both quantities. The dashed line represents the predictions of Roberts & Millar (2002b).

According to the available chemical models, high deuterium fractionations can only be obtained if the gas phase is strongly depleted in CO and O. Figure 3 shows the $[D_2CO]/[H_2CO]$ ratio versus the CO depletion factor for each core in the sample. Cores with high depletion factors tend to have high deuterium fractionations, in agreement with theoretical models. Quantitatively, however, active grain chemistry models cannot reproduce such high deuterium fractionations without assuming unrealistically high D/H ratios ($\sim$ 1; Roberts & Millar predict gas-phase D/H no higher than 0.01). The model of Roberts & Millar (2000b), shown as a dashed line in Fig. 3 can account relatively well for the deuterium fractionation of most cores in our sample. New rate coefficients (Roberts et al., this volume) could decrease the theoretical fractionations obtained, however. The models also assume large atomic oxygen depletion to have high deuterations, which is not consistent with observations (e.g. Caux et al. 1999; Vastel et al. 2000) and detailed modelling in prestellar cores (Caselli 2002).

4. Conclusions

We studied CO depletion and deuteration (through D_2CO) in a sample of seven prestellar cores. Our main results are:

1. All prestellar cores in the sample are depleted of CO by various factors (4 to 15, depending on the cores).

2. The relation between the CO depletion factor and the core density shows an increasing trend. Comparison with a simple model not taking desorption into account yields core ages of 10^4 yr; i.e., 2 orders of magnitude smaller than the estimated core ages. This suggests that desorption mechanisms play a role even at these low temperatures ($\sim$ 10 K).

3. The doubly deuterated form of formaldehyde, D_2CO, was searched for and detected in five of the seven sample cores. The deuterium fractionation was found to be extremely high, with $[D_2CO]/[H_2CO]$ ratios between 1 and 10 %. This is similar to values found in low-mass protostars, and provides convincing evidence that deuteration occurs during the cold, prestellar stage.

4. An increasing relation between CO depletion and deuteration, as predicted by present chemical models, is consistent with our data.

5. Such high deuterium fractionations as those found in prestellar cores cannot be accounted for if the gas phase is not also depleted of O. Observations of other cold cores suggest that O is, on the contrary, very abundant in the gas phase, which calls into question presently available chemical models.

References

Bacmann, A., et al. 2002a, A&A, 389, L6
Bacmann, A., et al. 2002b, submitted
Beichman, C.A., et al. 1986, ApJ, 307, 337
Caselli, P. 2002, Plan. Sp. Sci., in press
Caselli, P., et al. 2002, ApJ, 565, 344
Caux, E., et al. 1999, A&A, 347, L1
Ceccarelli, C., et al. 2001, A&A, 372, 998
Ceccarelli, C., et al. 2002, A&A, 381, L17
Ciolek, G.E., & Basu, S. 2001, ApJ, 247, 572
Frerking, M.A., Langer, W.D., & Wilson, R.W. 1982, ApJ, 262, 590
Loinard, L, et al. 2002, Plan. Sp. Sci., in press
Roberts, H., & Millar, T.J. 2000a, A&A, 361, 388
Roberts, H., & Millar, T.J. 2000b, A&A, 364, 784
Roueff, E., et al. 2000, A&A, 354, L63
Tielens, A.G.G.M. 1983, A&A, 119, 177
Turner, B.E. 1990, ApJ, 362, L29
Vastel, C., et al. 2000, A&A, 357, 994

WHERE IS THE EMISSION RESULTING FROM THE DECAY OF MHD TURBULENCE IN MOLECULAR CLOUDS?

Shantanu Basu

University of Western Ontario

ABSTRACT

Several numerical models of MHD turbulence reveal that the dissipation rate is comparable to that of hydrodynamic turbulence, so that the decay time is essentially the crossing time over the driving scale of the turbulence. We explore whether this is compatible with the observed characteristics of molecular clouds, including the measured ^{12}CO (J=1–0) luminosity.

New MHD simulations of turbulence which do not assume periodic boundary conditions and allow for density stratification within the cloud are also used to constrain our ideas about molecular clouds.

Key words: ISM: kinematics and dynamics – ISM: magnetic fields – turbulence

---

Chemistry as a Diagnostic of Star Formation (2003). *Edited by Charles L. Curry and Michel Fich.* Published in 2003 by NRC Research Press, Ottawa ON, K1A 0R6, Canada. ISBN 0-660-19089-3.

WATER ICE ABSORPTION AS A PROBE OF THE ENVIRONMENTS OF YOUNG STARS

Tracy L. Beck[1], **M. Simon**[2], **and L. Prato**[3]

[1]Gemini Observatory - Northern Operations, 670 N. A'ohoku Pl. Hilo, HI 92670
[2]Department of Physics & Astronomy, SUNY Stony Brook, Stony Brook, NY 11794-3800
[3]Department of Physics & Astronomy, UCLA, Los Angeles, CA 90095-1562

Abstract

We present the results of our study of the 3 μm water-ice feature observed in circumstellar disks and envelopes around obscured young stars. We compare the 2-4 μm absorption spectra of several stars with those of field sources obscured by the Ophiuchus molecular cloud from which the young stars formed. Comparison of the ice-band profile shapes with the presence of accretion indicators, such as hydrogen Br γ (2.16 μm) or H_2 (2.12 μm) emission, provides the opportunity to compare the temperature and structure of the material in the circumstellar and interstellar environments.

Key words: ISM: dust, extinction — stars: pre-main sequence

1. Introduction

Molecular absorption features which arise from the icy mantles of interstellar dust grains are useful in probing the composition and structure of material in cold, dense regions of molecular clouds. Teixeira & Emerson (1999) showed that a linear correlation exists between the 3.1 μm water ice column density, N_{ice}, and visual extinction A_V through the Ophiuchus molecular cloud. At temperatures less than ~ 110 K, water ice forms an amorphous structure and at higher temperatures, a crystalline lattice forms. The absorption profiles of water ice absorption features are dependent upon the temperatures of the dust grains from which they arise; amorphous water ice tends to have a broader, rounded profile, while crystalline ices at higher temperatures show a more narrow, distinct peak (Smith, Sellgren & Tokunaga 1988, 1989; Smith, Sellgren & Brook 1993). The dust grains in circumstellar disks which are best fit by warmer temperature models can be heated by the accretion flux from a young star. We were thus motivated to observe the 2-4 μm spectra of young stars forming in the Ophiuchus dark clouds, and compare the results and the presence of accretion indicators such as hydrogen Br γ (2.16 μm), Br α (4.05 μm) and H_2 (2.12 μm) with the spectra of background stars observed through the Ophiuchus molecular cloud.

2. Observations

The observations for this project were made on UT 26 & 27 May 2002 at NASA's Infrared Telescope Facility (IRTF) at the summit of Mauna Kea using the facility IR spectrograph SpeX. We observed using the long wavelength cross-dispersed mode to obtain the 2-4 μm spectra simultaneously in one setting. We observed 16 target stars with standards interspersed for correction of telluric features. The data were reduced using the SpexTool software available from the IRTF.

3. Results

We found that the spectra of the 16 stars for this project naturally group into four distinct classes:

Group 1: K and M giant stars designated as background from the presence of deep photospheric CO $\Delta v = 2$ bandhead absorption at $\sim 2.3 - 2.4\mu$m (Elias 1978). The four background stars we observed show tentative evidence for broad water ice absorption, with no distinct peak (see Fig. 1*a*). In some cases, the ice-band profiles are difficult to distinguish, though all of these stars have visual extinctions in the range of 8 – 12 magnitudes. In all of these sources, the 4 μm flux is much less than that at 2 μm. These stars show weak Br γ emission that cannot be attributed to poor correction of absorption features in the telluric standards.
Group 1 Members: Elias 15, Elias 35, Elias 44 and Elias 47.

Group 2: Four young stars have spectra that look very similar to the background sources (see Fig. 1*b*). The 3 μm

Chemistry as a Diagnostic of Star Formation (2003). *Edited by Charles L. Curry and Michel Fich.* Published in 2003 by NRC Research Press, Ottawa ON, K1A 0R6, Canada. ISBN 0-660-19089-3.

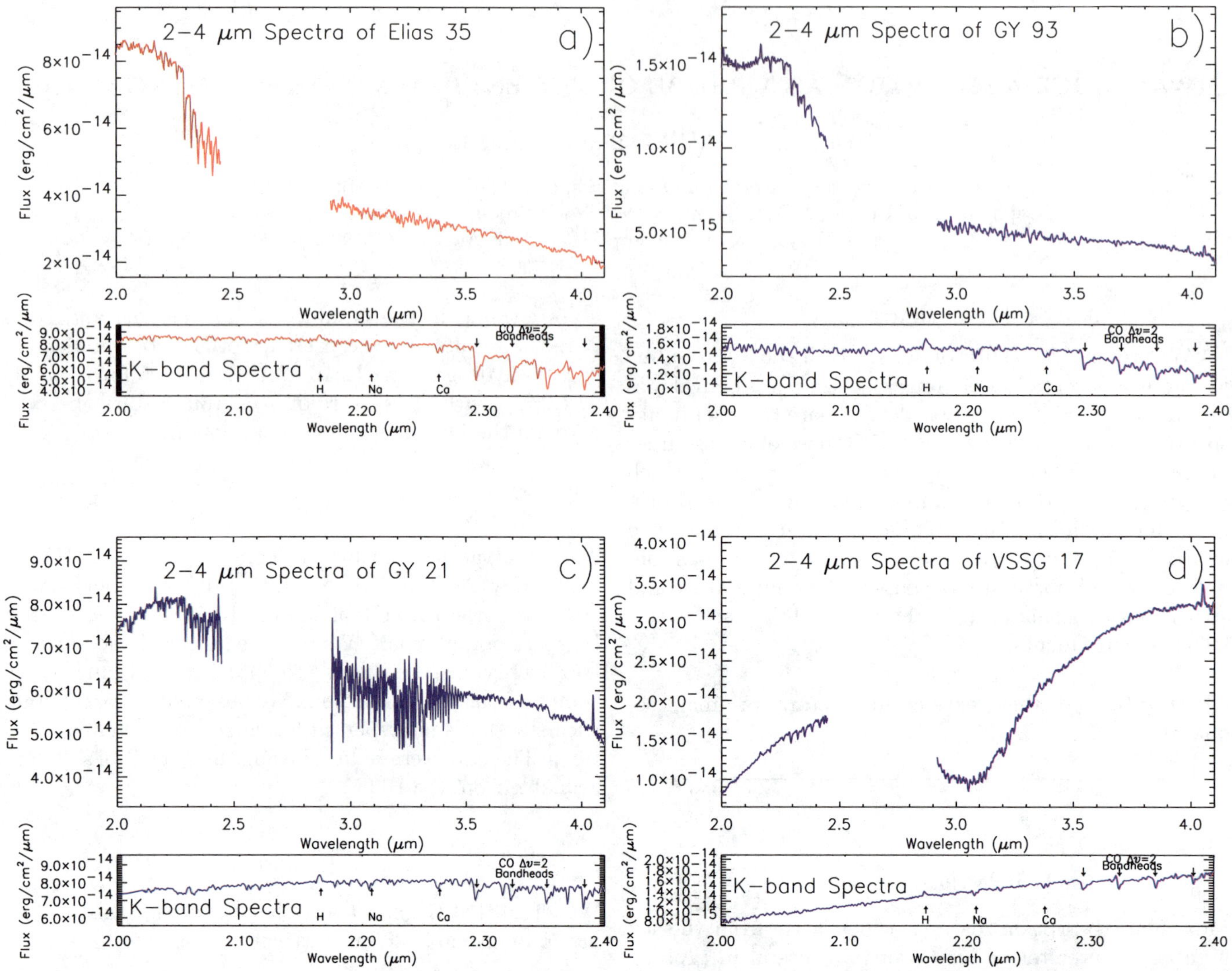

Figure 1. The spectra of the four groups from data obtained in May 2002. They were extracted using the SpeXTool data reduction software with a 1″ aperture. *(a) Group 1:* Background stars. *(b) Group 2:* Young stars, apparently not accreting material. *(c) Group 3:* Young stars with some accretion and evidence for water ice absorption. *(d) Group 4:* Young stars with high accretion activity and deep, pronounced water ice absorption features.

water ice absorption is hard to distinguish unambiguously. These sources have weak Br α and Br γ emission, and deep photospheric absorption features characteristic of young solar-type stars. This group of sources has an average $A_V \sim 14$ magnitudes and also shows a broad, shallow absorption feature centered at ~ 2.1 μm.

Group 2 Members: VSSG 5, GY 93, GSS 29, VSSG 14.

Group 3: Two stars show photospheric features typical of young solar type stars and have water ice absorption features that are slightly more apparent, although our spectra are contaminated by imperfect telluric correction (see Fig. 1*c*). The average visual extinction to these two sources is ~ 17 magnitudes. These stars also exhibit significant emission of hydrogen Br γ and Br α, indicating active accretion of material from a circumstellar disk. The hydrogen emission line ratios are close to 1. The 4 μm flux is closer to the 2 μm flux than in the Group 1 or 2 sources. This long wavelength IR excess indicates the presence of a circumstellar disk heated by accretion flux from the young star.

Group 3 Members: GSS 39, GY 21.

Group 4: Six young stars have spectra with deep and pronounced water ice absorption features and smooth, rounded profiles (see Fig. 1*d*). These stars have a wide range of visual extinctions (all greater than 15 magnitudes), with an average value of $\sim$ 28 magnitudes. All of these sources show significant Br γ and Br α emission, with F (Br γ)/F (Br α) < 1. In all of the cases, the 4 μm flux exceeds the 2 μm flux, also suggesting strong accretion activity from the stars' circumstellar disks. In two cases, the CO $\Delta v = 2$ bandhead absorption is undetectable because of significant 2 μm infrared excess.
Group 4 Members: WL 7, VSSG 27, VSSG 17, YLW 10b, WL 2, YLW 5a.

4. Interpretation

The young stars embedded within the Ophiuchus dark clouds that do not appear to be actively accreting material from a circumstellar disk (Group 2 stars) have characteristic spectra which appear similar to the background K & M giants. A unique characteristic observed only in the spectra of these stars is the broad, shallow absorption at 2.1 μm. Such a feature is observed in the reflectance spectra of icy solar system bodies (Bauer et al. 2002).

Stars that show evidence for active accretion of material from a circumstellar disk have more pronounced water ice absorption and, on average, have significantly higher visual extinctions (Group 3 and 4 stars). The enhanced extinction likely arises from the circumstellar disk and envelope of the young star. At the present time, it is unclear if the difference in water ice profile shapes is attributable to real temperature variations in the dust grains, or to the significant IR accretion excess which alters the continuum shape at wavelengths longward of 3 μm.

Acknowledgements

We are grateful to the staff at the IRTF for making the observations for this project possible. We also thank M. Cushing and J. Rayner for their assistance using the SpeXTool data reduction software. This research was supported in part by the Gemini Observatory, which is operated by the Association of Universities for Research in Astronomy, Inc., on behalf of the international Gemini partnership of Argentina, Australia, Brazil, Canada, Chile, the United Kingdom, and the United States of America.

References

Bauer, J.M. et al. 2002, Icarus, 158, 178
Elias, J.H. 1978, ApJ 224, 453
Smith, R.G., Sellgren, K., & Tokunaga, A.T. 1988, 334, 209
Smith, R.G., Sellgren, K., & Tokunaga, A.T. 1989, 344, 413
Smith, R.G., Sellgren, K., & Brook, T.Y. 1993, MNRAS, 263, 749
Teixeira, T.C., & Emerson, J.P. 1999, A&A, 351, 292

WATER EMISSION IN OUTFLOWS: A COMPARISON BETWEEN *ISO* AND *SWAS*

Milena Benedettini[1], Serena Viti[2], Teresa Giannini[3], Brunella Nisini[3], Paul F. Goldsmith[4], and Paolo Saraceno[1]

[1]CNR-IFSI, Area di Ricerca di Tor Vergata, via del Fosso del Cavaliere 100, I-00133, Roma, Italy
[2]Department of Physics and Astronomy, UCL, Gower St., London, WC1E 6BT, UK
[3]INAF-Osservatorio Astronomico di Roma, via Frascati 33, I-00040, Monte Porzio, Italy
[4]National Astronomy and Ionosphere Center, Department of Astronomy, Cornell University, Ithaca, NY 14853-6801

ABSTRACT

We compare the *SWAS* and *ISO* observations of water vapor emission in a sample of outflows. By taking into consideration the different methods used to derive the water abundance, we find that there is satisfactory agreement between *ISO* and *SWAS* results. Such an agreement is reached by assuming that most of the detected water comes from a single region, smaller than both the *ISO* and *SWAS* beam apertures, and considering temperatures higher ($T > 100$ K) and density lower ($n[H_2] < 2 \times 10^6$ cm^{-3}) than previously adopted. The ratio between the integrated flux of the o-H_2O 2_{12}–1_{01} line at 179.5 μm observed by *ISO* and the fundamental o-H_2O 1_{10}–1_{01} line at 538.3 μm observed by *SWAS* is always smaller (never by more than a factor of 7) than the expected ratio. We suggest that the fundamental o-H_2O line has a contribution from a cooler gas component to which *SWAS* is sensitive but which is not traced by *ISO*.

Key words: ISM: jets and outflows – ISM: abundances

1. INTRODUCTION

The direct detection of water vapor has become possible in the recent past thanks to the advent of two satellites: the Infrared Space Observatory (*ISO*) and the Submillimeter Wave Satellite (*SWAS*). The Long Wavelength Spectrometer (LWS) on board *ISO* has detected a variety of transitions of water vapor in the range 45-197 μm but without spectrally resolving the lines ($\Delta v =$ 1500 km s^{-1}). *SWAS*, on the other hand, has conducted pointed observations in the fundamental line of the o-H_2O at 538.3 μm, as well as in the low–lying transitions of four other molecules. The high spectral resolution of *SWAS* (Δv = 0.6 km s^{-1}) allows measurement of line profiles sufficiently well to distinguish between the emission from outflows and that from more quiescent regions. The *SWAS* beam is quite large (3.3′× 4.5′) and typically encompasses the entire outflow; the *ISO*-LWS beam is smaller ($\sim$ 1.3′), and for the extended outflows, different pointings were required to cover the entire outflow region.

Both *ISO* and *SWAS* observations of warm gas in outflow regions show an enhancement of the water abundance with respect to quiescent regions. From the *ISO* data a fractional water abundance with respect to H_2 of 10^{-5}–10^{-4} has been derived in outflows originating from low-mass protostars (Giannini et al. 2001 and references therein) and $\sim 5 \times 10^{-4}$ for the high-mass star forming region Orion BN/KL (Harwit et al. 1998). A high value (X[o-H_2O]$=3.5 \times 10^{-4}$) toward Orion BN/KL has also been found by *SWAS* (Melnick et al. 2000), while for low-mass outflows *SWAS* observations indicate an abundance $\lesssim 10^{-6}$ (Neufeld et al. 2000, hereafter NSA). The latter value, although it represents a mean value averaged over the large *SWAS* beam, is lower than what is expected in outflow regions, where the presence of shocks should enhance the fractional water abundance to values of more than 10^{-4} (Kaufman & Neufeld 1996).

In this poster we examine a sample of outflows observed by both the *ISO* and *SWAS* satellites in order to compare their results in more detail.

2. *ISO* AND *SWAS* WATER ABUNDANCE DERIVATION

ISO-LWS has detected a large number of water emission lines in several outflows, always in conjunction with high$-J$ ($J_{up} > 14$) CO rotational lines. Given the relatively large LWS beam ($\sim$75″) we cannot be sure where these lines originate. The LWS molecular lines observed in the on-source pointing have been widely interpreted by a non-dissociative, low velocity C-shock (Nisini et al. 1999; Molinari et al. 2000; Giannini et al. 2001) associated with the base of the outflow, although some of the emission could be also produced in the inner warm part of the circumstellar envelope for the most luminous sources (Ceccarelli et al. 1999). The molecular lines detected by LWS are usually fitted by means of a radiative transfer model and statistical equilibrium equations in the large

Chemistry as a Diagnostic of Star Formation (2003). *Edited by Charles L. Curry and Michel Fich.* Published in 2003 by NRC Research Press, Ottawa ON, K1A 0R6, Canada. ISBN 0-660-19089-3.

velocity gradient (LVG) approximation. This widely used method to derive the water abundance assumes that CO and H_2O come from the same component of the gas. The density and temperature derived from the high$-J$ CO lines are then used to fit the water lines and to derive the fractional water abundance.

SWAS observations and analyses of three outflows are reported by NSA. They assume that the water emission comes from the large-scale molecular outflow and derive the X(o-H_2O) through the following equation:

$$\frac{n(\mathrm{o-H_2O})}{n(\mathrm{H_2})} = 2.0 \times 10^{-5} \frac{d^2}{n_5 M q_{-11}} \int T^*_{\mathrm{A}} dv \ , \qquad (1)$$

where d is the distance in kpc, n_5 is the H_2 density in units of 10^5 cm^{-3}, M is the outflow mass in solar masses, and q_{-11} is the rate coefficient of the collisional excitation with H_2 for the transition 1_{10}-1_{01} in units of 10^{-11} cm^3 s^{-1}. NSA used the low$-J$ CO lines, tracing the large scale molecular outflow, to derive the emitting mass and molecules such as CS and $H^{13}CO^+$, tracing the denser gas closer to the central protostar, to derive the temperature and density. The shortcoming of this method, as pointed out by NSA, is that the density and the temperature derived from particular molecules are applied uniformly over the entire outflow mass traced by low$-J$ CO line wings. Equation (1) can be applied in the case of an optically thin transition. The 538 μm line has a high critical density ($\sim 10^9$ cm^{-3}), but if the density of the medium is much smaller than the effective critical density, as it is in the outflow regions, the transition may be optically thick but effectively thin.

3. Analysis & Results

We consider four outflows observed by both satellites. The *SWAS* spectra were published by NSA and Benedettini et al. (2002), the LWS spectra by Nisini et al. (2000), Giannini et al. (2001) and Saraceno et al. (2001). The *SWAS* lines exhibit a large line width (FWHM$\sim 15-25$ km s^{-1}) indicating that the o-H_2O fundamental transition in outflows is emitted from a shocked region; this is likely to be the same region emitting the high$-J$ CO and H_2O lines detected by *ISO*. We therefore assume in our analysis that the water observed by *SWAS* comes from the same gas component observed by *ISO*. To verify this hypothesis we compare the fractional water abundance derived by *ISO* with the one calculated from the line measured by *SWAS*. We use equation (1) and assume the density, temperature and mass derived from the high$-J$ CO lines observed by *ISO* and the excitation rate coefficients of Green et al. (1993). We also consider

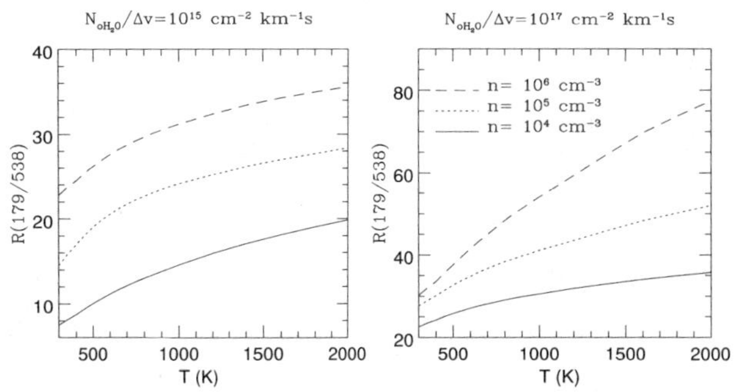

Figure 1. Theoretical ratio of the intensity of the 179.5 μm and 538.3 μm lines versus temperature for different densities (10^4 cm^{-3} [*solid curves*], 10^5 cm^{-3} [*dotted*], and 10^6 cm^{-3} [*dashed*] and N(o-H_2O)/Δv (10^{15} cm^{-2}km^{-1}s [*left*] and 10^{17} cm^{-2}km^{-1}s [*right*]).

the ratio $R(179/538)$ of the intensity of the *SWAS* o-H_2O 1_{10}-1_{01} line at 538.3 μm to the *ISO* o-H_2O 2_{12}-1_{01} line at 179.5 μm. Both are low–lying lines and they are the two channels through which the excited molecule can decay to the ground level. We compare the observed ratio with that calculated using the LVG model (Nisini et al. 1999) and assuming the same physical parameters of the gas derived by *ISO* and the same filling factor for the two lines. In Figure 1 we show the theoretical ratio for the range of the physical parameters typically derived from *ISO* in outflows. $R(179/538)$ increases with temperature, from $\sim$10 to $\sim$80, indicating that for the warm gas the 179.5 μm line is the preferred channel to decay to the ground state. In Table 1 we summarize the data and the results: the distance of the source (column 2); the density, mass and temperature, averaged along the outflow, derived from the *ISO* CO lines (columns 3-5); the observed and predicted $R(179/538)$ (columns 6 and 7, respectively); the fractional abundance of o-H_2O derived from the *SWAS* transition (column 9); and the H_2O (ortho+para) fractional abundance derived from *ISO* (column 10). For IC 1396 N we report the values of the two gas components detected by *ISO*.

4. Discussion

The X(o-H_2O) we derive from the *SWAS* data (see Table 1) is higher than previously found by NSA ($\lesssim 10^{-6}$). This is because while we assume that the water is emitted from the same gas emitting the high$-J$ CO lines, NSA assume that the fundamental o-H_2O line comes from the large-scale molecular outflow traced by the low$-J$ CO lines and attribute to this region density and temperature derived

Table 1. Values for the physical quantities of the observed outflows

Source	d	$n(H_2)$	M	T	$R(179.5/538.3)$		$X(\text{o-}H_2O)$	$X(H_2O)$
	kpc	$10^5 cm^{-3}$	$M_\odot$	K	obs.	model	*SWAS*	*ISO*-LWS
L1448-mm	0.35	0.4	0.015	900	10	25	1×10^{-5}	2×10^{-4}
NGC 1333-IRAS 4	0.35	0.8	0.013	1300	4	27	2×10^{-5}	1×10^{-4}
L1157-mm	0.44	3	0.004	600	4	6	5×10^{-5}	5×10^{-5}
IC1396N	0.75	20	0.003	1400	10	37	1×10^{-5}	1×10^{-4}
IC1396N	0.75	20	0.030	300	10	27	3×10^{-6}	3×10^{-5}

using other species tracing denser gas (CS, e.g.). However, these values cannot be representative of the mean density and temperature of the overall flow. On the contrary, the high$-J$ CO lines detected by *ISO*-LWS along the outflow are a better tracer of the shocked gas.

The $R(179/538)$ derived from the LVG model with the physical conditions deduced from the high excitation *ISO* lines is always higher by a factor of 1.5–7 than the observed ratio (see Table 1). A higher than predicted intensity of the ground state transition of o-H_2O detected by *SWAS* may indeed be explained if one takes into account that this line can have a contribution from a cooler water component not traced by the transitions available to *ISO*. These transitions have upper states at energies equivalent to $T \geq 100$ K above the ground state, and consequently they can be significantly populated only at relatively high temperatures.

The possible presence of a cooler gas component can also explain why the fractional abundance derived from *SWAS* is slightly lower than that derived from *ISO*. Indeed, if the *SWAS* transition has a contribution in emission from a cooler gas component, this should affect the respective estimated $X(\text{o-}H_2O)$ by a factor similar to the ratio between the observed and theoretical $R(179/538)$ and thus ≤ 7.

5. Conclusions

We have compared the observations of water vapor for four outflows observed by both *ISO*-LWS and *SWAS*. We assume that most of the water detected by *SWAS* is emitted from the same gas observed by *ISO* and, from the *SWAS* data, we derive higher $X(\text{o-}H_2O)$ than previously reported, although the abundances are still slightly lower than those derived from the *ISO* data. Also, the ratio between the integrated flux of the o-H_2O 2_{12}–1_{01} line at 179.5 μm observed by *ISO* and the fundamental o-H_2O 1_{10}–1_{01} line at 538.3 μm observed by *SWAS* is always smaller (never by more than a factor of 7) than the expected ratio calculated by means of a LVG model. Both discrepancies may be reconciled if one takes into account that the fundamental o-H_2O line can have a contribution from a cooler water component not traced by the higher excitation transitions available to *ISO*. We conclude that in general *ISO* and *SWAS* results are consistent with chemical models which predict a significant enhancement of water vapor abundance in outflow regions.

The high spectral and spatial resolution submillimeter data that will be collected by the instruments on board the *Herschel* satellite will undoubtedly shed light on the complex structure of the outflow regions and will help constrain their physical conditions, leading to a more accurate estimate of the water content.

References

Benedettini, M., et al. 2002, A&A, in press
Ceccarelli, C., et al. 1999, A&A, 342, L21
Giannini, T., Nisini, B., & Lorenzetti, D. 2001, ApJ, 555, 40
Green, S., Maluendes, S., & McLean, A.D. 1993, ApJ, 85, 181
Harwit, M., Neufeuld, D.A., Melnick, G.J., & Kaufman, M.J. 1998, ApJ, 497, 105
Kaufman, M.J., & Neufeld, D.A. 1996, ApJ, 456, 611
Melnick, G.J., et al. 2000, ApJ, 539, L87
Molinari, S., et al. 2000, ApJ, 538, 698
Neufeld, D.A., et al. 2000, ApJ, 539, L107 (NSA)
Nisini, B., et al. 1999, A&A, 350, 529
Nisini, B., et al. 2000, A&A, 360, 297
Saraceno, P., et al. 2001, in *The Promise of the Herschel Space Observatory*, ed. G.L. Pilbratt, J. Cernicharo, T. Prusti, & R. Harris (ESA SP-460), 203

THE LOW-J CO AND [C I] $^3P_1 \rightarrow {}^3P_0$ EMISSION OF TRANSLUCENT MOLECULAR CLOUDS: OBSERVATIONS AND PDR MODELS

Frank Bensch[1,2], Jürgen Stutzki[1], Uwe Leuenhagen[1,3], and Rudolf Schieder[1]

[1]I. Physikalisches Institut, Universität zu Köln, Germany
[2]Harvard-Smithsonian Center for Astrophysics, Cambridge, MA, USA
[3]LB Kiel, Martensdamm 6, 24103 Kiel, Germany

ABSTRACT

We present observations of the [C I] $^3P_1 \rightarrow {}^3P_0$ transition of neutral carbon made with the Submillimeter Wave Astronomy Satellite (*SWAS*) toward the high latitude translucent cloud MCLD 123.5+24.9. The [C I] data set is supplemented by ground-based observations of the three lowest rotational transitions of ^{12}CO and ^{13}CO. We find that the [C I] emission is spatially extended, following the gas traced by bright ^{12}CO emission. The [C I] line intensity is relatively weak with a [C I]/^{12}CO J=3→2 and [C I]/^{12}CO J=2→1 line flux density ratio between 0.7-2.0. We compare the observed line ratios to current photo-dominated region (PDR) models, which fail to reproduce the observed line ratios in a few key instances. This raises the question to what extent current PDR models can be applied to model the [C I] emission of clouds irradiated by weak interstellar radiation fields.

Key words: ISM: abundances – astrochemistry – ISM: clouds – ISM: individual (MCLD 123.5+24.9) – radio lines: ISM

1. INTRODUCTION

Translucent molecular clouds are characterized by low visual extinction, typically on the order of a few magnitudes. They are not gravitationally bound and show no signposts of star formation (cf. Heithausen 1999). The chemistry is governed by the interstellar FUV field, and as transition objects with only a fraction of their carbon being converted to CO they possibly represent early stages of molecular cloud formation. Chemical models predict significant amounts of neutral carbon in translucent clouds and detectable [C I] fine-structure emission at submillimeter wavelengths (e.g., van Dishoeck & Black 1988).

Here we present observations of the [C I] $^3P_1 \rightarrow {}^3P_0$ transition made towards the high latitude translucent cloud MCLD 123.5+24.9, located in the Polaris Flare. These observations are compared to [C I] and CO data of other translucent clouds available from the literature. We examine whether the observed emission is consistent with a cloud irradiated by a weak, external FUV field using a spherical photo-dominated region (PDR) code and published PDR models.

2. TRANSLUCENT CLOUD MCLD 123.5+24.9

MCLD 123.5+24.9 is visible as a local maximum in the ^{12}CO J=1→0 map of the Polaris Flare made with the 1.2 m "Mini" telescope at the CfA (Heithausen & Thaddeus 1990). It is one of the best studied translucent clouds with a large observational database constraining cloud parameters. This makes MCLD 123.5+24.9 an ideal test case for a translucent cloud PDR model and allows us to check whether the formation and excitation of neutral carbon is governed by the weak external FUV field.

The visual extinction determined by Zagury et al. (1999) ranges between 0.5-2.0 mag, characteristic of a translucent cloud. The strength of the external FUV is close to the mean interstellar radiation field in the solar neighborhood (Bernard et al. 1999). Following Heithausen (1999), we adopt a distance of 150 pc. This places the cloud well into the Galactic molecular disk, despite its high Galactic latitude of $l \sim 25^o$. The total cloud mass is $M_{\rm cloud} \sim 18\,M_\odot$ and the average cloud volume density is $\langle n \rangle \sim 500\,{\rm cm}^{-3}$, determined from the observations of CO, ^{13}CO and C^{18}O rotational transitions (Grossmann et al. 1990). Typical CO excitation temperatures range between 7-11 K (Bensch et al. 2002; Falgarone et al. 1998).

3. OBSERVATIONS

The [C I] $^3P_1 \rightarrow {}^3P_0$ emission was mapped with the Submillimeter Wave Astronomy Satellite (see Melnick et al. 2000 for a description of the *SWAS* satellite). Complementary observations of low-J ^{12}CO and ^{13}CO transitions were made with the KOSMA 3 m and the FCRAO

Chemistry as a Diagnostic of Star Formation (2003). *Edited by Charles L. Curry and Michel Fich.* Published in 2003 by NRC Research Press, Ottawa ON, K1A 0R6, Canada. ISBN 0-660-19089-3.

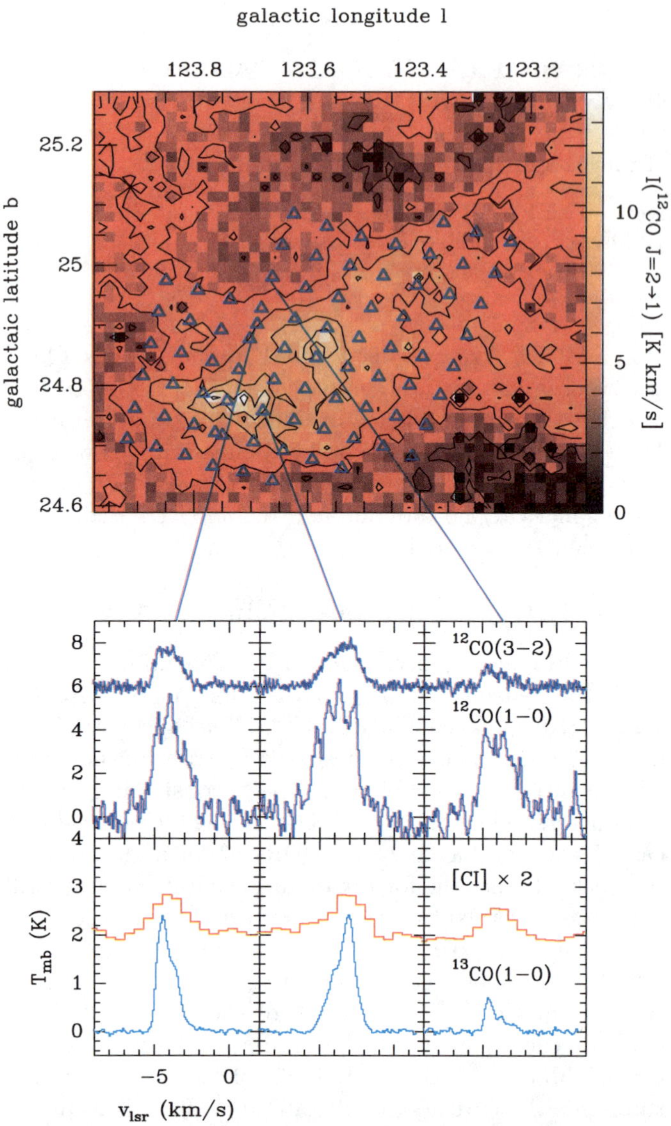

Figure 1. Upper panel: ^{12}CO J=2→1, velocity-integrated intensity (angular resolution: 2.2′). The positions observed with *SWAS* are marked by triangles (*filled, empty*). *Lower panel:* ^{12}CO, ^{13}CO and [C I] line profiles at the three indicated positions. All spectra have been smoothed to the angular resolution of the *SWAS* observations (4.3′).

14 m telescope. Figure 1 shows the spectra for three representative positions, together with the KOSMA ^{12}CO J=2→1 map.

The [C I] line is relatively weak, with typical line flux densities between 0.7 and 1.7 $\times 10^{-7}$erg s^{-1} cm^{-2} sr^{-1}. The [C I] to CO flux density ratios for the different CO transitions are summarized in Table 1. They are within the range of ratios observed towards other diffuse and translucent clouds, albeit at the low end of that range (see Bensch et al. 2002, and references therein). Similar line ratios are found for translucent clouds observed with *SWAS* toward the Galactic midplane (Bensch et al., in preparation) and for a number of galaxies (Gerin & Phillips 2000).

Table 1. [C I] to CO line flux density ratios for MCLD 123.5+24.9 and published translucent cloud observations.

Ratio	Average ± rms	Literature
R(C I/^{12}CO 1−0)	7.00±2.26	7.8-27
R(C I/^{12}CO 2−1)	1.37±0.41	1.0-5.4
R(C I/^{12}CO 3−2)	1.06±0.37	1.0-2.3
R(C I/^{13}CO 1−0)	42.7±22.3	70-620
R(C I/^{13}CO 2−1)	10.5±3.9	7.8-14.5

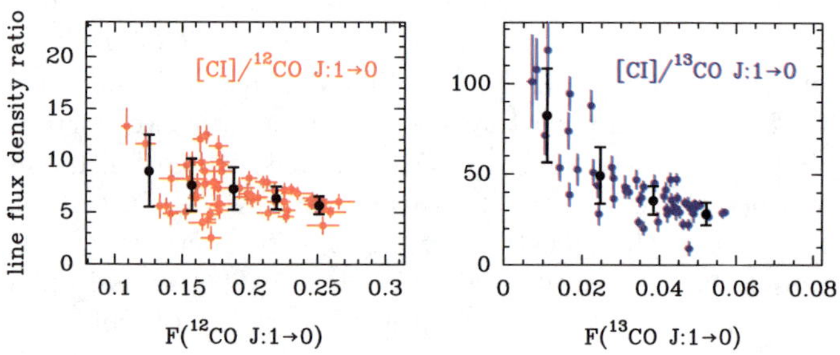

Figure 2. [C I] to CO line flux density ratio as a function of the CO line flux density F for the ^{12}CO and ^{13}CO J=1→0 transitions. F is given in units of 10^{-7} erg s^{-1} cm^{-2} sr^{-1}. The small error bars give the ratio determined for the individual positions, while the large black dots with error bars give the average and rms of the ratios in equally spaced bins.

4. PDR MODELLING

In order to model the observed line ratios we employ the PDR code for spherically symmetric geometry by Störzer, Stutzki & Sternberg (1996). The spherical clumps have a power-law density profile, $n(r) \propto r^{-1.5}$ for $r_c \leq r \leq 0.2\,r_c$ and a constant density for $r \leq 0.2\,r_c$, where r_c is the clump radius. The chemical network includes ions, neutral atoms and molecules containing H, He, C, ^{13}C, O and S. We use the UMIST 95 set of reaction rates, supplemented by the relevant fractionation reactions of ^{12}CO and ^{13}CO and taking into account the isotope-selective photo-dissociation process for both CO isotopomers. The external FUV field is assumed to be the mean local ISRF

($\chi = \chi_0$; Draine 1978). We ran a grid of 8×8 PDR models with the clump mass M_c between $10^{-5}\,M_\odot$ and $100\,M_\odot$, and the average clump density $\langle n_{\rm cl}\rangle$ between $600\,{\rm cm}^{-3}$ and $2 \times 10^6\,{\rm cm}^{-3}$. Figure 3 shows the clump-averaged line flux density ratios predicted by the PDR model as a function of $M_{\rm cl}$ and $\langle n_{\rm cl}\rangle$.

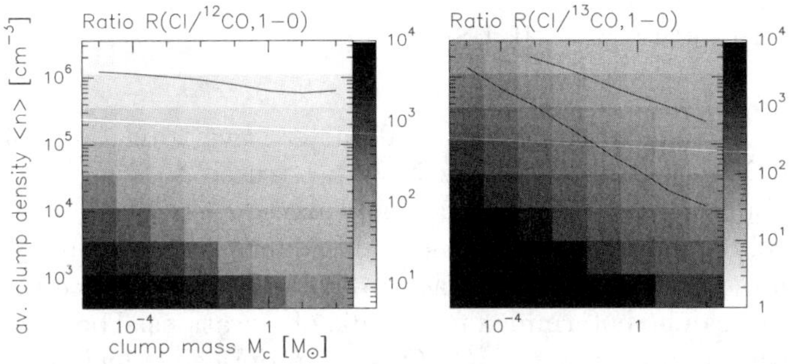

Figure 3. Results from the PDR model for an external FUV field with $\chi = \chi_0$. The clump-averaged [C I] to CO line flux density ratio (coded in grey-scale) is shown for the ^{12}CO and ^{13}CO, J=1→0 transitions as a function of the total clump mass and the average clump density. The lines bracket the $\pm 1\,\sigma$ range of the average line ratio in MCLD 123.5+24.9. (only the lower limit is shown for [C I]/^{12}CO J=1→0). Similar results are obtained for the corresponding ratios of the other CO rotational transitions (Bensch et al. 2002).

The comparison of the PDR model results and the observed line ratios for MCLD 123.5+24.9 can be summarized as follows:

1. Local gas densities close to the cloud average value of $500\,{\rm cm}^{-2}$ are inconsistent with the observations for translucent clouds such as MCLD 123.5+24.9. For such densities CO is expected to be photo-dissociated ($N_{\rm C^+}, N_{\rm C} \gg N_{\rm CO}$) and the [C I] to ^{12}CO line flux density ratio is 1-2 orders of magnitude larger than the observed ratios.

2. The [C I] to ^{12}CO line ratios imply that the emission is dominated by gas with a high volume density. Even for PDR models with a reduced FUV intensity of $\chi = 0.5\,\chi_0$, high densities exceeding $10^5\,{\rm cm}^{-3}$ are suggested.

High densities are indeed found for one of the cores in MCLD 123.5+24.9 (l = 123.69, b = 24.9; e.g., Heithausen, Bertoldi & Bensch 2002). The angular extent of the core is, however, smaller than the *SWAS* beam ($\lesssim 1'$, corresponding to $l \lesssim 0.04$ pc), whereas the constant line ratio across the map argues for similarly high densities *throughout* the whole cloud. This would require the gas to be confined to small clumps, and it is hard to see how the resulting high density contrast could be maintained. Additional evidence for high densities in the spatially extended part of the cloud is not available.

A detailed comparison of the observed line ratios to published PDR models cannot be easily done because of the few present simulations for low FUV fields. [C I] and low-J CO line flux densities from plane-parallel models are presented by Kaufman et al. (1999), and by Gerin & Phillips (2000) using the model of Le Bourlot et al. (1993). The comparison of the observed line ratios with the plots presented in these publications also argues for densities larger than $10^{4.5}\,{\rm cm}^{-3}$. For some of the transitions and line ratios, the models give even higher densities, exceeding $10^5\,{\rm cm}^{-3}$.

The PDR studies show that they consistently would require uncomfortably high values for the gas volume density in order to reproduce the low [C I] to CO line ratios observed for translucent molecular clouds. This raises the question of whether current PDR models sufficiently describe the relevant physical and chemical processes in order to model the [C I] emission of clouds in weak interstellar radiation fields.

ACKNOWLEDGEMENTS

We acknowledge the support provided by the DLR through grants 50 0090 090 and 50 0099 011. F.B. is supported by the Deutsche Forschungsgemeinschaft. DFG (research stipend Be 2412/1-1 and 2412/1-2).

REFERENCES

Bensch, F., et al. 2002, ApJ, submitted
Bernard, J.P., et al. 1999, A&A, 347, 640
Draine, B.T. 1978, ApJS, 36, 595
Falgarone, E., et al. 1998, A&A, 331, 669
Gerin, M., & Phillips, T.G. 2000, ApJ, 537, 644
Grossmann, V., et al. 1990, A&A, 240, 400
Heithausen, A., & Thaddeus P. 1990, ApJ, 353, L49
Heithausen, A. 1999, A&A, 349, L53
Heithausen, A., Bertoldi, F., & Bensch F. 2002, A&A, 383, 591
Kaufman, M.J., et al. ApJ, 527, 795
Le Bourlot, J., et al. 1993, A&A, 267, 233
Melnick, G., et al. 2000, ApJ, 539, L77
Störzer, H., Stutzki, J., & Sternberg, A. 1996, A&A, 310, 592
van Dishoeck, E.F., & Black, J.H. 1988, ApJ, 334, 771
Zagury, F., Boulanger, F., Banchet, V. 1999, A&A, 352, 645

MEASUREMENT OF RARE CO ISOTOPOMERS TOWARD THE ρ OPHIUCHI MOLECULAR CLOUD

Frank Bensch[1,2], Igor Pak[1,3], Jan G.A. Wouterloot[4,5], Gabriele Klapper[1], and Gisbert Winnewisser[1]

[1]I. Physikalisches Institut, Universität zu Köln, Germany
[2]Harvard-Smithsonian Center for Astrophysics, Cambridge, MA, USA
[3]Institute of Spectroscopy, Russian Academy of Sciences, Troitsk, Russia
[4]RAIUB, Universität Bonn, Germany
[5]Joint Astronomy Centre, University Park, Hilo, HI, USA

Abstract

We present observations of the J=1→0 and J=2→1 transitions for $C^{18}O$, $C^{17}O$ and $^{13}C^{17}O$, and the J=1→0 transition for $^{13}C^{18}O$, made toward core C of the ρ Oph molecular cloud.[1] Using these measurements we derive isotopic abundance ratios of $[^{12}C]/[^{13}C] = 65.0 \pm 6.3(stat) \pm 9.2(syst)$ and $[^{18}O]/[^{17}O] = 4.14 \pm 0.52(stat) \pm 0.59(syst)$. They are consistent with recent results from observations of CO rotational transitions, and absorption measurements of CO,^{13}CO and CH^+,$^{13}CH^+$ at visible/UV wavelengths. Our results support the idea that these isotope ratios are smaller in the local interstellar medium than the terrestrial values of 89 and 5.5.

Key words: ISM: abundances – ISM: evolution – ISM: individual (ρ Ophiuchi – ISM: molecules – nuclear reactions, nucleosynthesis, abundances

1. Introduction

The chemical evolution of the ISM is a function of stellar nucleosynthesis, the history of star formation, and the mixing of the gas in the interstellar medium. Therefore, measurements of the key isotopic ratios such as $^{18}O/^{16}O$, $^{17}O/^{16}O$ and $^{13}C/^{12}C$ provide sensitive tests of Galactic chemical evolution models. The $^{12}C/^{13}C$ ratio gives, in principle, a direct measure of primary to secondary nuclear processing ratio: ^{12}C is expected to be produced in early Galactic evolution of metal poor stars, while subsequent nuclear processing of the material in stars by the CNO cycle or non-equilibrium H burning increases the $^{13}C/^{12}C$ ratio. Similarly, repeated processing of material in stars by the CNO cycle effectively reduces the $^{17}O/^{16}O$ ratio (see Wilson & Matteucci 1992; Wilson & Rood 1994 and references therein).

[1] This paper is based on observations collected at the European Southern Observatory, Chile (ESO 66.C-0143).

Among other tracers, millimeter-wave transitions of CO and its isotopomers have been used to study the abundance ratio of carbon and oxygen isotopes. Optical depth effects of the most abundant CO isotopomers, however, limit the accuracy to which the abundance ratios can be determined from radio observations. The mm-wave emission from $^{12}C^{16}O$ is often optically thick, and optical depth effects have been noted to play a role for $^{13}C^{16}O$, and even for $^{12}C^{18}O$ toward regions with high column densities. Observations of the more rare $C^{17}O$ and $^{13}C^{17}O$ isotopomers would allow a more reliable measurement of the isotope ratios; however, $^{13}C^{17}O$ has not been detected in the ISM.

2. Observations

Using the 15 m SEST telescope, we conducted a survey of the J=2→1 and J=1→0 emission from rare CO isotopomers towards the centre position of core C in the ρ Oph molecular cloud in order to derive carbon and oxygen isotopic ratios.

The choice of ρ Oph C as a target is motivated by the relatively large CO column density ($N_{CO} \sim 7 \times 10^{19}\,cm^{-2}$) and small linewidths ($\Delta v[C^{18}O] \lesssim 1.5\,km\,s^{-1}$), derived from the $C^{18}O$ spectral lines. This allows us to detect the emission of even the doubly rare CO isotopomers, and the (partial) resolution of the hyperfine (hf) splitting for $C^{17}O$ and $^{13}C^{17}O$. The $C^{18}O$ J=1→0 map of core C made by Wouterloot et al. (2003, in prep.) shows a relatively broad emission maximum. Targeting core C therefore minimizes a possible systematic error of the detected emission and thus the derived abundance ratios due to small (residual) pointing errors for observations made on different days.

Simultaneous observations were made for both lower rotational transitions of all rare CO isotopomers, except for $^{13}C^{18}O$ J=2→1, which lies outside the tuning range of the receiver. We used the 115/230 GHz receiver as the frontend, and the high resolution acousto-optical spectrometer (HRS) with a channel spacing of 43 kHz as the

Chemistry as a Diagnostic of Star Formation (2003). *Edited by Charles L. Curry and Michel Fich.* Published in 2003 by NRC Research Press, Ottawa ON, K1A 0R6, Canada. ISBN 0-660-19089-3.

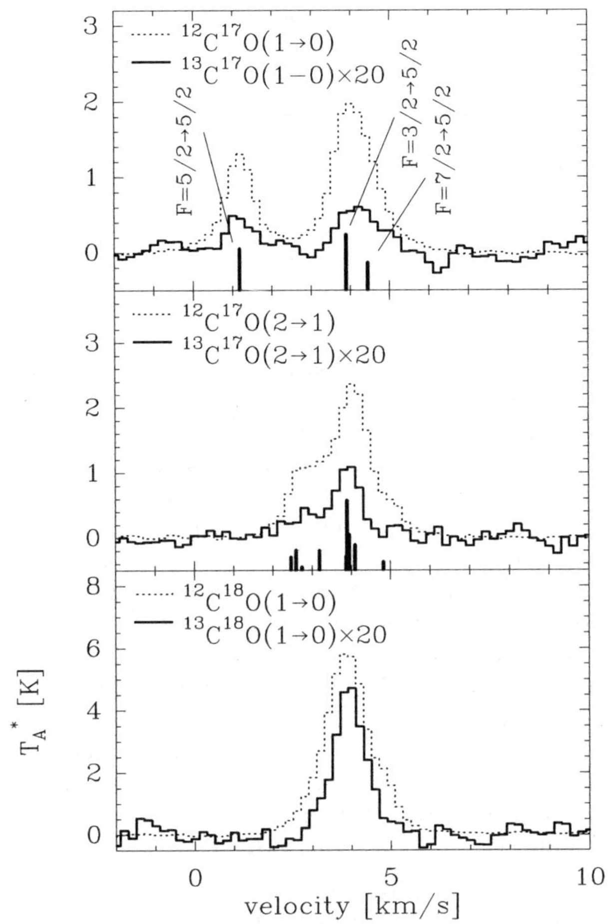

Figure 1. Spectral lines of rare CO isotopomers observed toward ρ Oph core C. For the $C^{17}O$ transitions, the position and relative strength of the hf transitions are indicated.

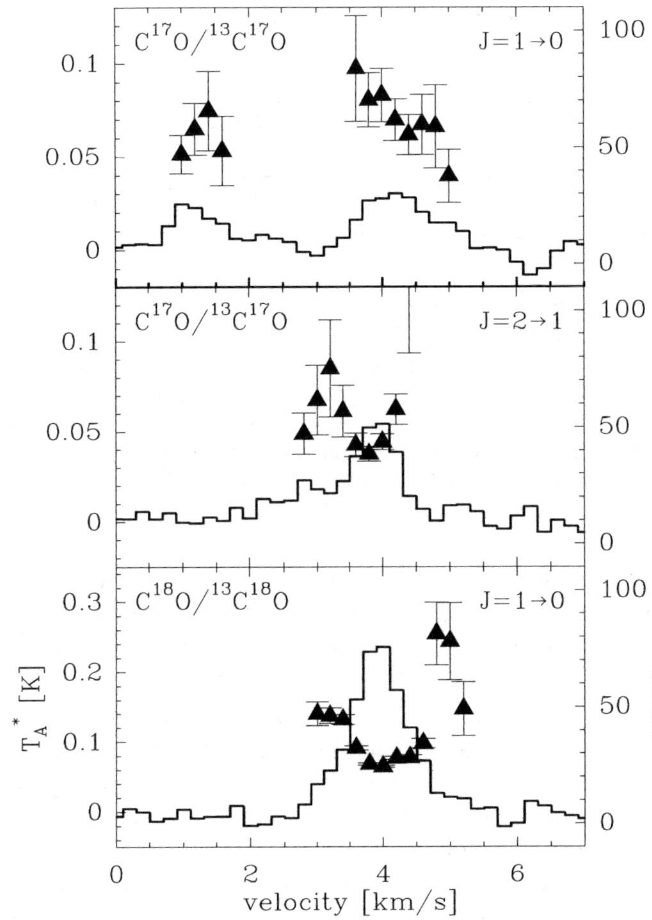

Figure 2. Ratios of the line profiles shown in Figure 1 (*points with error bars; right-hand axis*). For comparison, the line profile of the doubly rare isotopomer is indicated by the histogram (*left-hand axis*).

backend. All observations were re-sampled to a channel spacing of $\Delta v = 0.2\,\mathrm{km\,s^{-1}}$ for further data reduction.

We detected emission from all CO isotopomers and observed transitions, including the doubly rare $^{13}C^{18}O$ and $^{13}C^{17}O$. The line profiles are shown in Figure 1 and the velocity integrated intensities listed in Table 1. A more detailed description of the observations is presented by Bensch et al. (2001).

In addition to the rotational transitions of the CO isotopomers, we detected the emission from other species which fall in the bandpass of the HRS, and the bandpass of the low resolution spectrometer (LRS), which was connected to the receiver in addition to the HRS. We identified a spectral line of C_4H at 104.705 GHz in the HRS spectrum of $^{13}C^{18}O$ $J{=}1\to0$, and of cyclic C_3HD at 107.423 GHz in the LRS spectrum of $^{13}C^{17}O$ $J{=}1\to0$. The line peak intensities (in units of T_A^*) are 96 mK (C_4H) and 25 mK (C_3HD). We also note the detection of emission at 107.376 GHz, 112.355 GHz and 224.705 GHz, which we could not assign to a known transition. The line peak intensities are 19, 52 and 67 mK, respectively.

3. Results

3.1. Column Densities and Line Ratios

Both $C^{18}O$ transitions, $J{=}2\to1$ and $J{=}1\to0$, have a significant optical depth. This is obvious from the systematic variation of the $C^{18}O$ to $^{13}C^{18}O$ line ratio across the line profile shown in the bottom panel of Figure 2. The line ratio is much smaller than the abundance ratio of the corresponding isotopes for velocity channels close to the systemic velocity of the cloud, and increases from the line core to the line wings. Using the escape probability

Table 1. Observations of rare CO isotopomers toward ρ Oph core C

Isotopomer & transition	Frequency[a] (MHz)	$I \pm \Delta I_{\rm stat.} \pm \Delta I_{\rm syst.}$ $I = \int_{v_1}^{v_2} T_A^* \, dv$ (K km s^{-1})
$C^{18}O$ J=1→0	109782.173	8.96±0.03±0.90
$C^{18}O$ J=2→1	219560.357	7.33±0.04±0.73
$^{13}C^{18}O$ J=1→0	104711.404	0.25±0.02±0.03
$C^{17}O$ J=1→0	112358.983	4.26±0.02±0.43
$C^{17}O$ J=2→1	224714.284	4.07±0.03±0.41
$^{13}C^{17}O$ J=1→0	107288.681	0.062±0.006±0.006
$^{13}C^{17}O$ J=2→1	214573.940	0.071±0.006±0.007

[a] For transitions with hf splitting, the frequency of the strongest component is given (see Klapper [2000] for details).

Table 2. Isotopic abundance ratios

Isotope ratio	ρ Oph core C (ratio ± stat. ± syst.)	Terrestrial ratio
$^{12}C/^{13}C$	$65.0 \pm 6.3 \pm 9.2$	89
$^{18}O/^{17}O$	$4.15 \pm 0.52 \pm 0.59$	5.5

model of Stutzki & Winnewisser (1985) we derive a line centre optical depth of 7.5 and 3.5 for the $C^{18}O$ J=2→1, and J=1→0 transitions, respectively, with $n \sim 10^5\,\mathrm{cm}^{-3}$, $T_{\rm ex} = 13.5\,\mathrm{K}$ and $dN/dv = 8.9 \times 10^{18}\,\mathrm{cm}^{-2}$.

Optical depth effects are noticeable even for the $C^{17}O$ J=2→1 transition. A fit to the line profile with the known splitting and relative strength of the hf components gives a peak optical depth of $\tau_{\rm main} = 0.941 \pm 0.036$ for J=2→1 and 0.100 ± 0.006 for J=1→0 .

We use the $C^{17}O$ and $^{13}C^{17}O$ J=1→0 observations in order to derive the column density ratio because of the lower optical depth correction (4%) for the lower rotational transition. The correction for the slightly different rest frequencies (Linke et al. 1977) is $\sim$ 9%. This gives a $C^{17}O$ to $^{13}C^{17}O$ column density ratio of 65.0 ± 6.3. Similar corrections are applied to the integrated intensity ratio of both doubly rare isotopomers, giving $[^{13}C^{18}O]/[^{13}C^{17}O] = 4.15 \pm 0.52$. The contribution of systematic error is smaller than 10% for the integrated line intensities, estimated from $C^{17}O$ and $C^{18}O$ observations made with the SEST telescope at a different epoch (Wouterloot et al. 2003, in prep.). The corresponding systematic error for the column density ratios is listed in Table 3.2. The statistical error is determined from the rms noise in the spectra.

3.2. Isotopic Abundance Ratios

The FUV field in the vicinity of ρ Oph is relatively weak, and the detected emission probes well shielded material. Therefore, the influence of isotope-selective photo dissociation and chemical fractionation can be neglected, and the column densities can be used to derive the isotopic abundance ratios (Table 3.2).

The $[^{12}C]/[^{13}C]$ ratio is consistent with published results from observations of $C^{18}O$ and $^{13}C^{18}O$ rotational transitions (75 ± 8, Wilson, Langer & Goldsmith 1981; 62 ± 4 and 57, Langer & Penzias 1990; Langer & Penzias 1993), UV absorption measurements of CH^+ and $^{13}CH^+$ (66, Savage & Sembach 1990) and recent ground-based observations of CH^+ and $^{13}CH^+$ (67.6±4.5, Crane, Hegyi & Lambert 1991; 71 ± 3, Stahl & Wilson 1992; 63 ± 8, Hawkins, Craig & Meyer 1993). Our $[^{18}O]/[^{17}O]$ ratio is at the low end of the range published in the literature ($\sim$ 4, Wannier et al. 1976; 2.8 ± 0.2, Wilson, Langer & Goldsmith 1981).

Both isotope ratios for ρ Oph core C are significantly smaller than the terrestrial values. This suggests that the local interstellar medium has undergone some processing since the formation of the Solar system 4.5 Gyr ago.

Acknowledgements

We like to thank Jan Brand and Christian Henkel for discussions, and the SEST staff for their assistance during the observing run at La Silla. This research is supported in part by the Deutsche Forschungsgemeinschaft through the grant SFB-494. F.B. is supported by the DLR through grant 50 0099 011.

References

Bensch, F., et al. 2001, ApJ, 562, L185
Crane, P., Hegyi, D.J., & Lambert, D.L. 1991, ApJ, 378, 181
Hawkins, I., Craig, N., & Meyer, D.M. 1993, ApJ 407, 185
Klapper, G. 2000, Dissertation, Universität zu Köln
Langer, W.D., & Penzias, A.A. 1990, ApJ, 357, 477
Langer, W.D., & Penzias, A.A. 1993, ApJ, 408, 539
Linke, R.A., et al. 1977, ApJ, 214, 50
Savage, B.D., & Sembach, K.R. 1996, ARA&A, 34,279
Stahl, O., & Wilson, T.L. 1992, A&A 254, 327
Stutzki, J., & Winnewisser, G. 1985, A&A 144, 13
Wannier, P.G., et al. 1976, ApJ, 205, L169
Wilson, R.W., Langer, W.D., & Goldsmith, P.F. 1981, ApJ, 243, L47
Wilson, T.L., & Matteucci, F. 1992, A&A Rev., 4, 1
Wilson, T.L., & Rood, R.T. 1994, ARA&A, 32, 191

OBSERVATION AND INTERPRETATION OF A TEMPERATURE-SPECTRAL INDEX ANTICORRELATION

N. Boudet[1], X. Dupac[1], B. Stepnik[2], C. Nayral[1], and C. Mény[1]

[1]CESR, 9 av. du Colonel Roche, BP 4346-31028 Toulouse, Cedex 4, France
[2]IAS, UMR-CNRS, Université Paris Sud, 91405 Orsay, Cedex, France

Abstract

A synthesis of PRONAOS (Serra et al. 2001) observations of the extended submillimeter emission in various sites of the interstellar medium shows an anticorrelation between the temperature of the dust component radiating at thermal equilibrium and the emissivity spectral index of the dust emission. Such an anticorrelation, with spectral index values ranging from 1.0 to 2.5, cannot simply be explained by instrumental artifacts, grain models or a distribution of temperatures along the line of sight. This suggests variations of the intrinsic optical properties of the grains with the temperature, as observed in some laboratory experiments on materials of astrophysical interest. We test here an interpretation based on a distribution of two-level systems in a charge disorder structure, in order to model an amorphous material.

Key words: ISM: clouds, dust, extinction—infrared: ISM: continuum

1. Observations

1.1. The PRONAOS experiment

PRONAOS is a French balloon-borne experiment based on 2m telescope dedicated to submillimeter astronomy (Lamarre et al. 1994). Three flights have been performed (in 1994, 1996, and 1999) observing a few tens of astronomical objects in various sites of the interstellar medium (cirrus and molecular clouds to star-forming regions). PRONAOS uses a 2-meter telescope, pointed with a relative accuracy of 5″. The focal instrument performs simultaneous measurements in four broad filters extending from 180 to 1100 μm. The bolometers, cooled at 0.3 K using ^{3}He refrigerators, have an intrinsic noise equivalent power of about 10^{-15}W Hz$^{-1/2}$, close to the photon noise. The warm optics ensures the beam switching on the sky with an amplitude of 5′, and the in-flight calibration by means of two blackbodies. The accuracy of this calibration is better than 10% in absolute value, and 5% in relative value between bands. The PRONAOS instrument is well suited to detect cold ($T < 30$ K) extended condensations, and to determine both dust temperature and emissivity.

The large grain component (20–100 nm) of the interstellar dust radiates at thermal equilibrium . Its emission is deduced from measurements in the four PRONAOS photometric bands by fitting a modified blackbody emission law (for an optically thin medium):
$I_\nu(\lambda) = CB_\nu(\lambda, T)\lambda^{-\beta}$.
The best fit is determined using a least-squares method, considering 5% uncertainty in the relative values between the PRONAOS photometric bands. Fluxes are given at equivalent wavelengths taking into account the transmission profile and the source spectrum. When possible and useful, PRONAOS-simulated 100 μm emission from *IRAS* has been taken into account, with a 30% uncertainty.

1.2. Synthesis of Observations

Figure 1 presents a first synthesis of the observations. The measurements reveal values of the spectral index ranging from 1.1 up to 2.5, to be compared to the conventionally adopted $\beta = 2$ spectral index value for the dust component radiating in the submillimeter at thermal equilibrium. The $\beta(T)$ plot shows an apparent anticorrelation between the spectral index β and the dust temperature T.

1.3. Discussion of the β–T anticorrelation

The single temperature approximation is certainly not a very good one for warm regions ($T > 50$ K). If we consider a mixture of grains with a standard index value $\beta = 2.0$ and different temperatures along the line of sight, the flatness of the spectrum requires very high cold dust column densities (assuming the Désert et al. [1990] model), incompatible with those deduced from ^{13}CO and $C^{18}O$ measurements (Ristorcelli et al. 1998). While the

Chemistry as a Diagnostic of Star Formation (2003). *Edited by Charles L. Curry and Michel Fich.* Published in 2003 by NRC Research Press, Ottawa ON, K1A 0R6, Canada. ISBN 0-660-19089-3.

fitting procedure itself could induce an amount of inverse correlation between temperature and spectral index values, fit studies on noisy simulated data with no intrinsic $\beta - T$ correlation show that this cannot explain the observed anticorrelation (Dupac et al. 2001). Miyake et al. (1993) proposed that a grain size distribution extending up to 1 cm (instead of the usual 0.1 mm cutoff) can create some spectral index dependence of particle opacities, which could explain observations in circumstellar disks around T Tauri stars. Assuming a correlation between column density and temperature, Goldsmith et al. (1997) suggest that the spectral index-dust opacity anticorrelation induces a trend similar to our anticorrelation. If such an extended size distribution hypothesis has to be considered, however, its existence over the whole PRONAOS instrumental beam is unlikely. We believe that such an anticorrelation cannot result from data processing artifacts, nor simply from a temperature distribution.

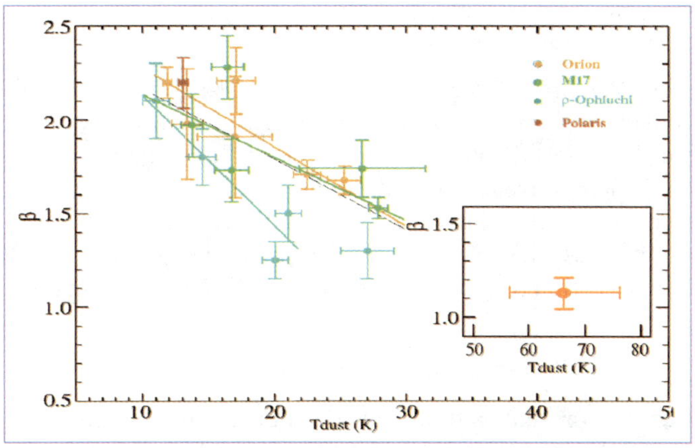

Figure 1. First synthesis over the PRONAOS observations of the spectral index β versus temperature dependence.

2. Grain Model

Recent far-infrared to submillimeter experimental spectroscopic studies on materials of astrophysical interest (Agladze et al. 1996; Mennella et al. 1998; Henning et al. 1997; Fitzgerald et al. 2001) have revealed various behaviors of temperature-dependent absorption for different materials and wavelength ranges. Several mechanisms can induce a temperature dependence of the absorption at long wavelengths, such as:

- two-phonon difference processes in ionic crystals (in which one phonon is created, and one of lower energy destroyed);
- free charge-carrier processes, mainly in semiconductors;
- absorption mechanisms in two-level systems (TLS) (Phillips 1987; Sievers 1998).

Such intrinsic grain properties could be the origin of our observed $\beta - T$ anticorrelation. It will be interesting to model these mechanisms to see if, with a set of plausible model parameters, it is possible to reproduce a $\beta(T)$ dependence similar to that observed by PRONOAS. We propose below a simple physical grain model, based on the TLS picture.

2.1. Optical absorption processes

This study is inspired by the work of Bösch (1978), who carried out a systematic study of the far-infrared properties of a number of different glasses. The overall absorption is composed of a temperature-independent vibrational contribution and a temperature-dependent contribution.

2.1.1. Temperature-independent contribution

In order to account for spectral index values up to 2.5, we follow Schlömann(1964) who calculated the dielectric absorption in a lattice with a disordered charge distribution (charge defects which model amorphous structure), assuming charge neutrality on a scale of a correlation length l. The corresponding absorption coefficient,

$$\alpha_{\text{back}} \approx \frac{2(\epsilon+2)^2 \langle q \rangle^2 \omega^2}{(3c_t)^3 c \sqrt{\epsilon m}} \left[1 - \left(1 + \frac{\omega^2}{\omega_0^2} \right)^{-2} \right],$$

gives two asymptotic behaviors:

$\alpha \propto \omega^2$ for $\nu l / c_t \gg 1$, and

$\alpha \propto \omega^4$ for $\nu l / c_t \ll 1$,

with c_t the sound velocity.

2.1.2. Temperature-dependent contribution

Following Bösch, we describe the temperature-dependent absorption using the TLS tunneling model. The physical origin of the TLS is still unclear: it could be due to motions of atoms or groups of atoms in a double-well potential present in the disorder structure, or motions of localized vibrational energies. In this model, the low temperature dielectric absorption at frequency ω comes from resonant absorption between the TLS with $\hbar\omega$ energy splitting. At higher temperatures, the absorption is driven by two relaxation processes: phonon-assisted tunneling and phonon-assisted hopping over the potential barrier between the two wells. Considering a uniform distribution of TLS, we use the following expressions for the absorption coefficients corresponding to the previous three processes:

– resonant tunneling (for $T \lesssim 10\,\mathrm{K}$):

$\alpha_{\rm res} = 4\pi^2 n\mu^2\omega \tanh(\hbar\omega/2kT)/\sqrt{\epsilon}c$

– phonon-assisted tunneling (for $10 \lesssim T \lesssim 30\,\mathrm{K}$):

$$\alpha_{\rm ph} = A \int_0^\infty \frac{\exp(E/kT)}{kT[1+\exp(E/kT)]^2} \frac{\omega^2\tau_r}{1+\omega^2\tau_r^2} dE$$

– thermally activated transitions (for $T \gtrsim 30\,\mathrm{K}$):

$$\alpha_{\rm therm} = (h/T) \int_0^\infty P(V)\omega^2\tau/(1+\omega^2\tau^2) d\tau$$

with $P(V)$ the distribution function of potential barrier heights.

2.2. Emission spectrum and $\beta(T)$ analysis

Assuming proportionality between α and $Q_{\rm abs}$ in the submillimeter range ($\epsilon'' \ll \epsilon'$), the emission spectrum is calculated from (for spherical particles of radius a):

$I_\nu(\lambda, T) = 4a^2\pi^2 Q_{\rm abs} B_\nu(\lambda, T) \propto \alpha B_\nu(\lambda, T)$.

For each temperature, the corresponding spectrum has been treated using the full PRONAOS data processing. A first study of the β versus T dependence with different sets of model parameters is presented below. The relative differences between theoretical and fitted temperature values are ~5% in the range 5–20 K. The best agreement with observations requires correlation lengths in the range 1–5 nm (typically the size of the smallest grains, going well with the aggregation process for the big grains), and high values of the hopping relaxation strength ($h \approx 10^{-10}\,\mathrm{cm^{-1}s}$) assuming $P(V)$ Bösch parameters (but strongly dependent on this $P(V)$ distribution function of potential barrier heights).

3. Perspectives

A deeper and more systematic study is in progress, as many questions are still unresolved. In TLS-induced absorption, what should be the influence of the TLS density profile and cutoff on the PRONAOS-simulated β versus T dependence? In what range of values should the model parameters governing the relaxation processes lie? Sievers et al. (1998) associated temperature-dependent features at frequencies above $20\,\mathrm{cm^{-1}}$ with excited states of the TLS in some mixed crystals. These excited states of TLS are

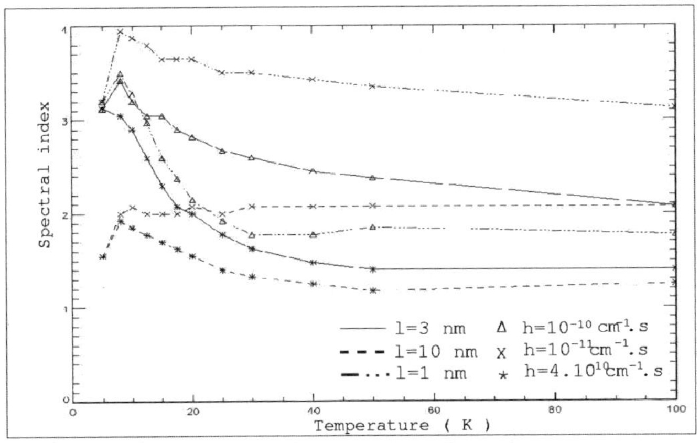

Figure 2. Calculated β versus T dependence with the following parameter values for the fit:
$\epsilon_r = 11.5$,
$P(V) = \exp[-(V - V_m)/V_0^2]$ if $V > 50\,\mathrm{K}$,
$P(V) = 0$ if $V < 50\,\mathrm{K}$,
$V_m = 550\,\mathrm{K}$, $V_0 = 410\,\mathrm{K}$,
$\tau = \tau_0 \exp(V/kT)\ \mathrm{sec}^{-1}$, $\tau_0 = 10^{-13}\ \mathrm{sec}^{-1}$,
$\tau_1 = a \tanh(E/2kT)/E^3\ \mathrm{sec}^{-1}$,
$A = 1.3 \times 10^{-12}\ \mathrm{cm}^{-1}$, $a = 4.2 \times 10^{-56}$, $\mathrm{erg}^3\ \mathrm{sec}$,
$n_e\mu^2 = 1.4 \times 10^{-3}$ in cgs units.

not considered in the Bosch model. The PRONAOS observations show evidence of a strong spectral index versus temperature dependence in the submillimeter emission of the dust. A full understanding of what is observed requires both solid state theoretical work and laboratory experiments.

References

Agladze, N. I., et al. 1996, ApJ, 462, 1026
Bösch, M. A. 1978, Phys. Rev. Lett., 40, 879
Desert, F.-X., et al. 1990, A&A, 237, 215
Dupac, X., et al. 2001, ApJ, 553, 604
Goldsmith, P. F. et al. 1997, ApJ, 491, 615
Fitzgerald, S. A., et al. 2001, J. Phys. Cond. Mat. 13, 2177
Henning, Th. & Mutscke, H. 1997, A&A, 327, 743
Lamarre, J. M., et al. 1994, Infrared Phys. Technol. 35, 277
Mennella, V., et al. 1998, ApJ, 496, 1058
Miyake, K. & Nakagawa, Y., 1993, Icarus, 106, 20
Phillips, W.A., 1987, Phys. Rev. Lett., 50, 1657
Ristorcelli, I., et al. 1998, ApJ, 496, 267
Schlömann, E., 1964, Phys. Rev. 135, A413
Serra, G., et al. 2001, C.R. Acad. Sci. Ser., IV, 1215
Sievers, A. J., et al. 1998, Physica B, 244, 159

A MULTI-SPECIES STUDY OF MOLECULAR DEPLETION IN *MSX* INFRARED-DARK CLOUDS

S. Carey[1], S. Charnley[2], M. Egan[3], R. Redman[4], and P. Feldman[4]

[1]SIRTF Science Center, Caltech, MS 220-6, Pasadena, CA, 91125, USA
[2]NASA/Ames Research Center, MS 245-3, Moffet Field, CA, 94035, USA
[3]Air Force Research Laboratory, MDT/AT, 7100 Defense Pentagon, Washington, DC 20301-7100, USA
[4]HIA, NRC of Canada, 5071 W. Saanich Rd., Victoria, BC, V9E 2E7, Canada

ABSTRACT

We have observed molecular species of high, intermediate and low volatility towards the mid-infrared extinction maxima of three infrared-dark clouds (IRDCs) using the NRAO/UASO 12m telescope and the JCMT. IRDCs are large molecular cores with significant (1–3) magnitudes of extinction from 8–25 microns. They are characterized by large columns of dense ($N[H_2] \sim 10^6\,cm^{-3}$), cold ($T \sim$ 7–20 K) molecular gas. Significant depletions of H_2CO in IRDCs have been previously reported in Carey et al. (1998). Isotopomers of CO, HCO^+, CS, SO, N_2H^+, $HCO_2{}^+$, SO_2, NH_2D and HCN are being used to measure molecular depletion and grain desorption rates in the interiors of IRDCs. Preliminary results indicate that molecules of lower volatility are more depleted than the higher volatility species. Sulphur-bearing molecules with the exception of CS are significantly depleted.

Key words: astrochemistry – ISM: molecules – ISM: clouds

Chemistry as a Diagnostic of Star Formation (2003). *Edited by Charles L. Curry and Michel Fich.* Published in 2003 by NRC Research Press, Ottawa ON, K1A 0R6, Canada. ISBN 0-660-19089-3.

INFRARED SPECTRAL VARIATIONS OF CRYSTALLINE PYROXENES DEPENDING ON CHEMICAL COMPOSITION

H. Chihara[1,2], C. Koike[2], A. Tsuchiyama[1], S. Tachibana[3], and D. Sakamoto[1,2]

[1]Department of Earth and Space Science, Osaka University, Machikaneyama, Toyonaka, Osaka 560-0043, Japan
[2]Laboratory of Physics, Kyoto Pharmaceutical University, Misasagi, Yamashina, Kyoto 607-8414, Japan
[3]Department of Geological Sciences, Arizona State University, Tempe, AZ 85287-1404, U.S.A.

Abstract

Crystalline Mg–Fe pyroxene solid solutions with different Mg/(Mg + Fe) ratio ($MgSiO_3 \sim Mg_{0.5}Fe_{0.5}SiO_3$ and $FeSiO_3$) were synthesized in the laboratory and their absorption properties were investigated in the IR region. We find that the absorption spectra depend on the chemistry and crystal structure of the material: (1) The far-IR features which are prominent in the Mg end members, vanish when the iron concentration is increased. However, even in the iron bearing pyroxenes, the variation of the mid-IR features with Fe concentration is less significant in comparison to that of the far-IR features. (2) Peak positions shifted to longer wavelength as increase in iron concentration; the dependence of the shift is approximately linear in wavenumber (cm^{-1}). (3) Bandwidths of the far-IR bands for the Mg and Fe end members are significantly smaller than those of solid solutions. These results suggest that the far-IR features of pyroxenes are very sensitive to chemical composition and crystal structure. Therefore, far-IR features are a very useful constraint on the the chemical composition and crystal structure of circumstellar pyroxenes.

Key words: ISM: dust, lines and bands – infrared: ISM – crystalline silicate – pyroxenes

1. Introduction

After the *ISO* (Infrared Space Observatory) mission, crystalline solid materials were found to exist in circumstellar environments (Waters et al. 1996; Waelkens et al. 1996; Whittet et al. 1996). Laboratory spectroscopic studies of circumstellar dust candidates are required to understand not only observational data correctly but also the physical and chemical properties of circumstellar environments. The existence of crystalline silicates such as olivine $(Mg, Fe)_2SiO_4$ and Mg-Fe pyroxene $(Mg, Fe)SiO_3$, in circumstellar dust shell of oxygen rich evolved and young stars, are considered particularly important. Therefore, optical properties of these minerals have been studied extensively in these years. The spectral variations of olivine depending on its chemical composition was first investigated by Koike et al. (1993). As for the olivines, optical data of pyroxenes is very important. To our knowledge the only compositional study on pyroxenes in the astronomical literature is that by Jäger et al. (1998) who measured absorption spectra of a synthetic clinoenstatite and some natural orthopyroxenes. However, unfortunately their natural samples were contaminated with weathering alteration products such as talc and metal oxides. Indeed, it is natural that the cosmic dust grain have diversity of chemical composition. However, measuring synthetic samples with simple chemical composition has the advantage of estimating the spectral properties depend only on the major chemical composition. Therefore, we tried to synthesize crystalline pyroxenes in a wider compositional range and finer steps including the Mg and Fe end members ($MgSiO_3 \sim Mg_{0.5}Fe_{0.5}SiO_3$ and $FeSiO_3$), and measured the absorption spectra in the mid- and far-infrared regions.

2. Experiment

The chemical formula of Mg–Fe pyroxene is described as $[Mg_x \cdot Fe_{(1-x)}]SiO_3$, $(0 \leq x \leq 1)$. The Mg end member ($x = 1$) is called enstatite (En), while the Fe end member ($x = 0$) is ferrosilite (Fs). Hereafter, following the conventional description, the chemical composition of $[Mg_x, Fe_{(1-x)}]SiO_3$ is described as En_X, where X is the mole percentage of the enstatite component ($X = 100\,x$).

In this work, pyroxene crystals of En_{100} (orthoenstatite; Oen, and clinoenstatite; Cen), En_{90}, En_{80}, En_{70}, En_{60}, En_{50} and ferrosilite (Fs) were synthesized. The orthoenstatite (Oen) single crystal was synthesized with the flux method, and the clinoenstatite (Cen) was obtained by annealing the orthoenstatite at 1200 °C for 9 hours in the atmosphere. For the synthesis of Mg–Fe pyroxenes, MgO, FeO and SiO_2 powders were mixed in appropriate

Chemistry as a Diagnostic of Star Formation (2003). *Edited by Charles L. Curry and Michel Fich.* Published in 2003 by NRC Research Press, Ottawa ON, K1A 0R6, Canada. ISBN 0-660-19089-3.

Table 1. Properties of synthesized samples

Sample	Nominal composition	Analytical composition	Crystal structure
Oen	$MgSiO_3$	—	ortho
Cen	$MgSiO_3$	—	clino
En_{90}	$Mg_{0.90}Fe_{0.10}SiO_3$	$Mg_{0.91}Fe_{0.08}SiO_3$	clino
En_{80}	$Mg_{0.80}Fe_{0.20}SiO_3$	$Mg_{0.82}Fe_{0.18}SiO_3$	ortho
En_{70}	$Mg_{0.70}Fe_{0.30}SiO_3$	$Mg_{0.72}Fe_{0.28}SiO_3$	ortho
En_{60}	$Mg_{0.60}Fe_{0.40}SiO_3$	$Mg_{0.63}Fe_{0.36}SiO_3$	ortho
En_{50}	$Mg_{0.50}Fe_{0.50}SiO_3$	$Mg_{0.52}Fe_{0.48}SiO_3$	ortho
Fs	$FeSiO_3$	—	ortho

mole ratios, and the mixtures were pressed into pellets with 5 mm radius and 2–3 mm thick. These pellets were hung by platinum wire and then heated in a furnace at 1200 °C for 96 hours at atmospheric pressure. O_2 fugacity was controlled by H_2–CO_2 gas flow. The value was estimated approximately 3×10^{-11} atm. The products were sintered aggregations of polycrystalline pyroxene. In general, ferrosilite can be produced only under high pressure environment. Therefore, orthoferrosilite was synthesized at 4GPa and 1000 °C using a multi-anvil press (Kobelco DIA-6 at Osaka University). Chemical compositions determined by electron micro probe (EPMA) and crystal systems of the synthesized samples are listed in Table 1.

To prepare the absorption measurements, the samples were crushed and ground in an agate mortar for about 1 hour until the grains were approximately less than 1 μm in size. These fine powders were dispersed and embedded in appropriate matrix; KBr for the mid–IR measurements (1600–300 cm^{-1}) or polyethylene for the far–IR measurements (650–50 cm^{-1}). The measurements were performed by Nicolet Nexus 670 FT–IR spectrometer (Kyoto Pharmaceutical University) at room temperature. The measured wavenumber range was 1600–50 cm^{-1}. The resolutions of the measurements were 1.0 cm^{-1} in the mid-IR region (1600–300 cm^{-1}) for all samples. In the far-IR region (300–50 cm^{-1}), the lower resolutions were used (4.0 cm^{-1} for Fs–En_{90}, and 2.0 cm^{-1} for the Mg end members) to avoid interference fringes.

3. Results and Discussion

Mass absorption coefficients, $\kappa(cm^2/g)$, of all samples are plotted in Figure 1. As a whole, the spectral feature of pyroxenes are very complex. The uppermost spectrum is orthoferrosilite (Fs). Below Fs the Mg/(Mg+Fe) ratio increases from $\sim$ 50% in En_{50} to 100% in En_{100} in increments of 10% for subsequent spectra. Crystal structures of En_{100}(Cen) and En_{90} were monoclinic; other samples were orthorhombic. It is noted that, for only En_{90}, some small absorption bands appeared at 12.9 and 16.5 μm which are probably due to forsterite, although the chemical and SEM-image analyses did not show any inhomogeneity and phase separation. Other than En_{90}, however, the spectral variation with chemical composition is almost continuous, and no discontinuous jump was observed.

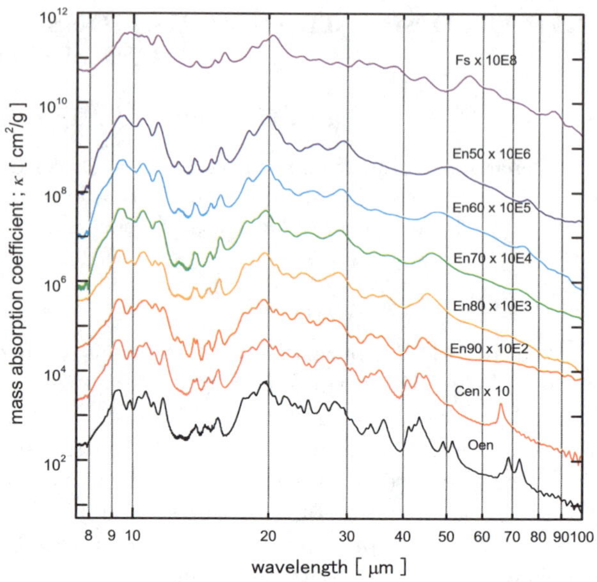

Figure 1. Mass absorption spectra of pyroxenes.

Some absorption bands newly appeared and/or disappeared, particularly near the chemical composition close to the Mg and Fe end members. For the Mg end members, the features were almost consistent with our previous results (Koike et al. 2000; Chihara et al. 2001). The two crystal structures of enstatite (Oen and Cen) show almost the same spectral features in the mid-IR region up to about 40 μm, while in the far-IR region, they can be distinguished easily, because both enstatites have small but sharp and characteristic features in this region, such as two double bands at around 50 and 70 μm for orthoenstatite, and a single band at 66 μm for clinoenstatite. Although, these far-IR characteristic features of the Mg end members are very sharp and prominent, En_{90} no longer exhibits these features clearly. In the spectrum of En_{90}, the 67 μm band of clinoenstatite vanished, and the triple bands at around 43 μm became a double structure. In the spectrum of En_{80}, both of the double structures of

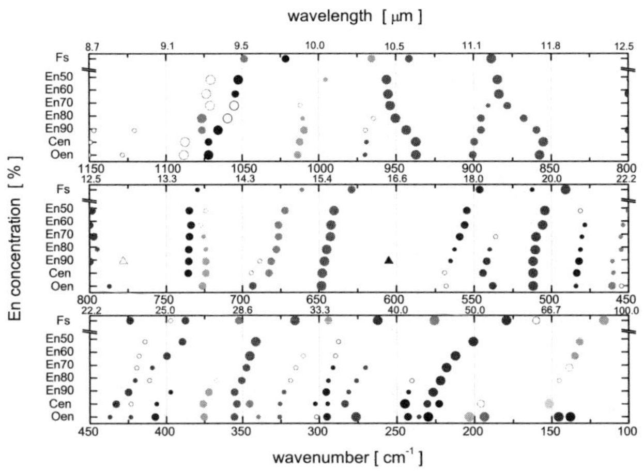

Figure 2. Peak shift as a function of chemical composition. The size of each symbol represents the intensity of the absorption peak. Open symbols represent shoulders.

the band of orthoenstatite at around 50 μm and 70 μm disappeared completely. Instead, a small shoulder began to appear at around 70 μm. This 70 μm shoulder become clearer and formed into a single band from about En_{60}, and shifted with the iron concentration to 86 μm for Fs. In contrast, in the mid-IR region, the variations of absorption features were not so drastic compared to those of the far-IR features.

Figure 2 shows the peak positions of absorption bands as a function of their chemical compositions. The horizontal axis indicates the band position by wavenumber (cm^{-1}) and the vertical axis indicates the enstatite content (%). Most of the absorption bands shifted to longer wavelength systematically as the iron concentration is increasing. Its tendency of variation against wavenumber is approximately linear. Especially the shift of 43 micron features on both enstatites are prominent. Their peak positions shifted from about 43 μm (233 cm^{-1}) on En_{100} to 56 μm (179 cm^{-1}) on Fs.

In addition to the band shift, the variation of bandwidth is observed, particularly in the sequence of the band shift from 44 to 56 micron (Table 2). It seems that the bandwidths of both the Mg and Fe end members are smaller than that of solid solutions. By analogy with olivine (see Jäger et al. 1998), the vibrational modes of SiO_4 tetrahedra and divalent cations can be considered as the origin of the far-IR features. The broadening of the bandwidth can be explained qualitatively by the overlap or unharmonicity of the potentials of mixed phonon modes due to $Mg^{2+} - SiO_4$ and $Fe^{2+} - SiO_4$. In contrast,

Table 2. Variation of FWHM for 44–56 micron bands

Sample	Peak center (μm)	FWHM (μm)
En_{90}	44.1	3.1
En_{80}	45.1	4.6
En_{70}	45.9	5.4
En_{60}	47.1	6.1
En_{50}	49.9	7.8
Fs	55.8	5.6

no significant variations of bandwidth in the mid-IR region are detected. These results suggest that the far-IR features are very sensitive to chemical composition and crystal structure.

Previously, from *ISO*-SWS observations the main constituent of circumstellar pyroxenes was identified as orthoenstatite due to the existence of large 33 μm band similar to the laboratory band reported by Koike & Shibai (1998). However, this band was not detected so strongly by other works. Recently, Molster et al. (2002) pointed out that the identification of orthoenstatite by this band should be reconsidered. And also in this study, we find that it is rather difficult to distinguish between ortho- and clinoenstatite in the *ISO*-SWS spectral range. However, results of this work suggest that the far-IR features are much better indicators of the differences between ortho- and clinoenstatite. Clinoenstatite is preferentially produced when the melt is cooled rapidly via phase transition from protoenstatite. While, if protoenstatite is cooled very slowly or annealed at moderate temperature (below melting point), it transforms to orthoenstatite. Hence, if the crystal structure of circumstellar pyroxenes is determined by observations in the future, we can obtain more information about the conditions of grain formation such as the time scale of annealing/cooling, temperature of crystallization, and so on.

References

Chihara, H., et al., 2001, PASJ 53, 243
Chihara, H., et al., 2002, A&A 391, 267
Jäger, C., et al., 1998, A&A 339, 904
Koike, C., et al., 1993, MNRAS 264, 654
Koike, C. & Shibai, H., 1998, ISAS Report, No. 671
Koike, C., et al., 2000, A&A 363, 1115
Molster, F. J., et al., 2002, A&A 382, 241
Waelkens, C., et al., 1996, A&A 315, L245
Waters, L. B. F. M., et al., 1996, A&A 315, L361
Whittet, D. C. B., et al., 1996, A&A 315, L357

OBSERVATIONS OF THE NGC 1333 IRAS 4A OUTFLOW IN HCN AND H_2CO

Minho Choi[1], Takeshi Kamazaki[2], and Ken'ichi Tatematsu[3]

[1]Institute of Astronomy and Astrophysics, Academia Sinica, P.O. Box 23-141, Taipei 106, Taiwan
[2]Nobeyama Radio Observatory, Minamimaki, Minamisaku, Nagano 384-1305, Japan
[3]National Astronomical Observatory of Japan, 2-21-1 Osawa, Mitaka, Tokyo 181-8588, Japan

Abstract

The molecular outflow driven by the Class 0 object NGC 1333 IRAS 4A is one of the best-collimated protostellar outflows, and the CO outflow map by Blake et al. (1995) shows an interesting directional variability. In this paper, we present interferometric maps of the IRAS 4A outflow observed in the HCN 1-0 and H_2CO 2_{12}-1_{11} lines. These species were chosen as outflow tracers in our study because the line survey by Blake et al. showed that they are highly enhanced in the outflow. The HCN and H_2CO maps show more detailed structure of the outflow than CO maps. The line wing emissions are highly enhanced along the outflow axis, and several knot-like outflow peaks can be seen. The northeastern outflow lobe shows a possible sign of periodic outflow variability.

Key words: ISM: jets and outflows — ISM: molecules — ISM: structure — stars: formation

1. Introduction

NGC 1333 IRAS 4 is a star formation region in the L1450 dark cloud at the distance of 220 pc from the Sun (Černis 1990). IRAS 4A is a Class 0 protobinary system and is driving a well collimated bipolar outflow (Liseau, Sandell, & Knee 1988; Blake et al. 1995; Hodapp & Ladd 1995). The IRAS 4A outflow shows an interesting directional variability. The outflow maps by Blake et al. (1995) show that the position angle of the flow changes from about 45° in the large-scale (~4′) to about 0° near IRAS 4A. Interferometric observations in the CO $J = 2 \rightarrow 1$ line by Girart, Crutcher, & Rao (1999) show that the position angle of the outflow changes by ~30° about 25″ away from IRAS 4A. Variability may be an important ingredient in understanding how molecular outflows are driven by the protostellar jet/wind (Masson & Chernin 1993; also see the review by Cabrit, Raga, & Gueth 1997 and references therein).

2. Observations

The NGC 1333 IRAS 4 region was observed in the HCN, HCO^+, ^{13}CO, and $C^{18}O$ $J = 1 \rightarrow 0$ lines using the BIMA array in the B and C-array configurations (Choi, Panis, & Evans 1999; Choi 2001) and in the H_2CO $J_{K_{-1}K_{+1}}=2_{12} \rightarrow 1_{11}$ line using the NMA in the C and D-array configurations. The systemic velocity of the IRAS 4A core, determined from the $C^{18}O$ line, is $V_{LSR} = 6.7$ km s^{-1}.

HCN and H_2CO were chosen as outflow tracers in our study because the line survey by Blake et al. (1995) showed that they are highly enhanced in the outflow. Our line spectra toward IRAS 4A show that the HCN and H_2CO lines have wing components highly enhanced relative to the core (Fig. 1).

3. Molecular Outflows in the IRAS 4 Region

Maps of the molecular outflows are shown in Figures 2 and 3. Position of the outflow peaks (HCNOP) are numbered and marked in the HCN map (Fig. 2). Several outflows can be identified in these maps. To the northeast of IRAS 4A, there is a flow mostly redshifted (NE outflow; containing HCNOP 1/2/3/4). To the south of IRAS 4A, the outflow morphology is more complicated. Close inspection of the outflow maps suggests that there are two blueshifted outflows driven by IRAS 4A: one flowing to the southwest (SW outflow; containing HCNOP 6/8) and the other flowing to the south (S outflow; containing HCNOP 5). Considering the outflow directions and the length of the outflow lobes, the SW outflow seems to be the counterpart of the NE outflow. Counterpart of the S outflow has not been found. The HCN map also shows a relatively compact bipolar outflow driven by IRAS 4BI.

The apparent 45° change of the outflow direction, reported by Blake et al. (1995), can be explained by the existence of the two blueshifted outflows. With a low angular resolution ($\gtrsim$10″), a blueshifted outflow map near IRAS 4A would be dominated by the S outflow (OP 5), but a large-scale map would show only the SW outflow.

Chemistry as a Diagnostic of Star Formation (2003). *Edited by Charles L. Curry and Michel Fich.* Published in 2003 by NRC Research Press, Ottawa ON, K1A 0R6, Canada. ISBN 0-660-19089-3.

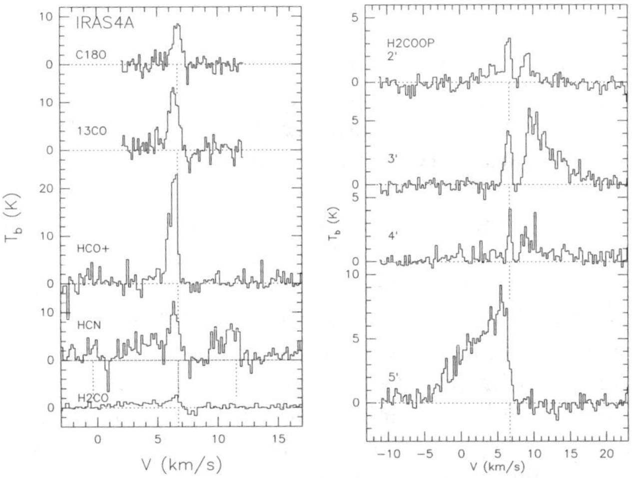

Figure 1. Left panel: Molecular line spectra toward IRAS 4A. *Right panel:* The H_2CO line spectra toward the outflow peaks. The vertical dotted line marks the systemic velocity.

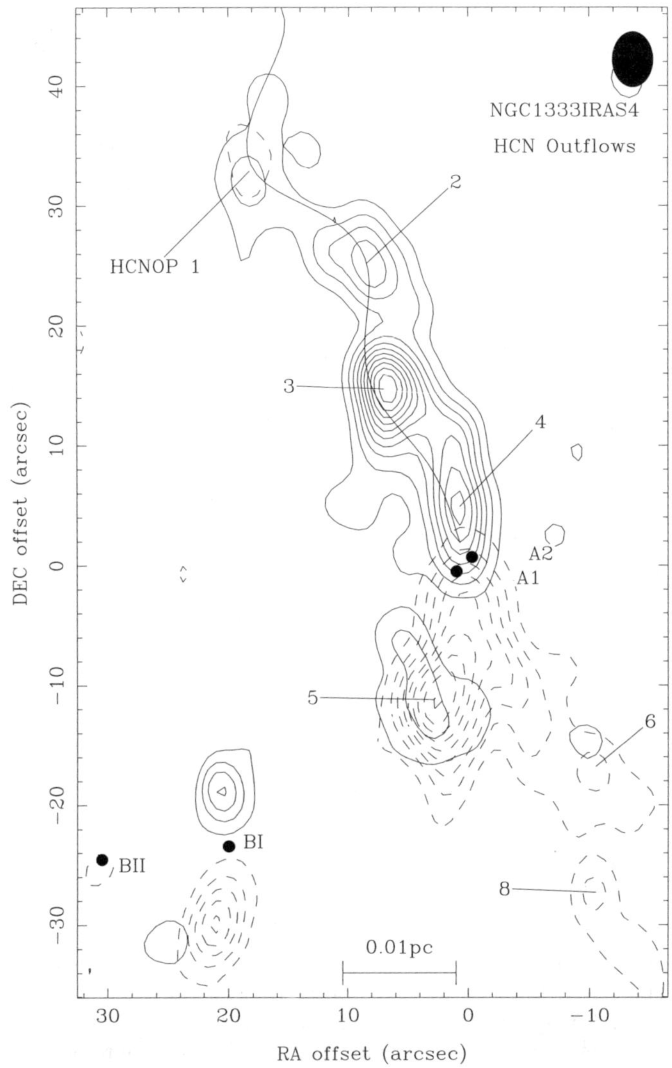

Figure 2. Map of the HCN $J = 1 \rightarrow 0$ line wings (see Choi 2001 for details.) The HCN line was averaged over the $V_{\rm LSR}$ (km s^{-1}) intervals of (–11.7, –1.3) and (2.1, 5.8) for dashed (blue) contours and over (9.2, 10.6) and (14.0, 21.6) for solid (red) contours. The lowest contour level and the contour interval are 48 mJy beam^{-1}. Shown at the top right-hand corner is the synthesized beam. The filled circles mark the millimeter continuum sources (Looney, Mundy, & Welch 2000).

On the other hand, the directional variability of the NE outflow turned out to be more complicated than what was reported by Girart et al. (1999). For the rest of this paper, we will concentrate our discussion on the variability of the NE outflow.

4. Absorption by Foreground Cloud

A characteristic feature of the interferometric maps of this region is the lack of emission in a ~2 km s^{-1} wide velocity interval centered at $V_{\rm LSR} \approx 8$ km s^{-1}. This feature makes unfortunate complications in the analysis of redshifted velocity components in this region. As a result, a significant fraction of the emission from redshifted outflows cannot be seen. Since this absorption feature can be seen over a region much larger than the IRAS 4A core (Fig. 1), it is most likely caused by a foreground cold cloud, not by infalling gas.

5. IRAS 4A NE Outflow

5.1. HCN Map

The NE outflow traced by the HCN line consists of a series of compact knots (Fig. 2). The compact nature of the outflow knots and the high degree of collimation suggest that these knots are intimately connected with the jet/wind driving the molecular outflow. A close examination of the spatial distribution of the knots reveals two interesting facts: a nearly regular spacing between knots and alternating changes of position angle. To show this periodic "wiggle" pattern clearly, a sinusoidal curve is drawn in Figure 2. The wavelength of the wiggle is $22{\pm}1''$. Assuming that the outflow is close to the plane of the sky and that the jet/wind speed is ~100 km s^{-1}, the period of the wiggle is $P_w \approx 220$ yr.

Outflow variability can be caused by many different physical mechanisms: intrinsic variability of the driving source, instability in the driving jet/wind, or influence of

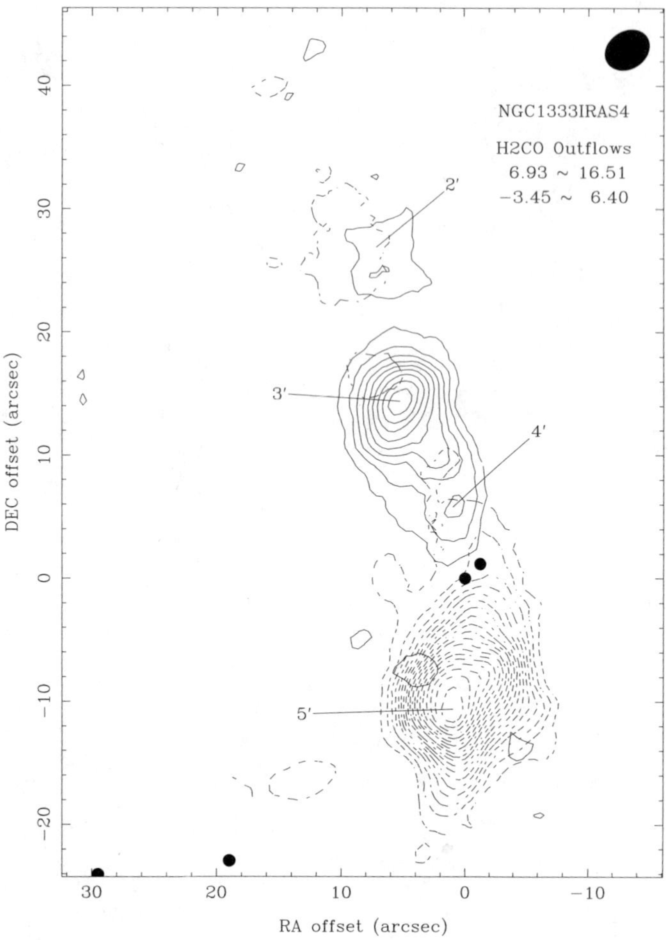

Figure 3. Map of the H_2CO $J_{K_{-1}K_{+1}}=2_{12} \rightarrow 1_{11}$ line wings. The H_2CO line was averaged over the V_{LSR} (km s^{-1}) interval of (−3.4, 6.4) for dashed contours and over (6.9, 16.5) for solid contours. The lowest contour level and the contour interval are 42 mJy $beam^{-1}$.

external environment. The periodic nature of the wiggle seems to rule out external perturbations. Among the previously proposed outflow variability mechanisms, several models can produce periodic variability of the outflow direction/intensity: outflow driven by a precessing disk, outflow driven by a wobbling disk, periodic outbursts of the central source, and Kelvin-Helmholtz instability in the outflow (Papaloizou & Terquem 1995; Stone, Xu, & Hardee 1997; Terquem et al. 1999; Bate et al. 2000). In principle, the models listed above can be tested observationally. Possible methods would be very high angular resolution imaging of the driving source, high resolution imaging of the outflow peaks (or shocked regions), and high resolution study of the outflow kinematics.

5.2. H_2CO Map

The H_2CO emission is dominated by the outflow (Fig. 3), even toward IRAS 4A (Fig. 1). The NE outflow traced by the H_2CO line also shows the periodic variability. The intensity variability is clearly visible, but the directional variability is not as obvious as in the HCN map. The H_2CO outflow peak positions coincide with the corresponding HCN peak positions, within the beam size.

While the HCN line wing is more highly enhanced than the H_2CO line wing, the H_2CO line is a better tracer of the outflow kinematics as it has no complication from hyperfine structure. In interpreting the kinematics, it should be kept in mind that the redshifted outflow is underrepresented relative to the blueshifted outflow because of the absorption by the foreground cloud.

In the H_2CO first moment map (not shown here), the velocity structure is more complicated than what are expected from simple models; the velocity peaks neither coincide with the intensity peaks nor are located between the intensity peaks. Instead, the velocity peaks are located several arcseconds away from the intensity peaks in the direction away from the driving source. More detailed analysis is needed to draw a firm conclusion.

ACKNOWLEDGEMENTS

We acknowledge partial support from NSC grants 89-2112-M-001-061 and 91-2112-M-001-045.

REFERENCES

Bate, M. R., Bonnell, I. A., Clarke, C. J., Lubow, S. H., Ogilvie, G. I., Pringle J. E., & Tout, C. A. 2000, MNRAS, 317, 773

Blake, G. A., Sandell, G., van Dishoeck, E. F., Groesbeck, T. D., Mundy, L. G., & Aspin, C. 1995, ApJ, 441, 689

Cabrit, S., Raga, A., & Gueth, F. 1997, IAU Symp. 182, 163

Černis, K. 1990, Ap&SS, 166, 315

Choi, M. 2001, ApJ, 553, 219

Choi, M., Panis, J.-F., & Evans, N. J., II 1999, ApJS, 122, 519

Girart, J. M., Crutcher, R. M., & Rao, R. 1999, ApJ, 525, L109

Hodapp, K.-W., & Ladd, E. F. 1995, ApJ, 453, 715

Liseau, R., Sandell, G., & Knee, L. B. G. 1988, A&A, 192, 153

Looney, L. W., Mundy, L. G., & Welch, W. J. 2000, ApJ, 529, 477

Masson, C. R., & Chernin, L. M. 1993, ApJ, 414, 230

Papaloizou, J. C. B., & Terquem, C. 1995, MNRAS, 274, 987

Stone, J. M., Xu, J., & Hardee, P. E. 1997, ApJ, 483, 136

Terquem, C., Eislöffel, J., Papaloizou, J. C. B., & Nelson, R. P. 1999, ApJ, 512, L131

WATER IN THE ENVELOPE OF SGR B2

Claudia Comito[1], Peter Schilke[1], Maryvonne Gerin[2], Tom G. Phillips[3], and Jonas Zmuidzinas[3]

[1]Max-Planck-Institut für Radioastronomie, Auf dem Hügel 69, D-53121 Bonn, Germany
[2]Laboratoire de Radioastronomie Millimétrique, Observatoire de Paris and Ecole Normale Supérieure, 24 rue Lhomond, F-75231 Paris, CEDEX 05, France
[3]California Institute of Technology, Downs Laboratory of Physics 320-47, Pasadena, CA 91125, USA

Abstract

We report the detection, with the Caltech Submillimeter Observatory[1], of the 894 GHz HDO($1_{1,1} - 0_{0,0}$) transition, observed in absorption against the background continuum emission of the Sgr B2 main cores (M and N). Radiative transfer modeling of this feature, together with the published data set of mm and submm HDO and $H_2^{18}O$ transitions, allows us to solve the controversy as to whether the $H_2^{18}O$ ground-state absorption features detected with *SWAS*, *ISO* and KAO are produced by the large column densities of gas located in the Sgr B2 warm envelope, or if they are instead a product of the hot, diffuse, thin gas layer lying in the foreground of the Sgr B2 complex.

Key words: ISM: molecules – ISM: structure – ISM: clouds

1. H_2O in Sgr B2

The Sgr B2 cloud is one of the most massive star-forming regions in our Galaxy. It has many unique characteristics, among them an exceptional chemistry. Several species (FeO, Walmsley et al. 2002; NH_2, van Dishoeck et al. 1993; HF, Neufeld et al. 1997; C_3, Giesen et al. 2001) have, in spite of searches elsewhere, been detected only toward this source. One possible explanation for this enigmatic chemistry is the existence of a layer of hot gas, which is thought to be produced by a shock. Since this layer has the same velocity as the ambient gas in Sgr B2, it has been proven difficult to assess its importance for the chemistry of many species, with a few exceptions. In particular its importance for the water chemistry has been a matter of recent debate (Ceccarelli et al. 2002, hereafter C02; Neufeld, Bergin & Goldsmith, in preparation). While its temperature and density are well known (cf. Hüttemeister et al. 1995), its column density and width have remained elusive. In this paper we are able to determine its importance for water chemistry and give constraints for column density and spatial width, by modeling the HDO and $H_2^{18}O$ emission and absorption.

Water is known to be a fundamental ingredient of the interstellar medium. An ubiquitous tracer of shock-heated gas, it dramatically influences the chemistry in shocked regions (cf. Neufeld & Melnick 1987; Bergin, Melnick & Neufeld 1998). In general, it acts as a major coolant in star-forming clouds (Ceccarelli, Hollenbach & Tielens, 1996). Although direct observation of non-masing water lines from ground-based observatories is made extremely difficult by atmospheric absorption, H_2O abundances can be estimated via observations of isotopomers, such as $H_2^{18}O$ (e.g., Phillips et al. 1978; Jacq et al. 1988, 1990; Gensheimer et al. 1996), and, in recent years, by the availability of satellites such as *ISO* (e.g., Cernicharo et al. 1997) and notably *SWAS* (cf. Neufeld et al. 2000, hereafter N00), but also by airborne observations of $H_2^{18}O$ (e.g., Zmuidzinas et al. 1995a, hereafter Z95a). However, the deuterated counterpart of water, HDO, presents many features observable from ground in the cm, mm and submm wavelength atmospheric windows (cf. Henkel et al. 1987; Jacq et al. 1990, 1999; Pardo et al. 2001), and is also used as a tracer of water abundance under the assumption that both the deuterated and non-deuterated species be spatially coexistent, and that their abundance ratio remain constant throughout the studied region.

We report the detection of the 894 GHz ground-state $1_{1,1} - 0_{0,0}$ transition of HDO *in absorption* against the background continuum sources Sgr B2(M) and Sgr B2(N). The radial velocity of the observed HDO features suggests that the absorption is directly connected to the Sgr B2 complex. The fact that i) we observed a ground-state transition, and ii) we observed it in absorption, allowed us to estimate, independently of assumption on the physical state of the gas, the HDO column density, which we found to be of 1.2×10^{13} cm^{-2} towards Sgr B2(M), and of 2.4×10^{13} and 1.2×10^{13} cm^{-2} towards the 65 and

[1] The CSO is operated by the California Institute of Technology under funding from the National Science Foundation, Grant No. AST-9980846.

Chemistry as a Diagnostic of Star Formation (2003). *Edited by Charles L. Curry and Michel Fich.* Published in 2003 by NRC Research Press, Ottawa ON, K1A 0R6, Canada. ISBN 0-660-19089-3.

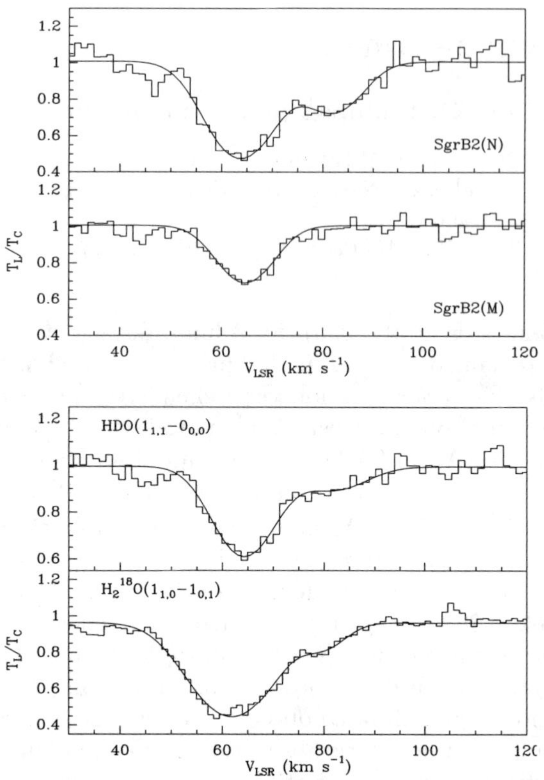

Figure 1. Upper panel: The 894 GHz HDO $(1_{1,1} - 0_{0,0})$ transition, observed in absorption against the continuum emission of Sgr B2(M) and of Sgr B2(N). *Lower panel:* Comparison between the 894 GHz HDO$(1_{1,1} - 0_{0,0})$ and the 547.7 GHz o-$H_2^{18}O(1_{1,0} - 1_{0,1})$ absorption features observed towards Sgr B2. The y-scale gives the line-to-continuum (T_L/T_C) ratio.

the 81 km s^{-1} velocity components of Sgr B2(N) respectively. Moreover, using the $H_2^{18}O$ *SWAS* data (N00), we were able to estimate a lower limit of 5×10^{-4} for the [HDO]/[H_2O] ratio towards the Sgr B2 complex.

The above mentioned results hold under the assumption that all the absorbing water lies in the warm envelope. In a recent paper, C02 propose that H_2O might instead be located in the hot gas layer that surrounds the whole Sgr B2 complex, implying that gas-phase water in the warm envelope does not contribute at all to the observed absorption features. In fact, while this hypothesis is reasonable for H_2O (the gas-phase production of water is triggered by high temperatures), it would not be applicable to HDO (the same high temperatures inhibit its enhancement). We therefore think it possible that the gas observed in HDO and that seen in H_2O absorption be located in distinct regions.

2. Modeling the distribution of water

Sgr B2 is often adopted as a template region for the study of sites of massive star formation. For this purpose, it is important to evaluate to what extent the warm envelope and the hot layer separately contribute to the observed HDO and H_2O absorption. In particular, in order to better understand the nature of the hot layer and possibly the mechanism that created it, it is vital to learn more about its chemical and physical characteristics, and especially about its water content. A precise determination of the column density of water would set a tight constraint on the physical models that attempt to identify the heating mechanism responsible for the high temperatures in the hot layer. Unfortunately, since both gas components show roughly the same radial velocity, it is not possible to separate, by purely observational means, the ground-state water absorption produced in the warm envelope from that produced in the hot layer. However, a number of HDO and $H_2^{18}O$ transitions have been observed towards the Sgr B2 cores with a variety of instruments.

Our 893 GHz HDO data, together with the published HDO and $H_2^{18}O$ observations carried out at mm and submm wavelengths, provides strong observational constraints and enables us to model, in a self-consistent manner, the HDO and H_2O abundance in all three components of the Sgr B2 cloud. In detail:

1. The observed *emission* lines of p-$H_2^{18}O$ at 203 GHz (Sgr B2(N), Gensheimer et al. 1996), and of HDO at 143, 226, 241 GHz (Sgr B2(M) and (N), Jacq et al. 1990) constrain the HDO and water abundance and the H_2 density in the hot-core-type components.

2. The 893 GHz HDO *absorption* feature observed towards Sgr B2(M) and N, together with the [HDO]/[H_2O] ratio estimated at point *a)*, help us set the HDO and H_2O abundance in the warm envelope.

3. Finally, having estimated the column density of water in the warm envelope, it is possible to assess whether or not the quantity of non-deuterated water in the warm envelope can be responsible for the $H_2^{18}O$ absorption observed at 548 GHz (N00, Z95a). The 180 μm o-$H_2^{18}O$ line observed by Cernicharo et al. (1997) towards the Sgr B2 complex is blended with the $H_3O^+(1_1^- - 1_1^+)$ line (Goicoechea & Cernicharo 2001), and cannot therefore be adopted as a reference for our purposes.

We use the radiative transfer code described by Z95b, with a few modifications (see C03 for details), to reproduce the intensities observed for the features. The model consists of two spherically symmetric components representing the hot core and the warm envelope (see Sec. 1). Most of the spectral line data are reproduced within errors $\sim 30\%$, with two exceptions:

1. The 226 GHz HDO emission line observed towards Sgr B2(M) is a factor of 2 weaker, and the 143 GHz HDO line in Sgr B2(N) is 4 times stronger than predicted by the model. Such discrepancies are likely to be due to the intrinsic chemical difference between the molecular gas components of the two hot cores (cf. Miao & Snyder [1997] and references therein), which has not been taken into account in our model.

2. Most interestingly, the predicted ground-state o-$H_2^{18}O$ absorption at 548 GHz is much shallower than observed (N00, cf. Fig. 1): *the abundance of non-deuterated water in the warm envelope is not sufficient to produce the absorption feature observed at 548 GHz.*

This is a very important result: our predictions confirm, at least from a qualitative point of view, the hypothesis of C02 that the most important contribution to the observed $H_2^{18}O$ absorption come from the foreground hot gas layer. Having predicted the contribution of the warm envelope to the $H_2^{18}O$ absorption, it is possible to estimate the spatial width and H_2O abundance necessary for the hot layer to produce the observed feature, given the observational constraints mentioned in Sec. 1.

We use the radiative transport equation to calculate the total intensity emerging from the hot layer, given the background emission produced by the complex made up of hot core and warm envelope. The data are best reproduced when the o-$H_2^{18}O$ column density in the hot layer is $\sim 10^{14}\,\mathrm{cm}^{-2}$. Assuming a $[^{16}O]/[^{18}O]$ ratio of 261 ± 20 (Whiteoak & Gardner 1981) and an ortho/para ratio of 3, we can estimate the H_2O column density in the hot gas component to be of order $\sim 3.5 \times 10^{16}\ \mathrm{cm}^{-2}$. If all the atomic oxygen not locked in CO is bound in gas-phase water, then the H_2O abundance in this region is about $5 \times 10^{-4}\ \mathrm{cm}^{-3}$, and the spatial width of the hot layer is only 0.02 pc.

ACKNOWLEDGEMENTS

The authors are grateful to David Neufeld for providing the 548 GHz *SWAS* data in digital format.

REFERENCES

Bergin, E.A., Melnick, G.J., Neufeld, D.A. 1998, ApJ, 499, 777

Ceccarelli, C., Hollenbach, D.J., Tielens, A.G.G.M. 1996, ApJ, 471, 400

Ceccarelli, C., Baluteau, J.-P., Walmsley, C.M., et al. 2002, A&A, 383, 603 (C02)

Cernicharo, J., Lim, T., Cox, P. et al. 1997, A&A, 323, 25

Comito, C., Schilke, P., Gerin, M., Phillips, T.G., Zmuidzinas, J., Lis, D.C., in preparation (C03)

Flower, D.R., Pineau des Forêts, G., Walmsley, C.M. 1995, A&A, 294, 815

Gensheimer, P.D., Mauersberger, R., Wilson, T.L. 1996, A&A, 314, 281

Giesen, T.F., Van Orden, A.O., Cruzan, J.D., et al. 2001, ApJ, 551, 181

Goicoechea, J.R., Cernicharo, J. 2001, ApJ, 554, 213

Henkel, C., Mauersberger, R., Wilson, T.L., Snyder, L.E., Menten, K.M., Wouterloot, J.G.A. 1987, A&A, 182, 299

Hüttemeister, S., Wilson, T.L., Mauersberger, R., et al. 1995, A&A, 294, 667

Jacq, T., Jewell, P.R., Henkel, C., Walmsley, C.M., Baudry, A. 1988, A&A, 199, 5

Jacq, T., Walmsley, C.M., Henkel, C., Baudry, A., Mauersberger, R., Jewell, P.R. 1990, A&A, 228, 447

Jacq, T., Baudry, A., Walmsley, C.M., Caselli, P. 1999, A&A, 347, 957

Martín-Pintado, J., de Vicente, P., Wilson, T.L., Johnston, K.J. 1990, A&A, 236, 193

Melnick, G.J., Ashby, M.L.N., Plume, R. et al. 2000, ApJ, 539, 87

Miao, Y., Snyder, L.E. 1997, ApJ, 480, L67

Neufeld, D.A., Melnick, G.J. 1987, ApJ, 322, 266

Neufeld, D. A., Zmuidzinas, J., Schilke, P., Phillips, T.G. 1997 ApJ, 488, 141

Neufeld, D.A., Ashby, M.L.N., Bergin, N.R. et al. 2000, ApJ, 539, 111

Pardo, J.R., Cernicharo, J., Herpin, F., Kawamura, J., Kooi, J., Phillips, T.G. 2001, ApJ, 562, 799

Phillips, T.G., Scoville, N.Z., Kwan, J., Huggins, P.J., Wannier, P.G. 1978, ApJ, 222, 59

van Dishoeck, E.F., Jansen, D.J., Schilke, P., Phillips, T.G. 1993, 1993 ApJ, 416, 83

Walmsley, C.M., Bachiller, R., Pineau des Forêts, G., Schilke, P. 2002 ApJ, 566, 109

Whiteoak, J.B., Gardner, F.F. 1979, MNRAS, 188, 445

Whiteoak, J.B., Gardner, F.F. 1981, MNRAS, 197, 39

Wilson, T.L., Ruf, K., Walmsley, C.M., et al. 1982, A&A, 115, 185

Zmuidzinas, J., Blake, G.A., Carlstrom, J., Keene, J., Miller, D., Schilke, P., Ugras, N.G. 1995b, in *Proceedings of the Airborne Astronomy Symposium on the Galactic Ecosystem: From Gas to Stars to Dust*, ed. M.R. Haas, J.A. Davidson and E.F. Erickson (San Francisco: ASP) (Z95a)

Zmuidzinas, J., Blake, G.A., Carlstrom, J., Keene, J., Miller, D. 1995, ApJ, 447, L125 (Z95b)

LINE SURVEY OF ORION-KL IN THE 350 MICRON BAND

Claudia Comito[1], Peter Schilke[1], Tom G. Phillips[2], Darek C. Lis[2], Frédérique Motte[2], and Dave M. Mehringer[3]

[1]Max-Planck-Institut für Radioastronomie, Auf dem Hügel 69, D-53121 Bonn, Germany
[2]California Institute of Technology, Downs Laboratory of Physics 320-47, Pasadena, CA 91125, USA
[3]University of Illinois, National Center for Supercomputing Applications, 605 E. Springfield Ave., Champaign, IL 61820, USA

Abstract

We report results of an unbiased spectral line survey of Orion-KL in the 350 μm band. The data have been acquired with the Caltech Submillimeter Observatory. Even at these high frequencies, the Orion Hot Core shows an extremely rich spectrum, and about 1000 features have been detected, most of which provide an insight on the densest, hottest region of this object. Moreover, the analysis of this survey will allow to predict the molecular emission of Orion-KL at THz frequencies, thus clearing the path for future astrochemical observations with APEX, ALMA and *Herschel*.

Key words: ISM: molecules – ISM: structure – ISM: clouds

The Orion Hot Core is considered to be the most prominent object of its kind, and is often adopted as a prototype for the study of hot-core-type sources. Partly due to its proximity ($\sim$ 450 pc), it is one of the richest known astronomical sources of molecular lines, which has made this object the ideal target for a number of line surveys at millimeter and submillimeter wavelengths, with the goal to create a complete inventory of its molecular emission within the frequency windows accessible from ground-based observatories (cf. Schilke et al. 2001 and references therein).

In the last two winters, we have completed the coverage of the 790-to-900-GHz band with the Caltech Submillimeter Observatory[1] (Comito et al., in preparation; see Schilke et al. 1999 for a report on the first half). The latter survey gives us an insight on the chemistry of the densest, hottest region of the Hot Core that is so far accessible from ground. In fact, although most molecules have already been well covered by lower frequency surveys, for almost all species the emission lines in the submillimeter band originate from higher energy levels than at lower frequencies, which, together with the decrease of beam size associated to the increase in frequency, allows to trace emitting gas from very small, high-density hot sources within the hot core.

1. Reconstruction and analysis of the data

The reconstruction of the double-sideband data has been carried out using the Maximum Entropy Method (Sutton et al. 1995; Schilke et al. 2001; Comito & Schilke 2002). The result is the whole-band single-sideband spectrum shown in Figure 1.

The first results of this study can be summarized as follows:

1. Even at these high frequencies, the spectrum of Orion-KL is extremely rich, with an average of 3 to 6 strong (> 5 K) lines per GHz.

2. The strongest emission lines are, as expected, produced by CO and ^{13}CO, HCN, SO_2, SO, CH_3OH.

3. A total of about 1000 lines has been assigned, based on the data available on the JPL Molecular Catalog[2] and on the Cologne Database for Molecular Spectroscopy[3]. Some interesting examples are shown in Figures 2 and 3. At this stage, 80% of them have been, at least tentatively, identified. We believe we will be able to further reduce the percentage of U-lines thanks to recent spectroscopic data, up to THz frequencies, of CH_3OH and H_2CO (Xu & Lovas 1997; Brünken et al., in preparation).

4. The spectral contribution to the total flux of the Orion Hot Core is about 15%. This result is in agreement with the percentage of line contamination, $\sim$ 50%, measured for this source at mm frequencies.

The analysis of the data is presently under way using the XCLASS extension to the GILDAS CLASS package.

[1] The CSO is operated by the California Institute of Technology under funding from the National Science Foundation, Grant No. AST-9980846.

[2] http://spec.jpl.nasa.gov/ftp/pub/catalog/catform.html

[3] http://www.ph1.uni-koeln.de/vorhersagen/index.html

Chemistry as a Diagnostic of Star Formation (2003). *Edited by Charles L. Curry and Michel Fich.* Published in 2003 by NRC Research Press, Ottawa ON, K1A 0R6, Canada. ISBN 0-660-19089-3.

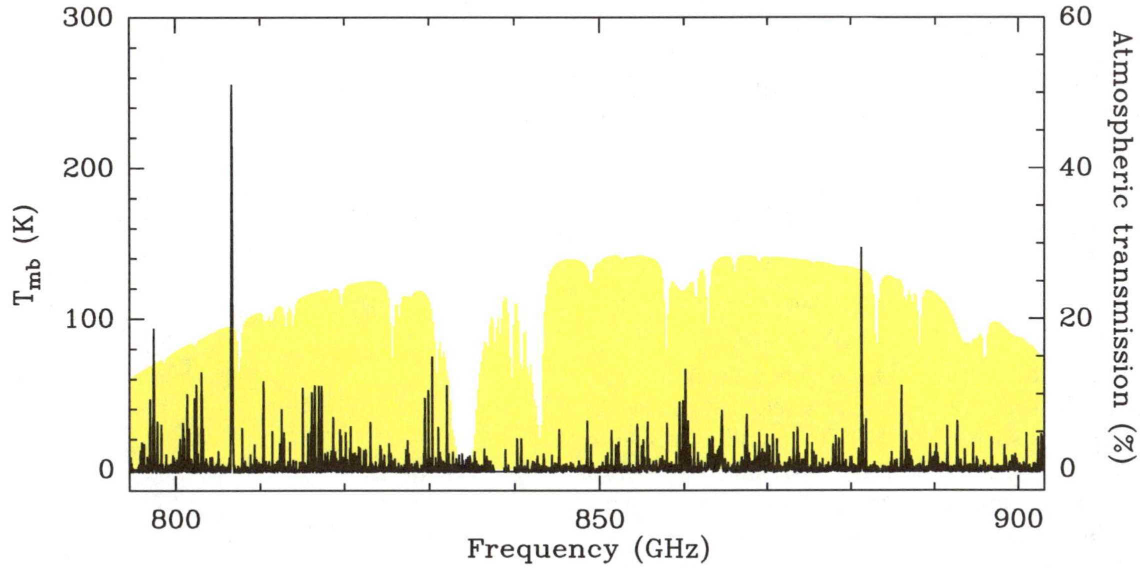

Figure 1. Line survey of Orion-KL in the 350 μm band. The spectrum has been reconstructed from the double-sideband data by using the Maximum Entropy Method. Note that, at these high frequencies, the atmospheric transmission (right-hand $y-$scale) drops to about 30% for good weather conditions typical of the Mauna Kea summit (1 mm precipitable water vapor). By comparison, at mm wavelengths the same weather conditions would allow an atmospheric transmission of more than 90%.

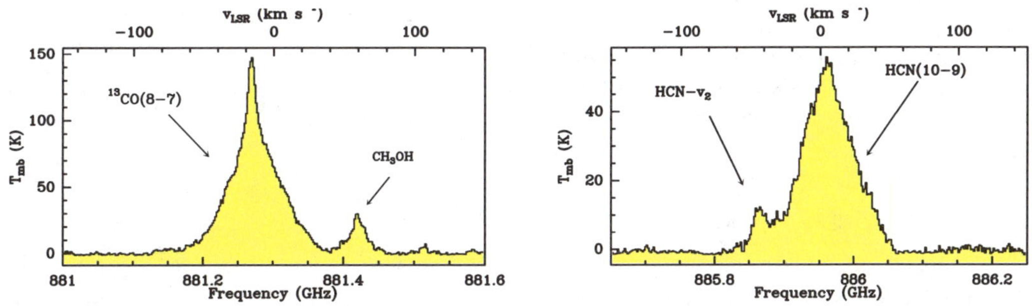

Figure 2. Left: ^{13}CO (8–7) transition at 881.3 GHz. The wings tracing the outflow of Orion-KL are clearly visible: the central narrow component has a width of $\sim$ 14 km s^{-1}, whereas the wing component is $\sim$ 90 km s^{-1} wide. *Right:* the HCN (10–9) rotational transition with its vibrationally excited companion. The HCN (10–9) line has a critical density $\geq 10^8$ cm^{-3}.

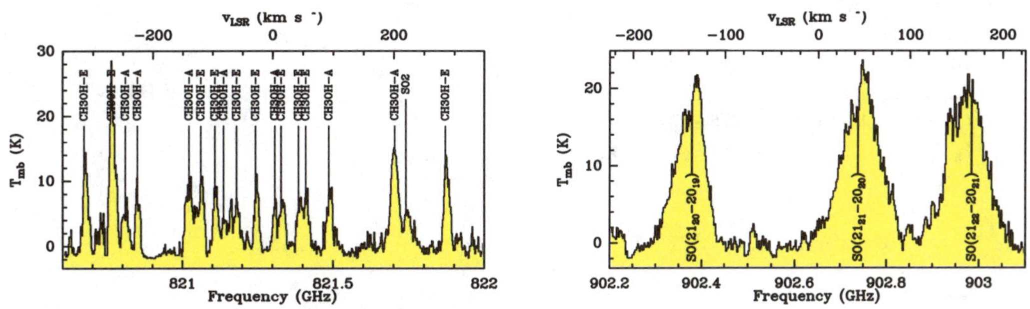

Figure 3. Left: Part of the CH_3OH (17–16) band. *Right:* The SO (21–20) triplet at 902 GHz is the highest frequency transition detected with the Caltech Submillimeter Observatory.

LTE fitting of rotational temperatures, column densities, source sizes, radial velocities and line widths is carried out for every chemical species and relative isotopomers. Note that *the whole spectrum* is fitted *at once*, which allows to correct for:

– optical depth effects and intra-species blending, since all the features radiated from the same molecule are fitted at once;

– inter-species blending, since *all* the lines from *all* species are fitted altogether.

The collected data set of Orion spectra will be extremely useful to spectroscopic studies carried out with future instruments such as APEX (cf. Schilke, this volume), SMA, ALMA, and the HIFI instrument on *Herschel.* Careful analysis of the data will also allow to predict spectra in the ranges, inaccessible from the ground, that *Herschel* will be able to observe, thereby allowing a detailed planning of the extremely precious satellite time. For this, the 350 μm data set is particularly valuable, because it will permit predictions in the THz bands, which are the most interesting bands of HIFI.

Acknowledgements

The authors are grateful to Sandra Brünken for providing her new measurements of high frequency H_2CO transitions prior to publication. This work would be impossible without the Cologne Database for Molecular Spectroscopy and the JPL Molecular Catalog.

References

Comito, C., & Schilke, P. 2002, A&A, 395, 357

Schilke, P., Groesbeck, T.D., Blake, & G.A., Phillips, T.G. 1997, ApJS, 108, 301

Schilke, P., Phillips, T.G., & Mehringer, D.M. 1999, in *The Physics and Chemistry of the Interstellar Medium*, Proceedings of the 3rd Cologne-Zermatt Symposium, held in Zermatt, September 22-25, 1998, ed. V. Ossenkopf, J. Stutzki, & G. Winnewisser (Herdecke: GCA-Verlag)

Schilke, P., Benford, D.J., Hunter, T.R., Lis, D.C., & Phillips, T.G. 2001, ApJS, 132, 281

Sutton, E.C., Peng, R., Danchi, W.C., Jaminet, P.A., Sandell, G., & Russell, A.P.G. 1995, ApJS, 97, 455 (S95)

Xu, L.-H. & Lovas, F. L. 1997, J. Phys. Chem. Ref. Data, 26, 17

A SEARCH FOR H_2 OUTFLOW SIGNATURES FROM MASSIVE STAR FORMATION REGIONS CONTAINING LINEARLY DISTRIBUTED METHANOL MASERS

James M. De Buizer

Cerro Tololo Inter-American Observatory, Casilla 603, La Serena, Chile

ABSTRACT

I present here the summary of results from a survey by De Buizer (2003) searching for outflows using near-infrared imaging. Targets were massive young stellar objects associated with methanol masers in linear distributions. Presently, it is a widely held belief that these methanol masers are found in (and delineate) circumstellar accretion disks around massive stars. A way to test the disk hypothesis is to search for outflow signatures perpendicular to the methanol maser distributions. The main objective of the survey was to obtain wide-field near-infrared images of the sites of linearly distributed methanol masers using a narrow-band 2.12 μm filter which is centered on the H_2 v=1–0 S(1) line. This line is a shock diagnostic that has been shown to successfully trace CO outflows from young stellar objects. Twenty-eight sources in total were imaged with eighteen sources displaying H_2 emission. Of these, only *two* sources showed emission found to be dominantly perpendicular to the methanol maser distribution. These results seriously question the hypothesis that methanol masers exist in circumstellar disks.

Key words: masers – stars: formation – ISM: lines and bands – ISM: molecules – infrared: ISM

1. INTRODUCTION

Methanol maser emission has been well studied in the last decade and is thought to be a good indicator of recent massive star formation. These masers occur in spatially localized regions or *spots* and serve as powerful probes of the small-scale structure, dynamics, and physical conditions of the environments near forming stars. Radio observations by Norris et al. (1993) first showed that methanol (CH_3OH) maser spots are frequently distributed in linear structures, with projected dimensions typically spanning 2500 AU. Furthermore, some of these linearly distributed methanol masers exhibit a gradient in velocity along the spot distribution. Norris et al. (1993) argue that such a velocity gradient is indicative of orbital motion, and suggest that these methanol maser spots occur in, and directly delineate, rotating circumstellar disks. Several authors have written papers on radio studies of these methanol masers and their motions (e.g. Norris et al. 1998; Phillips et al. 1998; Minier, Booth & Conway 2000) trying to decipher from this limited data the mass of the central protostars and sizes of circumstellar disks in which these methanol masers reside.

But do methanol masers really trace disks? More proof is needed than a line of masers displaying (perhaps) orbital motion. Direct detection of these disks would certainly be conclusive, however observations have proven to be problematic. Observations in the near-infrared suffer from the great extinction towards these regions. It has also been argued that in the mid-infrared it is difficult to tell if one is observing dust emission from a circumstellar accretion disk or dust emission from the placental envelope (Vinković et al. 2000). The far-infrared and sub-millimeter can probe the cool thermal emission from a dust disk with minimal contamination from the placental envelope, however no far-infrared facility exists that has the sub-arcsecond resolution to observe disks around these sources. Angular resolution presently is a problem for the sub-millimeter as well, however the construction of SMA (Submillimeter Array) and ALMA (Atacama Large Millimeter Array) are underway, and are presently 5 to 10 years away from carrying out these types of observations. Corroborative evidence that linearly distributed methanol masers exist in circumstellar disks, therefore, needs to come from something other than direct detection of the accretion disks.

Fortunately, there is an *indirect* way of testing whether or not linearly distributed methanol masers exist in accretion disks. According to the standard model of accretion, during the phase of stellar formation where the star is being fed by an accretion disk, it is also undergoing mass loss through a bipolar outflow. This bipo-

Chemistry as a Diagnostic of Star Formation (2003). *Edited by Charles L. Curry and Michel Fich.* Published in 2003 by NRC Research Press, Ottawa ON, K1A 0R6, Canada. ISBN 0-660-19089-3.

Table 1. List of targets observed in the De Buizer (2003) survey. The first column is the name of the target in galactic coordinates and the second column is the *IRAS* name of the target. The third and fourth columns give the source coordinates. The fifth column gives the methanol maser distribution position angle, and the sixth column describes the H_2 emission found in the target field. A '?' denotes that there may be some confusion associated with the nature or morphology of the H_2 emission.

Target	*IRAS* Name	Right Ascension (J2000)	Declination (J2000)	Maser p.a.	H_2
G305.21+0.21	13079-6218	13 11 13.72	-62 34 41.6	25°	parallel?
G308.918+0.123	13395-6153	13 43 01.75	-62 08 51.3	137°	parallel
G309.92+0.48	13471-6120	13 50 41.76	-61 35 10.1	30°	no detection
G312.11+0.26	14050-6056	14 08 49.30	-61 13 26.0	166°	no detection
G313.77-0.86	14212-6131	14 25 01.62	-61 44 58.1	135°	parallel
G316.81-0.06	14416-5937	14 45 26.44	-59 49 16.4	1°	not outflow
G318.95-0.20		15 00 55.40	-58 58 53.0	151°	parallel
G320.23-0.28	15061-5814	15 09 51.95	-58 25 38.1	86°	parallel
G321.031-0.484	15122-5801	15 15 51.64	-58 11 17.4	0°	parallel?
G321.034-0.483	15122-5801	15 15 52.52	-58 11 07.2	85°	parallel?
G327.120+0.511	15437-5343	15 47 32.71	-53 52 38.5	150°	no detection
G327.402+0.445	15454-5335	15 49 19.50	-53 45 13.9	62°	no detection
G328.81+0.63	15520-5234	15 55 48.61	-52 43 06.2	86°	parallel
G331.132-0.244	16071-5142	16 10 59.74	-51 50 22.7	90°	parallel
G331.28-0.19	16076-5134	16 11 26.60	-51 41 56.6	166°	perpendicular
G335.789+0.174		16 29 47.33	-48 15 52.4	136°	parallel?
G336.43-0.26	16303-4758	16 34 20.34	-48 05 32.5	163°	no detection
G337.705-0.053	16348-4654	16 38 29.61	-47 00 35.7	137°	no detection
G339.88-1.26	16484-4603	16 52 04.66	-46 08 34.2	137°	?
G339.95-0.54	16455-4531	16 49 07.99	-45 37 58.5	122°	no detection
G344.23-0.57	17006-4215	17 04 07.70	-42 18 39.1	117°	no detection
G345.01+1.79	16533-4009	16 56 47.56	-40 14 26.2	78°	parallel?
G345.01+1.80	16533-4009	16 56 46.80	-40 14 09.1	30°	parallel?
G348.71-1.04	17167-3854	17 20 04.02	-38 58 30.0	152°	not outflow
G353.410-0.360	17271-3439	17 30 26.17	-34 41 45.6	153°	not outflow
G00.70-0.04	17441-2822	17 47 24.74	-28 21 43.7	51°	no detection
G10.47-0.03	18056-1952	18 08 38.21	-19 51 49.6	98°	no detection
G11.50-1.49	18134-1942	18 16 22.13	-19 41 27.3	174°	perpendicular

lar outflow is perpendicular to the plane of the accretion disk, along the axis of rotation. Therefore, one can search these sources of linearly distributed methanol masers for evidence of outflow perpendicular to the methanol maser position angle. Such evidence would create an extremely solid case for the hypothesis that these methanol masers exist in circumstellar disks, without the need for their direct detection.

I present here the results of the survey of De Buizer (2003), the purpose of which was to image the regions around sites of linearly distributed methanol masers in the near-infrared with a narrow-band 2.12 μm filter. This filter is centered on the H_2 v=1–0 S(1) line, which is a shock indicator, and was convincingly shown to successfully trace CO outflows from young stellar objects by Davis & Eislöffel (1995). Each site was also observed with a continuum filter which yields a measure of the continuum-only flux from all the sources in the field. By subtracting the continuum image from the image with the hydrogen line, structures associated only with hydrogen emission emanating from the locations of the methanol masers were found.

2. Summary of Results

The full results of this survey are presented in De Buizer (2003). The target list consisted of 28 sites compiled mostly from the articles by Walsh et al. (1998), Phillips et al. (1998), and Norris et al. (1998). The coordinates for all of these sources are shown in the Table 1. These sites all contain groups of linearly distributed methanol masers, many with velocity gradients along their distributions indicative of rotation, and therefore represent the best circumstellar disk candidates.

Of the 28 maser targets observed by De Buizer (2003), H_2 emission was detected from 18 sites (64%). The distribution of the H_2 emission from these sites takes on three forms: 1) extended diffuse areas of H_2-dominated emission; 2) individual knots or blobs of H_2-only emission that range in number, sometimes cometary-shaped; and 3) some combination of extended H_2-dominated emission with knots of H_2-only sources.

The purpose of these observations was to try to confirm the hypothesis that linearly distributed methanol masers exist in, and directly delineate, circumstellar disks by searching for shock excited H_2 outflow signatures perpendicular to the methanol maser position angle. Of the 18 maser locations observed to have H_2 emission, only *two* displayed this morphology. The majority had H_2 emission found to be dominantly distributed within 45° of being *parallel* to their maser position angle (see Figure 1 for an example). H_2 emission can also be due to UV florescence, so for any one source in this survey, follow-up observations (in say, ^{12}CO or HCO+) would be needed to confirm the outflow nature. However, regardless of the nature of the H_2 emission, these results are unequivocally contrary to the circumstellar disk hypothesis. If the H_2 emission is indeed from outflows, then a likely explanation is that at least some linearly distributed methanol masers may be directly associated with outflows. The methanol masers appear to be located coincident with a stellar source at the center of the outflows in most cases. Perhaps the masers trace the jets or outflow surfaces near the central (proto-)stellar source.

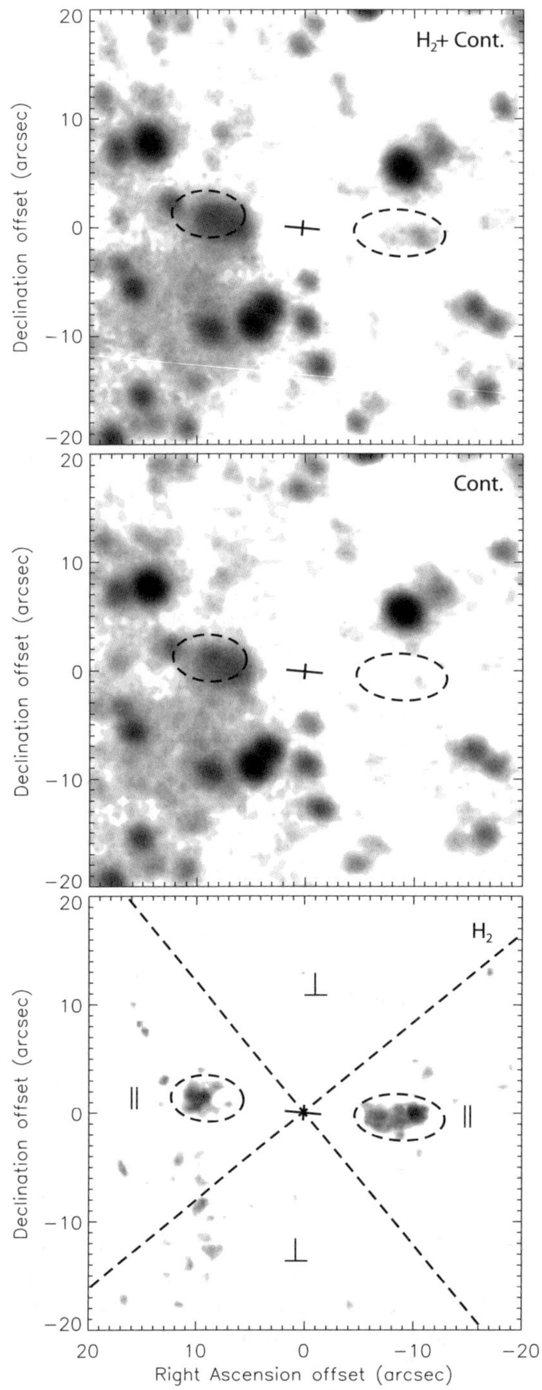

Figure 1. G320.23-0.28 (IRAS 15061-5814) H_2+continuum, continuum, and residual H_2 images. The cross represents the maser group location with the elongated axis showing the position angle of linear maser distribution. Dashed ellipses encompass areas of H_2 emission. Dashed lines in the H_2 image divide the frame into quadrants parallel and perpendicular to the maser position angle.

References

Davis, C. J. & Eisloeffel, J. 1995, A&A 300, 851

De Buizer, J. M. 2003, MNRAS, in press

Minier, V., Booth, R., Conway, J. E. 2000, A&A 362, 1093

Norris, R. P., Whiteoak, J. B., Caswell, J. L., Wieringa, M. H., Gough, R. G. 1993, ApJ 412, 222

Norris, R. P., et al. 1998, ApJ 508, 275

Phillips, C. J., Norris, R. P., Ellingsen, S. P., McCulloch, P. M. 1998, MNRAS 300, 1131

Vinković, D., Miroshnichenko, A., Ivezić, Ž., Elitzur, M. 2000, in ASP Conf. Ser. 196, Thermal Emission Spectroscopy and Analysis of Dust, Disks, and Regoliths, ed. M. L. Sitko, A. L. Sprague, & D. K. Lynch (ASP: San Francisco), 71

Walsh, A. J., Burton, M. G., Hyland, A. R., Robinson, G. 1998, MNRAS 301, 640

INTERSTELLAR TURBULENT VELOCITY FIELDS: SCALING PROPERTIES, SYNTHESIS AND EFFECTS ON CHEMISTRY

Nicolas Décamp and Jacques Le Bourlot

LUTH, Observatoire de Paris et Université Paris 7, France

ABSTRACT

Turbulence is thought to play a key role in the dynamics of interstellar clouds. Here, we do not seek to explain its origin or decipher the mechanisms that maintain it, but we start from the observational fact that it is present. Arnéodo et al. (1997) have developed a method based on wavelet analysis to study incompressible turbulence experiments. We propose to use this method with the same propagator to derive quantitative information on the structure of a turbulent field. Then we build a synthetic velocity field with the same statistical properties and we show that a reactive fluid subject to turbulent forcing exhibits self-organised structures that depend on the chemical species considered. Such effects could explain why observational evidence shows that the bulk of the mass is distributed smoothly whereas some chemical species are extremely patchy. See Décamp and Le Bourlot (2002) for more details.

Key words: turbulence – methods: numerical – ISM: clouds – ISM: molecules – ISM: structure

1. ANALYSIS AND GENERATION OF A VELOCITY FIELD

The observational map is a ^{12}CO 1 $\rightarrow$ 0 fully sampled 48 $\times$ 64 map of the Polaris cloud (Falgarone et al. 1998). The centroid velocity increment at scale Δ is $\delta C_\Delta = C(r+\Delta) - C(r)$, where C is the centroid velocity (mean radial velocity) and r is a position on the plane of the sky. Its probability density function (PDF) in a turbulent velocity field is indistinguishable from a Gaussian at integral scale and develops more and more non-Gaussian wings as the lag decreases (see Fig. 1).

Arnéodo et al. have shown that the PDF of velocity increments at one scale (Δ) (or of the wavelet coefficients T_Δ at this scale) can be expressed as a weighted sum of

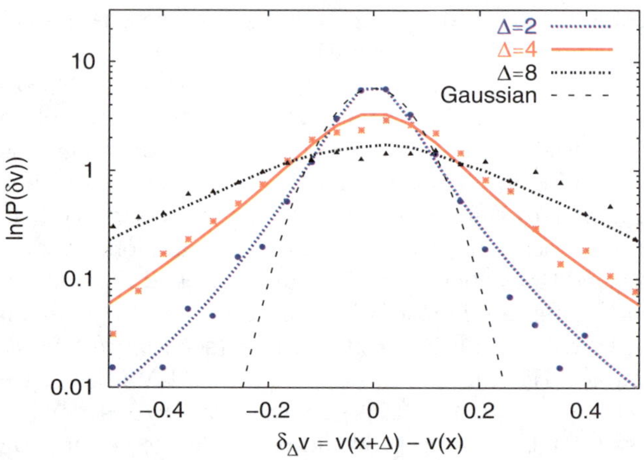

Figure 1. Velocity increment PDFs. Points are from the Polaris map and lines from the reconstructed field.

dilated PDFs at a larger scale (Δ'). This leads to a simple convolution equation for its logarithm:

$$Pdf(\ln|T_\Delta|) = \int G_{\Delta\Delta'}(x)\, Pdf(\ln|T_{\Delta'}| - x)\, dx \qquad (1)$$

$G_{\Delta\Delta'}$ is called a propagator. For a log-normal cascade, the propagator is Gaussian:

$$G_{\Delta\Delta'}(x) = \frac{1}{\sqrt{2\pi\sigma^2_{\Delta\Delta'}}} \exp\left(-\frac{(x-m_{\Delta\Delta'})^2}{2\sigma^2_{\Delta\Delta'}}\right) \qquad (2)$$

The only free parameters are $m_{\Delta\Delta'} = m\ln(\Delta'/\Delta)$ and $\sigma^2_{\Delta\Delta'} = \sigma^2 \ln(\Delta'/\Delta)$. From a fit to the PDFs' ratios, we found : $m \simeq -1 \pm 0.1$ and $\sigma \simeq 0.25 \pm 0.05$.

We generate a velocity field by constructing its wavelet coefficients decomposition: with f the velocity field, we have:

$$\forall f \in L^2(\Re),\ f = \sum_j \sum d_{j,k}\, \psi_{j,k} \qquad (3)$$

where $\psi_{j,k}(t) = 2^{-\frac{j}{2}}\psi(\frac{t-2^j k}{2^j})$ is an orthonormal basis of $L^2(R)$ and $d_{j,k} =< f, \psi_{j,k} >$ are the wavelet coefficients.

Chemistry as a Diagnostic of Star Formation (2003). *Edited by Charles L. Curry and Michel Fich.* Published in 2003 by NRC Research Press, Ottawa ON, K1A 0R6, Canada. ISBN 0-660-19089-3.

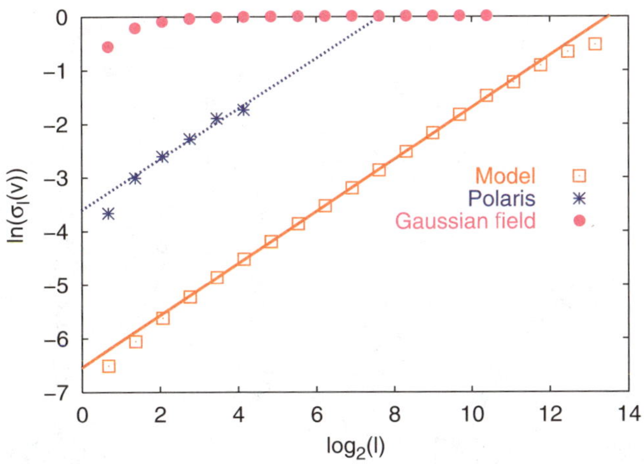

Figure 2. Standard deviation of the velocity field σ_l as a function of size l in pixel units (arbitrary vertical offset) Straight lines are "Larson" power laws $\sigma_l v \propto l^\beta$ with $\beta = 0.5$.

All the coefficients $d_{j+1,k}$ are generated from the $d_{j,k}$ by a multiplicative log-normal process such that:

$$\begin{cases} d_{j+1,2k} &= M_j^{(2k)} d_{j,k} \\ d_{j+1,2k+1} &= M_j^{(2k+1)} d_{j,k} \end{cases} \tag{4}$$

where $\left|M_j^{(k)}\right|$ are realisations of a random variable M_j that follows Eq. 2 with $\Delta = 2^{-j}$ and $\Delta' = 2^{-j-1}$.

We generate a velocity field in good agreement with the observed one (Fig. 1). It is correlated at all scales unlike the Gaussian field used for comparisons (Fig. 2).

2. Density field

Under the conditions of homogeneity, isotropy, and stationarity, Taylor hypothesis applies. Using a 2D field, the statistical properties of $v(X,t)$ along an axis X as t varies are the same as those of a collection of lines at a given time t_0. Time evolution of the density follows from the known evolution of $v(X,t)$ and mass conservation:

$$\frac{\partial \rho}{\partial t} + \vec{\nabla}.(\rho \vec{v}) = 0 \tag{5}$$

We use uniform initial conditions and assume periodic boundary conditions. After a transitory period, the density and velocity fields "couple" together and the standard deviation stabilises (the mean density is constant because periodic boundary conditions ensure conservation of the total mass). A typical example of a density field is given in the left panel of Figure 3. We see that large fluctuations of ρ are possible within a few cells and dense cores develop over a low-density background. The contrast is high with results from a Gaussian velocity field with the same mean and standard deviation.

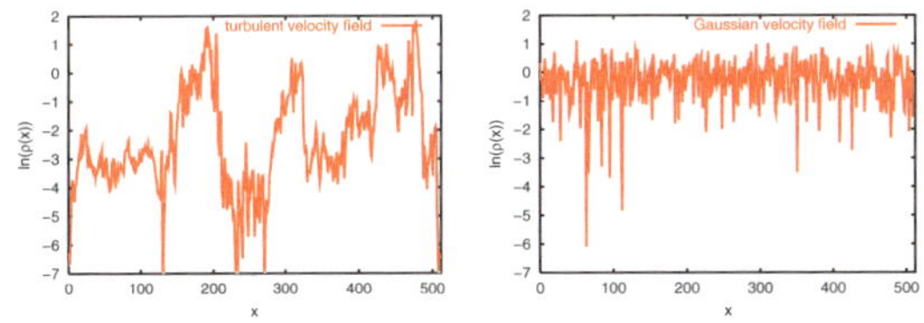

Figure 3. Log of the density field for a turbulent or a Gaussian velocity field (same mean and standard deviation).

Figure 7 shows a power law relation between size and density with an index of -0.32 ± 0.03. This is much lower than the $n \propto l^{-1}$ law of self-gravitating clouds. Such a flat index is probably a consequence of our 1D model and further discussions should wait for 2D results.

3. A toy model of interstellar chemistry

A minimal local dynamic requires 3 variables to exhibit bistability as may be found in some dark clouds (Le Bourlot et al. 1995). We use the following test model:

$$\begin{cases} R & \rightarrow & A & (k_1) \\ A & \rightarrow & B & (k_2) \\ A + 2B & \rightarrow & 3B & (k_3) \\ B & \rightarrow & P & (k_4) \end{cases} \tag{6}$$

Such a model can be seen as an excerpt from a larger chemical network with $R = R'\rho$, a production term, and product(s) P returning to the rest of the gas. With $k_1 = k_{10} \exp(-E_a/k_b T)$ and an exothermic reaction (4), thermal balance is governed by: $\Delta U = k_4 \Delta t \Delta H n_B - k_5 k_b \Delta t (n_A + n_B) T$ (with $U = \rho c_{p,\rho} T = n_R c_{p,R} T$). The cooling term mimics radiative cooling by both A and B after collisional excitation.

The local dynamics reduces to:

$$\begin{cases} d\alpha/d\tau &= r \exp(-\varepsilon r/u) - k\alpha - \alpha\beta^2 \\ d\beta/d\tau &= k\alpha + \alpha\beta^2 - \beta \\ du/d\tau &= \beta - \gamma(\alpha + \beta)u/r \end{cases} \tag{7}$$

with $\alpha \propto n_A$, $\beta \propto n_B$, $u \propto U/\Delta H$ and four parameters $(r, \varepsilon, k, \gamma)$.

Different parameters lead to various behaviours: bistability, limit cycle with Hopf bifurcation (see Fig. 4), . . . etc.

It is then possible to couple chemistry and turbulent mixing. A, B, U and R are advected and follow Eq. (7). Structures appear again at all scales. These structures are not identical for the different variables (A, B, R) neither in position nor in size (Fig. 5 and 7).

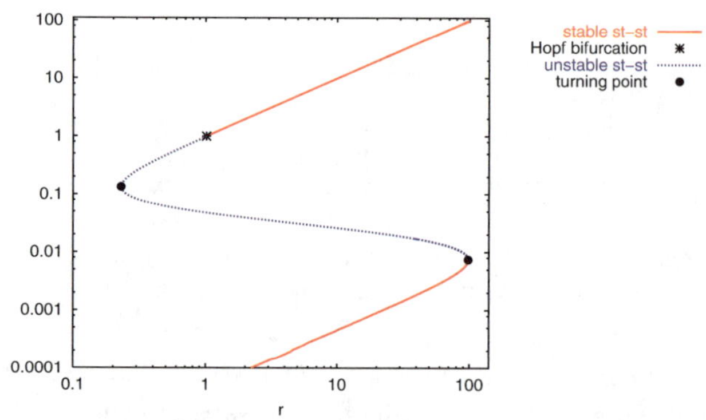

Figure 4. Steady states $\beta(r)$ with bistability and limit cycle with Hopf bifurcation for $k = 0.001$, $\gamma = 1$, $\varepsilon = 0.01$.

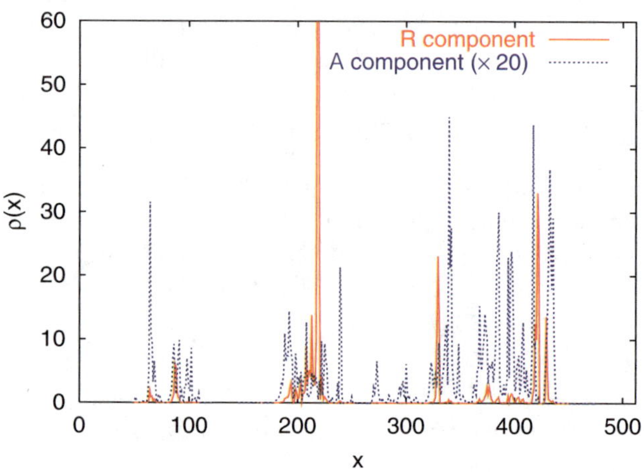

Figure 5. Density of A and R as a function of position at a given time (horizontal cut of Fig. 6).

4. Conclusion and perspectives

1. Interstellar turbulence properties are characterised by a small number of parameters with wavelet analysis (2 parameters for a pure log-normal cascade).

2. These parameters can be used to build synthetic velocity fields at a low numerical cost.

3. Turbulence leads to different structures for different variables, without requiring any external mechanism. The mixing properties of turbulence ensure that on any line of sight a fraction of the gas is in a far from equilibrium state.

4. Extension to 2D is required as a 1D field precludes vorticity and leads to excessive density contrasts.

5. Extension of the chemical set is underway.

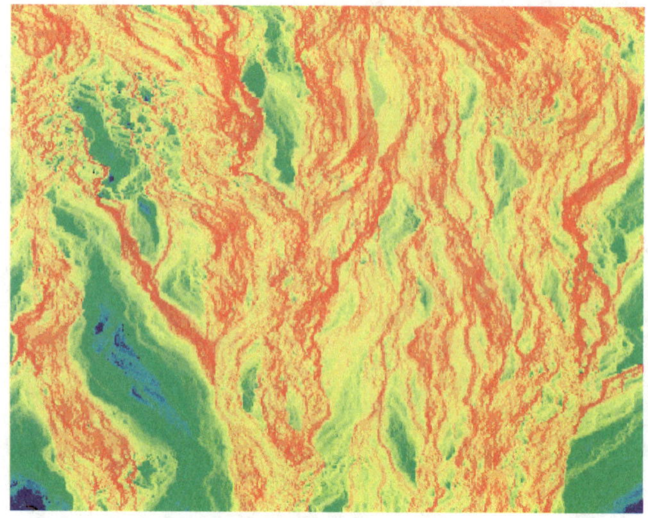

Figure 6. Density of A in a bistable case. The horizontal axis represents position and time flows from top to bottom. Density varies from deep blue to light red.

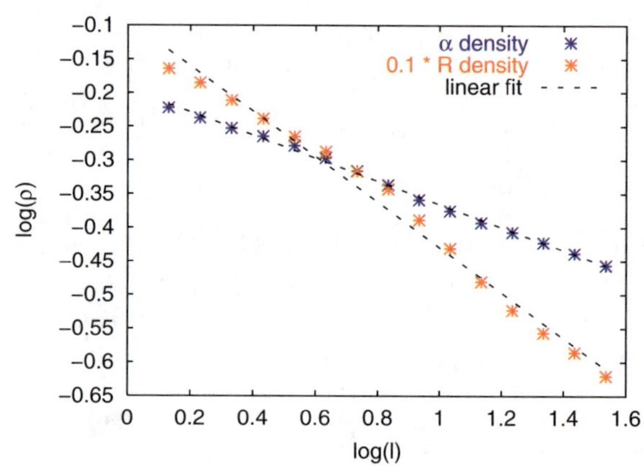

Figure 7. Decimal logarithm of α and R as a function of scale. The slopes are -0.17 ± 0.03 for α and -0.32 ± 0.03 for R.

6. Deviation from the log-normal cascade could be checked with larger maps.

Acknowledgements

We thank J. Pety for the map of Polaris, D. Pelat, A. Arnéodo and E. Falgarone for many discussions.

References

Arnéodo A., Muzy J. F. & Roux S. G. 1997, J. Phys. II (France), 7, 363

Décamp N. & Le Bourlot J. 2002, A&A, 389, 1055

Falgarone E., Panis J.-F., Heithausen A. et al. 1998, A&A, 331, 669

Le Bourlot J., Pineau des Forêts G., Roueff E. 1995, A&A, 297, 251

MOLECULAR ABUNDANCES IN BARNARD 68

James Di Francesco[1,2], Michiel R. Hogerheijde[1,3], William J. Welch[1], and Edwin A. Bergin[4]

[1]Radio Astronomy Laboratory, 601 Campbell Hall, University of California, Berkeley, Berkeley, CA, 94705-3411, U.S.A.
[2]National Research Council, Herzberg Institute of Astrophysics, 5071 West Saanich Rd., Victoria, BC V9E 2E7 Canada
[3]Steward Observatory, The University of Arizona, 933 N. Cherry Ave., Tucson, AZ, 85721-0065, U.S.A.
[4]Harvard-Smithsonian Center for Astrophysics, 60 Garden St. MS 42, Cambridge, MA, 02138-1516, U.S.A.

Abstract

Abundances for $C^{18}O$, CS, NH_3, H_2CO, C_3H_2, and N_2H^+ and upper limits for the abundances of ^{13}CO and HCO^+ are derived for gas within the Bok globule B68 using our own BIMA array data, single-dish data from the literature, and Monte Carlo radiative transfer models. B68 has had its density structure well determined, removing a major uncertainty from abundance determinations. All abundances for B68 are lower than those derived for translucent and cold dense clouds, but perhaps only significantly for N_2H^+, NH_3, and C_3H_2. CS depletion toward the extinction peak of B68 is hinted at by the large offset between the extinction peak and the position of maximum CS line brightness. $C^{18}O$ and N_2H^+ abundances derived here are consistent with recently determined values at positions observed in common.

Key words: ISM: molecules — ISM: abundances — ISM: globules

1. Introduction

We examine the abundances of several molecular species in the Bok globule Barnard 68 (B68; Barnard 1919), located at 125 pc in Ophiuchus (de Geus, de Zeeuw, & Lub 1989) with no evidence for internal star formation. Figure 1 shows an R-band image of B68 obtained from the Aladin sky atlas, revealing its compact morphology. Using the extinction of the background Galactic bulge population, Alves, Lada, & Lada (2001; ALL01) found a radial column density profile for B68 that was well matched with that of an isothermal Bonnor-Ebert sphere, specifically one with a near-critical center-to-edge density contrast of 16.5 (Bonnor 1956; Ebert 1955). This robust density determination removes a major uncertainty towards determining the molecular abundances in B68. Recently, Bergin et al. (2002) used single-dish maps of B68 in conjunction with models incorporating the ALL01 Bonnor-Ebert density structure to ascertain radial variations in $C^{18}O$ and N_2H^+ abundance, with the lowest values occurring toward the highest extinctions. We explore further abundances of these and other species in B68, also incorporating the ALL01 density structure into models.

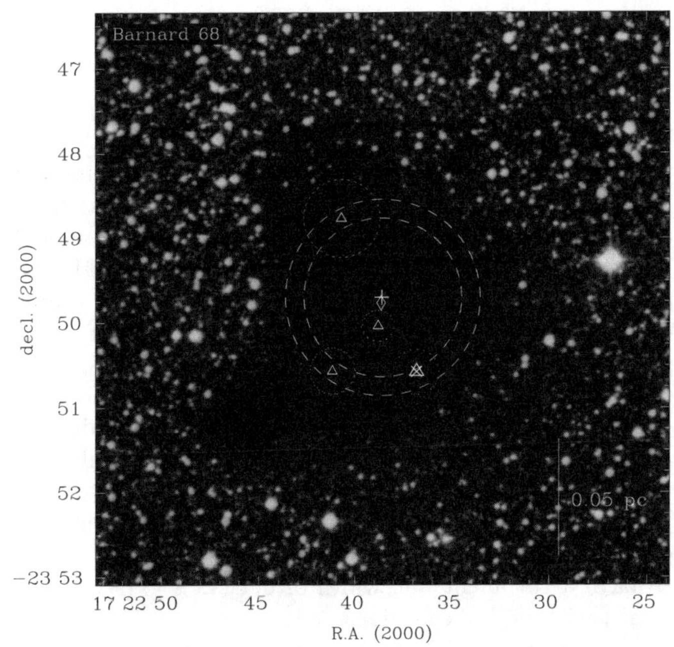

Figure 1. ESO MAMA R-band image of B68 from Aladin. The extinction peak is denoted by a diamond. The central pointing position of the BIMA data is shown as a cross and the primary beam FWHMs at 89.2 GHz and 110.2 GHz are shown respectively as outer and inner dashed circles. The positions and beam FWHMs of single-dish line observations from the literature are shown respectively as triangles and dotted circles. The "X" denotes the position of B68 from Clemens & Barvainis (1988).

Chemistry as a Diagnostic of Star Formation (2003). *Edited by Charles L. Curry and Michel Fich.* Published in 2003 by NRC Research Press, Ottawa ON, K1A 0R6, Canada. ISBN 0-660-19089-3.

2. Observations

Table 1 summarizes the data used to determine molecular abundances in B68, listing the lines used, the source observatory, the angular resolution, and the angular offset of the observation from the extinction peak. Figure 1 shows the position observed using the BIMA millimeter interferometer, approximately 4″ north of the extinction peak. Figure 1 also shows the minimum and maximum FWHM sizes of the BIMA primary beams at the frequencies of ^{13}CO 1–0 and HCO^+ 1–0. Line emission was not detected with the BIMA array at any observed frequency or position toward B68. Table 1 also lists studies where detections of lines were made toward B68 from single-dish observations having angular resolutions <60″. Figure 1 shows the resolutions of these data and the positions where line characteristics were reported. Note that these positions, typically the locations of peak line intensity, vary widely across B68.

Table 1. Summary of Line Observations of B68

Line	Site	Beam (″ × ″)	Offset (″)	Ref.
^{13}CO 1–0	BIMA	17.1 × 4.7	4.03	1
$C^{18}O$ 1–0	BIMA	17.7 × 4.5	4.03	1
$C^{18}O$ 2–1	CSO	30 × 30	59.3	2
CS 2–1	FCRAO	46 × 46	66.2	3
HCO^+ 1–0	BIMA	18.5 × 7.0	4.03	1
N_2H^+ 1–0	BIMA	17.7 × 6.6	4.03	1
N_2H^+ 1–0	Haystack	18 × 18	16.3	4
NH_3 (1,1)	Effelsberg	40 × 40	54.5	5
H_2CO 3_{12}–2_{11}	CSO	30 × 30	59.3	2
C_3H_2 2_{12}–1_{01}	Haystack	18 × 18	16.3	4

References— (1) This work; (2) Wang et al. (1995); (3) Launhardt et al. (1998); (4) Benson, Caselli & Myers (1998); (5) Lemme et al. (1996).

3. Abundances

Abundances were estimated using the 1D Monte Carlo code of Hogerheijde & van der Tak (2000) to calculate radiative transfer, molecular excitation, and line emission through a model of B68, and varying abundances until the output matched the observed data. Our B68 model had the thermal and density structures from ALL01, i.e., an isothermal sphere of T = 16 K with an outer radius of 12 500 AU and a "Bonnor-Ebert parameter" ξ_{max} of 6.9 (i.e., center-to-edge density ratio of 16.5). The central density is $\sim 2.5 \times 10^5$ cm^{-3}. Constant radial abundances were assumed. Although abundance gradients within B68 are quite possible (see Bergin et al.), they should not produce large differences in the derived abundances, given the small sizes of the beam widths relative to the spatial extent of B68. The velocity field of the gas was assumed to have a turbulent line width of 0.4 km s^{-1} (see Wang et al.) with no bulk motions.

To compare models with interferometer data, "visibility" datasets were produced from the model cubes using the baselines from the original data. These datasets were inverted, cleaned and restored in the same manner as the actual data, and velocity-integrated line intensities calculated. By properly accounting for spatial filtering, we estimated BIMA would have recovered only ~20% of the flux emitted from an ALL01 Bonnor-Ebert sphere in B68. To compare models with single-dish data, the model cubes were convolved with Gaussians of widths appropriately representing the resolutions of the respective observations.

Table 2 lists the fractional abundances of the sampled molecular species at various positions toward B68. The BIMA data yield the first upper limits to the abundances of HCO^+ toward the extinction peak of B68, as well as upper limits to the abundances of ^{13}CO, $C^{18}O$ and N_2H^+ at the same location. In addition, single-dish data of B68 from the literature yield abundance values of H_2CO, CS, NH_3, and C_3H_2 toward B68 for the first time, but at positions offset from the extinction peak. Finally, other single-dish data of B68 from the literature yield new abundance values of $C^{18}O$ and N_2H^+ at some of these latter locations.

Table 2 lists fractional abundances from the literature for the species studied here, in clouds similar in character to B68. The first set, for "translucent clouds," are those listed by Turner (2000) for round Clemens & Barvainis clouds with edge-to-center visual extinctions of 2.0. The second set, denoted by "cold dense clouds," are those compiled by Ohishi, Irvine, & Kaifu (1992) toward TMC-1 or L134N. Turner notes that these abundances should not be regarded as universal as they can vary by an order of magnitude both between different translucent clouds and within larger, dense clouds. Every value of molecular abundance we derive for B68 is less than the lowest value derived for other clouds. However, only the N_2H^+, NH_3, and C_3H_2 abundances differ by over an order of magnitude, and so only these species may be arguably depleted in B68, at least at the locations observed. However, Figure 1 and Table 2 show that less-discrepant abundances are found at positions relatively far from the extinction peak but more-discrepant abundances are found at posi-

Table 2. Abundances in Various Molecular Clouds[1]

Species	B68	Translucent Clouds	TMC-1	L134N
^{13}CO	<1.3(−7)	–	8.9(−7)	8.9(−7)
$C^{18}O$	<1.3(−7)	–	1.6(−7)	1.6(−7)
$C^{18}O$	3.0(−8)	–	1.6(−7)	1.6(−7)
CS	4.0(−10)	1.1(−9)	1(−8)	1(−9)
HCO^+	<1.4(−10)	2(−9)	8(−9)	8(−9)
N_2H^+	<1.3(−10)	1(−9)	5(−10)	5(−10)
N_2H^+	2.0(−11)	1(−9)	5(−10)	5(−10)
NH_3	7.0(−10)	2.1(−8)	2(−8)	2(−7)
H_2CO	4.0(−10)	6.3(−9)	2(−8)	2(−8)
C_3H_2	1.2(−11)	3.6(−8)	3(−8)	2(−9)

[1] Derived from ^{12}CO abundance assuming $^{12}C/^{13}C = 90$ or $^{16}O/^{18}O = 500$. $a(-b)$ denotes $a \times 10^{-b}$.

tions closer to the extinction peak (except notably NH_3). This pattern suggests C_3H_2 may be also depleted by some process related to extinction, e.g., the sublimation of gas-phase molecules onto grains, as suggested by Bergin et al. for $C^{18}O$ and N_2H^+. This same idea may also explain how CS and H_2CO appear relatively undepleted in the outer, less-extincted radii of B68. Moreover, the lack of bright emission in CS near the extinction peak hints that CS may be also depleted at high extinction in B68.

For their abundance estimates for B68, Bergin et al. assumed a Zucconi et al. temperature profile. They found the $C^{18}O$ abundance in B68 rises from very low values at $A_V << 1$ to a peak of 1×10^{-7} at $A_V = 2$, and decreases to 1×10^{-9} at $A_V > 20$, a contrast of ~100. Also, they found the N_2H^+ abundance rises from very low values at $A_V << 1$ to a peak of 6×10^{-11} at $A_V = 3$, and decreases to 3×10^{-11} at $A_V > 20$, a contrast of ~2. Any evidence of N_2H^+ depletion is remarkable, given its oft-described utility as a non-depleting probe of dense core interiors (e.g., see Tafalla et al. 2002.)

Adopting the same temperature profile in our models, derived abundances are increased by factors of 2-3 from those listed in Table 2. Our interferometer data of $C^{18}O$ and N_2H^+ do not provide much additional support for the abundance models of Bergin et al., except that the non-detection of compact line emission with BIMA suggests the low abundances found by Bergin et al. are not due to the beam dilution of small clumps of relatively abundant material. Our derived upper limits for $C^{18}O$ and N_2H^+ from the BIMA data are not particularly low, but they remain consistent with the still-lower values found by Bergin et al. throughout the core.

The single-dish $C^{18}O$ and N_2H^+ data from the literature only pertain to one line-of-sight per transition toward B68, but provide strong support for the abundances derived by Bergin et al., again assuming the same Zucconi et al. temperature profile. The $C^{18}O$ literature data, from a position 59.″4 offset from the extinction peak, yield an apparent abundance of 6–9 × 10^{-8}, quite consistent with their 3–9 × 10^{-8} abundances at this position. Also, the N_2H^+ literature data, from a position 16.″3 offset from the extinction peak, yield an apparent value of 4–6 × 10^{-11}, slightly larger than but still consistent with their 3–4 × 10^{-11} abundance at this position.

Acknowledgements

JD was supported by the Radio Astronomy Laboratory. MRH was supported by the Miller Institute for Basic Research in Science. We thank Charles J. Lada, João Alves, Tracy Huard, and Jon Swift for valued discussions. This research made use of the SIMBAD database and Aladin, both operated by CDS, Strasbourg, France.

References

Alves, J. F., Lada, C. J., & Lada, E. A. 2001, Nature, 409, 159 (ALL01)
Barnard, E. E. 1919, ApJ, 49, 1
Benson, P. J., Caselli, P, & Myers, P. C. 1998, ApJ, 506, 743
Bergin, E. A., Alves, J., Huard, T., & Lada, C. J. 2002, ApJ, 570, L101
Bonnor, W. B. 1956, MNRAS, 116, 351
Clemens, D. P., & Barvainis, R. 1988, ApJS, 6
de Geus, E. J., de Zeeuw, P. T., & Lub, J. 1989, A&A, 216, 44
Ebert, R. 1955, Z. Astrophys., 37, 217
Hogerheijde, M. R. & van der Tak, F. F. S. 2000, A&A, 362, 697
Launhardt, R., Evans, N. J., II, Wang, Y., Clemens, D. P., Henning, Th., & Yun, J. L. 1998, ApJS, 119, 59
Lemme, C., Wilson, T. L., Tieftrunk, A. R., & Henkel, C. 1996, A&A, 312, 585
Ohishi, M., Irvine, W. M., & Kaifu, N. 1992, in *Astrochemistry of Cosmic Phenomena*, Proceedings of the 150th Symposium of the International Astronomical Union, ed. P. D. Singh (Dordrecht: Reidel), 171
Tafalla, M., Myers, P. C., Caselli, P., Walmsley, C. M., & Comito, C. 2002, ApJ, 569, 815
Turner, B. E. 2000, ApJ, 542, 837
Wang, Y., Evans, N. J., II, Zhou, S., & Clemens, D. P. 1995, ApJ, 454, 217
Zucconi, A., Walmsley, C. M., & Galli, D. 2001, A&A, 376, 650

MULTIPLE SPECIES CHEMICAL MODELING AS A STRUCTURAL AND EVOLUTIONARY PROBE OF STAR-FORMING REGIONS

S. D. Doty[1], E. F. van Dishoeck[2], F. Schoier[2], J. Jørgensen[2], F. F. S. van der Tak[3] and A. M. S. Boonman[2]

[1]Department of Physics & Astronomy, Denison University, Ohio, USA
[2]Leiden Observatory, P.O. Box 9513, 2300 RA Leiden, The Netherlands
[3]MPIfR, Bonn, Germany

ABSTRACT

We present detailed models of the time- and space-dependent gas-phase chemistry for the envelopes of star forming regions. We consider both the hot, high-mass region AFGL 2591 as well as the low-mass region IRAS 16293-2422, which may contain a hot core. The chemical and thermal balance models are used to probe the structure inferred from observational studies. We compare the resulting model predictions with observations, which cover a wide range of physical conditions within the source. With this comparison, we study the chemistry and evolution of the regions.

Key words: ISM: molecules – ISM: structure – ISM: clouds

1. INTRODUCTION

Chemical reactions, by their nature, depend upon the temperature and density of the region in which they occur. As a result, chemical abundances should (and do) vary significantly with position within astrophysical sources. As a result, the abundance and excitation of different chemical species has the potential to provide a detailed picture of the physical structure and age of star-forming regions. Utilizing an observationally-driven underlying physical model, we have approached this problem for two sources: the relatively isolated massive star-forming region AFGL 2591, and the low-mass source IRAS 16293 which may contain a hot core.

2. AFGL 2591

2.1. PHYSICAL MODEL

We adopt the physical model for AFGL 2591 determined by van der Tak et al (1999). Fits to the continuum spectrum and line data, suggest a dust density distribution of $n(r) \sim r^{-1}$.

Given the density distribution, we compute the temperature structure via thermal balance in the gas. The resulting density and temperature distributions are shown in Figure 1. Notice how the gas temperature tracks the dust temperature due to collisional coupling.

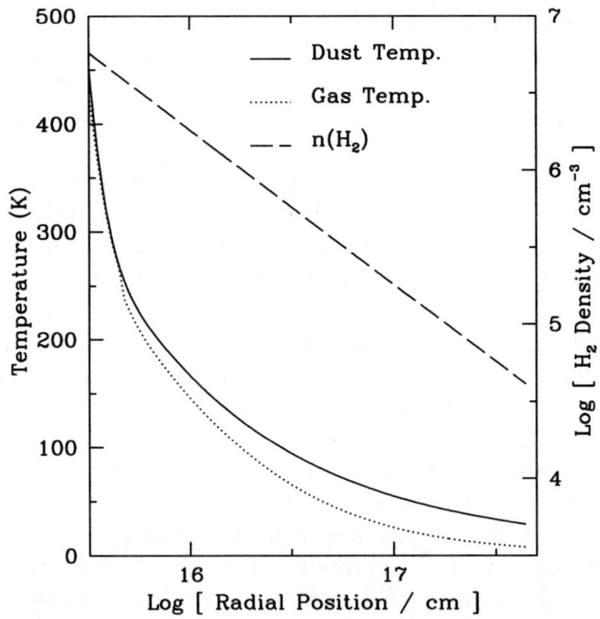

Figure 1. Density and temperature distribution in the model.

From $n(r)$ and $T(r)$, we can solve the time- dependent chemistry as a function of position. As can be seen in Figure 2, we allow for ice mantle evaporation at $T \sim 100$ K.

2.2. IMPORTANCE OF PHYSICAL-CHEMICAL MODELING

With the advancements in the quality and quantity of observations of molecular transition, and with the

Chemistry as a Diagnostic of Star Formation (2003). *Edited by Charles L. Curry and Michel Fich.* Published in 2003 by NRC Research Press, Ottawa ON, K1A 0R6, Canada. ISBN 0-660-19089-3.

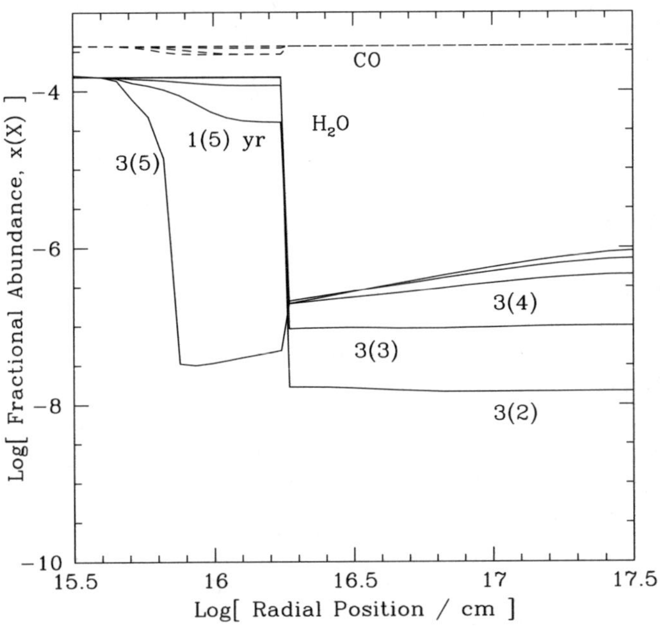

Figure 2. The time and position-dependent abundances of H_2O and CO for AFGL 2591.

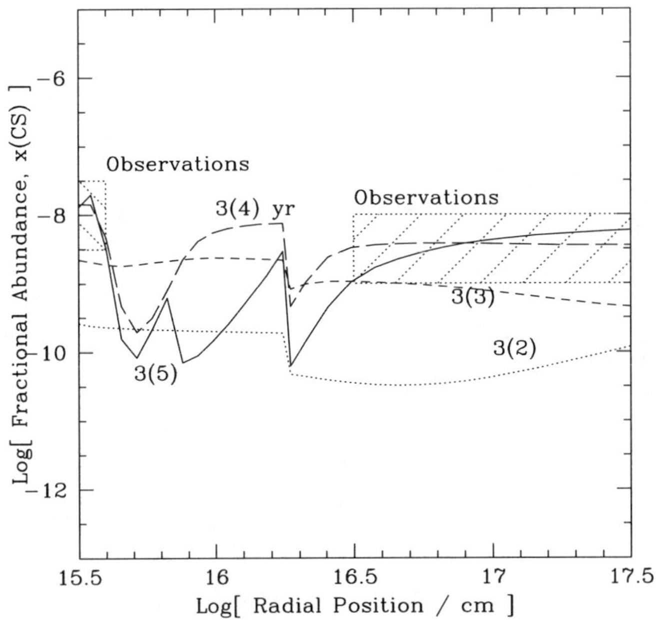

Figure 3. The time and position-dependent abundances of CS. Note the agreement with observations.

increased use of sophisticated line radiative transfer models to understand these data, it is possible to constrain abundance profiles significantly.

We take the physical-chemical structure above as input to our gas-phase chemical modeling. The chemistry is based upon the UMIST network. We construct time-dependent chemical models at 30 radial positions within the source to determine a time- and space-dependent chemistry.

One example is CS (Fig. 3). While there is structure in the abundance profile, there is agreement with the observations both in the interior and exterior regions. In particular, the models are consistent with an age of $> 3\times10^3$ years.

2.3. Structure and Age

A check on our understanding of the structure and processes of the source, then, is a detailed comparison of the predictions with observations.

In Figure 4 below we plot the chi-square between the model predictions and observational data. As can be seen, molecules observed in both infrared (absorption along the line of sight) and submillimeter (emission from the entire beam) can be simultaneously fit. These data suggest that the inferred structure is reasonable, and that AFGL 2591 is approximately 3×10^4 years old since turn-on of the central source.

3. IRAS 16293

IRAS 16293 is a low-mass region that may have a potential hot core. As a result, it is a good testbed for the range of applicability of our results. As a test, we have applied this same chemical model to IRAS 16293. We have also directly computed the line strengths and profiles for comparison with the observations in Figure 5, where we plot the percentage difference for the line strengths between the models and observations. Here the time ranges are $3 \times 10^3 - 3 \times 10^4$ years. While the agreement is not exceptionally poor, the range of values suggest that differences in initial abundances, effects from the internal source, and infall may play important roles in the interior of this source.

4. Conclusions

We have constructed detailed physical-chemical models of both high- and low-mass star-forming regions. This approach uses multiple chemical species and transitions

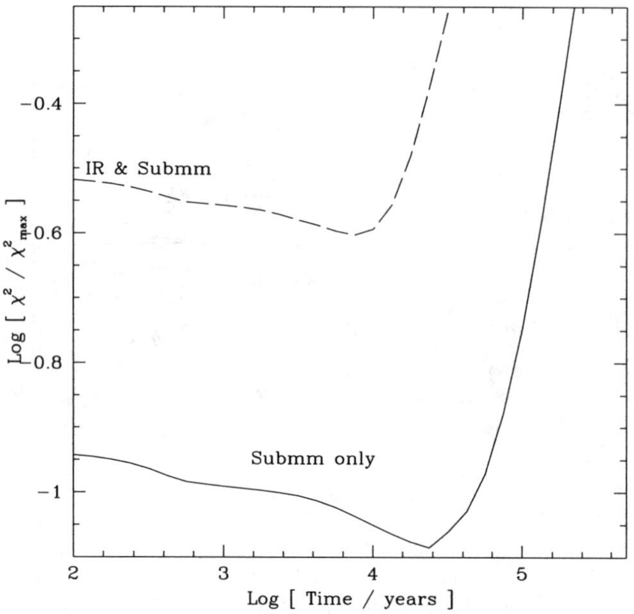

Figure 4. A comparison of the model predictions for abundances and column densities with the observational data are quantified as the chi-squared. Notice that both the IR and submm data give good fits for $t \sim 3 \times 10^4$ years since turn-on.

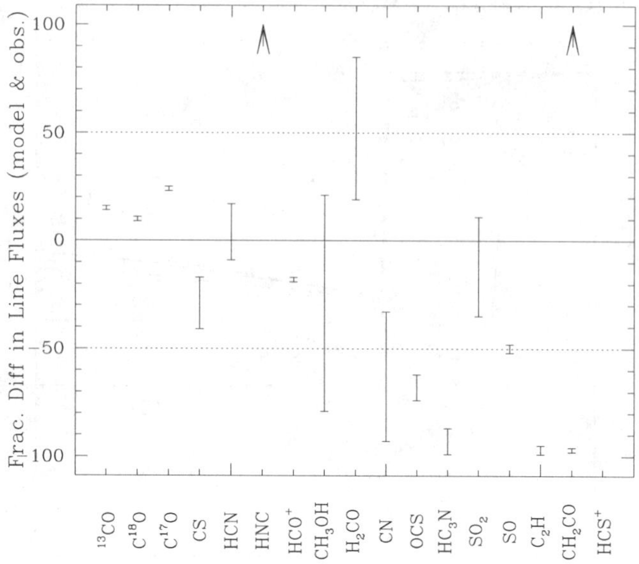

Figure 5. A percentage difference comparison for predicted and observed line strengths for species observed toward IRAS 16923.

to probe the temperature and density structure of these regions, with the goal of confirming a proposed physical structure and potentially age-dating the source. Based upon our results above, we find that:

1. It is possible to construct meaningful, detailed physical-chemical models for comparison with both inferred quantities (N, x) and line profiles.

2. The comparison of the model results with observations seems to provide further confirmation of proposed physical structure models.

3. The simultaneous comparison of model predictions with observations for many species may allow for the age-dating of these sources since turn-on.

Acknowledgements

This work was partially supported by support from the Research Corporation (SDD), and NWO grant 614-41-003 (AB, FvdT), and an NWO Spinoza award (EvD).

References

Doty, S. D., van Dishoeck, E. F., van der Tak, F. F. S., & Boonman, A. M. S., 2002, A&A, 389, 446

van der Tak, F. F. S., van Dishoeck, E. F., Evans, N. J., Bakker, E. J., & Blake, G. A., 1999, ApJ, 522, 991

INFERRING CLOUD STRUCTURE BY MODELING CONTINUUM DUST EMISSION

S. D. Doty[1], M. Palotti[1], M. Moore[1], Y. Shirley[2], C. Young[2], and N. J. Evans[2]

[1]Department of Physics & Astronomy, Denison University, Ohio, USA
[2]University of Texas, Texas, USA

Abstract

We present results from a study of the density structures of star-forming clouds. In 1-D modeling, we consider the robustness of existing semi-analytic methods of data analysis for the purpose of inferring source properties, as well as the need for detailed modeling to infer the density structure. We also adopt a new 3-D radiative transfer model to study non-spherical sources, with the goal of inferring the source structure from the SEDs and sky maps. We find that both the SEDs and surface brightness contours are necessary to properly constrain the models from observations.

Key words: ISM: molecules – ISM: structure – ISM: clouds

1. Introduction

The physical structure (density distribution) and source properties (grain properties, existence of a central heat source, etc.) are central to an understanding of star-forming regions. Here we report on a two-pronged program taken to study this problem. In one part, we consider the ability to invert the radiative transfer problem to infer source properties. In the second, we apply detailed radiative transfer model to astrophysical data in order to infer these very properties.

2. General Approach

In order to accurately infer the dust density and source structure, we construct a series of detailed, self-consistent, continuum radiative transfer models. The output from these models is then convolved with the appropriate beam response function, with the results treated as "simulated observations." We then analyze these "observations" to infer the source properties and compare the results with those adopted to study the reliability of the analysis methods.

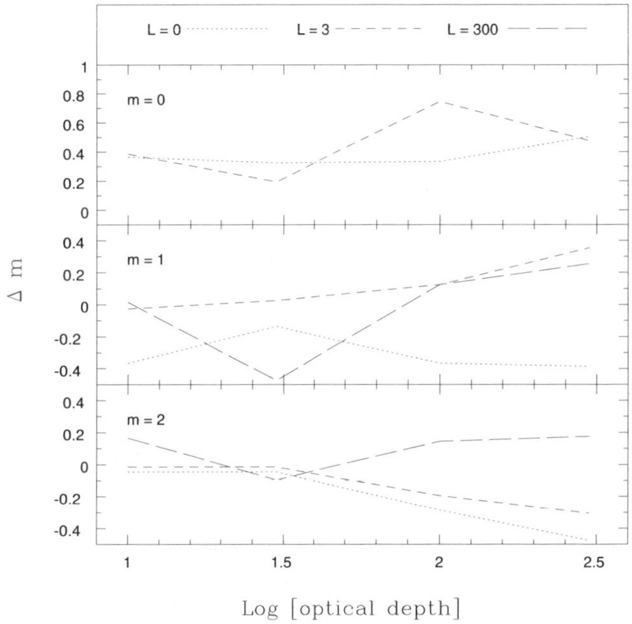

Figure 1. Density and temperature distribution in the model.

In cases where we compare to actual observational data (e.g., the 3-D models), the "simulated observations" are compared with the actual data, and the models are modified to best fit the data.

3. Cloud Structure

3.1. One Dimension

Question: Is it possible to accurately infer the dust density distribution from the emergent radiation?
Approach: Construct a grid of models, and compare the inferred results to the actual values used.

In the optically thin and isothermal limit, the dust density distribution may be inferred from the intensity: $I \sim N$. Results are shown in Figure 1. Note that the density power is poorly fit—due to the role of a small

Chemistry as a Diagnostic of Star Formation (2003). *Edited by Charles L. Curry and Michel Fich.* Published in 2003 by NRC Research Press, Ottawa ON, K1A 0R6, Canada. ISBN 0-660-19089-3.

temperature gradient and the breakdown of the R-J approximation.

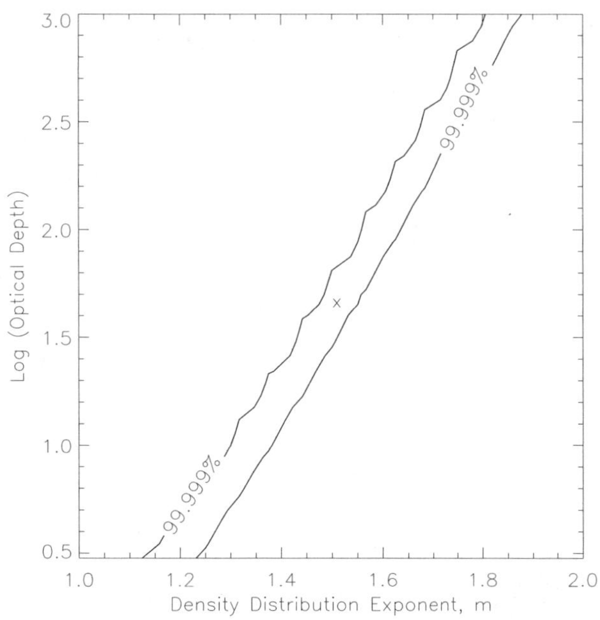

Figure 2. Density and temperature distribution in the model.

To test the ability of detailed radiative transfer models to constrain the density, we add noise to the simulated observations and compare to a large grid (> 2000) of models. The results are shown in Figure 2. The expected result is shown by the "x". Note the much better fits, even though the source is unresolved here.

Based upon these results, detailed radiative transfer modeling is superior to semi-analytic methods of inferring the dust density distribution.

3.2. Three-Dimensional Sources

Question: Is it possible to model non-spherical sources?
Approach: Construct detailed 3-D models, attempting to fit the observed data.

In Figure 3, we plot the observed (*points*) and predicted (*lines*) SED for the PPC L1527. Note that it is possible to simultaneously fit the observed data points—in this case, both with a disk surrounding the central source (*dashed line*) and with only an extended envelope

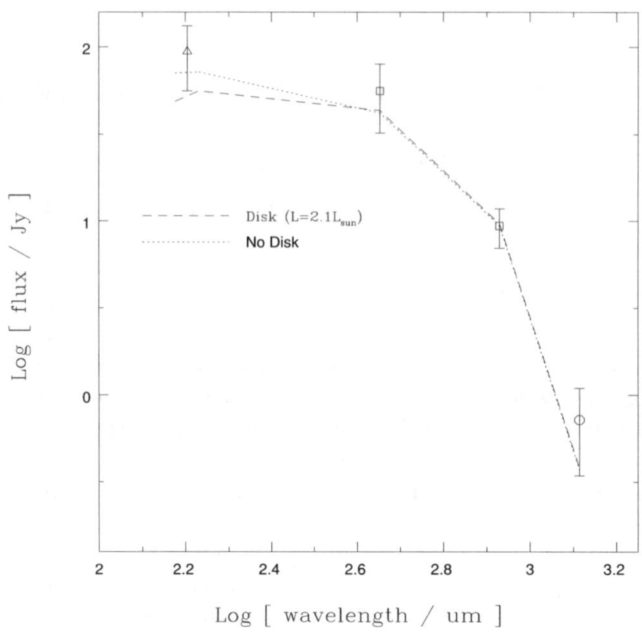

Figure 3. Observed (*points*) and modeled (*lines*) SED for L1527.

(*solid line*). This shows that any submm disk contribution to this source is probably due to free-free emission.

In Figure 4, we plot observed (*left*) and modeled (*right*) surface brightness contours for the PPC L1544. Note that it is possible to simultaneously fit the surface brightness.

4. Dust Properties

The dust opacity power law index is often used to infer the history (e.g., annealing), composition (e.g., amorphous vs. crystalline), and/or size of dust grains. In principle, the information is encoded in the SED. To test the ability to invert this problem, we have constructed models with an adopted $\beta = 1.5$, and then attempted to infer the dust opacity power law from the model results.

In Figure 5, we plot the inferred dust opacity index from the slope of the FIR flux spectrum. Notice that the results are poor until one reaches very long wavelengths, due to the breakdown of the R-J approximation and lack of true isothermality in the source.

To test another approach, we consider fitting the FIR flux spectrum with two modified blackbodies. The results are shown in Figure 6. Notice the much better agreement due to the removal of the isothermal and R-J assumptions.

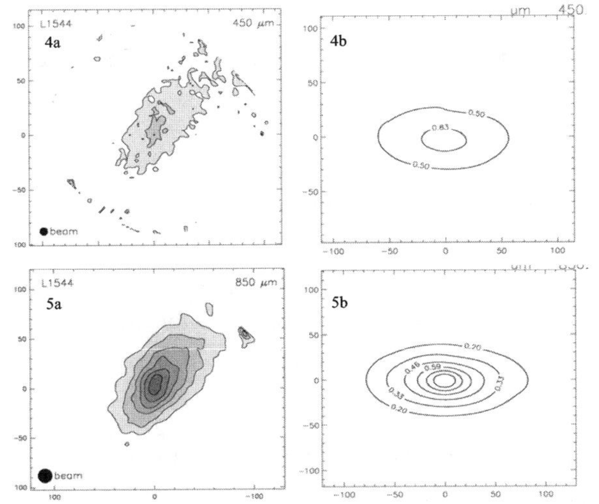

Figure 4. Observed (*left*) and modeled (*right*) brightness contours for L1544.

5. Conclusions

We have studied the ability to infer source properties from observations. By considering both "simulated" and real observations, we find that:

1. While semi-analytic methods of inferring source properties can work, the greater the sophistication of the modeling, the more reliable the results.

2. It is possible to infer source properties (such as density structure, geometry, and internal source properties) by considering 3-D models of non-spherical cores.

3. Even when using sophisticated modeling, it is important to use as much data as possible to constrain the models.

Acknowledgements

This work was partially supported by support from the Research Corporation (SDD).

References

Doty, S. D., & Palotti, M. L., 2002, MNRAS, 335, 993

Doty, S. D., Palotti, M. L., Shirley, Y., Young, C., & Evans, N. J., 2003, in preparation

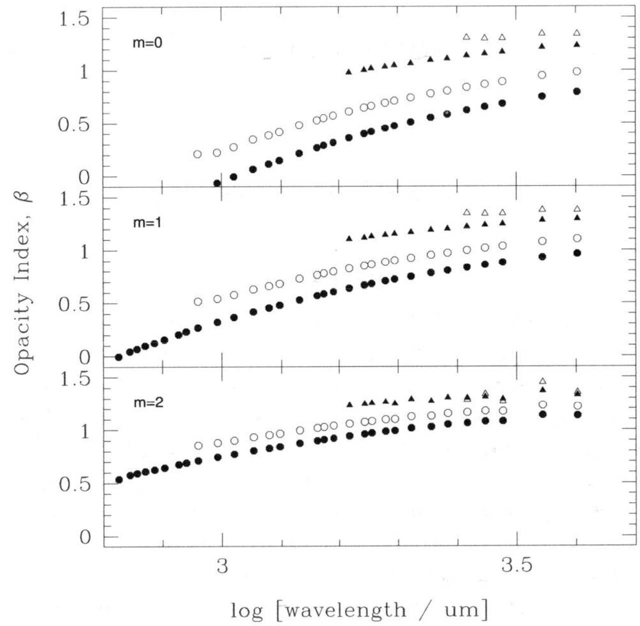

Figure 5. Opacity index inferred from the slope of the FIR spectrum. The adopted value is 1.5.

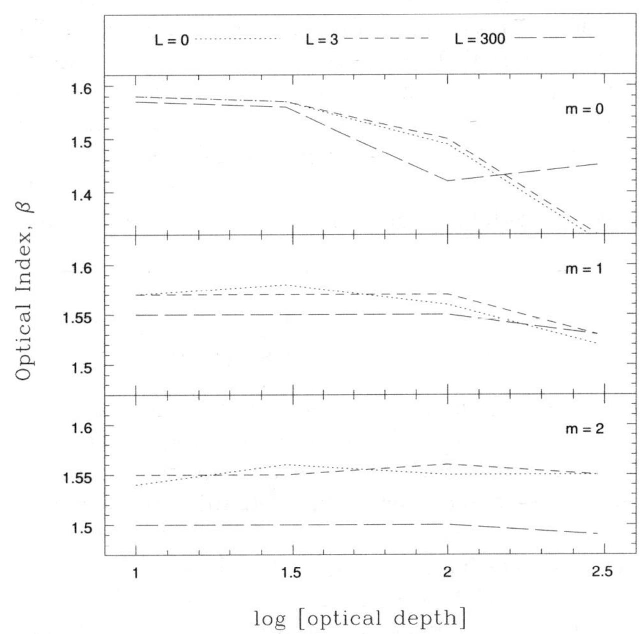

Figure 6. Opacity index inferred by fitting the FIR spectrum. The adopted value is 1.5.

DIFFUSE INFRARED EXCESS FEATURES IN THE CANADIAN GALACTIC PLANE SURVEY

Kevin A. Douglas and A. Russell Taylor

Department of Physics and Astronomy, University of Calgary Calgary, AB, Canada T2N 1N4

Abstract

As part of a multiwavelength study of the main constituents of the interstellar medium of the Milky Way, we investigate the possibility of tracing diffuse molecular hydrogen (H_2) through an excess in dust opacity. The dust column density inferred from infrared (*IRAS*) observations is found in some regions to be greater than predicted by correlations with neutral atomic hydrogen (H I) and carbon monoxide (CO). The Five College Radio Astronomy Observatory's entire CO $J = 1 - 0$ Outer Galaxy Survey (OGS) region is studied in the context of the Canadian Galactic Plane Survey (CGPS) datasets, to provide arcminute-scale intercomparisons of the H I , CO and dust features. After converting dust emission into optical depth, we compare the relative contributions of these species and identify areas where the dust opacity is greater than expected if a constant dust-to-gas ratio is assumed. An excess of dust that is not accounted for by either H I or CO may provide evidence of diffuse H_2 that is not detectable by the standard CO proxy. Alternatively, the excess may show evidence of variations in the assumed correlation between gas and dust.

Key words: ISM: molecules – ISM: structure – ISM: clouds

1. Infrared Excess Features

Observations of the gas and dust content of high-latitude gas clouds have shown interesting relationships between the emission by the different constituents of these structures. Several studies comparing *IRAS* dust emission with H I and CO data in these cirrus clouds have yielded evidence of a molecular component which is not traced by the emission of CO molecules (Reach, Koo & Heiles 1994, Meyerdierks & Heithausen 1996, Boulanger *et al.* 1998). Excess infrared emission from *IRAS* points to the possibility of a diffuse molecular gas component that may be comparable in H_2 mass to that of the atomic hydrogen content. This analysis is being extended to the plane of the Galaxy to search for sources of molecular hydrogen not traced by the CO surrogate.

2. Analysis of CGPS Datasets

The Canadian Galactic Plane Survey (Taylor *et al.* 2002) includes high resolution observations of dust, CO and H I for a 660 square-degree section of the Milky Way, bounded by galactic longitudes of 74 and 147 degrees, with a latitude coverage of $-3.5 < b < +5.5$. The *IRAS* datasets provide infrared dust emission maps, while the CO data come from the FCRAO OGS (Heyer 1998) and have been reprocessed to remove noise patterns caused by spatial correlations in the reference positions during the observations (Brunt et al., in preparation). The H I maps are obtained from 21 cm observations at the Dominion Radio Astrophysical Observatory.

2.1. Methodology

The FCRAO CO maps comprise the smallest of the CGPS datasets and therefore determine our spatial range of study. There are 20 CGPS 'mosaics,' each of 25 square degrees in area, which together contain the entire OGS region. From these 20 we analyse the region in five subsets, each containing four of the mosaics and each measuring $9° \times 9°$. Though all the datasets have similar resolutions, we convolve all maps to an identical resolution of three arcminutes to allow pixel-to-pixel comparisons to be made.

Since dust emissivity is temperature-dependent, it is essential to convert the infrared maps into dust *opacity*, to obtain the dust column density in the region of interest. Hence we derive a 'colour temperature' T_c for the dust using a ratio of the *IRAS* intensities at 60 and 100 μm. Once T_c is determined for each pixel, an opacity (τ) map is produced at each wavelength (λ) according to the relation $\tau_\lambda = I_\lambda / B_\lambda(T_c)$. In this equation, I_λ is the *IRAS* intensity and B_λ is the Planck blackbody intensity.

Chemistry as a Diagnostic of Star Formation (2003). *Edited by Charles L. Curry and Michel Fich.* Published in 2003 by NRC Research Press, Ottawa ON, K1A 0R6, Canada. ISBN 0-660-19089-3.

From our derived T_c we also identify pixels with a high inferred dust temperature and mask them out of the τ maps, because such regions can be attributed to warm dust rather than cold, diffuse molecular gas. From the masked image of 100-μm opacity, we look for spatial correlations between the dust and gas components of the interstellar medium in the CGPS region. Assuming a constant gas-to-dust ratio, the atomic and molecular gas (derived from the H I and CO emission respectively) in the region should account for essentially all of the observed dust. Since the *IRAS* images are two-dimensional, we integrate the CO and H I datacubes to compare the velocity-integrated intensities of these gas components.

We first look for correlations between the dust opacity and the molecular gas, since both have a rather clumpy structure, and the CO data is our spatially limiting dataset. A good correlation between τ_{100} and the CO intensity was observed in all five of the subsets. Four of the five subsets had the same slope, and the fifth was very close to the others. From this observed correlation we subtracted the appropriately weighted contribution of the CO from the dust opacity. It was found that many of the clumpy dust features in the τ_{100} map were removed following the subtraction. See Figure 1 for an example.

Following the CO subtraction, we seek a correlation between the (CO-subtracted) dust opacity and the H I line emission. The H I is much more widespread than the CO, and the integration over all velocities blends features in the local and Perseus spiral arms. Nevertheless, wide-scale variations between the dust and H I intensities seem to follow the some general spatial relationship. Indeed, a good pixel-by-pixel correlation was determined, allowing us to subtract the H I contribution to the dust opacity using the same slope and intercept in all five subsets.

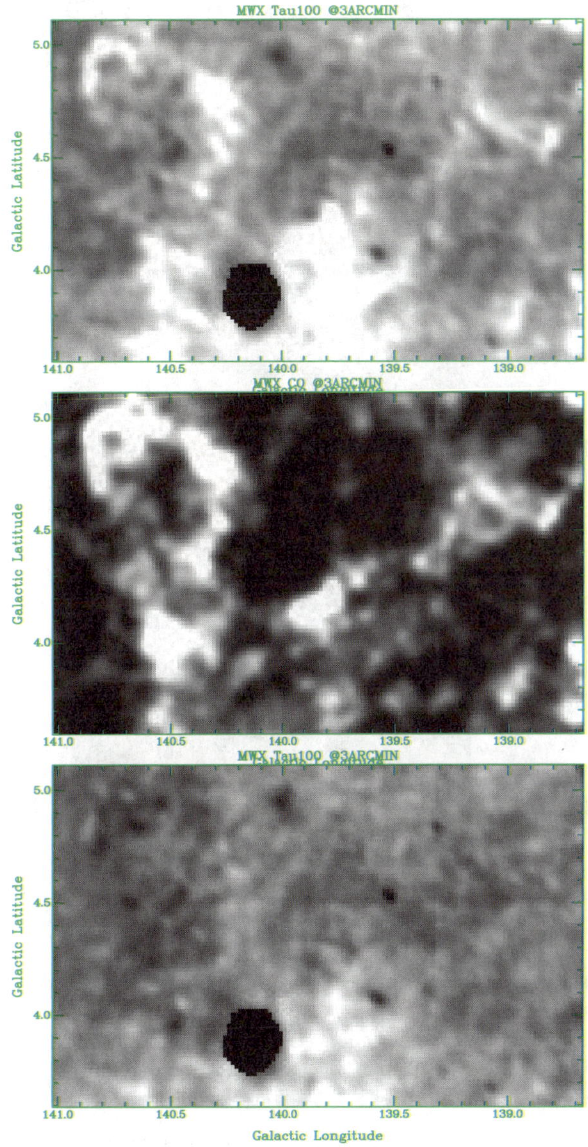

Figure 1. Example of CO subtraction from dust opacity maps. The first and third panels are dust opacity, before and after the subtraction. The middle panel shows the CO present in the region.

2.2. The Maps

While it is not feasible to present the images of the CGPS region of study at its many wavelengths in this paper, Figure 2 shows an interesting infrared feature that displays considerable excess dust opacity following the removal of its CO and H I contribution. The CO clouds near the feature seem to surround the dust feature, yet there is a noticeable different in their morphologies. The H I intensity is low compared to other areas within the region, but still much higher than in previous studies performed at higher galactic latitudes. This feature demonstrates the potential of finding diffuse features in the Galactic Plane that are not associated with atomic hydrogen, nor by the most common tracer of H_2, CO.

The infrared excess sources remaining after the subtractions of CO and H I are in need of intense scrutiny. Approximately 50 infrared excess features have been identified by eye, and we will examine methods to determine the validity of these identifications. The nature of the sources will be investigated; some are diffuse, others are

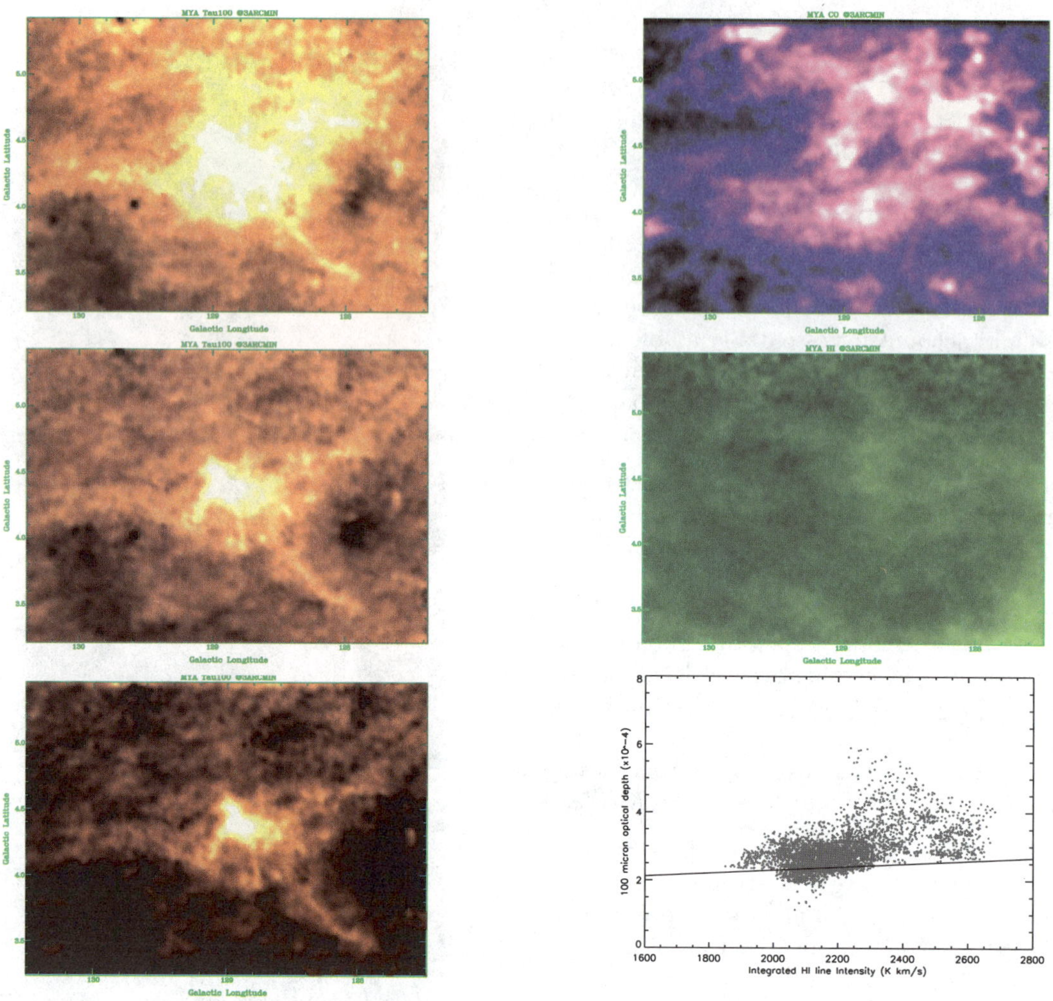

Figure 2. An infrared excess feature in the CGPS region. The left side panels show the dust opacity at various stages of the analysis. The top left panel is the original τ_{100} image; in the middle the CO has been subtracted, and at the bottom left the H I has also been removed. The top right image is the CO emission. The middle right is the neutral hydrogen image. The plot in the bottom right shows the correlation between the CO-subtracted τ_{100} and the H I intensity, with considerable dust column above the general trend.

filamentary, and others are quite compact. The challenge of unravelling the origin of these infrared excess sources will possibly lead to some conclusions about the amount of diffuse molecular gas (H_2) not traced by standard proxies.

The Star Formation and Chemistry conference gave insight to the possible manners in which H_2 can exist without an appreciable amount of CO to trace it. These include the depletion of CO onto grains or the self-shielding of H_2 in the outer layers of photodissociation regions. Also, the formation mechanisms of H_2 and the importance of the molecule to the early stages of star formation may be studied in greater detail if a population of cold molecular clouds that are deficient in heavier molecules can be found.

Acknowledgements

KAD thanks the SFChem2002 organizing committee for hosting a fantastic conference. KAD is grateful for financial support from the conference as well as from ART.

References

Boulanger, F., Bronfman, L., Dame, T. M. & Thaddeus, P., 1998, A&A, 332, 273

Heyer, M., 1998, *New Perspectives on the Interstellar Medium* (San Francisco: ASP Press)

Meyerdierks, H. & Heithausen, A., 1996, A&A, 313, 929

Reach, W. T., Koo, B.-C., & Heiles, C., 1994, ApJ, 429, 672

Taylor, A. R. et al. 2002, AJ, in press

NON-EQUILIBRIUM CHEMISTRY IN THE DISSIPATIVE STRUCTURES OF INTERSTELLAR TURBULENCE

Edith Falgarone[1], Guillaume Pineau des Forêts[2], Pierre Hily-Blant[1], Peter Schilke[3], and Jérôme Pety[4]

[1]LERMA, Ecole Normale Supérieure & Observatoire de Paris, France
[2]IAS, Orsay, France
[3]MPIfR, Bonn, Germany
[4]IRAM, Grenoble, France

Abstract

The cold diffuse medium (T <100 K) harbors molecules known to form only through endothermic reactions or by overcoming activation barriers of several thousand Kelvins. This suggests that tiny pockets of warm gas exist in this medium. We propose that they are generated by the bursts of dissipation of interstellar turbulence. The space-time intermittency of dissipation in incompressible turbulence being ascribed to the existence of small-scale but strong coherent vortices, we model the interstellar dissipative structures as Burgers vortices threaded by helical magnetic fields, an alternative to magneto-hydrodynamical shocks. The chemical evolution of the gas trapped in such a vortex for several hundred years is controlled by the temperature rise due to viscous dissipation at the vortex boundary and by the ion-neutral drift within the vortex. The fact that these two processes are closely associated in space and time explains, without fine-tuning the parameters of the model, the salient features of the molecular species observed in the diffuse medium.

It is enough to have less than one percent of the gas in these warm and chemically active regions to reproduce the observations, and the turbulent energy dissipated on average in these structures is smaller than that available in the turbulent cascade. We follow the isobaric cooling and chemical evolution of a parcel of diffuse gas heated and chemically enriched after its trapping in such a coherent vortex, up to the densities characteristic of molecular clouds. We confront the model results with observations of a high-latitude cloud: weak HCO^+(1–0) line emission and a mosaic in the ^{12}CO(1–0) line from the IRAM-PdB interferometer. The latter reveals a small-scale velocity shear $\sim 500\,\mathrm{km\,s^{-1}\,pc^{-1}}$.

Key words: ISM: kinematics and dynamics – ISM: magnetic fields – turbulence

Chemistry as a Diagnostic of Star Formation (2003). *Edited by Charles L. Curry and Michel Fich.* Published in 2003 by NRC Research Press, Ottawa ON, K1A 0R6, Canada. ISBN 0-660-19089-3.

DEPLETION OF CO IN *MSX* INFRARED-DARK CLOUDS

P. A. Feldman[1], R. O. Redman[1], S. J. Carey[2], and F. Wyrowski[3]

[1]HIA, National Research Council of Canada, 5071 West Saanich Road, Victoria, B.C., Canada V9E 2E7
[2]IPAC, California Institute of Technology, 770 South Wilson Avenue, Pasadena, CA, 91125, USA
[3]Max-Planck-Institut-für-Radioastronomie, Auf den Hügel 69, Bonn, D-53125, Germany

Abstract

Some regions of high dust column density in *MSX* Infrared-Dark Clouds (IRDCs) show evidence of CO depletion. In particular, we find that the bright, compact submm sources G28.34+0.06 P5 and G81.50+0.14 P3 are CO-depleted by factors of $\gtrsim 10$ and ≈ 6, respectively, compared to the other bright, compact sources in their IRDCs. Both of these sources appear to be unusually cold.

Key words: ISM: abundances – ISM: molecules – ISM: clouds

1. Introduction

The infrared-dark clouds (IRDCs) are a population of several thousand molecular clouds which have been identified by their substantial mid-infrared (8 – 25 μm) extinctions in *MSX* images (Egan et al. 1998). Egan et al. concluded that IRDCs possess hundreds of magnitudes of visual extinction and contain large column densities of cold dust. Spectral-line observations of a selection of these clouds by Carey et al. (1998) revealed that much of the gas inside the IRDCs has $T_K \approx$ 10-15 K, $n(H_2) \gtrsim 10^6$ cm^{-3}, and $N(H_2) \lesssim 10^{23}$ cm^{-2}. Their kinematic distances range between 1 and 8.5 kpc. Subsequent observations (Carey et al. 2000; Redman et al. 2000, 2002, 2003) found that the IRDCs contain many bright, compact submillimetre sources which appear to be molecular cores in a variety of the early stages of star formation.

We have obtained JCMT and BIMA images of several *MSX* Infrared-Dark Clouds (IRDCs) in both dust continuum emission and in several spectral-line tracers of molecular gas. In the IRDCs observed to date, N_2H^+ 1–0 emission seems to follow both the SCUBA 850 μm emission (see Fig. 1) and the *MSX* 8 μm absorption rather well, but CO appears to be depleted in some regions of high dust column density. In this paper, we will report initial results of our observations of CO depletion.

2. Observations of G28.34+0.06

Our observations of N_2H^+ 1–0, shown in Figure 1, demonstrate that the gas and dust are mixed uniformly throughout almost all of this IRDC. The exception is a region close to the active source G28.34+0.06 P2 (Redman et al. 2002), which will not be considered further here. In particular, the ratio of the integrated N_2H^+ $1-0$ line strength to the SCUBA 850 μm emission is nearly constant in the sources P1, P5, P6, and P7 that are under discussion here. To the extent that CO is also a constant fraction of the molecular gas, it would be expected that the line strengths in all four of these sources should be comparable (assuming similar gas and dust temperatures). However, our observations, shown in Figure 2, reveal that the CO and the dust are distributed quite differently in this cloud.

To the extent that the CO emission is optically thin and the dust and gas have uniform properties throughout the cloud, we can determine the (relative) ratio of the CO abundance to the dust fraction by plotting the integrated CO line intensity against the flux density of the dust emission. We have smoothed the 850 μm image with a Gaussian kernel to the same angular resolutions as the CO data, and regridded the resulting image onto the same set of points used in the CO images. Figure 2 shows the ^{13}CO 2–1 and C^{18}O 2–1 images, with contours of the smoothed 850 μm emission.

It will be seen immediately from this figure that CO is more widely distributed than the dust in this IRDC, with CO emission present over the whole area at about a quarter of the intensity of the peak near P1. For the current purpose we are not particularly interested in this widely distributed component of the gas, but instead want to know how much CO is associated with the dust in each of the compact sources. Small regions were selected for analysis around each source, shown as the white quadrangles in Figure 2 around P1, P5, P6, and P7.

Figure 3 shows the integrated ^{13}CO 2–1 line intensity plotted against the smoothed 850 μm flux density

Chemistry as a Diagnostic of Star Formation (2003). *Edited by Charles L. Curry and Michel Fich.* Published in 2003 by NRC Research Press, Ottawa ON, K1A 0R6, Canada. ISBN 0-660-19089-3.

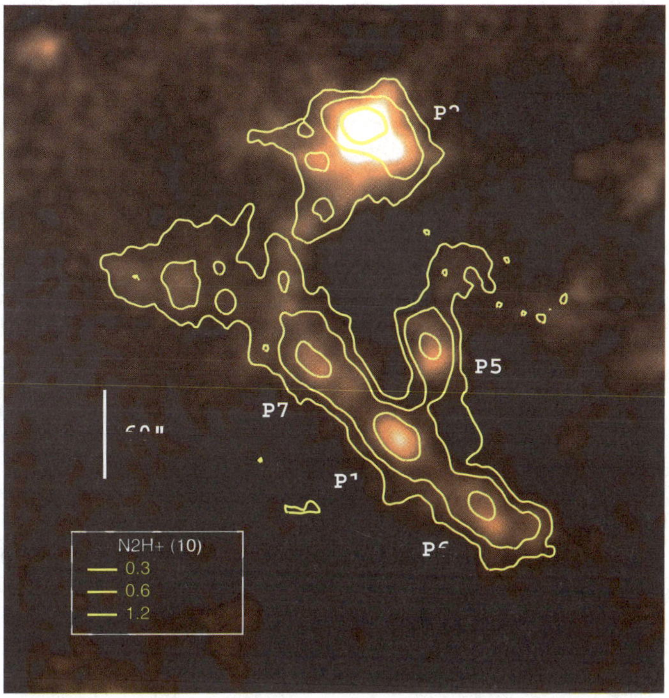

Figure 1. JCMT SCUBA 850 μm image of G28.34+0.06 with contours of BIMA N_2H^+ 1–0 integrated emission over the LSR velocity range 75–87 km/s. At the kinematic distance of 4.8 kpc, 60″ corresponds to a linear scale of 1.4 pc. The bright, compact SCUBA sources referred to in the text are labelled. The units for the BIMA contours are Jy/beam.

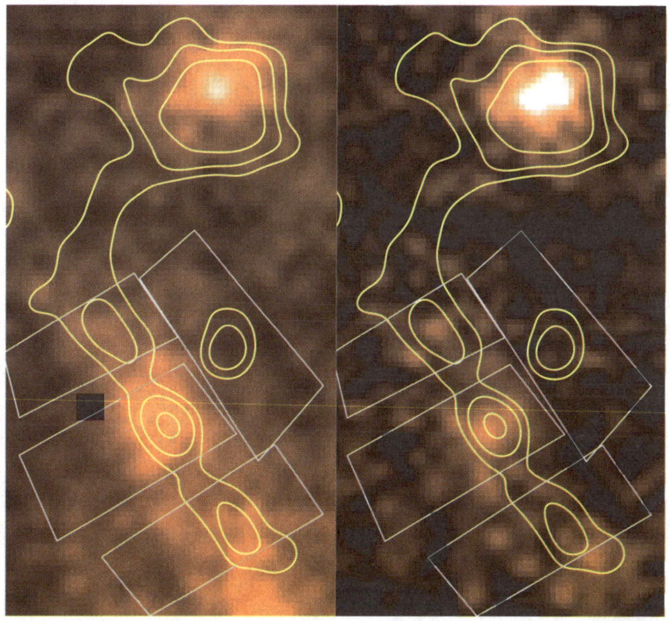

Figure 2. Integrated emission from G28.34+0.06 in the LSR velocity range 75–79 km/s from ^{13}CO 2–1 (left) and C^{18}O 2–1 (right). SCUBA 850 μm emission smoothed to match the angular resolution of the CO observations is shown with yellow contours in Jy/beam (0.15, 0.3, 0.6). The regions used to compare the CO emission with the 850 μm dust emission are shown as white quadrangles.

for the two sources P1 (Fig. 3a) and P5 (Fig. 3b). Two aspects of these data are notable. Firstly, the origins of the curves traced by the data points do not lie near (0,0) because there is extended line and continuum emission surrounding the compact sources. Secondly, the curve in Figure 3a flattens at high flux densities because the ^{13}CO line becomes optically thick.

The shape of the curve in the optically thick regime is a complicated function of the density, temperature and velocity structure of the cloud. The only part of the curve that is relatively simple to interpret is the linear portion in the optically thin regime near the origin. The slope of this part of the curve is a temperature-dependent measure of the ratio of CO to dust in the cloud. Accordingly, regression lines were fitted to each of the curves including only the data points in the optically thin parts of the curve. The lines are shown in Figure 3 as thick grey lines. Table 1 summarizes the measured ratios R of the integrated CO line emission to the 850 μm dust emission for all of the bright, compact sources in the two IRDCs. The estimated errors in Table 1 are 1 σ internal errors and probably underestimate the true errors by factors of a few because of correlated noise in the images.

The source G28.34+0.06 P5 stands out in both Figure 3 and in Table 1 because of its absence of detectable CO. For the other sources in this IRDC (P1, P6 and P7), the small variations in the measured values of R might be accounted for by variations in temperature, and are consistent between the ^{13}CO and C^{18}O observations. On theoretical grounds, we expect that the ratio R will vary inversely with temperature by a factor of about two over the range 10–20 K. The measured values of R for P5, by contrast, are consistent with zero for both ^{13}CO and C^{18}O, when allowance is made for systematic errors. This suggests that that CO is depleted in G28.34+0.06 P5 by at least a factor of 10 relative to the other bright, compact sources in the IRDC.

3. Observations of G81.50+0.04

The bright, compact source G81.50+0.14 P3 appears to show significant CO depletion by comparison with the other three bright, compact sources in the IRDC. The measured values of R for these four sources are recorded

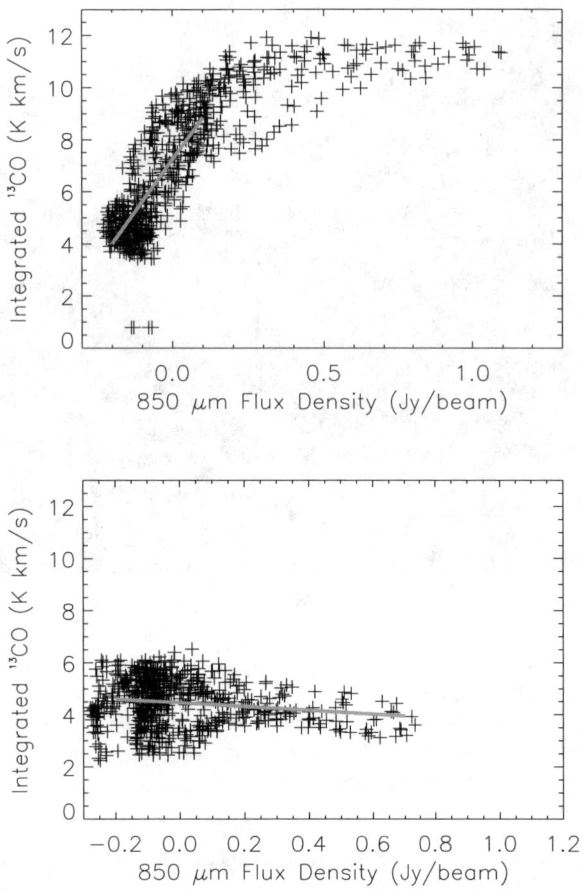

Figure 3. Integrated intensity of the ^{13}CO 2−1 line versus 850 μm flux density for G28.34+0.06 P1 (*upper panel*) and G28.34+0.06 P5 (*lower panel*). The thick grey lines were fitted to the optically thin portions of the data as discussed in the text.

in Table 1. The ^{13}CO measurement indicates that CO is depleted in P3 by a factor of about 6 compared to the nearby sources P1, P2, and P4. Within the errors, the measured value for CO depletion in P3 using $C^{18}O$ as the tracer was consistent with this value.

Based on the ratio of the SCUBA flux densities at 450 and 850 μm, we estimate temperatures for the CO-depleted sources G28.34+0.06 P5 and G81.50+0.14 P3 in the range $7-9$ K, depending on the dust emissivity index β, which we assume to be in the range of 2.5 to 1.5. The non-depleted sources appear to be significantly warmer.

Table 1. Ratio R of CO integrated to 850 μm dust emission

Source	^{13}CO			$C^{18}O$		
G28.34+0.06						
P1	16.1	±	0.7	2.8	±	0.15
P5	−0.7	±	0.2	0.5	±	0.11
P6	8.5	±	0.6	1.1	±	0.14
P7	14.4	±	1.1	2.0	±	0.16
G81.50+0.14						
P1	58.7	±	1.9	18.5	±	2.1
P2	40.3	±	1.6	21.3	±	2.4
P3	8.7	±	0.7	2.2	±	1.6
P4	66.6	±	3.4	33.0	±	5.1

Acknowledgements

We thank Brie Hoffman for her work in helping us analyse the data. The JCMT is operated by the JAC on behalf of the Particle Physics and Astronomy Research Council of the UK, the Netherlands Organisation for Scientific Research, and the National Research Council of Canada. The Berkeley-Illinois-Maryland Association (BIMA) is a consortium consisting of the Radio Astronomy Laboratory at U. California (Berkeley), the Laboratory for Astronomical Imaging at U. Illinois (Urbana) and the Laboratory for Millimeter-Wave Astronomy at U. Maryland (College Park), which operates a millimeter-wave radio interferometer at Hat Creek, California. The BIMA array is operated with support from the National Science Foundation under grants AST-9981308, AST-9981363, and AST-9981289.

References

Carey, S. J., Clark, F. O., Egan, M. P., Price, S. D., Shipman, R. F., & Kuchar, T. A., 1998, ApJL 508, 721

Carey, S. J., Feldman, P. A., Redman, R. O., Egan, M. P., MacLeod, J. M., & Price, S. D., 2000, ApJL 543, L157

Egan, M. P., Shipman, R. F., Price, S. D., Carey, S. J., Clark, F. O., & Cohen, M., 1998, ApJL 494, L199

Redman, R. O., Feldman, P. A., Carey, S. J., & Egan, M. P., 2000, in *Star Formation from the Small to the Large Scale*, ed. F. Favata, A.A. Kaas, & A. Wilson, ESA Special Publ. Ser. 445, 499

Redman, R. O., Feldman, P. A., Côté S., Wyrowski, F., Carey, S. J., & Egan, M. P., 2002, in *The Earliest Phases of Massive Star Birth*, ASP Conf. Ser. 267, ed. P. Crowther (San Francisco: ASP), 409

Redman, R. O., Feldman, P. A., Wyrowski, F., Côté S., Carey, S. J., & Egan, M. P., 2003, ApJ, 586, April 1 issue.

THE PERSISTENCE OF MOLECULAR EXCITED STATES UNDER ASTROPHYSICAL CONDITIONS

Robert J. Glinski and Clark R. Downum

Department of Chemistry, Tennessee Technological University, Cookeville, TN, USA 38505

Abstract

We are interested in the nature of energy distributions of molecules under astrophysical conditions. We have found that under certain conditions the populations are maintained by the interplay of radiative, collisional, and reactive processes. Chemical reactions are usually not included in standard fluorescence equilibrium or general radiative transfer models. The basic idea is that chemical reactions in space are generally exoergic and form products that carry the reaction energy in their internal degrees of freedom. Polar molecules will relax their rotational and vibrational populations via radiation, but non-polar molecules can only do this quite slowly. In addition, all molecules can only lose their excess translational energy by collisions with the relaxed gases. We have investigated these effects in the infrared spectra of homonuclear diatomics (MNRAS, 288, 286 [1997]) and in the high-resolution line profiles of CH^+, CH, and CN (MNRAS, 308, 29 [1999]). We examined these effects using detailed kinetic models that include radiative transfer and chemistry—i.e., solving the "master equation"—for both steady-state and time-evolving systems. It is becoming apparent that non-LTE energy distributions can be simply modelled by a combination of two Boltzmann-like populations. The low-temperature distribution is reasonably regarded as the local kinetic temperature, while the high-temperature distribution (often termed T_{exc}) represents the process that perturbs the system from the LTE. We seek the meaning of T_{exc} in terms of the rate of the non-thermal up-pumping process. Hence, the master equation may need only be solved once for any particular non-equilibrium process. Star-forming regions are particularly chemically active, and highly non-LTE populations are expected to be persistent. This approach can be useful in deducing the nature of the chemical processes in these regions—e.g., H_2 formation, maser activity, etc.—by careful examination of the remnant energy distributions.

Key words: astrochemistry – ISM: molecules

Chemistry as a Diagnostic of Star Formation (2003). *Edited by Charles L. Curry and Michel Fich.* Published in 2003 by NRC Research Press, Ottawa ON, K1A 0R6, Canada. ISBN 0-660-19089-3.

MODELLING THE FORMATION OF MOLECULAR CLOUDS WITH ZEUS-MP

Simon Glover[1], Mordecai-Mark Mac Low[1], Michael D. Smith[2], Alex Rosen[2], and Georgi Pavlovski[2]

[1]American Museum of Natural History, Central Park West at 79th Street, New York, NY 10024, USA
[2]Armagh Observatory, College Hill, Armagh, BT61 9DG, Northern Ireland, UK

ABSTRACT

The interstellar medium displays the signatures of supersonic turbulence over a wide range of scales and turbulent motions are believed to play a central role in the formation and support of molecular clouds. Large magnetohydrodynamical (MHD) simulations are important tools for studying this turbulence, but much of the work done with them has concentrated on idealized isothermal gas. In the past few years, however, it has become possible to extend these simulations to tackle more realistic problems. In this contribution, I report on the progress of our group in implementing an accurate treatment of molecular chemistry and cooling within ZEUS-MP, a modern high-performance MHD code, and discuss some of the main difficulties involved.

Key words: ISM: kinematics and dynamics – ISM: magnetic fields – turbulence

Chemistry as a Diagnostic of Star Formation (2003). *Edited by Charles L. Curry and Michel Fich.* Published in 2003 by NRC Research Press, Ottawa ON, K1A 0R6, Canada. ISBN 0-660-19089-3.

HCO^+ OBSERVATION OF STARLESS CORES

Erik M. Gregersen[1], Gerald Moriarty-Schieven[2], Ralph Pudritz[1], and Christine Wilson[1]

[1]McMaster University, Hamilton, Ontario, Canada
[2]Joint Astronomy Centre, Hilo, Hawaii, U.S.A.

Abstract

Star formation is a highly inefficient process. Some patches of a GMC form stars; most do not. Is there a necessary condition for star formation a core must fulfill? Large-area molecular line surveys indicate that starless cores all have column densities less than $8 \times 10^2\,\mathrm{cm}^{-2}$. We have observed nearly one hundred starless cores in HCO^+ J=3–2, a line particularly sensitive to collapse motions, to determine if a physically meaningful threshold separating quiescent from pre-protostellar cores exists.

Key words: stars: formation – ISM: clouds

Chemistry as a Diagnostic of Star Formation (2003). *Edited by Charles L. Curry and Michel Fich.* Published in 2003 by NRC Research Press, Ottawa ON, K1A 0R6, Canada. ISBN 0-660-19089-3.

DENSITY STRUCTURE AT THE EDGE OF THE HORSEHEAD NEBULA

E. Habart[1], A. Abergel[2], F. Boulanger[2], D. Teyssier[3], and L. Verstraete[2]

[1]Obs. Arcetri, Firenze, Italia
[2]Institut d'Astrophysique Spatiale, Orsay, France
[3]LERMA, Paris Obs., France

Abstract

We present near- and mid-infrared imaging observations of gas and dust emission of the Horsehead nebula emerging from the molecular cloud L1630. The 5–8.5 μm band of the ISOCAM camera has revealed that the edge of the nebula, illuminated by an O-star (σ Ori, χ ~100), presents a remarkably sharp bright filament (Abergel et al. 2000, 2001). This photodissociation region (PDR), seen edge-on, is an ideal target to study the gas density structure and the penetration of UV radiation across typical PDRs. Since much of the energy transfer that takes place between stars and gas occurs in PDRs, understanding their physical conditions is important in determining their role in regulating the star formation process.

Using SOFI (Son of ISAAC) at the NTT, we have obtained an image of the Horsehead nebula in the 1–0 S(1) emission line of H_2 at 2.12 μm revealing narrow filaments (width around 1 arcsec) spatially coincident with the aromatic emission. We have used the recent developments of the PDR model of Le Bourlot et al. (1993) to analyze our data. The adjustment of the brightness profiles with the model strongly constrains the geometry and the density structure of individual structures. Moreover, the combination of H_2 and ISOCAM data allows independent measurements of the local density, using the ratio I_{H_2}/I_{IR} which depends very strongly on the local density, but not on the total column density nor on the geometry.

Key words: ISM: structure – ISM: clouds – H II regions

Chemistry as a Diagnostic of Star Formation (2003). *Edited by Charles L. Curry and Michel Fich.* Published in 2003 by NRC Research Press, Ottawa ON, K1A 0R6, Canada. ISBN 0-660-19089-3.

SHOCK PROCESSING AND ICE MANTLE CRYSTALLIZATION

George E. Hassel[1,2], Wayne G. Roberge[1,2], Douglas C. B. Whittet[1,2], and Sachindev S. Shenoy[1,2]

[1]New York Center for Studies on the Origins of Life (NSCORT), Rensselaer Polytechnic Institute, Troy, NY 12180, U.S.A.
[2]Department of Physics, Applied Physics, and Astronomy, Rensselaer Polytechnic Institute, Troy, NY 12180, U.S.A.

Abstract

The water ice mantles on interstellar grains trap volatile molecules, such as CO and CH_3OH, with an efficiency that depends on the amorphous or crystalline structure of the ice. The ice structure therefore affects the composition of comets formed from the icy grains. We are studying the processing of mantled grains by shock waves in protoplanetary disks. The grains suffer a sudden increase in temperature which can evaporate the mantles. This is followed by an extended cooling time during which the mantles recondense on timescales comparable to the crystallization timescale for hydrodynamic parameters consistent with the Jupiter-Saturn region of the solar nebula. The crystallization of ice and the exclusion of volatiles from the matrix may explain the volatile-depleted composition observed recently in Comet C/1999 S4 (LINEAR), an Oort-cloud comet originating from the Jupiter-Saturn region.

Key words: hydrodynamics – shock waves – disks

1. Introduction

Recent advances in the study of star formation have enhanced the understanding of the incorporation of icy material from the parent molecular cloud into primitive bodies. We are investigating the processing of H_2O-rich ice which occurs in nebular shocks where interstellar gas and dust interact within protoplanetary disks. H_2O is the predominant component of ices in the molecular cloud and young stellar environments. On the basis of observations of the 3.05 μm, 6.00 μm, and 13.3 μm absorption features, Whittet (1993, 2003) finds that approximately 70% of the composition of ices in these shielded environments is H_2O. These observations have been used to interpret the spectra of hot cores surrounding high mass protostars (Whittet et al. 1996; Gibb et al. 2000). More recently, observations the low mass protostar Elias 29 provide evidence of the compositional predominance of H_2O ice surrounding a low mass protostar (Boogert et al. 2000).

Frictional heating in nebular shocks near low mass protostars is expected to sublimate icy dust mantles at heliocentric distances $R < 30\,\mathrm{AU}$ (Lunine et al. 1991; Neufeld & Hollenbach 1994). Protoplanetary disk models such as the viscous accretion disk (Aikawa et al. 1998) allow us to associate heliocentric position with hydrodynamic parameters. For $R \sim 3$–$4\,\mathrm{AU}$, the disks are cool enough for ices to reform in the post-shock region (Chick & Cassen 1997), but the volatile-trapping capacity of the ice depends on the degree (or lack) of crystalline structure (Notesco & Bar-Nun 1997). On the basis of a simplified treatment of the hydrodynamics, Lunine et al. (1991) concluded that reformed ices are amorphous.

Here, we carry out a more realistic calculation, and conclude that crystallization is efficient over a broad range of conditions. This result may account for the depletions of volatile organic molecules observed in Comet C/1999 S4 (LINEAR) with respect to other Oort-cloud comets (Mumma et al. 2001).

2. Physics

2.1. Shock Simulations

The structure of the gas is determined by solving the hydro equations for a steady, 1D shock:

$$\frac{d}{dz}[\rho v] = 0, \tag{1}$$

$$\frac{d}{dz}\left[\rho v^2 + P\right] = 0, \tag{2}$$

and

$$\frac{d}{dz}\left[\rho v\left(\frac{1}{2}v^2 + \frac{\gamma}{\gamma - 1}\frac{kT}{m}\right)\right] = -\Lambda, \tag{3}$$

where z is distance along the direction of the shock and ρ, m, v, P and T are the gas density, mean mass per particle, velocity, pressure, and temperature. The cooling rate, Λ, is the net rate per volume at which thermal energy is removed from the gas; we include the cooling due

Chemistry as a Diagnostic of Star Formation (2003). *Edited by Charles L. Curry and Michel Fich.* Published in 2003 by NRC Research Press, Ottawa ON, K1A 0R6, Canada. ISBN 0-660-19089-3.

to gas-grain collisions, rotational and vibrational emission from H_2O, CO, and H_2, evaporation of water molecules from grain mantles, and dissociation of H_2. The gas includes only those species - H, H_2, He, H_2O and CO - needed to compute the cooling accurately. Their abundances are governed by solving a simple chemical network. The "adiabatic exponent", γ, is computed from the time-dependent composition assuming LTE for the rotational degrees of freedom and neglecting vibrational excitation.

Our shocks include silicate grains with ice mantles. The silicate cores are allowed to have a range of radii as a discrete number of size bins. Because the gas cools primarily by collisional coupling with the dust, it is essential to compute the dust density, velocity, and temperature realistically. The density of cores in the kth size bin is determined by conserving the grain flux,

$$\frac{d}{dz}[n_k v_k] = 0, \tag{4}$$

where n_k and v_k are the number density and velocity of grains in bin k. (Note that we do not include grain growth or destruction at present.) The dust velocity is determined by Newton's Second Law,

$$m_k v_k \frac{dv_k}{dz} = F_{\rm drag}, \tag{5}$$

where m_k is the grain mass and $F_{\rm drag}$ is the drag force on a sphere (Draine & Salpeter 1979). At the densities of interest to us, the grains reach equilibrium temperatures in a time that is always much smaller than the flow time scale. We therefore calculate the dust temperature at each point in the flow by requiring the grains to be in thermal balance. Our models include heating/cooling of the grains by collisional coupling with gas, heating by nebular radiation, and cooling by thermal emission. The nebular radiation is represented as a dilute blackbody. Thermal emission is determined using Planck-averaged emissivities for bare silicate spheres. We calculate the emissivities using Mie theory for coated spheres with optical constants for "smoothed UV silicates" (Laor & Draine 1993; Weingartner & Draine 2001) and pure H_2O ice (Hudgins et al. 1993).

This preliminary calculation includes a single population of 1 μm radius silicate grains. Also, we do not consider the effects of radiative transfer at this time.

2.2. Crystallization of Mantle

The kinetics of H_2O ice crystallization are described in an astrophysical context in Seki & Hasegawa (1981), Kouchi et al. (1994), and Jenniskens & Blake (1996). In the temperature range of interest to this study, the phase of the ice can be amorphous or cubic-crystalline, depending on the deposition rate and grain surface conditions. Here, we consider crystallization through the process of nucleation and growth derived from classical nucleation theory.

The fraction crystallized, θ, is determined by solution of the differential equation:

$$\frac{d\theta}{dt} = \frac{1-\theta}{t_c}. \tag{6}$$

Amorphous ice anneals on the timescale t_c,

$$t_c = \left(\alpha J w^3\right)^{-\frac{1}{4}}, \tag{7}$$

where J is the nucleation rate of critical nuclei, w is the growth rate of crystals from nuclei, and α is the geometrical growth factor of crystals from nuclei, using the formulations of Kouchi et al. (1994).

2.3. Parameter Space

The structure of a disk shock depends primarily on the shock speed, v_s, and preshock density, $n_{\rm H0}$. We adopt the minimum mass solar nebula model for a viscous accretion disk which Aikawa et al. (1998) employ for chemical evolution. The preshock density is uncertain but unlikely to exceed the local disk density. We estimate the preshock temperature, T_0, (on which the Mach number depends) as the temperature of an α-disk at the corresponding locations. The shock speed is assumed to be equal to the local Keplerian speed. The calculations in the paper used an α-disk model with $M = 1M_\odot$, $dM/dt = 10^{-8}M_\odot$/yr, and $\alpha = 0.01$.

The shock also depends on the preshock fractional abundances, x_0, of H_2O, CO, and grains of various sizes. The preliminary calculations reported here assumed $x_0(H_2O) = 1.1\times10^{-4}$, $x_0(CO) = 7\times10^{-5}$, with H_2O initially all in solid form and CO all in the gas phase. We consider silicate grains with a single radius, $a = 1\,\mu$m, and abundance such that the dust-to-gas mass ratio is 1%.

3. Preliminary Results

Results describing the fate of the icy mantles for a shock with $v_s = 7$ km/s and $n_{\rm H0} = 10^{12}$ cm^{-3} are displayed. Figure 1 shows the temperature structure, in which the ice mantles sublimate almost instantaneously just downstream from the shock front (at $z = 0$) and recondense far downstream when the dust temperature, T_d drops below 140 K. At this density, the mantle is completely removed for $v_s > 2$ km/s, after which the ice recondenses onto the grain. The mantles are retained for smaller speeds.

The annealing process is most efficient when t_c is

comparable with the condensation timescale, $\sim 10^4$ s at $T_d \sim 140$ K. The final degree of crystallization of the recondensed ice increases with increasing shock speed. In the example case, we calculate $\theta = 0.75$. Slight annealing occurs below 2 km/s, to a maximum value of $\theta \sim 0.1$.

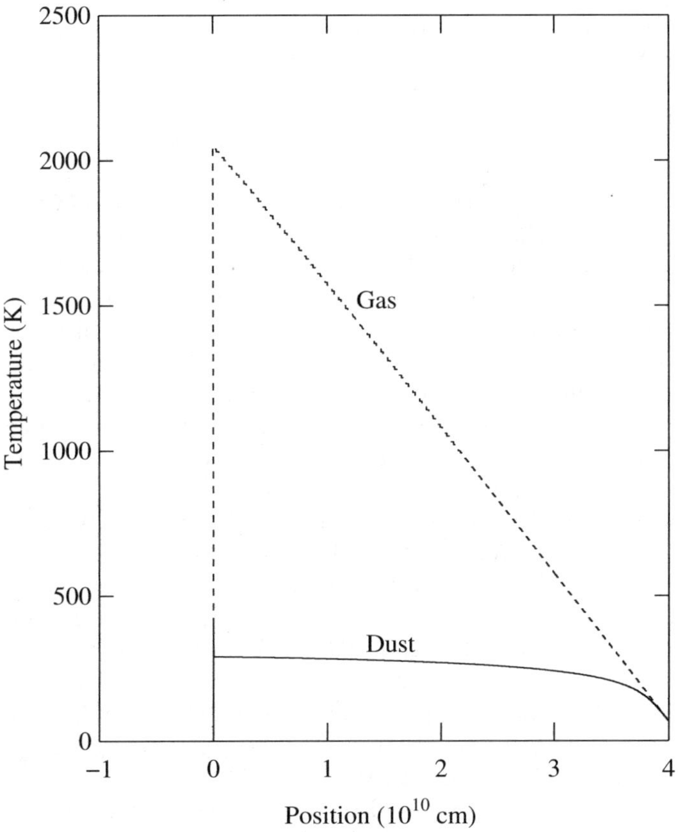

Figure 1. Temperatures of gas and dust populations versus position in a shock with $v_s = 7$ km/s and $n_{H0} = 10^{12}$ cm^{-3}.

4. Conclusions

The hydrodynamic shock code formulated for this work is applicable for a large range of parameters relevant to protoplanetary disks. The main mechanism for gas cooling is collisional coupling to the dust grains, and the dust radiates away excess energy as thermal emission. This process extends the cooling time of the dust, until the populations reach equilibrium, as shown in Figure 1.

When applied to representative parameters for the Jupiter-Saturn region of the solar nebula, we find that the mantles either slightly anneal or evaporate, recondense, and possibly undergo nearly total annealing. For the latter cases, the grain temperature remains sufficiently warm such that the crystallization time becomes comparable with the cooling time. A series of laboratory studies shows that amorphous H_2O ice can trap more volatile gaseous species, but that these molecules are excluded from the cubic-crystalline state of H_2O ice (Notesco & Bar-Nun 1997). This result may account for the volatile-depleted composition observed in Comet C/1999 S4 (LINEAR), an Oort-Cloud comet considered to be from the Jupiter-Saturn region (Mumma et al. 2001).

We are currently expanding this project by including the effects of radiative transfer in the disk. The ice model is being extended to evaluate crystallization of recondensing ice by means of a flux condition (Kouchi et al. 1994), to include a distribution of grain sizes, and to explore different disk environments. The results presented here underestimate the crystallization of the H_2O ice, but the revised models will provide a more thorough study of the problem and will yield a greater degree of crystallization in the mantles.

Acknowledgements

We are grateful to David Neufeld for providing a subroutine which calculates the cooling by H_2O, H_2, and CO molecules as described in Neufeld & Kaufman (1993). This work is supported by the New York Center for Studies on the Origins of Life (NSCORT) at RPI, under NASA grant NAG5-7598.

References

Aikawa, Y., et al. 1998, Faraday Discuss., 109, 281
Boogert, A. C. A., et al. 2000, A&A 360, 683
Chick, K. M., & Cassen, P. 1997, ApJ 477, 398
Draine, B. T., & Salpeter, E. E. 1979, ApJ 231, 77
Gibb, E. L., et al. 2000, ApJ 536, 347
Hudgins, D. M., et al. 1993, ApJS 86, 713
Jenniskens, P., & Blake, D. F. 1996, ApJ 473, 1104
Kouchi, A., et al., A&A 290, 1109
Laor, A., & Draine, B. T., 1993, ApJ 402, 441
Lunine, J. I., et al. 1991, Icarus, 94, 333
Mumma, M. J., et al. 2001, Science 292, 1334
Neufeld, D. A., & Hollenbach, D. J. 1994, ApJ 428, 170
Neufeld, D. A., & Kaufman, M. J. 1993, ApJ 418, 263
Notesco, G. & Bar-Nun, A. 1997, Icarus 126, 336
Seki, J., & Hasegawa, H. 1981, Prog. Theor. Phys. 66, 903
Weingartner, J. C., & Draine, B.T. 2001, ApJ 548, 296
Whittet, D. C. B. 1993, in *Dust and Chemistry in Astronomy*, ed. T.J. Millar & D.A. Williams (Bristol: Institute of Physics Publishing)
Whittet, D. C. B., et al. 1996, A & A, 315, L357
Whittet, D. C. B. 2003, *Dust in the Galactic Environment*, 2nd ed. (Bristol: Institute of Physics Publishing)

A SEARCH FOR CARBON CHAIN-RICH CORES IN DARK CLOUDS

Tomoya Hirota[1], Miho Ikeda[2], Masatoshi Ohishi[2], and Satoshi Yamamoto[3]

[1]Department of Physics, Faculty of Science, Kagoshima University, Korimoto 1-21-35, Kagoshima 890-0065, Japan
[2]National Astronomical Observatory of Japan, Mitaka, Tokyo 181-8588, Japan
[3]Department of Physics and Research Center for the Early Universe, University of Tokyo, Bunkyo-ku, Tokyo 113-0033, Japan

Abstract

We present the results of a survey of CCS, HC_3N, and HC_5N toward 31 dark cloud cores in order to search for carbon-chain-rich cores such as TMC-1 and L1521E. Because the abundances of carbon-chain molecules are expected to be anti-correlated with that of NH_3, most of the sources were selected from a list of the survey by Benson & Myers (1989), where the NH_3 lines were found to be weak or not detected. We observed the CCS, HC_3N, and HC_5N lines in the 45 GHz region using the 45 m radio telescope at Nobeyama Radio Observatory. The CCS, HC_3N, and HC_5N lines were detected toward 11, 9, and 4 sources, respectively, and some of them have very intense spectra, which are comparable to those in TMC-1 and L1521E, suggesting that they are possible candidates for new carbon-chain-rich cores.

We compared the abundances of carbon-chain molecules and NH_3 in the Taurus and Ophiuchus regions, combining our results with those of previous surveys. We found that the abundances of carbon-chain molecules relative to NH_3 tend to be lower in the Ophiuchus cores than in Taurus. This systematic abundance variation may reflect the difference in physical conditions such as cosmic-ray ionization rate, degree of depletion, and/or evolutionary stage between these regions.

Key words: ISM: abundances – ISM: molecules

1. Introduction

In order to investigate low-mass star formation processes, various observations have been carried out toward dark clouds mainly with molecular lines in radio wavelength. According to previous works, inversion transitions of the NH_3 molecule is known to be a good tracer of dense cores in dark clouds (Benson & Myers 1989). On the other hand, Suzuki et al. (1992) pointed out that NH_3 is not always a good tracer of dense cores because of chemical abundance variation from core to core. They carried out survey observations of CCS, HC_3N, HC_5N, and NH_3 toward 49 dense cores and found that the abundance of CCS is anti-correlated with that of NH_3, which is interpreted in terms of chemical evolution of dark cloud cores; CCS and NH_3 are abundant in early and late stage of chemical evolution, respectively (Suzuki et al. 1992). In their survey, four "carbon-chain-producing regions" were identified where the CCS and other carbon-chain molecules are extremely abundant while NH_3 is deficient, suggesting that these cores are chemically less evolved (Suzuki et al. 1992). Although extensive studies on physical and chemical properties have been carried out for TMC-1 (Olano, Walmsley, & Wilson 1988; Hirahara et al. 1992; Pratap et al. 1997), systematic studies for such carbon-chain-rich cores have never been carried out except for the recent observational study on L1521E (Hirota, Ito, & Yamamoto 2002).

This is partly because most of previous studies on dark cloud cores have been based on the pioneering survey of the NH_3 lines (Benson & Myers 1989) and the samples have been strongly biased to the NH_3 cores. In order to reveal the chemical evolutionary scheme of dark cloud cores, we need to search for more carbon-chain-rich cores. For this purpose, we have carried out a survey of the CCS, HC_3N, and HC_5N lines toward dark cloud cores where the NH_3 lines are weak, not detected, or not observed by the previous surveys, because the abundances of carbon-chain molecules are anti-correlated with that of NH_3 as mentioned above (Suzuki et al. 1992).

2. Observations

Observations were carried out with the 45 m radio telescope at Nobeyama Radio Observatory (NRO) in 2002 March. The CCS (J_N=4_3–3_2), HC_3N (J=5–4), and HC_5N (J=17–16) lines in the 45 GHz region were observed with an SIS mixer receiver whose system temperature was 200-300 K. We observed these lines simultaneously to minimize the pointing and calibration errors.

Chemistry as a Diagnostic of Star Formation (2003). *Edited by Charles L. Curry and Michel Fich.* Published in 2003 by NRC Research Press, Ottawa ON, K1A 0R6, Canada. ISBN 0-660-19089-3.

The beam size (HPBW) was 37″. Acousto-optical radio spectrometers with the frequency resolution of 37 kHz were used for the backend. All the observations were performed with the position-switching mode, in which the off position was 10′ away from the source position.

We selected observed sources mainly from a list of NH_3 survey by Benson & Myers (1989). We selected the sources where the NH_3 lines are not detected or not intense while the $C^{18}O$ lines are detected (Myers, Linke, & Benson 1983), indicating an evidence of a dense core. In addition, we selected the sources where the NH_3 lines are not observed by Benson & Myers (1989), while the CS are detected (Lee, Myers, & Tafalla 2001). For most of the sources, we observed 5 positions around a reference position for each source; (0″, 0″), (80″, 80″), (80″, -80″), (-80″, 80″), and (-80″, -80″) in order not to miss the spectral lines because the center position of the core might be shifted from the reference position by larger than the beam size. If one of the observed lines were detected, we further searched for the lines around the detected positions with a grid spacing of 40″. In total, we observed 31 dark cloud cores both with and without *IRAS* point sources. In our samples, 7 cores are in the Taurus molecular cloud, 8 in Ophiuchus, 5 in Cepheus, and 11 in other regions including isolated Bok globules.

3. Results

Examples of the observed spectra are shown in Figure 1. We detected the CCS, HC_3N, and HC_5N lines toward 11, 9, and 4 sources, respectively, indicating that there exist dense cores that have not been detected by previous NH_3 surveys. A few of them show intense spectra of carbon-chain molecules. For example, we found remarkably intense spectra of CCS, HC_3N, and HC_5N lines toward L492. As shown in Figure 1, intensities of these lines toward L492 are comparable to those of L1521E, one of the known carbon-chain-producing regions (Hirota et al. 2002). This result strongly suggests that L492 would be rich in carbon-chain molecules and a possible candidate for a new carbon-chain-producing region. Although the NH_3 lines have never been observed toward L492, the N_2H^+ lines are detected (Lee et al. 2001), indicating that the NH_3 would be also abundant (e.g. Caselli et al. 2002) unlike TMC-1 and L1521E. Further observations are needed to understand the chemical composition of L492.

For several sources where the CCS and HC_3N lines are intense, we have mapped these lines. Examples of integrated intensity maps are shown in Figure 2. Some of

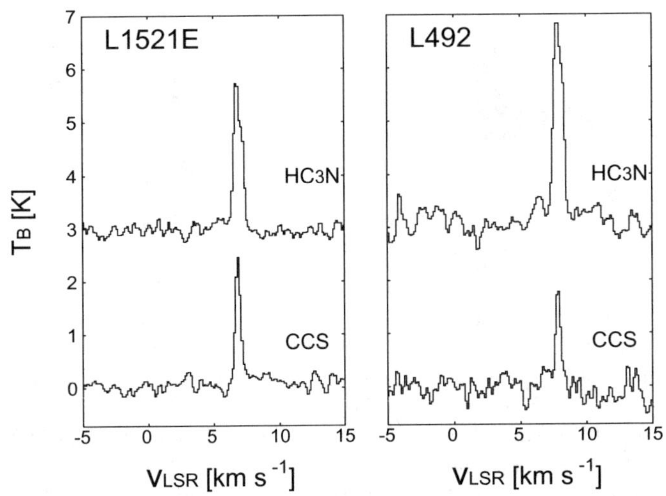

Figure 1. Observed spectra of CCS and HC_3N toward L1521E, one of the carbon-chain-producing regions (Hirota et al. 2002), and L492, a newly-found carbon-chain-rich core.

the mapped sources show different distributions between CCS and HC_3N, as can be seen in Figure 2. The distributions of HC_3N generally show simple spherical structure, and it is quite similar to that of NH_3 for L1172D (L1172A) (Benson & Myers 1989). On the other hand, CCS reside outside the HC_3N core. These difference in molecular distributions suggest chemical abundance variation within the cores. Since the CCS molecule is easily depleted in the center of evolved cores where the ionization degree decreases and freeze out of molecules onto grain surface occurs (Bergin & Langer 1997; Aikawa et al. 2001). On the other hand, the NH_3 molecule increases as the core evolves (Bergin & Langer 1997; Aikawa et al. 2001). Similar to CCS, the HC_3N molecule is also abundant in the early stage of chemical evolution (Suzuki et al. 1992). However, it is also expected that the HC_3N molecule would increase as the depletion of molecules occurs significantly in the late stage of chemical evolution (Ruffle et al. 1997). Therefore, the difference in distributions of CCS, HC_3N, and NH_3 would reflect the evolutionary stage of dark cloud cores.

4. Discussions

As suggested by Suzuki et al. (1992) and Ohishi & Kaifu (1998), the abundance ratio of CCS/NH_3 can be used as a good indicator of chemical evolutionary stage of dark cloud cores; CCS and NH_3 are abundant in early and late stage of chemical evolution, respectively, and the CCS/NH_3 ratio decreases as the core evolves. Based on

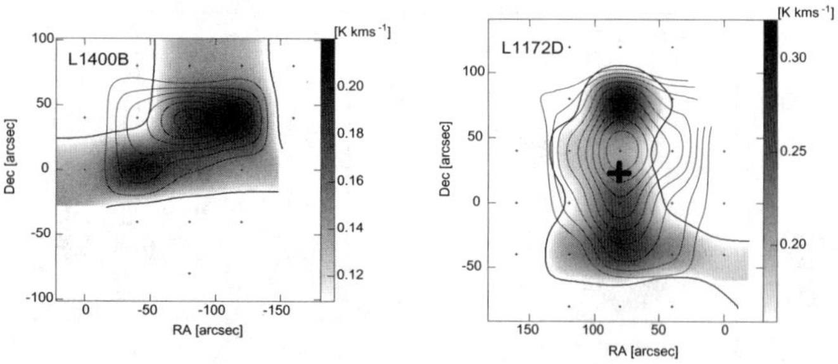

Figure 2. Integrated intensity maps of L1400B and L1172D. Contour and gray scale represent the integrated intensity maps of HC_3N and CCS, respectively. The interval of the contours are 0.05 K km s^{-1} and 0.1 K km s^{-1}, and the lowest ones are 0.15 K km s^{-1} and 0.6 K km s^{-1} for L1400B and L1172D, respectively. The half maximum contours of CCS are shown in thick lines. A cross in the map of L1172D indicates the *IRAS* point source, IRAS 21017+6742.

Table 1. Detection rates for CCS, HC_3N, HC_5N, and NH_3[a]

Cloud	n	CCS	HC_3N	HC_5N	NH_3
Tau	29(11)	19/29	19/29	12/29	20/29
Oph	24(9)	6/24	9/24	0/22	18/24
Total	92(39)	50/92	53/89	20/82	68/89

[a] Numbers in parentheses represent no. of *IRAS* sources.

our observations, we calculated the column densities of CCS, HC_3N, and HC_5N by the similar method described by Suzuki et al. (1992), and briefly discuss about their abundances using our results along with the previous surveys of CCS and NH_3 (Benson & Myers 1989; Suzuki et al. 1992; Benson, Caselli, & Myers 1998).

In the present survey, detection rates of carbon-chain molecules are significantly lower than those of Suzuki et al. (1992). Although we selectively observed the cores with low NH_3 abundance, we detected a few sources where the carbon-chain molecules are abundant such as L492. This result is consistent with the idea that the carbon-chain-producing regions are less evolved cores, and hence, rare objects. It is noteworthy that the carbon-chain-producing regions have been found only in the Taurus molecular cloud as reported by Suzuki et al. (1992), and that no possible candidate has been found in other regions such as Ophiuchus molecular cloud in both the present study and previous ones, except for L492. In addition, we found that the detection rate of carbon-chain molecules are significantly lower in the Ophiuchus cores than in the Taurus cores, while the total number of samples, the fraction of cores associated with *IRAS* sources, namely the fraction of star forming regions in our samples, and the detection rate of NH_3 are comparable to each other, as shown in Table 1. These results would suggest that carbon-chain molecules are generally less abundant in the Ophiuchus cores than in Taurus, and hence, the CCS/NH_3 ratios tend to be smaller in the Ophiuchus cores than in Taurus. Such a systematic abundance variation from cloud to cloud has been found in the deuterium fractionation ratio of DNC/$HN^{13}C$ (Hirota, Ikeda, & Yamamoto 2001). This may suggest the difference in physical conditions such as cosmic-ray ionization rate, degree of depletion, and/or evolutionary stage between these regions. To confirm this, further unbiased surveys would be needed.

Acknowledgements

We are grateful to all the staff of Nobeyama Radio Observatory for their assistance in observations. Nobeyama Radio Observatory is a branch of the National Astronomical Observatory of Japan, an inter-university research institute operated by the Ministry of Education, Science, Sports and Culture of Japan. TH thanks to the Inoue Foundation for Science for the financial support (Research Aid of Inoue Foundation for Science). This study is partly supported by Grant-in-Aid from Ministry of Education, Science, Sports and Culture of Japan (No. 07CE2002).

References

Aikawa, Y. et al. 2001, ApJ, 552, 639
Benson, P. J., Caselli, P., & Myers, P. C. 1998, ApJ, 506, 743
Benson, P. J., & Myers, P. C. 1989, ApJS, 71, 89
Bergin, E. A. & Langer, W. D. 1997, ApJ, 486, 316
Caselli, P. et al. 2002, ApJ, 572, 238
Hirahara, Y. et al. 1992, ApJ, 394, 539
Hirota, T., Ikeda, M., & Yamamoto, S. 2001, ApJ, 547, 814
Hirota, T., Ito, T., & Yamamoto, S. 2002, ApJ, 565, 359
Lee, C. W., Myers, P. C., & Tafalla, M. 2001, ApJS, 136, 703
Myers, P. C., Linke, R. A., & Benson, P. J. 1983, ApJ, 264, 517
Ohishi, M. & Kaifu, N. 1998, Faraday Discuss., 109, 205
Olano, C. A., Walmsley, C. M., & Wilson, T. L. 1988, A&A, 196, 194
Pratap, P. et al. 1997, ApJ, 486, 862
Ruffle, D. P. et al. 1997, MNRAS, 291, 235
Suzuki, H. et al. 1992, ApJ, 392, 551

THE SERPENS STAR-FORMING REGION IN HCO^+, HCN, AND N_2H^+

Michiel R. Hogerheijde

Steward Observatory, The University of Arizona, 933 N. Cherry Ave, Tucson, AZ 85721, U.S.A.

Abstract

This poster presents single-dish and aperture-synthesis observations of the J=1–0 ($\lambda \approx 3$ mm) transitions of HCO^+, HCN, and N_2H^+ towards the Serpens star-forming region. Jets driven by young stars affect the structure and the chemistry of their surrounding cloud, and this work aims to assess the extent to which the emission of these three molecular lines is dominated by such processes. In Serpens I find that N_2H^+ 1–0 traces the total amount of material, except in two regions slightly ahead of shocks. In contrast, the HCO^+ and, especially, HCN emission is dominated by regions impacted by outflows. One previously unknown, strongly shocked region is located $\sim$ 0.1 pc northwest of the young stellar object SMM 4. There is a marked spatial offset between the peaks in HCN and the N_2H^+ emission associated with shocked regions. I construct a simple, qualitative chemical model where the N_2H^+ emission increases in the magnetic precursor of a C-type shock, while N_2H^+ is destroyed deeper in the shock as the neutrals heat up and species like HCN and water are released from icy grain mantles. I conclude that N_2H^+ is a reliable tracer of cloud material, and that unresolved observations of HCO^+ and HCN will be dominated by material impacted by outflows.

Key words: ISM: molecules – ISM: clouds – ISM: jets and outflows – stars: formation

1. Introduction

The formation of stars is accompanied by energetic activity such as jets and outflows. This can affect the structure and the chemical composition of the surrounding cloud, possibly influencing its continued star formation. A relevant question therefore is, to what the extent the emission of commonly used tracers of dense gas reflects the energetics of star formation rather than the underlying structure of the cloud. This is especially interesting for unresolved observations of clouds and cloud complexes, at large distances in our Galaxy or in other galaxies (e.g., Aalto et al. 1995; Helfer & Blitz 1997; Wild & Eckart 2000; Kuno et al. 2002). This poster contribution investigates the emission lines of three molecules (HCO^+, HCN, and N_2H^+) in the Serpens star-forming region, with the specific question in mind of the relation between the emission and activity of young stars.

The Serpens star-forming region is ideally suited for the stated aim. It is relatively nearby at $\sim 310 \pm 40$ pc (de Lara et al. 1991) and harbors several deeply embedded protostars as well as apparently starless dust condensations (Casali et al. 1993; Davis et al. 1999). The molecular cloud consists of two sub-condensations, northeast (NE) and southwest (SW), each of which are broken up into numerous sub-clumps and filaments. Many outflows emanate from the Serpens region (Davis et al. 1999), apparently driven by the embedded protostars. The high density of young stellar objects (YSOs) and the complex structure of the region precludes identification of the driving sources of many of these flows. It is likely that the course of some of the flows are altered by collisions with dense cloud material, further confusing the picture.

2. Single-dish observations

In this poster I present single-dish observations of the J=1–0 lines of HCO^+, HCN, and N_2H^+ obtained with the Kitt Peak 12-meter (KP12m) telescope[1]; aperture-synthesis observations are discussed in Sec. 3. The KP12m maps cover a $420'' \times 420''$ area encompassing the NE and SW condensations (Fig. 1). The beam size of the telescope is $60''$. The N_2H^+ emission traces the NE and SW condensations equally, resembling the SCUBA 850 μm emission that traces cold dust (Davis et al. 1999). The HCN and HCO^+ emission is dominated by the SE

[1] The 12m Telescope is a facility of the National Science Foundation currently operated by the University of Arizona Steward Observatory under a loan agreement with the National Radio Astronomy Observatory.

Chemistry as a Diagnostic of Star Formation (2003). *Edited by Charles L. Curry and Michel Fich.* Published in 2003 by NRC Research Press, Ottawa ON, K1A 0R6, Canada. ISBN 0-660-19089-3.

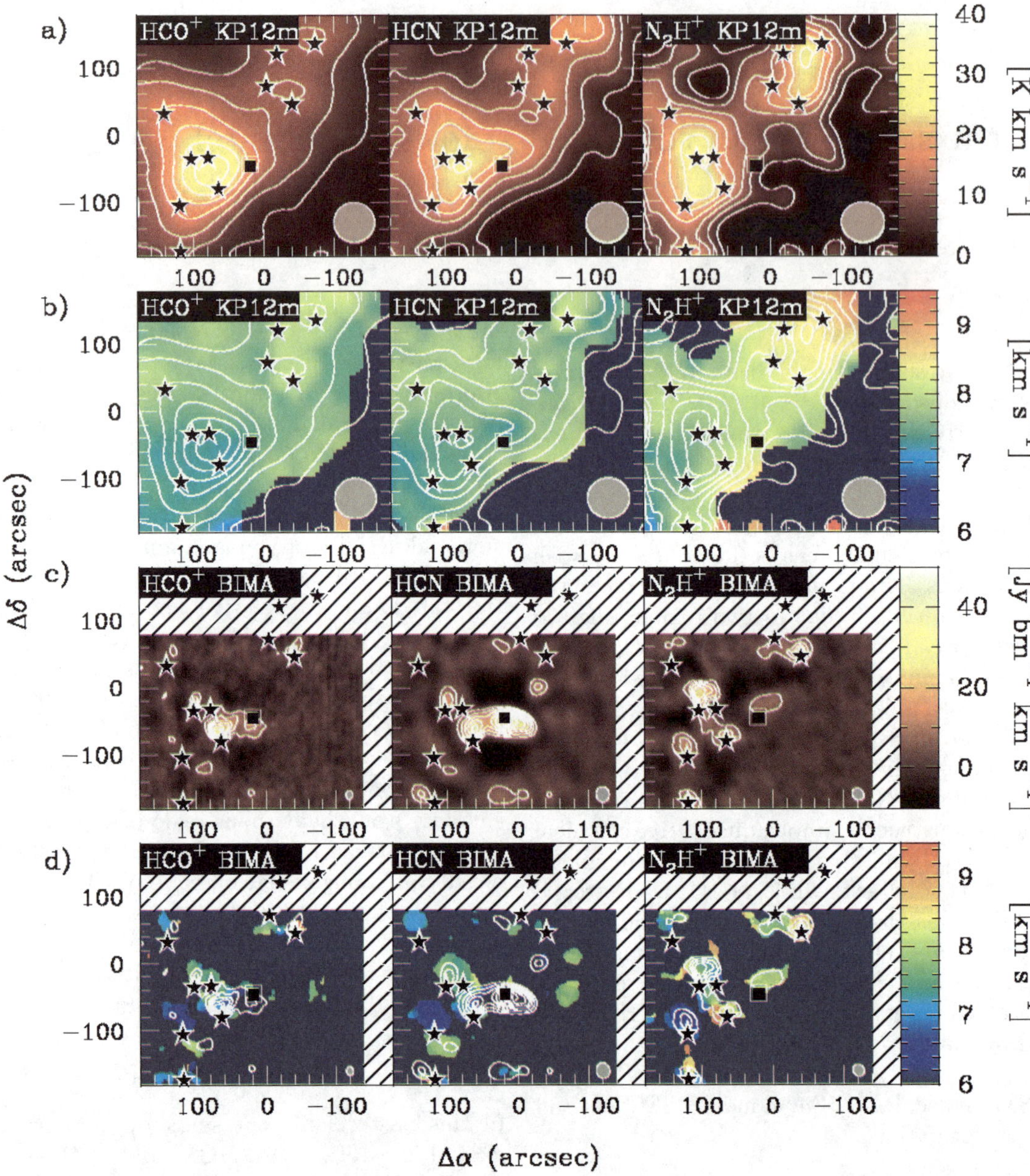

Figure 1. *(a)* KP12m maps of integrated intensity. *(b)* KP12m maps of the velocity centroid, with contours of integrated intensity superposed. *(c)* BIMA maps of integrated intensity. *(d)* BIMA maps of velocity centroid, with contours of integrated intensity superposed. The stars mark the locations of the YSOs; the square marks the location of the 'shock front'. Beam sizes are represented by the grey ellipses in the lower right-hand corner of each panel. The BIMA maps only cover the SE region.

condensation, where most of the YSOs are located and where most of the outflows originate. This indicates that N_2H^+ reliably reflects the distribution of the cloud material, while HCO^+ and HCN emission is enhanced near outflows. The velocity centroids of the emission lines show an east–west gradient of 2.5 km s^{-1}, most clearly seen in N_2H^+, indicating solid-body rotation of the entire cloud, as reported by Olmi & Testi (2002).

3. Interferometer observations

To investigate if the close match between N_2H^+ and dust continuum, and the association of HCN and HCO^+ with outflows holds on smaller scales, aperture-synthesis images are shown in Fig. 1. These interferometer observations where obtained at the BIMA[2] millimeter array, at resolutions of 12″–21″. Mosaics of 13 array pointings cover a 320″ × 260″ region around the SE condensation; the NW condensation has already been studied at high angular resolution by Williams & Myers (2000).

There exists an almost one-to-one correspondence between the peaks in the N_2H^+ BIMA map and the 850 μm emission. The only exceptions are two N_2H^+ peaks, one north of SMM 3 and one $\sim$ 60″ northeast of SMM 4. As discussed below, both locations probably are shocks. The association of HCO^+ and HCN with outflows become very clear at the small scales traced by BIMA. Emission is found north of SMM 3 and SMM 4, along the directions of the jets of these sources. An emission peak, especially prominent in HCN, is located $\sim$ 60″ northwest of SMM 4. The velocity-centroid of the emission around this peak stands out as extremely blue, at > 2 km s^{-1} from systemic. The likely interpretation of this peak is a 'shock front' of a jet impacting a cloud fragment, although no obvious driving source is present. Possible candidates are a jet from (the vicinity of) SMM 1 or from (the vicinity of) SMM 4. In the latter case it may be argued that at the 'shock front' this jet deflects off a cloud condensation and later connects up with the outflow lobe marked 'Wr' in the CO map of Davis et al. (1999).

Interferometers are only sensitive to small-scale emission, and filter out emission on spatial frequencies below the smallest antenna separation. Compared to the KP12m data that do cover these scales, the BIMA data contain $\sim$ 30% of the flux. While this is only a small fraction of the emission, the similarity in distribution of the emission in the KP12m and BIMA maps suggests that outflows contribute significantly to the total emission of HCO^+ and HCN. Hogerheijde (in preparation) analyses the emission and the abundances quantitatively.

4. A shock model for HCN and N_2H^+

Close inspection of the HCN and N_2H^+ BIMA maps near the tip of SMM 3's jet and near the 'shock front' reveals that N_2H^+ peaks $\sim$ 20″ ahead of HCN. The structure of C-type shocks (Draine & McKee 1993) may offer an explanation for this offset. C-type shocks have a magnetic precursor where charged particles accelerate and warm up ahead of neutral particles, enhancing the intensity of N_2H^+ emission lines. Deeper into the shock the neutrals accelerate and warm up. If the pre-shock density is sufficiently high, dust grains will be predominately neutral, and it is in this region that they will release the contents of their ice mantles. These include HCN, as seen toward other outflow (Lahuis & van Dishoeck 2000; Hogerheijde 2001), water, and CO. The latter two effectively destroy N_2H^+ through chemical reactions, causing the N_2H^+ emission to disappear where the HCN peaks. A quantitative model has to include time scales, length scales, and an adequate chemical network, and is discussed in Hogerheijde (in prep.). High-angular resolution observations with, e.g., the Smithsonian Millimeter Array, can help test this model.

5. Conclusions

Aperture-synthesis and single-dish maps of the Serpens star-forming region show that N_2H^+ accurately traces the distribution of the cloud material, while HCO^+ and especially HCN probe material that is shocked by outflows. Unresolved observations of the latter two species therefore would reflect star-forming activity rather than cloud column density. A difference map of N_2H^+ and HCN may be a useful tool to identify regions in cloud complexes that are actively star forming, as opposed to dense but quiescent. On small scales, however, N_2H^+ emission may be enhanced in magnetic precursors of C-type shocks, but appears effectively destroyed in the warmest regions of shocks where HCN peaks.

[2] The BIMA array is operated by the Universities of California (Berkeley), Illinois, and Maryland, with support from the National Science Foundation.

References

Aalto, S., Booth, R. S., Black, J. H., & Johansson, L. E. B. 1995, A&A, 300, 369
Casali, M. M., Eiroa, C., & Duncan, W. D. 1993, A&A, 275, 195
Davis, C. J., Matthews, H. E., Ray, T. P., Dent, W. R. F., & Richer, J. S. 1999, MNRAS, 309, 141
de Lara, E., Chavarria-K., C., & Lopez-Molina, G. 1991, A&A, 243, 139
Draine, B. T. & McKee, C. F. 1993, ARA&A, 31, 373
Helfer, T. T. & Blitz, L. 1997, ApJ, 478, 162
Hogerheijde, M. R. 2001, ApJ, 553, 618
Kuno, N., Nakai, N., Sorai, K., Nishiyama, K., & Vila-Vilaró, B. 2002, PASJ, 54, 555
Lahuis, F. & van Dishoeck, E. F. 2000, A&A, 355, 699
Olmi, L. & Testi, L. 2002, A&A, 392, 1053
Wild, W. & Eckart, A. 2000, A&A, 359, 483
Williams, J. P. & Myers, P. C. 2000, ApJ, 537, 891

MID-INFRARED SPECTROSCOPIC OBSERVATIONS OF OLD CTTSs

Mitsuhiko Honda[1,2], Takuya Yamashita[2], Hirokazu Kataza[3], Takashi Miyata[4], Yoshiko K. Okamoto[5], Shigeyuki Sako[1,2], Shinya Takubo[1,2], and Takashi Onaka[1]

[1]Department of Astronomy, School of Science, University of Tokyo, Bunkyo-ku, Tokyo 113-0033,Japan
[2]Subaru Telescope, National Astronomical Observatory of Japan, 650 North A'ohoku Place, Hilo, Hawaii 96720, U.S.A.
[3]Center for Advanced Spacecraft Technology, Institute of Space and Astronautical Science, 3-1-1 Yoshinodai, Sagamihara, Kanagawa 229-8510, Japan
[4]Kiso Observatory, Institute of Astronomy, School of Science, University of Tokyo, Mitake, Nagano 397-0101, Japan
[5]Institute of Physics, Center for Natural Science, Kitasato University 1-15-1 Kitasato, Sagamihara, Kanagawa 229-8555, Japan

Abstract

We have carried out mid-infrared N-band $R \sim 250$ spectroscopic observations of 12 old (5–10 Myr) low-mass YSOs (mostly classical T Tauri stars) with the COMICS (COoled Mid-Infrared Camera and Spectrometer) mounted on the 8.2m Subaru Telescope, and detected crystalline silicates features in two of them (Hen 3-600A, HD98800). In particular, the spectrum of Hen 3-600A shows clear forsterite and possible enstatite band features in its spectrum, which is the first clear evidence for crystalline silicates in low-mass YSOs. Crystalline silicates are already detected by *ISO* and other ground-based instruments in some Herbig Ae/Be stars, but few in low-mass YSOs such as T Tauri stars. Our detection of crystalline silicates in old T Tauri stars in the TW Hya association implies that the crystallization event occurs even in protoplanetary disks of low-mass YSOs in the T Tauri phase, though regional dependence cannot be ruled out.

Key words: astrochemistry – stars: formation

Chemistry as a Diagnostic of Star Formation (2003). *Edited by Charles L. Curry and Michel Fich.* Published in 2003 by NRC Research Press, Ottawa ON, K1A 0R6, Canada. ISBN 0-660-19089-3.

METALLICITY EFFECTS IN PHOTON DOMINATED REGIONS: CLUMPY CLOUDS

S. Jeyakumar and J. Stutzki

I. Physikalisches Institut, Universität zu Köln, Zülpicher Strasse 77, 50937 Köln, Germany.

Abstract

Several galaxies, such as dwarfs and irregulars as well as outer galactic clouds, have low metallicity. At low metallicities, a reduction in the amount of dust and heavy elements plays a significant role in the chemistry as well as the heating and cooling of the gas in the molecular regions called Photon Dominated Regions (PDRs). We present here the effects of reduced metallicity in PDRs and study the important PDR cooling lines ([C II], [C I] and CO). Moreover, much observational evidence suggests that molecular clouds are clumpy. We model the molecular emission from galaxies incorporating a mass spectrum of clumps. We also compare our results with the semi-analytical results obtained by Bolatto et al. (1999).

Key words: ISM: molecules – ISM: structure – ISM: clouds

1. Introduction

Photon Dominated Regions (PDRs) are predominantly molecular and atomic regions where the physical and chemical processes are dominated by UV radiation (Hollenbach & Tielens 1997, 1999). The molecular clouds in the vicinity of the newly-formed stars are heated by the FUV photons in the energy range from about 6 to 13.6 eV. These clouds cool through the atomic and molecular spectral lines, such as [C II] 158 μm, [O I] 63, 146 μm, [C I] 609, 370 μm, and the millimetric and sub-mm CO rotational lines. Plane-parallel and spherical models of PDRs have been constructed to understand the physical and chemical characteristics of these regions (Kaufman et al. 1999; Köster et al. 1994; Le Bourlot et al. 1993; Sternberg & Dalgarno 1995; Störzer et al. 1996; Tielens & Hollenbach 1985). However there are other important factors which affect the UV absorption and scattering as well as the basic heating and cooling processes in PDRs.

Several galaxies such as dwarf galaxies, irregular galaxies, and the Large and Small Magellanic Clouds have low metallicity (Wilson 1995). A radial gradient in metallicity of molecular clouds is found in the Milky Way and several other nearby galaxies (Arimoto et al. 1996). These low metallicity systems have much higher [C II]/CO and [C I]/CO line ratios as compared to the galactic value (Bolatto et al. 2000; Madden 2000). This suggests that the effects of metallicity should be considered while interpreting the molecular and atomic spectral line observations of these sources. Since the important surface coolant of the PDRs, the [CII]158 μm emission, is strongly affected by the metallicity factor, we study the effects of metallicity in PDRs. In addition, low-metallicity PDR models would also help us to understand the star-forming regions in dwarf galaxies which resemble primordial galaxies.

In addition, observations of edge-on PDR have suggested that the molecular clouds are clumpy, and that UV radiation can penetrate deep inside the clouds (Stutzki et al. 1988; Boissé 1990). This suggests that a simple single-component model may not be sufficient to explain the observed features. We also include a mass spectrum of clumps to understand the cooling lines of PDRs from low metallicity galaxies.

2. Metallicity effects

Low metallicity systems have lower dust-to-gas ratios and heavy element abundances compared to the galactic ISM. This reduction in the amount of dust grains affects the absorption of UV radiation, heating of gas by photo-electric emission from dust, formation of H_2 on the dust grains and the cooling of the gas through atomic and molecular lines. In addition, the chemistry is also affected by the reduction in dust and heavy elements (van Dishoeck & Black 1988; Lequeux et al. 1994). We use the self-consistent spherical PDR model of Störzer et al. (1996) to study the metallicity effects. We scale the dust-dependent parameters and the abundance of heavy elements with the metallicity factor, Z, in our PDR calculations.

Chemistry as a Diagnostic of Star Formation (2003). *Edited by Charles L. Curry and Michel Fich.* Published in 2003 by NRC Research Press, Ottawa ON, K1A 0R6, Canada. ISBN 0-660-19089-3.

The variation of the temperature at the surface of the PDR clumps with the incident UV field is plotted in Figure 1, for a clump of mass $M = 1\,M_\odot$ and density $n=10^3 \mathrm{cm}^{-3}$. The UV field, χ, is expressed in units of the mean UV field of Draine (1978). It is seen from the figure that at high UV fields the temperature is proportional to metallicity, whereas at low UV fields there is no significant change in the surface temperature.

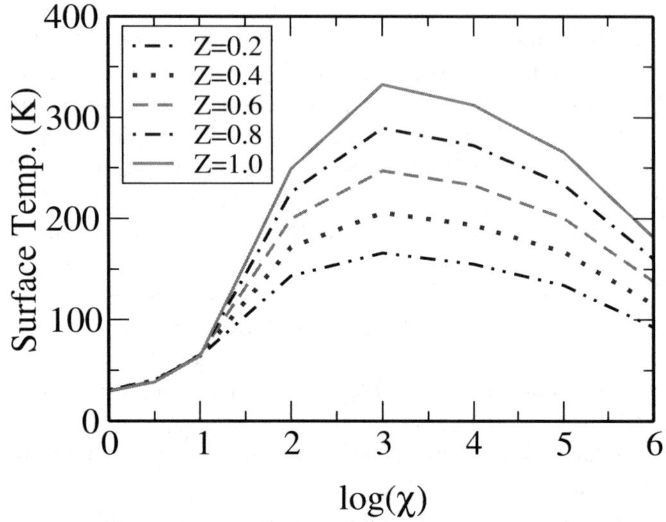

Figure 1. The temperature at the surface of the PDR clump of mass $M = 1\ M_\odot$ and density $n = 10^3$ cm^{-3} is plotted against the UV field for different metallicities.

This correlation can be understood analytically by balancing the dominant heating and cooling processes. In the case of PDRs exposed to high UV fields, grain photo-electric emission (PE) dominates the heating. The photo-electric heating rate given by Bakes & Tielens (1994) is

$$\Gamma_{PE} = 10^{-24}\chi n Z \frac{3\mathrm{e}^{-2}}{1 + 2\times 10^{-4}\chi T^{1/2}/n_e}, \tag{1}$$

where Z is the metallicity factor.

The cooling is dominated by fine structure [O I] emission, gas-grain collisions and fine structure [C II] emission. The cooling rate can be written as $L_{ij} = n_i A_{ij} h\nu_{ij}\beta(\tau_{ij})$ where β is the escape probability, A_{ij} is the transition probability, n_i the number of atoms at level i, and $h\nu_{ij}$ is the corresponding frequency of radiation.

Following Hollenbach & McKee (1979), for the population of the first level of the O I atom, the cooling rate of [O I] 63μm can be written as

$$L_{OI} = \frac{8.45\times 10^{-22}\beta(\tau) n_H Z}{1 + 5\mathrm{e}^{228/T} + 1/3\mathrm{e}^{-98/T}}. \tag{2}$$

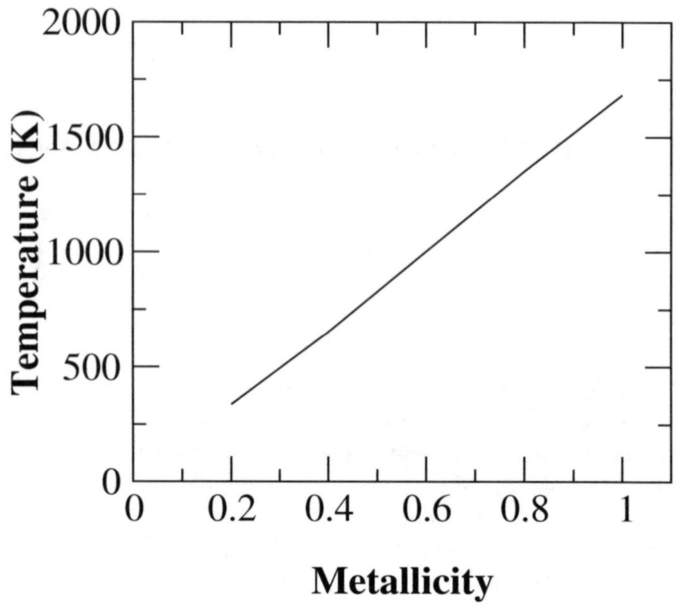

Figure 2. The equilibrium temperature for the balance between grain photo-electric heating rate and [O I] cooling rate is plotted against the metallicity for a density $n = 10^{5.5}$ cm^{-3} and UV field $\chi = 10^6$.

Although equations (1) and (2) show a similar dependence with metallicity, the grain heating rate has an additional dependence on metallicity through the charge state of the grains. The latter is expressed as $\chi T^{1/2}/n_e$, which depends on the availability of electrons for recombination. The main source of electrons at high UV fields is the ionization of C I . Since at the surface almost all of the C I is ionized, the electron density $n_e \sim n_C \propto n_{C,Z=1}Z$. With this assumption the heating rate, Γ_{PE}/Z, decreases with decreasing Z whereas the cooling rate, L_{OI}/Z, remains constant. The equilibrium temperature obtained by solving equations (1) and (2) decreases as the metallicity decreases as seen in our PDR calculations (Fig. 2).

3. Clumpy PDR model

In our clumpy model, the molecular cloud is composed of many spherical clumps of mass spectrum of the form, $dN/dM = AM^{-\alpha}$, with $\alpha = 1.8$ (e.g., Kramer et al. 1998). We assume that the turbulent velocity dispersion of the cloud is larger than the intrinsic line width of each clump. Thus the clumps do not interact radiatively and the total intensity of a spectral line from the cloud can be written as

$$I_T = \int_{M_{\min}}^{M_{\max}} I(M)\left(\frac{dN}{dM}\right)_{\text{cloud}} f_{b,c}\, f_{M,b} dM.$$

The beam filling factor of each clump is $f_{M,b}(= \Omega_M/\Omega_b)$ where Ω_M is the solid angle of the clump of mass M and Ω_b is the beam solid angle. The fraction of the clumps within the beam is given by $f_{b,c} = \Omega_b/\Omega_c$.

By scaling the mass with $M_{\max}$ ($x = M/M_{\max}$),

$I_T = AM_{\max}^{1-\alpha}\ f_{b,c}\ f_{M_{\max},b}\ \int_{M_{\min}/M_{\max}}^{1} I_x x^{-\alpha} f_{M,M_{\max}} dx$

$A \approx M_{tot}(2-\alpha)M_{\max}^{-(2-\alpha)}$,

where f_b is the beam filling factor, $x_0 = x(M_{\min})$ and $f_x = R^2(x)/R^2_{cloud}$. Although the total intensity depends on the scaling constant and the beam filling factors, the line ratio between any two spectral lines depends only on $M_{\min}/M_{\max}$ and α.

4. Clumpy clouds at low metallicity

The observations of [C II] and [C I] emission from low metal galaxies have been modelled by Bolatto et al. (1999), assuming that the size of the C II region scales inversely with metallicity. It is also assumed that the size of the C I region remains between two limiting scenarios of an inverse dependence and no variation with metallicity. Our spherical PDR model calculations based on the model of Störzer et al. (1996) show that the size of the C II layer is indeed inversely proportional to the metallicity factor Z. However, the size of the C I region shows very weak dependence at low Z and a roughly inverse dependence at high Z for a typical spherical clump of density $10^4\,\mathrm{cm}^{-3}$.

The intensity ratio $I_{[C\,ii]158\,\mu m}/I_{CO(1-0)}$, observed in many nearby low metallicity galaxies, shows a power law dependence with metallicity. This dependence has been predicted using a semi-analytical clumpy model by Bolatto et al. (1999). We use our clumpy model to study the metallicity dependence of this line ratio. The observed ratios of nearby galaxies are well-represented by a clumpy model of density $10^4\,\mathrm{cm}^{-3}$ exposed to a UV field of $\chi = 10^2$, as shown in Figure 3. The higher observed ratio for the 30 Doradus region can be explained by a similar clumpy model, but exposed to an UV field of $\chi = 10^5$. These results compare well with the semi-analytical model of Bolatto et al. (1999). This trend suggests that at low metallicities, C II is a tracer of molecular hydrogen rather than CO. However, the observed variation of [C I]/CO line ratio with metallicity is steeper than the model prediction. This is most likely due to the large [C I] line intensities predicted by the PDR models, and requires further investigation.

Acknowledgements

The research presented here is supported by the Deutsche Forschungsgemeinschaft (DFG) via Grant SFB 494.

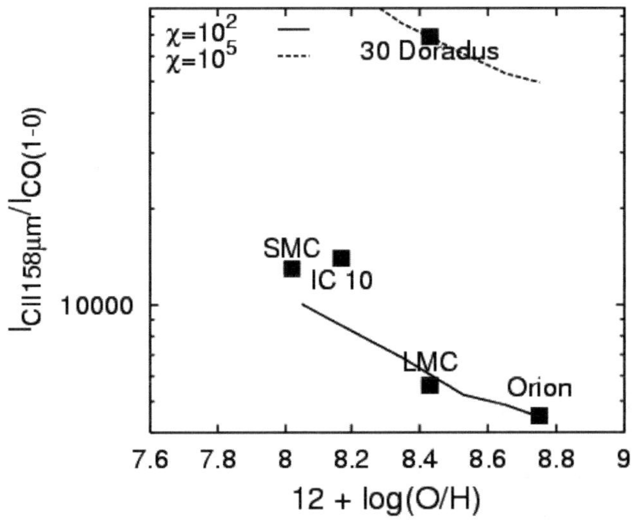

Figure 3. The intensity ratio $I_{[CII]158\,\mu m}/I_{CO(1-0)}$ is plotted against the metallicity. The maximum and minimum masses of the clump spectrum are 10^3 and 10^{-2} $M_\odot$. The density at the surface of the clumps is 10^4 cm^{-3}. The observed ratios of nearby galaxies are plotted as filled squares. The model result, plotted as a solid curve, compares well with that of the trend shown in Fig. 5 of Bolatto et al. (1999).

References

Arimoto, N., Sofue, Y., & Tsujimoto, T. 1996, PASJ, 48, 275
Bakes, E.L.O., & Tielens, A.G.G.M. 1994, ApJ, 427, 822
Boissé, P. 1990, A&A, 228, 483
Bolatto, A.D., Jackson, J.M., & Ingalls, J.G. 1999, ApJ, 513, 275
Bolatto, A.D., et al. 2000, ApJ, 541, 17
Draine, B.T. 1978, ApJS, 36, 595
Hollenbach, D., & McKee, C.F. 1979, ApJS, 41, 555
Hollenbach, D.J., Tielens A.G.G.M. 1999, Rev. Mod. Phys., 71, 173
Hollenbach, D.J., Tielens A.G.G.M. 1997, ARA&A, 35, 179
Kaufman, M.J., et al. 1999, ApJ, 527, 795
Köster et al. 1994, A&A, 284, 545
Kramer, C., et al. 1998, A&A, 329, 249
Le Bourlot, J., et al. 1993, A&A, 267, 233
Lequeux, J., et al. 1994, A&A, 292, 371
Madden, S.C. 2000, NewAR, 44, 249
Sternberg, A., & Dalgarno, A. 1995, ApJS, 99, 565
Störzer, H., Stutzki, J., & Sternberg, A. 1996, A&A, 310, 592
Stutzki, J., et al. 1988, ApJ, 332, 379
Tielens, A.G.G.M., Hollenbach, D. 1985, ApJ, 291, 722
van Dishoeck, E. F. & Black, J. H. 1988, ApJ, 334, 771
Wilson, C.D. 1995, ApJ, 448, L97

A MULTI-LINE STUDY OF ATOMIC CARBON AND CARBON MONOXIDE IN THE GALACTIC STAR-FORMING REGION W3

H. Jakob, C. Kramer, B. Mookerjea, S. Jeyakumar, and J. Stutzki

Kölner Observatorium für Submillimeter-Astronomie (KOSMA), I. Physikalisches Institut, Universität zu Köln, Zülpicher Strasse 77, 50937 Köln, Germany

Abstract

We present results from observations of both fine structure lines of neutral carbon at 492 and 809 GHz and two mid-J transitions of CO from the Galactic star-forming region W3 and compare these with the results in the literature. The observations were conducted with the newly installed submillimeter array receiver SMART on the KOSMA telescope. The region observed covers the embedded infrared cluster, masers and outflow sources, as well as the more extended photon-dominated regions. A preliminary analysis of [C I] line ratios indicates warm gas of $T_{\rm kin} > 50\,{\rm K}$ in the central core. Further analysis will include [C II] intensities (Howe et al. 1991) to derive coherent models of the physical and chemical structure within the photon-dominated regions applying models of Störzer et al. (1996) and Kaufman et al. (1999).

Key words: ISM: molecules – ISM: structure – ISM: clouds

1. Introduction

Molecular clouds are the sites of star formation. The newly formed massive stars in particular, do strongly influence their molecular environment due to their strong UV radiation and stellar winds. Due to the complex, fractal nature of the embedding clouds, the UV radiation scatters deep into the clouds. photon-dominated regions (PDRs) where the physical and chemical processes are governed by UV radiation are thus formed on the surfaces of clumps embedded in the molecular cloud. A large fraction of the cloud volume therefore consists of PDRs. Within the PDR interface layer, the dominant carbon bearing species changes from C^+ to C^0 and to CO. The dominant cooling lines of the neutral & molecular gas in the PDRs are the [C II] and [C I] fine structure lines, and the FIR-, submm- and mm-rotational lines of CO. These lines trace different density and temperature regimes as shown in Table 1. To understand the properties of the C^0 layer, we have been observing several rotational transitions of CO as well as both fine-structure lines of atomic carbon. In combination with appropriate modelling (e.g. Störzer et al. 1996, Kaufman et al. 1999) these lines provide a unique diagnostic of the physical conditions (UV-field, density, mass distribution) of the clouds.

Table 1. Major cooling lines of the cold, dense ISM. The critical density is defined by $n_{\rm crit} = A_{ij}/\gamma$, where A_{ij} is the Einstein A-coefficient and γ is the collision rate. Densities higher than $n_{\rm crit}$ signify thermalized line emission. The population of the upper energy level E_u depends on the kinetic temperature of the gas.

	Transition	$n_{\rm crit}\,[{\rm cm}^{-3}]$	$E_{\rm u}\,[{\rm K}]$
CO	$J \rightarrow J-1$	$4 \times 10^3 \times J^3$	$5.53 \times J(J+1)$
C^0	$^3P_1 \rightarrow {}^3P_0$	1.0×10^3	23.8
C^0	$^3P_2 \rightarrow {}^3P_1$	1.4×10^3	62.4
C^+	$^2P_{3/2} \rightarrow {}^2P_{1/2}$	5×10^3	91.2

The W3 complex, at a distance of 2.3 kpc, is a massive star-forming region in the Perseus arm at the rim of a Giant Molecular Cloud complex of 7×10^4 $M_\odot$ (Lada et al. 1978). This region shows many evidences of massive star formation, such as maser and outflow sources, associated NH_3 emission, and dense molecular cores (Tieftrunk et al. 1998 and references therein). Observations of the core region W3 Main show high ^{13}CO and [C I] column densities (Plume et al. 1999) corresponding to a total mass of $\sim 5 \times 10^3$ $M_\odot$ (Thronson 1986). W3 Main is divided into an East and a West cloud each containing several strong embedded UV sources which provide an extended radiation field. The stellar cluster in W3 East near IRS 5 has a luminosity of $\sim 2 \times 10^5$ $L_\odot$. It consists of an association of seven B stars (Claussen et al. 1994). IRS 4 in W3 West shows a similarly high luminosity. Both IRS sources show outflow activity. The FUV field of this association impinges on the surrounding dense gas and warms it. Due to the strong embedded sources, the total FUV

Chemistry as a Diagnostic of Star Formation (2003). *Edited by Charles L. Curry and Michel Fich.* Published in 2003 by NRC Research Press, Ottawa ON, K1A 0R6, Canada. ISBN 0-660-19089-3.

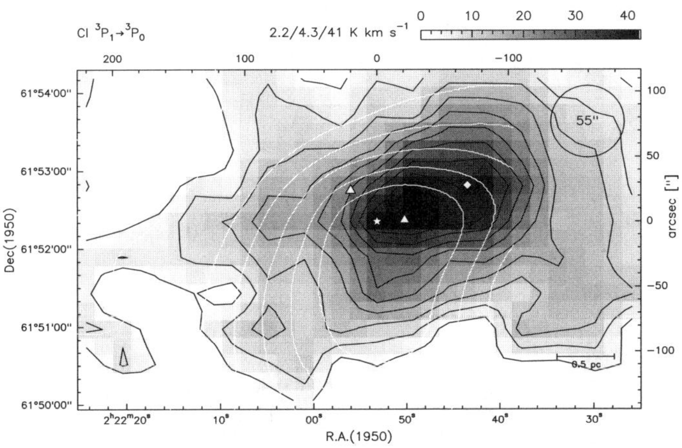

Figure 1. Map of integrated [C I] 1–0 intensities of W3. Four IR sources are marked (Wynn-Williams et al. 1972). The star marks the position of IRS5. The square marks IRS4. The two triangles mark IRS1 and 3. White contours mark [C II] integrated intensities observed by Howe et al. (1991).

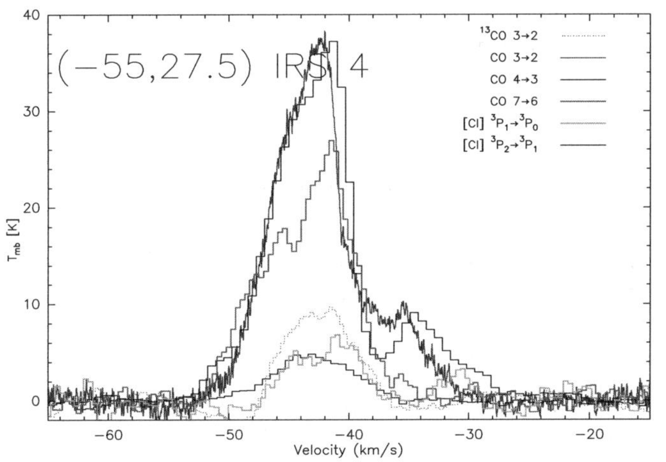

Figure 2. Observed spectra at the position of the embedded protostar IRS4.

flux of the core region is $\sim 10^4$ times stronger than the average interstellar radiation field.

2. Observations

The observations of W3 Main presented here, cover a region of $360'' \times 220''$ (4.2 pc × 2.6 pc). We observed the two [C I] fine structure lines and two mid-J rotational transitions of CO with the SMART/KOSMA[1] receiver on the Gornergrat/Switzerland. The observations were done in Dual-Beam-Switch mode with a $6'$ chop throw in azimuth. Additional larger-scale observations in low-J CO transitions were done in the on-the-fly mode with the dual channel 230/345 GHz SIS receiver at the second Nasmyth port. Some parameters of the KOSMA observations are listed in Table 2. In addition, we compare with [C II] data at 1.9 THz taken on board the KAO by Howe et al. (1991) with a beamwidth of $55''$.

[1] SMART: SubMillimeter Array Receiver for Two frequencies; KOSMA: Kölner Observatorium für Submillimeter Astronomie.

Table 2. KOSMA observations of the W3 region. Frequencies marked with a * were observed simultaneously with SMART.

Species	Transition	ν [GHz]	Positions	HPBW [$''$]
^{13}CO	$J = 3 \to 2$	330.588	1091	85
CO	$J = 3 \to 2$	345.796	1091	85
CO	$J = 4 \to 3$	461.041	192	57
CO	$J = 7 \to 6$	806.652*	192	42
C^0	$^3P_1 \to {}^3P_0$	492.1607*	192	55
C^0	$^3P_2 \to {}^3P_1$	809.3420*	192	42

3. Results

Figure 1 shows a map of integrated [C I] 1-0 emission and Figure 2 shows all spectra taken at IRS4, an embedded protostar. The CO 3-2 transition traces the cooler molecular material while [C I] is found in the thin interface regions at the surfaces of PDRs. The spatial extent especially of [C II] emission strongly depends on how far the UV-field can penetrate through the clumpy molecular medium.

The spectral resolution of the data set permits the analysis of the line wing emission at outflow positions and the absorption of foreground material which is known to affect the line intensities of W3. A coherent analysis has to take into account the origins for all these features.

Figure 3 shows the observed line ratios of integrated intensities. The line ratios are to a first order independent of beam filling factors and thus easy to compare with model results. In addition, some of the line ratios stem from simultaneous observations and are thus largely unaffected by the atmospheric calibration accuracy.

The observed line ratios tend to be constant over a wide intensity range. Mean values found in W3 are compared with the values from the limited high-frequency observations found in the literature (Table 3).

Of particular interest is the ratio of both [C I] transitions. In the optically thin limit, it measures the excitation temperature of the 2 → 1 transition: $T_{\rm ex}$ =

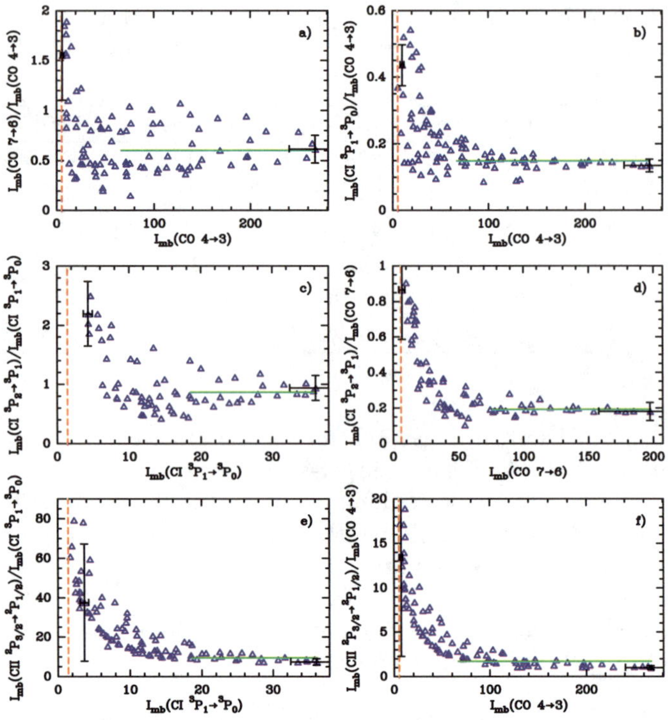

Figure 3. Observed ratios of line integrated intensities. Horizontal lines show average ratios calculated for intensities above 40σ (vertical lines denote the 3σ level).

$38.8\,\mathrm{K}/\ln[2.11/R]$ with the ratio of integrated intensities I in $\mathrm{K\,km\,s^{-1}}$: $R = I(2 \rightarrow 1)/I(1 \rightarrow 0)$. In this limit the low [C I] ratios of 0.16 observed with the *COBE* satellite in the inner Galactic disk, correspond to excitation temperatures of $\sim 15\,\mathrm{K}$. In contrast, the high ratios observed in massive star forming regions and in the archetypal starburst galaxy M82, correspond to 50 K. The kinetic temperatures of the emitting regions will thus be 50 K or more.

4. Outlook

In this study, we aim to observe and interpret the emission due to the major coolants of photon-dominated regions in Galactic star forming regions. We have started to compare the present data set of W3 Main with PDR models of Störzer et al. (1996) and Kaufman et al. (1999). The aim is to derive important physical parameters such as the UV flux and its spatial variation, the density, and the underlying unresolved mass distribution of clumps. Such a comparison will also eventually show whether all important physical processes are taken into account in the current models to reproduce the observed line ratios.

Table 3. Ratios of integrated intensities observed in W3 and other Galactic and extragalactic star-forming regions

Ratio	W3	Other regions
[C I] $2 \rightarrow 1$/[C I] $1 \rightarrow 0$	0.9 ± 0.1	0.22[a], 0.16[b], 1.0[c]
[C I] $2 \rightarrow 1$/CO $7 \rightarrow 6$	0.22 ± 0.03	1.1[a], 0.25[d]
CO $7 \rightarrow 6$/CO $4 \rightarrow 3$	0.62 ± 0.06	0.11[a], 1.2-3.5[e]
[C I] $1 \rightarrow 0$/CO $4 \rightarrow 3$	0.16 ± 0.01	0.35-0.4[f], 0.2-0.6[g]
[C II]/[C I] $1 \rightarrow 0$	10 ± 2	1.38[a], 3.54[b]
[C II]/CO $4 \rightarrow 3$	1.8 ± 0.2	0.71[a], 4.29[b]

References— [a]Galactic center (Fixsen et al. 1999, *COBE*);
[b]Inner Galactic disk (Fixsen et al. 1999, *COBE*);
[c]M82 (Stutzki et al. 1997) and Orion KL (Yamamoto et al. 2001);
[d]W3 (Krügel et al. 1989);
[e]Orion (Wilson et al. 2001);
[f]NGC 6946 and M83 (Israel et al. 2001);
[g]Sample of 15 spiral galaxies (Israel et al. 2002).

Also, it allows the derivation of a consistent picture of the photon-dominated regions in W3 Main in the vicinity of massive star formation.

References

Claussen et al. 1994, ApJ, 424, L41
Hasegawa et al. 1994, ApJ, 426, 215
Howe et al. 1991, ApJ, 373, 158
Kaufman et al. 1999, ApJ, 527, 795
Köster et al. 1994, A&A, 284, 545
Lada et al. 1978, ApJ, 226, L39
Megeath et al. 1996, A&A, 307, 755
Plume et al. 1999, ApJ, 512, 768
Störzer et al. 1996, A&A, 310, 592
Tieftrunk et al. 1998, A&A, 340, 232
Thronson, 1986, ApJ, 306, 160
Wynn-Williams et al. 1972, MNRAS, 160, 1

Acknowledgements

The KOSMA 3m submillimeter telescope at the Gornergrat-Süd is operated by the University of Cologne in collaboration with Bonn University, and supported by special funding from the Land NRW. The observatory is administrated by the International Foundation Gornergrat & Jungfraujoch.

THE GAS TEMPERATURE IN CIRCUMSTELLAR DISKS

B.J. Jonkheid, G.J. van Zadelhoff, F. Faas, and E.F. van Dishoeck

Leiden Observatory, P.O. Box 9513, 2300 RA Leiden, The Netherlands

Abstract

The gas temperature in circumstellar disks is calculated in a self-consistent way using a quasi-2-dimensional model. A simple chemical network is adopted to calculate the densities of species relevant to the heating and cooling processes of the gas, and special attention is given to the effects of gas-dust coupling on the temperature. This coupling has a noticeable effect, but is not strong enough to validate the assumption that the gas temperature equals the dust temperature throughout the disk. The difference is greatest in the upper layers of the disk, where T_{gas} exceeds T_{dust} by up to 50 K.

Key words: ISM: molecules – circumstellar material – protoplanetary disks

1. Introduction

The gas temperature T_{gas} plays an important role in the study of circumstellar disks: both the chemistry, the population of molecular energy levels, the overall density structure of a disk in hydrostatic equilibrium, and the dynamics of dust grains depend on this temperature. It is often assumed that gas and dust are strongly coupled in disks by collisions, so that T_{gas} equals the dust temperature T_{dust}. T_{dust} is determined by radiative equilibrium and can be inferred from the continuum radiation coming from the disk. We present here explicit calculations of T_{gas} to determine if this assumption is generally valid in circumstellar disks.

2. Disk model

A quasi-2-dimensional model (hereafter called the 1+1-D model) was used to calculate, at different radii, the radiative transfer and gas temperature structure in the vertical direction (i.e., perpendicular to the midplane of the disk; see Fig. 1). The code used for 1-dimensional photon-dominated regions (PDRs) by Jansen at al. (1995) was adopted to calculate the chemical and temperature structure. The intensity of the incident stellar radiation

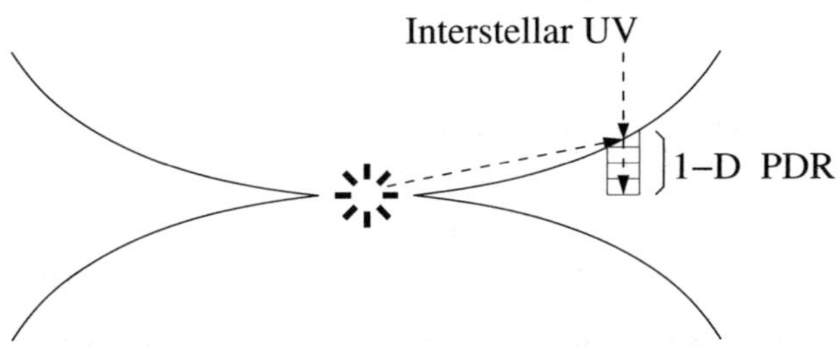

Figure 1. In our model, the intensity of the stellar radiation is calculated at the surface of the disk and added to the interstellar radiation field. This intensity is then used in a 1-D PDR code to calculate the chemistry and temperature structure in the vertical direction.

on the disk's surface was computed separately using the 2-D UV radiative transfer code of van Zadelhoff et al. (2002). The densities and dust temperatures are taken from the flared disk model of D'Alessio et al. (1999) (see Fig. 2), using a central star of $0.5\,M_\odot$, a radius of $2\,R_\odot$ and an effective temperature of 4000 K. The disk is assumed to have a viscosity parameter $\alpha = 0.01$, and an accretion rate of $10^{-8}\,M_\odot\,\mathrm{yr}^{-1}$; the radiation from accretion onto the central star is assumed to have a luminosity of $0.04 - 0.04\,L_\odot$ and a temperature of 10,000 K.

The cooling of the gas occurs through line emission from C^+, C, O and CO. The cooling rate is determined by calculating the populations of the fine structure levels of C^+, C and O, assuming statistical equilibrium of those levels, and of the rotational levels of CO, assuming a thermal distribution. Heating terms include photoelectric heating (by both grains and PAHs), the dissociation, formation and excitation (and subsequent collisional deexcitation) of H_2, photoionisation of C, cosmic ray heat-

Chemistry as a Diagnostic of Star Formation (2003). *Edited by Charles L. Curry and Michel Fich.* Published in 2003 by NRC Research Press, Ottawa ON, K1A 0R6, Canada. ISBN 0-660-19089-3.

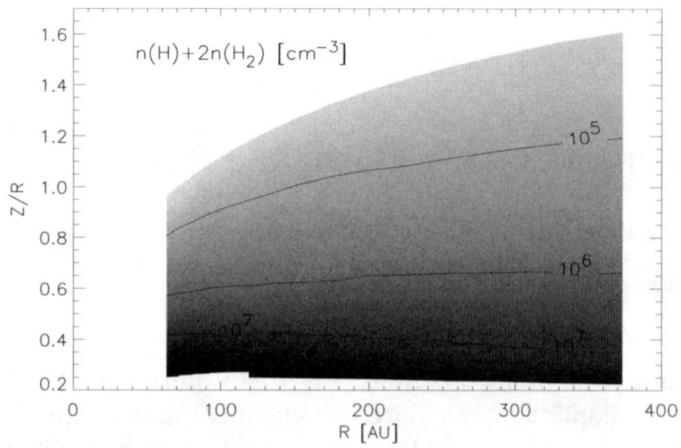

Figure 2. The density structure of one quadrant of the flared disk calculated by D'Alessio et al. (1999), used as input in our model.

ing and chemical heating. For the heating rate resulting from gas-dust coupling, the formula

$$\Gamma_{\rm GD} = n_{\rm H}\, n_{\rm dust}\, \sigma_{\rm dust} \left(\frac{8kT_{\rm gas}}{\pi m_{\rm H}}\right)^{1/2} \bar{\alpha}_T\, 2k\,(T_{\rm dust} - T_{\rm gas})$$

is used, taken from Burke & Hollenbach (1983). Here $n_{\rm H}$ is the total density of hydrogen nuclei, $n_{\rm dust}$ the density of dust grains, $\sigma_{\rm dust}$ the geometrical cross section of the grains and $\bar{\alpha}_T$ the thermal accommodation coefficient (≈ 0.3 for graphite and silicate grains). A simple chemical network is used to calculate the abundances of species relevant to the heating and cooling rates.

The main disadvantage of the 1+1-D model is its poor treatment of radiative transfer: only transfer in the vertical direction is included. As a result, radiation coming from the central star effectively changes direction when it hits the disk's surface, and radiation from cooling lines is also forced to travel in the vertical direction instead of escaping via the shortest route to the surface. This means that the UV radiation from the star penetrates too deep into the disk, and that our escape probabilities for line cooling radiation are too low. This in turn introduces uncertainties in the determination of the temperature structure. In the upper layers of the disk this effect is small, however, because the optical depths there do not change much if the geometry is changed to 2-D. In the regions near the midplane, the optical depths due to the dust are so large that little radiation reaches them, and the results of our temperature calculations are correct there; furthermore, the effects of gas-dust coupling are strongest in these dense parts of the disk, so that $T_{\rm gas}$ is close to $T_{\rm dust}$. This indicates that the largest uncertainty in the determination of the temperature is in the regions between the direct surface and the midplane.

3. Results

The results of our calculations are shown in Figures 3 and 4. It can be seen that in the absence of gas-dust coupling $T_{\rm gas} > T_{\rm dust}$ in the outer parts of the disk. If gas-dust coupling is included in the calculations, $T_{\rm gas}$ is generally still greater than $T_{\rm dust}$, even up to a factor of 2 in the outer layers and at large radii. At larger depths $T_{\rm gas}$ decreases, falling below $T_{\rm dust}$ until the gas-dust coupling becomes strong enough to set $T_{\rm gas}$ equal to $T_{\rm dust}$. The assumption of strong coupling is therefore only correct in the high density region near the midplane ($Z/R < 0.5$). The effects of gas-dust coupling can be seen in the surface region: its cooling effect is strong enough to have a noticeable effect on the gas temperature, but the coupling is weak in those parts. The coupling becomes strong (i.e., the assumption $T_{\rm gas} = T_{\rm dust}$ is correct) at densities of $n > 10^6 - 10^7\,{\rm cm}^{-3}$, dependent on local conditions.

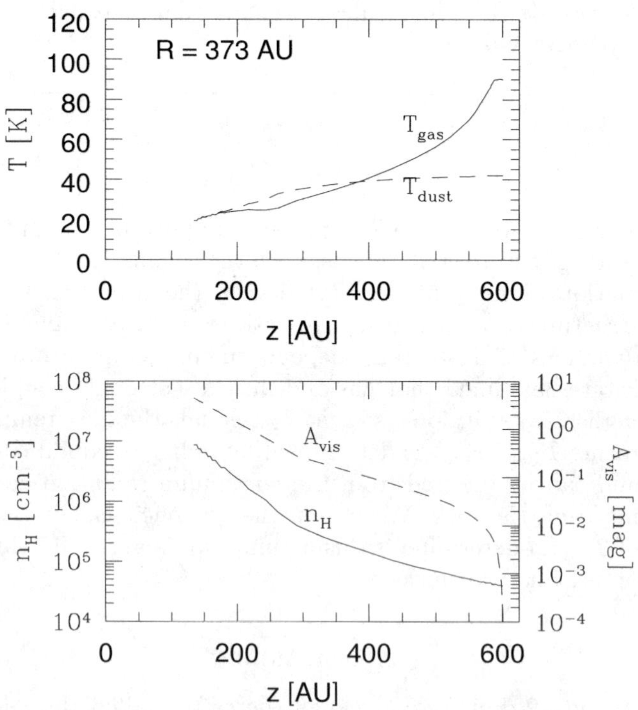

Figure 3. In the upper panel the vertical structure of $T_{\rm gas}$ and $T_{\rm dust}$ are shown at a radius of 373 AU. In the lower panel the density and visual extinction are shown at the same radius. Gas-dust coupling was included here.

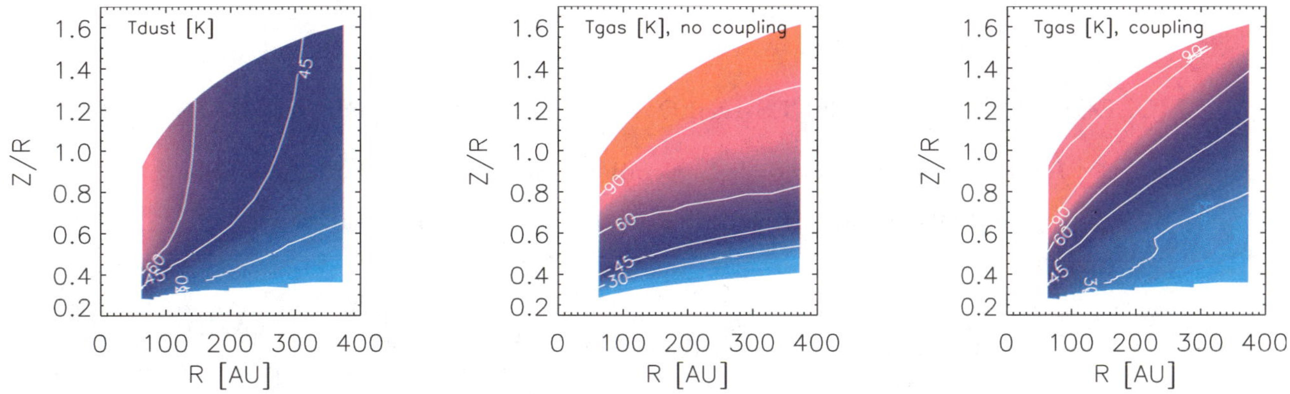

Figure 4. Temperature distributions for one quadrant of a flaring disk. The left panel shows $T_{\rm dust}$. The middle panel shows the results of $T_{\rm gas}$ calculations where gas-dust coupling is ignored. The right panel shows the results for $T_{\rm gas}$ when the effects of gas-dust coupling are calculated explicitly. The color scales and contours trace the same temperatures in all panels.

4. Conclusions

Our calculations indicate that the assumption $T_{\rm gas} = T_{\rm dust}$ is generally invalid in the surface layers of circumstellar disks. This assumption is only valid at small heights, where the coupling term is the dominant term in the heating-cooling balance. At larger heights the effects of gas-dust coupling are still noticeable in the determination of $T_{\rm gas}$, as can be seen in Figure 4, but not strong enough to validate the assumption of strong coupling. In these parts of the disk explicit calculations of $T_{\rm gas}$ are necessary, meaning that models that calculate the density structure of disks using the strong coupling assumption will give incorrect results. As can be seen in Figure 4, $T_{\rm gas}$ is higher than $T_{\rm dust}$ in the upper layers, which may lead to an effective outgassing of the disk.

5. Future work

One of the motivations for this work is to investigate the effects of dust settling and coagulation, and future models will include these effects. Eventually, a true 2-D formalism to calculate the heating-cooling balance in disks will be included to get a more accurate temperature distribution. This is expected to result in a steeper temperature profile near the disk's surface: UV radiation will penetrate less deeply into the disk, while radiation from cooling lines can more easily escape in a 2-D model. Other plans involve using our temperature calculations in the code by Dullemond, van Zadelhoff & Natta (2002) to iteratively calculate the disk's density structure in hydrostatic equilibrium.

References

Burke, J.R. & Hollenbach, D.J. 1983, ApJ 265, 223

D'Alessio, P., Calvet, N., Hartmann, L., Lizano, S. & Cantō, J. 1999, ApJ 527, 893

Dullemond, C.P., van Zadelhoff, G.J. & Natta, A. 2002, A&A 389, 464

Jansen, D.J., van Dishoeck, E.F., Black, J.H., Spaans, M & Sosin, C. 1995, A&A 302, 223

van Zadelhoff, G.J. et al. 2002, A&A 395, 373

MOLECULAR ABUNDANCES IN LOW-MASS PROTOSTELLAR ENVELOPES

Jes K. Jørgensen, Fredrik L. Schöier, and Ewine F. van Dishoeck

Leiden Observatory, P.O. Box 9513, 2300 RA Leiden, The Netherlands

Abstract

A study of the chemical structure of the envelopes around a sample of protostars is introduced. Physical models for the envelopes derived using 1-D radiative transfer modeling of their dust continuum emission are used as input for Monte Carlo modeling of single-dish line observations in order to establish the chemical inventories for the different sources. CO and HCO^+ abundances are found to be correlated with envelope mass: the sources with the most massive envelopes show the lowest abundances, supporting the idea that the depletion of these molecules is most efficient in the colder, denser environments. Other molecules like CS and HCN do not show a similar trend. Deuterated species like DCO^+ and DCN also show signs of correlations with envelope mass and temperature: the deuterium fractionation of HCO^+ is most prominent for the coldest and most massive envelopes in the sample and at the same time anti-correlated with the fractionation of HCN. This puts constraints on the low-temperature gas-phase deuterium chemistry.

Key words: ISM: molecules – ISM: abundances – stars: formation – radiative transfer – astrochemistry

1. Introduction

Understanding the chemistry of protostellar environments is of great importance for addressing topics of star formation. It is directly related to processes regulating star formation, for example through the ionization and the chemistry also potentially serves as an evolutionary clock against which physical models—e.g., for the protostellar collapse—can be compared (see, e.g., van Dishoeck & Blake (1998) for an overview). Recent evidence suggests that many of the traditional molecular tracers of density and temperature may be subject to significant gradients in abundances. Examples are the freeze-out of molecules like CO onto dust grains both in pre-stellar cores (Caselli et al. 1999; Bacmann et al. 2002) and in the cold and dense parts of low-mass protostellar envelopes (Jørgensen et al. 2002) and enhancement of molecules due to liberation of ice mantles in regions of higher temperatures similar to "hot cores" or in regions subject to shocks driven by protostellar outflows (Bachiller & Pérez Gutiérrez 1997; Ceccarelli et al. 2000; Schöier et al. 2002).

In this paper, a preliminary overview of the results from a large survey of the chemical properties of a sample of low-mass protostars is presented. For further details, see Jørgensen et al. (2002) and Jørgensen et al. (2003, in preparation).

2. Sample, observations and modeling

The sample of objects in this study comprises 18 pre- and protostellar objects, including 11 Class 0 objects, 5 Class I objects and 2 pre-stellar cores; the full sample is presented in Jørgensen et al. (2002). These objects have been observed in a range of molecular transitions using the 15 m James Clerk Maxwell Telescope (JCMT), the 20 m telescope at Onsala Space Observatory and the IRAM Pico Valeta 30 m telescope between 2000 and 2002.

The physical structure of each envelope is derived using 1-D radiative transfer calculations. Power-law density profiles are assumed for a range of envelope parameters (e.g., density slope and optical thickness) and the corresponding temperature profiles calculated using the radiative transfer code, DUSTY (Ivezić et al. 1999) for a given central source of heating. For each model, images are constructed and compared to the observed SCUBA images (450 and 850 μm) and SEDs over the range from 60 μm to 1.3 mm. It is found that each protostellar envelope can successfully be modeled under these assumptions, whereas the pre-stellar cores cannot, which is not unexpected since these sources by definition are characterized as not having central sources of heating.

The derived physical models are subsequently used as a basis for calculating the molecular excitation and ra-

Chemistry as a Diagnostic of Star Formation (2003). *Edited by Charles L. Curry and Michel Fich.* Published in 2003 by NRC Research Press, Ottawa ON, K1A 0R6, Canada. ISBN 0-660-19089-3.

diative transfer for the molecular lines using the Monte Carlo code of Hogerheijde & van der Tak (2000). The fractional abundances (assumed to be constant) are constrained through comparisons with the integrated line intensities and it is found that the majority of the lines can successfully be accounted for with the presented models and assumptions. In Figure 1, an overview of the average abundances for the Class 0 and I objects are compared to the molecular cloud L134N (Dickens et al. 2000) and the abundances in the steady state or pre-collapse phase of the models of Bergin & Langer (1997). The abundances are, where possible, derived from the minor isotopic lines. The main isotopic lines of, e.g., CS and HCO^+ are more sensitive to departures from spherical symmetry or the interaction of molecular outflows with the envelope material.

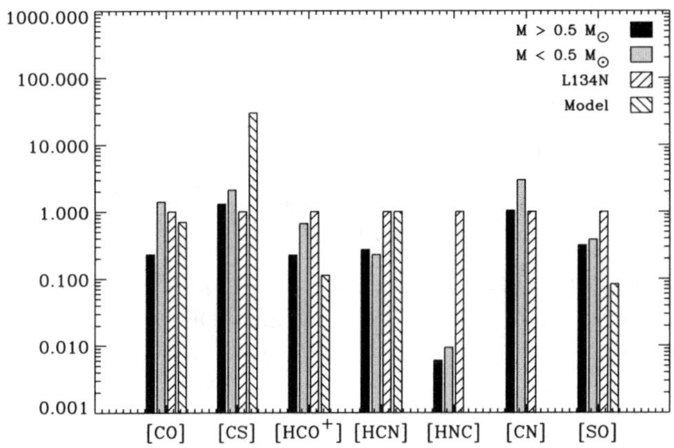

Figure 1. Comparisons between average abundances with respect to H_2 for different molecules for objects with envelope masses larger or smaller than 0.5 $M_\odot$, and with the abundances towards the "C" position in the dark cloud L134N (Dickens et al. 2000) and the abundances in the stage before the collapse in the chemical models of Bergin & Langer (1997). All abundances are normalized to the abundances of L134N. (From Jørgensen et al. 2003, in prep.).

3. Molecular depletion

As can be seen from Figures 1 and 1, significant differences exist between the various molecules relative to the two comparison sets of abundances and also between the samples of Class 0 and I objects for some molecules—in particular CO and HCO^+. CO is thought to be frozen out onto dust grains at low temperatures ($T \leq 30\,$K; Jørgensen et al. 2002) and CO is indeed most depleted for the sources with the most massive envelopes (Fig. 1); i.e., the envelopes with a region in which the temperatures are low enough and densities high enough that the depletion can be effective. The primary formation route of HCO^+ is connected to the presence of CO through the reaction $CO + H_3^+ \rightarrow HCO^+ + H_2$, as is indeed illustrated by the data in Figure 3.

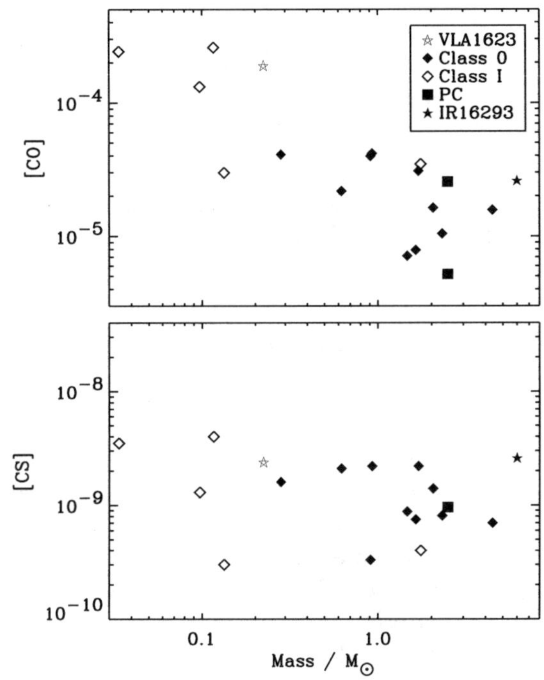

Figure 2. CO *(upper panel)* and CS *(lower panel)* abundances as functions of mass. (Jørgensen et al. 2002; 2003, in prep.).

Other molecules like CS and HCN do not show a similar trend with mass, which at first may seem unexpected (lower panel of Fig. 1): CS depletion is expected to occur at even lower densities than for CO and CS should stay bound to the dust grains to even higher temperatures than CO due to its larger binding energy to the dust grains. Comparison to the models of Bergin & Langer (1997) shows that the CS abundances observed in the protostellar environments are indeed lower by more than an order of magnitude, which could indicate that the observed lines only probe the region where CS is depleted. A detailed comparison with other sulfur-bearing species like SO will be needed to fully address this, and similarly for the nitrogen-bearing species that also show interest-

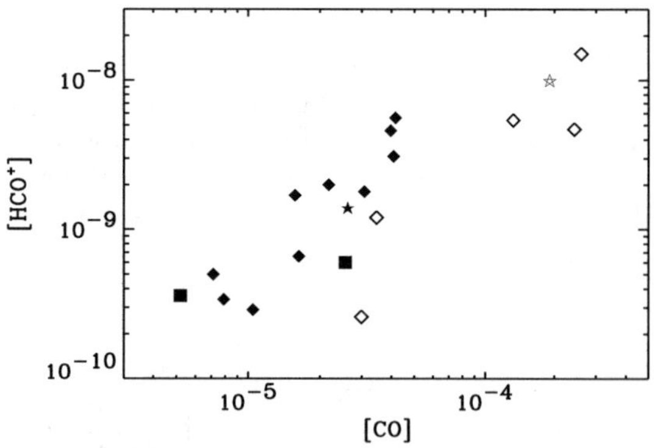

Figure 3. HCO^+ abundance as a function of CO abundance. Symbols as in Fig. 1. (Jørgensen et al. 2003; in prep.)

ing features—in particular, the strikingly low HNC abundances for the protostars compared to other star-forming environments.

4. Deuterium fractionation

Another interesting aspect of the chemistry in the envelopes around protostars is the degree of deuterium fractionation. Gas-phase deuterium fractionation of molecules like HCO^+ is expected to be particularly efficient at low temperatures (e.g., Roberts & Millar 2000) due to small zero-point energy differences of reactions like $HD + H_3^+ \rightarrow H_2D^+ + H_2$. As illustrated in Fig. 4, the highest degree of deuterium fractionation is seen to occur in the pre-stellar cores and Class 0 objects with low temperatures ($T \leq 20$ K) and high densities. Interestingly, the HCN deuterium fractionation seems to be anti-correlated with that of HCO^+ as shown in the lower panel of Fig. 4: for HCN deuterium fractionation through other species like CH_3^+ (i.e. forming CH_2D^+) could become increasingly important in producing DCN at temperatures $T \sim 30$ K (e.g., Turner 2001). Such a scenario could explain the observed anti-correlation between the deuteration of HCO^+ and HCN.

Acknowledgements

The authors are grateful to the Leids Kerkhoven-Bosscha Fond for supporting their participation in the conference financially.

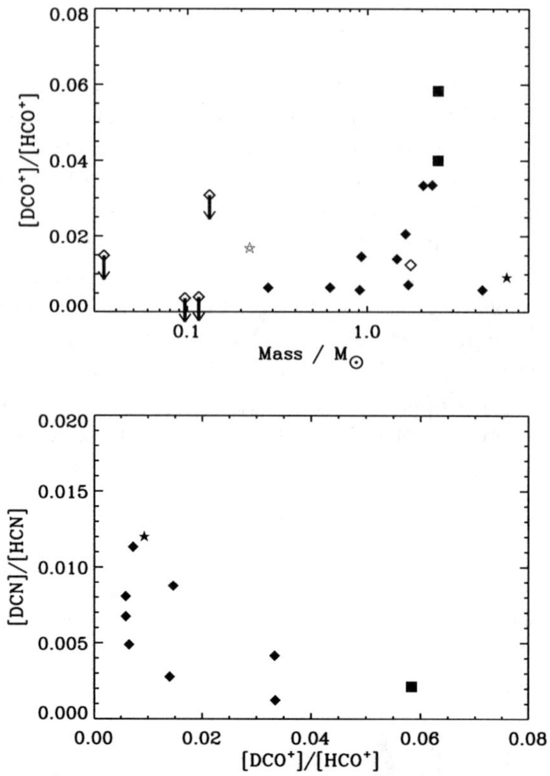

Figure 4. $[DCO^+]/[HCO^+]$ ratio plotted against mass *(upper panel)* and [DCN]/[HCN] plotted against $[DCO^+]/[HCO^+]$ ratios *(lower panel)*. Symbols as in Fig. 1. (Jørgensen et al. 2003, in prep.).

References

Bachiller, R. & Pérez Gutiérrez, M. 1997, ApJ 487, L93

Bacmann, A., Lefloch, B., Ceccarelli, C., Castets, A., Steinacker, J., & Loinard, L. 2002, A&A 389, L6

Bergin, E. A. & Langer, W. D. 1997, ApJ 486, 316

Caselli, P., Walmsley, C. M., Tafalla, M., Dore, L., & Myers, P. C. 1999, ApJ 523, L165

Ceccarelli, C., Castets, A., Caux, E., Hollenbach, D., Loinard, L., Molinari, S., & Tielens, A. G. G. M. 2000, A&A 355, 1129

Dickens, J. E., Irvine, W. M., Snell, R. L., Bergin, E. A., Schloerb, F. P., Pratap, P., & Miralles, M. P. 2000, ApJ 542, 870

Hogerheijde, M. R. & van der Tak, F. F. S. 2000, A&A 362, 697

Ivezić, Ž, Nenkova, M., & Elitzur, M. 1999, *User Manual for DUSTY*, University of Kentucky Internal Report

Jørgensen, J. K., Schöier, F. L., & van Dishoeck, E. F. 2002, A&A 389, 908

Roberts, H. & Millar, T. J. 2000, A&A 361, 388

Schöier, F. L., Jørgensen, J. K., van Dishoeck, E. F., & Blake, G. A. 2002, A&A 390, 1001

Turner, B. E. 2001, ApJS 136, 579

van Dishoeck, E. F. & Blake, G. A. 1998, ARA&A 36, 317

A STUDY OF WARM CLOUDS IN THE LINES OF COMPLEX MOLECULES

Sergei V. Kalenskii, Alexei V. Alakoz, and Vitaly G. Promyslov

Astro Space Center of P.N. Lebedev Physical Institute, Moscow, Russia

Abstract

We observed a number of molecular clouds in the lines of CH_3OH, CH_3CCH, CH_3CN, and HC_3N. Temperatures, densities, and molecular abundances were determined both analytically (applying rotational diagrams) and by means of statistical equilibrium calculations. The obtained gas parameters are typical for warm clouds: temperatures 20–60 K, densities $10^5 - 10^7$ cm^{-3}, and molecular abundances that match early-time chemistry.

Key words: ISM: molecules – ISM: structure – ISM: clouds

1. Introduction

Massive stars are known to form in warm molecular clouds, which have temperatures 30–50 K, masses from hundreds to thousands $M_{\odot}$, and sizes 0.1–3 pc. Hence, the exploration of these clouds is an important part of the exploration of star formation. Therefore, for a number of years our team have observed warm clouds in different molecular lines. Here we present the results on methyl acetylene (CH_3CCH), methanol (CH_3OH), methyl cyanide (CH_3CN), and cyanoacetylene (HC_3N). In addition, we observed the 1–0 ^{13}CO line to derive H_2 column densities necessary for determining molecular abundances.

Methyl acetylene, like ammonia, is a symmetric top molecule and can be used as a probe of interstellar gas temperature. Moreover, we believe that methyl acetylene may be a better "thermometer" for warm clouds than ammonia. The reason is as follows: NH_3 emission may be strongly affected by the contribution from compact regions of hot gas (hot cores), where the abundance of ammonia is enhanced by several orders of magnitude owing to grain-mantle evaporation. On the contrary, the methyl acetylene abundance is not enhanced in the Orion Hot Core (Wang et al. 1993), and probably, it is not enhanced in other hot cores either. Since hot cores, which have sizes of approximately 0.05 pc, are much smaller than warm clouds, their contribution to the CH_3CCH emission is probably negligible.

We observed the $6_K - 5_K$ lines at 102 GHz and $5_K - 4_K$ lines at 85 GHz. Five sources were mapped at 102 GHz. Gas kinetic temperatures and CH_3CCH column densities were derived applying rotational diagrams, built according to Turner (1991).

Methanol is one of the most abundant complex molecules in Galactic molecular clouds. Some rotational lines of methanol are grouped in series of lines with closely spaced frequencies. These series can be used for estimating gas temperature and density, the estimates being not affected by calibration errors. To derive the interstellar gas parameters from the observations of methanol lines one should perform statistical equilibrium (SE) calculations for a number of parameter sets and choose the sets that are in agreement with observational data. Previously the usage of methanol for interstellar gas probing could lead to erroneous results due to poorly known collisional decay rates, but recently the decay rates were calculated by Pottage et al. (2001). Using the new decay rates, we analyzed available observational data on the series of $J_0 - J_{-1}E$ lines near 157 GHz (Slysh et al. 1999), and the series of $2_K - 1_K$ lines near 97 GHz (Minier & Booth 2001), observed with similar spatial resolution. The agreement between the observational data and SE models was evaluated by comparing the model line intensity ratios at 157 and 97 GHz with the corresponding observed ratios.

Methyl cyanide (CH_3CN) is another symmetric top molecule, which can be used as a probe of interstellar gas temperature. Having a high dipole moment (3.91 Debye) CH_3CN is a tracer of dense gas. CH_3CN have been studied in a number of hot cores, where its abundance is increased due to grain-mantle evaporation. Much less is known about CH_3CN in warm clouds. Therefore we made a survey of Galactic molecular clouds in the $6_K - 5_K$ and $5_K - 4_K$ series of methyl cyanide lines near 110 and 92 GHz, respectively, most sources from our list be-

Chemistry as a Diagnostic of Star Formation (2003). *Edited by Charles L. Curry and Michel Fich.* Published in 2003 by NRC Research Press, Ottawa ON, K1A 0R6, Canada. ISBN 0-660-19089-3.

Table 1. Gas parameters derived from observations of complex molecules

Source	T_{kin} (K)			Molecular abundance $\times 10^9$				n_{H_2} (10^5 cm^{-3})	
	CH_3CCH	CH_3OH	CH_3CN	CH_3CCH	CH_3OH	CH_3CN	HC_3N	CH_3OH	virial
W 3(OH)	35	30	70[a]	2.3	5.3			100	
ORI S6	41	35	40	5.2	10.0	0.2		100	
OMC 2	<30	20		4.1	2.6			100	
S 231	38	20	55	4.8	6.5			18	
S 235	<61	25		1.1	1.5			18	
S 252	27	25		2.1	4.0			10	
NGC 2264	20	30	45	3.9	4.3	0.05	2.1	5.6	4.8
G345.01+1.79		40						18	
NGC 6334B		20							
M 8E		30						5.6	
G11.94–0.62		15						32	
W 33Met		30						5.6	
GGD 27		20						5.6	
M 17		20						$\geq$ 100	
G29.95–0.02	43	20		2.0	2.05			100	
G34.26+0.15	31	30	140[a]	< 16				32	1.9
G35.19–0.74	32			8.5	190	0.1			
W 51E1/E2	49		160[a]	3.6			0.5		
W 51 MET1	<26	10						32	
Onsala 1	39	25		6.1	2.3	0.06		10	
W 75N	38	30	40	< 24	13	0.08		10	
DR 21 West	26			2.6			2.5		
DR 21 (OH)	39	25	45	4.5	3.9	0.09	3.2	56	1.0
S 140	26	30		4.5	1.3	0.1	1.6	18	5.5
Cep A	33	35		5.7	1.4	0.02	1.0	32	
NGC 7538	31	60	50	1.9	4.3	0.04		18	

[a] Hot core.

ing warm clouds. Methyl cyanide column densities were derived applying rotational diagrams; gas temperatures were determined utilizing SE calculations (Kalenskii et al. 2000).

Cyanoacetylene (HC_3N) has already been observed in different molecular clouds (e.g., Suzuki et al. 1992) Nevertheless, no systematic surveys of warm clouds in HC_3N lines have been performed so far. Therefore, we observed a number of sources in the 4–3 HC_3N line at 36 GHz with the 20-m radio telescope of the Pushchino Radio Astronomy Observatory. Most of the observed sources are warm clouds.

2. Results and discussion

Figure 1 shows the $6_K - 5_K$ methyl acetylene and methyl cyanide lines towards W 3(OH), G34.26+0.15, and NGC 7538S. The NGC 7538S spectra look alike: i.e., in both spectra there are no lines with $K > 3$, and the LSR velocities are the same. Therefore, we believe that the emission of both molecules arises in the same region. On the contrary, the W 3(OH) and G34.26+0.15 spectra are different: high energy lines with $K > 3$ are present only in the methyl cyanide spectrum. Since the $K = 4$ and $K = 5$ methyl cyanide lines appear due to the contribution from hot cores (Kalenskii et al. 2000), the non-detection of analogous methyl acetylene lines suggests that the contribution from hot cores to the CH_3CCH emission is negligible. High-energy CH_3CCH lines are either weak or not detected in all the observed sources. Since many of these harbor hot cores, it is reasonably to assume that the contribution from hot cores to the methyl acetylene emission is negligible.

The derived temperatures and molecular abundances are presented in Table 1. The temperatures fall within the range 20–60 K, typical for warm clouds, and turned out to be in a reasonably good agreement with each other.

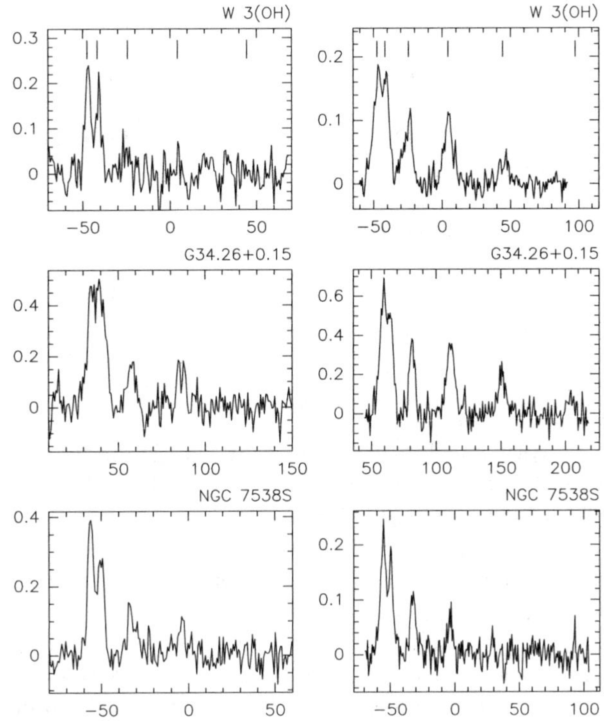

Figure 1. Spectra of three methyl acetylene (*left column*) and methyl cyanide (*right column*) sources. The vertical lines in the upper row indicate the positions of the 0 to 5 K components, with the K values increasing to the right.

Molecular abundances $N_{\rm mol}/N_{\rm H_2}$ were determined employing H_2 column densities that were derived using the 1–0 ^{13}CO line intensities. Methyl acetylene and methanol abundances appeared to be $10^{-9} - 10^{-8}$, similar to the abundances in dark clouds and the Orion Extended Ridge (Friberg et al. 1988; Blake et al. 1987). Note that in some sources the abundances may be overestimated as a result of ^{13}CO depletion or high optical depth. CH_3CN abundances in most of the sources are several $\times 10^{-11}$, much lower than that in TMC-1 (10^{-9}; Irvine et al. 1987). However, since methyl cyanide has a high dipole moment and the collisional excitation of CH_3CN requires high density, in the case of density inhomogeneities CH_3CN sources may appear more compact than ^{13}CO sources. Hence, our estimates may represent only the lower limits of CH_3CN abundances.

Table 2. Angular diameters, linear diameters, virial masses and densities of the mapped sources

Source	θ ($''$)	D (pc)	$M_{\rm vir}$ ($M_\odot$)	n_{H_2} (cm^{-3})
NGC 2264	57	0.20	89	4.8×10^5
G30.8-0.1	53	1.66	6208	5.6×10^4
G34.26+0.15	42	0.80	2332	1.9×10^5
DR 21 (OH)	56	0.83	1336	9.6×10^4
S 140	49	0.12	135	5.5×10^5

The abundances of cyanoacetylene are enclosed in the range 0.5–3.2 $\times 10^{-9}$. They are similar to that in, e.g., L 1544, but much smaller than that in the TMC-1 cyanopolyyne peak or L 1544 ($\approx 10^{-8}$, see Table 5 in Walmsley et al. 1980). Note that the abundances of all molecules, including CH_3CN, match early-time chemistry.

The parameters of the sources, mapped in CH_3CCH lines, are presented in Table 2. The densities were derived dividing virial masses by source volumes. These densities fall in the range $1-5.5\times10^5$ cm^{-3}. The densities, derived by methanol modeling, appeared to be $5\times10^5-10^7$ cm^{-3}. For G34.26, DR 21(OH), and S 140, the latter densities are much higher than the former, probably indicating that the objects are clumpy.

Acknowledgements

The authors are grateful to Prof. D.R. Flower, who presented the methanol collisional decay rates, and Dr. V.I. Slysh for helpful discussions. The work was partially supported by the Russian Foundation for Basic Research (grants no. 98-02-16916 and 01-02-16902), project no. 315 "Radio Astronomy Educational and Scientific Center," and the scientific program of the RAS Presidium "Non-stationary Processes in Astronomy."

References

Blake, G. A., et al. 1987, ApJ 315, 621
Friberg, P., et al. 1988, A&A, 195, 281
Irvine, W. M., Goldsmith P.F., & Hjalmarson, Å1987, in *Interstellar Processes*, ed. D.J. Hollenbach & H.A.Thronson Jr. (Dordrecht: Reidel), 561
Kalenskii, S. V., et al. 2000, Astronomy Reports, 44, 725
Minier, V., & Booth, R. S. 2001, A&A, 387, 179
Pottage, J. T., Flower, D. R., & Davis, S. L. 2001, J. Phys. B. 34, 3313
Slysh, V. I., et al. 1999, ApJS, 123, 515
Suzuki, H., et al. 1992, ApJ, 392, 551
Turner, B. E. 1991, ApJS, 76, 617
Walmsley, C. M., Winnewisser, G., & Toelle, F. 1980, A&A, 81, 245
Wang, T. Y., Wouterloot, J. G. A., & Wilson, T. L. 1993, A&A, 277, 205

AST/RO OBSERVATIONS OF CO $J = 7 \rightarrow 6$ AND $J = 4 \rightarrow 3$ EMISSION TOWARD THE GALACTIC CENTER REGION

S. Kim, C.L. Martin, A.A. Stark, and A.P. Lane

Harvard-Smithsonian Astrophysical Observatory

Abstract

We present position-velocity strip maps of the Galactic Center region in the CO $J = 7 \rightarrow 6$ and $J = 4 \rightarrow 3$ observed with the Antarctic Submillimeter Telescope and Remote Observatory (AST/RO) located at the Amundsen-Scott South Pole Station. Emission from the two rotational transitions of ^{12}CO was mapped at $b = 0°$ for $3.5° > \ell > -1.5°$, on a $1'$ grid with a FWHM beamsize of $58''$ at 806 GHz and $105''$ at 461 GHz. CO $J = 4 \rightarrow 3$, and [C I] (Ojha et al. 2001) emission from this region show that these lines are distributed in a manner similar to CO $J = 1 \rightarrow 0$ (Stark et al. 1987); the CO($J = 4 \rightarrow 3$)/CO($J = 1 \rightarrow 0$) line ratio map is almost featureless across the entire Galactic Center region.

In contrast, the CO $J = 7 \rightarrow 6$ emission from the Galactic Center is strongly peaked toward the Sgr A and Sgr B molecular complexes. A Large Velocity Gradient analysis shows that aside from the two special regions Sgr A and Sgr B, the photon-dominated regions within a few hundred parsecs of the Galactic Center are remarkably uniform in mean density and kinetic temperature at $n = 2500$ to $4000\,\mathrm{cm}^{-3}$ and $T = 30$–$45\,\mathrm{K}$. The CO($J = 7 \rightarrow 6$)/CO($J = 4 \rightarrow 3$) line temperature ratios near Sgr B are a factor of two higher than those observed in the nuclear region of the starburst galaxy M82 (Mao et al. 2000), while the CO($J = 7 \rightarrow 6$)/CO($J = 4 \rightarrow 3$) line temperature ratios around Sgr A are similar to M82. The line ratio on large scales from the Galactic Center region is an order of magnitude less than that from M82.

Key words: Galaxy: center – ISM: clouds

Chemistry as a Diagnostic of Star Formation (2003). *Edited by Charles L. Curry and Michel Fich.* Published in 2003 by NRC Research Press, Ottawa ON, K1A 0R6, Canada. ISBN 0-660-19089-3.

MID-INFRARED ABSORPTION SPECTROSCOPY TOWARD NGC 7538 IRS 1

Claudia Knez[1], John H. Lacy[1], Neal J. Evans, II[1], Annemieke M.S. Boonman[2], Ewine F. van Dishoeck[2], and Matthew J. Richter[3]

[1]Department of Astronomy, 1 University Station, C1400, Austin, TX 78712
[2]Leiden University, P.O. Box 9513, 2300 RA Leiden, The Netherlands
[3]Physics Department, University of California Davis, Davis, CA 95616, U.S.A.

Abstract

We present high-resolution observations of acetylene (C_2H_2) and hydrogen cyanide (HCN) ro-vibrational absorption near 13 μm toward NGC 7538 IRS 1 taken with the Texas Echelon Cross Echelle Spectrograph (TEXES). Both the C_2H_2 and HCN velocity offsets indicate that the probed gas is quiescent gas associated with IRS 1. For HCN, the data were best fitted by two temperature components: 110 K and 600 K. C_2H_2 also shows multiple temperature components: 110 K and 800 K. The higher temperatures for both molecules were consistent with results obtained by Lahuis & van Dishoeck (2000) using *ISO* data. We have detected the methyl radical (CH_3). The identification of CH_3 marks the first detection of this radical toward dense molecular gas.

Key words: ISM: molecules – ISM: abundances – ISM: clouds

1. Introduction

Regions of high-mass star formation provide laboratories in which to study dust grain chemistry at cold temperatures. As protostars heat the surrounding gas, the ice mantles on dust grains evaporate. By studying molecules which evaporate from grains, we can learn about the low-temperature, chemical processes that occur on the surface of grains (van Dishoeck & van der Tak 2000). In addition, the chemistry in star forming regions can be a useful tool in determining a sequence of massive star formation (van Dishoeck & van der Tak 2000). In the cold prestellar and collapse phases, molecules are frozen onto dust grains forming icy mantles. Surface chemistry on dust grains can alter the mantle composition. Then, as the protostar begins to warm the surrounding gas, the ices evaporate and return to the gas phase. A rich chemistry is driven by these newly evaporated molecules in compact, optically thick hot cores. Hot cores have densities $\geq 10^7$ cm^{-3}, diameters ≤ 0.1 pc, and temperatures ≥ 100 K (Kurtz et al. 2000). High abundances of fully hydrogenated molecules such as NH_3, H_2S, OH, and H_2O and saturated complex organic molecules such as CH_3OH, CH_3CN, CH_3OCH_3 are characteristic of hot cores.

Mid-infrared absorption spectroscopy, along with millimeter or radio (hereafter simply radio) spectroscopy, is a powerful probe of the physical conditions and chemical abundances in molecular clouds. Most radio spectroscopy involves emission spectra, where the angular resolution is set by the size of the beam, which varies with frequency and with telescope. However, in infrared absorption spectroscopy, the spatial resolution is set by the size of the background source. For this case, embedded protostars serve as the background source, and small scales in star forming regions are probed for all excitation levels. In addition, mid-infrared spectroscopy can be used to study symmetric molecules like acetylene, C_2H_2 (Lacy et al. 1989b; Evans et al. 1991; Carr et al. 1995), which do not have a permanent dipole moment and, therefore, no pure rotational transitions. The study of ro-vibrational transitions allows for the sampling of populations for many rotational levels with a single instrument, often simultaneously, therefore reducing the calibration problems encountered in radio observations. Previous infrared spectroscopy has been limited by poor spectral resolution (i.e., $R \sim 2000$ for the Short Wavelength Spectrometer (SWS) on board the Infrared Space Observatory (*ISO*) and $R \sim 10^4$ for Irshell, a previous ground-based high resolution instrument; see Lacy et al. 1989a). We now have a new instrument, the Texas Echelon Cross Echelle Spectrograph (TEXES) (Lacy et al. 2002), that achieves $R \approx$ 80,000 at 13 μm, or $\delta v \sim 4$ km s^{-1}, allowing resolution of long-standing questions and a new class of observations.

The NGC 7538 (S158) molecular cloud, located in the Perseus arm at a distance of 2.8 kpc (Ungerechts et al. 2000), contains many star-forming objects, making it a good candidate to study different stages of star formation. Furthermore, there seems to be a progression in evolution

Chemistry as a Diagnostic of Star Formation (2003). *Edited by Charles L. Curry and Michel Fich.* Published in 2003 by NRC Research Press, Ottawa ON, K1A 0R6, Canada. ISBN 0-660-19089-3.

from NW (IRS 2, a visible H II region) to SE (IRS 9, a deeply embedded core) (Elmegreen & Lada 1977; Campbell & Thompson 1984). IRS 1, 2, and 3, located in the high-density core, have H II regions associated with them. IRS 1 appears to be the youngest of these three sources since it is the brightest infrared source (Bloomer et al. 1998) with a luminosity of 1.3×10^5 $L_\odot$.

Lahuis & van Dishoeck (2000) observed C_2H_2 and HCN towards massive young stellar objects using *ISO*-SWS. With a resolving power of $R \approx 2000$, they were unable to resolve the individual lines; instead they observed the Q-branch, in which the lines are blended together. From the observations, excitation temperatures for C_2H_2 and HCN were $800\pm^{250}_{150}$ K and $600\pm^{50}_{40}$ K, respectively, for IRS 1. Complex organics (e.g., CH_3OCH_3 and CH_3COOH) have been detected toward IRS 1 (van Dishoeck & van der Tak 2000). In this paper, we present high resolution observations of HCN, C_2H_2 and CH_3 absorption towards IRS 1.

2. Observations

We observed R-branch and Q-branch lines of the ν_5 band of C_2H_2 and R-branch lines of the ν_2 band of HCN and the ν_2 band of CH_3 near 13 μm(see Fig. 1 for sample spectra). In addition, hot band lines were observed for both HCN and C_2H_2, including the combination band $\nu_4+\nu_5-\nu_4$ of C_2H_2. Observations were made with TEXES (Lacy et al. 2002) at the NASA Infrared Telescope Facility in 2001 June and 2001 November. The data reduction procedure, described in Lacy et al. (2002), involved flat-fielding using the difference of frames of a black chopper blade and of the sky. The black-sky difference corrects for the absorption of light by the telescope and the sky. The removal of telluric lines by division by a comparison source. For these data, asteroids were used as comparisons because stars bright in the infrared have photospheric lines that may interfere with the lines being observed. The wavelength was calibrated from narrow telluric absorption lines in the spectrum.

3. Results for HCN

Using a curve of growth analysis to find the populations for the rotational levels (N_J), we found that the data were fitted by two temperature components for HCN, $T_{\rm cold} = 110$ K and $T_{\rm hot} = 685$ K. The corresponding column densities were $N_{\rm cold} = 1 \times 10^{16}$ cm^{-2}and $N_{\rm hot} = 1 \times 10^{16}$ cm^{-2}. Subsequently, radiative transfer models agreed with the initial findings. A single temperature could not fit the

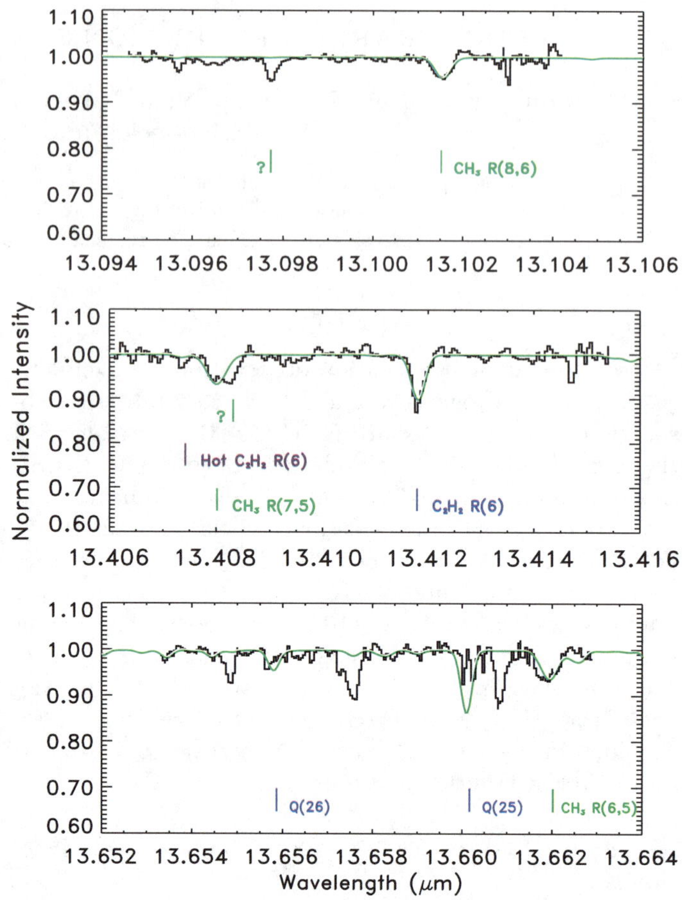

Figure 1. The observed lines of the R-branch of CH_3 are shown. Overplotted in green is a model spectrum corresponding to absorption from gas at two temperatures, $T_{\rm cold} = 100$ K and $T_{\rm hot} = 200$ K. The light blue and dark blue labels indicate C_2H_2 and hot C_2H_2, respectively. Features not marked are due to telluric lines.

data. Further modeling showed that a temperature gradient also fits the data. The presence of a temperature gradient seems the more plausible explanation for the observed spectra. We are currently working on characterizing the temperature gradient using power-law temperature and density gradients.

4. Results for C_2H_2

For C_2H_2, the interpretation was more complicated. We first analyzed the R-branch lines using a curve of growth to determine the populations. In a rotational energy diagram, the populations should be decreasing linearly (on a semi-log plot). However, the fit to the ortho (odd) populations gives a lower temperature than does the fit to the para (even) populations. One possibility is that the statistical weight ratio (ortho:para) is not the assumed 3:1. The data show that the ratio between neighboring lines is not 3:1; this may be a result of saturation. We are still exploring this and other possible explanations for this phenomenon.

We also ran radiative transfer models in order to derive the temperature and column density. Several hot band lines were seen throughout our spectra. The lower-J lines (up to J=6) of the R-branch of C_2H_2 are explained by a two-temperature model ($T_{\rm cold}$ = 110 K and $T_{\rm hot}$ = 800 K). The corresponding column densities were $N_{\rm cold} = 4 \times 10^{15}$ cm^{-2} and $N_{\rm hot} = 6 \times 10^{15}$ cm^{-2}. However, middle and high-J para lines are underpredicted. One possible explanation is that emission is filling in the absorption. Another puzzle for this molecule is that the model overpredicts the ortho lines. Around the Q-branch, there are also hot band Q-branches which lie on top of the fundamental Q-branch. This results in the blending of lines and a poor determination of the continuum level. More analysis and modeling are required to interpret the results from the C_2H_2 spectra.

5. Results for CH_3

Among our spectra we found several lines that were not due to either C_2H_2 or HCN. Some of them have been identified as CH_3. CH_3 has been detected towards diffuse interstellar gas (Feuchtgruber et al. 2000). Our possible detection of CH_3 marks the first detection of the radical towards dense interstellar gas. For this molecule, we ran simple radiative transfer models and found two temperature components. The two temperatures, $T_{\rm cold}$ = 100 K and $T_{\rm hot}$ = 200 K, are not as different as for HCN and C_2H_2. The corresponding column densities for CH_3 are $N_{\rm cold} = 7.5 \times 10^{16}$ cm^{-2} and $N_{\rm hot} = 2 \times 10^{15}$ cm^{-2}. These results are preliminary and further modeling is necessary. The abundance of CH_3 will help constrain chemical models of regions of massive star formation.

6. Conclusions

Both HCN and C_2H_2 show at least two temperature components. The temperature of the hot component agrees with the results obtained by Lahuis & van Dishoeck (2000). More detailed analysis is needed to determine the temperature structure as probed by these two molecules. The hot band lines will provide constraints on the temperature.

Some unidentified lines are due to CH_3, which will be an important constraint for chemical models for regions of massive star formation. From the first-order results for CH_3, CH_3 seems to be probing different gas than the hot components of HCN and C_2H_2. We will supplement these observations with submillimeter observations of HCN in order to get a more complete model of the region.

Acknowledgements

C. Knez is grateful to F. Lahuis for his help with the models.

References

Campbell, B., & Thompson, R. I. 1984, ApJ, 279, 650

Carr, J. S., Evans, N. J. II, Lacy, J. H., & Zhou, S. 1995, ApJ, 450, 667

Elmegreen, B. G., & Lada, C. J. 1977, ApJ, 214, 725

Evans, N. J., II, Lacy, J. H., & Carr, J. S. 1991, ApJ, 383, 674

Kurtz, S., et al. 2000, in *Protostars and Planets IV*, ed. V. Mannings, A. P. Boss, & S. S. Russell (Tucson: University of Arizona Press), 299

Lacy, J. H., et al. 1989a, PASP, 101 1166L

Lacy, J. H., et al. 1989b, ApJ, 342, L43

Lacy, J. H., et al. 2002, PASP, 114, 153

Lahuis F. & van Dishoeck, E. F. 2000, A&A, 355, 699

van Dishoeck, E. F. & van der Tak, F. F. S. 2000, in *Proceedings of IAU Symposium 197*, ed. Y. C. Mihn & E. F. van Dishoeck, 97

COMPOSITIONAL DEPENDENCE OF INFRARED ABSORPTION OF OLIVINE

C. Koike[1], H. Chihara[1,2], A. Tsuchiyama[2], H. Suto[3], H. Sogawa[1], and H. Okuda[4]

[1]Kyoto Pharmaceutical University, Yamashina, Kyoto 607-8412, Japan
[2]Department of Earth and Space Science, Osaka University, Toyonaka, Osaka 560-0043, Japan
[3]Subaru Telescope, National Astronomical Observatory of Japan, Hilo, Hawaii, USA
[4]Gunma Astronomical Observatory, Gunma 377-0702, Japan

ABSTRACT

The mass absorption spectra of crystalline olivine particles with different Mg/(Mg+Fe) ratios, between forsterite Mg_2SiO_4 (Fo) and fayalite Fe_2SiO_4 (Fa), were measured for synthetic and natural samples in the mid- and far-infrared regions. The positions of main strong peaks shift to longer wavelength systematically as the Mg/(Mg+Fe) ratio decreases. In the 10–17 μm region, these trends are very clear and the positions are a good indicator to know the chemistry of dust grains. In the 20–100 μm region, the trends are also rather clear, but a little complicated, because the intensity and width of each band become weak and broad, or bands disappear as the concentration of Fo decreases. However, the trends are clear only in the limited composition range at near Fo and Fa: the peak positions of 33, 50, and 69 μm bands of forsterite shift linearly as the Fo concentration decreases to about Fo_{70}. Those of double band at 50 μm region for Fe-rich olivine shift linearly as the Fa concentration increases from Fo_{20} to Fa. Only when the spectral signature of Mg-rich or Fe-rich olivine might be observed, are the positions in the 20–100 μm region a good indicator to know the chemistry of dust grains.

Key words: ISM: circumstellar dust – dust, extinction – astrochemistry

1. INTRODUCTION

Many characteristic features in the far-infrared region were observed around circumstellar environment of both evolved and young stars and Comet Hale-Bopp. The majority of these features were identified with Mg-rich crystalline olivine and pyroxene (Molster et al. 2002a,b). Further, the detected 69 μm feature in the *ISO* spectra led to the conclusion that the crystalline olivine are also very Mg-rich compared with laboratory data depending on the composition and temperature (Molster et al. 2002c). But no evidence for the presence of iron-rich silicates has been obtained to date. The absence of Fe-rich silicates is very curious as the most abundant elements forming refractory compounds are Mg, Si, and Fe together with oxygen. The spectra of crystalline silicates depending on the Mg/(Mg + Fe) ratio is one of the keys to understanding the structure, origin, and evolution of circumstellar and interstellar dust, and to identify the chemical condensation path for circumstellar minerals.

As for laboratory data depending on composition, we already measured the spectra of olivine with different Mg/(Mg + Fe) ratio, which are Fo_{100}, Fo_{90}, Fo_{80}, Fo_{60}, and Fo_{40} (Koike et al. 1993). We made clear that the peak position of main bands shifts to longer wavelength as this ratio decreases. Chihara et al. (2002) (Paper I and this volume) investigated and confirmed this trend in more detail for Mg–Fe pyroxenes with different Mg/(Mg + Fe) ratio. In order to examine in more detail how these peak positions shift with the Mg/(Mg + Fe) ratios at near forsterite and fayalite, we obtained especially Mg-rich olivine and Fe-rich olivine samples. Following Paper I, we measured the absorption spectra of the whole range of olivine solid solution from forsterite to fayalite in the mid-infrared region and far-infrared region (Koike et al. 2002). We report the spectra of olivine series, and made clear the correlation between the peak positions and the Mg/(Mg + Fe) ratio in detail.

2. EXPERIMENTS

Olivine is a solid solution having two end members of forsterite Mg_2SiO_4 (Fo) and fayalite Fe_2SiO_4 (Fa), and is expressed as $(Mg_x, Fe_{1-x})_2SiO_4$ $(0 \leq x \leq 1)$. The composition of olivine is expressed as Fo_X, where X is the mole percentage of forsterite (X = 100 x). In this study, both of synthetic and natural olivines are prepared. Synthetic Fo was made with Czochralski (CZ) method and Fa was made with floating zone (FZ) method. Other synthetic olivine samples - Fo80, Fo60, Fo40, Fo20, Fo15, and Fo10 - were produced at 1200 °C by heating a mixture of

Chemistry as a Diagnostic of Star Formation (2003). *Edited by Charles L. Curry and Michel Fich.* Published in 2003 by NRC Research Press, Ottawa ON, K1A 0R6, Canada. ISBN 0-660-19089-3.

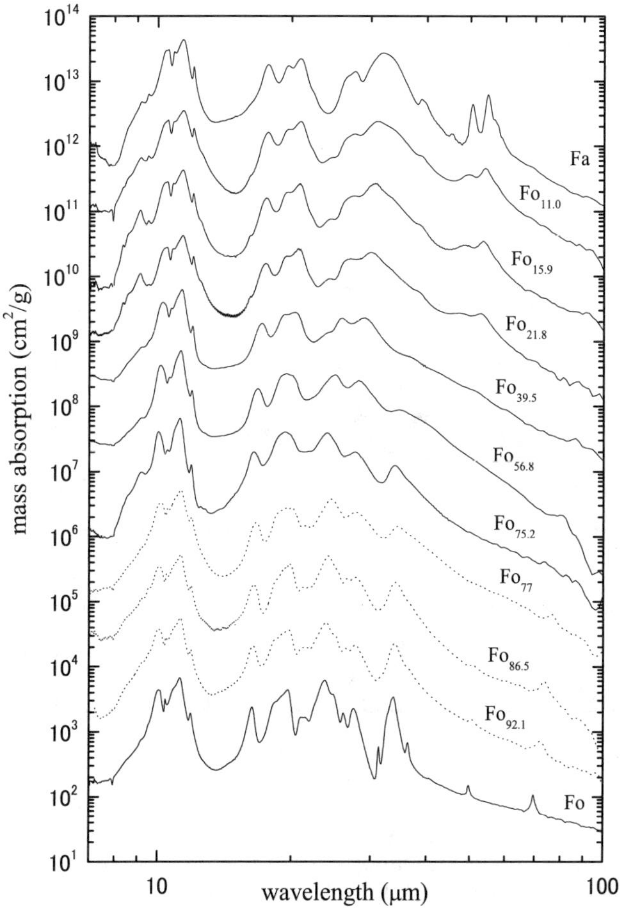

Figure 1. Spectra of synthetic olivine and natural olivine. Each spectrum, indicating composition, is multiplicative lower spectrum by 10. Thin lines and dotted lines are synthetic and natural olivine, respectively. The compositions are expressed on each spectrum.

MgO, SiO_2 and FeO powder under controlled O_2 fugacity at atmospheric pressure. This method is same as for synthesizing pyroxenes mentioned in Paper I. The composition of olivine samples has been determined by an EPMA (electron probe micro analyzer) at Osaka University (JEOL733). With the EPMA investigations, chemical composition of natural 8 samples are from $Fo_{92.1}$ to Fo_{77}. As for samples of natural olivine from Akita, Fo80, Fo60, and Fo40 samples, which are same samples as in Koike et al. (1993). The synthetic Fo, Fa and natural samples are single crystal, whereas other synthetic olivines are polycrystal. The synthesized and natural samples were crushed and ground in an agate mortar. The transmittances T of each KBr and polyethylene pellet were measured from 7000 cm^{-1} to 50 cm^{-1}. The spectrometers are Jasco FT/IR-350 and Nicolet FTIR (Nexus 670) at resolution of 1 cm^{-1}.

3. Results

The spectra of samples are shown as mass absorption coefficients κ (cm^2/g) in Figure 1, where only typical spectra of natural olivine are shown among 8 samples. The previous data for natural olivine of Ichinomegata Akita, $Fo_{89.3}$ in the far infrared region (Koike et al. 1993) are also included. The spectra of the synthetic olivine gradually changed from forsterite to fayalite as the Fe content increases, and this change became distinct especially in the far infrared region as shown in Fig. 1. As for natural olivine, those chemical compositions were limited to a narrow range from $Fo_{92.1}$ to Fo_{77}. The spectra of natural samples were resemble each other in the mid- and far-infrared region compared with the spectra of synthetic olivine. As for 69 μm band, the peak positions of natural olivine were definitely shifted to longer wavelength as the Fe content increases. For all bands, the correlation between the peak positions of each band and concentration of Fo (mol %), X, is plotted in Figure 2. The trend for each band seems no different between synthetic and natural olivine. The sizes of symbols are indicated the strength of peaks. From Fig. 2, the following trends have been found:

– all peaks are shifted toward longer wavelengths as the Fo content decreases (the Fa content increases);

– for several strong bands at 11.2, 16.4, and 23.7 μm of Fo, peak positions are strongly correlated with the Fo content, X. The peak positions in wavenumber, ν, are nearly linear with X:

$$\nu = 0.15\ X + 876 \quad \text{for the 11 } \mu\text{m band}$$
$$\nu = 0.12\ X + 829 \quad \text{for the 12 } \mu\text{m band}$$
$$\nu = 0.51\ X + 562 \quad \text{for the 16 } \mu\text{m band}$$
$$\nu = 0.65\ X + 359 \quad \text{for the 24 } \mu\text{m band}$$
$$\nu = 0.37\ X + 104 \quad \text{for the 69 } \mu\text{m band.}$$

For olivine samples, the peak positions in the mid-infrared region shift linearly as Fo concentration decreases, whereas those in the far-infrared region this trend appears clearly only near the end members: The peak positions of 33 , 50 , and 69 μm bands of Fo clearly shift linearly from Fo to about Fo_{70}, and the peak positions of Fa at 50 μm bands clearly shift linearly as Fa concentration decreases to about Fo_{20}. In contrast, these trends appear

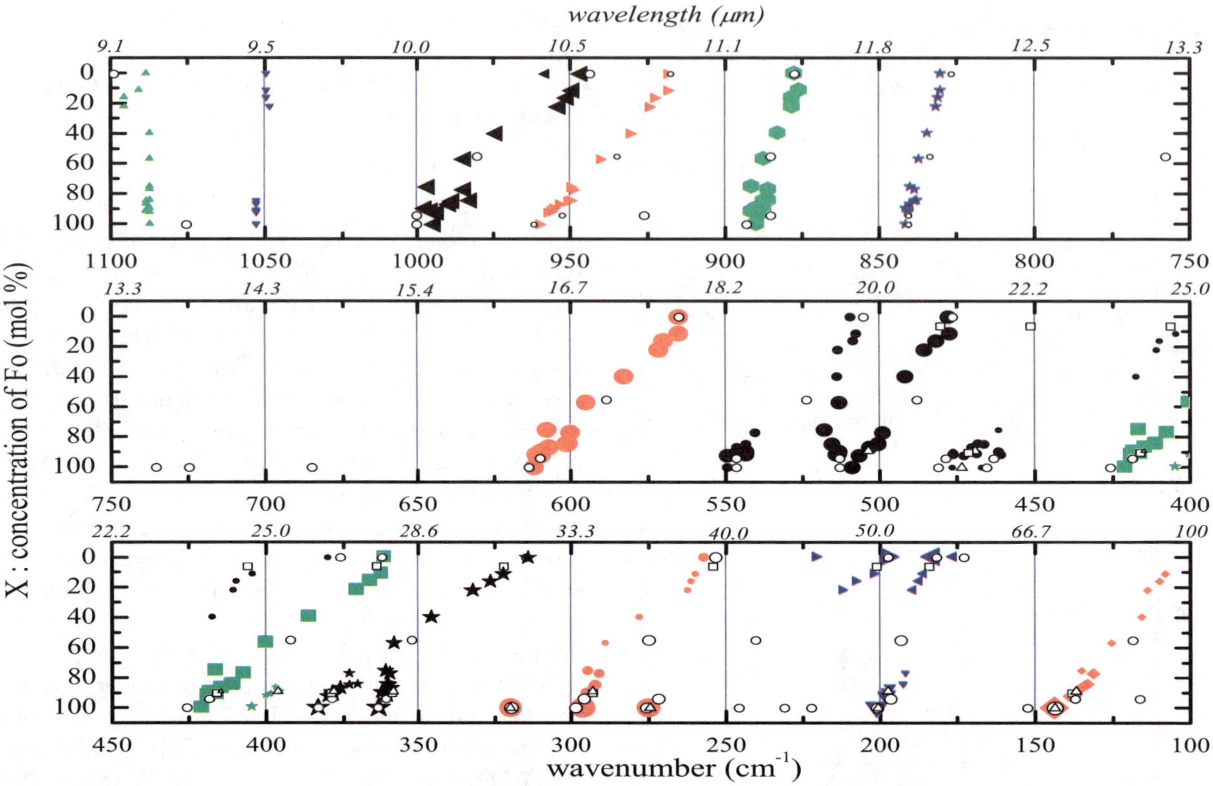

Figure 2. Peak shifts of absorption bands depending on chemical composition. Full symbols are synthetic and natural samples. Open circles, open squares, and open triangles are from Jäger (1998), Mennela (1998), and Bowey (2001), respectively. The sizes of symbols indicate the strength of peaks.

a little different from pyroxene (Paper I). Although the compositions of pyroxene are limited to En – En50 and Fs, the peak shifts do not so clearly appear in the mid-infrared region as olivine, whereas the peak shifts clearly appear in the far-infrared region, especially at around 40–50 and 70 μm region.

Up to now, the crystalline olivine and pyroxene were detected in *ISO* observations. This trend of peak shifts of olivine and pyroxene will become help to identify the bands and to know the chemical composition on the whole. Especially, from the peak shifts of olivine in the mid-infrared region and those of pyroxene in the far-infrared region, we can easily know the chemical composition of crystalline silicates. High resolution spectra around circumstellar sources will become clues to clarify the growth of circumstellar dust.

Acknowledgements

We thank Professor H. Takei of Osaka University for preparing Fayalite crystal. This work was supported by the Grant-in-Aid of Japanese Ministry of Education, Science, and Culture (12440054), and the Research Fellowship of the Japan Society for Promotion of Science for Young Scientists (Chihara).

References

Bowey J.E., et al., 2001, MNRAS, 325, 886
Chihara, C. et al., 2002, A&A, 76, 56 (Paper I)
Jager C., et al., 1998, A&A, 339, 904
Koike, C., et al., 1993, MNRAS, 264, 654
Koike, C., et al., 2002, A&A, submitted
Mennela S. S., et al., 1998, ApJ, 496, 1058
Molster, F., et al. 2002a, A&A, 382, 184
Molster, F., et al. 2002b, A&A, 382, 222
Molster, F., et al. 2002c, A&A, 382, 241

DUST AND MOLECULES IN THE R CORONAE AUSTRALIS CLOUD

Samu Kontinen[1], Jorma Harju[1], Malcolm Walmsley[2], Paola Caselli[2], and Arto Heikkilä[3]

[1]Observatory, P.O. Box 14, 00014 University of Helsinki, Finland
[2]Osservatorio Astrofisico di Arcetri, Largo E. Fermi 5, 50125 Firenze, Italy
[3]Laboratory of Electronics and Physics, Halmstad University, P.O. Box 823, 301 18 Halmstad, Sweden

Abstract

We discuss the dust distribution and the abundances of some simple molecules in the R CrA cloud. $C^{18}O$ is found to be undepleted (and optically thick) in a quiescent dense core, whereas moderate degrees of depletion can be measured in cores in the vicinity of newly born stars. We suggest that the undepleted core represents an early stage of chemical evolution where freezing of molecules has not yet taken effect. NH_3 and N_2H^+ maps are seen to follow closely the dust emission, and these molecules are useful for the interpretation of the continuum map. A cold condensation standing out in DCO^+ is missed in the 1.2 mm dust map, showing that different tracers are needed to study the structure of molecular clouds.

Key words: ISM: clouds – ISM: molecules – ISM: dust – ISM: depletion – ISM: star formation

1. Introduction

Quiescent dense cores, cores associated with protostars and outflows, and regions which seem to be left-overs of the star formation process often coexist in molecular clouds. The determination of chemical abundances and the identification of characteristic molecules in cores at different stages of dynamical evolution can lead to a better understanding of chemical evolution and its interplay with star formation.

One of the objects which appear particularly favourable for studies of evolutionary effects is the nearby R Coronae Australis (R CrA) cloud. It has a simple, elongated structure with well separated active and quiescent parts. The cloud is located at a distance of 170 pc (Knude & Hg 1998), about 20° off the Galactic plane. The northwestern end of the cloud is a region of low- to intermediate-mass star formation (Wilking et al. 1999), whereas the southeastern end contains a massive, starless core.

In this paper we compare the distributions of dust and some molecular species in the R CrA cloud. The dust emission map at 1.2 mm was obtained with the SIMBA bolometer array at the SEST. The observations in the spectral lines of $C^{18}O$, $C^{17}O$, DCO^+, $H^{13}CO^+$, N_2H^+ (among others) were made with the SEST and the NH_3 $(1,1)$ and $(2,2)$ observations were performed with the 64-m Parkes telescope. The purpose of these observations was to study if the apparent gradient in the star formation activity along the cloud axis is reflected in the physical conditions, the gas phase chemical abundances and the degree of CO depletion.

2. The SIMBA map

The 1.2 mm thermal dust emission map obtained with the SIMBA bolometer array at the SEST is shown in Figure 1, superposed on a $C^{18}O$ map from Harju et al. (1993). The brightest spot in the continuum map is a dense clump designated here as CrA A2, which lies next to a cluster young stars. The gas kinetic temperature in CrA A2 is 21-24 K. Further to the northwest there is another dense clump with a temperature of about 15-18 K (called here CrA A1). The latter clump is more clearly discernible in the molecular line maps below. The core which is located in the southeastern tail of the cloud (designated as CrA C) is weak in the dust continuum because of its low temperature.

3. Molecular line maps of selected regions

The $H^{13}CO^+$, DCO^+ and $NH_3(1,1)$ maps of the northwestern star-forming region are shown in the right of Fig. 1. The $H^{13}CO^+$ and DCO^+ maps are taken from Anderson et al. (1999) and they are smoothed to the Parkes resolution. In these maps, a third dense clump (designated here as CrA SE) turns up on the southeastern side of the continuum peak. Of the three molecules

Chemistry as a Diagnostic of Star Formation (2003). *Edited by Charles L. Curry and Michel Fich.* Published in 2003 by NRC Research Press, Ottawa ON, K1A 0R6, Canada. ISBN 0-660-19089-3.

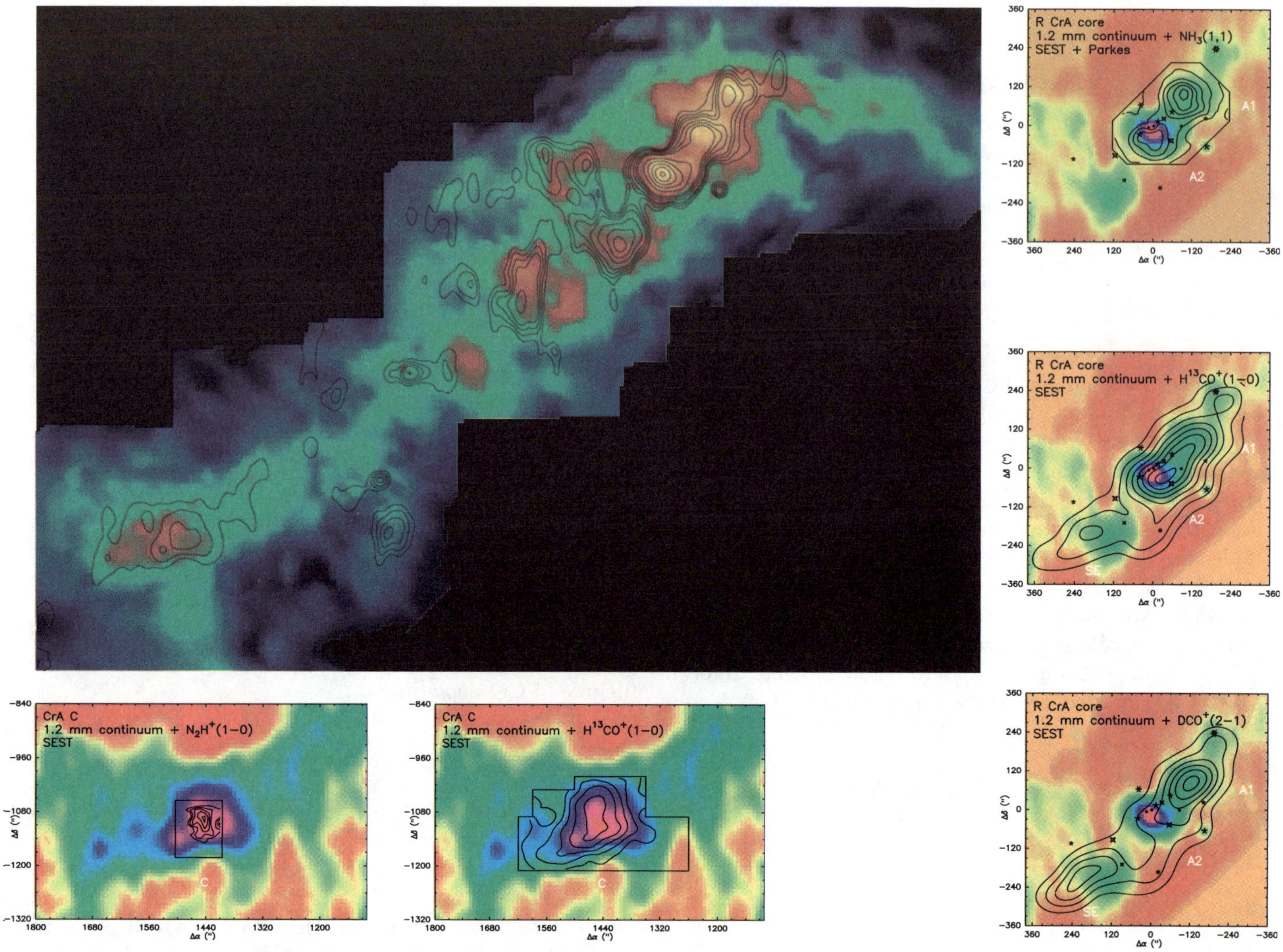

Figure 1. Top Left: The 1.2 mm dust emission map of the R CrA cloud (*contours*) superposed over a $C^{18}O$ map (*colour*). Besides dense cores, the SIMBA map shows warm, dusty envelopes around young stars; e.g., VV CrA and IRAS 18595-3712 in the south, and IRS2 in the north. The projected extent of the filament is about 30′, which corresponds to 1.5 pc at the distance 170 pc. *Right*: NH_3, $H^{13}CO^+$ and DCO^+ maps of the northwestern part of CrA (*contours*), superposed over the 1.2 mm continuum map (*colour*). The locations of the clumps CrA A1, A2 and SE are indicated. The young stars associated with the cloud are marked with filled circles (Class I protostars), or asterisks (Class II protostars, T Tauri stars or Herbig Ae/Be stars). *Bottom*: N_2H^+ and $H^{13}CO^+$ contour maps of CrA C overlaid on the continuum map.

shown here, NH_3 correlates best with the dust emission. The cooler clumps A1 and SE stand out in DCO^+. The N_2H^+ peak (not shown here) in CrA A1 coincides with that of the 1.2 mm dust continuum.

The $N_2H^+(J = 1-0)$ and $H^{13}CO^+(J = 1-0)$ integrated intensity maps of the CrA C core are shown in the bottom of Fig. 1 (the $H^{13}CO^+$ and $H^{13}CO^+$ data of CrA C are provided by Lauri Haikala). Both molecules peak towards the dust continuum maximum. This is the case also for SO, which is very strong in this core (Kontinen et al. 2000). The $C^{18}O/C^{17}O$ intensity ratio shows that $C^{18}O$ is optically thick here (Harju et al. 1993).

4. Abundances

In Table 1 we list estimates for the fractional abundances of some simple molecules in CrA C, SE, A1 and A2, to-

Table 1. Fractional abundances, $\chi[\mathrm{A}] \equiv N[\mathrm{A}]/N[\mathrm{H_2}]$, in selected clumps

		C	SE	A1	A2
$N(\mathrm{H_2})$	$(10^{23}\,\mathrm{cm^{-2}})$	0.4	0.5	1	2
T_{kin}	(K)	≤ 13	13	15–18	21–24
$\chi(\mathrm{C^{18}O})$	(10^{-8})	9	8	5	4
$\chi(\mathrm{NH_3})$	(10^{-9})	3	4	6	2
$\chi(\mathrm{N_2H^+})$	(10^{-11})	7	4	10	–
$\chi(\mathrm{H^{13}CO^+})$	(10^{-11})	3	2	1	1
$\chi(\mathrm{DCO^+})$	(10^{-11})	3	4	1	0.4

gether with the estimates of the total molecular hydrogen column density, $N(\mathrm{H_2})$, and the kinetic temperature, T_{kin}. The gas kinetic temperatures are derived from $\mathrm{NH_3}$ and $\mathrm{CH_3CCH}$, and the $N(\mathrm{H_2})$ estimates were derived using equation (1) of Motte, André, & Neri (1998), by assuming that the dust temperature is equal to the gas temperature.

The fractional $\mathrm{C^{18}O}$, $\mathrm{H^{13}CO^+}$ and $\mathrm{DCO^+}$ abundances are larger in the quiescent CrA C and SE than in A1 and A2. On the other hand, $\mathrm{NH_3}$ and $\mathrm{N_2H^+}$ peak towards A1. The variation in the $\mathrm{DCO^+}$ abundance is due to temperature dependence of deuterium fractionation and is discussed for this cloud in Anderson et al. (1999).

The core CrA C is close to virial equilibrium and may actually be the centre of a large-scale contraction (Harju et al. 1993). It can be therefore classified as a pre-stellar core. Based on the large SO abundance, Kontinen et al. (2000) suggested that CrA C, despite being pre-stellar, represents an advanced chemical state, probably associated with a high degree of molecular depletion due to freezing onto grain surfaces. Comparison with the $\mathrm{H_2}$ column densities derived from the new SIMBA map shows, however, that the depletion is *not* significant in CrA C. This is consistent with the fact that $\mathrm{H^{13}CO^+}$ peaks towards the dust column density maximum (Fig. 1). Since depletion requires, besides a high density, some time to develop (e.g., Caselli et al. 2002), the inference is that in CrA C we are witnessing a stage before depletion sets in. It is puzzling that this stage is linked with a large SO abundance, which is usually thought to be an indication of "late-type chemistry".

The clumps CrA A1 and A2 lie in the midst of newly-born stars. They are dense ($n[\mathrm{H_2}] > 10^5\,\mathrm{cm^{-3}}$) and have elevated kinetic temperatures. The clump A2 has the powerful outflow source IRS7 in its vicinity and may itself embrace a Class 0 protostar (van den Ancker 1999). The conditions in these northwestern clumps, especially A2, should be favourable for desorption, and we did not expect to see here lower $\mathrm{C^{18}O}$ abundances than in CrA C. The kinetic temperature in A1 is probably not high enough for effective thermal evaporation of CO. In accord with these conditions a moderate degree of CO and $\mathrm{HCO^+}$ depletion can be seen towards this clump. In the case of A2, the $\mathrm{C^{18}O}$ abundance may be lowered by selective photodissociation due to radiation from the surrounding stars.

5. Conclusions

The observations discussed here show that CO and related molecules in the quiescent dense core CrA C are less depleted than in cores associated with star formation in the same cloud. We suggest that this is an evolutionary effect. Our previous experience with SO in CrA C suggests that the degree of CO depletion may actually be a better indicator of chemical age than a particular molecular abundance ratio. Varying initial conditions might be a source of confusion when abundance ratios are used.

Temperature estimates are of vital importance when studying molecular cloud structure with the aid of dust continuum mapping. In dense cores, where the dust and gas temperatures are expected to be similar (Galli, Walmsley, & Goncalves 2002), ammonia observations are useful because this molecule does not seem to freeze out. In agreement with the results of Tafalla et al. 2002, the nitrogen-containing species $\mathrm{NH_3}$ and $\mathrm{N_2H^+}$ follow closely the dust distribution. On the other hand, it is once again demonstrated that $\mathrm{DCO^+}$ can reveal cold condensations, which do not coincide with column density maxima traced by millimetre continuum.

Acknowledgements

We thank Lauri Haikala for providing us with the $\mathrm{H^{13}CO^+}$ and $\mathrm{DCO^+}$ maps of CrA C.

References

Anderson, I. M., et al., 1999, A&A 347, 983
Caselli, P., et al., 2002, ApJ 565, 344
Galli, D., Walmsley, M., & Goncalves, J., 2002, A&A 394, 275
Harju, J., et al., 1993, A&A 278, 569
Knude, J. & Hog, E., 1998, A&A 338, 897
Kontinen, S., et al., 2000, A&A 361, 704
Motte, F., André, P., & Neri, R., 1998, A&A 336, 150
Tafalla, M., et al., 2002, ApJ 569, 815
van den Ancker, M., 1999, Ph.D. Thesis
Wilking, B. A., et al., 1997, AJ 114, 2029

THE ORIGIN OF INTERSTELLAR DUST: THE SYNTHESIS OF INORGANIC AND ORGANIC GRAINS IN THE CIRCUMSTELLAR ENVIRONMENT

Sun Kwok

University of Calgary

Abstract

Infrared spectroscopic observations of the circumstellar envelopes of AGB and post-AGB stars have identified a variety of inorganic grains, including amorphous silicates, crystalline silicates, silicon carbide, carbonates, corundum, spinels, and possibly rutiles. Isotopic studies of meteorites have also identified similar species of pre-solar origin (Bernatowicz et al. 1987, Nature, 330, 728; Nittler et al. 1997, ApJ, 483, 475). These grains therefore represent an important link between stars and the solar system.

The existence of aromatic and aliphatic features and their associated plateau continuum suggests that organic compounds in solid state form are also made during the post-AGB phase of stellar evolution (Kwok et al. 1999, A&A, 350, L35; Kwok et al. 2001, ApJ, 554, L87). These features show similarity with the infrared spectra of kerogen (Papoular 2001, A&A, 378, 597), which are also found in meteorites.

The origin of the strong emission feature at 20 microns remains a mystery. Since its initial discovery by the *IRAS* satellite in 1989 (Kwok et al. 1989, ApJ, 345, L51), now 12 carbon-rich post-AGB stars have been found to show this feature. The broad and smooth nature of this feature suggests that it is due to a solid (Volk et al. 1999, ApJ, 516, L99). Although the 20-micron feature has been suggested to be due to hydrogenated fullerenes or titanium carbide nanoclusters, no firm identification has been made.

In this paper, we will summarize the spectroscopic evidence for inorganic and organic grains in the circumstellar environment and discuss the role they play in the chemical enrichment of the ISM, and possible pathways of their delivery to the solar system.

Key words: astrochemistry – ISM: lines and bands – stars: AGB and post-AGB

Chemistry as a Diagnostic of Star Formation (2003). *Edited by Charles L. Curry and Michel Fich.* Published in 2003 by NRC Research Press, Ottawa ON, K1A 0R6, Canada. ISBN 0-660-19089-3.

HOW TO GET RID OF FRINGES IN *SIRTF*/IRS DATA

Fred Lahuis[1,2] **and Adwin Boogert**[3]

[1]Leiden Observatory, P.O. Box 9513, 2300 RA Leiden, The Netherlands
[2]SRON National Institute for Space Research, P.O. Box 800, 9700 AV Groningen, The Netherlands
[3]California Institute of Technology, Department of Astronomy 105-24, Pasadena, CA 91125, USA

Abstract

The InfraRed Spectrograph (IRS) is one of three instruments on board the Space Infrared Telescope Facility (*SIRTF*) scheduled for launch in January 2003. It will take low and high resolution spectra from $\sim 5 - 40\,\mu$m. Interference of light waves reflected on the surfaces of filters, detectors, and other components in the *SIRTF*/IRS light path is expected to result in significant fringing in the measured spectra of astrophysical sources. Fringes associated with the IRS detectors have been observed in spectra measured in a laboratory set-up. Although precise fringe characteristics remain to be measured in orbit, it is clear that reliable defringing methods are essential in order to warrant an accurate analysis of interstellar features.

Key words: Instrumentation: spectrographs – methods: data analysis – infrared: general

1. Introduction

Initial defringing algorithms have been developed for the *SIRTF*/IRS instrument. The InfraRed Spectrograph IRS (Roellig 1998) is one of three instruments on board the Space Infrared Telescope Facility *SIRTF* (Fanson 1998) scheduled for launch in January 2003. The IRS is an echelle spectrograph composed of four modules. Two modules will provide low resolution (R$= \lambda/\delta\lambda = 60 - 120$) spectra from $5.3 - 40\,\mu$m and another two modules will take high resolution (R $\sim$ 600) spectra from $10 - 37\,\mu$m.

In the IRS wavelength range many spectral features of solid ices, silicates, PAHs, molecular bands and unresolved molecular and atomic emission and absorption lines can be observed. The presence of fringes in the spectra will hinder the detection and/or analysis of these often very weak spectral features.

The development effort for the IRS defringing forms part of the *SIRTF*-Legacy program "From Molecular Cores to Planet-Forming Disks" (PI: N.J. Evans), which will obtain IRS spectra of ~200 young stellar objects in Perseus, Ophiuchus, Chamaeleon, Lupus and Serpens (co-I's: E.F. van Dishoeck and G.A. Blake). The work is done in collaboration with the IRS Instrument Team and the *SIRTF* Science Center. The developed algorithms form an integral part of the *SIRTF*/IRS data analysis environment SMART[1] and have also been made available to the SSC for distribution to the general observer.

2. Instrumental Fringes

Fringes originate from interferences on plane-parallel surfaces in the light path of the instrument. These surfaces act as Fabry-Pérot etalons, each of which can add unique fringe components to the source signal. In the infrared wavelength range, surfaces separated by a few mm up to a few cm form the most efficient FP's. Fringes are therefore expected to be seen in the *SIRTF*/IRS spectra and have indeed been observed in spectra measured in a laboratory set-up.

In principle the instrumental fringes are removed from the observed spectrum by calibrating out the response of the instrument. However, due to the nature of the fringes this calibration will for most observations be imperfect. The detailed fringe pattern is mainly determined by the wavelength scale and the resolution. Small offsets in wavelength (offsets much less than a resolution element) can change the position and spacing of the fringe peaks. These offsets can result from an imperfect wavelength calibration or from an off-center location of the source in the slit. The extent of the source determines the effective resolution and thus the amplitude of the observed fringes.

Reliable defringing routines will therefore be needed to achieve the highest possible signal to noise. Such

[1] SMART is a code developed at Cornell University for the analysis of *SIRTF*/IRS data. Additional plug-in modules for peak-up photometry have been developed at the University of Arizona and for defringing at Leiden University, SRON Groningen and Caltech.

Chemistry as a Diagnostic of Star Formation (2003). *Edited by Charles L. Curry and Michel Fich.* Published in 2003 by NRC Research Press, Ottawa ON, K1A 0R6, Canada. ISBN 0-660-19089-3.

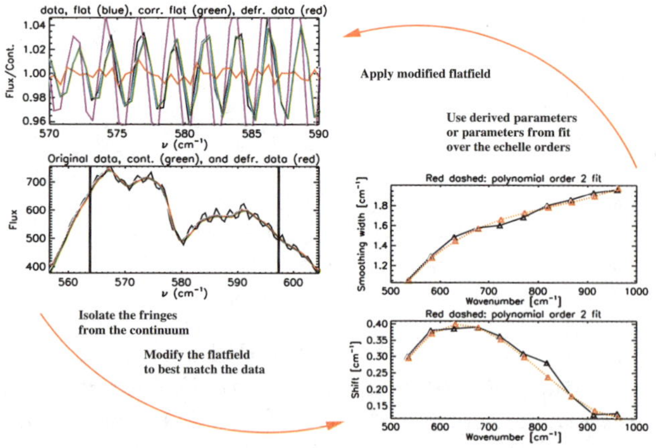

Figure 1. Modifying a flat field in practice. For each echelle order the shift and resolution correction is determined. Shown in the left plot are the corrections for one Short High (SH) order. Subsequently, there is the choice to either use the derived correction values or an interpolated version of these to correct the flat field before applying it to the observed spectrum. This is shown in the right plot.

defringing methods have also proven to be essential in extracting weak features from data from the Short Wavelength Spectrometer (SWS) on board the Infrared Space Observatory (*ISO*) (Lahuis & van Dishoeck (2000); Lahuis, this volume)

3. Defringing Algorithms

Two algorithms have been developed for the defringing of the IRS spectra. The first method minimizes fringe residuals by correcting the flat field to best match the data. The second method uses a robust method of iteratively fitting sine functions. The two methods can be used either separately or in combination. In the latter case the fringes are first removed by optimizing the flat field to the data and then the sine-wave fitting is used to remove residual fringes in the data. In-orbit data are needed to test both methods to the extreme and it may well depend on the specific observation which method or combination of methods will work best.

In order to analyze the fringes it is essential to first isolate the fringes from the continuum for both the flat field and the observed spectrum. In addition, any unresolved spectral features present in the spectrum need to be identified. These features will influence the correlation of the fringes of the flat field and the observation, as well as the parameter estimates in the sine-wave fitting. The masking of the unresolved spectral features is fully automated. In the unlikely event this should not work, the user can of course define these by hand.

4. Flat Field Modification

The first method minimizes the fringe residuals in the data by using a correlation method between the IRS flat field data and the observation. The fringe phase is a function of source position in the slit, while the spectral resolution depends on the filling factor of the source in the slit. Straight division of observation and flat field is thus expected to give residual fringes in a nominal observation. The fringe phase (i.e., wavelength scale) is therefore shifted and the flat field spectral resolution is matched to that of the observation by smoothing or artificially enhancing the fringe amplitudes. The modified flat field fringe spectrum is then multiplied back into the flat field 'continuum'. The astronomical spectrum is finally divided through the modified flat field, giving a spectrum with minimal fringe residuals. Figure 1 gives a short overview of the whole flat field modification procedure.

5. Sine-Wave Fitting

The second method uses robust sine-wave fitting to remove the fringes. The starting point is the optical theory of Fabry-Pérot interference from which an approximation

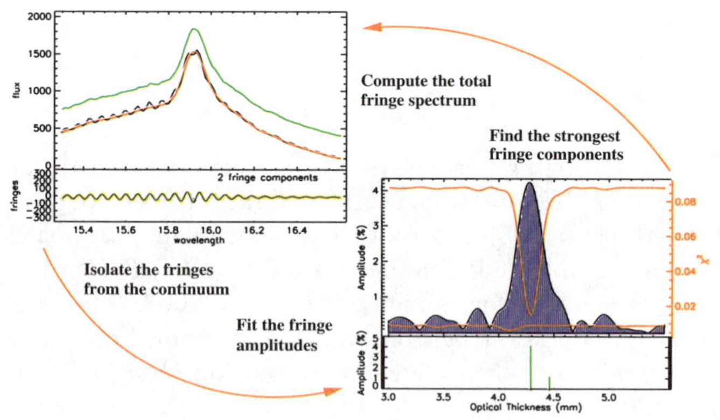

Figure 2. Sine-wave fitting in practice. In the left figure, the original data with continuum are plotted. Offset the final defringed spectrum is shown. In the lower panel the continuum subtracted data and the computed fringe spectrum are plotted. In the right figure, the derived fringe amplitude and the effective decrease in chi-square are plotted for the selected range of optical thickness. In the lower panel the selected fringe components are shown.

is derived which is computationally stable and easy to implement.

5.1. Fabry-Pérot interference

The transmission function of a Fabry-Pérot element is given by the Airy function,

$$I_t = \frac{T^2}{(1-R)^2} \frac{I_o}{1 + \frac{4R}{(1-R)^2}\sin^2(wD\pi)}, \tag{1}$$

with:	R	reflectance [%]	$R + T + A = 1$
	T	transmission [%]	
	A	absorption [%]	$A \simeq 0$ on the surface
	D	optical thickness	$= 2\,nd$
	d	mechanical thickness	$\sim 500\,\mu$m
	n	refractive index	~ 4.3 for the SH detectors
	w	wavenumber [cm^{-1}]	

where R, T, A and n are functions of w and d is a function of position on the detector wafer. For a reflectance of a few percent the transmission function may be approximated by $I_t = A\cos(2\pi wD + \alpha)$. This can be rewritten in a computationally more handy form as:

$$I_t = A_f \cos(2\pi wD) + B_f \sin(2\pi wD). \tag{2}$$

This is the basic formula which was used in the *ISO*-SWS defringing algorithm (Kester 2001; Lahuis & van Dishoeck 2000) which has been adapted for use with the *SIRTF*/IRS data.

5.2. Fringe characteristics

The detailed fringe spectrum is determined by the reflectance and the optical thickness. The reflectance determines the modulation depth given by $(1-R)^2/(1+R)^2$. The optical thickness (i.e., the mechanical thickness and the reflective index) determines the position of the transmission maxima ($w_{\max} = N/D$ with N the FP-order which ranges from approximately 250 to 400).

It is clear that any change in one of the parameters will modify the spectral shape of the fringe system. The reflectance and the reflective index are known to change with wavelength and the mechanical thickness of the detector wafer can vary with detector position.

Once enough high quality in-orbit data are received and more about the IRS instrument and its calibration is learned, the fringe model will be improved. For example, the definition of an accurate model for the wavelength dependence of the reflective index and reflectivity will be attempted.

5.3. Implementation

In the current implementation, all parameters are assumed to be constant. This is a reasonable assumption when data over a small wavelength range are considered like those from one IRS echelle order. Limiting ourselves to this small wavelength range, the fringes in Short-High (SH) data can be fitted with one main component.

Figure 2 gives an overview of how this is currently implemented. First the fringes are isolated from the continuum. Then A_f and B_f in formula (2) are fitted for a limited range of optical thicknesses. The strongest fringe components are selected and the fringe spectrum is computed and divided out of the spectrum. It is possible to use this procedure on both the astronomical observation and the flat field to divide a defringed observation with a defringed flat field. It is also possible to use this procedure to remove residual fringes from the standard flat-fielded spectrum. High quality in-orbit data are needed to test which option is best.

6. Conclusions

It has been possible to have a working set of defringing tools available before the launch of *SIRTF*. This will allow us to improve the reduction of the IRS spectra from the Evans Legacy program in an early phase of the mission. It will also help us and the IRS calibration experts to better understand the IRS instrument in orbit. With the arrival of in-orbit data, the IRS fringe characteristics can be further determined. The defringing algorithms will also be improved, e.g., by improving the physical fringe model and extending the model from 1-D to 2-D. It is our and the SSC's intention to make a stable version of the defringing algorithms available to the general observer as soon as possible.

Acknowledgements

The authors would like to thank the *SIRTF*/IRS instrument team and the *SIRTF*-SSC, in particular Sarah Highdon, Pat Morris and Martin Burgdorf, for their help and collaboration.

References

Fanson, J.L. et al. 1998, in *Space Telescopes and Instruments V*, Proc. SPIE Vol. 3356, 478

Kester, D. et al. 2001, in *The Calibration Legacy of the ISO Mission* (ESA-SP481)

Lahuis, F. & van Dishoeck, E.F. 2000, A&A 355, 699

Roellig, T.L. et al. 1998, in *Infrared Astronomical Instrumentation*, Proc. SPIE Vol. 3354, 1192

GAS-PHASE CO_2, C_2H_2, AND HCN TOWARD ORION-KL

F. Lahuis[1,2], A.M.S. Boonman[1], E.F. van Dishoeck[1], S.D. Doty[3], C.M. Wright[4], and D. Rosenthal[5]

[1]Leiden Observatory, P.O. Box 9513, 2300 RA Leiden, The Netherlands
[2]SRON National Institute for Space Research, P.O. Box 800, 9700 AV Groningen, The Netherlands
[3]Department of Physics and Astronomy, Denison University, Granville, Ohio 43023, USA
[4]School of Physics, University College, ADFA, UNSW, Canberra ACT 2600, Australia
[5]Max-Planck-Institut für Extraterrestrische Physik, Giessenbachstrasse, D–85741 Garching, Germany

ABSTRACT

The infrared spectra toward Orion-IRc2 Peak 1 and Peak 2 in the 13.5-15.5 μm wavelength range are summarized. The spectra are obtained with the Short Wavelength Spectrometer on board the *Infrared Space Observatory* and show absorption and emission features of the vibration-rotation bands of gas-phase CO_2, HCN, and C_2H_2. The CO_2, HCN, and C_2H_2 absorption toward the deeply embedded massive protostar IRc2 probably originates in the warm plateau gas.

The CO_2 emission toward the shocked regions Peak 1 and Peak 2 is best explained by excitation in the warm component of the shock. The HCN and C_2H_2 emission toward Peak 2 suggests a similar explanation. A summary of the full analysis by Boonman et al. (2002) of these absorption and emission bands toward Orion-IRc2 Peak 1 and Peak 2 is presented here.

Key words: Stars: formation – ISM: individual objects: Orion IRc2, Orion Peak 1, Orion Peak 2 – ISM: lines and bands

1. INTRODUCTION

The Orion-IRc2/KL region is probably the best studied source in astrochemistry. A multitude of millimeter, submillimeter single-dish and interferometer observations of nearly a hundred different molecules exists. However, molecules such as CO_2 and C_2H_2 cannot be observed through rotational transitions at millimeter wavelengths due to the lack of a dipole moment. Mid-infrared vibration-rotation spectra allow us to study these and other species. These observations can be used to further constrain the physical structure of this complex region and to study the chemical processes involving these molecules.

Figure 1 gives an overview of the complex environment of the Orion-IRc2/KL region. The observations are centered on IRc2, which is surrounded by a clumpy hot core with the clump exteriors currently being evaporated and/or ablated by the winds from the embedded massive young star(s). This hot core is contained in a cavity, surrounded by a dense, quiescent ridge in the NE-SW direction. To the NW and SE, two shocked regions – Peak 1 and Peak 2 – are revealed by bright H_2 2 μm emission, indicating the positions where the high-velocity plateau or outflow runs into the ambient molecular cloud. Peak 1 is located $\sim 25''$ NW and Peak 2 $\sim 24''$ SE of IRc2.

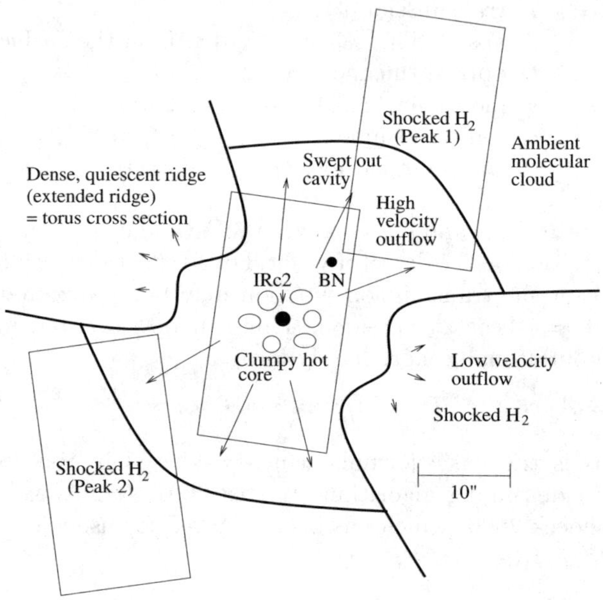

Figure 1. Cartoon of the Orion-KL region. The figure represents a cross section in the plane of the sky. The size and orientation of the *ISO*-SWS beam around 15 μm at the approximate positions of IRc2, Peak 1 and Peak 2 are indicated by the rectangular boxes (adapted from van Dishoeck et al. 1998).

Chemistry as a Diagnostic of Star Formation (2003). *Edited by Charles L. Curry and Michel Fich.* Published in 2003 by NRC Research Press, Ottawa ON, K1A 0R6, Canada. ISBN 0-660-19089-3.

2. Data reduction

The observations have been performed with the Short Wavelength Spectrometer (SWS) on board the Infrared Space Observatory (*ISO*). The spectra are reduced using the Off Line Processing software and calibration files (version 10). Instrumental fringes have been removed by a combination of optimized response correction and a robust sine wave fitting method (see Lahuis & van Dishoeck [2000] for a detailed description). The reduced and continuum subtracted spectra are shown in Figure 2. The vibration-rotation bands of gas-phase CO_2, HCN, and C_2H_2 are detected in absorption toward IRc2 and in emission toward the shocked region Peak 2. Toward Peak 1 only CO_2 is seen in emission, but this could be due to the somewhat higher noise level in this spectrum.

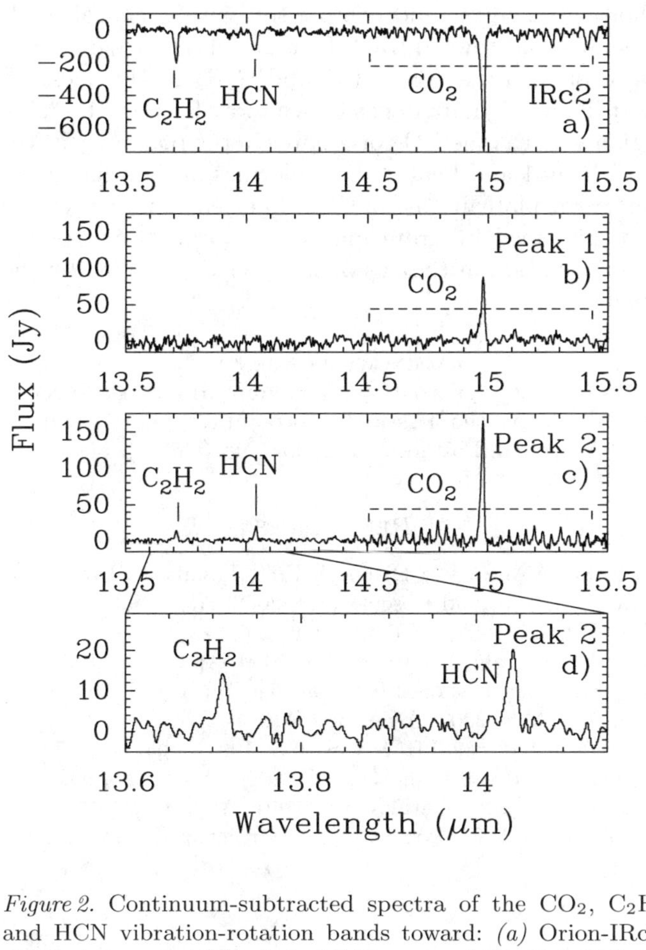

Figure 2. Continuum-subtracted spectra of the CO_2, C_2H_2 and HCN vibration-rotation bands toward: *(a)* Orion-IRc2, *(b)* Peak 1, and *(c)* Peak 2. *(d)* Blow-up of the 13.6-14.5 μm region toward Peak 2. (From Boonman et al. 2002).

3. Absorption bands toward IRc2

The CO_2, HCN and C_2H_2 absorption bands toward IRc2 have been analyzed by computing synthetic spectra, adopting a homogeneous source with a constant temperature and column density (Lahuis & van Dishoeck 2000). It is assumed that only absorption takes place. All three bands are characterized by excitation temperatures of ~175–275 K, indicating that they originate in the plateau gas. The inferred abundances are listed in Table 1. Emission filling in the absorption changes the derived abundances by less than ~25% for C_2H_2 and CO_2, but becomes important for HCN for $T_{ex} \gtrsim 250$ K.

HCN and C_2H_2 are only seen in absorption in the plateau gas in the direction of IRc2 (Evans et al. 1991), whereas the CO_2 absorption is probably more widespread. The CO_2 abundances are best explained by both grain mantle evaporation and reformation in the gas phase on time scales of $t \sim 10^5$ yr after destruction by the shock or a heating event.

The HCN and C_2H_2 absorption bands on the other hand probably originate in gas that has been blown away from the hot core and now resides in the plateau gas close to IRc2. Their high abundances can be explained by a combination of high-temperature gas-phase chemistry and grain mantle evaporation.

4. Emission bands toward Peak 1 and 2

4.1. CO_2

The CO_2 emission toward Peak 1 and Peak 2 does not show evidence for a hot component at $T \sim 3000$ K. The derived excitation temperatures are ~150–225 K, suggesting that the CO_2 emission probably originates in the warm component of the shock, like the CO_2 absorption toward IRc2/BN.

The question arises as to what excitation mechanism is responsible for the observed CO_2 emission bands toward Peak 1 and Peak 2. Therefore three possible scenarios have been investigated and compared with chemical models:

1. Radiative excitation by IRc2/BN;
2. Collisional excitation;
3. Excitation by warm dust mixed with the shocked gas.

In the case of radiative excitation by IRc2/BN, CO_2 abundances of ~10^{-4} are needed to explain the observed emission, more than two orders of magnitude higher than predicted by chemical models (Charnley & Kaufman 2000; Doty et al. 2002). If the CO_2 molecules are excited

Table 1. IRc2 versus Peak 1 and 2 abundances compared with chemical models (from Boonman et al. 2002)

	IRc2/BN complex	Peak 1	Peak 2	Shock model[a]	Envelope model[b]	Hot core model[c]
	10^{-7}	10^{-7}	10^{-7}	10^{-7}	10^{-7}	10^{-7}
CO_2/H_2	$1.6-8.9$	$0.06-0.3$	$0.03-0.6$	$\sim 1-7$	$\sim 0.6-30$	~ 10
HCN/H_2	$0.9-9.8$	< 0.06	$0.004-0.1$	–	$\lesssim 0.1$	$\sim 7-10$
C_2H_2/H_2	$0.4-4.9$	< 0.04	$0.002-0.07$	–	$\lesssim 0.1$	$\sim 0.3-2$

[a] Charnley & Kaufman (2000) for reformation through gas-phase reactions at $T \sim$200 K and $t \sim 10^4$ yr after destruction by a shock.
[b] Doty et al (2002) at $T \sim$150–200 K and $t \sim 10^4$ yr for CO_2 and at $T <$200 K, $t \lesssim 10^5$ yr for HCN and C_2H_2. It is assumed that these molecules are not originally present in ices.
[c] HCN and C_2H_2 from Rodgers & Charnley (2001) for $t \sim 10^5$ yr at T=300 K for HCN and T=100-300 K for C_2H_2. CO_2 from Charnley (1997) at T=200 K and $t \sim 10^5$ yr. It is assumed that HCN, C_2H_2, and CO_2 are not present originally in ices.

by collisions, abundances of $\sim 10^{-6}$ are inferred, which agree with the highest abundances predicted by chemical models. However, since the color temperature of the dust around 15 μm is comparable to the CO_2 excitation temperature, it is likely that dust is mixed with the gas in the warm component of the shock at $T \sim 150-200$ K. Therefore, the last scenario is the preferred scenario. This scenario gives CO_2 abundances of $\sim 10^{-8}$ for both Peak 1 and Peak 2 (Table 1). These abundances are in agreement with predictions from chemical models including grain mantle evaporation and/or little reformation in the gas-phase after destruction by the shock.

4.2. HCN and C_2H_2

The HCN and C_2H_2 emission profiles detected toward Peak 2 are narrower than the CO_2 profile, resulting in lower excitation temperatures of $T_{ex} \sim$50–150 K. This suggests an origin in colder gas. In order to investigate the excitation of the HCN and C_2H_2 emission, the same three scenarios have been investigated as for CO_2.

In the cases of radiative excitation by IRc2/BN and collisional excitation, the HCN and C_2H_2 abundances are about an order of magnitude higher than predicted by chemical models (Doty et al. 2002). In the preferred scenario of excitation by dust in the warm component of the shock their low abundances of $\sim 10^{-9}$ suggest that they are destroyed by the shock or have only been in the warm gas for a short time ($t \lesssim 10^4$ yr).

In case they reside in the cooler foreground gas of the extended ridge, the inferred abundances are much higher and do not agree with predictions from current chemical models at low temperatures.

5. Conclusions

Absorption and emission bands of CO_2, HCN and C_2H_2 toward Orion-IRc2, Peak 1 and Peak 2 are analyzed. Thorough analysis and comparison with chemical models shows that toward IRc2 the molecules all reside in the warm plateau gas. The HCN and C_2H_2 at this location most likely originate in gas blown away from the hot core region whereas the CO_2 originates in the plateau gas. Toward Peak 1 and Peak 2 the molecules likely originate in the warm plateau gas and the inferred abundances are best explained by grain mantle evaporation and/or little reformation in the gas-phase after destruction by the shock.

Acknowledgements
This work was supported by NWO grant 614-41-003, a NWO Spinoza grant, the Research Corporation (SDD), and an ARC Australian Postdoctoral Fellowship (CMW).

References

Boonman A.M.S., van Dishoeck E.F., Lahuis F., Doty S.D., Wright C.M. and Rosenthal D. 2002, A&A, submitted
Charnley S.B. 1997, ApJ, 481, 396
Charnley S.B. and Kaufman M.J. 2000, ApJ, 529, L111
Doty S.D., van Dishoeck E.F, van der Tak F.F.S. and Boonman A.M.S. 2002, A&A, 389, 446
Evans N.J. II, Lacy J.H. & Carr J.S. 1991, ApJ, 383, 674
Lahuis, F. and van Dishoeck, E.F. 2000, A&A, 355, 699
Rodgers S.D. and Charnley S.B. 2001, ApJ, 546, 324
van Dishoeck E.F., Wright C.M., Cernicharo J., et al. 1998, ApJ, 502, L173

THE CHEMICAL STRUCTURE AND DYNAMICAL STATUS OF THE PRE-PROTOSTELLAR CORE B68

Shih-Ping Lai, T. Velusamy, W. D. Langer, and T. B. H. Kuiper

Jet Propulsion Laboratory, California Institute of Technology, Pasadena, CA 91109, U.S.A.

Abstract

A previous extinction study suggested that the density profile of B68 is remarkably consistent with a Bonnor-Ebert sphere with 2.1 $M_{\odot}$ at 16 K. We mapped B68 in C_3H_2, CCS, and NH_3 with the Deep Space Network (DSN) 70 m and 34 m telescopes at Goldstone. Our results show strong chemical differentiation, suggesting that B68 is indeed a cold, dense prestellar core. We measured the kinetic temperature of 11 K in B68 with NH_3 (1,1) and (2,2) spectra, which is significantly lower than that previously assumed. We also study the turbulence in B68 using the molecular linewidths and show that B68 is thermally dominated with little contribution from turbulence support ($< 10\%$). We use these results to re-examine the dynamical status of B68. Considering the uncertainty in distance to B68, we conclude that a stable Bonnor-Ebert sphere is a possible solution, consistent with all observable parameters, only if the distance is $\sim$ 95 pc with a corresponding mass $\sim 1.2\ M_{\odot}$.

Key words: ISM: molecules – ISM: abundances – ISM: globules – ISM: individual (B68) – stars: formation

1. Introduction

B68 is an isolated quiescent Bok Globule near the Ophiuchus complex. Its nearly spherical geometry makes it a good candidate for comparison between observations and theoretical models (which are primarily spherically symmetric). The importance of B68 is that it is proposed to be a core on the verge of collapse (Alves, Lada, & Lada 2001, hereafter ALL2001). They derive the radial extinction profile of B68 from the optical and near infrared photometry of background stars, and the profile is remarkably consistent with a 16 K isothermal Bonnor-Ebert sphere with a mass of 2.1 $M_{\odot}$ (assuming a distance of 125 pc) which is slightly over the critical mass. Their result on the center-to-edge density contrast implies that B68 can adequately be supported by the thermal pressure (T=16 K) without support from turbulence and/or magnetic fields before the dynamical collapse takes place. As the best candidate of a Bonnor-Ebert sphere to date, B68 has attracted great attention recently. Here, we present spectral line maps of B68 in C_3H_2, CCS, and NH_3 to characterize its chemical status and to derive the temperature and turbulence in the core. Our results when combined with the density structure derived from extinction data (ALL2001) provide stringent constraints on the distance (linear size and mass) to B68 for its interpretation as a stable prestellar core.

2. Observations and Results

We made single-dish maps of B68 in C_3H_2, CCS, and NH_3 lines between January and July 2002 using NASA's Deep Space Network (DSN) 70 m antenna at Goldstone. We mapped B68 over a $\sim 4'\times4'$ region. We also used a DSN 34 m antenna to obtain the spectra of NH_3 (1,1) and (2,2) lines simultaneously in order to derive the kinematic temperature of B68. The NH_3 spectra were taken at (0.′25, 0.′25) offset from the nominal center of B68 so that the 34 m beam includes most of the NH_3 emission.

2.1. Chemical Differentiation

Figure 1 shows our DSN 70 m maps of C_3H_2, NH_3, and CCS. The NH_3 map shows a centrally peaked distribution with the maximum close to the nominal center. By contrast, C_3H_2 and CCS both have their strongest peaks to the northeast of the NH_3 peak and their intensity decreases towards the southwest of the NH_3 peak. The C_3H_2 intensity increases again at the southwest edge. CCS has another strong peak at the southeast extension of B68. The JCMT dust continuum maps of B68 at 850 μm also shows a distinct peak at the southeast extension (Visser, Richer, & Chandler 2002), indicating this extension could be a separate core.

Chemistry as a Diagnostic of Star Formation (2003). *Edited by Charles L. Curry and Michel Fich.* Published in 2003 by NRC Research Press, Ottawa ON, K1A 0R6, Canada. ISBN 0-660-19089-3.

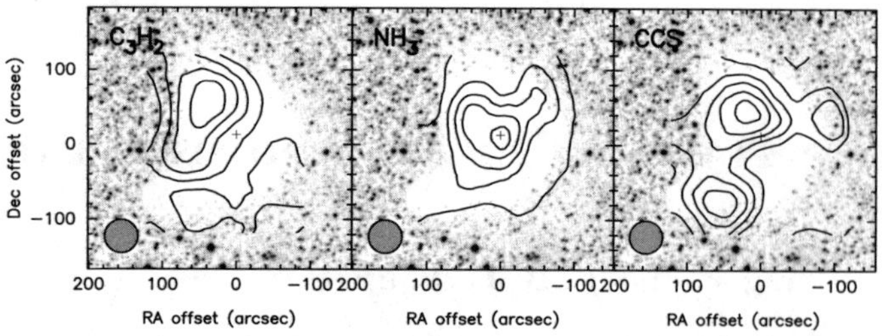

Figure 1. DSN 70 m maps of C_3H_2, NH_3, and CCS in B68 overlaid on the Palomar Optical Sky Survey image. The contours are 10, 30, 50, 70, and 90% of the peak integrated antenna temperature, which are 0.23, 0.25, and 0.084 K km s^{-1} for C_3H_2, NH_3, and CCS, respectively. The plus symbol indicates the peak of the NH_3 emission.

In Figure 2 we show an overlay of the NH_3 map with the N_2H^+ map obtained at the IRAM 30 m telescope (Bergin et al. 2002) and the 200 μm dust emission obtained with ISO (Langer & Willacy 2001). We found that the NH_3 peak coincides with both the dust continuum peak and is located between the two N_2H^+ peaks. This result is consistent with the expectation of the chemical models that NH_3 and N_2H^+ are relatively more abundant in high density regions than other molecules. The double peaks of N_2H^+ have been interpreted to indicate that even N_2H^+ depletes in the region with $A_V > 25$ mag. A similar depletion of NH_3 could take place in the center of B68, but may not be spatially resolved with our observations.

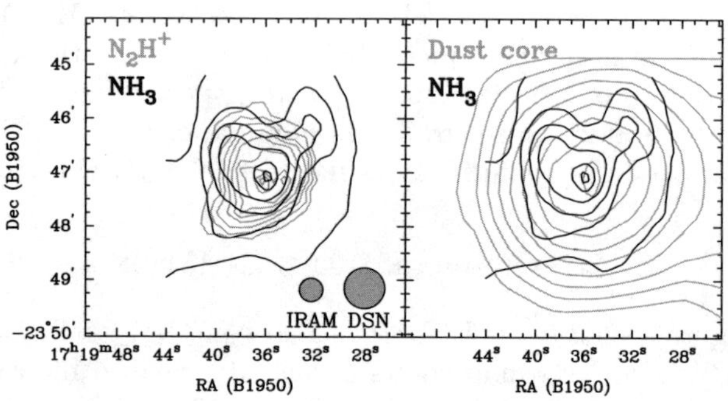

Figure 2. Comparison of DSN 70 m NH_3 map with IRAM 30 m N_2H^+ map (*left*; Bergin et al. 2002) and 200 μm dust emission (*right*; Langer & Willacy 2001).

2.2. Kinetic Temperature

At low temperatures where collisional excitation dominates, the rotation temperature of NH_3 can be estimated from the intensity ratio of the (1,1) and (2,2) lines T_{22}/T_{11} and the optical depth of the (1,1) main line τ_{11} (Ho & Townes 1983):

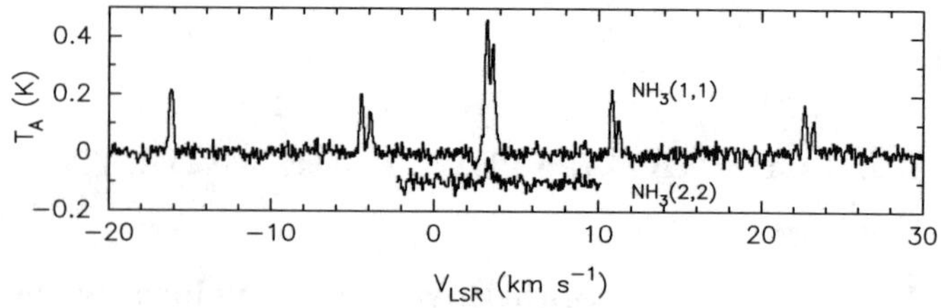

Figure 3. DSN 34 m NH_3 (1,1) and (2,2) spectra of B68.

$$T_r = -41.5/\ln\left\{\frac{-0.282}{\tau_{11}}\ln\left[1-\frac{T_{22}}{T_{11}}(1-e^{-\tau_{11}})\right]\right\}. \quad (1)$$

We fitted NH_3 (1,1) and (2,2) hyperfine lines using CLASS, and obtained $T_{22}/T_{11} = 0.20 \pm 0.04$K and $\tau_{11} = 2.5 \pm 0.2$ (Fig. 3). From equation (1), we derived $T_r = 10.9 \pm 0.8$K. Figure 4 shows how the uncertainty in the estimate of T_r varies with T_{22}/T_{11} and τ_{11}. The rotation temperature slightly underestimates the kinetic temperature T_k in dark clouds (Walmsley & Ungerechts 1983), but it approaches the kinetic temperature when $T_k \rightarrow 10K$. We adopt $T_k = 11K$ in further calculations.

2.3. Turbulence

The non-thermal linewidth (FWHM), $\Delta v_{\rm nth}$, can be estimated from $\Delta v_{\rm obs}^2 = \Delta v_{\rm th}^2 + \Delta v_{\rm nth}^2$. The observed linewidth $\Delta v_{\rm obs}$ can be obtained from the spectral line data. The thermal linewidth $\Delta v_{\rm th} = [(8\ln 2)kT_k/\mu m_H]^{1/2}$ (μ is the atomic weight and m_H the mass of a hydrogen atom) can be calculated with the kinetic temperature determined in Sec. 2.2. We derived $\Delta v_{\rm nth}$ from NH_3 and CCS separately. The multiple line fitting of NH_3 gives a reliable determination of the observed linewidth. The CCS molecule has a large atomic weight, and hence a narrow thermal linewidth. We derived $\Delta v_{\rm nth} \sim 0.14 \pm 0.01$ km s^{-1} from NH_3 and $\Delta v_{\rm nth} \sim 0.14 \pm 0.03$ km s^{-1} from CCS. The $\Delta v_{\rm nth}$, interpreted as that due to turbulence, is small in B68—of the order of 0.3 $\Delta v_{\rm th}(H_2)$.

3. Discussion

3.1. Chemistry of B68 vs. other prestellar cores

Since CCS is an early-time molecule and NH_3 is abundant at a later stage, the column density ratio of CCS and NH_3 has been suggested as a possible indicator for cloud evolution (Suzuki et al. 1992; Bergin & Langer 1997). As the evolutionary status of B68 has been established

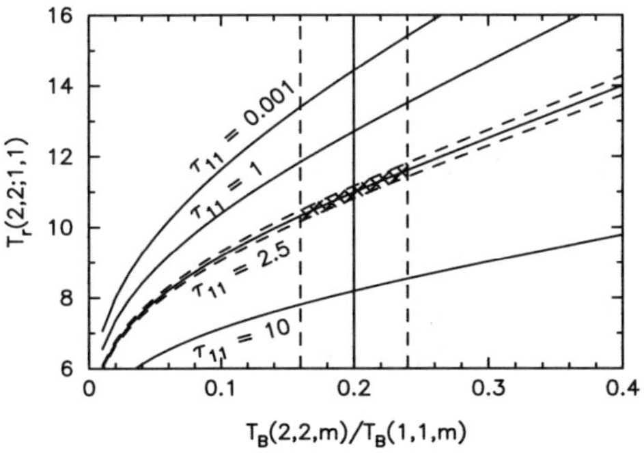

Figure 4. Determination of the rotational temperature, T_r.

to be a likely pre-protostellar core on the verge of collapse, we can compare its N(CCS)/N(NH_3) to two other well-studied pre-protostellar cores, L1498 and L1544. We find that N(CCS)/N(NH_3) for L1498, B68, and L1544 are ~0.008, 0.006, 0.002, respectively. Among these cores, L1544 is the most evolved, because the collapse motion has been directly observed (Tafalla et al. 1998). L1498 could be younger than B68, because CCS is more widely distributed in L1498 than in B68. Therefore, B68 appears to be in an intermediate stage between L1498 and L1544 in core evolution, as indicated by N(CCS)/N(NH_3).

3.2. How stable is B68?

ALL2001 demonstrate that the density profile of B68 inferred from extinction is remarkably consistent with that of an isothermal Bonnor-Ebert sphere at 16 K, for an assumed distance of 125 pc. However, our observations show that: (1) B68 is colder (T_k=11 K), and (2) thermal support dominates over turbulence, but that turbulent energy is present. Is the Bonnor-Ebert sphere still a good description for B68?

The physical parameters that determine the status of a core are: mass, density structure, temperature, rotation and the presence of turbulence and magnetic fields. The extinction data provides density structure which is described by: (i) the dimensionless radius parameter $\xi_{\max}$ = 6.9, (ii) the angular radius $\theta = 100''$, and (iii) the central column density, N_c. The dimensionless radius parameter $\xi_{\max}=(R/a)\sqrt{4\pi G\rho_c}$ (where $\rho_c = mn_c$ is the central density, $a = \sqrt{kT/m}$ is the sound speed, and m is the mean molecular weight) is the only parameter characterizing the shape of the density profile for a Bonnor-Ebert sphere. Thus the stability criterion for B68 as a Bonnor-Ebert sphere requires a choice of linear size ($R = \theta D$, where D is the distance to B68) and temperature (T) such that $\xi_{\max}$ is kept at the observed value. The three observed quantities ($\xi_{\max}$, θ, N_c) are invariant to T and D. We can express

$$\frac{\xi_{\max}}{\theta} = mD\sqrt{\frac{4\pi G n_c}{kT}}. \qquad (2)$$

Because $\xi_{\max}$ and θ are fixed by observations, the number density at the center $n_c \propto D^{-2}\,T$. Furthermore, n_c is also related to the column density $N_c = \int n\,ds = Kn_cD\theta$, where K is a constant which depends only on the shape of the density profile characterized by $\xi_{\max}$. Eliminating n_c, we find that $N_c \propto D^{-1}T$. Because N_c is fixed by observation, $D^{-1}T$ is a constant and $n_c \propto D^{-1}$. ALL2001 assumed D=125 pc and T=16 K, to derive a model consistent with $\xi_{\max} = 6.9$ and $M = 2.1\ M_\odot$. However, if we use an effective temperature of 12 K, which includes both thermal and non-thermal (turbulence) support, for the same $\xi_{\max}$, B68 would have to be at a closer distance $\sim$ 95 pc with a lower mass of $\sim 1.2\ M_\odot$.

The stability of the core depends on the distance assumed for B68 (since the mass is determined from the observed core size and densities). Since the distance to B68 itself is not well determined (ALL2001 use an average distance 125 pc of the Ophiuchus complex), a static Bonnor-Ebert sphere is consistent with the data only if the core is at 95 pc. It is also possible that magnetic fields provide additional support to the core; however, the role of magnetic fields in prestellar cores is poorly known. If B68 is at > 95 pc, B68 can be a stable core only with additional support from magnetic fields. Otherwise, it is not in equilibrium, but could be undergoing slow contraction in the center of the core within an angular scale which cannot be resolved with our observations.

Acknowledgements

This research was conducted at JPL, Caltech with support from NASA, while SPL held a National Research Council Research Associateship at JPL.

References

Alves, J., Lada, C. J., & Lada, E. A. 2001, Nature, 409, 159
Bergin, E. A., Alves, J., Huard, T., & Lada, C. J. 2002, ApJL, 570, L101
Bergin, E. A. & Langer, W. D. 1997, ApJ, 486, 316 de Zeeuw, P. T., & Lub, J. 1989, A&A, 216, 44
Ho, P. T. P. & Townes, C. H. 1983, ARA&A, 21, 239
Langer, W. D. & Willacy, K. 2001, ApJ, 557, 714
Suzuki, H., Yamamoto, S., Ohishi, M., Kaifu, N., Ishikawa, S., Hirahara, Y., & Takano, S. 1992, ApJ, 392, 551
Tafalla, M., Mardones, D., Myers, P. C., Caselli, P., Bachiller, R., & Benson, P. J. 1998, ApJ, 504, 900
Visser, A.E., Richer, J.S., & Chandler, C.J. 2002, AJ, in press
Walmsley, C. M. & Ungerechts, H. 1983, A&A, 122, 164

CHEMISTRY AND DYNAMICS IN PRE-PROTOSTELLAR CORES

Jeong-Eun Lee[1], Neal J. Evans II[1], Yancy L. Shirley[1], and Ken'ichi Tatematsu[2]

[1]Department of Astronomy, The University of Texas at Austin, 1 University Station C1400, Austin, Texas 78712–0259
[2]National Astronomical Observatory of Japan, 2-21-1 Osawa, Mitaka, Tokyo 181-8588, Japan

Abstract

We have compared the intensity distribution of molecular line emission with that of dust continuum emission, and modelled molecular line profiles in three different preprotostellar cores in order to test how dynamical evolution is related to chemical evolution, and whether we can use different chemical tracers to identify specific dynamical evolutionary stages. We used dust continuum emission to obtain the input density and temperature structures by calculating radiative transfer of dust emission. Our results show that chemical evolution is dependent on dynamical processes, which can give different evolutionary timescales, as well as the density structure of the core.

Key words: ISM: molecules – stars: formation

1. Introduction and Observations

Low-mass star formation has received more detailed study, both in theory and observation, compared to massive star formation. An evolutionary sequence of low mass star formation based on the spectral energy distribution has been developed (Adams et al. 1987). However, the initial conditions of collapse are still poorly understood. Preprotostellar cores (PPCs) are believed to be gravitational bound, but they have no central hydrostatic protostar (Ward-Thompson et al. 1994). Therefore, PPCs can be considered as the potential sites of future star formation and give us chances to probe the initial conditions of star formation. Evans et al. (2001) modelled dust continuum in three PPCs using Bonnor-Ebert spheres and showed that the dust temperature decreases toward the center because PPCs are heated only by the interstellar radiation field, and the radiation from the cores is optically thin. Therefore, they argued that the actual density structure of a preprotostellar core is more centrally condensed compared to the intensity distribution of the dust continuum emission.

Several studies on chemical evolution in PPCs have been done by Bergin & Langer (1997), Aikawa et al. (2001), Caselli et al. (2002), and Li et al. (2002). They tested different dynamical scenarios, but all included the interaction of gas phase molecules with the surfaces of dust grains as well as gas phase chemistry and showed that some molecules such as CS and CO are depleted substantially from the gas phase as the density increases. Those molecules have high binding energy onto the surfaces of dust grains, so their timescales for evaporation by thermal energy or cosmic rays are much longer than those of molecules that have lower binding energy. For instance, nitrogen-bearing molecules such as N_2H^+ and NH_3, which become abundant in later stages of dynamical evolution and have low binding energy, do not show as significant depletion as CO and CS. Comparison of the abundance change of a molecule with time and the differential distributions of molecules in a given time seems to be able to trace the dynamical evolutionary stages. We tested this idea by comparing the line emission from several molecules that have different chemical evolution timescales with dust emission, which can trace physical conditions, in three PPCs: L1512, L1544, and L1689B.

We observed the three PPCs in $C^{18}O$ ($J = 2-1$ and $J = 3-2$), $C^{17}O$ ($J = 2-1$), DCO^+ ($J = 3-2$), HCO^+ ($J = 3-2$), and $H^{13}CO^+$ ($J = 3-2$) with the 10.4 m telescope of the Caltech Submillimeter Observatory (CSO) at Mauna Kea, Hawaii from 1995 to 2002. We also observed these cores in $H^{13}CO^+$ ($J = 1-0$), N_2H^+ ($J = 1-0$), and CCS ($N_J = 4_3 - 3_2$) with the 45 m telescope of the Nobeyama Radio Observatory in Japan on January 2002. See Lee et al. (2003) for details about observations.

2. Simple Analysis

We first compare the H_2 column densities calculated from molecular lines ($C^{18}O$ J=2–1 and $C^{17}O$ J=2–1) and dust continuum, which can be used as the tracer of column density least affected by chemistry (Fig. 1*a*). We assume that temperature and abundance are constant along

Chemistry as a Diagnostic of Star Formation (2003). *Edited by Charles L. Curry and Michel Fich.* Published in 2003 by NRC Research Press, Ottawa ON, K1A 0R6, Canada. ISBN 0-660-19089-3.

the line of sight, that gas and dust have the same temperature, that all levels are in LTE, and that lines are optically thin. The H_2 column densities calculated from $C^{18}O$ J=2−1 are smaller than those calculated from 850 μm dust emission by a factor of 10 to 25. Also, the $C^{18}O$ emission shows no peak at the position of the peak of dust continuum emission. If this line is optically thin, so that it traces all material along the line of sight, the difference would be caused by the depletion of CO. The optical depth of $C^{17}O$ J=2−1 can be calculated by fitting the relative strength of its hyperfine components, and the calculated optical depths in three PPCs show that the optical depth of $C^{17}O$ J=2−1 is negligible in all three cores, but $C^{18}O$ J=2−1 is optically thick in L1544 and L1689B (see Sec. 4.2.1 of Lee et al. 2003). Therefore, if we use an optically thick line, the depletion factor can be overestimated. The comparison between the H_2 column densities calculated from $C^{17}O$ J=2−1 and 850 μm dust emission shows that the CO molecule is significantly depleted in L1512 and L1544 but not in L1689B.

We also compare the distributions of the intensity of molecular emission and dust continuum emission (Fig. 1*b*). L1512 is the weakest, and L1544 and L1689B are similar in dust continuum. However, the CCS intensity of L1689B is more similar to that of L1512. Even though the distribution of CCS is very different from that of the dust continuum, it does not show the central dip expected if the molecule is depleted significantly as some chemical modelings show. We suggest that this trend would be caused by the significant depletion of CO, helping C^+ to react more with molecules that are not oxygen-bearing species (Li et al. 2002). The intensity of N_2H^+ follows dust emission well in L1544 and L1689B, but it is weaker in L1689B than in L1544. The difference is possible because CO, which is a destroyer of N_2H^+, is much more depleted in L1544. On the other hand, in the central region of L1512, the N_2H^+ distribution is flatter than the dust emission indicating the possible depletion of N_2H^+.

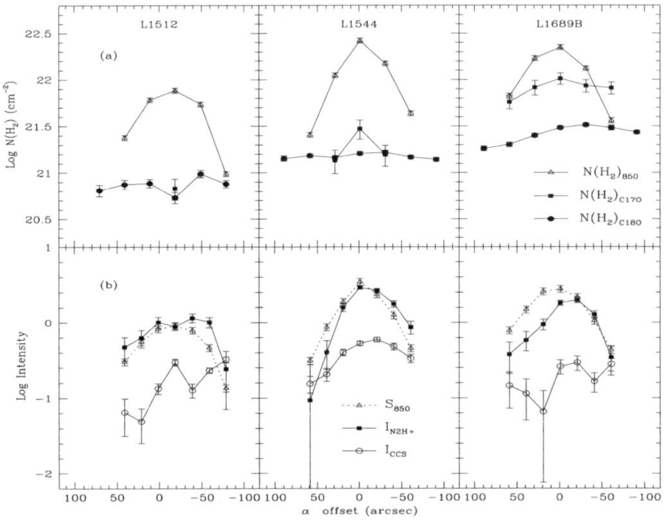

Figure 1. *(a)* Comparison between the H_2 column densities calculated from the 850 μm dust continuum emission, $C^{18}O$ $J = 2-1$ (without the correction of τ), and $C^{17}O$ $J = 2-1$ (with correction of τ) through the cuts marked in Fig. 2 of Lee et al. (2003). *(b)* Comparison of the integrated intensities of CCS $N_J = 4_3 - 3_2$ and N_2H^+ $J = 1-0$ lines with S_{850} through the same cuts in *(a)*. Here, S_{850} is shifted by 0.9.

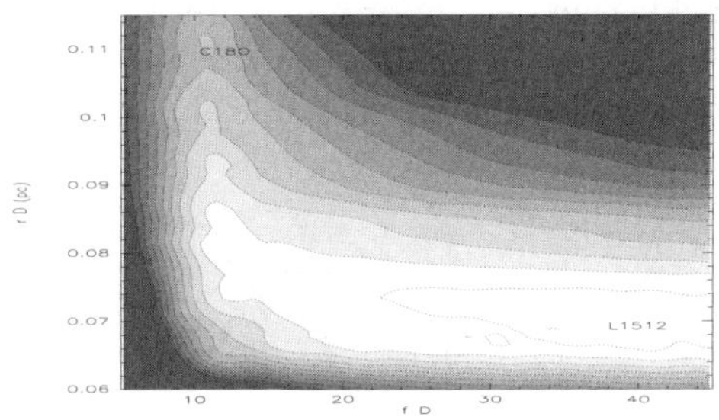

Figure 2. Distribution of the reduced χ^2 of models that have different r_D and f_D in $C^{18}O$ 2–1 and 3–2 toward L1512. The contour levels are from 5 to 95 with an interval of 10.

3. Detailed Models

The simple method used in the previous section has limitations for quantitative analysis because it cannot include the variation of the excitation temperature and the abundance of a molecule along the line of sight. Therefore, we used the Monte Carlo (MC) code and the virtual telescope simulation code developed by Choi et al. (1995) to calculate radiative transfer of molecular lines and to simulate specific molecular line profiles, respectively. For the simulation, we adopted two physical models: Bonnor-Ebert spheres (Bonnor 1956; Ebert 1955), which give only density structure; and Plummer-like spheres (Whitworth & Ward-Thompson 2001), which give density and velocity structures and can be used for optically thick lines to simulate their self-absorbed line profiles (see Sec. 5.2 of Lee et al. [2003] for details). We calculated the radiative transfer of dust emission to obtain the temperature structure (Evans et al. 2001). For the structure of molecular abundances, we tested several functional forms; the step function, which defines a depletion factor (f_D) and a depletion radius (r_D), was the best to fit line profiles from the center to the outer regions in the three PPCs (Fig. 5 in Lee et al. 2003). We modelled $C^{18}O$, $H^{13}CO^+$,

Table 1. Results of Monte Carlo models of molecular lines

		$C^{18}O$	$H^{13}CO^+$	DCO^+
	X_0	4.82×10^{-7}	3.0×10^{-10}	2.8×10^{-10}
L1512	r_D	0.075 pc	0.021 pc	[a]
	f_D	25	25	[a]
	X_0	4.82×10^{-7}	5.0×10^{-10}	3.5×10^{-10}
L1544	r_D	0.045 pc	0.026 pc	0.022 pc
	f_D	25	20	3
	X_0	4.82×10^{-7}	2.2×10^{-10}	3.5×10^{-10}
L1689B	r_D	0.030 pc	0.012 pc	0.011 pc
	f_D	3	5	4

[a] We could not calculate r_D and f_D of DCO^+ in L1512 because we have only one spectrum toward the center.

DCO^+, and HCO^+ lines using a step function of each abundance in order to test how much those molecules are depleted within what radii. Figure 2 shows that the depletion radius is well constrained around 0.075 pc, but the fractional depletion is constrained only to be greater than about 25 in the $C^{18}O$ lines of L1512. This trend occurs in every molecule that is significantly depleted, so we consider f_D to be the lower limit of the depletion factor. In other words, the data are consistent with complete depletion inside some r_D. Figure 3 shows the best-fit model of $C^{18}O$ $J = 2 - 1$ in L1512. The results of modelling in the three cores are summarized in Table 1.

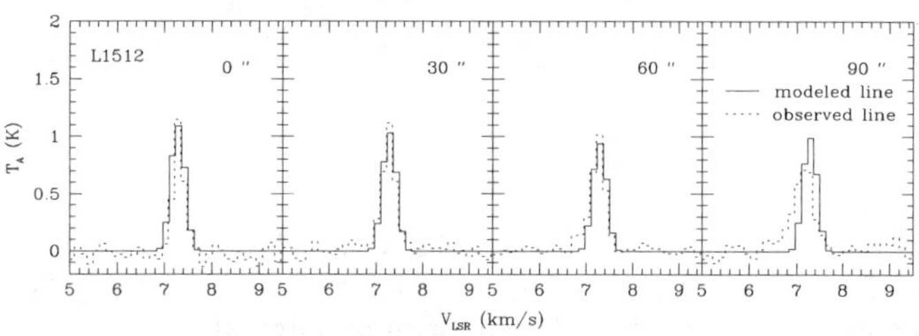

Figure 3. Best-fit model of $C^{18}O$ $J = 2 - 1$ in L1512. The arcseconds marked in panels represent the angular distance from the dust peak.

According to our results, CO is substantially depleted in L1512 and L1544 but not in L1689B. In addition, the depletion radius of CO in L1512 is greater than that in L1544. The $C^{18}O$ lines in L1689B might be affected by surrounding warm gas (see Sec. 5.3.2 of Lee et al. 2003). HCO^+ is also depleted significantly in L1512 and L1544, but the depletion radii are smaller than those of CO. On the other hand, DCO^+ does not show significant depletion in any of these three cores. We might not detect the depletion of DCO^+ due to poor observational resolution.

Table 2. Results of this study[a]

	L1512	L1544	L1689B
Den.	Young	Evolved	Evolved
Chem.	Evolved	Evolved	Young
Time	$\tau_{che} \ll \tau_{dyn}$	$\tau_{che} \approx \tau_{dyn}$	$\tau_{che} \gg \tau_{dyn}$
Dyn.	Stable (?)	AD (?)	Free-fall (?)

[a] In the leftmost column, "Den." ≡ density structure, "Chem." ≡ chemical evolution, "Time" ≡ timescale, and "Dyn." ≡ dynamical state.

4. Discussion

Our simple analysis and detailed modelling of molecular line profiles combined with dust continuum emission show that L1512 is least centrally condensed, but CCS, CO, and HCO^+ are significantly depleted. Even N_2H^+ is possibly depleted in L1512. On the other hand, L1689B has a similar density structure to L1544, which is quite centrally condensed and has substantial depletion of CCS, CO, and HCO^+, but no molecules except for CCS show significant depletion. This result may indicate that L1689B has evolved too fast to have time to evolve chemically. Therefore, we suggest that the stage of dynamical evolution of a core cannot be simply probed by its chemical status because chemical evolution depends on the size of dust grains (Caselli et al. 2002, in prep.) or the absolute dynamical timescale, which is different for each dynamical process, as well as its density structure that shows the relative dynamical evolutionary stages (Table 2).

References

Aikawa, Y., Ohashi, N., Inutsuka, S.-I., Herbst, E. & Takakuwa,S. 2001, ApJ, 552, 639
Adams, F.C., Lada, C.J., & Shu, F.H. 1987, ApJ, 312, 788
Bergin, E.A., & Langer, W.D., 1997, ApJ, 486, 316
Bonnor, W.B. 1956, MNRAS, 116, 351
Caselli, P., Walmsley, D.M., Zucconi, A., Tafalla, M., Dore, L., & Myers, P.C., 2002, ApJ, 565, 344
Choi, M., Evans, N.J.II, Gregersen, E.M., & Wang, Y., 1995, ApJ, 448, 742
Ebert, R. 1955, Z. Astrophys., 37, 217
Evans, N.J.II, Rawlings, J.M.C., Shirley, Y., & Mundy, L.G., 2001, ApJ, 557, 193
Lee, J.-E., Evans, N.J.II, Shirley, Y.L., & Tatematsu, K., 2003, ApJ, 583, in press
Li, Z.-Y., Shematovich, V.I., Wiebe, D.S., & Shustov, B.M., 2002, ApJ, 569, 792
Ward-Thompson, D., Scott, P.E., & André, P., 1994, MNRAS, 268, 276
Whitworth, A.P. & Ward-Thompson, D., 2001, ApJ, 547, 317

LARGE-SCALE MAPPING OF THE ρ OPHIUCHI REGION BY *SWAS*

Di Li[1], Paul F. Goldsmith[2,3], and Gary J. Melnick[1]

[1]Harvard-Smithsonian Center for Astrophysics, Cambridge, MA, U.S.A.
[2]Astronomy Department, Cornell University Ithaca, NY, U.S.A.
[3]National Astronomy and Ionosphere Center, Ithaca, NY, U.S.A.

Abstract

We have completed a 3 square degree C I $^3P_1$-$^3P_0$ map of the ρ Ophiuchi regions using *SWAS*. The remarkably stable receiver and back end systems of *SWAS* allow for uniformly calibrated data set on this scale. Combined with ^{12}CO, ^{13}CO, and C^{18}O maps made using FCRAO, this data set will facilitate a thorough study of the physics and chemistry in this nearby star-forming region.

Key words: Radio lines: ISM – ISM: individual (ρ Ophiuchi)

1. The Ophiuchi Region

At only 125 pc from the Earth (de Geus 1989), the ρ Ophiuchi molecular cloud is one of the closest regions of active star formation. This region has been the focus of numerous observational investigations, as witnessed by mapping efforts by Wilking & Lada (1983) and Loren et al. (1989), to give only some references with extensive mapping data. At the same time, it was recognized to be an interesting area for studying stars at the earliest phases of their evolution, and also properties of interstellar dust (Vrba, Strom, & Strom 1976). More recently, there has been a reawakening of interest, as witnessed by *ISO* observations in the infrared (Abergel et al. 1996; Liseau et al. 1999).

Its overall size of approximately 15 pc and its proximity present us a unique opportunity for studying both large scale structure and resolved dense cores. Our goal is to map the region in C I and three CO isotopologues (CO, ^{13}CO, C^{18}O). The optically thick J=1–0 line of ^{12}CO provides information on the kinetic temperature of clouds, and thus sheds light into the cloud thermal balance. The lines with modest opacity such as C I fine structure line ($^3P_1$-$^3P_0$) and ^{13}CO J=1–0, are ideal for studying the large-scale structure, particularly the effects of nearby ionizing sources, such as HD 147889. The largely optically thin tracer of C^{18}O 1–0 should provide a relatively complete census of cores from scale sizes of 1 pc to 0.03 pc.

We focus on the C I mapping of the ρ Ophiuchi region in this discussion.

2. Observations

The Submillimeter Astronomy Wave Satellite (*SWAS*) is particularly suitable to carry out large-scale mapping in C I . At 492 GHz, its 54×68 cm antenna gives a relatively large beam of $3.5'\times5.0'$ full width half-maximum (FWHM). The off-axis Cassegrain design provides a 90% main beam efficiency, minimizing the sidelobe pickup. Although C I can be observed from ground-based telescopes, the opacity and instability of the atmosphere at this frequency make it hard to perform position switching with large throws. Finally, the acousto-optical spectrometer (AOS) on board *SWAS* has proven to be extremely stable. This is important for maintaining high quality of the overall calibration for a data set taken over a time span of a couple of years.

From February 1999 to March 2002, *SWAS* has taken data toward 4345 individual positions with $1.6'$ spacing. The Nyquist sampled maps cover a region of 3 square degrees including the ρ Oph A cloud and Lynds 1689. The data are convolved with a Gaussian beam of $4.25'$ FWHM to mimic the true instrumental spatial resolution. The AOS channels has a nearly Gaussian response with FWHM channel width = 1.25 MHz, corresponding to a velocity resolution of 0.76 kms^{-1}. The smoothed data have a representative RMS noise equal to 0.2 K per channel and most of the spectra are spectrally resolved (Fig. 1).

Maps using the J=1–0 transitions of ^{12}CO, ^{13}CO, and C^{18}O were made at the Five College Radio Astronomy Observatory (FCRAO) during roughly the same period as the *SWAS* observations. ^{12}CO and ^{13}CO maps cover the same 3 square degree region. The C^{18}O data were only

Chemistry as a Diagnostic of Star Formation (2003). *Edited by Charles L. Curry and Michel Fich.* Published in 2003 by NRC Research Press, Ottawa ON, K1A 0R6, Canada. ISBN 0-660-19089-3.

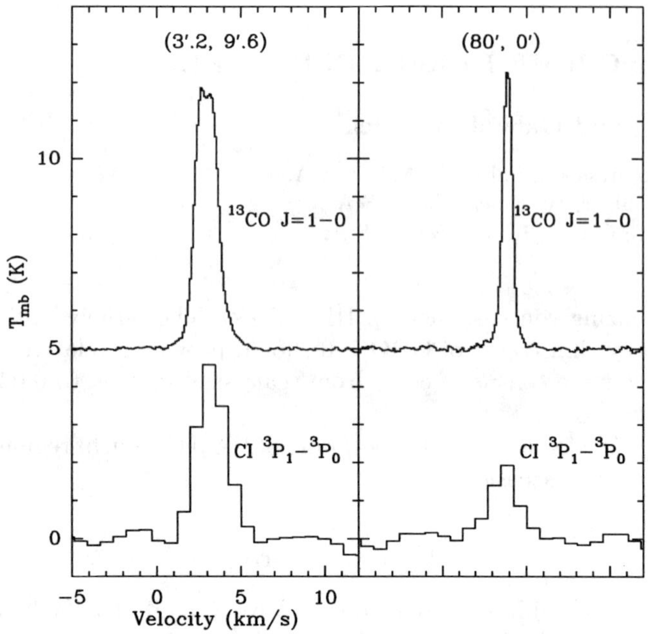

Figure 1. The C I and ^{13}CO spectra from the maps of the ρ Ophiuchi clouds. The map center is at ($16^h26^m23.4^s$, -24°23′02″) (J2000). Given in the top of each box is the angular offset (ΔRA, ΔDEC). All spectra have been convolved with a Gaussian beam of 4.25′ FWHM and corrected for the main beam efficiencies of the respective telescopes. The ^{13}CO line intensities have been reduced by a factor of 2 in order to be shown on the same scale with C I .

taken toward selected areas with high column density. The details of these data will be discussed elsewhere.

3. Cloud Temperature and C I Column Density

For a spectral line with modest opacity like C I , the column density derived from the line integrated intensity can be parameterized as

$$N_{C\,i} = N_1 F_\tau F_b F_u ¿¿, \quad (1)$$

where

$$N_1\ (\mathrm{cm}^{-2}) = 5.94 \times 10^{15} \int T_{\mathrm{mb}} dV\ (\mathrm{km\,s^{-1}}) \quad (2)$$

is the column density in the ^{3}P$_1$ level (Frerking et al. 1989),

$$F_\tau = \frac{\int \tau dv}{\int (1 - e^{-\tau}) dv} \quad (3)$$

accounts for the line opacity,

$$F_b = \left(1 - \frac{e^{\frac{h\nu}{kT_x}} - 1}{e^{\frac{h\nu}{kT_{\mathrm{bg}}}} - 1} \right)^{-1} \quad (4)$$

accounts for the non-zero background temperature T_b, and

$$F_u = \frac{1}{3} e^{23.6/T_x} + 1 + \frac{5}{3} e^{-38.8/T_x} \quad (5)$$

converts the population in the ^{3}P$_1$ level to the total population.

Both F_b and F_u diverge for excitation temperatures $T_x < 5$ and remains essentially constant for $T_x > 20$. For an average density of $n \sim 10^4$ cm^{-3} and a nominal carbon abundance [C^0]/[CO] ~ 0.1, the C I ^{3}P$_1$– ^{3}P$_0$ line is close to being thermalized based on Large Velocity Gradient calculations. The opacity of this line can then be estimated using the antenna temperature and the gas temperature under these conditions. It is, therefore, important to determine the gas temperature.

For cloud surfaces, the ^{12}CO 1–0 antenna temperatures provides a measure of the gas temperature due to its large opacity. The temperatures thus obtained show a relatively smooth distribution in the range 20 to 25 Kelvins for the bulk part of the clouds. Patches of temperature enhancement (30–40 K) appear around the center of ρ Oph A and along some cloud edges of both ρ Oph A and L1689. A comprehensive explanation for the temperature structure should include both internal and external heating. Particularly, the heating of the cloud edges, if proven exclusively due to external heating, could provide us a measure of the UV enhancement (Li & Goldsmith 2002). The change in the UV field must be tied with the gas density and with the [C^0]/[CO] ratio in a consistent manner in developing a completed physical and chemical model of the region.

Overall, the physical conditions of ρ Oph clouds produce thermalized C I emission at relatively warm temperatures ($T_x > 20$ K), which restricts the variation in the combined factor $F_u F_b$ to be no more than 5%. One major uncertainty in the derived $N(C\,i\,)$ lies in the significant opacity close to cloud centers, where the C I line may become optically thick and the CO line shows self-absorption. For cloud edges, the CO line may not be thick enough to allow an accurate determination of gas temperatures. At the cloud edges, the density also drops so that the C I line may no longer be thermalized.

The derived C I column density shows a striking correlation in morphology with that of the ^{13}CO (Fig. 2). In low extinction regions, ring-like structures around (120′, 0′) and (–10′, –15′′) are seen both in ^{13}CO and in C I . In high extinction regions, C I and ^{13}CO peak at the same locations both in L1689 and ρ Oph A.

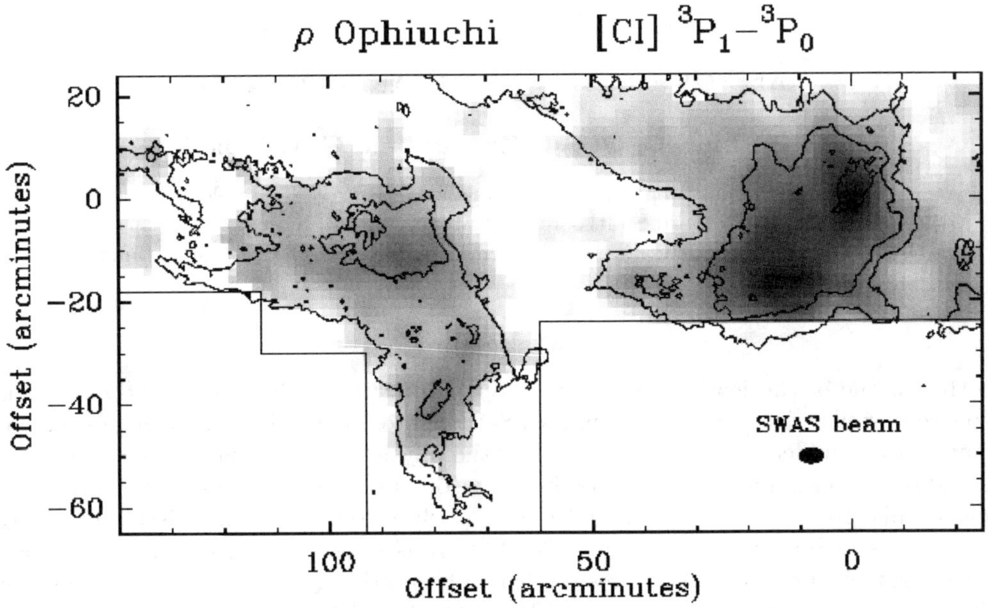

Figure 2. Image of derived C I column density overlaid with the ^{13}CO column density. The ^{13}CO contour levels are 9×10^{15}, 4×10^{16}, and 1.5×10^{17} cm^{-2}. The ρ Oph A cloud is around (0′, 0′) and L1689 is to the left. The peak C I column density is 2.8×10^{17} cm^{-2}.

4. Discussion

According to van Dishoeck and Black (1988), the fractional abundance of C I increases from very low visual extinction (A_V) and to a peak at about $A_V = 2$ for an average interstellar radiation field. C I disappears in highly extincted regions before A_V reaches 10. Apparently, a large portion of the clouds are darker than $A_V = 10$ with the central extinction of ρ Oph A being a couple of hundred. If only a skin of C I is seen toward the lines of sight with large A_V, we would expect a more uniform image of C I column density. Similar phenomena are also observed in other regions, particularly those with stronger UV fields, such as M17 (Howe et al. 2000) and Orion (Plume et al. 2000). To explain the presence of C I in high extinction regions, suggestions have been made in terms of clumpy structure and/or dynamics, which allow UV to penetrate further. We will study the spatial correlation scales of C I and CO, which could test the plausibility of these two types of models

As whole, C I and CO data form a valuable data set, which allow us to pursue a consistent picture of the large-scale structure and chemistry of the ρ Ophiuchi region.

Acknowledgements

This work was supported by NASA's *SWAS* contract NAS5-30702. The Five College Radio Astronomy Observatory is supported by NSF grant AST 97-25951. The National Astronomy and Ionosphere Center is operated by Cornell University under a Cooperative Agreement with the National Science Foundation.

References

Abergel, A., et al. 1996, A&A, 315, L329
de Geus, E., de Zeeuw, P., & Lub, J. 1989, A&A, 216, 44
Frerking, M. A., Keene, J., Blake, G. A., & Phillips, T. G. 1989, ApJ, 344, 311
Howe, J. et al. 2000, ApJ, 539, L137
Kenyon, S.J., Lada, E.A., & Barsony, M. 1998, ApJ, 115, 252
Li, D., & Goldsmith, P.F. 2002, ApJ, submitted
Liseau, R., et al. 1999, A&A, 344, 342
Loren, R.B. 1989, ApJ, 338, 902
Motte, F., André, P., & Neri, R. 1998, A&A, 336, 150
Plume, R. et al. 2000, ApJ, 539, L133
Wilking, B.A. & Lada, C.J. 1983, ApJ, 274, 698
van Dishoeck, E. F. , & Black, J. H. 1988, ApJ, 334, 771
Vrba, F.J., Strom, S.E., & Strom, K.M. 1976, AJ, 81, 317

AN "OVRO-VIEW" OF IRAS 20126+4104

S.-Y. Liu[1], C. Qi[2], Q. Zhang[2], and G.A. Blake[3]

[1]Astronomy Department, California Institute of Technology
[2]Harvard Smithsonian Center for Astrophysics
[3]Divisions of Geological & Planetary Sciences, California Institute of Technology

Abstract

IRAS 20126+4104 is one of the few massive young stellar objects which were recently found to possess dusty disks. Contrary to circumstellar disks around low-mass YSOs, where observations with single-element telescopes and interferometers have made significant progress in revealing the physical and chemical structures in these sources, the existence of disks and their relation with the surrounding envelope in massive YSO systems remain relatively elusive. Emission from species like HCN, CN, N_2H^+, and CH_3OH is often found to trace distinctly different components in both low and high-mass YSOs, suggesting their distribution as sensitive measures of photochemical and depletion/desorption processes. We present here the results from observations, carried out with the Owens Valley Radio Observatory Millimeter Array, of these species toward IRAS 20126+4104.

Key words: stars: formation – dust, extinction – ISM: molecules

Chemistry as a Diagnostic of Star Formation (2003). *Edited by Charles L. Curry and Michel Fich.* Published in 2003 by NRC Research Press, Ottawa ON, K1A 0R6, Canada. ISBN 0-660-19089-3.

DOUBLY DEUTERATED FORMALDEHYDE IN PROTOSTELLAR SOURCES

Laurent Loinard[1], A. Castets[2], C. Ceccarelli[2,3], B. Lefloch[3], Ch. Vastel[4,5], J.-J. Benayoun[3], E. Caux[5], A.G.G.M. Tielens[6], and E. Dartois[7]

[1]Instituto de Astronomía, Universidad Nacional Autónoma de México, Apartado Postal 72–3 (Xangari), 58089 Morelia, Michoacán, México
[2]Observatoire de l'Université de Bordeaux I, B.P. 89, F-33270 Floirac, France
[3]Laboratoire d'Astrophysique, Observatoire de Grenoble, B.P. 53, F-38041 Grenoble Cedex 09, France
[4]California Institute of Technology, Downs Laboratory of Physics 320-47, Pasadena, CA 91125, USA
[5]Centre d'Étude Spatiale des Rayonnements, Centre National de la Recherche Scientifique, Université Paul Sabatier, B.P. 4346, 31028 Toulouse Cedex 04, France
[6]Space Research Organization of the Netherlands, P.O. Box 800, 9700 AV Groningen, The Netherlands
[7]Institut d'Astrophysique Spatiale, Bat.121, Campus d'Orsay, F-91405 Orsay, France

Abstract

We report the results of a search for doubly deuterated formaldehyde (D_2CO) in a large sample of objects associated with low-mass star formation. D_2CO was detected towards the 16 low luminosity protostars (Class 0 and Class I) in our sample, with relative abundance ratios ([D_2CO]/[H_2CO]) of 5-40%. In contrast, it was not detected towards any of the high-luminosity objects observed, with upper limits on the [D_2CO]/[H_2CO] abundance ratio of about 0.2%. D_2CO was not detected either towards any of the more evolved Class II sources (classical T–Tauri stars: CTTS) in the sample, nor towards any of the shocks associated to the outflows powered by extremely young protostars.

Key words: ISM: abundances – ISM: molecules – astrochemistry – stars: formation – stars: winds, outflows

1. Introduction

The mono–deuterated forms of most hydrogen-bearing molecules have long been known to be significantly more abundant in space than the cosmic abundance of deuterium would suggest. For instance, [DCO^+]/[HCO^+] abundance ratios of the order of 10% are commonly found in cold molecular clouds (Guélin et al. 1977), and similar values for the deuteration levels of neutral molecular species (water, ammonia, formaldehyde, methanol, etc.) have been reported in protostars (van Dishoeck et al. 1995), and cold molecular cloud cores (Tiné et al. 2000, and references therein). It has recently been found that even stronger relative fractionation could be obtained for multiple deuteration (Turner 1990; Ceccarelli et al. 1998; Loinard et al. 2000, 2001; Roueff et al. 2000; Parise et al. 2002; Lis et al. 2002; van der Tak et al. 2002; Bacmann et al. 2003). For doubly or triply deuterated species, the relative enhancement can reach more than ten orders of magnitude.

Such a large fractionation is challenging for the chemistry models. Though it is fairly clear that the fractionation is initiated by the deuteration of the molecular ion H_3^+, and that the deuteration occurs in cold, dense cores, the exact processes remain unknown. To reach the observed levels of deuteration, the active grain surface chemistry schemes (Tielens et al. 1983) require atomic [D]/[H] ratios larger than inferred from the observations, while ion-molecule gas-phase scenarios (Roberts & Millar 2000a,b; Rodgers & Charnley 2001) fail to produce the relative observed isotopic ratios unless somewhat *ad hoc* prescriptions on the reaction speeds involving neutrons are included (van der Tak et al. 2002). Further constraints will only be obtained from systematic measurements of the isotopic ratios in larger samples of sources. As part of such an effort, we present here a search for D_2CO emission in a large sample of objects associated with star formation. A preliminary analysis of part of the present data was published in Loinard et al. (2002), and the full survey will be presented in Loinard et al. (2003, in preparation).

2. Data

The data reported here were obtained towards the sample of 34 objects associated with star–formation listed in Table 1. Including IRAS16293–2422 and 16293E previously observed (Loinard et al. 2000, 2001), it gathers 15 *bona fide* low-luminosity (L_{bol} = 0.1–20 $L_\odot$) Class 0 protostellar sources, 3 low–luminosity Class I objects, and 3 low–luminosity Class II sources. It is worth noting that our sample contains almost half of the total number of known

Chemistry as a Diagnostic of Star Formation (2003). *Edited by Charles L. Curry and Michel Fich.* Published in 2003 by NRC Research Press, Ottawa ON, K1A 0R6, Canada. ISBN 0-660-19089-3.

low–luminosity Class 0 sources (André et al. 2000), and covers the complete range of intrinsic properties (L_{bol}, L_{smm}, etc.) exhibited by those sources. It can thus be considered as a fully representative sample of those kind of sources. Because they live longer, there are many more Class I and Class II sources known than Class 0. Consequently, our sample of Class I and Class II sources is far from representative.

It has been noticed before (Loinard et al. 2002) that the $[D_2CO]/[H_2CO]$ ratio found in low-luminosity protostars (Loinard et al. 2000, 2001) is 1 to 2 orders of magnitude larger than that found in the high-mass star-forming site Orion KL (Turner et al. 1990). To decide whether this difference between low-luminosity and high-luminosity objects was statistically significant, we also included 6 high-luminosity protostars ($L > 5000\ L_\odot$), and 2 intermediate-luminosity protostars ($100 < L < 5000\ L_\odot$) in our sample.

The two schemes that can produce high levels of deuteration (active grain surface chemistry and gas-phase chemistry in a medium strongly depleted in CO) require that the deuteration occurred during the pre-stellar phase, and that the material enriched in deuterium is only presently released into the gaseous phase. While thermal evaporation is most likely to be at work in the protostellar envelopes themselves (Ceccarelli et al. 2001), sputtering or partial destruction of the grains by fast shocks (e.g. Schilke et al. 1997), in particular those associated with the molecular outflows powered by protostars, are known to also be efficient at injecting dust grain material in the gas phase. Our sample thus also includes 5 shock positions associated with molecular outflows powered by very young low-luminosity protostars.

The observations were performed on November 5–9, 1999, November 22–24, 1999, and March 21–26, 2001 with the IRAM 30–m telescope located near the summit of Pico Veleta near Granada in Southern Spain; and on October 4–5, 1999 with the 15-m Swedish ESO Submillimeter Telescope (SEST) located at La Silla in Chile. Details on the observing procedures can be found in Loinard et al. (2002, and in prep.).

To estimate the $[D_2CO]/[H_2CO]$ abundance ratio, spectral lines of both species were searched. The lines (not all of which have been searched for in all sources) correspond to the $2_{1,2}$–$1_{1,1}$, $3_{0,3}$–$2_{0,2}$, $3_{1,3}$–$2_{1,2}$, $4_{1,4}$–$3_{1,3}$, and $3_{2,2}$–$2_{2,1}$ transitions of H_2CO; and to the $2_{1,2}$–$1_{1,1}$, $3_{1,3}$–$2_{1,2}$, $4_{0,4}$–$3_{0,3}$, $5_{1,5}$–$4_{1,4}$, and $3_{2,2}$–$2_{2,1}$ transitions of D_2CO. They were chosen to provide a large span of upper level energies (from E_u = 10 to 70 K) to make the estimation of column densities possible with rotational diagrams (see below). In order to estimate the opacities involved, we also observed the $4_{1,4}$–$3_{1,3}$ transition of the rarer isotope $H_2{}^{13}CO$ of formaldehyde in three of our sources: IRAS4A, L1448–N, and HH211.

Table 1. $[D_2CO]/[H_2CO]$ ratio in all the sources of our sample

Name	D pc	Type	$[D_2CO]/[H_2CO]$ %
L 1448–C	300	Class 0	8.8
L 1448–N	300	Class 0	2.8
NGC1333/IRAS2	350	Class 0	8.1
NGC1333/IRAS4A	350	Class 0	7.3
NGC1333/IRAS4B	350	Class 0	4.6
IRAS 03282+3035	300	Class 0	14.7
IC348/HH211–mm	300	Class 0	11.9
IRAS16293–2422	160	Class 0	5.0
16293E	160	Class 0	40.0
VLA1623	160	Class 0	16.7
IRAM04191	140	Class 0	9.8
L1527	140	Class 0/I	11.7
L1157–mm	440	Class 0	6.2
BHR71–mm	200	Class 0	7.2
NGC2264G–VLA2	800	Class 0	37.6
NGC1333/IRAS6	350	Class I	4.7
NGC1333/IRAS7	350	Class I	5.9
NGC1333/SVS13	350	Class I	3.8
IC1396–N	740	Interm.	2.1
NGC 7129–FIRS2	1000	Interm.	< 2.7
NGC 7538–IRS1	2800	Massive	< 0.2
NGC 7538–IRS9	2800	Massive	< 0.4
W3(OH)–TW	2200	Massive	< 0.4
W3–IRS5	2200	Massive	< 0.4
S140–IRS1	910	Massive	< 0.3
RAFGL7009S	3100	Massive	< 0.5
DGTau	140	Class II	< 7.4
HLTau	140	Class II	—
GGTau	140	Class II	—
L 1448–C(R3)	300	Outflow	—
L 1448–C(R4)	300	Outflow	—
NGC1333/IRAS2(R2)	350	Outflow	< 0.7
IRAS16293–2422(W1)	160	Outflow	< 0.9
L1157(S2)	440	Outflow	< 0.7

To derive the column densities of D_2CO and H_2CO from the observed line intensities, we use rotational diagrams. The method is explained in detail in Goldsmith & Langer (1999), and its application to formaldehyde lines was summarized by Loinard et al. (2000). It rests on two assumptions: first that the lines are optically thin, and second that the medium is in local thermodynamic equilibrium (LTE). While the second hypothesis is probably roughly valid here, the first one might be violated for the H_2CO lines, particularly in the more massive sources. In the low–luminosity source IRAS 16293–2422, Loinard et al. (2000) showed that the opacity of the H_2CO transitions was only very moderate using observations of the less abundant isotopic form $H_2{}^{13}CO$. The situation is even more favourable in the three low–luminosity sources where $H_2{}^{13}CO$ has been observed here, since we find a $^{12}C/^{13}C$ intensity ratio is between 40 and 70. Assuming an intrinsic $^{12}C/^{13}C$ abundance ratio of 60, this implies that the determinations based on the $H_2{}^{12}CO$ observations underestimate the H_2CO column densities by at most 30%.

3. Results and Conclusions

All 15 low-luminosity Class 0 sources and 3 low-luminosity Class I sources in our sample have been detected in D_2CO. The abundance ratio $[D_2CO]/[H_2CO]$ in these protostars varies from about 3 to 40%. On the other hand, no D_2CO was detected in any of the 3 Class II sources. However, in two of the three, H_2CO is not detected either, so the estimation of abundance ratio is only possible in one Class II source, yielding $[D_2CO]/[H_2CO] < 7\%$.

The H_2CO spectral line profiles obtained towards the Class 0 and Class I sources fall in two categories. In most cases, they appear to be well fitted by gaussians, but in a few instances, they exhibit strong wings, a result of contamination by the associated outflows. It is found that the D_2CO line profiles are never affected by that outflow contamination. This already suggests that less D_2CO emission is associated with the shocks along the molecular outflows than with the protostellar sources themselves. The most prominent shock features in outflows are, however, usually not located in the close vicinity of the central protostars, but some tens of arc-seconds away, towards the terminal bowshocks. At those positions, the H_2CO line profiles are strongly dominated by emission associated with the outflows, but no D_2CO emission is found at the velocity of the shocked material. The upper limit on the value of the $[D_2CO]/[H_2CO]$ abundance ratio in shocks is found to be about 0.5–1%. Although this still leaves ample room for some over–abundance compared to the cosmic abundance of deuterium, it is at least one order of magnitude lower than in the corresponding powering sources. Both gas-phase and active grain surface chemistry schemes predict that large levels of double deuteration can only be attained in cold and dense media. The lack of D_2CO detection in the shocks is quite naturally explained in this scheme, because the material found at high velocities in molecular outflows mostly corresponds to previously ambient gas that never went through the cold and dense phase required for large enrichments in D_2CO to occur.

While D_2CO was clearly detected in the intermediate luminosity source IC1396N (with a $[D_2CO]/[H_2CO]$ ratio of 2%), but only marginally in the other intermediate luminosity object (NGC7129–FIRS2; $[D_2CO]/[H_2CO] \sim 2\%$). None of the higher luminosity sources was detected in D_2CO, with $[D_2CO]/[H_2CO]$ typically less than 0.2 to 0.5 %. Again, while this leaves ample room for some over-abundance compared to the cosmic abundance of deuterium, it is one to two orders of magnitude lower that in lower-mass protostars. This difference between low- and high-luminosity protostars might be a result of a different temperature during the pre-stellar phase.

References

Bacmann, A., Ceccarelli, C., Lefloch, B., Steinacker, J., Castets, A., & Loinard, L. 2003, A&A, in press
Ceccarelli, C., Castets, A., Loinard, L., Caux, E., & Tielens, A.G.G.M. 1998, A&A, 338, L43
van Dishoeck, E.F., Blake, G.A., Jansen, D.J., & Groesbeck, T.D. 1995, ApJ, 447, 760
Goldsmith, P.F., & Langer, W.D. 1999, ApJ, 517, 209
Guélin, M., Langer, W.D., Snell, R.L., & Wootten, H.A. 1977, ApJ, 217, 165
Lis, D.C., Roueff, E., Gerin, M., et al. 2002, ApJ, 571, L55.
Loinard, L., Castets, A., Ceccarelli, C., Caux, E., & Tielens, A.G.G.M. 2001, ApJL, 552, L163
Loinard, L., Castets, A., Ceccarelli, C., et al. 2002, Planetary & Space Science, 50, 1205
Loinard, L., Ceccarelli, C., Castets, A., Tielens, A.G.G.M., Faure, A., Caux, E., & Duvert, G. 2000, A&A, 357, 9
Roberts, H., & Millar, T.J. 2000a, A&A, 361, 388
Roberts, H., & Millar, T.J. 2000b, A&A, 364, 780
Rodgers, S.D., & Charnley, S.B. 2001, ApJ, in press
Roueff, E., Tiné, S., Coudert, L.H., Pineau des Forêts, G., Falgarone, E., & Gérin, M. 2000, A&A, 354, L63
Schilke, P., Walmsley, C.M., Pineau des Forêts, G., & Flower. D.R. 1997, A&A 321, 293
van der Tak, F.F.S., Schilke, P., Müller, H.S.P. et al. 2002, A&A, 388, L53
Tielens, A.G.G.M. 1983, A&A, 119, 177
Tiné, S., Roueff, E., Falgarone, E., Gérin, M., & Pineau des Forêts, G. 2000, A&A, 356, 1039
Turner, B.E. 1990, ApJ, 362, L29

ENERGY TRANSFER IN MOLECULAR HYDROGEN: DOES THE COLLIDER MATTER?

Margot E. Mandy

Program in Chemistry, University of Northern British Columbia, Prince George, BC, Canada V2N 4Z9

Abstract

An understanding of collisional energy transfer processes in molecular hydrogen is essential to the understanding of the relaxation of interstellar shocks and the associated quadrupole emission spectrum. Accurate ab initio potentials exist for the interaction of molecular hydrogen with each of atomic hydrogen, atomic helium, and molecular hydrogen. Using the method of quasi-classical trajectories, energy transfer in molecular hydrogen as the result of each of these colliders has been examined. The similarities and differences cannot simply be accounted for by whether the collider is open or closed shelled. These similarities and differences are discussed in terms of the role of certain features of the interaction potential. In the case of molecular hydrogen as the collider, the role of internal energy in a collider is elucidated, especially with regard to energetic and dynamic thresholds to energy transfer and dissociation. Comparison of quantum and classical calculations is also considered. The shortcomings and adequacies of earlier approximations are discussed.

Key words: ISM: molecules – ISM: hydrogen

1. Introduction

Molecular hydrogen is the most abundant molecule in the interstellar medium. Energy transfer in molecular hydrogen affects the evolution of shock fronts in the interstellar medium. Shocks in the interstellar medium are linked to star formation. In these regimes, the density is low enough that competition of quadrupole radiative emission with the collisional processes must be explicitly considered. Under the conditions of the interstellar medium, a hydrogen molecule is most likely to be hit by an H atom, a He atom, or another hydrogen molecule.

Each of these colliders has distinct features. The H atom is a light collider with one unpaired electron. Since it is open shelled it has the possibility of undergoing a reactive interaction with molecular hydrogen. In contrast the He atom is a closed shelled species with two paired electrons. It can cause only nonreactive energy transfer or dissociation of the target molecule. Molecular hydrogen is also a light collider with a closed shell electronic conformation, but as a diatomic, it is nonspherical and there is the possibility that internal vibrational and rotational energy of the collider may affect the energy transfer behaviour in the target molecule.

Chemists and physicists interested in reaction dynamics have carried out extensive calculations on the systems $\{H,He,H_2\}$ + H_2. The three systems are real systems with light nuclei and few electrons which render them amenable to calculations that are tractable with a high level of precision. Since light atoms are involved quantum effects are potentially important and the 348 (v,j) energy levels of H_2 (Fig. 1) cannot be treated as a continuum. Chemical accurate potential energy surfaces (PESs) are available (Boothroyd et al. 1996; Wilson et al. 1974; Boothroyd et al. 2002) and insights gained will lead to a better understanding of other systems.

In Figure 1 of Mandy et al. (1998) were shown collinear cuts of the potential energy surfaces for He + H_2, H + H_2, and H_2 + H_2. The qualitative similarity of the He + H_2 and H_2 + H_2 PESs was noted when the H_2 collider was at its equilibrium situation. In contrast, when the H_2 collider was vibrationally excited and at its outer turning point, the H_2 + H_2 PES was qualitatively similar to the H + H_2 PES. On this basis, it may be argued that the energy transfer behaviour due to collisions with H_2 is expected to differ from that due to He or H.

2. Method

Calculations were carried out with the quasi-classical trajectory (QCT) method, which is tractable in situations where quantum calculations are not. QCT does not increase in difficulty with energy of the system making it possible to consider all 348 (v,j) states of H_2. With a

Chemistry as a Diagnostic of Star Formation (2003). *Edited by Charles L. Curry and Michel Fich.* Published in 2003 by NRC Research Press, Ottawa ON, K1A 0R6, Canada. ISBN 0-660-19089-3.

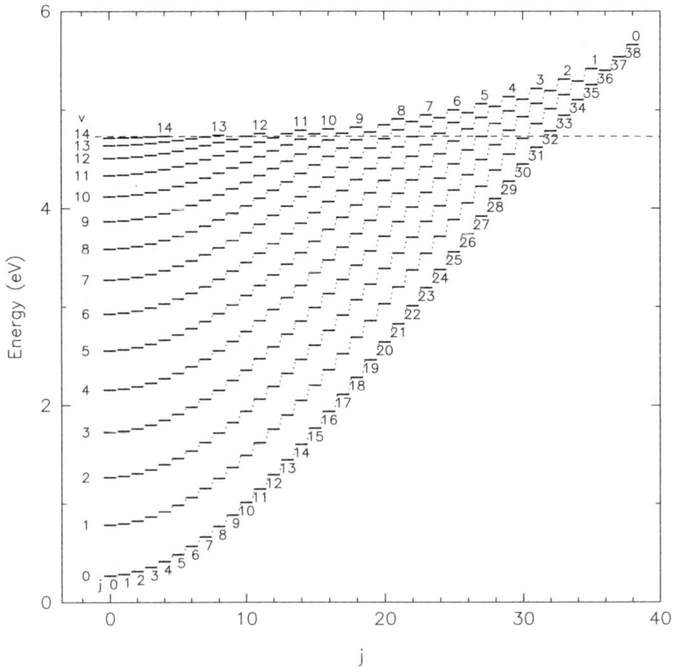

Figure 1. (v, j) levels of the ground state hydrogen molecule, $H_2(\Sigma_g^+)$. The dotted line is the minimum energy required for dissociation. Energy levels above this line are quasi-bound.

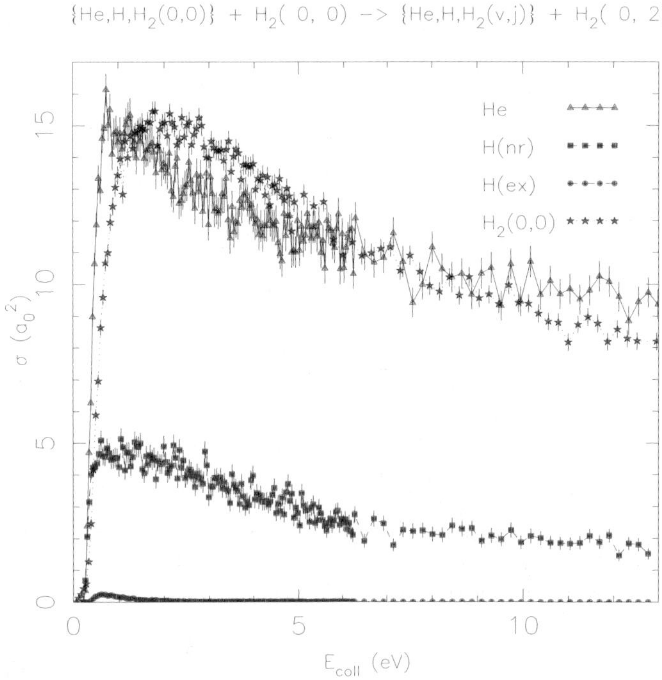

Figure 2. Integral cross sections for rotational excitation from $H_2(0{,}0) \to H_2(0{,}2)$.

good potential energy surface, chemical accuracy is possible. The calculations were carried out as described in Mandy et al. (1998). In that study (see Fig. 2, 3, and 4 therein) we found significant differences in energy transfer behaviour with the different colliders. The H atom was the most efficient at causing large changes in (v, j), while the He atom led to rotational excitation or interchange of vibrational and rotational energy. The energy transfer due to H_2 in low (v, j) states was dampened compared to that due to He. If the H_2 molecular collider was excited, some large changes in (v, j), such as are typical of an open-shell atomic collider, were observed. We concluded that results from He + H_2 cannot be readily extrapolated to approximate results from H_2 + H_2 and thus a full survey of energy transfer in H_2 + H_2 is required.

3. Results and Discussion

In Figure 2 are shown cross sections for rotational excitation from $H_2(0{,}0) \to H_2(0{,}2)$ as the result of collisions with H, He, or $H_2(0{,}0)$. The cross sections for the latter two colliders are similar with those for He rising more rapidly from threshold and are considerably larger than either the reactive or nonreactive for the atomic H collider.

In the case of vibrational excitation from $H_2(0{,}0) \to H_2(1{,}0)$ (Fig. 3), it is the nonreactive collisions with H atoms that give the largest cross sections while reactive collisions yield small cross sections. The cross sections for He or H_2 as colliders are similar at higher energies but there is a substantial difference in the threshold behaviour. There is appreciable dynamic elevation of the threshold in the case of He while the threshold for $H_2(0{,}0)$ as a collider is not elevated significantly above the energetic requirement. Since the threshold region makes the dominant contribution to the corresponding rate coefficients, this means that rate coefficients for $H_2(0{,}0)$ as a collider will be considerably larger than those for He.

Cross sections for collisional dissociation from $H_2(0{,}0)$ are shown in Figure 4. The H atom is the most efficient collider for dissociation of $H_2(0{,}0)$. The $H_2(0{,}0)$ molecule as a collider, while less efficient that atomic H, is much more efficient than He, since there is no dynamic elevation of the threshold. Further elaboration of this may be found in Martin et al. (1998).

These three representative cases indicate clearly that energy transfer behaviour in H_2 due to collisions H_2 is distinct from that due to H or He. Caution must be exercised in extrapolating data for He + H_2 or H + H_2 to H_2 + H_2.

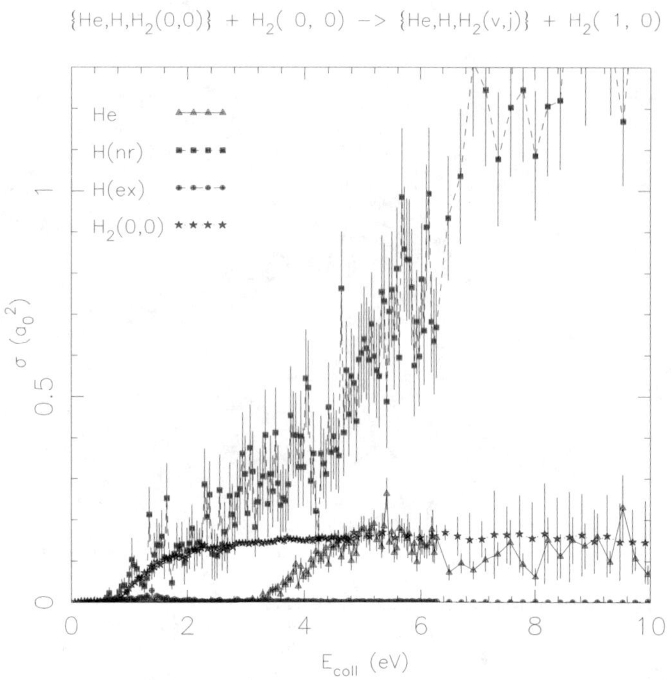

Figure 3. Integral cross sections for vibrational excitation from $H_2(0,0) \rightarrow H_2(1,0)$.

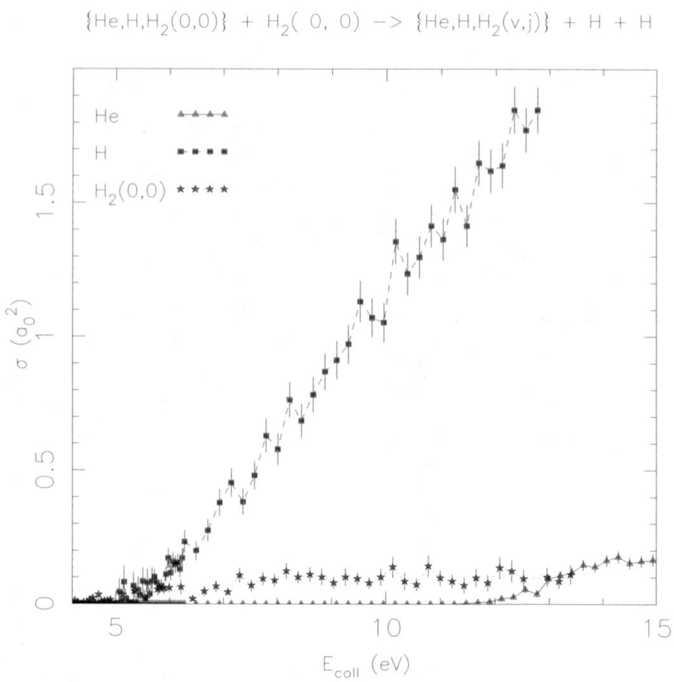

Figure 4. Integral cross sections for dissociation from $H_2(0,0)$.

4. Future Directions

Quasi-classical calculations are in progress to examine energy transfer for all $H_2(v,j)$ $+H_2(v',j')$ with the objective of calculating a complete set of state-to-state rate coefficients. In addition, in collaboration with D. C. Clary and S. K. Pogrebnya, (Pogrebnya et al. 2002; Mandy et al., in preparation) full dimensional quantum calculations are underway to benchmark the quasi-classical calculations. Recently, Boothroyd & Martin have calculated a new ab initio potential energy surface for the He+H_2 system which is a significant improvement upon PES used in the reported calculations. A new calculation of state-to-state rate coefficients for He+H_2 is being carried out by Mandy and collaborators.

Acknowledgements

The author thanks the Visitors' Programme of the Canadian Institute for Theoretical Astrophysics and the Centre for Theoretical and Computational Chemistry at University College London. This work was supported by grants from the Natural Sciences and Engineering Research Council of Canada. Some of the calculations were carried out using the high performance computing facility jointly supported by the Canadian Foundation for Innovation, the British Columbia Knowledge Development Fund, and Silicon Graphics at the University of Northern British Columbia.

References

Boothroyd, A. I., Martin, P. G., Keogh, W. J., & Peterson M. J., 2002, J. Chem. Phys. , 116, 666

Boothroyd, A. I., Keogh, W. J., Martin, P. G.,& Peterson M. J., 1996, J. Chem. Phys. , 104, 7139

Mandy, M. E., Martin, P. G., & Keogh, W. J. 1998, J. Chem. Phys. , 108, 492

Martin, P. G., Keogh, W. J., & Mandy, M. E. 1998, ApJ, 499, 793

Pogrebnya, S. K., Mandy, M., & Clary, D. C. 2002, Int. J. Mass Spectrom. , in press

Wilson, C. W. Jr., Kapral, R., & Burns, G. 1974, Chem. Phys. Lett. , 24, 488

WATER EMISSION IN NGC1333 IRAS 4: PROBING THE SURROUNDING ENVELOPE

S. Maret[1], C. Ceccarelli[2,3], E. Caux[1], A.G.G.M. Tielens[4], and A. Castets[2]

[1]Centre d'Etude Spatiale des Rayonnements, CESR/CNRS-UPS, BP 4346, F-31028 Toulouse Cedex 04, France
[2]Observatoire de Bordeaux, BP 89, F-33270 Floirac, France
[3]Laboratoire d'Astrophysique, Observatoire de Grenoble, B.P. 53, F-38041 Grenoble Cedex 09, France
[4]Space Research Organization of the Netherlands, P.O. Box 800, 9700 AV Groningen, The Netherlands

Abstract

Sun-like stars are born from the collapse of fragments of molecular clouds. During the first embedded phase (the so called Class 0/1), the protostar is surrounded by a collapsing envelope, whose physical and chemical structure would set up the initial conditions of the protostellar disk which may eventually form planets. In this contribution, we show how spectroscopic far infrared to millimeter observations can be a powerful diagnostic to derive the chemical and physical structure of the innermost regions of envelopes collapsing around solar type protostars, where disk are believe to form. Our claim is based on the simultaneous modelling of the dynamics, chemistry and radiative transfer in a protostellar envelope. Here we present a study of a Class 0 protostar NGC1333 IRAS 4, obtained using *ISO*-LWS observations of water far-infrared lines. We show the derived physical and chemical structure of the protostar, namely the density and temperature profiles, as well as the water abundance across the envelope. Besides, the mass of the central protostar and its accretion rate can be constrained by these observations and modelling, yielding an alternative method to measure these two poorly known parameters.

Key words: Stars: formation – circumstellar matter – ISM: abundances – ISM: individual (NGC1333 IRAS 4)

1. Introduction

NGC1333 IRAS 4 is a binary protostellar system. located in the south part of the Perseus cloud. The relatively large mass of the surrounding envelopes, together with the relative submillimeter to bolometric luminosity ratio suggest that both source are highly embedded and probably very young. They have been classified as *Class 0* protostars (André et al. 1993). IRAS4 is associated with two molecular outflows, observed in the CO, CS (Blake et al. 1995) and millimeters transitions (Lefloch et al. 1998). The dynamical age of these outflow is a few thousand of years. In the past years, several studies have focused on the molecular emission of this source. In particular, Blake et al. carried out a study of several molecules in NGC1333 IRAS 4. Their major results were large depletion factors of all molecules in the envelopes, and the presence of a region with an increased abundance of CS, SiO and CH_3OH that the authors attribute to grain mantle desorption caused by grain-grain collisions induced by the powerful outflow originating from this source. More recently, Di Francesco et al. (2001) reported the detection of inverse P-Cygni profiles of several lines toward NGC1333 IRAS 4A, providing the least ambiguous detection of infall motion towards a protostar ever. Here we present a study of the far infrared water line spectrum of NGC1333 IRAS 4 observed with the Long Wavelength Spectrometer (Clegg et al. 1996; hereafter LWS) on board *ISO* (Kessler et al. 1996). Using an accurate model of the thermal emission of the envelope, we show how the physical and chemical structure of the protostar can be derived (see also Maret et al. 2002).

2. Observations

A full range spectral survey of (43-196 μm) of NGC1333 IRAS 4 was performed using LWS in the low resolution ($R \sim 200$) grating mode. The observations were obtained on three positions, the first one being centered between NGC1333 IRAS 4A and NGC1333 IRAS 4B, and the two other were taken along the outflow powered by NGC1333 IRAS 4, NW and SW of it respectively. Figure 1 shows the spectra observed on the three positions. We detected fourteen H_2O lines, seven CO lines, three OH lines, the two [O I] and one [C II] fine structure line towards the central position. No significant H_2O, CO and OH emission was detected on the NE and SW positions, whereas the [O I] and [C II] lines were detected towards the three position at the same intensity.

Chemistry as a Diagnostic of Star Formation (2003). *Edited by Charles L. Curry and Michel Fich.* Published in 2003 by NRC Research Press, Ottawa ON, K1A 0R6, Canada. ISBN 0-660-19089-3.

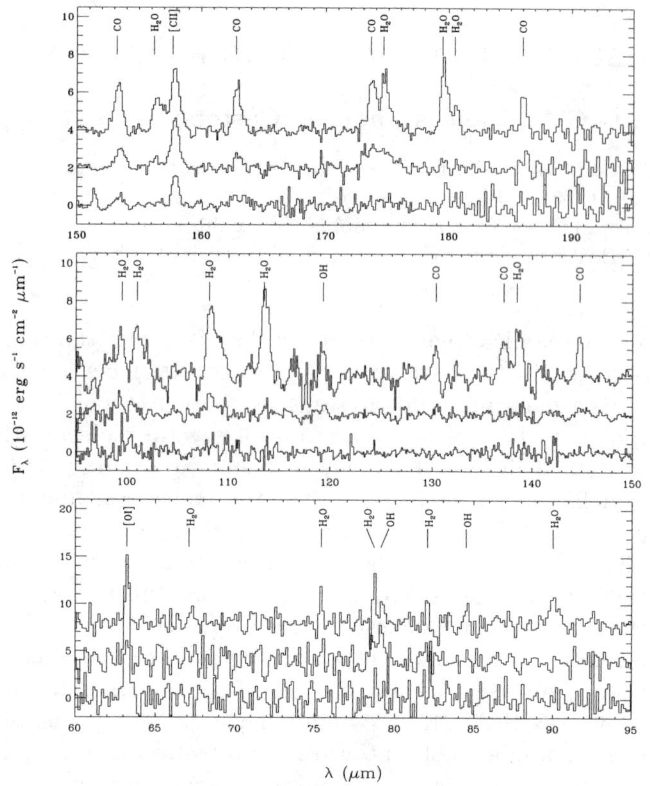

Figure 1. *ISO*-LWS spectra obtained towards NGC1333 IRAS 4A at the central (*top*), NE (*middle*) and SW (*bottom*) positions.

3. Line modelling

The observed emission observed towards NGC1333 IRAS 4 can have a priori three different origin: the outflow powered by this source, the PDR at the surface of the cloud, or the thermal emission of the protostellar envelopes. Origin of molecular emission can be usually determined if a sufficient enough spatial or spectral resolution is achievable. In the case of *ISO*-LWS observations, all the gas components are mixed in the 80" beam of *ISO*. Moreover, the relatively low spectral resolution does not allow to spectrally disentangle these components. The outflows powered by protostars has been invoked in the literature as a possible origin of the water line emission observed towards them (Giannini et al. 2001), but the absence of detection of water emission on the outflow lobes does not play in favor of this hypothesis. Moreover, Ceccarelli et al. (1999) noted the absence of the correlation between the H_2O and SiO emission[1] toward seven protostars, whereas they noted a correlation between the 1.3 mm envelope emission. This suggest that the FIR water line emission can be due to the thermal emission of the envelopes surrounding the two protostars.

In order to test this hypothesis, we modelled the water emission by an accurate model which simultaneously compute the thermal balance, chemistry and radiative transfer in the envelope (Ceccarelli, Hollenbach & Tielens 1996). The model has four free parameters: the central mass of the protostar M_*, its accretion rate $\dot{M}$, and the water abundance across the envelope. In order to take into account the increase of water abundance in the inner parts of the envelope – due to the evaporation of grain mantle when the gas reach the sublimation temperature of water ices – we used a different abundance in the inner ($X_{\rm in}$) and outer parts ($X_{\rm out}$) of the envelope. We run several models, varying the central mass between 0.3 and 1 $M_\odot$, the accretion rate between 10^{-5} and 10^{-4} $M_\odot{\rm yr}^{-1}$, and the water abundance in the outer and inner parts between 10^{-7}–10^{-6} and 10^{-6}–10^{-5} respectively. The best fit parameters were obtained by minimizing the χ^2 between the observations and the model results. The minimum χ^2 value is obtained for $M_*= 0.5\ M_\odot$, $\dot{M}= 5 \times 10^{-5}\ M_\odot{\rm yr}^{-1}$, $X_{\rm out}= 5 \times 10^{-7}$ and $X_{\rm in}= 5\times 10^{-6}$. The model reproduce relatively well the observed water emission, with the exception of two lines which seems to be underestimated by our model.

4. Discussion

The first result of this study is the capability to reproduce the observed water line emission with an envelope thermal emission model. Our model also allowed us to derive the density and temperature profile in the envelope. Figure 2 shows the density and temperature profile derived. The observations are consistent with a Shu (1977) "inside-out" density profile, with a transition region starting at 1500 AU. This profile is in good agreement with the one derived by Jørgensen et al. (2002), using continuum observations. The dust temperature closely track the dust one, with the exception of a small region at $R \sim 50$ AU. This is causes at the radius where the temperature reach 100 K, and the grain mantle evaporates, releasing large amount of water in the gas phase (about a factor ten more), increasing the cooling efficiency of

[1] SiO is usually associated with outflow strong shocks (e.g., Bachiller & Perez Gutierrez 1997) and it is believed to be a product of grain mantle desorption (Caselli et al. 1997; Schilke et al. 1997).

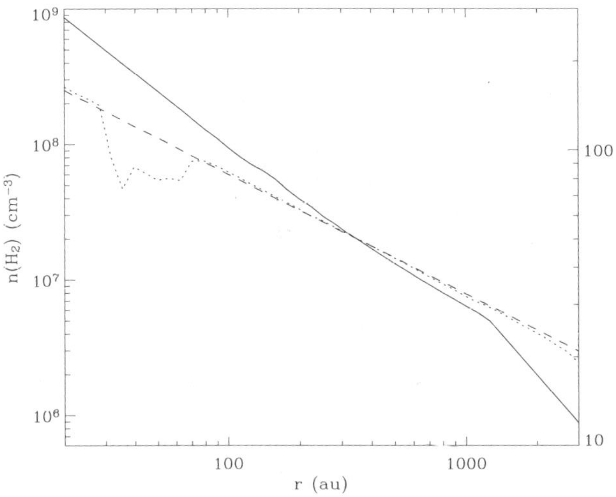

Figure 2. Density and temperature derived from the model. The solid line is the density profile in the envelope, the dashed line is the dust temperature, and the dotted line is the gas temperature.

the gas. This *hot core-like region*, similar to the ones in massive star-forming regions—here indirectly probed by the jump in the water abundance needed to reproduce the observed emission—has been already observed in IRAS 16293-2422, where a similar study was performed (Ceccarelli et al. 2000a,b). This claim was confirmed by a recent study by Schöier et al. (2002).

The central mass of the protostar is also constrained by our model, giving an alternative measure to this poorly-known parameter. We derived a central mass of 0.5 $M_{\odot}$and an accretion rate of 5×10^{-5} $M_{\odot}\mathrm{yr}^{-1}$. This gives an age of 10^4 year, close to the dynamical age of the outflow. This values compares well with the values derived recently by Di Francesco et al. The water abundance we found for the outer envelope is in good agreement with recent *SWAS* observations of the Perseus cloud (Bergin et al. 2002). As inner water abundance is dominated by the grain mantle evaporation, the derived inner abundance should be a indirect measure of the water abundance in the grain mantle. the abundance we derive is somewhat lower than the water ice abundance observed towards massive star forming region (Tielens et al. 1984). Finally the abundances derived for NGC1333 IRAS 4 are in good agreement with the one derived by Ceccarelli et al (2000a) for IRAS 16293-2422. This study show that NGC1333 IRAS 4 is probably younger than IRAS 16293-2422.

5. Conclusion

Using an accurate model of the thermal emission of a protostellar envelope, we successfully modelled the water emission observed with *ISO*-LWS. We derived density and temperature profiles in the envelopes, as well as the water abundance. The water abundance is enhanced by a factor ten in the inner parts, in a hot core-like region. We also derived the central mass of the protostar, which gives us an estimate of its age, if the accretion rate remains constant. The mass, accretion rate, and water abundances compare well with recent studies. This work also emphasis the need of ground based millimeter observation, where higher spatial and spectral resolutions are achievable. From this point of view, H_2CO and CH_3OH are of particular interest because of their large abundances in grain mantle. H_2CO and CH_3OH abundances should therefore be increased in the central parts of envelopes. This is indeed confirmed by very recent observations of these molecules towards several Class 0 protostars (Maret et al. in preparation).

References

André, P. Ward-Thompson, D. & Barsony, M. 1993, ApJ 406, 122
Bachiller, R. & Perez Gutierrez, M. 1997, ApJ 487, L93
Bergin, E. A., et al. 2002, ApJ, in press
Blake, G. A., et al., 1995, ApJ 441, 689
Caselli, P. Hartquist, T. W. & Havnes, O. 1997, A&A 322, 296
Ceccarelli, C. et al., 1999, A&A 342, L21
Ceccarelli, C., et al., 2000a, A&A 355, 1129
Ceccarelli, C., et al., 2000b, A&A 357, L9
Ceccarelli, C. Hollenbach, D. J. & Tielens, A. G. G. M. 1996, ApJ 471, 400
Clegg, P. E., et al., 1996, A&A 315, L38
Di Francesco, J., et al., 2001, ApJ 562, 770
Giannini, T. Nissini, B. & Lorenzetti, D. 2001, ApJ, 555, 40
Jørgensen, J. K. Schöier, F. L. & van Dishoeck, E. F. 2002, A&A, 389, 908
Kessler, M. F., et al, 1996, A&A 315, L27
Lefloch, B., et al, 1998, A&A 334, 269
Maret, S., et al, 2002, A&A 395, 573
Schilkem, P., et al, 1997, A&A 321, 293
Schöier, F. L. Jørgensen, J. K.& van Dishoeck, E. F. 2002, A&A, 390, 1001
Schu, F. H. 1977, ApJ 214, 488
Tielens, A. G. G. M., et al, 1984, ApJ 287, 697

MOLECULES IN THE NEARBY STARBURST GALAXY IC 342

David S. Meier[1] **and Jean L. Turner**[2]

[1]Department of Astronomy, University of Illinois, Urbana-Champaign, Urbana, IL 61801
[2]Department of Physics and Astronomy, UCLA, Los Angeles, CA 90095-1562

Abstract

We present high resolution ($\sim 5''$) images in the emission lines of several important molecules in the nearby starburst galaxy, IC 342. The data were taken with the OVRO millimeter array. The morphology of the species vary strongly across the nucleus, indicating widespread chemical differentiation or strong variations in the excitation of their dense gas components across the starburst region. Peak molecular abundances are similar to those found in the envelope of Sgr B2. The carbon chain molecules are confined to the central starburst region, with HC_3N tracing the densest gas associated with the star formation, and C_2H lining the inner "ring" of molecular gas presumably associated with the cloud PDR surfaces. Widespread, bright emission in CH_3OH and HNCO demonstrate that gas-grain chemistry is important across the nucleus. Surprisingly, the dense, quiescent gas tracer N_2H^+ is ubiquitous.

Key words: galaxies: individual(IC 342) – ISM: molecules

1. Introduction

For external galaxies, the large distances prohibit resolving the inner workings of GMCs, but observations of trace molecules may afford the opportunity to investigate the structure and physical conditions on GMC size scale. External galaxies provide three principle benefits for studying the large-scale chemical structure of the ISM: (i) the overall geometry of a star-forming region is often easier to understand than in our Galaxy, (ii) one can simultaneously study many GMCs in one observation, and (iii) external galaxies can have more extreme conditions than found anywhere in the Galaxy. These combine to make external galaxies potentially powerful complimentary tools for studying the global interplay between astrochemistry and star formation or dynamics.

2. Observations and Results

We surveyed, at resolutions high enough to resolve individual GMCs, the molecules HNC, HC_3N, C_2H, $C^{34}S$, HNCO, CH_3OH, N_2H^+ and SO in the nearby starburst nucleus, IC 342, with the OVRO array. IC 342 is the nearest ($D \sim 2$ Mpc), gas-rich spiral with active star formation. It is nearly face-on with widespread molecular gas in both dense GMCs (about the mass of Sgr B2) and a diffuse medium (comparable to the Galaxy's "central molecular zone"). A multitude of CO isotopomers have been observed providing a detailed knowledge of the distribution of H_2 column density, temperature, and density (e.g., Meier, Turner & Hurt 2000; Meier & Turner 2001). Molecular gas is distributed along x_1 and x_2 bar orbitals, in response to a small central bar of recently formed stars ($\sim 10^7$yrs). Inflow along the arms are driving two younger, optically obscured starbursts (~1000 O7 stars) at the orbit intersections (GMC B and C; Böker, Forster-Schrieber & Genzel 1997; Meier & Turner 2001).

2.1. Dense, Quiescent Gas

Dense gas is confined largely to the leading edges of the arms and to the central ring, and is collected into five major GMCs (see Fig. 1) with similar HCN(1–0) and HNC(1–0) intensities (I[HNC]/I[HCN] $\sim$ 0.5). HC_3N has the largest dipole moment and E_u of any observed transition, and its emission is confined to the two young starbursts seen in 3 mm continuum (Meier & Turner 2001). Considering that N_2H^+ avoids active regions, and in general traces dense quiescent gas in the Galaxy (Womack, Ziurys & Wyckoff 1992), it is surprising N_2H^+ is ubiquitous ($N_2H^+/H_2 \sim 3 \times 10^{-10}$, assuming $T_r = 10$ K and H_2 column densities estimated from $C^{18}O$) and follows the CH_3OH morphology closely.

These observations show that most of the dense gas is confined to the GMCs along the central molecular ring, with the warmest, densest gas occurring at the starbursts. However, the fact that N_2H^+ also peaks here suggests

Chemistry as a Diagnostic of Star Formation (2003). *Edited by Charles L. Curry and Michel Fich.* Published in 2003 by NRC Research Press, Ottawa ON, K1A 0R6, Canada. ISBN 0-660-19089-3.

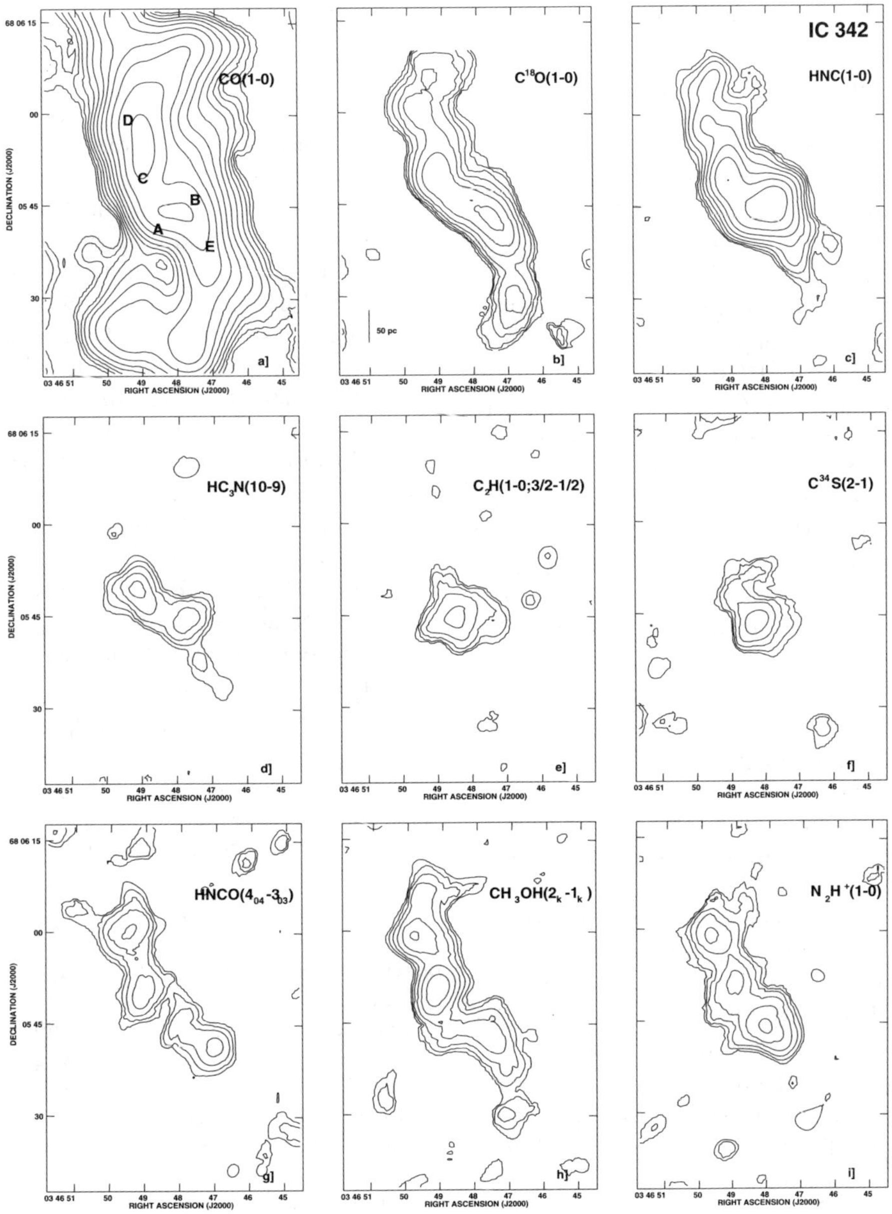

Figure 1. The integrated intensities of nine tracer molecules observed towards the nucleus of IC 342. Each map is contoured at $2^{i/2}$, i = 0,1,2,... times the 2σ times the RMS intensity in each map, except for ^{12}CO(1–0). *(a)* The ^{12}CO(1–0) transition contoured in steps of 7.7 K km s^{-1} for $6.''5 \times 5.''5$ resolution. Locations of the five major GMCs based on HCN are labeled A-E (Downes et al. 1992). *(b)* The C^{18}O(1–0) transition contoured in steps of 1.4 K km s^{-1} for $6.''0$ resolution. *(c)* the HNC(1–0) transition in steps of 2.0 K km s^{-1} for $5.''9 \times 5.''1$ resolution. *(d)* The HC_3N(10–9) transition in steps of 2.0 K km s^{-1} for $5.''9 \times 5.''1$ resolution. *(e)* The C_2H $N(J,F)$= 1(3/2, 2)–0(1/2, 1) transition in steps of 3.6 K km s^{-1} for $5.''5 \times 4.''9$ resolution. *(f)* The C^{34}S(2–1) transition in steps of 2.5 K km s^{-1} for $6.''6 \times 5.''6$ resolution. *(g)* The HNCO($4_{04} - -3_{03}$) transition in steps of 2.4 K km s^{-1} for $5.''9 \times 5.''1$ resolution. *(h)* The CH_3OH($2_k - -1_k$) transition in steps of 2.0 K km s^{-1} for $6.''0 \times 4.''8$ resolution. *(i)* The N_2H^+(1–0) transition in steps of 1.8 K km s^{-1} for $6.''2 \times 5.''1$ resolution. SO($2_3 - -1_2$) was observed but not detected at a 2σ RMS equal to that of the C^{18}O(1–0) transition.

that despite the strong star formation, either significant amounts of sterile, quiescent gas must be present, or that N_2H^+ is not exclusively a tracer of quiescent gas.

2.2. PDRs

The radical C_2H is observed exclusively towards the central star formation complex, peaking towards GMC A and inside the central ring, with fractional abundances of $C_2H/H_2 \sim 5 \times 10^{-8}$. $C^{34}S$ is the only other species as confined as C_2H, also peaking at GMC A.

From the gas-phase chemistry, C_2H is expected to be abundant where high abundances of C^+ and FUV photons are present, namely PDRs (Sternberg 1999). Likewise, the enhancements of CS (23 × $C^{34}S$) likely arise from the commensurate increase in S^+ in the PDR (Jansen et al. 1995). $N(SO)/N(CS)$ as low as < 0.05 also favors early-time C-rich conditions. These results provide excellent confirmation that the gas physical conditions of the central ring are strongly influenced by PDRs. C_2H and $C^{34}S$ also demonstrates that GMC A has different physical and chemical conditions from the other GMCs.

2.3. Grain Chemistry

CH_3OH is widespread and bright, and traces the leading edges of the molecular bar arms and the cusps in the x_1 orbits (GMC D). It has a high fractional abundance, $CH_3OH/H_2 \sim 2 \times 10^{-8}$. An additional peak in methanol is seen towards GMC C, one of the locations of obscured massive star formation. HNCO emission displays a qualitatively similar behavior, being especially strong towards GMC D. A similar enhancement of HNCO near the bar ends is seen in two other nearby barred galaxies, Maffei 2 and NGC 6946 (Meier et al., in preparation).

The high abundance of CH_3OH seen in IC 342 probably requires liberation of grain mantles to explain the observed abundances (Millar, Herbst & Charnley 1991), suggesting that grain processing is important on nucleus-wide scales. Moreover, the fact that HNCO (and CH_3OH) exhibit enhancements at the sites of dynamical resonances, leads to the speculation that on large scales (coherent enhancements over at least 50 pc), disruption of grain mantles is through cloud-cloud collisions induced by non-circular motions present in barred galaxies. Future studies of more galaxies may reveal the exciting prospect that chemistry can even be used to understand galaxy dynamics. The CH_3OH peak towards GMC C is most likely due to mantle evaporation due to the massive star formation there. This GMC, which is bright in species like HCN, HNC, HC_3N, HNCO, CH_3OH and NH_3 (Ho et al. 1990), is an excellent extragalactic analogue of a hot core like Sgr B2.

3. The Chemical Structure of IC 342

A significant numbers of molecules are detected in IC 342, opening up the possibility of studying the localized effects of star formation and dynamical shocks on gas chemistry. Morphological variations are more dramatic than seen in the CO isotopomers, suggesting that the dense gas component of starbursts either exhibits significant variations in gas excitation or that there is widespread chemical differentiation across the nucleus. We see for the first time, gas with M 82-like *and* NGC 253-like chemistries (Mauersberger & Henkel 1993) in the same galaxy. Based on the location for each in IC 342, it is tempting to conclude that the ISM in M 82 is dominated by GMC A-like chemistries, or PDR molecular gas, while NGC 253 is dominated by gas-grain chemistries caused by grain mantles liberated in bar driven shocks. These observations suggest that the prospects for high resolution extragalactic astrochemistry with future arrays, such as CARMA and ALMA, are excellent.

Acknowledgements

We are grateful to L. E. Snyder, M. Morris and M. Jura for useful discussions. We acknowledge support from the Laboratory for Astronomical Imaging at the University of Illinois and from NSF grant AST-0071276 to J.L.T.

References

Böker, T., Forster-Schrieber, N. M. & Genzel, R. 1997, AJ, 114, 1883

Downes, D., Radford, S. J. E., Guilloteau, S., Guelin, M., Greve, A. & Morris, D. 1992, A&A, 262, 424

Ho, P. T. P., Martin, R. N., Turner, J. L. & Jackson, J. M. 1990, ApJL, 355, L19

Jansen, D. J., van Dishoeck E. F., Black, J. H., Spaans, M. & Sosin, C. 1995, A&A, 302, 223

Mauersberger, R. & Henkel, C. 1993, RvMA, 6, 69

Meier, D. S. & Turner, J. L. 2001, ApJ, 551, 687

Meier, D. S., Turner, J. L. & Hurt, R. L. 2001, ApJ, 531, 200

Millar, T. J., Herbst, E. & Charnley, S. B. 1991, ApJ, 369, 147

Sternberg, A. 1998, in *The Molecular Astrophysics of Stars and Galaxies*, ed. T.W. Hartquist & D.A. Williams (Clarendon Press: Oxford), 201

Womack, M., Ziurys, L. M. & Wyckoff, S. 1992, ApJ, 387, 417

DEPLETION OF SMALL DUST GRAINS IN THE EARLY STAGES OF MOLECULE FORMATION

M.-A. Miville-Deschênes

Canadian Institute for Theoretical Astrophysics (CITA), Toronto, Canada

ABSTRACT

ISOCAM allowed the observation of diffuse interstellar regions where the transition from atomic to molecular gas takes place. In this contribution we report on spectacular variations of small dust grains abundance in such regions observed with ISOCAM. In the four cirrus-like clouds observed, we note a systematic disappearance of small dust grains in regions where CO emission starts to appear. We suggest that the variations of the small dust grain abundance observed here are related to the damping of the gas turbulent motions that favors the coagulation of dust through grain-grain collisions.

Key words: ISM: dust – turbulence

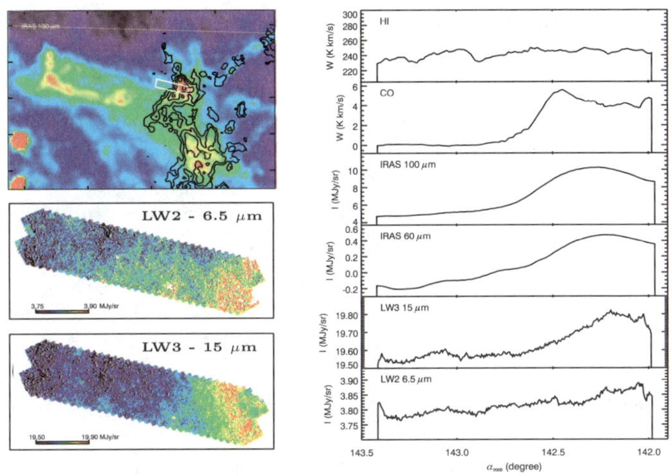

Figure 1. Left: IRAS 100 μm *(top)* emission of the Ursa Major cirrus (contours are the ^{13}CO integrated emission of Pound & Goodman 1997) and 6.5 μm *(middle)* and 15.0 μm *(bottom)* ISOCAM observations of the H I-H_2 interface (shown as a white rectangle). *Right:* Average emission as a function of right ascension through the H I-H_2 interface. Note that the mid-infrared emission gets significantly flatter than the 100 μm emission in the molecular region, indicating a decrease by a factor 3 of the small dust grain abundance with respect to the big grains.

1. INTRODUCTION

We know that interstellar dust plays an important role in the physics of the interstellar medium. For instance, dust grains heat the gas via the photo-electric effect and provide the surfaces necessary to the formation of H_2. These processes are sensitive to the size distribution, the structure and composition of dust grains and, especially, to the abundance of the smallest particles. The *IRAS* images of nearby molecular complexes and bright cirrus clouds ($A_V > 1$) have revealed remarkably that the abundance ratio between small and large dust grains varies widely from cloud to cloud and within clouds. The gain in brightness sensitivity and angular resolution provided since then by dedicated observations carried out with ISOCAM, the camera on board the *Infrared Space Observatory (ISO)*, now allows us to carry on these *IRAS* investigations and in particular to extend them to the range of column densities ($N_{\rm H} \sim 5 \times 10^{20}$ cm^{-2}) where the transition from atomic to molecular gas takes place. It is the motivation of this work to shed further light on the evolution of small dust grains within the diffuse interstellar medium in relation to gas physical conditions.

In this contribution we present mid-IR imaging observations of four high Galactic latitude cirrus obtained with ISOCAM at 6″angular resolution. The observations were done with two filters LW2 (5-8.5 μm) and LW3 (12-18 μm), that measure respectively the aromatic carbon bands and the underlying continuum emission from small dust particles. The comparison of these observations with 21 cm, CO and *IRAS* data allows us to measure the mid-IR emissivity per hydrogen which, in such clouds transparent to stellar light, is related to the abundance and the optical properties of small dust particles independently of any modelling of the penetration of the dust heating radiation.

Chemistry as a Diagnostic of Star Formation (2003). *Edited by Charles L. Curry and Michel Fich.* Published in 2003 by NRC Research Press, Ottawa ON, K1A 0R6, Canada. ISBN 0-660-19089-3.

2. New results on infrared cirrus

The ISOCAM database contains several observations of high-latitude interstellar regions that can be used to study the evolution of dust in the diffuse interstellar medium. In this section, I present ISOCAM observations of four diffuse clouds that confirm the very fast evolution of small dust grains in the early stages of molecular gas formation.

2.1. Ursa Major

The first diffuse interstellar region we have studied was the Ursa Major cirrus. The details of this analysis are described in Miville-Deschênes et al. (2002). We have observed three 0.05 square degree fields in the Ursa Major Galactic cirrus with ISOCAM in the LW2 and LW3 filters. These observations of a weakly illuminated and diffuse atomic cloud revealed striking variations of the properties of small dust particles, responsible for the emission in the 4-18 μm range. First, the combination of ISOCAM images and 21 cm interferometric data allows us to highlight a rotating filament with a small dust particles abundance $\sim$ 5 times higher than its surrounding. Second, an unexpected variation of the mid-infrared color was found, at an interface between a fully atomic region and a denser part of the cloud where molecular hydrogen is formed and survives (see Fig. 1). The comparison of the ISOCAM observations with *IRAS* data suggests a spectacular decrease of the small dust grain abundance in the CO region, suggestive of grain coagulation.

2.2. Polaris

Figure 2 presents a region of the Polaris flare observed at 100 μm (*IRAS*), 21 cm (DRAO) and in the mid-infrared (ISOCAM). This region of the Polaris flare shows very different structure at *IRAS* 100 μm and in the 21 cm integrated emission map. In particular, the brightest filament observed at 100 μm is not seen at all at 21 cm, indicating that this diffuse filamentary structure is mainly composed of molecular gas. What is most interesting here is that this filament does not show any evidence of mid-infrared emission attributed to PAH-like particles (ISOCAM observations; Fig. 2, *top right*). In fact, in the restricted region observed with ISOCAM, there is a very clear spatial correlation between the integrated 21 cm and the mid-infrared emissions, indicating that most of the mid-infrared emission comes from the atomic gas and that the abundance of small dust grains in the molecular gas is considerably decreased.

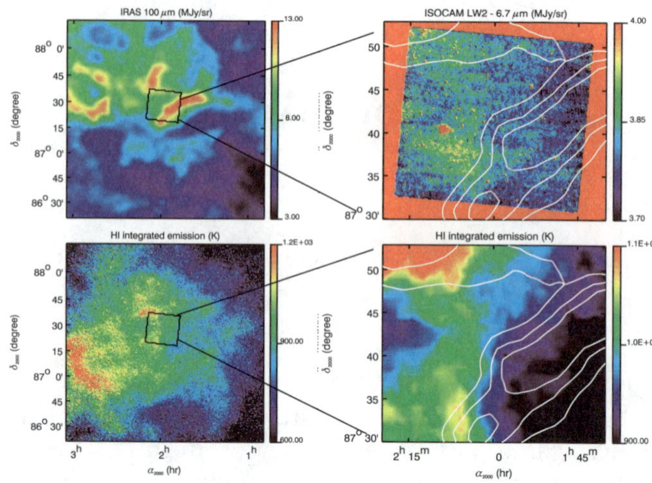

Figure 2. The Polaris flare. *Left: IRAS* 100 μm *(top)* and 21 cm integrated emission *(bottom)* images of the Polaris cirrus cloud. The 21 cm observations were obtained at the Dominion Radio Astrophysical Observatory (DRAO) in Penticton. *Right:* A smaller region has been observed with ISOCAM at 6.5 μm *(top)*. An enlargement of the same region seen at 21 cm is also shown *(bottom)*. On these last two images, the *IRAS* 100 μm emission is shown as contours.

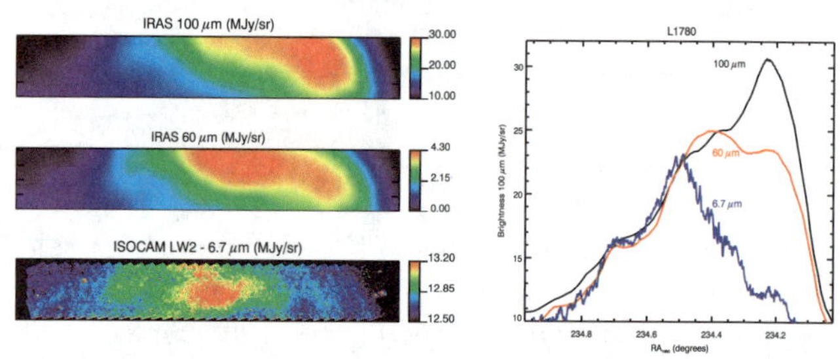

Figure 3. The L1780 cloud. *Left: IRAS* 100 *(top)* and 60 μm *(middle)*, and ISOCAM 6.5 μm *(bottom)* of a cut across the L1780 cloud. *Right:* Average emission of the three tracers as a function of right ascension. The 6.5 and 60 μm were multiplied by standard factors to be compared to the 100 μm emission. There is a spectacular decrease of the 6.5/100 μm ratio inside the L1780 filament. This is in accordance with a decrease of the smallest dust grain abundance, spatially correlated with the apparition of CO emission.

2.3. L1780

Based on *IRAS* observations, Laureijs et al. (1991) showed that the central part of the L1780 cloud, which has a moderate opacity ($A_V < 4$), is characterized by a low 60/100 μm ratio. This has been interpreted as the sign of a low dust temperature ($T \sim 15\,$K) and a change in the Very Small Grains (VSG) properties. The mid-

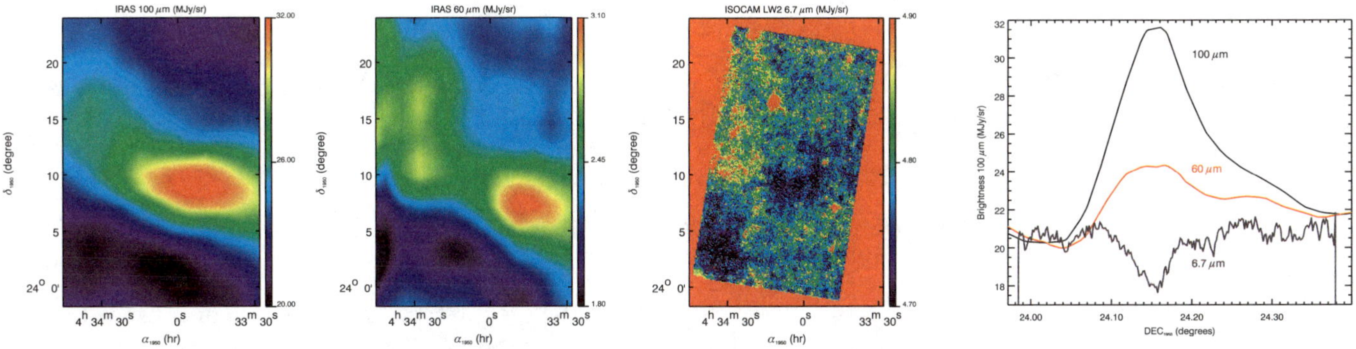

Figure 4. The Taurus filament. *IRAS* 100 μm, *IRAS* 60 μm and ISOCAM 6.5 μm observations of a filament in Taurus. On the right, we show a cut across the filament (average emission of the three tracers as a function of declination).

infrared observations of the L1780 cloud done with ISOCAM show that the decrease of the 6.5/100 μm ratio in the central part of the filament is even more spectacular (Fig. 3). These observations are a strong indication that the mid-infrared emitters (PAH-like particles) are rapidly coagulating on bigger grains in the denser part of the cloud, where CO emission is detected. This scenario is in accordance with the hypothesis of Laureijs et al. (1991) who state that the change in the VSGs optical properties may be due to the formation of dirty-ice mantle.

2.4. Taurus

Non-star-forming regions with low 60/100 μm ratio values (often referred as *cold* infrared emission) are generally associated with CO emission. A tight correlation between these two quantities has been remarkably shown in the Taurus complex by Abergel et al. (1994). We extend this analysis with mid-infrared ISOCAM observations of a particular filament in this complex (Fig. 4). Like in the three other regions observed in this study, the Taurus filament observed with ISOCAM has very low 6.5 μm emission associated with a low 60/100 μm and CO emission. Again, these observations are in accordance with a spectacular decrease of the abundance of the mid-infrared emitters in diffuse molecular gas. On the other hand, as the optical opacity are moderate in this Taurus filament, like in the L1780 filament, we cannot completely rule out the effect of absorption.

3. Discussion

Thanks to ISOCAM's high sensitivity and to the data processing techniques that allow us to reach it (Miville-Deschênes et al. 2000) we are now able to study the properties of VSGs even in diffuse cirrus clouds. From the comparison with gas phase observations (H I and CO), such observations are strong evidence that interstellar turbulence has an impact on the dust size distribution. As dust grains are coupled to the gas motions, turbulent motions of great amplitude may produce grain-grain collision strong enough to fragment big grains and to enhance the abundance of small grains. On the other hand, when turbulent motions are damped, like in the molecular gas, grain-grain collisions may lead to the disappearance of small grains through coagulation.

The ISOCAM observations of such regions also raise interesting questions regarding the evolution of the interstellar medium in general. As most of the reactive surface for the photo-electric effect and for the formation of molecules is on small grains, the strong variations of small dust grains abundance reported here suggest significant spatial variations of the chemical and thermal balance in the diffuse ISM. In particular, the rapid disappearance of small dust grains in molecular regions raises the question of the H_2 formation rate in such environment. In addition, if small dust grains coagulate that rapidly, we can wonder if there are any of them in condensed regions. It also raises the question of the origin of small dust grains at the surface of molecular clouds and in PDRs.

References

Abergel, A., et al. 1994, ApJ, 423, L59

Laureijs, R. J., Clark, F. O., Prusti, T. 1991, ApJ, 372, 185

Miville-Deschênes, M.-A., et al. 2002, A&A, 381, 209

Miville-Deschênes, M.-A., et al. 2000, A&AS, 146, 519

Pound, M. W., Goodman, A. A. 1997, ApJ, 482, 334

LABORATORY SPECTRA OF HOT WATER AND METHANE FOR MODELING LATE-TYPE STARS AND BROWN DWARFS

Ray Nassar and Peter Bernath

Department of Chemistry, University of Waterloo, Waterloo, ON, Canada N2L 3G1

Abstract

Absorption bands from water and methane are prominent in the infrared spectra of many late-type stars and brown dwarfs, and are expected to be observed in the infrared spectra of hot extrasolar giant planets. In a continuing project, laboratory high resolution Fourier transform spectra of water emission were recorded from 400-13,000 cm^{-1} using two sources, a tube furnace at 1800 K and an oxyacetylene torch at 3000 K. These spectra have been used for the identification of water lines in sunspots. We have also recorded laboratory high resolution Fourier transform spectra of methane emission using a tube furnace at 1273, 1000, and 800 K, spanning the region of 2000-6400 cm^{-1}. Spectra of water and methane at high temperatures differ significantly from room temperature spectra; therefore, modelers have shown an interest in our laboratory data for interpreting the spectra of late-type stars and brown dwarfs, and modeling the atmospheres of hot extrasolar giant planets.

Key words: Stars: low-mass, brown dwarfs – infrared: stars – molecular data – methods: laboratory – techniques: spectroscopic

1. Introduction

Water and methane have been detected in numerous astronomical objects over a wide range of temperatures. The main problem with the astronomical spectroscopy of water is the massive blanket of telluric water vapor that blocks the detection of cool water from the ground. In spite of this, observations are still made in a number of atmospheric windows. The matching of laboratory spectra to that of a sunspot, first near 10 μm, but later in many other regions, proved that there was *water on the Sun* (Wallace et al. 1995). M dwarfs are too hot to contain methane, but their spectra have an abundance of water lines and closely resemble the spectra of sunspot umbrae (3200 K).

L dwarf infrared spectra are dominated by metal hydrides and CO, with prominent water lines and a weak 3.3 μm fundamental methane band that appears for dwarfs later than about L5 (Noll et al. 2000). The infrared spectra of T dwarfs are dominated by numerous hot water and methane bands (Strauss et al. 1999) and their subclassification is based on their water and methane features in the 1.0-2.5 μm region (Burgasser et al. 2002; Geballe et al. 2002).

Due to the temperature-sensitive equilibrium reaction $CO + 3H_2 = CH_4 + H_2O$, the abundance ratio between methane and CO at a given pressure may be used as an indicator of the temperature of a brown dwarf (Lodders & Fegley 2002) and possibly hot extrasolar giant planets. Over 100 extrasolar planets have been discovered within the past few years, and differences in proximity to their parent stars result in a wide range of temperatures for these objects. Many of those discovered so far are gas giants with orbits very close to their parent stars and are believed to contain appreciable amounts of hot methane and water near the temperatures of our spectra. We intend to make all of our line lists available to modelers, who have shown an interest in using them for modeling the atmospheres of late-type stars, brown dwarfs and hot extrasolar giant planets.

2. Experimental Work

In our laboratory tube furnace experiments, gases were heated in alumina or mullite tubes, with the ends water-cooled and sealed with CaF_2 windows. Emission was measured with a Bruker IFS 120 HR Fourier transform spectrometer (FTS) utilizing a CaF_2 beam splitter and liquid-nitrogen-cooled InSb detector.

We have recorded spectra of methane in a tube furnace at 1273, 1000, and 800 K with 0.020 cm^{-1} resolution for the 2000-5000 cm^{-1} region and 0.040 cm^{-1} resolution for 5000-6400 cm^{-1}. This resulted in a total of over

Chemistry as a Diagnostic of Star Formation (2003). *Edited by Charles L. Curry and Michel Fich.* Published in 2003 by NRC Research Press, Ottawa ON, K1A 0R6, Canada. ISBN 0-660-19089-3.

65,000 lines, which were calibrated in position and intensity using the HITRAN line lists (Rothman et al. 1998). Bands observed include the prominent ν_3 fundamental, the $2\nu_3$ and $2\nu_4$ overtones, numerous combination bands and many hot bands, but actual line assignments for our methane spectra have not yet been carried out.

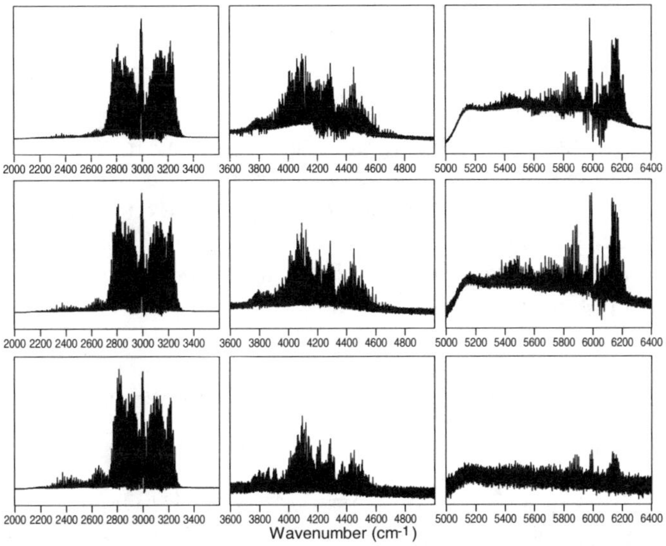

Figure 1. Our laboratory hot methane spectra: 1273 K *(top row)*, 1000 K *(middle)*, and 800 K *(bottom)*. Each panel is plotted on a different intensity scale, but the line lists have been calibrated to give absolute intensities.

We have recorded spectra of water in a tube furnace at approximately 1800 K over the region of 400-7552 cm^{-1} as an ongoing project already described in several publications (Wallace et al. 1995; Zobov et al. 2000; Tennyson et al. 2001; Bernath 2002). The most recent is an analysis of the 4878-7552 cm^{-1} region at a resolution of 0.020 cm^{-1} (Tereszchuk et al. 2002). Most of the lines have been assigned by comparison with sunspot spectra and theoretical line lists from our collaborators.

Very recently, we have obtained water spectra using an oxyacetylene torch at 3000 K (with our collaborators in Brussels, Belgium) to cover the 1200-13000 cm^{-1} region. These spectra match more closely the spectra of M dwarfs and sunspots, since they are at essentially the same temperature. Line assignments and a publication on this work are in progress.

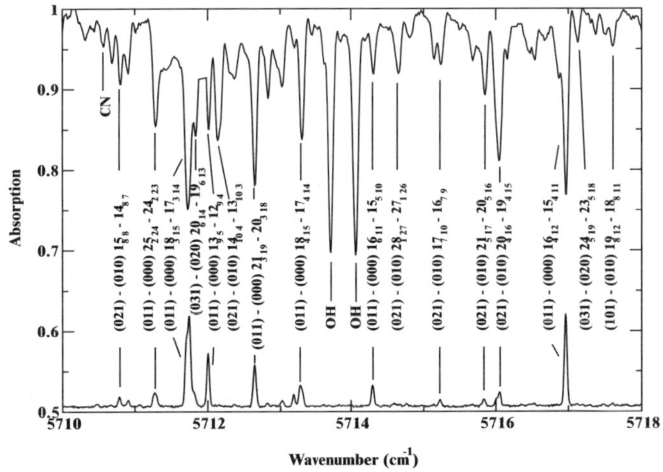

Figure 2. Laboratory hot water spectrum *(bottom trace)* and the sunspot spectrum *(top trace)*.

ACKNOWLEDGEMENTS

Funding for this work was provided by the Natural Sciences and Engineering Research Council of Canada with partial support provided by the NASA Laboratory Astrophysics program.

REFERENCES

Bernath, P.F. 2002, Phys. Chem. Chem. Phys., 4, 1501
Burgasser, A.J., et al. 2002, ApJ, 564, 421
Geballe, T.R., et al. 2002, ApJ, 564, 461
Lodders, K. & Fegley, B. Jr. 2002, Icarus, 155, 393
Noll, K.S., et al. 2000, ApJ, 541, L75
Rothman, L.S. et al. 1998, JQSRT, 60, 665
Strauss, M.A., et al. 1999, ApJ, 522, L61
Tennyson, J., et al. 2001, J. Phys. Chem. Ref. Data, 30, 735
Tereszchuk, K., et al. 2002, ApJ, 577, 496
Wallace, L., et al. 1995, Science, 268, 1155
Zobov, N.F., et al. 2000, ApJ, 530, 994

H_2 FORMATION AND EXCITATION IN CIRRUS CLOUDS

Cyrine A. Nehmé[1], Franois Boulanger[2], Edith Falgarone[3], Cécile Gry[4,5], Emilie Habart[2], Jacques Le Bourlot[1] and Guillaume Pineau des Forêts[2]

[1]LUTH, Observatoire de Paris, Université Paris 7, France
[2]Institut d'Astrophysique Spatiale,Université Paris Sud,Bat. 121,91405 Cedex, France
[3]Ecole Normale Supérieure, 24 Rue Lhomond,75005 Paris, France
[4]*ISO* Data Center, ESA,PO Box 50727,28080 Madrid, Spain
[5]Laboratoire d'Astrophysique de Marseille, BP 8, 13376, Marseille Cedex12, France

Abstract

We present *FUSE* observations of three late B stars located behind the Chameleon clouds. These data combined with a model relying on chemical and thermal balance in photon-illuminated gas allow us to calculate the H_2 formation rate R for the three lines of sight. We find similar H_2 formation rates for the three stars, about $R = 4 \times 10^{-17}\,cm^{-3}\,s^{-1}$. We also show that the H_2 excitation in the $J>2$ levels cannot be accounted for by UV pumping of the cold H_2 alone. For more details see Gry et al. (2002).

Key words: ISM: molecules – ISM: clouds – ISM: individual: HD 102065, HD 108927, HD 96675

1. Introduction

FUSE observations of H_2 UV absorption lines towards HD 102065, HD 108927 and HD 96675 are used to derive the effective rate of formation of H_2 on grains in diffuse clouds. These three stars are taken out of a sample of 8 stars initially observed with *IUE* to correlate the UV extinction curve with changes in the dust size distribution traced by the *IRAS* mid-IR to far-IR colors (Boulanger et al. 1994). The *IRAS* maps show that the target stars do not interact with matter responsible for the absorption. Characteristics of the lines of sight are given in Table 2. *FUSE* observations are used together with past observations to investigate H_2 formation and excitation. The data interpretation relies on a model of the chemical and thermal balance in photon-illuminated gas.

2. Data Analysis

The J=0 and 1 H_2 lines are blended and heavily saturated. Lorentzian wings allow to derive column densities independently of the gas velocity distribution. This is done by line fitting with Martin Lemoine's code "Owens" (see Fig 1). Column densities are shown in Table 1. For higher excited levels, the lines are less saturated and depend heavily on the gas velocity distribution. We use the curve of growth (taking for H_2 the gas velocity distribution fitted on CH and CH^+ high resolution [10^5] observations) to calculate the column densities of excited H_2. Results are gathered in Table 2.

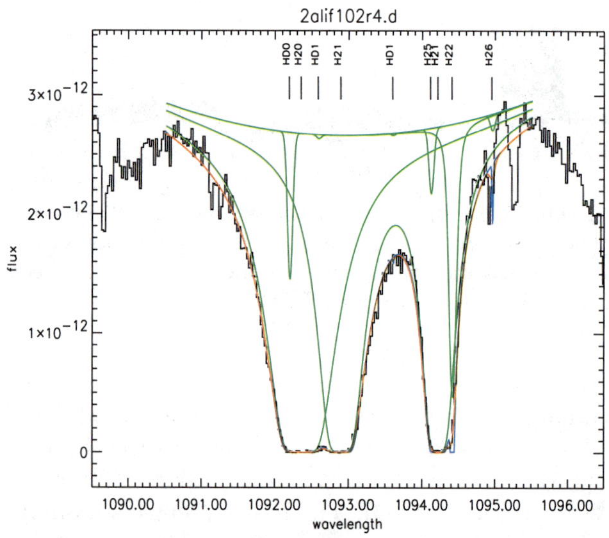

Figure 1. Sample fit of H_2 J=0 and J=1 lines towards HD 102065. *FUSE* observations with synthetic absorption components due to individual transitions and resulting profile.

3. Illuminated gas model

Observations are interpreted through a model of chemical and thermal balance of a 1D slab of gas illuminated by the solar neighborhood mean radiation field. Transfer in the H_2 lines is taken into account. This model is derived from that introduced by Le Bourlot et al. (1993). For all models we assume a constant gas density. For dust

Chemistry as a Diagnostic of Star Formation (2003). *Edited by Charles L. Curry and Michel Fich.* Published in 2003 by NRC Research Press, Ottawa ON, K1A 0R6, Canada. ISBN 0-660-19089-3.

Table 1. Interstellar absorption toward the stars

	HD 102065	HD 108927	HD 96675
Spec. Type	B9IV	B5V	B7V
d (pc)	170	390	250
$E(B-V)$	0.17 ± 0.04	0.23 ± 0.04	0.31 ± 0.03
A_V (mag)	0.67 ± 0.12	0.68 ± 0.10	1.1 ± 0.15
R_V	3.9 ± 0.4	3.0 ± 0.2	3.5 ± 0.2
N(CH) (cm^{-2})[a,b]	6.3 (12)	1.4 (13)	2.2 (13)
b(CH) ($km\,s^{-1}$)	1.9	2.2	1.6
$N(CH^+)$	1.2 (13)[a]	–	2.8 (12)
$b(CH^+)$	3.0	–	2.2
N(CO)	(7 ± 3) (13)	–	>(15)
N(C I)[c]	(4 ± 1) (14)	–	>1.7 (15)
N_{total}[d]	9.9 (20)	1.3 (21)	1.8 (21)
$N(H_2,J{=}0)$	(2.0 ± 0.2) (20)	(2.0 ± 0.2) (20)	(4.9 ± 0.9) (20)
$N(H_2,J{=}1)$	(1.4 ± 0.1) (20)	(1.2 ± 0.1) (20)	(2.1 ± 0.6) (20)
$2\,N_{H_2}/N_H$	0.69 ± 0.12	0.49 ± 0.09	0.76 ± 0.15

NOTES:
[a] The notation $n(m)$ denotes $n \times 10^m$.
[b] Derived from high resolution (R = 110,000) optical spectra obtained with the ESO 3.6 m telescope.
[c] Derived from *HST*-GHRS observations.
[d] Derived from E(B-V), $N_{total} = 5.8 \times 10^{21}\ E(B\text{-}V)$.

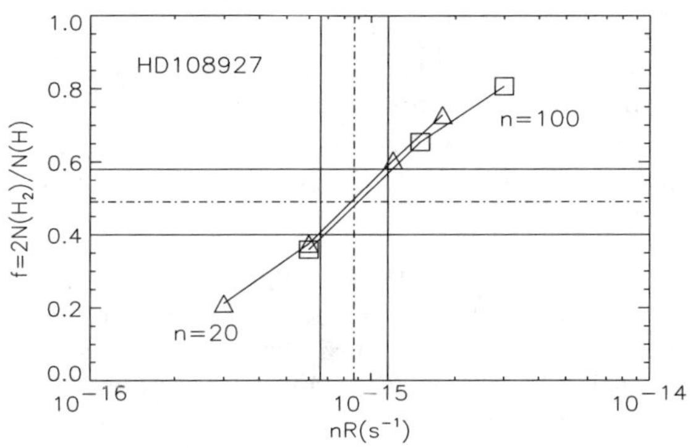

Figure 2. Determination of nR from H_2 fraction f for HD 108927. Symbols represent model calculations.

Table 2. H_2 column densities (cm^{-2}) in the excited ($J{\geq}2$) levels, derived from a curve of growth analysis by assuming successively $b{=}b_{CH}$ and $b{=}b_{CH^+}$ ($km\,s^{-1}$). The "true" $N(H_2)$ should be close to one of these two estimates.

	HD 102065		HD 108927	HD 96675	
b	1.9	3.0	2.2	1.6	2.2
$J{=}2$	2.6(18)	2.5(18)	1.7(17)	3.3(18)	3.4(18)
$J{=}3$	3.1(17)	1.1(17)	9.3(16)	1.7(17)	8.7(16)
$J{=}4$	5.6(16)	6.0(15)	1.5(15)	4.0(15)	1.4(15)
$J{=}5$	1.2(15)	4.5(14)	$\leq$ 6(13)	$\leq$ 2(14)	

extinction the specific extinction curve measured for each star is introduced in the model.

4. Results

By assuming chemical and thermal balance of the gas, the model allows us to derive the product nR of the gas density n and the H_2 formation rate R from the molecular fraction $f{=}2\ N_{H_2}/N_H$, as illustrated in Figure 2. The results for the product nR for the three stars are gathered in Table 3. Our results are in agreement with Jura (1975a) and Jura (1975b) values from *Copernicus* observations; however, our uncertainties are an order of magnitude lower. A comparison of data and model also shows that the H_2 excitation for $J-$levels larger than 2 is well above what is expected from UV pumping alone. This could arise from collisional excitation regions where the gas is much warmer. Verstraete et al. (1999) have studied a long ($A_V{=}18$) line of sight through the Galaxy

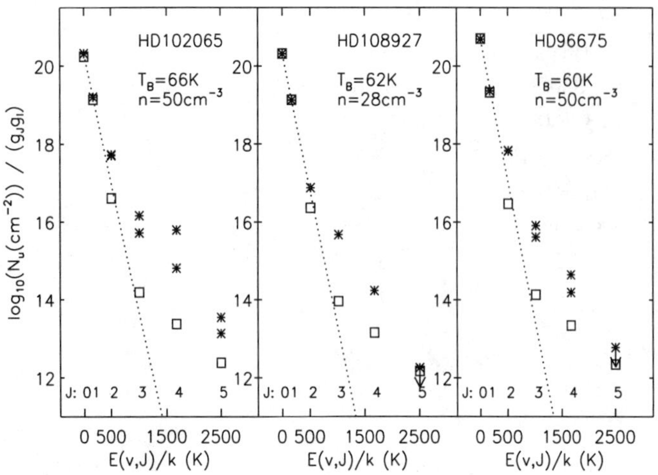

Figure 3. H_2 excitation diagrams. Asterisks are measured column densities, squares represent the best model fit to J=0 and J=1 observations. The dashed line is the Boltzmann distribution at the $N(J{=}1)/N(J{=}0)$ excitation temperature T_B.

Table 3. H_2 formation rate R from the product nR and the density n estimated from H_2 excitation diagrams (see Fig. 3).

Etoile	HD 102065	HD 108927	HD 96675
$nR\,(s^{-1})$	2.3(−15)	0.87(−15)	2.0(−15)
$n\,(cm^{-3})$	50	28	50
$R\,(cm^3s^{-1})$	4.5(−17)	3.1(−17)	4.0(−17)

Table 4. Excited H_2 column densities ($\times\ 10^{17}\,cm^{-2}$) and extinction A_V (mag) for our three lines of sight and a deep *ISO* line (see Vertraete et al. 1999).

	HD 102065	HD 108927	HD 96675	*ISO*
$N(H_2)_{ex}$	1.7–3.7	1.0	0.9–1.7	77.0
A_V	0.67	0.68	1.1	18
$N(H_2)_{ex}/A_V$	2.5–5.4	1.5	0.8–1.6	4.3

devoid of strong star formation. The ratio $N(H_2)_{ex}/A_V$ for that line of sight is comparable to our estimates for these three close lines of sight (see Table 4). Warm H_2 thus seems to be present throughout the diffuse ISM.

5. Soon to come

Far-UV spectra from the STIS spectroghraph aboard the *HST* (*Hubble Space Telescope*), towards HD 102065 are

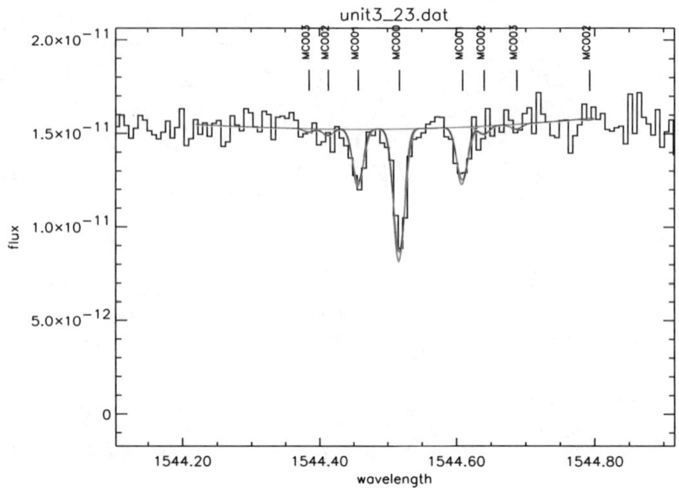

Figure 4. Example of fit of CO absorption lines from v=0 and (J=0, J=1 and J=2) towards HD 102065. *HST*/STIS observations.

currently being analysed (Nehmé, C. et al. 2003, in preparation). Preliminary results include the measurements of the column density of many elements co-existing with the H_2 (C I, C I*, C I**, S I, Fe I, O I, S I* and CO).Those lines are used to determine physical conditions of the gas (density, temperature and pressure). Three ro-vibrational bands of CO are observed at 1544.54 ρA, 1509.4 ρA and 1477.57 ρA (Fig. 4). CO was also detected in emission towards this line of sight, with column densities $(2.1{\pm}1) \times 10^{13}\,cm^{-2}\,(J{=}0)$, $(1.6{\pm}0.7) \times 10^{13}\,cm^{-2}\,(J{=}1)$, and $<2.5 \times 10^{12}\,cm^{-2}\,(J{=}2)$. We combine those two observations to derive physical parameters. We find that $N(CO)/N(H_2) \simeq 6 \times 10^{-8}$.

Acknowledgements

We are grateful to Martin Lemoine for his profile-fitting software Owens, to Vincent Lebrun for reprocessing the data, and to Michèle Eidelsberg for providing UV CO line strengths in machine readable format.

References

Boulanger, F., Prévot, M.L., & Gry, C. 1994, A&A, 284, 956

Gry, C., Boulanger, F., Falgarone, E., Pineau des Forêts, G., & Lequeux, J. 1998, A&A, 331, 1070

Gry, C., Boulanger, F., Nehmé, C., Pineau des Forêts, G., Habart, E., & Falgarone, E. 2002, A&A, 391, 675

Jura, M. 1975a, ApJ, 197, 575

Jura, M. 1975b, ApJ, 197, 581

Le Bourlot, J., Pineau des Forêts, G., Roueff, E. & Flower, D.R. 1993, A&A, 267, 233

Verstraete, L., et al. 1999, in *The Universe as seen by ISO*, ESA SP-427, 779

presence of T Tau stars, compact and extended outflows, respectively. Five out of six proposed embedded YSOs by Kun & Prusti (1993) in the region we surveyed are probably still deeply embedded in the dense gas of the cloud: sources 4, 5 and 6 in the tail area, source 8—i.e., IRAS 22343+7501—in the northern area (core N 1) and source 14—i.e., IRAS 22376+7455, embedded in the H 2b core. Additionally, based on their projection onto the surveyed dense areas, two more T Tauri stars are probably associated with cores: sources 16 and 17 in the topmost part of the head area are both coincident with detected H α emission stars (Kun et al. 2000). Although marked as embedded YSOs in Fig. 1, sources 13 and 15 (both blended by the strong IR emission from IRAS 22376+7455) were originally classified as T Tauri candidate stars based on their IR colors (Kun & Prusti 1993).

3. Molecular abundances

We calculated fractional abundances (Table 2) from the data of optically thin rarer isotopomers, assuming LTE and local ISM relative abundances of atoms. For the CO species we used T_{ex} = 10 K, while for other molecules values of 30 K and 20 K were used where appropriate. The HNC fractional abundances were derived both from the ^{13}C and ^{15}N isotopomers. The H_2 column density was determined from ^{13}CO using a conversion factor of $N(H_2)/N(^{13}CO) = 4.8 \times 10^5$ (Dickman & Clemens 1983). NH_3 is from Tóth & Walmsley (1996).

Using methyl acetylene and the HCN/HNC ratios as thermometers, we find a "temperature gradient" in the cloud. The highest temperature is detected in the head region, in a core which has already formed one or more T Tauri stars (H1 a).

The branching ratio in the main production route for HCN and HNC was estimated to be 0.2–0.4 for HCN in all cores but the core with the highest detected gas kinetic temperature. In this core (H 1a) the formal branching ratio for the HCN production is as high as 0.8, probably a reflection of an increased HNC destruction rate.

Table 2. Fractional abundances of the main isotopomers relative to H_2

Molecule		H 1a	H 2a	H 2b	N 1a	T 1a
CS	($\times 10^{-9}$)	19	5.3	8.1	3.2	3.9
HCO^+	($\times 10^{-9}$)	14	1.8	11	10	7.9
SO	($\times 10^{-9}$)	9.1	1.8	2.7	0.82	1.7
HCN	($\times 10^{-8}$)	6.6	0.14	0.26	0.2	1.1
HNC	($\times 10^{-8}$)	1.0	2.4	1.2	0.6	2.6
CH_3CCH	($\times 10^{-9}$)	16	3.5	4.7	2.8	2.0
NH_3	($\times 10^{-7}$)	1.2	–	0.68	–	5.7

3.1. Star formation influence on chemistry

The presence of a T Tauri star in H1 a would indicate an age of $10^6 - 10^7$ years for this core, while the embedded YSOs in the remaining 4 cores point to ages of $10^4 - 10^5$ years. The fractional abundances given in Table 2 lack any clear trends with respect to the adopted dynamical ages of the cores. In an attempt to isolate trends suggested by chemical modelling, we have taken the naive approach to form the abundance ratios of "late-time" (i.e., SO, HCO^+, NH_3) relative to "early-time" molecules, that is CS, HCN and HNC. If anything, the ratios point in general to trends opposite to those expected. We think that, if present chemical networks of dense and cold clouds predict the evolution of molecular abundances correctly, such trends are masked by the onset of processes like, e.g., star formation and external shocks.

References

Berkhuijsen, E.M. 1971, A&A, 14, 252
Dickman, R.L., Clemens, D.P. 1983, ApJ, 271, 143
Grenier, I. A., et al. 1989, ApJ, 347, 231
Kun, M. & Prusti, T. 1993, A&A, 272, 235
Kun, M., Vinkó, J., Szabados, L. 2000, MNRAS, 319, 777
Kun, M. 1998, ApJS, 115, 59
Sato, F. et al. 1994, ApJ, 435, 279
Tóth, L.V. & Walmsley C.M. 1996, A&A, 311, 981

MORPHOLOGY AND DYNAMICS OF THE ENVIRONMENT OF THE YOUNG MASSIVE STAR S106-IR REVEALED BY BEAR SPECTRO-IMAGING IN H AND H_2

B. Noel[1], C. Joblin[1], and J.P. Maillard[2]

[1]Centre d'Etude Spatiale des Rayonnements, 9 avenue du colonel Roche, 31028 Toulouse Cedex 4, France
[2]Institut d'Astrophysique de Paris, 98 bis boulevard Arago, 75014 Paris, France

Abstract

Data from S106 were obtained in the H and H_2 lines (Br γ, H_2 2-1 S[1] and H_2 1-0 S[1]) with BEAR, an imaging Fourier Transform Spectrometer, at the CFH Telescope. S106 is a star formation region where a bipolar H II region embedded in a molecular cloud is driven by the young massive star S106-IR. Thanks to the high spatial and spectral resolution of BEAR, the morphology and the dynamics of the close environment of S106-IR have been investigated. We confirm a population of nearby low-mass stars around S106-IR. The interpretation of the observations shows evidence for the disruption of the parent cloud through the combined action of winds, jets, magnetic fields and UV radiation. A new geometrical model of the region is proposed.

Key words: Stars: formation – ISM: jets and outflows – ISM: individual (S106)

Chemistry as a Diagnostic of Star Formation (2003). *Edited by Charles L. Curry and Michel Fich.* Published in 2003 by NRC Research Press, Ottawa ON, K1A 0R6, Canada. ISBN 0-660-19089-3.

LOW UPPER LIMITS ON THE O_2 ABUNDANCE FROM THE *ODIN* SATELLITE

Laurent Pagani[1] **and the *Odin* Team**[2]

[1]LERMA & FRE 2460 du CNRS, Observatoire de Paris, 61, A. de l'Observatoire, 75140 Paris, France
[2]http://www.snsb.se/Odin/Odin.html

Abstract

For the first time, a search has been conducted in our Galaxy for the 119 GHz transition of O_2, using the *Odin* satellite. *Odin* has reached unprecedented upper limits on the abundance of O_2. Here we report upper limits for a dozen sources. In cold dark clouds we improve upon the published *SWAS* upper limits by more than an order of magnitude, and we reach $N\ (O_2)/N(H_2) \leq 10^{-7}$ in half of the sources. While standard chemical models are definitively ruled out by these new limits, our results are compatible with several recent studies that derive lower O_2 abundances. Goldsmith et al. (2002) recently reported a *SWAS* tentative detection of the 487 GHz transition of O_2 towards ρ Oph A. In a brief and partial covering of the *SWAS* region ($\approx$ 65%), we did not detect the corresponding 119 GHz line. Our 3-sigma upper limit on the O_2 column density is 7.3×10^{15} cm^{-2}. We presently cannot exclude the possibility that the *SWAS* signal lies mostly outside of the 9′ *Odin* beam.

Key words: ISM: molecules – Galaxy: abundances

1. Introduction

Oxygen, the third most abundant cosmic element, is a key species in the chemistry of the interstellar medium (ISM). Our knowledge of its prevalence and distribution depends on the ability to measure the abundance of the main oxygen-bearing species, including H_2O and O_2. The task is difficult due to telluric blocking. Standard chemical models predicted that the abundance of O_2 should be comparable to that of CO inside well-shielded cloud cores (Herbst & Klamperer 1973; Graedel et al. 1982; Maréchal et al. 1997). Consequently, O_2 was considered an important cooling species for clouds (Goldsmith & Langer 1978). It has been actively searched for from the ground (Liszt 1985; Goldsmith et al. 1985; Maréchal et al. 1997), by balloon (Olofsson et al. 1998), and by satellite (*SWAS*, Goldsmith et al. 2000). None of these efforts has produced a convincing detection of O_2. *SWAS* reported the lowest upper limits yet: $N(O_2)/N(C^{18}O) \leq 1.5\ (3\sigma)$ and $N(O_2)/N(H_2) \leq 2.6 \times 10^{-7}\ (3\sigma)$ in sources with star-forming activity and $N(O_2)/N(H_2) \leq 3 \times 10^{-6}\ (3\sigma)$ in cold dark clouds (TMC-1 and L183/L134N). More recently, the *SWAS* team reported a possible detection of the 487 GHz O_2 line in the ρ Oph A outflow (Goldsmith et al., 2002) with $N(O_2)/N(H_2) \approx 10^{-5}$. Despite its diminished importance as a cloud coolant, O_2 still attracts interest because it provides strong constraints on chemical models.

Odin, a 1.1m space-borne millimeter and submillimeter telescope, is equipped with a cryogenically cooled 3mm HEMT preamplifier (fixed-) tuned to the 118.750343 GHz ($N_J = 1_1 - -1_0$) transition of O_2 for which the beam diameter is 9′. Since the 119 GHz line is easy to thermalize, stronger than any of the submm transitions at temperatures below 100 K (Maréchal et al. 1997), it is probably the best line to search for, especially in cold dark clouds. At 10 K, the 119 GHz line is 20 times stronger than the 487 GHz line. With a system temperature 8 times lower than that of *SWAS* (T_{sys}[SSB] = 600 K at 119 GHz), we gain 2 orders of magnitude in sensitivity in extended cold sources.

2. Observations and Data Reduction

The first useful *Odin* observations were obtained towards Orion on 2001-08-19 and all reported sources but TMC-1 were subsequently observed until 2001-10-04. At that time, the phase-lock system became unstable and was lost after 2001-10-14. The phase-lock was recovered no later than 2002-03-09. Data taken towards TMC-1 (2002-03-23 to 2002-03-25) were correctly phase-locked. The absolute pointing uncertainty is within 30″, which is negligible for our beam size of 9′. The autocorrelator spectral sampling was 0.125 MHz in a 100 MHz bandwidth for all sources except G0.26-0.01, for which we used 1 MHz sampling.

Chemistry as a Diagnostic of Star Formation (2003). *Edited by Charles L. Curry and Michel Fich.* Published in 2003 by NRC Research Press, Ottawa ON, K1A 0R6, Canada. ISBN 0-660-19089-3.

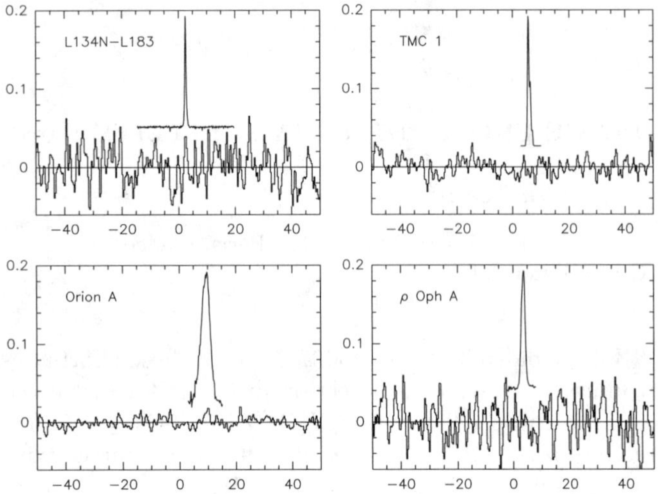

Figure 1. O_2 spectra with unscaled $C^{18}O$ spectra. The noise peaks at source velocity are 1.5σ (L134N) and 2.5σ (Orion A). The horizontal axis is LSR velocity in km s^{-1}, the vertical axis, T_A in K (uncorrected for beam efficiency).

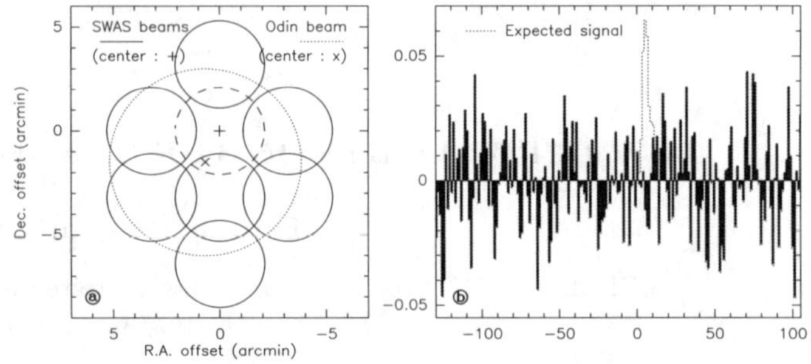

Figure 2. ab. **a** The *SWAS* and *Odin* beams towards ρ Oph A. The dashed circle represents the *SWAS* central beam, not taken into account in their detection (Goldsmith et al. 2002). **b** Smoothed ρ Oph A O_2 spectrum, with the signal expected at 119 GHz from the *SWAS* O_2 column density estimate superimposed. Axes are the same as in Fig.1

We adopt $\eta_{MB} = 0.9$, as measured at 557 GHz towards Jupiter.

Our first concern was to establish the exact center frequency of the O_2 receiver band at all times and check its stability. The receiver tuning is checked by measuring the telluric O_2 line position in the backend during the 40% of each orbit *Odin* spends pointing at Earth's atmosphere. The tuning was extremely stable during the two periods we report here. The observing mode (switching against blank sky via an off-axis reference beam of 4.4° size) gives reasonably good baselines at high resolution but rather bad ones for large bandwidth. In the latter case (G0.26-0.01), we subtracted an off-source observation, itself switched against blank sky and devoid of $C^{18}O$ signal, from the on-source observation to improve the baselines. Some of our observational results are displayed in Figure 1.

3. Results and Discussion

Table 1 gives the list of sources with their final noise limit, the $C^{18}O$ ($J = 1\text{–}0$) integrated intensity convolved in a $9'$ gaussian beam taken from literature, and the final O_2 relative abundance upper limit (3σ). To derive O_2 column density upper limits, we assumed that the 119 GHz transition is thermalized (LTE). To estimate the O_2 integrated intensity upper limit, we took $I \leq 3\sigma\,(\delta v \Delta v)^{1/2}$ where Δv is the $C^{18}O$ linewidth.The $C^{18}O$ ($J = 1\text{–}0$) line is rather easy to thermalize and is relatively insensitive to the density and temperature ranges spanned by the observed sources. For a $9'$ beam size, depletion and saturation effects should be of limited importance. To derive $N(C^{18}O)$ we also assumed LTE and we took $N(C^{18}O)/N(H_2) = 1.7 \times 10^{-7}$ (Frerking et al. 1982) for local clouds and modulated this value as a function of galactocentric distance following the $^{16}O/^{18}O$ variation described by Wilson & Rood (1994). The $C^{18}O$ data are all in the T_R^* scale as defined by Kutner & Ulich (1981), appropriate for extended sources.

In a recent paper (Goldsmith et al. 2002), the *SWAS* team reported a possible detection of 487 GHz O_2 in the red wing of the ρ Oph A outflow. Their tentative detection was made by adding 7 different *SWAS* positions surrounding, but excluding, the source itself (Fig. 2a). Our larger beam, with a slightly different target position (ρ Oph A-VLA) covers about 65% of the *SWAS*-mapped region, not including their central beam. With the same assumptions as in Goldsmith et al. (2002), we find a 3σ upper limit of $X(O_2) \leq 4.4 \times 10^{-6}$, a factor of 3 below their estimate (1.5×10^{-5}). Using their $N(O_2)$, we have computed the expected gaussian line at 119 GHz and overlaid it on our spectrum in Figure 2b (the gaussian line has been combined with our own noise spectrum to simulate real observing conditions). A clear signal should have been detected if O_2 is evenly distributed in the *SWAS* mapped region. Because the *Odin* beam does not completely cover the region, we cannot exclude the possibility that most of the emission lies outside our beam and thus escaped detection at 119 GHz. Further *Odin* observations are underway to settle this point.

Odin has improved upon the *SWAS* upper limits by a factor of a few in star-forming regions, and by 20 for L134N and 40 for TMC-1, yet still has not detected O_2.

Table 1. Derived upper limits on O_2 column density relative to H_2. Resolution is 0.3 km s^{-1}, except for G0.26-0.01 (2.5 km s^{-1}).

Source	α_{2000}	δ_{2000}	O_2rms (K)	$\int T_r^*(C^{18}O)dv$ (K km s^{-1})	$\Delta v(C^{18}O)$ (km s^{-1})	T_{kin} (K)	N(O_2)^a (cm^{-2})	N($C^{18}O$) (cm^{-2})	$X(O_2)$	Ref.
TMC1 − NH_3	$4^h41^m16^s$	25°48′44″	0.010	1.6	1.1	10	$\leq 6.8(14)$	1.5(15)	$\leq 7.7(-8)$	1
OrionA	$5^h35^m14\overset{s}{.}36$	−5°22′30″	0.006	2.7	3.8	50	$\leq 1.9(15)$	3.6(15)	$\leq 8.9(-8)$	2
NGC2071	$5^h47^m4\overset{s}{.}1$	0°21′42.″8	0.026	2.86	1.8	20	$\leq 2.6(15)$	2.8(15)	$\leq 1.5(-7)$	3
L134N − NH_3	$15^h54^m08\overset{s}{.}52$	−2°52′48″	0.024	1.16	0.6	10	$\leq 1.1(15)$	1.1(15)	$\leq 1.7(-7)$	4
ρOphA	$16^h26^m26\overset{s}{.}4$	−24°24′30.″5	0.027	5.17	1.53	30	$\leq 3.4(15)$	6.2(15)	$\leq 9.3(-8)$	5
IRAS16293	$16^h32^m22\overset{s}{.}8$	−24°28′35″	0.017	2.07	0.94	30	$\leq 1.7(15)$	2.5(15)	$\leq 1.2(-7)$	5
NGC6334I	$17^h20^m53\overset{s}{.}38$	−35°47′1.″5	0.017	30.^{b}	4.0	50	$\leq 5.0(15)$	^{d}	$\leq 7.1(-8)^e$	6 − 8
G0.26 − 0.01	$17^h46^m11^s$	−28°42′39″	0.021	11.75	164.0^c	20?	$\leq 5.6(16)$	1.2(16)	$\leq 7.6(-7)$	9
M17SW	$18^h20^m23\overset{s}{.}11$	−16°12′47.″2	0.024	2.84	4.3	50	$\leq 7.3(15)$	3.0(15)	$\leq 5.7(-7)$	10
S68FIRS1	$18^h29^m50\overset{s}{.}3$	1°15′18.″6	0.016	2.63	1.4	25	$\leq 1.6(15)$	2.8(15)	$\leq 9.7(-8)$	11
G34.3 + 0.2	$18^h53^m18\overset{s}{.}34$	1°14′58.″4	0.021	.−	6.0	30	$\leq 5.2(15)$	^{d}	$\leq 5.2(-8)$	12 − 14

NOTES— ^{a} 6.8(14) means 6.8×10^{14}. ^{b} Core only: 2′×3′. ^{c} Many clouds on the line of sight of this cloud in a large range of velocities. ^{d} From ^{13}CO, $N(H_2) \approx 10^{23}$ cm^{-3}. ^{e} $\leq 1.7 \times 10^{-7}$ $C^{18}O$ core only.

REFERENCES— (1) Onishi et al. (1996); (2) Dutrey et al. (1996); (3) Aoyama et al. (2001); (4) Pagani et al. (2001); (5) Tachihara et al. (2000); (6) Kraemer & Jackson (1999); (7) Schwartz et al. (1989); (8) Dickel et al. (1977); (9) Dahmen et al. (1997); (10) Ando et al. (2002); (11) McMullin et al. (2000); (12) Churchwell et al. (1992); (13) Matthews et al. (1987); (14) Little et al. (1994).

It is now well established (since the work of Olofsson et al. 1998) that standard, steady state, chemical models are ruled out by this absence of O_2. In the meantime, more sophisticated models have been developed which can accommodate O_2 abundances compatible with these new upper limits. Goldsmith et al. (2000) and Bergin et al. (2000) discuss those models which appeared up to the time of the *SWAS* observations. More recent work includes Casu et al. (2001), Spaan & van Dishoeck (2001), Viti et al. (2001), Roberts & Herbst (2002), and Willacy et al. (2002). All models are compatible with our upper limits. The detection of O_2 in one or more sources, even at a low level, would greatly help in this task.

ACKNOWLEDGEMENTS

We thank the NANTEN group, G. Dahmen, J. McMullin and J. Bally for kindly providing us with their $C^{18}O$ data. Based on observations with *Odin*, a Swedish-led satellite project funded jointly by the Swedish National Space Board (SNSB), the Canadian Space Agency (CSA), the National Technology Agency of Finland (Tekes) and Centre National d'Etudes Spatiales (CNES). The Swedish Space Corporation was the industrial prime contractor and is operating *Odin*.

REFERENCES

Ando M., Nagata T., Sato S., et al., 2002, ApJ, 574, 187
Aoyama H., et al., 2001, PASJ, 53, 1053
Bergin E.A., et al., 2000, ApJL, 539, L129
Casu S., et al., 2001, MNRAS, 325, 826
Churchwell E., 1992, A&A, 253, 541
Dahmen G., et al., 1997, A&AS, 126, 197
Dickel H.R., Dickel J.R., Wilson W.J. 1977, ApJ, 217, 56
Dutrey A., Duvert G., Castets A., et al. 1993, A&A, 270, 468
Frerking M.A., et al. 1982, ApJ, 262, 590
Goldsmith P.F. & Langer W.D., 1978, ApJ, 222, 881
Goldsmith P.F., et al., 1985, ApJ, 289, 613
Goldsmith P.F.,et al., 2000, ApJL, 539, L123
Goldsmith P.F., Li D., Bergin E.A.,et al., 2002, ApJ, 576, 814
Graedel T.E., et al., 1982, ApJS, 48, 321
Herbst E. & Klamperer, W., 1973, ApJ, 185, 505
Kraemer K.E. & Jackson J.M., 1999, ApJS, 124, 439
Kutner M.L. & Ulich B.L., 1981, ApJ, 250, 341
Liszt H.S., 1985, ApJ, 298, 281
Little L.T., et al., 1994, MNRAS, 271, 649
Maréchal P., et al., 1997a, A&A, 324, 221
Maréchal P., et al., 1997b, A&A, 318, 252
Matthews N., et al., 1987, A&A, 184, 284
McMullin J.P., et al., 2000, ApJ, 536, 845
Olofsson G., Pagani L., Tauber J., et al., 1998, A&A, 339, L81
Onishi T., et al., 1996, ApJ, 465, 815
Pagani L., Pardo J.R., Fich M., Motte F., Stepnik B. 2001, in *Infrared and Submillimeter Space Astronomy*, ed. M. Giard et al., EAS Publ. Ser. V.4, EdP Sciences, 145
Roberts H. & Herbst E., 2002, A&A, 395, 233
Spaans M. & van Dishoeck E.F., 2001, ApJL, 548, L517
Tachihara K., Mizuno A., Fukui Y., 2000, ApJ, 528, 817
Schwartz P.R., Snell R.L., Schloerb F.P., 1989, ApJ, 336, 519
Viti S., et al., 2001, A&A, 370, 557
Willacy K., Langer W.D., Allen M., 2002, ApJL, 573, L119
Wilson T.L. & Rood R.T., 1994, ARA&A, 32, 191

DETECTION OF DEUTERATED METHANOL IN THE LOW-MASS PROTOSTAR *IRAS* 16293-2422

B.Parise[1], C.Ceccarelli[2,3], A.G.G.M.Tielens[4], E.Herbst[5], B.Lefloch[3], E.Caux[1], A.Castets[2], I.Mukhopadhyay[6], L.Pagani[7], and L.Loinard[8]

[1]CESR CNRS-UPS, BP 4346, 31028 - Toulouse Cedex 04, France
[2]Observatoire de Bordeaux, BP 89, 33270 Floirac, France
[3]Laboratoire d'Astrophysique, Observatoire de Grenoble -BP 53, F-38041 Grenoble Cedex 09, France
[4]SRON, P.O. Box 800, NL-9700 AV Groningen, the Netherlands
[5]Department of Physics, The Ohio State University, 174 W. 18th Ave. Columbus, OH 43210-1106, USA
[6]Laser Spectroscopy Lab-II, Chemistry Department, University of Akron, Akron, OH 44333-3601, USA
[7]LERMA & FRE 2460 du CNRS, Observatoire de Paris, 61 Av. de l'Observatoire de Paris, 75014 Paris, France
[8]Instituto de Astronomía, UNAM, Apdo Postal 72-3 (Xangari), 58089 Morelia, Michoacán, México

Abstract

We present multi-frequency observations of singly deuterated methanol (CH_3OD and CH_2DOH) as well as the first detection of doubly deuterated methanol (CHD_2OH) towards the low-mass protostar *IRAS* 16293-2422. We derive the following fractionation ratios:
$[CH_2DOH]/[CH_3OH] = 0.9 \pm 0.3$,
$[CH_3OD]/[CH_3OH] = 0.04 \pm 0.02$, and
$[CHD_2OH]/[CH_3OH] = 0.2 \pm 0.1$.
These observations unveil the first case of a molecule whose deuterated forms are more abundant overall than the main hydrogenated isotopomer, and provide a severe challenge to both gas-phase and grain-surface models of deuteration.

Key words: ISM: molecules – stars: formation

1. Introduction

Despite the relatively low elemental abundance of deuterium in space ($\sim 1.5 \times 10^{-5}$ less abundant than H nuclei; Linsky 1998), extremely large amounts of doubly deuterated formaldehyde ($D_2CO/H_2CO \sim 10\%$) have been observed in the solar-type protostar *IRAS* 16293-2422 (Ceccarelli et al. 1998; Loinard et al. 2000; Ceccarelli et al. 2001). Subsequently, multiply deuterated formaldehyde and ammonia have been observed in other protostars and dark clouds from where protostars form (Roueff et al. 2000; Loinard et al. 2001; Ceccarelli et al. 2002; van der Tak et al. 2002; Lis et al. 2002). The observations suggest that the large deuteration of formaldehyde is produced during the cold and dense pre-collapse phase of low-mass protostars. Highly deuterated ices are very likely formed via active grain chemistry (Tielens 1983), stored on the grain mantles, and eventually released into the gas phase during the collapse phase, when the heating of newly-formed protostars evaporates the CO-rich ices (e.g., Ceccarelli et al. 2002). If, as claimed, formaldehyde and methanol are produced on grain surfaces by simple successive hydrogenations of CO, then the abundance ratios between deuterated isotopomers and their normal counterparts are a crucial test for the grain-surface theory of deuteration (Charnley et al. 1997; Caselli et al. 2002).

We report here the very first detection of a doubly deuterated isotopomer of methanol (CHD_2OH), with 15 detected lines, towards the low-mass protostar *IRAS* 16293-2422. We also report the detection of the two singly deuterated forms (CH_2DOH and CH_3OD) of methanol towards the same object. We compare the derived fractionation ratios as well as the formaldehyde fractionation (Loinard et al. 2000) with predictions based on active grain chemistry.

2. Observations and results

Using the IRAM 30-meter telescope (Pico Veleta, Spain), we detected 23 CH_2DOH lines, 6 CH_3OD lines, and 15 CHD_2OH lines with frequencies ranging from 83 to 226 GHz and upper level energies ranging from 4 to 80 cm^{-1}. The telescope was pointed at the coordinates $\alpha(2000) = 16^h32^m22.6^s$ and $\delta(2000) = -24^o28'33.0''$. The observations were performed in November 2001 and May 2002.

Examples of observed spectra are shown in Figure 1. We derived the total column density of the three species from the rotational diagrams presented in Figure 2. The observed scattering is likely due to non-LTE and opacity effects. Note that the level column densities are averaged over a $10''$ beam, the smallest beam of our observations, following the suggestion by van Dishoeck et al. (1995) of enhanced methanol emission in the central $\leq 10''$ re-

Chemistry as a Diagnostic of Star Formation (2003). *Edited by Charles L. Curry and Michel Fich.* Published in 2003 by NRC Research Press, Ottawa ON, K1A 0R6, Canada. ISBN 0-660-19089-3.

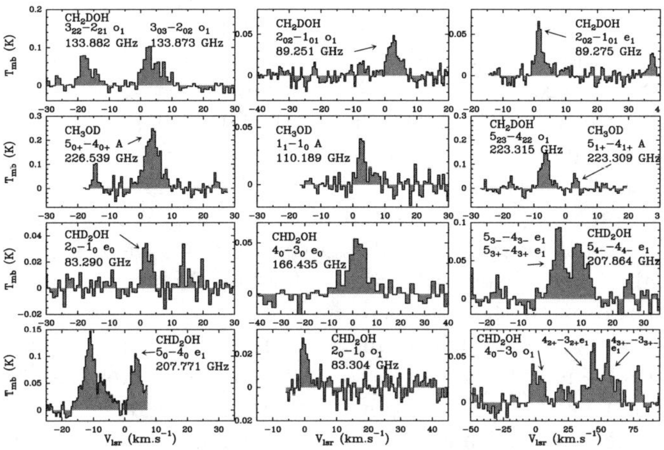

Figure 1. Examples of CH_2DOH, CH_3OD and CHD_2OH lines. Intensities are in main beam brightness temperature.

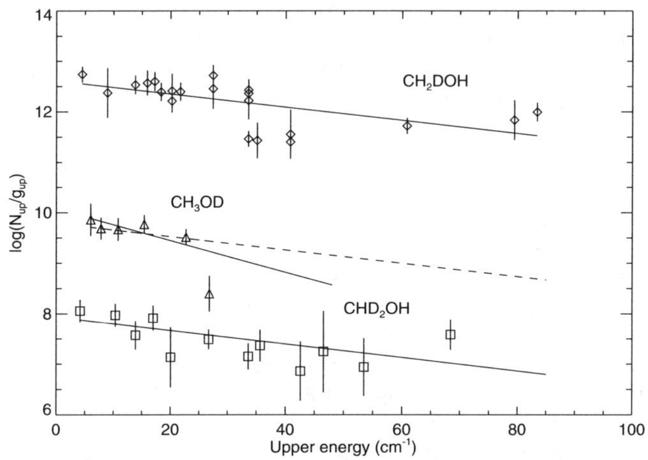

Figure 2. Rotational diagrams for CH_2DOH *(diamonds)*, CH_3OD *(triangles)*, and CHD_2OH *(squares)*. For the sake of clarity, the CH_3OD and CHD_2OH points have been shifted by -2 and -4, respectively. The dashed line is the result of the fit obtained by fixing the rotational temperature at 47.5 K.

gion of *IRAS* 16293. The derived rotational temperatures and overall column densities for the three molecules are reported in Table 1. The CH_3OD rotational temperature is significantly lower compared with the other two, very probably because the detected lines have lower upper level energies. We therefore estimated the CH_3OD column density assuming the same rotational temperature as the two other species, namely 47.5 K. Using the methanol column density derived by van Dishoeck et al. (1995) (3.5×10^{15} cm^{-2} with a source size of 10″), we finally obtain the following fractionation ratios:

$$[CH_2DOH]/[CH_3OH] = 0.9 \pm 0.3,$$
$$[CH_3OD]/[CH_3OH] = 0.04 \pm 0.02,$$
$$[CHD_2OH]/[CH_3OH] = 0.2 \pm 0.1,$$
$$[CH_2DOH]/[CH_3OD] = 20 \pm 14.$$

If the CH_3OD column density derived with T_{rot}=47.5 K is taken, the ratios involving this species change by a factor of two, reflecting the increased abundance of this molecule.

3. Discussion and conclusions

As discussed in the Introduction, the abundances of deuterated methanol and deuterated formaldehyde provide a strong test of models involving active grain chemistry. The basic hypothesis behind these models is that formaldehyde and methanol form by successive hydrogenations of CO accreted onto the grains via reactions with atomic hydrogen (Tielens & Hagen 1982). The enhanced deuteration is caused by an enhanced (atomic) D/H ratio in the gas during the era of mantle formation (Tielens 1983). With some simplifying assumptions, this leads to predictions for steady-state ratios of singly- and doubly deuterated formaldehyde and methanol to their normal isotopic forms in terms of the D/H atomic abundance ratio in the accreting gas. Figure 3 shows fractionation ratios predicted by the Charnley et al. (1997, hereafter CTR97 model. The CH_2DOH and CHD_2OH observations are compatible with an atomic D/H ratio of 0.2 in the accreting gas. The CH_3OD abundance falls short of the predicted value for a D/H ratio of 0.2. However, the gas phase abundance of CH_3OD in the hot core can be affected by gas phase ion-molecule reactions. Specifically, protonation of methanol by H_3^+ or H_3O^+ followed by dissociative electron recombination back to methanol will drive the CH_3OD/CH_3OH ratio to the deuterium fractionation of molecules in the warm gas, which is very low (CTR97). The timescale for this process is some $3{\times}10^4$ yrs which is comparable to the lifetime of *IRAS* 16293 ($\sim 2{\times}10^4$, Ceccarelli et al. 2000). We note that this chem-

Table 1. 10″-averaged molecular column density N and rotational temperature T_{rot}

Molecule	N (cm^{-2})	T_{rot} (K)
CH_2DOH	$(3.0 \pm 0.6) \times 10^{15}$	48 ± 3
CH_3OD	$(1.5 \pm 0.7) \times 10^{14}$	20 ± 4
CH_3OD^a	$(2.8 \pm 0.3) \times 10^{14}$	47.5
CHD_2OH	$(6.0 \pm 2.2) \times 10^{14}$	47 ± 7

[a] Assuming T_{rot}=47.5 K (see text).

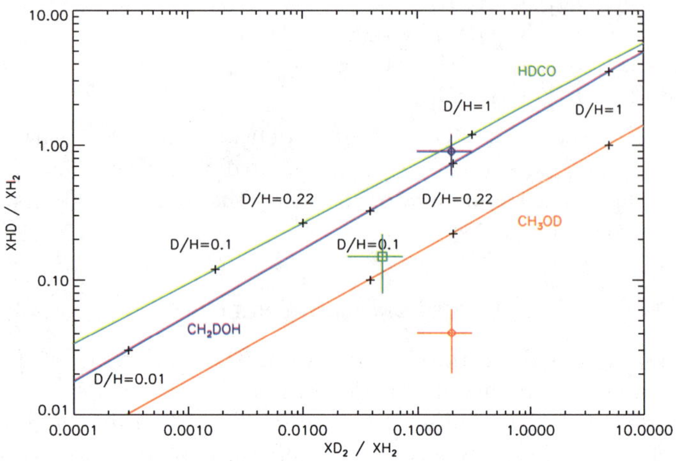

Figure 3. Singly deuterated vs. doubly deuterated fractionation ratios: CH_2DOH/CH_3OH *(blue)* or CH_3OD/CH_3OH *(red)* versus CHD_2OH/CH_3OH and $HDCO/H_2CO$ vs. D_2CO/H_2CO *(green)*. Lines show the model predictions with a varying $(D/H)_{at}$ ratio in the gas phase when accretion occurs on the grains. Observations are reported with error bars.

ical reshuffling of the deuterium will not affect the deuterium fractionation on the methyl group (e.g., CH_2DOH, CHD_2OH). The CTR97 model, however, has some difficulties in explaining the observed formaldehyde fractionation ratios towards *IRAS*16293 (Loinard et al. 2000). In particular, the value of atomic D/H compatible with CH_2DOH and CHD_2OH is reasonably compatible with HDCO but results in too low a fractionation ratio for D_2CO by a factor of 5 or so. Possible gas-phase alterations have not been considered. The formaldehyde discrepancy suggests that at least some of the assumptions in the CTR97 model may be too drastic. In particular, the CTR97 model only considers formaldehyde and methanol formation, ignoring other competing reactions.

Very recently, Caselli et al. (2002) [hereafter CSS02] proposed a somewhat more detailed but related model for the formation of formaldehyde, methanol, all of their deuterated isotopomers, and selected other species on grains. In their model, CSS02 consider accretion onto grains of H, D, CO and O, followed by a comprehensive set of reactions to form H_2O, H_2, CO_2, H_2CO, CH_3OH and all singly- and multiply deuterated isotopomers of these species. Their model differs slightly from that of CTR97 in that it specifically includes small differences in the barriers to reaction between non-deuterated and deuterated species. Predictions for methanol and formaldehyde fractionation ratios are similar to the CTR97 model and substantially the same discrepancies remain. In particular, for $D/H = 0.3$, a temperature of 10 K, and so-called high-density conditions, CSS02 agree approximately with our observed fractionation ratios for CH_2DOH and CHD_2OH, but produce approximately 5 times too much CH_3OD, a value similar to that of CTR97 with $D/H = 0.2$ and no subsequent gas-phase chemistry. For formaldehyde, CSS02 obtain a fractionation ratio for HDCO that is twice the observed value and a fractionation ratio for D_2CO that is a factor of two below the observed value.

In summary, neither model is in good agreement with all of our data. More importantly perhaps, the CTR97 and CSS02 models require a D/H atomic ratio in the range 0.2-0.3, which is a significantly larger value than can be produced by current gas-phase models, even in the presence of a large CO depletion (e.g., Roberts & Millar 2000a). Future progress will probably require more detailed chemical models in which gas-phase and surface chemistry occur simultaneously. Although such models are currently in existence, they do not yet contain fractionation processes.

References

Anderson, Crownover, Herbst, & de Lucia 1988, ApJS, 67, 135
Caselli, Stantcheva, Shalabeia, et al. 2002, PSS, in press
Ceccarelli, Vastel, Tielens, et al. 2002, A&A, 381, L17
Ceccarelli, Loinard, Castets, et al. 2001, A&A, 372, 998
Ceccarelli, Castets, Caux, et al. 2000, A&A, 355, 1129
Ceccarelli, Castets, Loinard, et al. 1998, A&A, 338, L43
Charnley, Tielens, & Rodgers 1997, ApJ, 482, L203
Gottlieb, Ball, Gottlieb, et al. 1979, ApJ, 227, 422
Jacq, Walmsley, Mauersberger, et al. 1993, A&A, 271, 276
Linsky, 1998, Sp. Sci. Rev., 84, 285
Lis, Roueff, Gerin, et al. 2002, ApJ, 571, L55
Loinard, Castets, Ceccarelli, et al. 2001, ApJ, 552, L163
Loinard, Castets, Ceccarelli, et al. 2000, A&A, 359, 1169
Loinard L., et al. 2002, PSS in press
Lubowich, Pasachoff, Balonek, et al. 2000, Nature, 405, 1025
Mauersberger, Henkel, Jacq, et al. 1988, A&A, 194, L1
Menten, Walmsley, Henkel, et al. 1988, A&A, 198, 253
Mukhopadhyay 1998, Spectrochimica Acta, 54, 927
Roberts & Millar 2000, A&A, 361, 388
Roberts & Millar 2000, A&A, 364, 780
Rodgers & Charnley 2001, ApJ, 553, 613
Roueff, Tiné, Coudert, et al. 2000, A&A, 354, L63
Tielens & Hagen 1982, A&A, 114, 245
Tielens 1983, A&A, 119, 177
Turner 1990, ApJ, 362, L29
van der Tak, Schilke, Müller, et al. 2002, A&A, 388, L53
van Dishoeck, Blake, Jansen, et al. 1995, ApJ, 447, 760
Williams, Bergin, Caselli, et al. 1998, ApJ, 503, 689

HCN ABUNDANCE CONTRAST IN WARM DENSE CORES OF G1.6-0.025

Ruisheng Peng[1], John B. Whiteoak[2], Martin Houde[1], and Hiroshige Yoshida[1]

[1]Caltech Submillimeter Observatory, 111 Nowelo St., Hilo, HI 96720
[2]Australia Telescope National Facility, Epping, NSW 2031, Australia

Abstract

We report a 5× contrast in HCN abundance between two warm dense cores of G1.6-0.025. The cores, at a LSR velocity of 145 km s^{-1} and 162 km s^{-1} respectively, are closely aligned along the line of sight with a size of ~2.5 pc and a typical FWHM line width of ~18 km s^{-1}. They also present very similar physical characteristics: core centers as probed by HCN lines have a gas density of $\sim 10^7$ cm^{-3} and a temperature of ≥ 80 K, while the outer peripheries seen in ^{13}CO lines are less dense ($\sim 10^3$ cm^{-3}) at a temperature of 70–120 K.

The contrast in HCN abundance between the two seemingly identical cores may indicate the 145 km s^{-1} core as a site of active star formation, whereby X-rays from the embedded protostars maintain a higher rate of ionization leading to an enhanced HCN abundance in comparison with the more quiescent 162 km s^{-1} core.

Key words: Astrochemistry – ISM: abundances – stars: formation

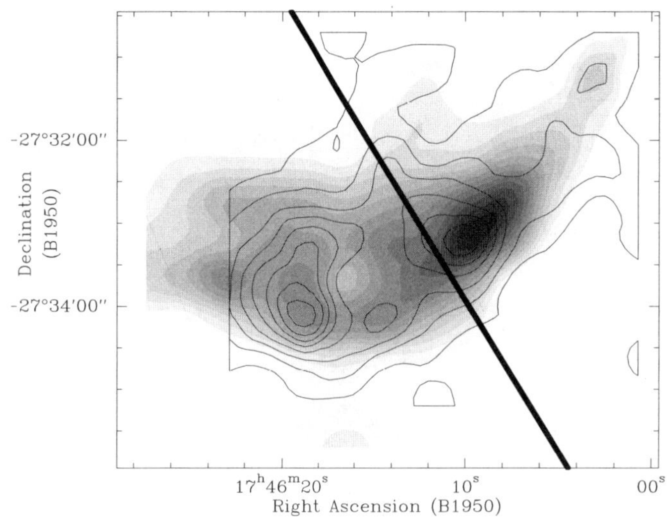

Figure 1. Integrated line emission maps of the 165 km s^{-1} cloud of G1.6-0.025, with HCN 3–2 map in contours superposed on the ^{13}CO 2–1 map in greyscale. The greyscale shade starts from 6 K km s^{-1} and the lowest contour is at 2 K km s^{-1}. Both have the same intensity increment of 4 K km s^{-1}. The thick solid line in the figure represents the Galactic plane with the Galactic center 1°.595 away to the southeast.

1. The Cloud

G1.6-0.025 is a GMC-type molecular cloud located within 300 pc of the Galactic center. Associated with no radio continuum and infrared point source, the cloud was first detected in the 4.8 GHz H_2CO absorption at ~50 km s^{-1} with a line width of ~20 km s^{-1} and spans ~10′ in extent(Whiteoak & Gardner 1979). NH_3 emission at 23 GHz reveals a warm (≥50 K), clumpy, and optically thick structure, typical of Galactic center molecular cloud (Gardner et al. 1985). An additional cloud at ~160 km s^{-1} was later discovered also in NH_3 inversion lines at 23 GHz(Gardner & Boes 1987). Compared with the 50 km s^{-1} cloud, the 160 km s^{-1} cloud is much warmer (≥120 K) and more spatially confined (~5′ in size).

We show in Figure 1 the integrated intensity maps of the 160 km s^{-1} cloud made with the CSO, with ^{13}CO 2–1 in gray scale and HCN 3–2 as contours. Line flux was integrated over a LSR velocity range of 135–195 km s^{-1} and 125–190 km s^{-1} for the ^{13}CO 2–1 and HCN 3–2 lines, respectively. One sees a well-confined, elongated structure oriented nearly perpendicular to the Galactic plane, as indicated by the thick solid line in the figure.

2. Intriguing Line Profiles

In Figure 2 we present the spectra of various molecular lines observed towards the NE peak of the 160 km s^{-1} cloud in Fig. 1. One sees two types of line profiles: the isotopic CO lines exhibit extended blue wings and the HCN as well as SiO lines show extended red wings, with the latter peaking at lower velocities than the former.

Chemistry as a Diagnostic of Star Formation (2003). *Edited by Charles L. Curry and Michel Fich.* Published in 2003 by NRC Research Press, Ottawa ON, K1A 0R6, Canada. ISBN 0-660-19089-3.

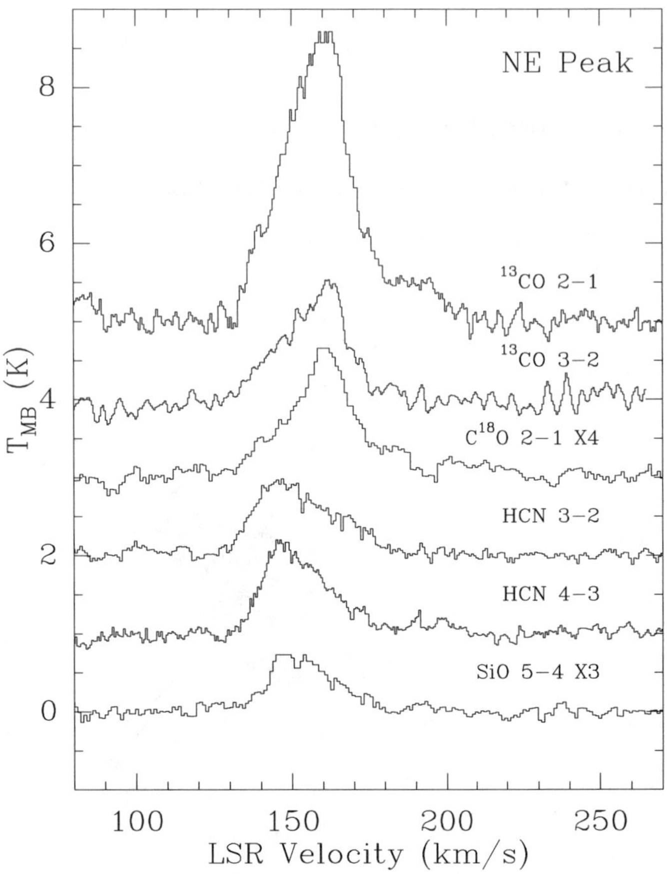

Figure 2. Spectra taken towards the NE peak in the 160 km s^{-1} cloud of G1.6-0.025.

Such a spectral contrast between isotopic CO and HCN lines is intriguing as one would expect that the isotopic CO line profiles follow that of the dense gas traced by the HCN lines, especially when the isotopic CO lines can be reasonably assured to be optically thin by the resemblance of their line profiles.

3. Two Different Worlds

Taking the cue from profiles of the isotopic CO lines and HCN lines, we decompose each profile into two Gaussian components. Table 1 lists the derived Gaussian parameters. The 145 km s^{-1} component has a line temperature ratio of ~0.5 between the ^{13}CO 3–2 and 2–1 lines, and ~1.2 between the HCN 4–3 and 3–2 lines. For the 162 km s^{-1} component, the ratio is ~0.4 and ~1.3 respectively.

LVG modelling unambiguously ties the origin of the HCN line emission to a warm, dense environment: $T_K \geq 80$ K, $n(H_2) \sim 10^7$ cm^{-3}, and N(HCN)$\sim 1.0 \times 10^{13}$ cm^{-2} and $\sim 6.3 \times 10^{12}$ cm^{-2} for the 145 km s^{-1} component and the 162 km s^{-1} component, respectively.

Table 1. Gaussian components at the NE peak

Line	$v_{\rm LSR}$ km s^{-1}	w km s^{-1}	$T_{\rm MB}$ K
^{13}CO (2–1)	146.2	18.0	1.36
	161.2	18.9	3.44
^{13}CO (3–2)	144.3	17.0	0.67
	161.2	17.4	1.36
HCN (3–2)	145.1	18.8	0.89
	162.8	20.9	0.42
HCN (4–3)	146.4	15.2	1.06
	161.5	18.0	0.55

^{13}CO lines, on the other hand, arise from relatively low density regions ($n[H_2] \sim 10^3$ cm^{-3}). The gas temperature is likely in the range of 70–120 K, with the upper limit constrained by the gas density needed to withstand the strong tidal shearing in the Galactic center region. The lower temperature limit is the typical gas temperature in the Galactic center region (Morris & Serabyn 1996). Over this temperature range, the ^{13}CO column density $N(^{13}$CO) remains fairly constant at $\sim 6.8 \times 10^{15}$ cm^{-2} and $\sim 2.2 \times 10^{16}$ cm^{-2} for the 145 km s^{-1} component and the 162 km s^{-1} component, respectively.

Given the detailed correspondence on the spectral features as well as the projected sky positions of the two molecular species, it would be reasonable to associate the region where ^{13}CO lines originates and the region where HCN lines originate to the same physical entity. This is to say that the 145 km s^{-1} component and the 162 km s^{-1} component each represent a centrally condensed core where HCN lines come from the core center and ^{13}CO lines originate in the outer peripheries of the core.

4. Where Is the Dense CO Gas?

Now that the isotopic CO line emission is largely attributed to the outer peripheries of the dense cores, one naturally wonders about the CO molecules in the core centers traced by the HCN lines. After all, CO depletion occurs in cold dense cloud cores through adsorption of CO molecules onto dust grains (e.g., Bergin et al. 2002). In warm gas, however, the chemical reaction network lacks any viable means to efficiently remove CO molecules from gas phase.

To gauge the extent of any CO depletion, we partition the observed ^{13}CO line temperatures into two parts, one attributable to the dense core center, and the other to the outer peripheries. By solving the equations that combine the two parts to the observed line temperature, the part attributable to the core center, $T_c(2-1)$, is simply

$$T_c(2-1) = \frac{\alpha_0 - \alpha_2}{\alpha_1 - \alpha_2}\, T(2-1),$$

where $T(2-1)$ is the observed ^{13}CO 2–1 line temperature, α_0 is the observed ^{13}CO 3–2 to 2–1 line ratio (0.4–0.5), α_1 that of the core center, and α_2 that of the peripheries. At core centers, LVG calculations yield an $\alpha_1 \sim 2$. For the peripheries, α_2 is not likely to be much lower than α_0 due to the constraint on the gas density by the strong tidal shearing in the Galactic center region. With $\alpha_0 - \alpha_2 \leq 0.1$, $T_c(2-1)$ would be less than 6% of $T(2-1)$. This leads to a ^{13}CO column density of $\leq 2 \times 10^{15}$ cm^{-2} and $\leq 5 \times 10^{15}$ cm^{-2} for the dense center of the 145 km s^{-1} and the 162 km s^{-1} core, respectively. With a typical ^{12}CO to ^{13}CO abundance ratio of 24 for the Galactic center region (Langer & Penzias 1990), this leads to a ^{12}CO to HCN abundance ratio of $\leq 4.8 \times 10^3$ and $\leq 1.9 \times 10^4$ for the two core centers, respectively, comparing favorably to a value of 10^3–10^4 often seen in star-forming regions in the Galaxy(van Dishoeck & Blake 1998). There is therefore no evidence of CO depletion in the warm dense core centers.

5. HCN Enhancement in the 145 km s^{-1} Core

The observed spectral contrast between the isotopic CO lines and the HCN lines is then truly a reflection of a higher HCN abundance in the 145 km s^{-1} core as compared to the 162 km s^{-1} core. By comparing the ^{13}CO to HCN column density ratios in the 145 km s^{-1} core to that of the 162 km s^{-1} core, one sees that HCN abundance in the former is enhanced by a factor of five.

Given the almost identical physical environment for the two cores as inferred from molecular line observations, such a contrast in HCN abundance is striking. It signals significant differences in their chemical environments. A higher rate of ionization has been found to enhance HCN abundance in protostellar disks(Aikawa & Herbst 1999; Markwick et al. 2002). In a cloud absent of any compact H II regions and near- and mid-infrared point sources, X-rays from protostars (Tsujimoto et al. 2002) can be a significant source of ionization, as shown by Glassgold, Najita, & Igea (1997). In relating to the two warm dense cores in G1.6-0.025, this would imply that the 145 km s^{-1} core harbors one or likely more deeply embedded protostars. The core, with a size of ~2.5 pc and a virial mass in excess of $8\times10^4 M_\odot$, is naturally a site of potential star formation. In addition, the attenuation length of X-rays, 10^{21}–10^{23} cm^{-2} hydrogen nuclei (Aikawa & Herbst 1999), is comparable to the H_2 column density of the 145 km s^{-1} core ($\sim 6 \times 10^{20}$ cm^{-2}, with a ^{12}CO to ^{13}CO abundance ratio of 24 and a ^{12}CO abundance of 2.7×10^{-4} [Lacy et al. 1994]). So ionization by X-rays would occur on a global scale in the 145 km s^{-1} core, resulting in the observed HCN abundance enhancement relative to the 162 km s^{-1} core which may have no or a lower level of star forming activity. The SiO line (see Fig. 2) also lends support for a heightened level of star formation in the 145 km s^{-1} core as it often signals shocks and outflows.

Future high spatial resolution observations should help to resolve the issue of whether the 145 km s^{-1} core harbors any protostars. In addition, observation of CN lines, preferably in the 3 mm range where the hyperfine lines are well separated, can determine whether the 145 km s^{-1} core has indeed a higher rate of ionization as CN is expected to be more abundant under such conditions in the models of Aikawa & Herbst (1999) and Markwick et al. (2002).

References

Aikawa, Y. & Herbst, E., 1999, A&A, 351, 233

Bergin, E. A., Alves, J., Huard, T. & Lada, C. J., 2002, ApJ, 507, L101

Gardner, F. F. & Boes, F., 1987, PASA, 7, 185

Gardner, F. F., Whiteoak, J. B., Forster, J. R., Peters, W. L. & Kuiper, T. B. H., 1985, PASA, 6, 176

Glassgold, A. E., Najita, J. & Igea, J., 1997, ApJ, 480, 334

Lacy, J. H., Knacke, R., Geballe, T. R. & Tokunaga, A. T., 1994, ApJ, 428, L69

Langer, W. D. & Penzias, A. A., 1990, ApJ, 357, 477

Morris, M. & Serabyn, E., 1996, ARA&A, 34, 645

Markwick, A. J., Ilgner, M., Millar, T. J. & Henning, Th., 2002, A&A, 385, 632

Tsujimoto, M., Koyama. K., Tsuboi, Y., Goto, M. & Kobayashi, N., 2002, ApJ, 566, 974

van Dishoeck, E. F., & Blake, G. A., 1998, ARA&A, 36, 317

Whiteoak, J. B. & Gardner, F. F., 1979, MNRAS, 188, 445

CHEMICAL DIAGNOSTICS FOR GAS-PHASE METAL DEPLETION IN COLD ASTROPHYSICAL ENVIRONMENTS

Simon Petrie

Department of Chemistry, the Faculties, Australian National University, Canberra, ACT 0200, Australia

ABSTRACT

Negative results from searches for metal-containing molecules (MCMs) in dense interstellar (IS) clouds, and current models of IS cloud chemistry, suggest severe depletion of gas-phase metals. To properly interpret the observational results, however, the true gas-phase reactivity of metal-containing species must be known: non-detection of a MCM is useful only if its chemistry is well-understood. Here we suggest that the MCMs sought to date in IS clouds either lack viable formation processes or have an intrinsically short lifetime, severely limiting both their expected abundance and their usefulness as "fingerprints" for gas-phase metals. We explore efforts to develop more detailed models of metal chemistry in IS clouds and circumstellar envelopes, with a view to predicting detectable gas-phase markers for metals in such environments.

Key words: ISM: molecules – ISM: clouds – astrochemistry

1. INTRODUCTION

Dense interstellar (IS) clouds are rich in unsaturated organic, and small inorganic, molecules, but no metal-containing molecules (MCMs) have been detected in such objects. In other molecular astrophysical environments, known MCMs are restricted to halides such as KCl and AlF (Cernicharo & Guelin 1987) and the cyanides Na(CN) (Turner et al. 1994), Mg(CN) (Kawaguchi et al. 1993; Ziurys et al. 1995) and AlNC (Ziurys et al. 2002). These compounds are found in the outflowing carbon-rich envelope of the post-AGB star IRC+10216; some have recently also been seen in the protoplanetary nebula CRL2688 (Highberger et al. 2001). A larger set of MCMs, including hydrides of most of the cosmically abundant metals, oxides, carbides, and sulfides, has been sought without success (Turner & Steimle 1985; Turner 1991; Sakamoto et al. 1998). The consensus is that, except in very "youthful" molecular environments such as stellar outflows, cumulative depletion of gas-phase metals onto IS dust grains effectively precludes the detection of MCMs.

The inferred large depletion of gas-phase metals is convenient for modelling cold IS cloud chemistry: it is widely assumed (see, e.g., Caselli et al. 1998) that any gas-phase metal M exists mainly as its positive ion M^+, produced by charge transfer from abundant molecular ions such as HCO^+. In most models the sole M^+ loss process is inefficient radiative recombination (RR):

$$M^+ + e \rightarrow M + h\nu$$

(a typical metal ion, Na^+, has a recombination coefficient of 9×10^{-12} cm^3 molecule^{-1} s^{-1}). If gas-phase metals are sufficiently abundant that M^+ becomes the predominant carrier of positive charge in IS clouds, the synthetic chemistry initiated by molecular ions is hindered, and chemical models are unable to account for the observed development of molecular complexity in clouds such as TMC-1.

Recent studies have, however, demonstrated that RR is not the main loss process for M^+ (Millar 1982; Petrie 1996a, 1999; Petrie & Dunbar 2000; Dunbar & Petrie 2002). Radiative association (RA) to polyatomic neutrals, e.g.,

$$Mg^+ + HC_7N \rightarrow MgNC_7H^+ + h\nu,$$

reduces the lifetime of a typical metal ion, Mg^+, in TMC-1 from 3×10^6 years to 3×10^4 years (Dunbar & Petrie 2002). Such RA reactions are also the only known viable pathways to the observed metal cyanides MgNC and MgCN (Petrie 1996a, 1999). Thus, gas-phase metals need not be highly depleted (if sufficient abundant neutrals exist to efficiently remove M^+), and MCMs amongst the reaction products of such processes are potentially detectable. In this Note, using Mg as a typical metal, I outline efforts to fill in some of the many gaps in our knowledge of the IS chemistry of metals.

Chemistry as a Diagnostic of Star Formation (2003). *Edited by Charles L. Curry and Michel Fich.* Published in 2003 by NRC Research Press, Ottawa ON, K1A 0R6, Canada. ISBN 0-660-19089-3.

2. Discussion of Recent Results

2.1. Mg^+ in cold clouds: radiative association

Magnesium is a useful 'prototype' with which to explore IS metal chemistry: Mg is among the ten most cosmically abundant elements (of metals, only Fe has a comparable cosmic abundance) and has been detected in circumstellar outflows in the isomeric MgNC (Kawaguchi et al. 1993) and MgCN (Ziurys et al. 1995) radicals. As a main-group metal, Mg is amenable to existing ab initio quantum chemical methodologies, which is advantageous given the difficulties inherent in laboratory study of gas-phase MCMs.

One important characteristic of Mg is a tendency to form weak bonds. For example, the Mg-NC bond strength of 321 kJ mol^{-1} (Petrie 1996b) is almost 190 kJ mol^{-1} lower than that of Al-NC (Petrie 1996c). Thus, while several pathways may exist for formation of circumstellar AlNC (Petrie 1999; Ziurys et al. 2002), the production of MgNC and MgCN from efficient exothermic reactions can only be satisfied by RA reactions of Mg^+ with the cyanopolyynes HC_nN (n = 5–11) (Petrie 1996a; Dunbar & Petrie 2002). This is a useful 'test case' for the IS chemistry of Mg: the efficiency of these RA reactions derives, first, from the comparatively large number of atoms in the collision complex $[MgNC_nH^+]^*$, ensuring a long complex lifetime and therefore a high probability of radiative stabilization, and second from the high exothermicity of Mg^+-ligand bond formation. As we shall see, other IS neutrals share these features of polyatomicity and large Mg^+ affinity; but it is, e.g., difficult to see how RA reactions could produce MgH, $MgCH_3$ or MgCCH, all of which have been proposed as potential MCMs in IS clouds.

From our studies of Mg^+ RA kinetics (Petrie & Dunbar 2000; Dunbar & Petrie 2002), we can establish "ground rules" for efficient RA reactions of metal ions under cold IS conditions. First, the reactant neutral should possess at least 7 atoms (the near-collision-rate RA of Mg^+ with HC_5N is over 1000 times more efficient than RA with HC_3N). Second, the neutral's metal-ion affinity should exceed 100 kJ mol^{-1}. Third, the neutral should be abundant, at least $10^{-10}\times$ $n(H_2)$ and preferably considerably higher. Some neutrals which come closest to meeting these criteria are listed in Table 1. Calculations to establish the efficiency of the RA processes in Table 1 are underway. Within TMC-1, high abundances of the cyanopolyynes HC_5N and HC_7N (though not as high as in IRC +10216) justify a search for MgNC. Few other apparent 'precursors' to MCMs are present in high abundance.

Table 1. Pertinent parameters for some metal-ligand complexes

X	$N(X)^a$	$n(X)^b$	$D_0(Na^+\text{-}X)^c$	$D_0(Mg^+\text{-}X)^c$
$CH_3{}^d$	4	2.2(−9)	31.8	201.3
$C_2H_3{}^d$	5	2(−11)	52.0	226.5
C_4H^e	5	8.6(−7)	50?	>350?
CH_3CN^f	6	1.9(−8)	128.5	163.0
HC_5N^h	7	1.4(−9)	134.3	184.5
$(CH_3)_2O^f$	9	4.4(−9)	102.5	154.8
$C_2H_5OH^f$	9	3.7(−12)	109.6	155.5
HC_8H^g	10	9.6(−12)	81.3	129.2

Notes: [a] Number of atoms in the ligand X. [b] Abundance ratio ("early time"), relative to $n(H_2)$, from UMIST95 rate file (Millar et al. 1997). Notation: $a(b) \equiv a \times 10^b$. [c] Bond energy in kJ mol^{-1}. [d] CP-dG2thaw calculation (Petrie 2002). [e] Estimated from the CP-dG2thaw value for C_2H. [f] CP-dG2thaw calculation (Petrie 2001). [g] B3-LYP/aug-TZ calculation (Dunbar & Petrie 2002).

Two of the most promising species listed here are C_4H, a highly abundant radical, and dimethyl ether CH_3OCH_3. (Note that, although C_4H fails to fulfil our requirement of at least 7 atoms, its exceptionally high expected Mg^+ affinity is likely to promote efficient RA). RA of Mg^+ with C_4H is a potential route to the MCMs MgC_2 and MgC_4:

$$\begin{aligned} MgCCCCH^+ + e &\rightarrow Mg + C_4H \\ &\rightarrow MgCC + C_2H \\ &\rightarrow MgCCCC + H. \end{aligned}$$

The reaction with dimethyl ether (a neutral not detected in TMC-1, but seen in high abundance in warmer giant molecular clouds) appears well suited to production of $MgOCH_3$, or of MgO, and searches for these MCMs in cold clouds known to contain dimethyl ether may be worthwhile.

The RA product of Mg^+ with a closed-shell neutral always has one unpaired electron. In the dissociative recombination of the RA product ion $MgXY^+$ with e, there is a 75% probability of recombination on a triplet surface and 25% on a singlet surface. This distinction is important because, although the Mg-X bond is always expected to be the weakest bond on the singlet surface, it is often one of the stronger bonds on the triplet surface. While bond strength is not the only factor influencing the dissociative recombination product ratio, it is reasonable to expect that the triplet surface recombination favours retention of the Mg-X bond to a greater degree than does the singlet-state recombination, and it has been suggested (Petrie 1999) that this effect particularly favours forma-

Table 2. Thermochemistry of some X^+ + Mg reactions

Reactants	Products	ΔH° / kJ mol^{-1} [a]
$HCNH^+$ + Mg	Mg^+ + HCN + H	+144
	Mg^+ + HNC + H	+203
	Mg^+ + HCNH	+73
	MgH^+ + HCN	**−53**
HCO^+ + Mg	Mg^+ + H + CO	+20
	Mg^+ + HCO	−44
	MgH^+ + CO	**−177**

[a] Enthalpy of reaction. Constituent thermochemical values taken from the NIST webbook, except for HNC, HCNH, and MgH^+ for which high-level computational values (Petrie 2002) have been used.

tion of detectable abundances of MCM radicals derived from doublet-state atomic metal ions.

2.2. Other aspects of Mg^+ chemistry

The RA chemistry discussed above requires a reservoir of Mg^+. In existing IS models, it is assumed that molecular ions react with Mg by charge transfer (i.e., $AB^+ + Mg \rightarrow Mg^+ + AB$) or by dissociative charge transfer (i.e., $AB^+ + Mg \rightarrow Mg^+ + A + B$). But Mg has both a high ionization energy (IE = 7.646 eV) and a high proton affinity (PA = 762 kJ mol^{-1}) favouring MgH^+ over Mg^+ formation. The thermochemistry for reactions of two abundant cold-cloud ions with Mg is summarized in Table 2.

From Table 2, reaction of $HCNH^+$ with Mg can only proceed via proton transfer, while proton transfer from HCO^+ may well also be preferred. The molecular ion MgH^+ has not been considered in IS models, but may be an important reservoir of gas-phase Mg. This ion can also form in reactions of Mg^+ with radicals such as C_2H_3 and C_2H_5 (Petrie 2002), which may be abundant in warmer giant molecular clouds: e.g.,

$$Mg^+ + C_2H_5 \rightarrow MgH^+ + C_2H_4.$$

Unless MgH^+ is observed directly, it will be difficult to infer its presence: while dissociative recombination of MgH^+ is presumably efficient, it can't lead to an MCM.

Reactions of neutral Mg may even be relevant to MCM formation. Consider the reaction

$$Mg + C_4H \rightarrow MgCCCCH + h\nu$$

(a highly exothermic reaction with an abundant polyatomic neutral). The calculated rotational constant for MgCCCCH is within 1% of the carrier of a prominent set of unidentified lines in IRC+10216, although another Mg-containing molecule is also a candidate for these U-lines (Petrie 1999).

3. Conclusions

Three general assumptions have been made concerning the gas-phase chemistry of metals in cold IS environments:

(1) Metal atoms don't react with anything except positive ions.

(2) Metal atom/positive ion reactions only produce atomic metal ions.

(3) Atomic metal ions don't react with anything except free electrons.

In this short Note, I have argued that all three of these assumptions are incorrect, and that the metal chemistry of such environments is much more complex than has previously been acknowledged. While I have described some aspects of IS metal chemistry currently under investigation, there is clearly much to be uncovered on this subject. In continuing collaborations with Rob Dunbar at Case Western Reserve, the Ziurys group in Arizona, Barry Turner at NRAO, and Kentarou Kawaguchi at Nobeyama, efforts are underway to enlarge our current 'knowledge base' of IS metal chemistry.

References

Caselli, P, et al. 1998, ApJ, 499, 234

Cernicharo, J. & Guelin, M. 1987, A&A, 183, L10

Dunbar, R. C. & Petrie, S. 2002, ApJ, 564, 792

Highberger, J. L., et al. 2001, ApJ, 562, 790

Kawaguchi, K., et al. 1993, ApJ, 406, L39

Millar, T. J. 1982, in *Galactic and Extragalactic Infrared Spectroscopy*, ed. Kessler, Phillips & Guyenne (ESA SP-192), 33

Millar, T. J., et al. 1997, A&ASS, 121, 139

Petrie, S. 1996a, MNRAS, 282, 807

——— 1996b, J. Chem. Soc., Far. Trans., 92, 1135

——— 1996c, J. Phys. Chem. A, 100, 11581

——— 1999, MNRAS, 302, 482

——— 2001, J. Phys. Chem. A, 105, 9931

——— 2002, J. Phys. Chem. A, 106, 5188

——— & Dunbar, R. C. 2000, J. Phys. Chem. A, 104, 4480

Sakamoto, S., et al. 1998, MNRAS, 301, 872

Turner, B. E. 1991, ApJ, 376, 573

Turner, B. E. & Steimle, T. C. 1985, ApJ, 299, 956

Turner, B. E., Steimle, T. C., & Meerts, L. 1994, ApJ, 426, L97

Ziurys, L. M., et al. 1995, ApJ, 445, L47

——— 2002, ApJ, 564, L45

PAH/DIAMOND HYBRIDS: A STUDY OF SOME MOLECULAR PROTOTYPES

Simon Petrie[1,2] and Walter W. Duley[3]

[1]Department of Chemistry, the Faculties, Australian National University, Canberra ACT 0200, Australia
[2]School of Chemistry, University College, University of New South Wales, Australian Defence Force Academy, Canberra ACT 2600, Australia
[3]Physics Department, University of Waterloo, Waterloo, ON, Canada N2L 3G1

ABSTRACT

We present results of density functional theory calculations on a novel class of polycyclic hydrocarbons, termed here "PAH/diamond hybrids," which are feasible candidates for production in carbon-rich molecular astrophysical environments.

Key words: ISM: molecules – astrochemistry

1. INTRODUCTION

Sharp IR emission features, seen in type C UIR sources (Tokunaga 1996), correlate with CH groups in a diamond-like framework. Conditions in these sources are very similar to those in type A and B objects, in which PAHs have been implicated in the carbon chemistry. It is known that graphite can act as a site for diamond nucleation (Jungnickel et al. 1996); PAHs, or layered assemblies of PAHs, may share this ability to nucleate. Thus, PAH/diamond hybrids are legitimate candidates for production in stellar atmospheres.

While research into the large molecular carriers of various interstellar (IS) spectral features has begun to consider PAHs modified either by hydrogenation (Bernstein et al. 1996; Wagner et al. 2000; Beegle et al. 2001; Pauzat & Ellinger 2001) or by aliphatic substitution (Geballe et al. 1992; Joblin et al. 1996; Sloan et al. 1997; Langhoff et al. 1998; Pauzat et al. 1999), the "hybrid" structures which have been investigated to date may be described as 'PAH-dominant' in that they generally feature only minor modifications to a planar PAH skeleton. In the present work, we consider structures which retain some aromatic (sp^2) features but which possess a considerably greater degree of aliphatic (sp^3) character than has featured in previous astrophysically motivated studies on PAH-derived hydrocarbons. In this regard, the structures investigated here (which are necessarily small, no larger than 40 carbon atoms at most, for reasons of computational expediency) can be considered molecular analogues to the particulate hydrogenated amorphous carbon (HAC) (Duley & Williams 1983; Duley 1994; Grishko & Duley 2000) which may be an important reservoir of IS carbon.

Astrochemical identification of the title compounds will require a detailed knowledge of their spectroscopic characteristics, which we (begin to) explore here. We have focussed principally on the far-IR spectroscopy (ν < 600 cm^{-1}), since modes in this region show the greatest dependence on molecular morphology (these are often "whole-body" vibrations) and therefore represent a potential "fingerprint region" for these structures.

2. THEORETICAL METHODS

Hydrocarbon structures investigated in the present work have been subjected to geometry optimization and frequency calculation (including determination of IR, but not Raman, intensities) at the B-LYP/6-31G* level of theory. This combination of the Becke exchange (Becke 1988) and Lee-Yang-Parr correlation (Lee et al. 1988) density functionals is one of the most widely-used density functional theory (DFT) approaches. In a detailed study by Scott and Radom (1996) of performance in (unscaled) frequency calculations using the standard 6-31G* basis set, B-LYP was found to outperform the seven other density functional and quantum chemical methods surveyed as assessed by its agreement with a "test set" of known vibrational frequencies for 122 small molecules.

All calculations reported here used the GAUSSIAN98 quantum chemical program suite (Frisch et al. 1998). While smaller species were studied via Linux-based workstations, calculations on the larger species investigated were executed on the APAC Compaq AlphaServer supercomputer housed at the ANU Supercomputer Facility.

Chemistry as a Diagnostic of Star Formation (2003). *Edited by Charles L. Curry and Michel Fich.* Published in 2003 by NRC Research Press, Ottawa ON, K1A 0R6, Canada. ISBN 0-660-19089-3.

3. Results and Discussion

3.1. Structural effects

Representative structures for which calculations were pursued are shown in Figure 1. In these images, sp^2 (aromatic) C atoms are shown in red while sp^3 (aliphatic, i.e., "diamond-like") carbons are grey. In three of the examples given, we have imposed the most compact feasible "layering arrangement" for the aromatic layers, in which aromatic subunits are anchored to every second row of sp^3 carbons. The spacing between such sp^3 C atoms is approximately 2.6 ρA, significantly less than the 3.4 ρA graphite interlayer separation. Repulsive interactions between the π clouds of the aromatic layers lead to a distortion of the structure around the points of sp^3-sp^2 attachment, such that these anchorage points splay apart; resulting interlayer distances, in structures such as $C_{28}H_{26}$ and $C_{36}H_{32}$ are, at 3.0 to 3.2 ρA, rather closer to the graphitic value. Similar distortion gives rise to the 'open-mouth' appearance of $C_{26}H_{24}$, which unlike the other structures shown has only one aliphatic "fragment" to which the aromatic subunits are attached. In contrast, the structure $C_{34}H_{36}$ in which the anchorage points are 5.2 ρA (4 sp^3 rows) apart features a more modest degree of distortion.

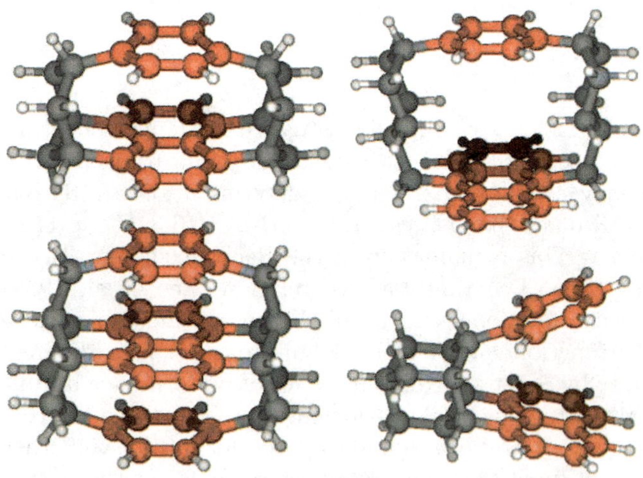

Figure 1. Geometries, optimized at the B-LYP/6-31G* level of theory, for the PAH/diamond hybrid molecules $C_{28}H_{26}$, $C_{36}H_{32}$ *(left)*, $C_{34}H_{36}$, and $C_{26}H_{24}$ *(right)*. Colours show the formal hybridization on each carbon, as noted in the text.

The evidence for substantial strain in structures such as $C_{28}H_{26}$ suggests that these species may not enjoy high stability. While this may appear to disfavour their formation in astrophysical environments, it is worth noting that significantly strained structures may well arise at high temperature, as is witnessed by the observed laboratory (and proposed circumstellar) formation of fullerenes such as C_{60}.

3.2. Spectral effects

Comparatively intense frequency modes obtained through our calculations on various structures are summarized in Table 1. Note that this table includes only a few representative structures; space does not permit a full tabulation of the species surveyed to date.

The calculations performed indicate that these hydrocarbons possess relatively little spectral activity in the low frequency region. In fact, preliminary calculations suggest that the activity of these structures below ν = 600 cm^{-1} is considerably enhanced by derivatization by -OH or -NH_2. It is therefore probable that transitions in the 'fingerprint' region will prove more useful in the identification of such derivatized structures than of the "pure" hydrocarbons which form the basis of this Note.

Another feature of the calculations which may warrant further investigation is the C-H stretching frequencies for the "border" regions of the molecules; i.e., those regions featuring adjacent sp^2 and sp^3 carbons. We might expect that proximity to aromatic subunits could distort the aliphatic C-H stretch frequencies away from their typical values, and this may prove useful in identifying IS spectral features arising from the types of structure under consideration here. However, at this stage the results of our study are inconclusive in this regard.

4. Conclusions

Density functional theory calculations on a series of PAH/diamond hybrids, containing up to 40 atoms, have for the first time provided some insight into the structural and spectral characteristics of these molecules, which might be considered as "hydrogenated amorphous carbon in miniature." Further study is needed to explore the prospects for formation and persistence of such structures in astrophysical environments, as well as to refine strategies by which such structures might be detected.

Acknowledgements

This research was supported by a grant from the NSERC of Canada.

Table 1. Notable vibrational modes obtained from B-LYP/6-31G* calculations on various PAH/diamond hybrid structures

Formula [a]	PAH subunits [b]	Aromatic layer spacing	$\nu < 600$ [c] $I > 5$ km mol^{-1}	$600 < \nu < 750$ [c] $I > 10$	$750 < \nu < 1000$ [c] $I > 20$	$1000 < \nu$ [c] I large
$C_{26}H_{32}$	bz,bz	4.9 ρA	475 (7)			2960 (213)
			519 (9)			3013 (193)
						3121 (51)
$C_{27}H_{34}$	bz,bz	4.9 ρA	520 (8)			2954 (154)
			546 (5)			3012 (142)
						3121 (48)
$C_{28}H_{26}$ *	bz,naph	3.2 ρA	514 (5)		796 (20)	2967 (328)
					821 (35)	3091 (76)
$C_{34}H_{36}$ *	bz,anth	5.3 ρA		654 (12)	758 (45)	2955 (152)
				690 (13)	786 (19)	3020 (158)
						3113 (67)
$C_{36}H_{32}$ *	bz,naph,bz	3.0 ρA, 3.0 ρA	525 (10)	669 (16)	807 (53)	2959 (504)
						3092 (58)
$C_{40}H_{30}$	naph,anth	3.0 ρA			794 (35)	2963 (256)
					811 (21)	3094 (87)
$C_{26}H_{24}$ *	bz,naph	— [d]		685 (17)	769 (30)	2965 (151)
				714 (12)	809 (21)	3114 (61)
				736 (10)		

NOTES:

[a] Asterisks denote structures shown in Fig. 1.
[b] Abbreviations used: bz (benzene), naph (naphthalene), anth (anthracene).
[c] Frequency in cm^{-1} (intensity in parentheses).
[d] Aromatic layers are not parallel in the optimized structure.

REFERENCES

Becke, A. D. 1988, Phys. Rev. A, 38, 3098
Beegle, L. W., Wdoniak, T. J., & Harrison, J. G. 2001, Spectrochim. Acta A, 57, 737
Bernstein, M. P., Sandford, S. A., & Allamandola, L. J. 1996, ApJ, 472, L127
Duley, W. W. 1994, ApJ, 430, L133
Duley, W. W., & Williams, D. A. 1983, MNRAS, 205, 67P
Frisch, M. J., et al. 1998, *GAUSSIAN98*, Gaussian Inc., Pittsburgh, PA
Geballe, T. R., et al. 1992, ApJ, 387, L89
Grishko, V. I., & Duley, W. W. 2000, ApJ, 543, L85
Joblin, C., et al. 1996, ApJ, 458, 610
Jungnickel, G., et al. 1996, Phys. Stat. Sol. (a), 154, 109
Langhoff, S. R., et al. 1998, J. Phys. Chem. A, 102, 1632
Lee, C., Yang, W., & Parr, R. G. 1988, Phys. Rev. B, 37, 785
Pauzat, F., & Ellinger, Y. 2001, MNRAS, 324, 355
———, Talbi, D., & Ellinger, Y. 1999, MNRAS, 304, 241
Scott, A. P. & Radom, L. 1996, J. Phys. Chem., 100, 16502
Sloan, G. C., et al. 1997, ApJ, 474, 735
Tokunaga, A. T. 1996, in *ASP Conf. Ser. 124, Diffuse Infrared Radiation and the IRTS*, eds., H. Okada, T. Matumoto & T. L. Roellig (San Francisco: ASP), 149
Wagner, D. R., Kim, H.-S., & Saykally, R. J. 2000, ApJ, 545, 854

RESULTS FROM A VLT/ISAAC SURVEY OF ICES AND GAS AROUND YOUNG STELLAR OBJECTS

K. M. Pontoppidan[1], H. Fraser[1], F. L. Schöier[1], E. Dartois[2], W.-F. Thi[1,3], and E. F. van Dishoeck[1]

[1]Leiden Observatory, P.O. Box 9513, 2300 RA Leiden, The Netherlands
[2]Insitut d'Astrophysique Spatiale, Bât. 121, Université Paris XI, 91405 Orsay Cedex, France
[3]Astronomical Institute "Anton Pannekoek," University of Amsterdam, Kruislaan 403, 1098 SJ Amsterdam, The Netherlands

Abstract

General results from a $3-5$ μm spectroscopic survey of nearby low-mass young stellar objects are presented. L and M-band spectra have been obtained of ~ 50 low-mass embedded young stars using the ISAAC spectrometer mounted on UT1-Antu at Paranal Observatory. For the first time, a consistent census of the CO, H_2O ices and the minor ice species CH_3OH and OCN^- and warm CO gas present around young stars is obtained, using large number statistics and resolving powers of up to R=10,000. The molecular structure of circumstellar CO ices, the depletion of gaseous CO onto grains in protoplanetary disks, the presence of hot gas in the inner parts of circumstellar disks and in outflows and infalls are studied. Furthermore, the importance of scattering effects for the interpretation of the spectra has been addressed.

Key words: Circumstellar matter – dust, extinction – ISM: molecules – infrared: ISM

1. Introduction

The $3-5$ μm region of young stellar objects contains a large range of spectroscopic signatures of both cold ices as well as warm and hot gas. The most important are the broad absorption band at 3.08 μm due to water ice, the narrow absorption band around 4.67 μm due to solid CO and the large number of lines from the fundamental ro-vibrational transitions from gaseous CO covering the entire M-band. The ice mantles are thought to be located on dust grains in the cold midplanes of optically thick circumstellar disks, in the cold and dense envelopes of collapsing protostars and to some extent in the more quiescent parts of dense molecular clouds. However, due to the nature of absorption line studies, it is often hard to disentangle contributions from different environments along the line of sight. Studying large samples statistically is one way of circumventing the problem. Alternatively absorption studies can be combined with observations of rotational transitions and submm continuum to map the foreground clouds (Boogert et al. 2000).

The fundamental ro-vibrational transitions of CO present in the $4.5-5.0$ μm region probe gas of temperatures from 20 K to more than 5000 K due to the large range of rotational and vibrational levels present in a narrow wavelength region and are thus sensitive temperature indicators. Ro-vibrational transitions of various molecules in the mid-infrared region are among the few effective probes of warm and hot gas located behind dense, dusty envelopes and are therefore a powerful probe of warm gas in the inner few AU of circumstellar disks and the hot gas in the region between the disk, the central star and the outflow.

2. Observations

The observations were obtained between January 2001 and May 2002 using the Infrared Spectrometer and Array Camera (ISAAC) mounted on UT1-Antu at Paranal Observatory. L-band spectra were taken at low resolving powers of $\lambda/\Delta\lambda = 600-1200$ and M-band spectra were obtained using the medium resolution mode, resulting in $\lambda/\Delta\lambda = 5\,000-10\,000$. An advantage of the ISAAC instrument is the ability to rotate the detector to include two sources in the slit. This, combined with the often excellent seeing in the $3-5$ μm region (typically $0\farcs3$ to $0\farcs5$), enabled us to obtain simultaneous spectra of several close binaries with separations of $1-5''$ to probe the difference in ice chemistry and structure along lines of sight separated by only a few hundred AU. A total of ~ 50 sources has been observed. The survey was designed to focus on low-mass ($L = 0.1-50$ $L_\odot$) Class I and borderline Class I-II sources in the major southern low-mass star-forming clouds ρ Ophiuchus, Corona Australis, Chameleon and Serpens. Some intermediate to high-mass sources as well as some background stars were observed for comparison. Spectra of high quality of sources as faint as $L = 10$ and $M = 8$ were obtained, thus probing the entire stellar mass

Chemistry as a Diagnostic of Star Formation (2003). *Edited by Charles L. Curry and Michel Fich.* Published in 2003 by NRC Research Press, Ottawa ON, K1A 0R6, Canada. ISBN 0-660-19089-3.

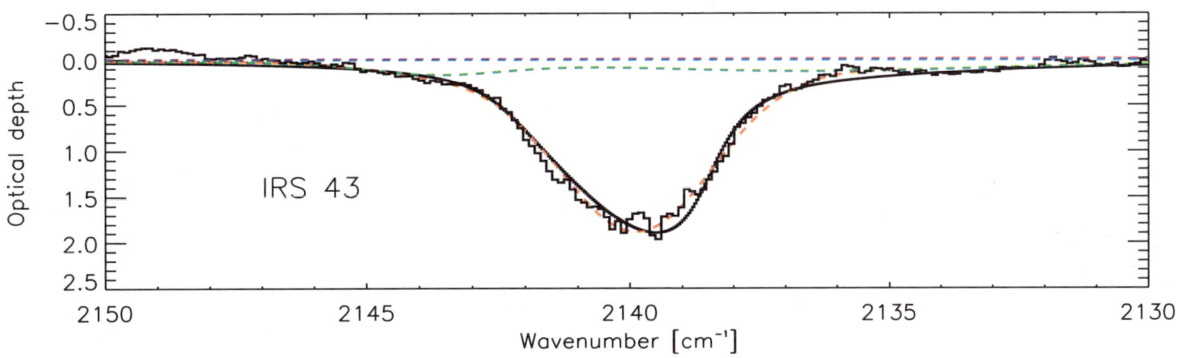

Figure 1. Example of a fit to the solid CO profile using a simple physical model (Pontoppidan et al. 2003). The red curve shows the total analytical model, the black solid curve shows the total physical model while the green curve shows the sum of the blue and red analytical components.

range. Here we present recent results from the analysis of the M-band ($4.5 - 5.0$ μm) spectra. An initial account is given by van Dishoeck et al. (2002), which includes the full list of team members.

3. Summary of results

3.1. The solid CO absorption band profile

The profiles of the solid CO stretching vibration mode around 4.67 μm are analysed in Pontoppidan et al. (2003). It is found that the structure of the profile is almost identical along all observed lines of sight. A phenomenological decomposition is used to show that the observed profiles can be split into three unique components. They have the same centers and widths, such that only their relative intensities need to be varied to reproduce all the observed circumstellar and interstellar CO bands. The three components are centered at 2143.7 cm^{-1}, 2139.9 cm^{-1} and 2136.5 cm^{-1}. Using a physical model of the CO ice, it is found that the two blue components can be fully explained by pure crystalline CO (α-CO). The model depends only on two free parameters, namely the number of CO oscillators per unit volume and the linear polarisation fraction of the background source.

The identification of the third red component is less secure. It may be associated with CO molecules embedded on the water ice known to exist in large quantities in all lines of sight also containing solid CO, although there are serious inconsistencies with the available laboratory spectra of water-rich ice mixtures. For instance, the 2152 cm^{-1} feature due to CO-OH bonds is not detected in any of the observed spectra, but strong upper limits can be derived. This feature, in addition to a red $2137-2135$ cm^{-1} feature, is always present in amorphous and unprocessed water-rich mixtures. Substantial thermal or energetic processing will remove the 2152 cm^{-1} feature but will also irreversibly change the shape of the 2136.5 cm^{-1} feature so that it no longer matches the astronomical spectra. The optical depth of a broad absorption band centered on 2175 cm^{-1} is found to correlate strongly with the depth of the 2136.5 cm^{-1} band for low-mass stars only. This feature may be an indication that another carrier is present in this region in addition to the 2165 cm^{-1} XCN band.

An example of a fit using the physical model of the CO ice is shown in Figure 1. The similarity between the CO ices observed along a large variety of lines of sight shows that the CO profile is not a good probe of chemical processing of circumstellar ices. The presence of secondary species in the ice also cannot be inferred from the profile of the CO stretching vibration, contrary to what has been previously reported in the literature. The abundance of species such as N_2, O_2 and CO_2 mixed with the CO ice is probably less than 10%. However, the presence of solid CO profiles which show only the red component suggests that the carrier of the red component is the least volatile of three carriers. The ratio of the middle and blue components to the red component may then be a good indicator thermal processing. This supports the idea that the CO responsible for the red component is trapped in the water ice or in another environment.

The presence of pure crystalline CO ice and the apparent low concentration of contaminants mixed with the bulk of the CO ice sets strong constraints on the condition under which CO freezes out onto dust grains.

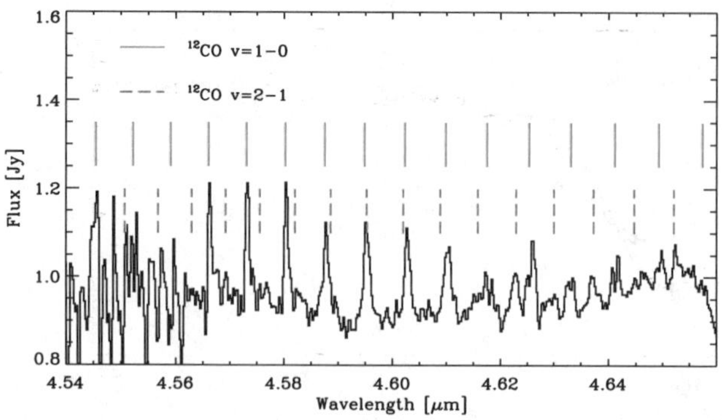

Figure 2. Example of broad gas-phase CO lines in the Class I source Elias 32 located in the ρ Ophiuchi cloud (Pontoppidan et al. 2003).

3.2. Gaseous CO ro-vibrational emission lines

Luminous sources often exhibit deep absorption lines sometimes showing signatures of infall or outflow in the gas. Low luminosity sources often show pure emission lines or self-absorbed emission lines. One extreme example which has been studied as an isolated case is GSS 30 IRS1 in the ρ Ophiuchi cloud core (Pontoppidan et al. 2002). This source shows narrow (< 30 km s^{-1}) and bright emission lines from CO gas. The emission lines are well-fitted by gas at a single temperature of 515 ± 10 K. The gas is not seen directly and the lines are probably scattered into the line of sight on the walls of a bipolar cavity. The fact that a range of temperatures does not seem to characterise the gas is surprising. The heating mechanism may be related to an accretion shock as circumstellar gas is incorporated into a circumstellar disk. A single temperature of ~ 500 K then arises naturally as the recombination of H_2 behind the shock front releases chemical energy which maintains a thermal equilibrium (Neufeld & Hollenbach 1994).

Other low-mass sources show very broad ($>$ 100 km s^{-1}) CO emission lines from hot ($T_{\rm gas} > 10^3$ K) CO gas. An example of this is shown in Figure 2 for Elias 32. The origin of the different classes of ro-vibrational lines (broad versus narrow, emission versus absorption) is presently unclear, although geometry and the presence or absence of foreground clouds likely plays a role. In general, studies employing full radiative transfer models are needed to constrain the detailed origin of the lines.

4. Conclusions and future Work

The combination of spectra of solid state features taken at the highest possible spectral resolution combined with physical models of the molecular structure and careful laboratory experimentation, also using high resolution spectrometers, is a powerful tool for the study of the microscopic structure and composition of interstellar solid state material. Furthermore, high resolution spectroscopy of the ro-vibrational transitions of molecules is well suited for penetrating the obscuring envelopes of young embedded stars to study the physics and relationships of the inner parts of circumstellar disks, accretion of circumstellar material and the inner engines of outflows. Detection of strong solid CO absorption through an edge-on disk has recently been published by Thi et al. (2002).

Work in progress includes:

1. Full radiative transfer modeling of the observed ro-vibrational gas-phase CO lines to constrain the origin of the different classes (narrow absorption lines, narrow emission lines and broad emission lines).
2. Using the statistics of the large sample to analyse the relationship between solid CO and water ice in interstellar grain mantles.
3. Performing laboratory experiments to determine the origin of the 2175 cm^{-1} feature and its relation to the red component of the CO ice profiles.
4. Further modeling of the formation of ice mantles to explain the observed icy features. Emphasis will be put on determining the detailed structure of the mantles, including the degree to which the ices are expected to be mixed, the composition as a function of the depth in a single grain mantle and the evolution of the mantle during the formation of a low-mass star.

Acknowledgements

The authors wish the thank the VLT staff for assistance and advice, in particular Chris Lidman for many helpful comments on the data reduction. This research was supported by the Netherlands Organization for Scientific Research (NWO) grant 614.041.004, the Netherlands Research School for Astronomy (NOVA) and a NWO Spinoza grant.

References

Boogert, A. C. A., et al. 2000, A&A, 360, 683

Neufeld, D. A. & Hollenbach, D. J. 1994, ApJ, 428,170

Pontoppidan, K. M., Schöier, F. L., van Dishoeck, E. F., & Dartois, E. 2002, A&A, 393, 585

Pontoppidan, K. M., Fraser, H., et al. 2003, A&A, submitted

Thi, W.-F., et al., 2002, A&A, 394, L27

van Dishoeck, E. F., et al. 2002, in *The Origins of Stars and Planets: The VLT View* (ESO Workshop), 238

INITIAL RESULTS FROM THE SUBMILLIMETER ARRAY

Chunhua Qi

Harvard-Smithsonian Center for Astrophysics

Abstract

We report on high resolution images obtained from the Submillimeter Array. The images were made in CO J=2–1 at 230 GHz and J=3–2 at 345 GHz. Observations were obtained of several astronomical sources including the evolved stars CRL 2688, V Hya, and the young stars LkCa 15, HD 163296 and TW Hya. The images are quite consistent with images from the IRAM interferometer at Plateau de Bure and the OVRO Millimeter Array. These preliminary observations demonstrate that the partially completed SMA is capable of producing scientific images at submillimeter wavelengths.

Key words: Submillimeter – instrumentation: interferometers – stars: general

Chemistry as a Diagnostic of Star Formation (2003). *Edited by Charles L. Curry and Michel Fich.* Published in 2003 by NRC Research Press, Ottawa ON, K1A 0R6, Canada. ISBN 0-660-19089-3.

STAR FORMATION: 3-D COLLAPSE OF TURBULENT CLOUD CORES

Michael A. Reid, Ralph E. Pudritz, and James Wadsley

Department of Physics and Astronomy, McMaster University, Hamilton, ON, Canada L8S 4R8

ABSTRACT

We have performed fully 3-D simulations of the collapse of molecular cloud cores which obey the logatropic equation of state. By following the collapse of these cores from states of near hydrodynamic equilibrium, we are able to produce accretion histories which closely resemble those of observed cores. The accretion proceeds in four distinct stages: an initial period of very slow accretion; a period of vigorous accretion following the development of a central density singularity, with $\dot{M} \propto t^3$, as predicted by self-similar models; a period of relatively stable, vigorous accretion; and finally a gradual decrease in the accretion rate once about 50% of the available mass of the molecular cloud core has been accreted. These results may explain the accretion histories of cores as they pass through the pre-protostellar, Class 0, and Class I stages.

Key words: Stars: formation – ISM: clouds

1. INTRODUCTION

Mounting evidence indicates that massive molecular cloud cores possess properties which are not adequately explained by models employing an isothermal equation of state. For example, surveys of prestellar cores in a variety of star-forming complexes show that intermediate and high mass cores have nonthermal line widths which are significantly greater than their thermal line widths (Caselli & Myers 1995). Models of the collapse of isothermal molecular cloud cores exhibit a number of features which are not borne out by observations, such as a time-invariant rate of accretion onto the central object (Shu 1977) and highly supersonic infall (Foster & Chevalier 1993).

To address some of the shortcomings of isothermal models, McLaughlin & Pudritz (1996) advanced a model for molecular cloud core collapse which accounts for some degree of turbulent support via a phenomenological equation of state. This logatropic equation of state reads $P/P_c = 1 + A \ln(\rho/\rho_c)$, where the subscript "c" denotes "central" values. The constant A is an adjustable parameter whose value for a range of molecular cloud core masses was determined empirically by McLaughlin & Pudritz (1996) to be $A = 0.2 \pm 0.02$.

The logatropic equation of state admits two equilibrium solutions. The features of each solution are developed in McLaughlin & Pudritz (1996). The "singular" equilibrium solution is analogous to the singular isothermal sphere (SIS) of Shu (1977), except that its density profile is $\rho \propto r^{-1}$. The "non-singular" solution is analogous to the Bonnor-Ebert sphere, in that it is a pressure-truncated equilibrium solution with a finite central density and density profile in its outer regions which matches that of the singular solution.

The non-singular logatrope has the potential to be as useful a model for massive star-forming cores as the Bonnor-Ebert sphere is for low mass star-forming cores (see, e.g., Alves, Lada, & Lada [2001], who showed that a Bonnor-Ebert sphere is an excellent fit to the structure of the dark cloud B68). Hence, we undertook a numerical study of the collapse of a non-singular logatropic sphere from an initially near-equilibrium state until most of its mass had been accreted onto the central condensation.

2. NUMERICAL METHODS AND TESTS

Our fully 3-D simulations were conducted using the ZEUS-MP parallelized hydrodynamics code. We used a 256^3 Cartesian grid with a "sink cell" at its centre. Mass falling into the sink cell was presumed to have accreted onto the central protostar and was represented thereafter by a point mass at the grid centre. The initial configuration for each of our simulations was a spherical molecular cloud core in near hydrodynamic equilibrium, surrounded by an external medium of low and constant pressure. We conducted extensive tests of our numerical method using the collapse of the singular logatrope (McLaughlin & Pudritz

Chemistry as a Diagnostic of Star Formation (2003). *Edited by Charles L. Curry and Michel Fich.* Published in 2003 by NRC Research Press, Ottawa ON, K1A 0R6, Canada. ISBN 0-660-19089-3.

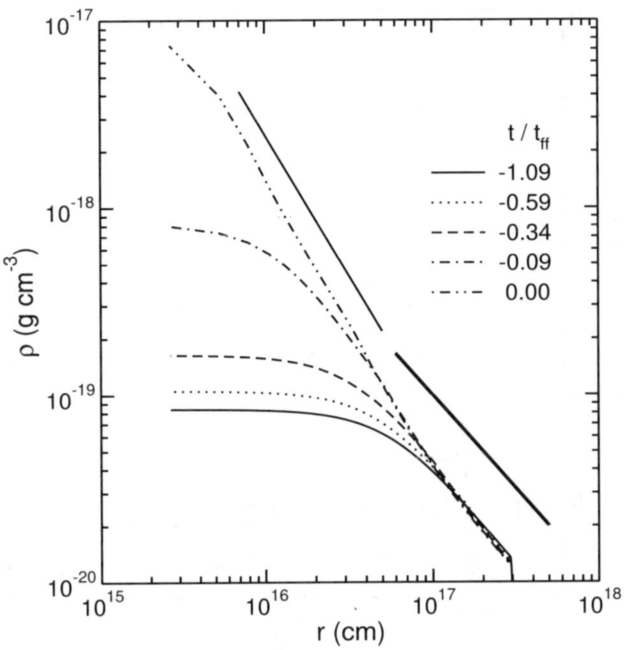

Figure 1. Density evolution of the 1 $M_\odot$ non-singular core. The singular density profile is reached at $t = 0$. The thick straight line indicates $\rho \propto r^{-1}$ and the thinner straight line indicates $\rho \propto r^{-3/2}$. The solid curve ($t = -1.09\,\bar{t}_{ff}$) represents the initial density profile of the non-singular core. Adapted from Reid, Pudritz, & Wadsley (2002).

1997) as our test case. Excellent agreement between the analytic and numerical solutions was found. A full description of our numerical method may be found in Reid, Pudritz, & Wadsley (2002).

3. Results

The most interesting of our results from the collapse of the non-singular logatrope are presented in Figures 1–3. Figures 1 and 2 show the evolution of the density and infall speed profiles of a collapsing non-singular logatrope prior to the formation of a singular density profile. The model has been scaled such that the total mass of the core is 1 $M_\odot$. The total core mass is determined by the choice of central temperature and external pressure—our choices of $T_c = 10$ K and $P_s = 1.3\times10^5$ k_B result in a core of total mass 1 $M_\odot$, but the results can be rescaled and applied to models of higher mass (for example, by assuming a higher confining pressure). Indeed, observational evidence has emerged which indicates that the confining pressures on molecular cloud cores are one or two orders of magnitude higher than we assumed (Johnstone et al. 2000).

Perhaps one of the most notable features of Fig. 1 is the behaviour of the density profile at late times. At the moment when the density becomes singular (labelled $t = 0$), the density in the inner parts of the core scales as $r^{-3/2}$. This is the density profile expected for a collapsing singular logatropic sphere, as well as for a collapsing SIS. This implies that, whereas observations of $r^{-3/2}$ density profiles in molecular cloud cores have traditionally been interpreted as evidence for isothermal collapse, this should no longer be the case. Even models which incorporate turbulent support will show such a density profile.

Because of their extra nonthermal support, non-singular logatropic spheres collapse more "gently" than do their isothermal counterparts. As can be seen by comparing our Figure 2 with Fig. 1*a* of Foster & Chevalier (1993), the logatropic collapse is more gentle than that of the Bonnor-Ebert sphere. Foster & Chevalier (1993) find that, at the time of singularity formation, 44% of the mass of a Bonnor-Ebert sphere is in supersonic infall. We find that less than 5.5% of a non-singular logatrope is in supersonic infall at the same point in its evolution.

Figure 3 shows the rate of accretion onto the central object in collapsing non-singular logatropes of masses 1, 2.5, and 5 $M_\odot$. The reader should bear in mind that these are fiducial masses only—the models can be rescaled to produce much more massive cores and hence higher accretion rates by increasing the confining pressure. Each curve shows the evolution of the accretion rate from the near-equilibrium starting point and until 95%, 91%, and 29% of the total available mass has been accreted (for the 1, 2.5, and 5 $M_\odot$ cases respectively).

4. Discussion

The accretion profiles shown in Figure 3 bear several important similarities to what is known observationally about accretion in young stellar objects. A simple picture of the evolution of an accreting molecular cloud core suggests an accretion history of the following form: accretion proceeds slowly through the pre-protostellar core (PPC) phase, increases markedly at the formation of a central source (to power the substantial outflows seen in Class 0 sources), and finally declines and dissipates as the core passes into the Class I stage. André, Ward-Thompson, & Barsony (2000) mark the transition between the Class 0 and Class I phase at the point where the mass accreted onto the central object is roughly equal to that remaining in the envelope. Our results show a similar transition at the same point. As indicated for the 1 $M_\odot$ curve in Fig. 3, there is an initial period of slow accretion, which we liken

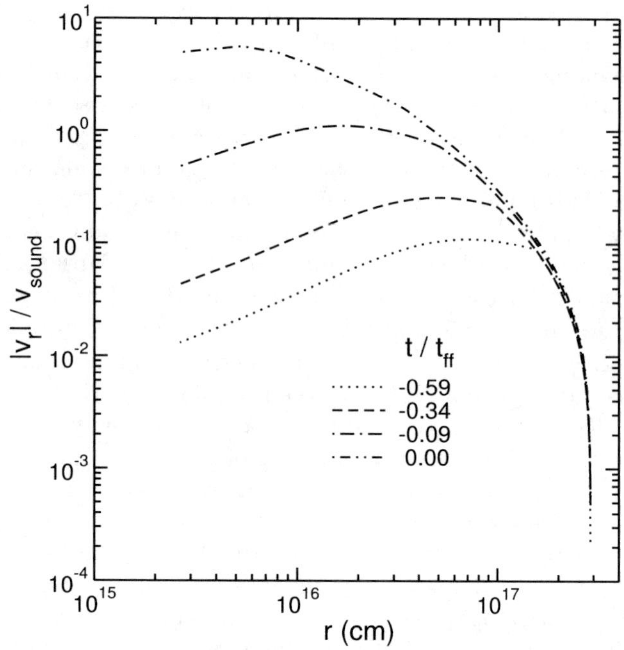

Figure 2. Evolution of the radial infall speed, $|v_r|$, of a 1 $M_\odot$ non-singular logatrope preceding the development of the singular density profile. Adapted from Reid, Pudritz, & Wadsley (2002).

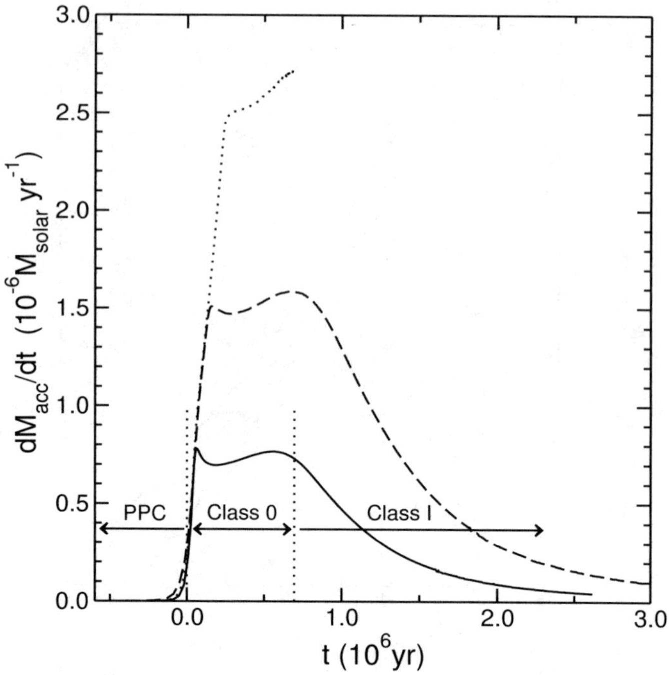

Figure 3. Time evolution of $M_{\rm acc}(t)$ and $\dot{M}_{\rm acc}(t)$ for the collapse of non-singular logatropic spheres with dimensionless radii of $R/r_o = 1.34$ *(solid curve)*, 2.21 *(dashed)*, and 3.26 *(dotted)*. The data are scaled such that each core has a central temperature of $T_c = 10$ K and surface pressure $P_s = 1.3 \times 10^5$ k_B, giving them total masses of 1, 2.5, and 5 $M_\odot$. The vertical dotted lines indicate suggested transitions between the PPC, Class 0, and Class I stages for the 1 $M_\odot$ model. Time $t = 0$ marks the moment at which the core's density profile becomes singular. Adapted from Reid, Pudritz, & Wadsley (2002).

to the PPC stage. As soon as the density profile of the core becomes singular ($t = 0$), which can be thought of as the moment of protostar formation, the accretion rate increases markedly, launching a phase of vigorous accretion sufficient to power an outflow. We liken this to the Class 0 phase. Once half of the available mass has been accreted, the accretion rate declines steadily to near-zero.

We suggest that our simulations of the collapse of logatropic molecular cloud cores bear significant similarities to the observations. They offer an alternative method for modelling the structure of intermediate and higher mass molecular cloud cores. Cores which exhibit density power-law indices of between 1 and 1.5 (see, for example, van der Tak et al. [2000]) should be pursued as possible candidate non-singular logatropes. Our results bolster the argument that models of collapsing protostellar cores *must* account for some degree of nonthermal support in order to reproduce the observations.

Acknowledgements

We are grateful to the organizers of SFCHEM 2002 for allowing us to present our work. The work of M.A.R. was supported in part by a scholarship from the Natural Sciences and Engineering Research Council of Canada (NSERC). The work of R.E.P. was supported by NSERC. J.W. is a SHARCNET Research Associate.

References

Alves, J. F. Lada, C. J. & Lada, E. A. 2001, Nature, 409, 159

André, P., Ward-Thompson, D., & Barsony, M. 2000, in *Protostars and Planets IV*, ed. V. Mannings, A. P. Boss, & S. S. Russell (Tucson: Univ. of Arizona Press), 59

Caselli, P., & Myers, P. C. 1995, ApJ, 446, 665

Foster, P. N. & Chevalier, R. A. 1993, ApJ, 416, 303

Johnstone, D., et al. 2000, ApJ, 545, 327

McLaughlin, D. E. & Pudritz, R. E. 1996, ApJ, 469, 194

McLaughlin, D. E. & Pudritz, R. E. 1997, ApJ, 476, 750

Reid, M. A. Pudritz, R. E. & Wadsley, J. 2002, ApJ, 570, 231

Shu, F. H. 1977, ApJ, 214, 488

van der Tak, F. S. et al. 2000, ApJ, 537, 283

NEW RATES FOR DEUTERIUM FRACTIONATION REACTIONS APPLIED TO INTERSTELLAR CHEMISTRY

H. Roberts[1], E. Herbst[2], and T.J. Millar[3]

[1]Department of Physics, The Ohio State University, Columbus, OH, USA
[2]Departments of Physics and Astronomy, The Ohio State University, Columbus, OH, USA
[3]Department of Physics, UMIST, Manchester, UK

ABSTRACT

In this paper, we look at the effects of incorporating rate coefficients for deuterium fractionation reactions, which have recently been studied at 10 K in the laboratory for the first time, on predicted deuterium fractionation from models of gas-phase interstellar chemistry. We compare our results with those obtained previously and with the dark clouds TMC-1, L134N, and L1544. In general, the new rates tend to reduce the amount of fractionation that can occur in the gas, which may present problems in regions where high molecular D/H ratios have been observed.

Key words: molecular processes – ISM: molecules – ISM: clouds

1. INTRODUCTION

Observations of molecular D/H ratios in dark clouds are currently used to probe physical conditions, such as temperature, cosmic-ray ionization rate and ionization fraction (e.g. Caselli et al. 1998; Williams et al. 1998), and to determine whether gas-grain interactions are significant (e.g. Caselli et al. 1999; Markwick et al. 2001). In star forming regions, where ice-mantles have evaporated and released the products of grain-surface chemistry, deuterium fractionation can provide valuable information on the temperatures under which the mantles formed and how species have been processed in the ice (e.g. Hatchell et al. 1998, 1999; Ceccarelli et al. 2001).

All of the above studies rely on our understanding of the most basic deuterium fractionation processes. Deuterium in the ISM is mostly locked up in the form of HD, with an abundance relative to H_2 of only $\sim10^{-5}$ (Linsky et al. 1995; Wood et al. 2002), but under certain conditions deuterium can be preferentially brought out of this interstellar reservoir and into an active chemistry, resulting in greatly enhanced abundances of deuterium bearing molecules.

Table 1. Comparison of rate coefficients for three fractionation reactions at 10 K

Reaction	Standard ($cm^3\,s^{-1}$)	Updated ($cm^3\,s^{-1}$)
H_3^+ +HD → H_2D^+ + H_2	1.7(-09)	3.5(-10)
H_2D^+ + H_2 → H_3^+ + HD	2.5(-18)	7.0(-15)
CH_3^+ + HD → CH_2D^+ + H_2	1.3(-09)	2.6(-10)
CH_2D^+ + H_2 → CH_3^+ + HD	7.4(-26)	2.2(-15)
$C_2H_2^+$ + HD→ C_2HD^+ + H_2	1.0(-09)	7.5(-10)
C_2HD^+ + H_2 → $C_2H_2^+$ + HD	3.2(-33)	<8(-16)

NOTES: $a(-b)$ implies $a \times 10^{-b}$; "standard rates" are those used in Millar et al. (1989) and Roberts & Millar (2000a,b); "new rates" are from Gerlich et al. (2002) and Roberts et al. (2002).

Table 1 lists the reactions which are primarily responsible for the deuterium fractionation in dark clouds (Millar et al. 1989; Roberts & Millar 2000a). In each case, the forward reactions are exothermic, and so the deuterated molecular ions are destroyed very slowly by endothermic reactions with H_2 at temperatures <100 K. H_2D^+ is the most important species for fractionation at typical dark cloud temperatures of 10 K or so because H_3^+ is more abundant than the two carbonaceous ions. As temperatures rise, however, the greater exothermicities of reactions (2) and (3) play an increasingly important role since CH_2D^+ and C_2HD^+ are destroyed far more slowly than H_2D^+.

The rates of the forward reactions were measured by Adams & Smith (1981), Smith et al. (1982) and Herbst et al. (1987) at temperatures down to 80 K. Chemical models using these rates have been fairly successful in explaining the levels of deuterium fractionation seen in various molecular clouds (e.g. Millar et al. 1989; Roberts & Millar 2000a,b). Recently, however, Gerlich et al. (2002) and Gerlich & Schlemmer (2002) have studied the reactions in

Chemistry as a Diagnostic of Star Formation (2003). *Edited by Charles L. Curry and Michel Fich.* Published in 2003 by NRC Research Press, Ottawa ON, K1A 0R6, Canada. ISBN 0-660-19089-3.

an ion trap at 10 K, finding that the rate coefficients for the forward reactions lie considerably below the Langevin value previously assumed. Table 1 compares these updated rates with the standard values. A fuller discussion of how these rate coefficients were estimated is given in Roberts et al. (2002).

2. Results

The updated rates have been incorporated into the chemical kinetic models of deuterium fractionation described in Roberts & Millar (2000a,b), for a dark cloud at 10 K, with constant physical conditions. As was discussed in those papers, and as shown in Table 2, a simple gas-phase chemistry does not account for the high fractionation seen in sources such as L134N and L1544, or for the most recent detections of multiply deuterated species in dark clouds and star forming regions. The steady-state fractionation in cold clouds is governed by the H_2D^+ fractionation. The H_2D^+ fractionation is, itself, limited by the abundance of neutral species, such as CO and O, in the gas-phase (Millar 2002; Roberts et al. 2002). In the centres of cold dense clouds species are freezing out onto the surfaces of dust grains, as evidenced by recent observations of CO depletion (e.g. Bacmann et al. 2002a) and so deuterium fractionation is expected to increase. This expectation has been confirmed by the detection of very large DCO^+/HCO^+ ratios in L1544 (Caselli et al. 1999) and D_2CO/H_2CO ratios in pre-stellar cores (Bacmann et al. 2002b) in clumps in which CO is significantly depleted.

Table 2 lists selected molecular D/H ratios that have been observed towards dark clouds in the local ISM. As mentioned above, L134N and L1544 both show enhanced D/H ratios, which indicate some level of depletion. Table 2 also compares selected molecular D/H ratios predicted by steady-state gas-phase and accretion models at selected times using the different rate coefficients shown in Table 1. Information at other times for the accretion models is shown in Figure 1. The specific times chosen for accretion models are defined so as to obtain the best agreement for the doubly deuterated isotopomer NHD_2, as observed in L134N.

The newly adopted rates for the formation of H_2D^+ and CH_2D^+ are 5 times lower than the Langevin value, and formation of C_2HD^+ is $\sim$1.3 times slower. Thus, the new rates lead to lower molecular D/H ratios for all species in dark interstellar clouds, whether we utilize purely gas-phase or accretion models for the chemistry.

The accretion results are generally too high for TMC-1, which is reasonable since this source is not known for

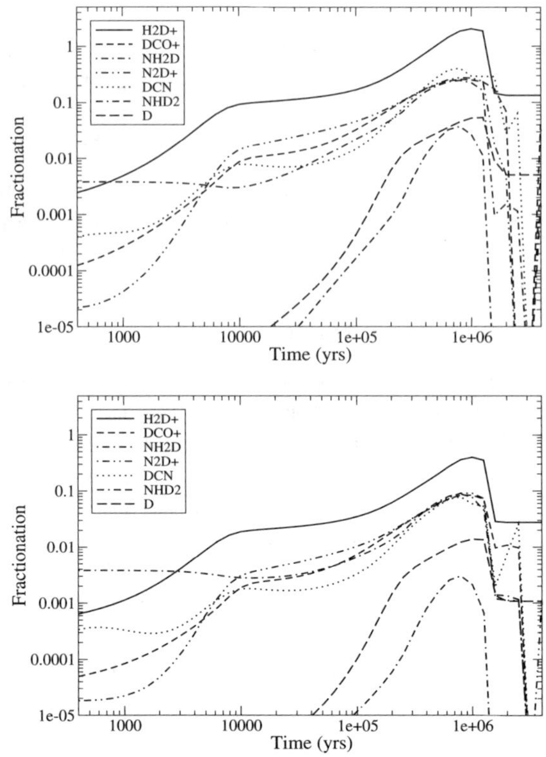

Figure 1. Evolution of selected molecular D/H ratios from accretion models using the standard (*top*) and updated (*bottom*) rates in from Table 1.

strong depletion of heavy neutral species. Using the new reaction rates, but including a modest depletion, where CO is between 2 and 5 times lower than its steady state abundance, would increase the ratios to the level observed in TMC-1.

The molecular D/H ratios predicted by the gas-phase models are generally lower than those observed towards L1544 and L134N, so that some level of depletion must be assumed. With the standard rates, the accretion model is in good agreement with the observations of L1544. For L134N, the accretion model which uses the standard rates at a time of 8×10^6 yr shows reasonable agreement with observed D/H ratios including that for doubly deuterated ammonia. Moreover, Roberts & Millar (2000b) showed that this can be achieved without the degree of depletion being too far advanced.

For the updated rates, however, the peak NHD_2/NH_3 abundance ratio (0.001, which occurs at $\sim10^5$ yrs) is lower than that observed in L134N. Also, because the rate of formation of H_2D^+ has been reduced by a factor of 5, a more severe depletion is now required to increase

Table 2. Molecular D/H ratios observed towards several sources, along with model predictions assuming $T_{\rm kin}$=10 K and $n(H_2)=10^4$ cm^{-3}

	Observations			Model predictions			
	TMC-1	L134N	L1544	Gas-phase models		Accretion models	
Rates used				standard	updated	standard	updated
Time (yr)				steady state		8×10^5	1.3×10^6
H_2D^+/H_3^+	—	—	—	0.105	0.024	0.580	0.228
DCO^+/HCO^+	0.02[a]	0.18[a]	0.12[d]	0.045	0.011	0.144	0.059
N_2D^+/N_2H^+	0.08[a]	0.35[a]	—	0.038	0.009	0.147	0.064
DCN/HCN	0.011[a]	0.05[b]	—	0.021	0.005	0.155	0.055
NH_2D/NH_3	0.011[a]	0.1[a]	0.13[e]	0.021	0.007	0.157	0.070
NHD_2/NH_3	—	0.005[c]	—	1.5(−4)	1.2(−5)	0.006	0.001

REFERENCES: [a] Tiné et al. 2000; [b] Turner 2001; [c] Roueff et al. 2000; [d] Caselli et al. 1999; [e] Shah & Wootten 2001.

the molecular D/H ratios to the levels which were observed. For CO, for example, the predicted value is more than 40 times lower than that observed.

3. DISCUSSION

We have examined the effects of new rate coefficients (Gerlich et al. 2002; Gerlich & Schlemmer 2002) for fractionation reactions which are important in modelling deuterium chemistry in interstellar clouds.

For TMC-1, the consequences of the calculated reduction in deuterium fractionation are not drastic; reasonable if somewhat worse agreement with observations was shown, and this could be improved by assuming slight depletion of heavy species. For L134N, the situation is far worse. Although the degree of deuteration and the absolute abundances in this source were previously well fit by an accretion model (Roberts & Millar 2000b), this can no longer be said to be the case. The situation regarding L1544 is nowhere near as dire since strong CO depletion is observed.

As the rate for the reaction of H_3^+ with HD is so important, it is now crucial that this reaction be studied over the entire temperature range from 10-300 K. It would also be useful if a theoretical understanding of the rate coefficients for this reaction system were available.

ACKNOWLEDGEMENTS

E.H. and H.R. acknowledge the support of the National Science Foundation (US) for their research in astrochemistry. Astrophysics at UMIST is supported by PPARC.

REFERENCES

Adams, N.G., & Smith, D. 1981, ApJ, 248, 373
Bacmann, A., et al. 2002, A&A, 389, L6
Bacmann, A., et al. 2002, A&A, submitted
Caselli, P., Walmsley, C.M., Terzieva, R., & Herbst, E. 1998, ApJ, 499, 234
Caselli, P., et al. 1999, ApJ, 523, 165
Caselli, P., et al. 2002, Plan. Sp. Sci., in press
Ceccarelli, C., et al. 2001, A&A, 372, 998
Charnley, S.B., Tielens, A.G.G.M., & Rodgers, S.D. 1997, ApJ, 482, L203
Gerlich, D., & Schlemmer, S. 2002, Plan. Sp. Sci., in press
Gerlich, D., Herbst, E., & Roueff, E. 2002, Plan. Sp. Sci., in press
Hatchell, J., Millar, T.J., & Rodgers, S.D. 1998, A&A, 332, 695
Hatchell, J., Roberts, H., & Millar, T.J. 1999, A&A, 346, 227
Herbst, E., Adams, N.G., Smith, D., & DeFrees, D.J. 1987, ApJ, 312, 351
Linsky, J.L., et al. 1995, ApJ, 451, 335
Loinard, L., et al. 2001, ApJ, 552, L163
Markwick, A.J., Charnley, S.B., & Millar, T.J. 2001, A&A, 376, 1054
Millar, T.J. 2002, Plan. Sp. Sci., in press
Millar, T.J., Bennett, A., & Herbst, E. 1989, ApJ, 340, 906
Roberts, H, Herbst, E., & Millar, T.J. 2002, MNRAS, 336, 283
Roberts, H., & Millar, T.J. 2000a, A&A, 361, 388
Roberts, H., & Millar, T.J. 2000b, A&A, 364, 780
Rodgers, S.D., & Charnley, S.B. 2001, ApJ, 553, 613
Roueff, E., et al. 2000, A&A, 354, L63
Shah, R.Y., & Wootten, A. 2001, ApJ, 554, 933
Smith, D., Adams, N.G., & Alge, E. 1982, ApJ, 263, 123
Tiné, S., et al. 2000, A&A, 356, 1039
Turner, B.E. 1989, ApJ, 347, L39
Turner, B.E. 2001, ApJS, 136, 579
Williams, J.P., et al. 1998, ApJ, 503, 689
Wood, B.E., Redfield, S., & Linsky, J.L. 2002, Plan. Sp. Sci., in press

ENERGETICS OF H_2 FORMATION ON ASTROPHYSICALLY RELEVANT SURFACES

Joseph E. Roser[1], Gianfranco Vidali[1], Giulio Manicò[2], and Valerio Pirronello[2]

[1]Syracuse University, 201 Physics Bldg., Syracuse, N.Y. 13244-1130, USA
[2]Dipartimento di Metodologie Fisiche e Chimiche per l'Ingegneria, Università di Catania, Viale Doria 6, 95125 Catania, Sicily, Italy

Abstract

The surface recombination of H atoms releases approximately 4.5 eV of energy, which may be distributed between the solid, the desorption of the newly-formed H_2 molecule and the translation, vibrational, and rotational degrees of freedom of the desorbed molecule. Computational studies by Takahashi and Uehara (2001) and Duley and Williams (1993), among others, suggest that formation pumping of excited H_2 may be observable in diffuse or dark clouds, and that it can be differentiated from UV pumping and be surface specific. In this poster, we present our experimental apparatus designed to perform the first laboratory measurements of the energy distribution of H_2 molecules forming on astrophysically relevant surfaces.

Key words: Astrochemistry – ISM: molecules

Chemistry as a Diagnostic of Star Formation (2003). *Edited by Charles L. Curry and Michel Fich.* Published in 2003 by NRC Research Press, Ottawa ON, K1A 0R6, Canada. ISBN 0-660-19089-3.

HCO$^+$ (J=1–0) INTERFEROMETRY OF THE SHOCKED OUTFLOW IN THE ORION-KL REGION

Tomoki Saito[1,2], M. Momose[3], R. Kawabe[4], S. Sakamoto[4], K. Sunada[2], T. Kamazaki[2], and N. Hirano[5]

[1]Department of Astronomy, University of Tokyo, 7-3-1 Hongo, Bunkyo, Tokyo, Japan
[2]Nobeyama Radio Observatory, Nobeyama, Minamimaki, Minamisaku, Nagano 384-1305, Japan
[3]Ibaraki University, Mito, Ibaraki, Japan
[4]National Astronomical Observatory of Japan, Osawa, Mitaka, Tokyo, Japan
[5]Academia Sinica Institute of Astronomy and Astrophysics, Taipei, Taiwan

Abstract

Millimetric interferometry has been carried out toward the Orion-KL region, the closest massive star-forming region, to unveil the detailed morphology and mechanism of the outflow from a massive protostar. We have used the Nobeyama Millimeter Array with molecular emission of HCO$^+$ (J=1–0) at 89 GHz. Our high resolution ($6'' \times 4''$, 1 km s^{-1}) observations have revealed the structure of bipolar outflow with both lobes having expanding shell-like structure, extending $\sim 50''$ toward NW and $\sim 30''$ toward SE direction. Another distinct result is that there are elongated clumps at $\sim 1'$ downstream of the blue side of the flow, associated with the "fingers" of shocked H_2 ($2\,\mu$m) observations. These "fingers" show contrast with "shells" not only in their morphology, but also in their velocity features, suggesting that these features are different components of the outflow. Analysis of these outflow components shows that they are ejected within quite a short time (several 10^2 to 10^3 yr).

Key words: ISM: individual (Orion-KL) – ISM: jets and outflows – ISM: molecules – radio lines: stars – stars: formation

1. Introduction

The object Orion-KL, lying at a distance of $\sim$ 480pc from us, is the most well-studied massive star-forming region. The prominent outflow is known, together with its highly complex morphology (e.g., shocked features called "fingers" identified in $2\,\mu$m H_2 observations; Kaifu et al. [2000]). The outflow has weak bipolarity along NW-SE direction, nearly perpendicular to the filamentary structure of the ambient gas cloud referred to as "the ridge". This region is thought to contain a radio continuum source, Source I, thought to be a massive protostar with mass of $\sim 25 M_\odot$, luminosity $\sim 10^{4-5} L_\odot$, in its center. Properties of outflows from such massive protostars are not very well studied yet, compared with well-known low-mass stars, and hence the case study is still of great importance to clarify the structure of the massive outflow and its underlying mechanisms.

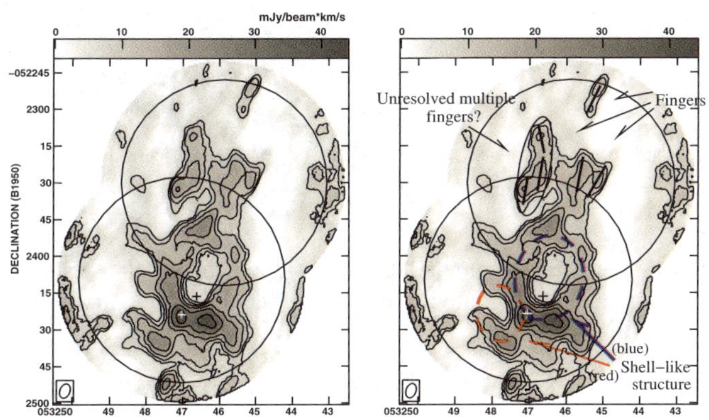

Figure 1. Integrated intensity map of HCO$^+$. Note *(1)* the shell-like region with clear bipolarity centered at the vicinity of Source I, and *(2)* the finger-like clumps at $\sim 1'$ N–NW from Source I. The lowest contour level is $\sim 3\sigma$, and each level is $\sim \sqrt{2}$ times the previous level. The white and black crosses show the position of Source I and BN, respectively.

The HCO$^+$ (1–0) line is capable of tracing denser regions compared with lines such as CO (1–0). It is experimentally known to be enhanced in shocked regions like outflows—actually enhanced abundance is suggested in the outflow of Orion-KL (Blake et al. 1987). In addition, this line has no hyperfine transitions like HCN, so that the line profiles are not smeared by its intrinsic linewidth. Therefore this line is suitable to investigate detailed morphology and mechanisms of the outflow from this source, together with the high angular- and velocity-resolution interferometer.

Chemistry as a Diagnostic of Star Formation (2003). *Edited by Charles L. Curry and Michel Fich.* Published in 2003 by NRC Research Press, Ottawa ON, K1A 0R6, Canada. ISBN 0-660-19089-3.

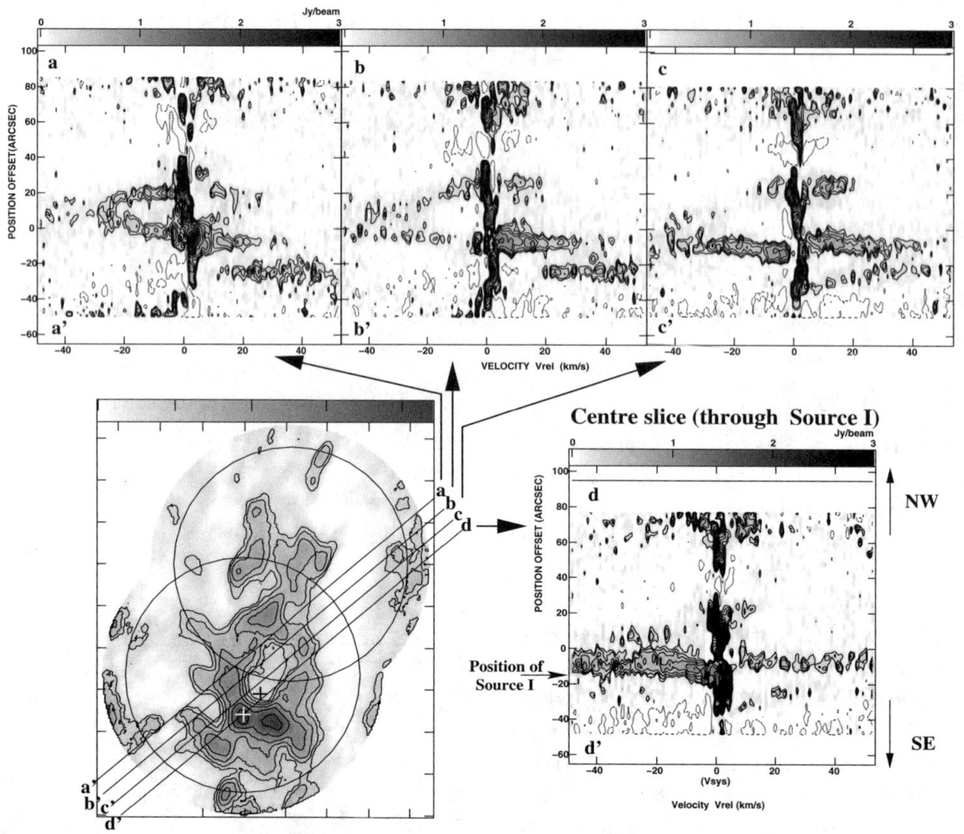

Figure 2. Position-velocity diagrams along the outflow axis direction. The double-ring stream is characteristic of the picture of "bipolar expanding shell-like outflow" centered on the vicinity of Source I.

2. Observations

Observations have made with the Nobeyama Millimeter Array, consists of six 10 m antennas. Two array configurations, C (intermediate)- and D (compact)-array, were used to achieve the angular resolution of $\sim 6'' \times 4''$, equivalent to $\sim$ 0.014 $\times$ 0.01 pc in linear scale at $d \sim 480$ pc. Two fields ($\sim$ 87″ FWHM each) were mosaiced to cover almost the whole outflow region with shocked H_2 emission. The HCO^+ (J=1–0) transition at 89.188518 GHz was observed with velocity resolution of $\sim$ 1 km s^{-1}, covering the extent of $\sim$ 100 km s^{-1}. High sensitivity ($\sim$ 80 mJy beam^{-1} within 1 km s^{-1} channel) is also a remarkable feature to detect faint emission from highly localized structures such as "fingers".

3. Results

Figure 1 shows the integrated intensity map of HCO^+, with the integrated range of $-50 \lesssim V_{\rm rel} \lesssim 50$km s^{-1}. On both sides of Source I, the shell-like structures can be identified, extending $\sim 50''$ toward NW and $\sim 30''$ toward SE direction. These shell-like components show clear bipolarity centered on the position of Source I. On the other hand, there are elongated clumpy structures at $\sim 1'$ N–NW from Source I (i.e., downstream the blue lobe of the outflow).

Detailed inspection of the velocity structure is shown in Figure 2. Here we are assuming the position angle of the bipolar outflow axis to be 130°. The position-velocity diagrams along the axis direction make a double-ring stream on the P–V plane, centered at the vicinity of Source I. The double-ring stream seems to be extended most in the slices near the bipolar axis, and gradually collapsing with shifting the slice toward NE. This is characteristic of the bipolar outflow, with the both lobes having an expanding shell-like structure. The spatial extent of each "shell" is $\sim$ 30–50″, with a thickness of $\sim 10''$. Fig. 2 shows that its expanding velocity is ~ 50 km s^{-1}.

Figure 3 shows an overlay of the integrated wing component of HCO^+ and H_2 image. Overall distributions of the HCO^+ and H_2 show good correlation, showing that both HCO^+ and H_2 are tracing the outline of bipolar flow. The shell-like structures are located on the verge of the major peaks of the H_2. These shells seems to be main component of the outflow since they show clear bipolarity centered on the vicinity of Source I and most of the flux comes from this bipolar shell-like region.

The second component, the region of "fingers", shows quite a good correlation with the H_2. These finger-like clumps are located on the tip of the shell-like bipolar outflow. We also identified H_2 bow-shocks, behind which there are finger-like clumps identified in the HCO^+ image. The finger-clumps of HCO^+ seem to be identical to, or at least associated with, the shocked H_2 "fingers."

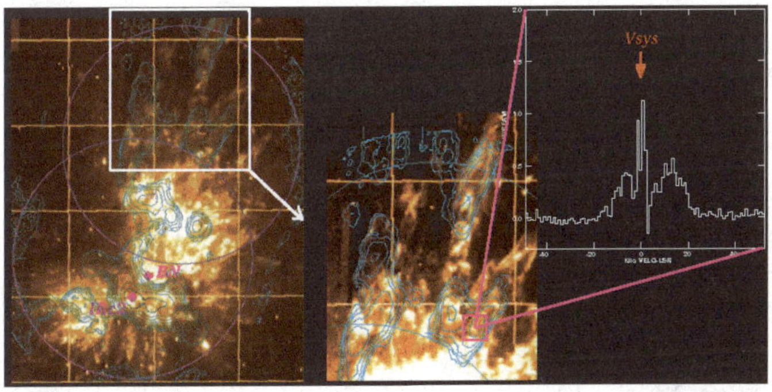

Figure 3. The left and center panels are overlays of the HCO^+ integrated intensity map of line wing component *(contours)* and shocked H_2 image (Kaifu et al.2000). The right panel shows the nearly blue-red symmetric line profile of the finger-like clump.

The finger-like clumps of HCO^+ show anomalously broad line profiles (Fig. 3). Their linewidth is more than 40 km s^{-1} (FWHM), with nearly blue-red symmetric spectra (note that channels around the systemic velocity are not reliable due to a large amount of missing flux), making it difficult to determine their motions. Each clump has a mass of $\sim 1 - 2M_{\odot}$ (LTE), and is thought to be transient structure associated with the outflow. They show contrast with the rest of the flow (i.e., shell-like components), not only in their elongated finger-like morphology, but also in their velocity features.

4. Discussion

Our high resolution observations have revealed the $\lesssim 0.1$ pc scale structure of the outflow from Orion-KL. It can be morphologically distinguished into two major components: (1) the main part of the outflow showing bipolarity with both lobe showing expanding shell-like features, and (2)elongated finger-like clumps on the tip of the outflow.

Detailed spectral and morphological analysis of the former component suggests its activity center is located in the vicinity of Source I. The shell-like structure has "broken shell" features, rather than "shell with smooth surface", suggesting that it was formed by plenty of collimated jets. The "shell" lies on the verge of the major peaks of shocked H_2, suggesting that the ISM is swept up to form the dense "cavity wall". The driving source of this "shell" should be the Source I, but in addition, there is a distinct peak at $\sim 10''$ west, possibly contributing to the asymmetry of the outflow.

The finger-like clumps seem to be molecular counterpart of the H_2 "fingers" seen as the bow-shocks at several arcsecs ahead. Shock-compressed gas is thought to be detected in HCO^+. The directions of these finger-clumps do not always point back to the position of Source I, but sometimes to the position of the peak near the IRc3. Most of these finger-clumps originate from the Source I, but some originate from, or are bent by some mechanism at, the position near the IRc3.

The shell-like component shows clear velocity feature, and the results shown in Fig. 2 lead to the dynamical age of a few 10^3 years. Whereas it is difficult to determine the dynamics of the finger-clumps component, one can estimate their motions by measuring the proper motions of HH objects on the tips of the "fingers" (Lee & Burton 2000), leading to the dynamical age of $\sim 10^3$ yrs. Both components have obviously different distance from the outflow source; however, the dynamical ages are not significantly different. This suggests that they are ejected within quite a short time (within $\sim 10^2 - 10^3$ yr), in contrast with well known low-mass protostars, in which multi-epoch ejection is seen with a time interval of $\sim 10^4$ yr (Bachiller et al. 1994).

Acknowledgements

The authors greatly appreciate Prof. S. K. Okumura, Prof. K. Tatematsu, and Dr. M. Saito for their helpful advice and suggestions, and also thank Dr. J. Koda for fruitful discussions. This work is based on observations at the Nobeyama Radio Observatory (NRO), a branch of the National Astronomical Observatory of Japan.

References

Blake, G. A., et al. 1987, ApJ, 315, 621
Bachiller, R., et al. 1994, ApJ, 425, L93
Kaifu, N., et al. 2000, PASJ, 52, 1
Lee, J.-K., & Burton, M. G. 2000, MNRAS, 315, 11

CARBON MONOXIDE DEPLETION IN ORION B MOLECULAR CLOUD CORES

D. Savva[1], A. G. Gibb[2], L. T. Little[1], and R. R. Phillips[3]

[1]Electronics Laboratory, University of Kent, Canterbury, Kent CT2 7NT, UK,
[2]Department of Astronomy, University of Maryland, College Park, MD 20742, USA
[3]Joint Astronomy Centre, 660 N. A'ohoku Place, University Park, Hilo, HI 96720, USA

Abstract

Carbon monoxide emission is compared to observations of dust continuum emission to deduce apparent molecular abundances in four star-forming cores. Relative to 'normal abundance' values, the CO appears to be depleted by about a factor of 10 at the core positions. CO condensation on dust grains provides a natural explanation for the apparent depletion both through gas-phase depletion of CO, and through a possible increase in dust emissivity in the cores. The high brightness of HCO^+ relative to CO is then naturally accounted for by time-dependent interstellar chemistry starting from 'evolved' initial conditions.

Key words: ISM: clouds – ISM: molecules – ISM: abundances

1. Introduction

L1630 (Orion B) lies at 400 pc and contains several star-forming regions, including NGC2024. Previous observations of a number of LBS cores revealed considerable substructure in the dust continuum and high$-J$ HCO^+ line emission. However, observations of $C^{18}O$ showed that in two cases, $C^{18}O$ did not appear to trace the mass distribution, leading to the conclusion that CO was probably depleted in the centres of these cores. Radiative transfer modelling on $C^{18}O$ emission towards LBS23 showed that the $C^{18}O$ abundance must decrease towards the centre of the clumps. Some caution must be exercised when making claims of depletion, especially of CO, as it only freezes out at $T < 20$ K. Also the $C^{18}O$ may be optically thick.

An alternative interpretation of the apparent low abundance is that there are compact 'sub-clumps' within the 22 arcsec JCMT beam which are highly optically thick in $C^{18}O$ emission. To check the optical depth and derive the relative abundance of CO we have made observations of rare isotopomers (i.e., $C^{18}O$, $C^{17}O$, and $^{13}C^{18}O$) at selected positions near and around cold cores in L1630.

2. Observations

Observations of the J=2→1 transitions of $C^{18}O$, $C^{17}O$ and $^{13}C^{18}O$ were carried out over the period of 1998 February 16–28 at the James Clerk Maxwell Telescope, using the common user SIS receiver A2 and the digital autocorrelation spectrometer DAS. Key positions were selected from previous maps of the cores, and, in addition, a new map of $C^{18}O$ in LBS18 was made. The positions of cores were observed, as well as a selection of references off-core where there was sign of continuum emission. The observations are described in Savva et al. (2002). An example of the spectra obtained is shown in Figure 1, where the three isotopomers have been observed at the core position of LBS18S.

The continuum observations were made using the submillimetre bolometer array receiver SCUBA and are described in Phillips, Gibb & Little (2001). For the purposes of this study we have smoothed their 850 μm maps to the same resolution as that of our CO data; i.e., 22 arcsec.

3. Results and analysis

3.1. Temperatures and masses

We use the 450 μm and 850 μm observations to derive the temperatures at the chosen positions. The results depend on the opacity law, which is uncertain. Rather than seeking to justify a law we chose to investigate the effect of varying it by comparing the different results obtained assuming two typical but reasonably extreme examples: the laws proposed by Hildebrand (1983) (β=2) and by Testi & Sargent (1998) (β=1.1).

The total number of hydrogen molecules $N(H_2)$ implied by observation of a flux F_ν, from dust assumed to be optically thin, is

$$N(H_2) = \frac{D^2c^2}{2h\nu^3\kappa_{H_2}}\left[\exp\left(\frac{h\nu}{kT}\right) - 1\right] F_\nu, \quad (1)$$

Chemistry as a Diagnostic of Star Formation (2003). *Edited by Charles L. Curry and Michel Fich.* Published in 2003 by NRC Research Press, Ottawa ON, K1A 0R6, Canada. ISBN 0-660-19089-3.

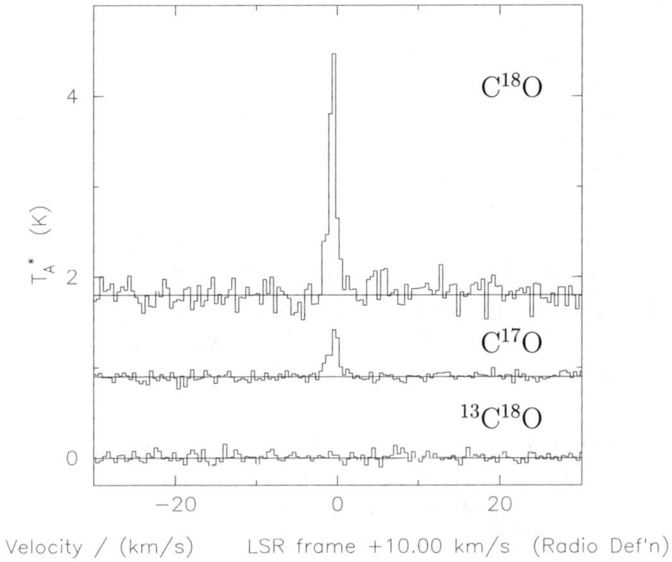

Figure 1. LBS18: Spectra at the core position (RA=$05^h43^m54.2^s$, DEC=$+00°18'23.0''$).

where D is the distance from Earth and κ_{H_2} is the opacity coefficient. The temperature of the gas was derived by forming the ratio of the submillimetre continuum fluxes at the two wavelengths and comparing it with that predicted for different temperatures using equation 1. We find that the core temperatures vary from 10 K to 38 K when using β=1.1, and from 7 K to 13 K when using β=2. The off-core temperatures vary from 16 K to 37 K when using β=1.1, and from 9 K to 13 K when using β=2.

It is of interest to compare our results with those from grey body fits to the spectra of dust continuum emission, which give T_d=17–35 K for HH25MMS (Gibb & Davis 1998), T_d=20±5 K for HH24MMS (Ward-Thompson et al. 1995) and T_d <20 K for LBS17H (Gibb & Little 2000). We can also compare our temperatures with those obtained from NH_3 inversion lines. Harju, Walmsley & Wouterloot (1993) derive temperatures of about 15K at the positions of HH24MMS and HH25MMS. Matthews & Little (1983) also mapped LBS23 and found a temperature of 14 K for HH25MMS, which is similar to that of Harju, Walmsley & Wouterloot (1993).

On the basis of the above comparisons we regard conservative limits for the dust temperatures to be 20±5 K for HH24MMS and HH25MMS and 15±5 K for LBS17H and LBS18S, while for the off–core envelope positions we take T=15±5 K.

Table 1. Temperatures and CO abundances

Source	T (K)	$\chi_{C^{18}O}$ ($\times10^{-8}$)	$\chi_{C^{17}O}$ ($\times10^{-8}$)	$\chi_{^{13}C^{18}O}$ ($\times10^{-10}$)
HH24MMS	20±5	$2.4^{+1.1}_{-0.8}$	$0.6^{+0.3}_{-0.2}$	$2.8^{+1.6}_{-1.4}$
HH25MMS	20±5	$1.9^{+0.9}_{-0.7}$	$0.5^{+0.2}_{-0.2}$	$<1.3^{+2.0}_{-2.0}$
LBS17H	15±5	$2.0^{+0.8}_{-0.7}$	$0.5^{+0.3}_{-0.2}$	$4.1^{+2.3}_{-2.0}$
LBS18S	15±5	$2.9^{+1.1}_{-0.9}$	$0.6^{+0.3}_{-0.3}$	$<1.5^{+3.0}_{-3.0}$
LBS23E1	15±5	$1.7^{+0.9}_{-0.8}$	$0.4^{+0.2}_{-0.2}$	—
LBS17F	15±5	$4.8^{+2.6}_{-2.2}$	$1.2^{+0.6}_{-0.5}$	—
LBS18A	15±5	$3.6^{+1.9}_{-1.6}$	$0.9^{+0.5}_{-0.4}$	—
LBS18B	15±5	$3.9^{+2.1}_{-1.8}$	$1.0^{+0.6}_{-0.5}$	—
LBS23A	15±5	$5.4^{+2.9}_{-2.4}$	—	$<10.8^{+6.9}_{-6.1}$

3.2. Carbon monoxide optical depths

The mean value of the ratio of $C^{18}O$ to $C^{17}O$ line intensities in the cores is 4.28; towards the other positions, 6.1. The latter value is in good agreement with the solar expectation of 5.4, suggesting that these positions are optically thin, and that the $[C^{17}O]/[C^{18}O]$ ratio is normal. The former suggests that $C^{18}O$ may be slightly optically thick in the cores, but not sufficiently to discredit a former conclusion that CO is depleted in several of the cores of LBS23 (Gibb & Little 1998). The $[C^{17}O]/[^{13}C^{18}O]$ ratios are consistent with the solar value.

3.3. Relative abundance of CO

The relative abundance of each CO isotopomer was found by dividing the column density of the isotopomer by the H_2 column density, which was derived from the dust data. The abundance of $C^{18}O$ was increased to correct for its slight optical thickness where necessary. The resulting relative abundances of $C^{17}O$, $C^{18}O$ and $^{13}C^{18}O$ at each position can be found in Table 1.

4. Discussion

4.1. Depletion

The canonical values for the abundances of $C^{18}O$, $C^{17}O$ and $^{13}C^{18}O$ may be taken to be 2×10^{-7}, 4.7×10^{-8} and, assuming simple proportion, 2.7×10^{-9} respectively, from Frerking, Langer & Wilson (1982). Relative to the canonical values, the isotopic species appear to be depleted by factors of about 10 on the cores, and approximately half that in the envelopes. The values are typically less than those quoted by Gibb & Little (1998), the differences being due to the assumption of slightly higher temperatures, careful matching of effective beam sizes and correction for the optical depths of $C^{18}O$.

A natural explanation for the depletion observed in these cores is the condensation of CO molecules on dust grains. This would explain both the depletion of CO in the gas phase and an increase in the dust emissivity as the grain size increases (Hildebrand 1983).

4.2. Implications for chemical models

Rawlings et al. (1992) show that as a protostar collapses it is plausible that the chemistry occurring while molecules deplete out onto dust grains leads to an increasing ratio of HCO^+/CO (by up to 50), although their absolute abundances finally fall. A requirement for this to happen is a high initial H_2O abundance such as might result from initial conditions following a shock through already processed material, but not from initial conditions which are of simple atomic form.

Other models have been developed which have different assumptions. Bergin & Langer (1997) followed chemistry during the collapse using a gas-grain chemical code, including desorption and depletion, but starting from an initial pristine state with elements only. The increase in HCO^+/CO suggested by observations was not produced by the model. Neither was it produced by the different type of model of Charnley (1997).

Since star formation has already taken place in L1630, the initial conditions for the chemistry will not be pristine. This supports our consideration of the model of Rawlings et al. (1992), which starts from 'evolved' conditions (including a high abundance of water). Only this model predicts an actual increase of HCO^+ while CO is declining. It thus provides a natural explanation for the brightness of high excitation HCO^+ emission relative to CO emission in cores. It also gives better estimate of HCO^+ abundance (5×10^{-10}, GLHL). It thus appears that the role of the initial conditions is very important in determining the later behaviour of the chemistry.

5. Conclusions

The main conclusions of this study are:

1. The emission from the CO isotopomers is optically thin, which rules out a model by which the cores contain optically thick sub-clumps to explain the low isotopic CO emission.

2. Although the dust observations do not allow the temperatures to be constrained for β=1.1, we believe it unlikely, from the NH_3 observations of Harju, Walmsley & Wouterloot (1993) and Matthews & Little (1983), that temperatures will be greater than 15–20 K. It seems unlikely that the temperatures exceed the dust-derived values in the sources.

3. If the dust emissivity falls within what are presently considered to be reasonable limits, CO is depleted in both the cores and the surrounding regions, but by a higher factor in the cores. Relative to the canonical abundances of, the CO appears to be depleted by about a factor of 10 at the core positions and a factor of 5 in the envelopes.

4. The low abundance of CO and the brightness of HCO^+ emission in the cores tends to support the model of Rawlings et al. (1992) which suggests that the amount of HCO^+ in the core increases as that of CO decreases.

Acknowledgements

The authors would like to thank the JCMT staff for their efforts in obtaining the observational data. D. Savva would like to thank PPARC for the award of a studentship.

References

Bergin, E. A. & Langer, W. D. 1997, ApJ, 486, 316
Charnley, S. B. 1997, MNRAS, 291, 455
Frerking, M. A., Langer, W. D. & Wilson, R. W. 1982, ApJ, 262, 590
Gibb, A. G., et al., 1995, MNRAS, 277, 341
Gibb, A. G. & Little, L. T. 1998, MNRAS, 295, 299
Gibb, A. G. & Davis, C. J. 1998, MNRAS, 298, 644
Gibb, A. G. & Little, L. T. 2000, MNRAS, 313, 663
Harju, J. Walmsley, C. M. & Wouterloot, J. G. A. 1993, A&AS 98, 51
Hildebrand, R. H. 1983, QJRAS 24, 267
Matthews, N. & Little, L. T. 1983, MNRAS, 205, 123
Phillips, R. R., Gibb, A. G. & Little, L. T. 2001, MNRAS, 326, 927
Rawlings, J. M. C., et al., 1992, MNRAS, 255, 471
Savva, D., et al., 2002, MNRAS, submitted
Testi, L. & Sargent, A. I. 1998, ApJ, 508L, 91
Ward–Thompson, D., et al., 1995, MNRAS, 274, 1219

A DATABASE OF MOLECULAR LINE DATA FOR ROTATIONAL TRANSITIONS FROM SELECTED SPECIES OF ASTROPHYSICAL INTEREST

Fredrik L. Schöier[1], Floris F. S. van der Tak[2], Ewine F. van Dishoeck[1], and John H. Black[3]

[1]Leiden Observatory, P.O. Box 9513, 2300 RA Leiden, The Netherlands
[2]Max-Planck-Institut für Radioastronomie, Auf dem Hügel 69, 53121 Bonn, Germany
[3]Onsala Space Observatory, SE-439 92 Onsala, Sweden

Abstract

Molecular line data for the rotational transitions of a number of astrophysically interesting species have been collected and are made publicly available on the World Wide Web. These data, including energy levels, statistical weights, Einstein A-coefficients and collisional rate coefficients, are essential input for non-LTE molecular radiative transfer programs. A computer program for performing statistical equilibrium calculations is made publicly available for use as well. This database should form an important tool in analyzing observations from current and future infrared and (sub)millimetre telescopes.

Key words: Molecular data – radiative transfer

1. Introduction

A wide variety of molecules has been detected in space to date, ranging from simple molecules like CO to more complex organic molecules including ethers and alcohols. Observations of molecular emission at millimetre and infrared wavelengths, supplemented by careful and detailed modelling, are powerful tools to investigate the physical and chemical conditions of astrophysical objects (e.g., Black 2000). For some objects, lines with a large range of critical densities and excitation temperatures are needed, since the densities can range from $\sim 10^2 - 10^9\,\mathrm{cm}^{-3}$ and the temperatures from $\sim 10 - 1000\,\mathrm{K}$ in the interstellar and circumstellar environments probed by current and future instrumentation.

To extract astrophysical parameters, the excitation and radiative transfer of the lines needs to be calculated. A number of radiative transfer codes have been developed for the interpretation of molecular line emission [see van Zadelhoff et al. (2002) for a review]. The radiative transfer analysis requires molecular data in the form of energy levels, statistical weights and transition frequencies as well as the spontaneous emission probabilities and collisional rate coefficients.

2. Molecular line data

Literature data on the rotational transitions of 23 different molecules are summarized and extrapolation of collisional rate coefficients to higher energy levels and temperatures are made. The data are made available for the community through the World Wide Web.[1] Many of the data files presented here were adopted by Schöier et al. (2002) to model the circumstellar environment of the protostar IRAS 16293–2422.

2.1. Energy levels and radiative rates

The energy levels and Einstein A-coefficients are obtained from the JPL[2] and CDMS[3] catalogues. In some cases, better or more recent laboratory data have been used instead. Generally, only the ground vibrational state is retained, although in a few cases, e.g. CO and HCN, vibrationally excited levels are included. More study of collisional rate coefficients for vibrational transitions is urgently needed, as there are certainly cases where vibrationally excited molecules are important. Examples include HC_3N (Wyrowski et al. 1999) and torsionally excited CH_3OH. In the future, data files including excited vibrational states will be added for more molecules.

Hyperfine splitting can be of importance in line transfer and introduce non-local effects for lines overlapping in frequency (Lindqvist et al. 2000). However, often the splitting between individual hyperfine components is small compared to the line-broadening, so that it can safely be neglected and treated as a single level for the purpose of excitation analysis. We present data files for some of the relevant molecules, such as HCN and CN, both with and without hyperfine splitting.

[1] `http://www.strw.leidenuniv.nl/~moldata`
[2] `http://spec.jpl.nasa.gov`
[3] `http://www.cdms.de`

Chemistry as a Diagnostic of Star Formation (2003). *Edited by Charles L. Curry and Michel Fich.* Published in 2003 by NRC Research Press, Ottawa ON, K1A 0R6, Canada. ISBN 0-660-19089-3.

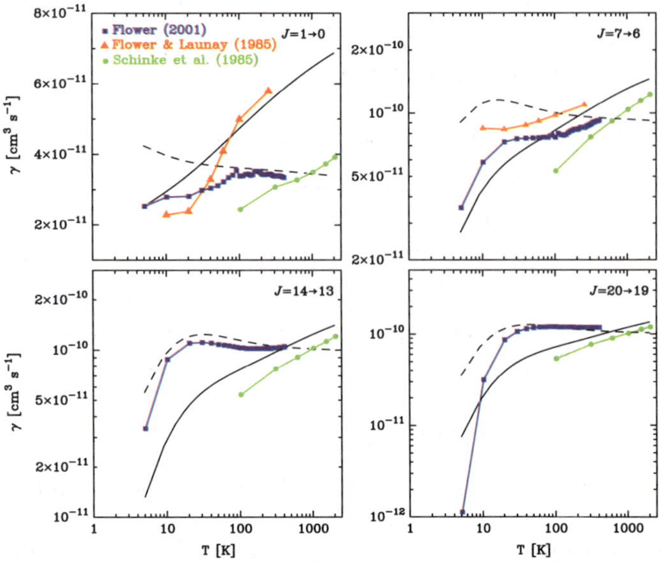

Figure 1. Calculated collisional de-excitation rate coefficients for CO in collisions with para-H_2 for selected transitions *(symbols)*. Black lines indicate fits to the rate coefficients of Flower & Launay (1985) *(solid curves)* and Flower (2001) *(dashed curves)*. Also shown are the rate coefficients presented by Schinke et al. (1985).

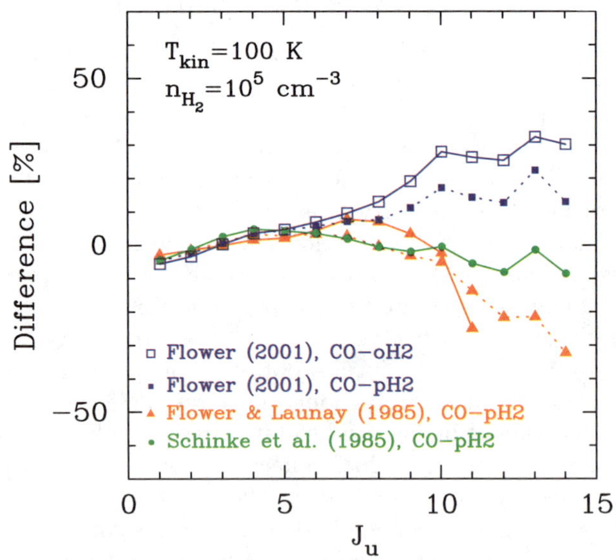

Figure 2. Predicted CO line intensities using different sets of collisional rates for an isothermal homogeneous sphere with $T_{\rm kin} = 100$ K and $n_{\rm H_2} = 10^5$ cm^{-3}, compared to the values obtained using the CO-pH_2 rate coefficients from Flower (2001). Points connected with full lines are based on the original sets of rates while the dotted and dashed line styles indicate values obtained from fitting and extrapolation of the original rates

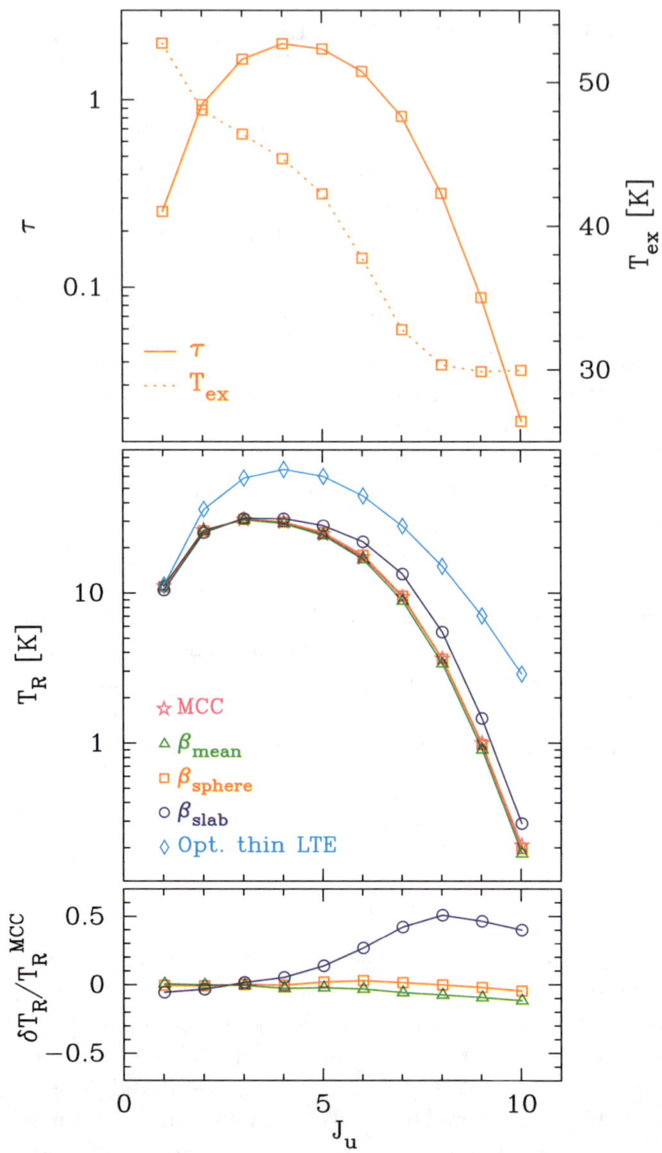

Figure 3. Comparison of the predicted line strengths for the 10 lowest rotational transitions of CO for a homogeneous isothermal sphere, with $n_{\rm H_2} = 10^5$ cm^{-3} and $T_{\rm kin} = 50$ K, using different methods. *Upper panel:* The total optical depth through the sphere at line centre τ and excitation temperature $T_{\rm ex}$ as a function of the upper rotational level J involved in the transition. *Middle panel:* The radiation temperature $T_{\rm R}$ obtained for each transition using RADEX with different prescriptions of the escape probability, β, and compared with the result from the Monte-Carlo code (MCC) of Schöier (2000). Also shown are the results for optically thin emission in LTE. *Lower panel:* $T_{\rm R}$ obtained from RADEX compared with the results from MCC, $\delta T_{\rm R} = T_{\rm R}^{MCC} - T_{\rm R}$.

2.2. Collisional rate coefficients

The adopted collisional rate coefficients usually pose the largest source of uncertainty of the molecular data input to the radiative transfer analysis. Detailed summaries of the theoretical methods and the uncertainties involved in determining collisional rate coefficients are given by Green (1975) and Flower (1990). Published collisional rate coefficients cover only a limited range of temperatures and energy levels, and extrapolations are often necessary (see Fig. 1) introducing additional uncertainties. Data files containing the originally calculated set of rates as well as files where an extrapolation has been carried out are available for all molecular species.

In Figure 2 the predicted line intensities for the 15 lowest rotational transitions of CO of an isothermal homogeneous sphere are shown using different sets of collisional rates. Deviations up to $\sim 50\%$ are found for the higher transitions which are sub-thermally excited. The lower transitions are close to LTE and less sensitive to the adopted set of collisional rate coefficients.

3. Radiative transfer

A radiative transfer code, *RADEX*[4], is made available for public use as part of the database. *RADEX* is a one-dimensional radiative transfer code aimed at solving the statistical equilibrium equations using the escape probability formulation. The code makes no assumption for the geometry or large scale velocity fields and its basic assumption is that of an isothermal and homogeneous medium. *RADEX* provides a useful tool in rapidly analysing a large set of observational data providing constraints on physical conditions, such as density and kinetic temperature. *RADEX* provides an alternative to the widely used rotation temperature diagram method (e.g., Blake et al. 1987; Goldsmith & Langer 1999) which relies upon the availability of many transitions of optically thin emission lines and is useful only in roughly constraining the excitation temperature in addition to the column density.

Benchmarking of the *RADEX* code is presented in Figure 3 where various expressions for the escape probability are used. The output is compared both to an optically thin LTE analysis (rotation diagram method) and a full radiative transfer analysis using a sophisticated non-LTE code based on the Monte-Carlo method (*MCC;* Schöier 2000). The molecular data used in this example are the collisional rate coefficients for CO collisions with para-H_2 from Flower (2001). Levels up to $J = 29$ are included. The excitation temperatures of the lines vary from nearly thermalized, for transitions involving low J-levels, to sub-thermally excited for the higher-lying lines. The optical depth in the lines is moderate ($\sim 1-2$) to low. It is seen that the expressions of the escape probability for a sphere and the mean $[(1-e^{-\tau})/\tau]$ give almost identical solutions and close to that obtained from the full radiative transfer. However, the slab geometry gives quite different results, in particular for high lying lines. The optically thin approximation, where the gas is assumed to be in LTE at 50 K produces the largest discrepancy and only gives the correct answer for the $J = 1 \rightarrow 0$ line, where the requirements are fulfilled.

[4] `http://www.strw.leidenuniv.nl/~moldata/radex.html`

Acknowledgements

FLS and EFvD are grateful to the Leids Kerkhoven-Bosscha Fond for financial support for their participation in the conference. This research was supported by the Netherlands Organization for Scientific Research (NWO) grant 614.041.004, the Netherlands Research School for Astronomy (NOVA) and a NWO Spinoza grant.

References

Black, J. H. 2000, in *Astrochemistry: From Molecular Clouds to Planetary Systems*, ed. Y. C. Minh & E. F. van Dishoeck, Vol. 197 (San Francisco: Astronomical Society of the Pacific), 81

Blake, G. A., Sutton, E. C., Masson, C. R., & Phillips, T. G. 1987, ApJ 315, 621

Flower, D. R. 1990, *Molecular collisions in the interstellar medium* (Cambridge: Cambridge: University Press)

Flower, D. R. 2001, MNRAS 328, 147

Flower, D. R. & Launay, J. M. 1985, MNRAS 214, 271

Goldsmith, P. F. & Langer, W. D. 1999, ApJ 517, 209

Green, S. 1975, in *Atomic and molecular physics and the interstellar matter* (Amsterdam: North-Holland Publishing Co.), 83

Lindqvist, M., Schöier, F. L., Lucas, R., & Olofsson, H. 2000, A&A 361, 1036

Schinke, R., Engel, V., Buck, U., Meyer, H., & Diercksen, G. H. F. 1985, ApJ 299, 939

Schöier, F. L. 2000, Ph.D. Thesis, Stockholm Observatory

Schöier, F. L., Jørgensen, J. K., van Dishoeck, E. F., & Blake, G. A. 2002, A&A 390, 1001

van Zadelhoff, G.-J., Dullemond, C. P., van der Tak, F. F. S., Yates, J. A., Doty, S. D., et al. 2002, A&A 395, 373

Wyrowski, F., Schilke, P., Walmsley, C. M. 1999, A&A 341, 882

DOES IRAS 16293–2422 CONTAIN A HOT CORE? NEW INTERFEROMETRIC RESULTS

Fredrik L. Schöier[1], **Jes K. Jørgensen**[1], **Ewine F. van Dishoeck**[1], **and Geoffrey A. Blake**[2]

[1]Leiden Observatory, P.O. Box 9513, 2300 RA Leiden, The Netherlands
[2]Division of Geological and Planetary Sciences, California Institute of Technology, Pasadena, CA 91125, USA

Abstract

High angular resolution H_2CO line observations have been carried out for the low-mass protobinary star IRAS 16293–2422 using the OVRO millimeter array. Simultaneous continuum observations reveal that the source most probably contains two accretion disks, less than ~ 250 AU in size, through which matter is fed onto the stars. The observations suggest that the binary has cleared most of the material in the inner part of the envelope, out to the binary separation. The H_2CO interferometer observations indicate the presence of large-scale rotation roughly perpendicular to the large scale CO outflow associated with this source. The H_2CO emission is dominated by the compact dense and hot ($T > 100$ K) gas close to the positions of the continuum peaks.

Key words: ISM: molecules – ISM: abundances – stars: formation – astrochemistry

1. Introduction

Recent studies (Jørgensen et al. 2002; Schöier et al. 2002; Shirley et al. 2002) have suggested that the inner envelopes of low-mass protostars are dense ($> 10^6\,\mathrm{cm}^{-3}$) and warm (> 80 K), as would be expected from scaling of high-mass protostars (Ceccarelli et al. 1996; Ivezić & Elitzur 1997). Detailed modelling of multi-transition single dish observations towards the deeply embedded low-mass protostellar object IRAS 16293–2422 (Ceccarelli et al. 2000; Schöier et al. 2002) shows that the abundances of some molecules, e.g. H_2CO and CH_3OH, are drastically increased in the warm and dense inner region of the circumstellar envelope. The location where this increase occurs is consistent with the radius at which ices are expected to thermally evaporate off the grains (~ 90 K). However, the small spatial scale of the region of hot gas (< 100 AU) and the infalling nature of the envelope lead to very different chemical time scales between low-mass and high-mass hot cores, which may prevent production of second generation complex molecules in low-mass protostars. Also, shocks are thought to be relatively more important in low-mass protostars. High angular resolution observations are needed to pinpoint the origin of the abundance enhancements and possibly distinguish these two scenarios. We present here observations of H_2CO at 1 mm using the Owens Valley Radio Observatory (OVRO) millimeter array at $\sim 3''$ resolution. The line ratio of the two H_2CO lines observed is a measure of the gas temperature.

2. Observations and data reduction

IRAS 16293–2422 was observed using the OVRO millimeter array between September 2000 and March 2002. The continuum emission at 1.37 mm was obtained simultaneously with the H_2CO $3_{22} \rightarrow 2_{21}$ and $3_{03} \rightarrow 2_{02}$ line emission at 218.222 and 218.475 GHz, respectively. IRAS 16293–2422 was observed in the L and E configurations. The natural weighted continuum observations has a 1σ noise of about 20 mJy beam^{-1} with a beam size of $3''\!.9 \times 1''\!.9$.

3. Results and discussion

3.1. Continuum emission

In Figure 1 the 221.7 GHz (1.37 mm) continuum emission observed toward IRAS 16293–2422 is presented. As expected for this protobinary star, two continuum sources are detected, separated by approximately $5''$ with faint extended emission also present. The total observed continuum flux density at 1.37 mm is about 3.5 Jy. This is roughly 50% of the flux obtained by Walker et al. (1990) from mapping with a single dish telescope, indicating that the interferometer resolves out some of the emission. The positions of the continuum sources are consistent with the two 3 mm sources MM1 (southeast) and MM2 (northwest) found by Mundy et al. (1992). At the distance of

Chemistry as a Diagnostic of Star Formation (2003). *Edited by Charles L. Curry and Michel Fich.* Published in 2003 by NRC Research Press, Ottawa ON, K1A 0R6, Canada. ISBN 0-660-19089-3.

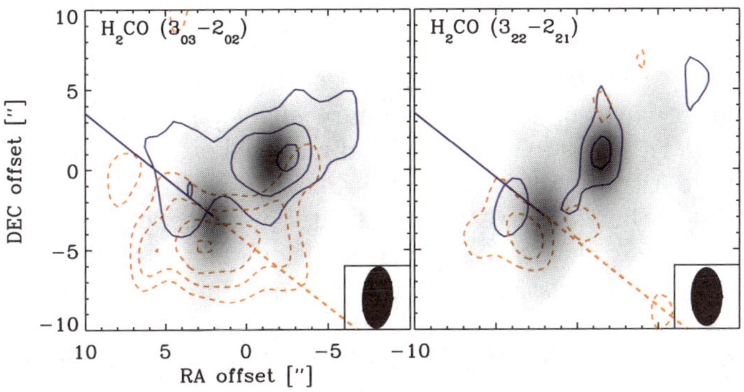

Figure 1. OVRO interferometer maps of H_2CO emission *(contours)* overlayed on the 1.37 mm continuum emission *(greyscale)* from IRAS 16293–2422. The H_2CO emission has been separated into blue (4–7 km s^{-1}; *solid curves*) and red (1–4 km s^{-1}; *dashed curves*) parts. Contours start at the 2σ level (1.0 Jy beam^{-1} km s^{-1}), spaced by 2σ. Also indicated is the direction of the large-scale CO outflow. The southeast continuum source is MM1 whereas MM2 is to the northwest.

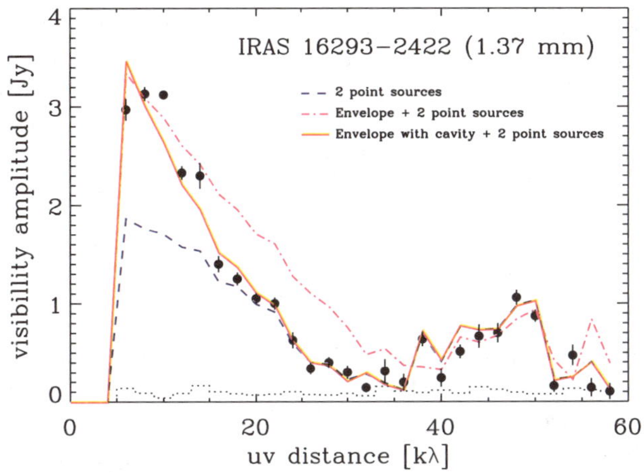

Figure 2. Visibility amplitudes of the OVRO 1.37 mm continuum emission towards IRAS 16293–2422 as a function of the projected baseline length. The central position is taken to be midway between MM1 and MM2. The observations are plotted as filled symbols with 1σ error bars. Also shown are the predictions from the continuum model (Schöier et al. 2002), with the same *uv* sampling as the observations *(dash-dotted magenta curve)* with unresolved compact emission *(dashed blue curve)* added to the model. A model envelope with a cavity *(solid red line)* best reproduces the observations.

IRAS 16293–2422 (160 pc) the projected separation of the sources is about ~ 800 AU. The sources are unresolved and the fitted sizes in the *uv*-plane of MM1 and MM2 provide upper limits to the sizes of these disks of ~ 250 AU in diameter. The total fluxes estimated for the compact components MM1 and MM2 are 0.7 Jy and 1.1 Jy, respectively. The remaining 1.7 Jy of emission is attributed to the contribution from an extended circumbinary envelope. Assuming the emission to be optically thin and adopting a dust temperature of 40 K sets lower limits to the masses of 0.25 and 0.4 $M_{\odot}$, respectively.

Schöier et al. (2002) determined the circumstellar structure of IRAS 16293–2422 from detailed modelling of the observed continuum emission. In addition to the SED, resolved images at 450 and 850 μm obtained with the SCUBA bolometer array at the JCMT were used to constrain the large-scale envelope structure. In order to investigate if the brightness distribution at 1.37 mm from the envelope model is consistent with the flux picked up by the interferometer, the same *uv* sampling was applied to the model output. Furthermore, the interferometer data constrain the envelope structure at smaller scales (~ 1″) than the JCMT single dish data (~ 10″). In Figure 2 the visibility amplitudes of the observed emission are compared to the model predictions. Fitting just two compact sources underestimates the flux at small baselines, indicating additional large-scale emission. Addition of the best-fit envelope model of Schöier et al. (2002) produces the correct amount of flux at the smaller baselines but provides too much emission at intermediate baselines

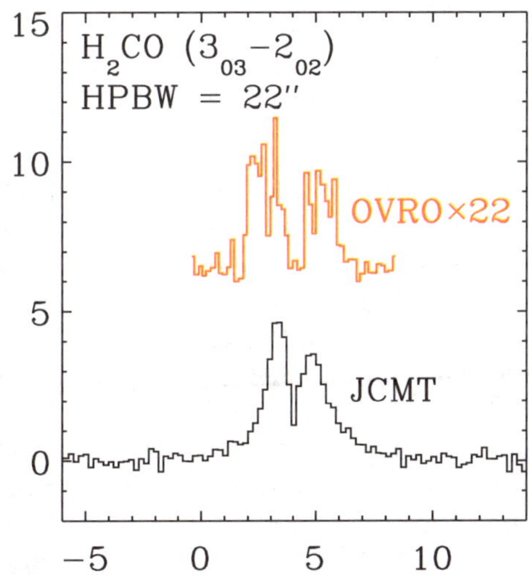

Figure 3. Comparison between the H_2CO ($3_{03} \rightarrow 2_{02}$) line emission towards IRAS 16293–2422 at the source position from JCMT single dish observations *(lower spectrum)* and OVRO interferometric observations *(upper spectrum)* restored with the JCMT beam (22″). The OVRO spectrum has been scaled by a factor of 22 in order to account for all the flux seen in the JCMT spectrum.

($\sim 5''$), i.e., at scales of the binary separation. It is found that an envelope model that is void of material on scales smaller than the binary separation best reproduces the observed visibilities. The emission from the two unresolved components, possibly accretion disks, is estimated to contribute $\sim 25\%$ to the total flux at 1.37 mm.

3.2. Molecular line emission

The observed H_2CO ($3_{22} \rightarrow 2_{21}$ and $3_{03} \rightarrow 2_{02}$) line emission obtained at OVRO is shown in Fig. 1. The emission is clearly resolved and shows a $\sim 6''$ separation between the red ($4-7\,\mathrm{km\,s^{-1}}$) and blue ($1-4\,\mathrm{km\,s^{-1}}$) emission peaks. The direction of the red-blue asymmetry is roughly perpendicular to the large-scale CO outflow associated with MM1 (Walker et al. 1988) and is indicative of an overall rotation of the material encompassing both MM1 and MM2. The emission from both lines peaks close to the positions of MM1 and MM2. The $(3_{22} \rightarrow 2_{21})/(3_{03} \rightarrow 2_{02})$ line ratio is sensitive to the temperature of the emitting gas (e.g., van Dishoeck et al. 1995), and is > 0.4 close to the positions of MM1 and MM2 indicating hot (> 100 K) gas. Single dish observations (van Dishoeck et al. 1995; Schöier et al. 2003, in prep.) show that the H_2CO line emission is extended to scales of $\sim 30''$. Also, the large-scale $(3_{22} \rightarrow 2_{21})/(3_{03} \rightarrow 2_{02})$ line ratio is ~ 0.25, much lower than for the interferometer data, suggesting that a cold (< 50 K) component dominates the single dish flux.

Care has to be taken when interpreting the interferometer maps due to the low sensitivity to weak large-scale emission. A direct comparison between the single dish spectrum and that obtained from the interferometer observations restored with the single dish beam is shown in Figure 3. The interferometer picks up only $\sim 5\%$ of the single dish flux, suggesting that the extended cold material is resolved out by the interferometer and that a hotter compact component is predominantly picked up. This is to be expected from the 'jump models' introduced by Schöier et al. (2002) where a steep gradient in abundance needed to be introduced, for a number of molecular species including H_2CO, in order to model emission lines probing a wide range of excitation conditions. Simulations where the *uv* sampling from the observations has been applied to the best fit H_2CO models from Schöier et al. (2002) with and without a jump in the abundance show that the 'hot core' component dominates the flux as seen by the interferometer ($> 50\%$).

4. Summary and outlook

Interferometric continuum observations suggest that the inner parts of the circumbinary envelope around IRAS 16293–2422 is relatively void of material on scales smaller than the binary separation ($\sim 5''$). The simultaneous H_2CO line observations indicate presence of hot and dense gas close to the peak positions of the continuum emission consistent with thermal evaporation of ices in 'hot core' regions close to the protostars. Based on the morphology and line widths little of the observed interferometric emission is thought to be associated with the known outflow(s) associated with this source.

Additional interferometric data have been obtained for IRAS 16293–2422 in a wide range of lines including $C^{18}O$, CH_3OH, HCN, N_2H^+ and SiO, that will provide additional constraints on the physical and chemical structure in the circumstellar envelope. Whatever their precise origin, the molecules located in the inner envelope of IRAS 16293–2422 can be incorporated into the growing circumstellar disk(s) and become part of the material from which planetary bodies are formed.

Acknowledgements

This research was supported by the Netherlands Organization for Scientific Research (NWO) grant 614.041.004, the Netherlands Research School for Astronomy (NOVA) and a NWO Spinoza grant. OVRO observations were supported by the National Science Foundation, grant AST-9981546.

References

Ceccarelli, C., Hollenbach, D. J., & Tielens, A. G. G. M. 1996, ApJ 471, 400

Ceccarelli, C., Loinard, L., Castets, A., Tielens, A. G. G. M., & Caux, E. 2000, A&A 357, L9

Ivezić, Ž. & Elitzur, M. 1997, MNRAS 287, 799

Jørgensen, J. K., Schöier, F. L., & van Dishoeck, E. F. 2002, A&A 389, 908

Mundy, L. G., Wootten, A., Wilking, B. A., Blake, G. A., & Sargent, A. I. 1992, ApJ 385, 306

Schöier, F. L., Jørgensen, J. K., van Dishoeck, E. F., & Blake, G. A. 2002, A&A 390, 1001

Shirley, Y. L., Evans, N. J., & Rawlings, J. M. C. 2002, ApJ 575, 337

van Dishoeck, E. F., Blake, G. A., Jansen, D. J., & Groesbeck, T. D. 1995, ApJ 447, 760

Walker, C. K., Lada, C. J., Young, E. T., & Margulis, M. 1988, ApJ 332, 335

Walker, C. K., Adams, F. C., & Lada, C. J. 1990, ApJ 349, 515

TUNABLE ETALONS FOR *NGST* NIRCAM

A. Scott, N. Rowlands, and C. Evans

EMS Technologies Canada, Ltd., Ottawa, ON, Canada K2C 0P9

Abstract

Tunable etalons will be an important component of the *NGST* NIRCam instrument. NASA recently selected the University of Arizona, Lockheed-Martin NIRCam design for implementation. In addition to broadband imager modules, this design uses two sets of tunable Fabry-Perot etalons to cover the spectral ranges of <1.2 to 2.5 μm and 2.5 to ≥ 4.5 μm in separate fields. These filter sets, with R=100, will allow extraction of critical spectral information to meet the NIRCam science goals. These include measuring the population of nearby sub-stellar objects and brown dwarfs, the extent of photodissociation regions in obscured sources, and gathering information about galaxy clusters and AGN at high redshift. EMS and the CSA are developing the technology for space-based cryogenic tunable filter modules. PZT actuator performance has been characterized down to 77 K. Reflective coatings with low phase dispersion are being designed to cover the <1.2 to 5 μm region for the NIRCam etalons. Materials tests have been performed in support of mechanical design concepts.

Key words: Instrumentation: detectors

1. Science Objectives

1.1. Galaxy Formation

Tunable filters will search for the first Lyman-α sources. A deep emission line survey will cover a wide range of redshifts and will also detect He II and either Hβ or Pβ to guard against misidentifications.

1.2. Formation of Stars and Brown Dwarfs

Tunable filters will help to constrain the low mass end of the IMF by providing spectral information on protostellar objects in nearby embedded clusters. The filters are able to effectively isolate these dim objects in the midst of a complex field containing bright young stars.

1.3. Planetary Systems from Birth to Maturity

Selected disks, in a range of evolutionary states, will be imaged using the coronagraphic mode of the tunable filter modules permitting composition measurements. The tunable filter coronagraph will also enable low resolution spectroscopy of planets around other stars.

2. Tunable Filter Requirements

Tunable filter requirements have been derived in order to best accomplish the science goals for the mission. These requirements are listed in Table 1.

Table 1. NIRCam Tunable Filter Requirements

Parameter	Requirement	Goal
Spectral resolution	>50	~100
Finesse	~30	50
Wavelength coverage	1.2 to 2.5 μm 2.5 to 4.5 μm	<1 to 2.5 μm 2.5 to 5 μm
Wavelength stability	<0.2 nm	<0.05 nm
Throughput	>80%	>85%
No. blocking filters	<11	few as possible
Order blocking	<2%	<0.05%
Clear aperture	>32 mm	

3. Opto-Mechanical Design

The actuators are low voltage piezo-electric stacks (PZTs) with the following features:

– similar to those extensively used in telecommunication fibre-optic networks;

– current drive circuit range is –30 to +100 V;

Chemistry as a Diagnostic of Star Formation (2003). *Edited by Charles L. Curry and Michel Fich.* Published in 2003 by NRC Research Press, Ottawa ON, K1A 0R6, Canada. ISBN 0-660-19089-3.

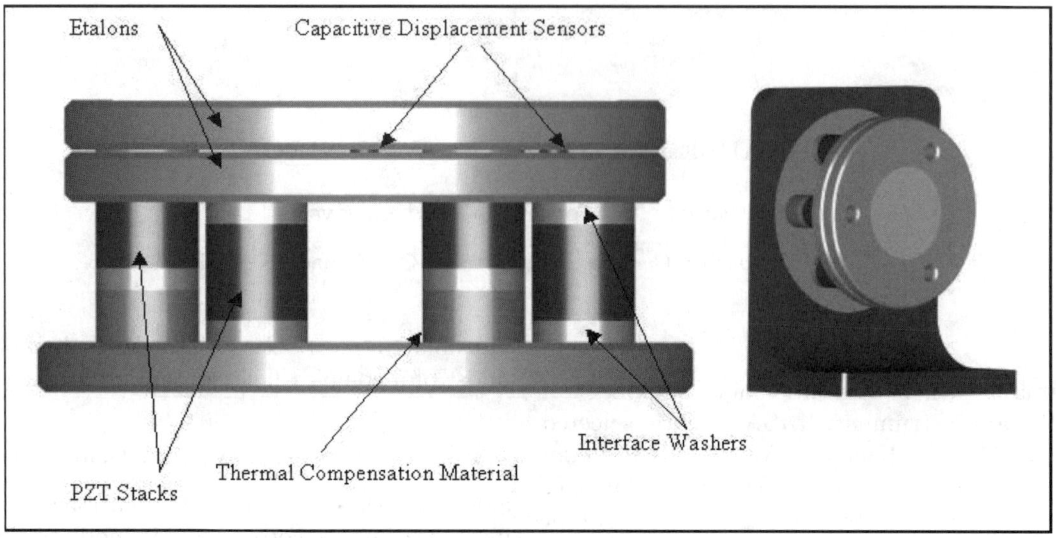

Figure 1. Solids model of the NIRCam tunable etalon.

– 4.2 μm displacement at 79 K (2.4 μm at 30 K);

– two sets of PZT's will be used (one set for each optical plate) to provide $\sim$5 μm of travel.

The *NGST* filter design is shown in Figure 1. The etalon optical substrates (Si & SiO_2 are candidates) are mounted on PZTs which in turn are mounted on a support ring of identical material (mounting interface). Low order operation requires an accurate gap near 2 μm be constantly maintained between plates. The etalon gap may need to be larger during launch vibrations to prevent damage. The PZT ceramic is also very weak under tensile load and is expected to require mechanical support. Thermal compensation material could be used to give a larger gap at ambient temperatures during launch. Kevlar straps mounted axially inside the PZTs can provide the necessary compressive support at ambient temperatures and will release this force upon cooling due to differential coefficient of thermal expansion. Overall etalon structural integrity and optical performance issues are being explored in an on-going development of a 9 μm tunable filter. These issues will be further explored in an *NGST* prototype development project.

4. Coating Design

The reflective coatings which bound the tunable air gap are challenging due to the need to span a full spectral order:

1. The coating should have $R < 91\%$ across the waveband, so that physical finesse does not degrade transmittance (depends on etalon flatness).

2. The coating should minimize phase variation ($\Delta\phi < 80°$) across the waveband (see Fig. 2). Large phase gradients can result in spurious transmittance peaks which scan at different rates.

3. Initial designs have considered 1st or 2nd order to minimize blocking filters. This results in small gaps (0.5 to 0.9 μm for the short wavelength channel).

4. Results show that silicon substrates (+Si/MgF2 layers) could be used $\lambda > 1.8$ μm.

5. Control Electronics Design

The etalon gap near 2 μm must be maintained parallel to a few nanometres in order to maintain peak throughput and finesse. This will be accomplished using capacitive displacement sensors in an FPGA-controlled feedback loop based on a capacitive bridge design as shown in Figure 4.

The circuit is directly based on a design for a Michelson interferometer phase stepping mirror currently under development for a separate flight program.

Digital interface inputs:

– input: etalon set point (wavelength to digital value conversion in flight software);

– input: digital filtering parameters, PZT control parameters (gains, offsets);

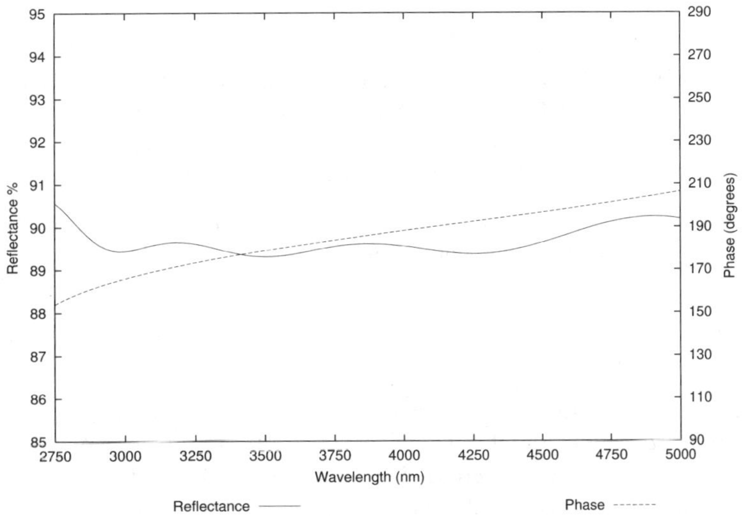

Figure 2. Preliminary coating performance for long wave etalon.

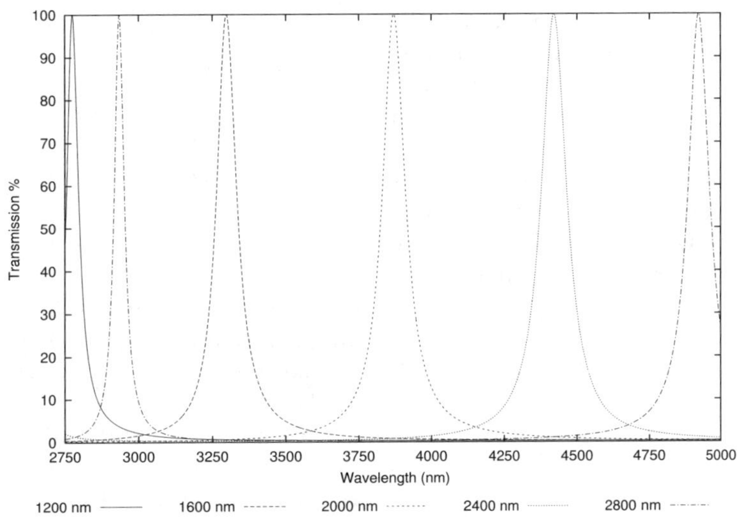

Figure 3. Spectral performance of coating design shown in Figure 2 for various gap spacings.

– output: sampled etalon position.

In-flight calibration (and ground interpretation) allows new set points, and filtering and control parameters to be revised in flight software.

ACKNOWLEDGEMENTS

This work was supported by EMS Technologies Canada, Ltd., and the Canadian Space Agency.

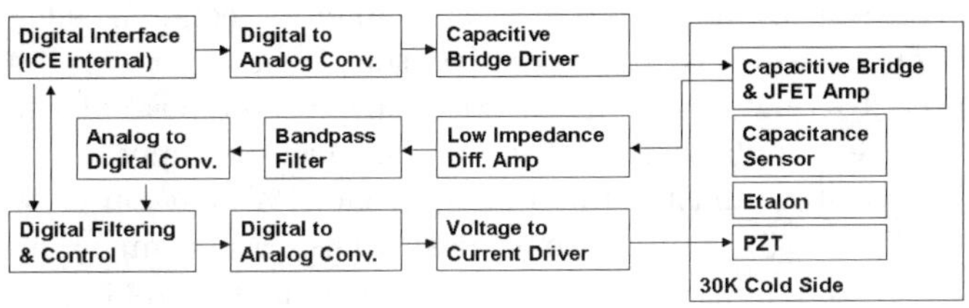

Figure 4. Tunable etalon control system block diagram.

DEUTERIUM FRACTIONATION ON INTERSTELLAR GRAINS STUDIED WITH THE MASTER EQUATION APPROACH

Tatiana Stantcheva[1] and Eric Herbst[1,2]

[1]Department of Physics, The Ohio State University, Columbus, OH, 43210, USA
[2]Department of Astronomy, The Ohio State University, Columbus, OH, 43210, USA

Abstract

We have applied the master equation approach to study deuterium fractionation on the surfaces of dust grains. The species considered are H, O, and D atoms, CO molecules accreting onto the dust grains, and their various products produced by surface reactions. We consider dark clouds at a late age when most of the hydrogen is already in a molecular form and the carbon has been converted into CO. We present results for the surface abundances of the stable surface species, as well as the fractionation ratios between deuterated and normal isotopomers. A comparison between the results and those obtained in Monte Carlo simulations shows excellent agreement between the two approaches. In most cases, the master equation approach proved to be more time efficient than the Monte Carlo method. We believe that it will be possible for the method to be used for much more complex systems.

Key words: ISM: abundances – ISM: molecules – ISM: molecular processes

1. Overview of the system

When the average abundances of some surface species become as low as unity, or even less, the rate equation method is not applicable to models of surface chemistry occurring by diffusion of weakly-bound species. In such circumstances, it is imperative that other methods be used to take into consideration the discreteness of the surface populations on the grains. One such method is the so-called master equation approach. We have already applied it to simple chemical systems (Wakelam et al. 2002) and now we report our success in its application to a relatively more complex one. Our current chemical network consists of H, O, and D atoms, and CO molecules, and the products of various surface reactions among them. Overall, there are 31 species that react with each other via 33 surface reactions. The reactions that require an activation barrier are listed in Table 1 along with the barrier. Our chemical network is almost identical to that of Caselli et al. (2002).

Table 1. Surface reactions in with barriers in the H,D,O,CO model

No.	Reaction			E_a (K)
1	H + CO	$\longrightarrow$	HCO	2000
2	H + H_2CO	$\longrightarrow$	H_3CO	2000
3	H + HDCO	$\longrightarrow$	H_2DCO	1965
4	H + D_2CO	$\longrightarrow$	HD_2CO	1925
5	O + CO	$\longrightarrow$	CO_2	1000
6	CO + D	$\longrightarrow$	DCO	1930
7	H_2CO + D	$\longrightarrow$	H_2DCO	1799
8	D + HDCO	$\longrightarrow$	HD_2CO	1758
9	D + D_2CO	$\longrightarrow$	D_3CO	1713

We have used the same values for grain parameters as in Hasegawa et al. (1992). We have also followed Hasegawa et al. (1992) closely in calculating the various accretion, evaporation, and reaction rates. All the calculations were performed for a duration of 10^4 years and a temperature of 10 K. The gas-phase abundances of the accreting species were assumed to be constant throughout the calculations. The models were performed for three different gas-phase densities—low, intermediate, and high—based on gas-phase steady-state solutions for dense quiescent clouds (Table 2). The results reported here, however, are only for the high density case.

2. Theory

The master equation approach to modelling the chemistry on grain surfaces was first proposed independently by Green et al. (2001) and Biham et al. (2001). The idea behind this approach is to follow the evolution of all sur-

Chemistry as a Diagnostic of Star Formation (2003). *Edited by Charles L. Curry and Michel Fich.* Published in 2003 by NRC Research Press, Ottawa ON, K1A 0R6, Canada. ISBN 0-660-19089-3.

Table 2. H, O, and CO gas-phase abundances (cm^{-3}) utilized

Abundance n	Low	Intermediate	High
H	1.15	1.15	1.10
O	0.09	0.75	7.0
CO	0.075	0.75	7.5
D	0.3	0.3	0.3

face populations simultaneously. Since we cannot know the exact population distribution of the surface species at a given time, it is necessary to consider all possible states of the system, defined as the set of populations for each species. For example, one possible state for the grain surface consists of 1 particle of species A, 0 particles of species B, etc. Another state consists of 2 particles of A, 1 of B, and so on. Each such state is associated with a probability that the grain surface exists in such a state, and a set of coupled differential equations is numerically solved to propagate the probabilities forward in time. The solution of these equations gives the probabilities at any given time for the grain surface to be in each state. The calculated values for the probabilities can then be used to calculate the average abundances of the surface species. Examples of the probability equations and the means of calculating the surface abundances of the species can be found in Wakelam et al. (2002) and Stantcheva and Herbst (2002).

It is important to note that not all the surface species need to be considered in the states of the system. We have restricted our consideration to only 11 species, all of low abundance, that readily react with other species without an activation barrier. We have called these species stochastic since they require a probabilistic treatment. The species—H, O, D, OH, OD, HCO, DCO, H_3CO, H_2DCO, HD_2CO, and D_3CO—are atoms and radicals. The time dependence of the abundance of stable surface species, which have large surface populations, can be treated by rate-like equations. In order to restrict the number of probability equations that we need to solve, we have imposed two requirements. They are based on the physical assumption that the abundance of any of the stochastic species is likely to be low. The first requirement is that no states be considered for which the abundance of any stochastic species exceeds a fixed number $\mathcal{N}_i$, which can be different for different stochastic species. Secondly, we ignore all the states for which the total surface abundance of the stochastic species is greater than another parameter, $\mathcal{N}_{\text{tot}}$.

Table 3. Populations in mono-layers

Species	High density	
	MC	ME
CO	5.0	4.9
H_2O	1.4	1.2
O_2	2.7	2.4
CO_2	0.67	0.71
H_2CO	0.44	0.44
CH_3OH	0.088	0.095
Total abundance	11.1	10.5

3. Results

Results for the high density case only are presented in this paper. The details of the individual and total population limits are discussed in Stantcheva and Herbst (2002) which also contains results for the other densities. Table 3 shows the surface abundances of selected non-deuterated stable molecules achieved after 10^4 yr. The abundances are given in mono-layers, with each layer considered to consist of 10^6 particles. The master equation (ME) results are compared with those from previous Monte Carlo simulations (MC). The comparison shows excellent but not perfect agreement between the two methods. With the population limits chosen, the master equation approach is far more efficient in terms of computing time than the Monte Carlo approach. As the necessary population limits get larger, the standard rate equation technique becomes more useful.

The abundances of the deuterated species depend strongly on the fraction of deuterium atoms to hydrogen atoms in the gas. We have utilized a very large value of 0.27 for this ratio since there is strong evidence that efficient deuterium fractionation occurs on grain surfaces in low-mass star formation regions. In Table 4, we list the abundance ratios f between deuterated and normal isotopomers for the major deuterated species in the model.

Once again, results for both the master equation and Monte Carlo simulation approaches are shown. Once again, the methods are in excellent agreement. Note that the D/H abundance ratios are very large. Since the high abundance ratios detected in low-mass star formation regions are found by radioastronomical methods in the gas phase, a theory in which deuterated species are formed on grain surfaces requires subsequent desorption of these species into the gas phase.

The calculations reported here prove that the master equation approach can be successfully applied to mod-

Table 4. Abundance ratios f

Species	High density	
	MC	ME
HDCO	0.33	0.34
D_2CO	0.026	0.027
CH_3OD	0.20	0.20
CH_2DOH	0.70	0.71
CH_2DOD	0.14	0.14
CHD_2OH	0.17	0.18
CHD_2OD	0.034	0.035
CHD_3OH	0.014	0.015
CHD_3OD	0.0025	0.0029
HDO	0.39	0.38
D_2O	0.038	0.036

erately complex grain surface models. Like the Monte Carlo method, the master equation approach is an exact method, but it is clearly less wasteful of computer time than the former in situations (low surface populations) where an exact approach is needed. Based on our results, we believe that the master equation method can be extended to study the surface chemistry of far more complex systems, especially in conjunction with the standard rate equations when the surface populations of reactive species become high. The latter method, which uses differential equations for average abundances, remains the most efficient method for modelling the chemistry both in the gas and on the grain surface. Since the solution of the master equation also utilizes differential equations, this approach can be effectively coupled with rate equations in complex models of gas-grain chemistry. Here the gas-phase chemistry is treated by rate equations while the surface chemistry is treated by the master equation or the rate equation approach depending upon the surface populations.

Acknowledgements

We thank V. Shematovich for fruitful discussions and for his help in the implementation of the master equation approach. The Astrochemistry Program at The Ohio State University is supported by The National Science Foundation (US). We thank the Ohio Supercomputer Center for time on their Cray SV1 machine.

References

Biham, O., Furman, I., Pirronello, V., & Vidali, G. 2001, ApJ, 553, 595

Caselli, P., Stantcheva, T., Shalabiea, O., Shematovich, V. I., & Herbst, E. 2002, Plan. Sp. Sci., in press

Green, N. J. B.,Toniazzo, T., Pilling, M. J., Ruffle, D. P., Bell, N.,& Hartquist, T. W.. 2001, A&A, 375, 1111

Hasegawa, T. I., Herbst, E., & Leung, C. M. 1992, ApJS, 82, 167

Stantcheva, T., Shematovich, V. I., & Herbst, E. 2002, A&A, 391, 1069

Stantcheva, T., & Herbst, E., 2002, MNRAS, submitted

DETECTABILITY OF INFRARED H_2 EMISSION SPECTRA WITH NEW FORMATION PUMPING MODELS

Junko Takahashi[1] **and Hideya Uehara**[2]

[1]Meiji Gakuin University, 1518 Kamikurata-cho, Totsuka-ku, Yokohama 244-8539, Japan
[2]SGI Japan, Ltd., 2-5-25 Umeda, Kita-ku, Osaka 530-0001, Japan

Abstract

Recently, we built up new formation pumping models of H_2 newly formed on various dust surfaces, based on quantitative computer simulations. Then, we calculated the H_2 emission spectra dominated by formation pumping and UV pumping, respectively, with our new formation pumping models, considering the early stage of an evolving region from H I clouds to H_2 clouds. It was found that the resulting infrared H_2 emission spectra dominated by formation pumping can be discriminated from those dominated by UV pumping. We also calculated the detectability of the infrared H_2 emission spectra arising from formation pumping, and found that the fluxes of the most intense emission lines should be strong enough to be detectable with Subaru-IRCS Echelle. The infrared H_2 emission spectra arising from formation pumping can be a novel probe which will directly show us the formation process of interstellar H_2 on dust surfaces as well as a new probe to observe the spots where the evolution from H I clouds to H_2 clouds occurs.

Key words: ISM: lines and bands – ISM: molecules – ISM: clouds – ISM: dust, extinction

1. Formation Pumping of Molecular Hydrogen

It has been generally considered that interstellar molecular hydrogen ought to be formed on the surface of cosmic dust grains. Recently, the energy distribution of the product H_2 on the formation process on various dust surfaces was quantitatively investigated, based on computer simulations, by several authors (e.g., Takahashi et al. 1999; Meijer et al. 2001). They found that H_2 newly formed on dust surfaces are in highly vibrationally and rotationally excited states of the ground electronic state via the formation pumping mechanism.

The formation pumping mechanism can be explained by the redistribution of the H–H bond energy (= 4.5 eV) via the H–surface interaction into the vibrational, rotational, and translational energies of the product H_2 and the energy absorbed by dust. The formation pumping levels of H_2 produced on various dust surfaces are summarized in Table 1 (Takahashi 2001; Takahashi & Uehara 2001).

The formation pumping to vibrationally and rotationally excited states of the ground electronic state is followed by spontaneous radiative cascade transitions to lower rovibrational energy levels. As was suggested by Duley & Williams (1993), this means that the infrared H_2 emission may be detectable even in regions without a source of UV pumping or dynamical excitation.

2. Calculation of H_2 Emission Spectra

The infrared H_2 emission spectra with formation pumping were previously calculated by Black & van Dishoeck (1987), Le Bourlot et al. (1995), and Draine & Bertoldi (1996). However, their formation pumping models were not based on any theoretical or experimental studies, and far from the results of the recent quantitative studies about H_2 formation on dust surfaces by Takahashi et al. (1999), Meijer et al. (2001), and so on. Thus, we attempted to build up our new formation pumping models for H_2 formed on icy mantles, carbonaceous dust, and silicate dust, based on the above studies (Takahashi & Uehara 2001).

Then, we calculated the H_2 emission spectra with our new formation pumping models, considering an evolving region from H I clouds to H_2 clouds dominated by the H_2 formation process on the surface of dust grains, of uniform density $n_H = 10^3$ cm^{-3} and gas temperature $T = 10$ K, in interstellar clouds under a weak UV radiation field. We solved the following master equations for the population, n_i, of vibration-rotation levels, i [$i = (v, \mathrm{J})$], of H_2:

$$\frac{dn_i}{dt} = \sum_{E_i<E_k} (A_{ki} + \beta_{ki})\, n_k - n_i \sum_{E_k<E_i} A_{ik} - n_i\beta_{\mathrm{diss},i} + f(i)\, R_{H_2} n_H, \qquad (1)$$

Chemistry as a Diagnostic of Star Formation (2003). *Edited by Charles L. Curry and Michel Fich.* Published in 2003 by NRC Research Press, Ottawa ON, K1A 0R6, Canada. ISBN 0-660-19089-3.

Table 1. Formation pumping levels of H_2 on various dust surfaces

	Icy mantles	Silicate dust	Carbonaceous dust
H–surface interaction	weak	middle	strong
H–surface binding energy (eV)	< 0.1	~ 0.4	1.7
Main vibrational energy level	$v = 7$–8	$v \leq 7$	$v = 1$–2
Vibrational energy [eV (%)]	3.1 (70)	2.7 (60)	0.9 (20)
Rotational energy [eV (%)]	0.6 (14)	0.7 (15)	0.7 (15)
Translational energy [eV (%)]	0.5 (12)	0.7 (15)	1.1 (25)
Energy absorbed by dust [eV (%)]	0.2 (4)	0.5 (10)	1.8 (40)
Mean vibrational temperature (K)	36200	31000	10200
Mean rotational temperature (K)	3660	4000	4000
Mean translational temperature (K)	2050	2600	4560

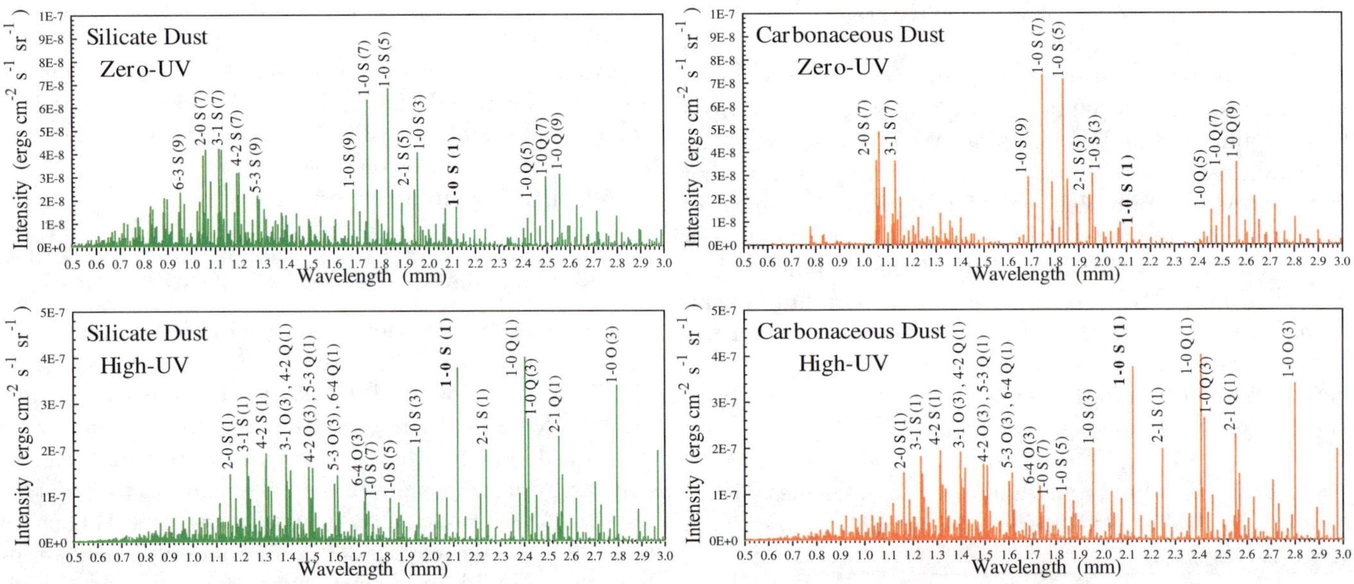

Figure 1. H_2 emission spectra dominated by formation pumping (Zero-UV) and UV pumping (High-UV) for silicate and carbonaceous dust (T=10 K, $n_H = 10^3$ cm^{-3}, $\Sigma_H = 10^{21}$ cm^{-2}).

where A_{ik} are the Einstein A-coefficients for spontaneous rovibrational transitions, β_{ki} are transition probabilities by UV pumping, $\beta_{\mathrm{diss},i}$ are photodissociation rate coefficients, E_i are the energy levels, R_{H_2} is the H_2 formation rate coefficient on dust grains taken from Tielens & Hollenbach (1985), and $f(i)$ are the vibration-rotation populations by formation pumping for which the functional formulas in our new formation pumping models were employed (Takahashi & Uehara 2001). Since the time scale of radiative transition($\sim 10^0$ years) is much shorter than that of H_2 formation ($\sim 10^7$ years), the population settles into the steady state rapidly. The intensities of the transitions from states i to k were given by

$$I_{ik} = (1/4\pi) E_{ik} A_{ik} n_i (\Sigma_H / n_H) \ \mathrm{erg\,cm^{-2}s^{-1}sr^{-1}}, \qquad (2)$$

where Σ_H is the column density of neutral H atoms.

In Figure 1, calculated H_2 emission spectra for silicate dust and carbonaceous dust are shown. The spectra for icy mantles are similar to those for silicate dust. Zero-UV and high-UV mean the pure formation pumping case without a UV field and the formation pumping case with a strong UV field, respectively. In the high-UV case, since the typical interstellar UV field of solar neighbor-

Table 2. Detectability of the intense H_2 emission lines with Subaru-IRCS *K*-band Echelle

Line	Wavelength	Dominant pumping	Zero UV		High UV	
			Magnitude	Exposure time	Magnitude	Exposure time
2–1 S(5)	1.9449 μm	formation	20.4	6.3 hrs	20.4	6.3 hr
1–0 S(3)	1.9577 μm	formation + UV	20.0	3.6 hrs	18.6	16.5 min
1–0 S(1)	2.1218 μm	UV	20.7	13.2 hrs	17.4	1.8 min
1–0 Q(5)	2.4545 μm	formation + UV	20.5	9.1 hrs	19.1	41.5 min

hood ($G_0 = 1$) is used but the effect of H_2 self-shielding is not taken into account, the spectra are considered to correspond to those with a UV field of 10^3–10^5 G_0. It is expected that the spectra observable in dark clouds with a weak UV field will be in the intermediate case between the zero-UV and high-UV cases.

The above proved that H_2 emission spectra dominated by formation pumping and UV pumping can be discriminated from each other. The line intensities of transitions from higher rotational states (e.g., 2–1 S[5]) are little enhanced by UV pumping but are mainly dominated by formation pumping. On the contrary, those from lower rotational states (e.g., 1–0 S[1]) are significantly enhanced by UV pumping. If the former are relatively stronger than the latter, it should be the evidence of H_2 emission arising from formation pumping.

It is also found that the H_2 emission spectrum for carbonaceous dust can be discriminated from those for icy mantles and silicate dust. There are many strong emission lines in the range of $\lambda < 1.6$ μm for the latter while they are lost or weakened for the former. These emission lines belong to the transitions from vibrationally excited states higher than $v = 3$, which are strongly pumped for icy mantles and silicate dust but little pumped for carbonaceous dust.

3. Calculation of Detectability

In Table 2, the calculated brightness and exposure times after the correction of the system efficiency and transmission to obtain $S/N = 5$ of the intense H_2 emission lines in the K-band are shown (T= 10 K, $n_H = 10^3$ cm^{-3}, $\Sigma_H = 10^{21}$ cm^{-2}), considering the observation with Subaru-IRCS Echelle, with the spectral resolution (R = 5000, 0.60″ slit, 10 and 96 pixels binned in the dispersion and spatial direction, respectively).

Since the exposure times of other lines decrease rapidly with the increasing UV field, 6.3 hrs of the 2–1 S(5) line, arising from formation pumping and independent of the UV field, is taken to be the standard exposure time. If the temperature increases from 10 K to 100 K, the brightness of 2–1 S(5) becomes about three times stronger, which means that the exposure time will go down to about 40 min.

If the UV field is relatively so strong, H_2 emission lines arising from formation pumping will be buried in those dominated by UV pumping. Then, we consider that one of possible targets to be observed for the detection of the H_2 emission spectra arising from formation pumping may be the boundary regions of dark clouds with weak FUV fields, where any one has never expected that H_2 emission would be detectable.

The infrared H_2 emission spectra arising from formation pumping can be a novel probe which will directly show us the formation process of interstellar H_2 on dust surfaces as well as a new probe to observe the spot where the evolution from H I clouds to H_2 clouds occurs.

Acknowledgements

The authors are grateful to Prof. B. T. Draine, Dr. P. G. Martin, and Dr. J. Le Bourlot for their having kindly sent us their data to calculate the H_2 emission spectra. The authors also thank Dr. Tomonori Usuda and Dr. Naoto Kobayash for their helpful discussions about the detectability of the infrared H_2 emission spectra arising from formation pumping.

References

Black, J. H. & van Dishoeck, E. F. 1987, ApJ 322, 412
Draine, B. T. & Bertoldi, F. 1996, ApJ 468, 269
Duley, W. W. & Williams, D. A. 1993, MNRAS 260, 37
Le Bourlot, J., et al., 1995, ApJ 449, 178
Meijer, A. J. H. M., et al., 2001, J. Phys. Chem. 105, 3359
Takahashi, J., Masuda, K., Nagaoka, M. 1999, MNRAS 306, 22
Takahashi, J., Masuda, K., Nagaoka, M. 1999, ApJ 520, 724
Takahashi, J. 2001, ApJ 561, 254
Takahashi, J. & Uehara, H. 2001, ApJ 561, 843
Tielens, A. G. G. M. & Hollenbach, D. 1985, ApJ 291, 722

CONNECTION BETWEEN PAHs AND SMALL HYDROCARBONS IN THE HORSEHEAD NEBULA PHOTODISSOCIATION REGION

David Teyssier[1], David Fossé[1], Maryvonne Gerin[1], Jérôme Pety[1,2], Alain Abergel[3], and Emilie Habart[4]

[1]LERMA, UMR 8540, CNRS/Observatoire de Paris, 24 rue Lhomond, 75231 Paris Cedex 5, France
[2]IRAM, 300 rue de la Piscine, Domaine Universitaire, 38406 Saint Martin d'Hères, France
[3]IAS, UMR-8617, Université Paris-Sud, bâtiment 121, 91405 Orsay, France
[4]Osservatorio Astrofisico di Arcetri, Largo E. Fermi 5, 50125 Florence, Italy

Abstract

We present recent observations of small hydrocarbons (C_3H_2, C_2H, C_4H) with high abundances in the photodissociation region of the Horsehead Nebula. Our results show for the first time observational indications that the small hydrocarbon distribution follows the Aromatic Infrared Bands emission traced by *ISO*-LW2 (5-8.5 μm), whereas it does not coincide with the CO and isotopes large-scale distribution. The derived abundances are significantly higher than in local clouds. This enhancement might be explained by an *in situ* formation assisted by the release of carbonaceous molecules from UV-irradiated aromatic particles.

Key words: ISM: molecules — ISM: clouds — ISM: individual (Horsehead Nebula)

1. Introduction

The spatial distribution of the Aromatic Infrared Bands emission features (AIBs: 3.3, 6.2, 7.7, 8.6, 11.3 and 12.7 μm features) has been mapped at high spatial resolution by ISOCAM in many photodissociation regions (PDRs). The identification of the AIBs carriers is still debated, however it is well established that they are complex carbon compounds heated transiently by the absorption of a single UV or visible photon. The best candidates are polycyclic aromatic hydrocarbons (PAHs; e.g., Léger & Puget 1984) or hydrogenated amorphous carbons (HACs; Duley & Williams 1988). These large-sized carbonaceous molecules ($\sim$ 50 carbon atoms) are likely to be chemically connected to the numerous small hydrocarbons detected in the ISM, including cyanopolyynes or acetylenic chains and rings.

Various theoretical works and laboratory experiments support this point: modelling of dark cloud chemistry (Herbst 1991; McEwan et al. 1999), calculation of the acetylene formation rate by photo-erosion of UV irradiated PAHs (Allain et al. 1996), experiments of carbon chains production during the irradiation of HACs by a UV laser pulse (Scott et al. 1997). Nevertheless we still miss a firm observational link between AIB carriers and small hydrocarbons. PDRs are the best places to look for such a link as they exhibit strong AIB emission features which have been mapped at arcsec resolution by *ISO* in several objects. They are moreover highly structured regions where, due to the penetration depth of UV irradiation, the aromatic emission, the H_2 ro-vibrational lines, and the molecular (CO) emission are spatially separated (see Hollenbach & Tielens 1997).

2. Observations

We have used the IRAM 30-m telescope to observe small carbon chains (C_2H, c-C_3H, l-C_3H, c-C_3H_2, l-C_3H_2, and C_4H) in the PDR of the Horsehead Nebula. We obtained maps of $2.5' \times 3.5'$ in three species. To our knowledge, these are the first maps of small hydrocarbons reported in PDRs. In complement, we mapped with unprecedented spatial resolution the emission of the optically thick ^{12}CO, and the likely optically thin $C^{18}O$ isotope. These molecular data were compared to the AIB emission mapped by ISOCAM in the 5.5–8.5 μm band (LW2 filter; Abergel et al. 2002). The considered data are compiled in Figure 1.

Figure 1 shows that the small carbon chains are present until the edge of the PDR. The maps reveal moreover that all three hydrocarbon integrated intensities peak very close to the *ISO*-LW2 one, which is not true for the CO and isotope emission. There is perfect correlation between C_2H and c-C_3H_2, indicating likely similar physical conditions. This behavior holds for C_4H, but is still limited by a poor S/N ratio. A shift towards the molecular cloud is however observed, suggesting beam dilution of larger-scale structures. Interferometric data obtained at IRAM PdBI (Pety et al., in preparation) are no more affected by this effect and show a perfect coincidence of the

Chemistry as a Diagnostic of Star Formation (2003). *Edited by Charles L. Curry and Michel Fich.* Published in 2003 by NRC Research Press, Ottawa ON, K1A 0R6, Canada. ISBN 0-660-19089-3.

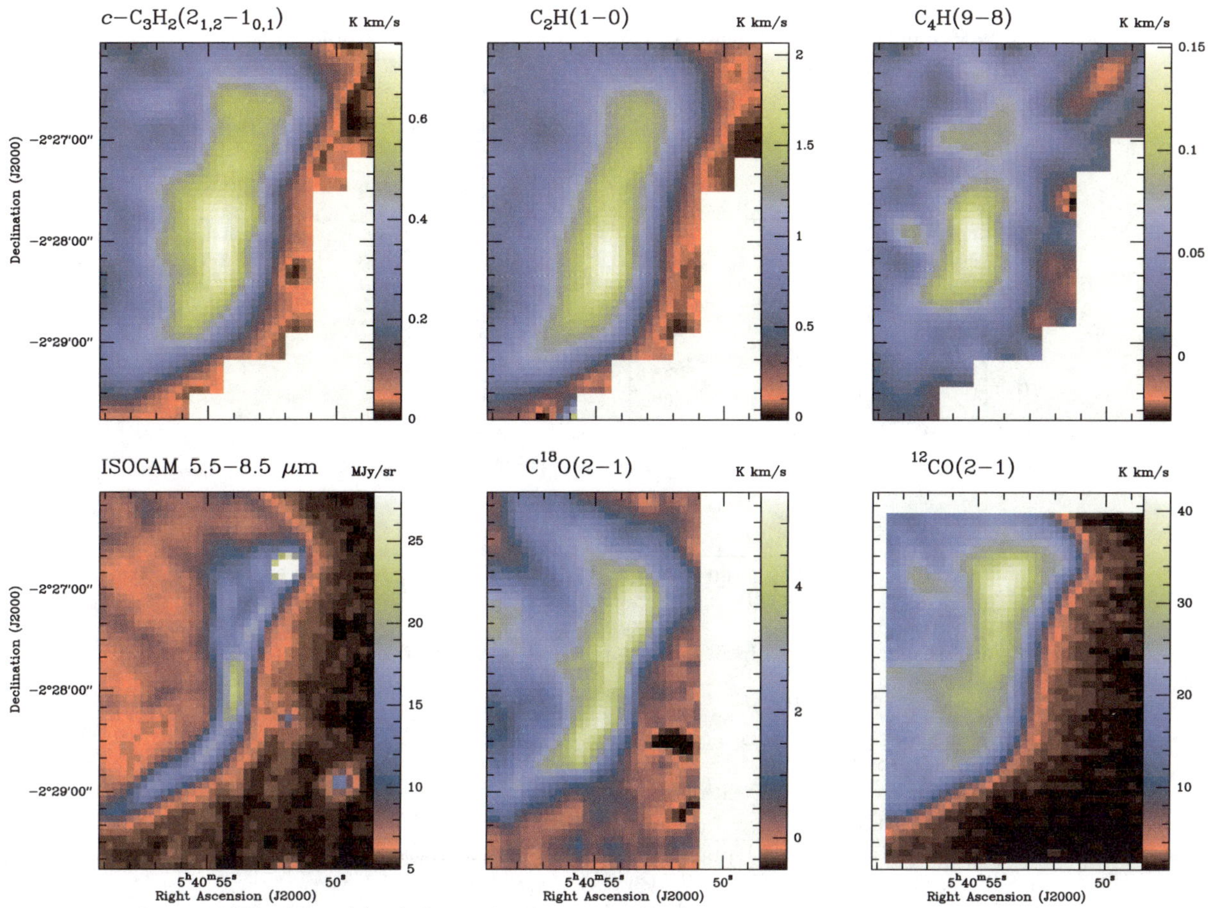

Figure 1. Integrated maps (T_A^*, K km/s) of the molecular emission for various species, as well as the aromatic emission as covered by *ISO*.

hydrocarbon and aromatic emission peaks. At first sight, there seems to be a spatial correlation between aromatic particles and small hydrocarbons.

3. Column densities

We have computed the molecular column densities in the areas corresponding to the AIB and the CO peaks respectively. Table 1 gathers the results and shows that the hydrocarbon column densities are maximum at the aromatic emission peak, which is not the case for $C^{18}O$. The values are of the same order as those observed in the TMC-1 dark cloud, known as an efficient hydrocarbon factory. However the illumination conditions ($G_0 = 100$) are different from the dark cloud ones and small hydrocarbons are expected to be rapidly destroyed by the intense radiation fields.

4. Discussion

We have compared our measurements to PDR models using several chemistry networks (Fig. 2) and assuming representative conditions ($G_0 = 100$, $n_{\rm H} = 2\times10^4\,{\rm cm}^{-3}$). We find the following results:

– the models reproduce well the abundance ratios of $[C_3H_2]/[C_4H]$ and $[C^{18}O]/[C_2H]$ ($\sim$ 1 and 20 resp.)

– C_3H_2 and C_4H are at least one order of magnitude more abundant than expected compared to C_2H.

Part of the explanation possibly lies in the underestimated H_2 column density. Kramer et al. (1996) reported A_V=3 inside the nebula (2′ resolution), translating into molecular column densities 10-100 times lower than observed. Recent continuum observations performed at 1.2 mm (11″ resolution) suggest visible extinctions more likely in the range A_V =10−20, which reconciles the observed $C^{18}O$ and C_2H results with the model cal-

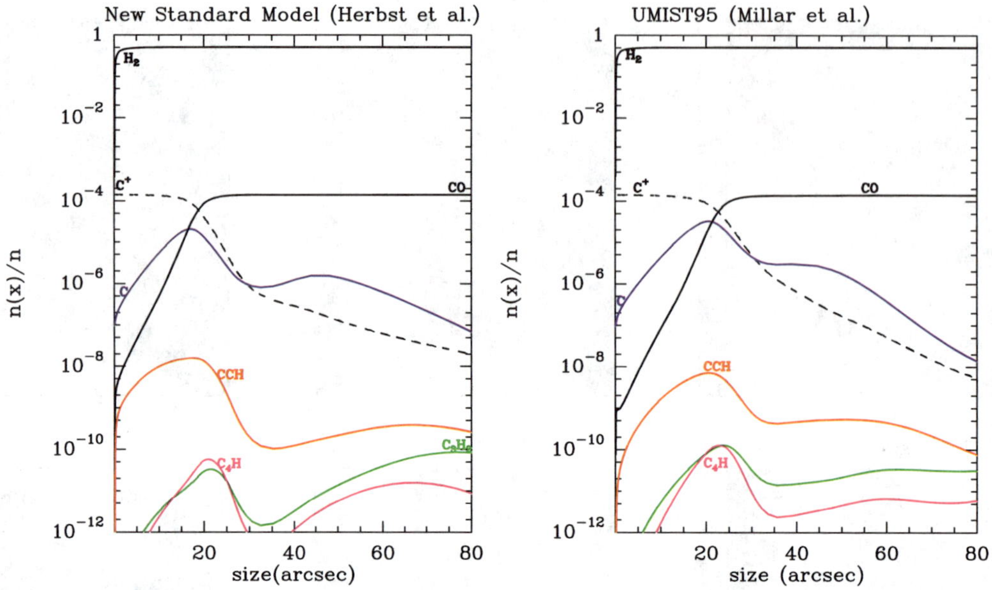

Figure 2. Abundances of various species derived from an updated Le Bourlot et al. PDR model, using two chemistry networks (calculations courtesy of E. Roueff).

Molecule	Observed Frequency (GHz)	Position	Column Density (cm^{-2}) Measured	Model $A_V^{peak} = 3$	Model $A_V^{peak} = 15$
C_2H	87.360	AIB peak	$1.8\pm0.2\times10^{14}$	1.9–5×10^{13}	1–2.5×10^{14}
		CO peak	$1.3\pm0.1\times10^{14}$		
C_3H_2	85.339	AIB peak	1-2×10^{13}	1–3×10^{11}	0.5–1.5×10^{12}
		CO peak	0.7-1.4×10^{13}		
C_4H	85.634	AIB peak	2×10^{13}	1.7–3×10^{11}	0.9–1.5×10^{12}
		CO peak	1.3×10^{13}		
$C^{18}O$		AIB peak	$4\pm0.2\times10^{15}$	5.4×10^{14}	2.7×10^{15}
		CO peak	$5\pm0.2\times10^{15}$		

Table 1. Molecular column densities measured in the PDR (use of LTE and LVG approximations), and inferred from models (Fig. 2) for two peak extinctions, in areas corresponding to the AIB and the CO emission peaks. Assumptions are: $n(H_2) = 10^4\,cm^{-3}$, $T_{ex} \sim 10\,K$ for all hydrocarbons, and $T_{kin} \sim 35\,K$ for $C^{18}O$.

culations. Nevertheless, a residual overabundance (factor $\sim$ 20) remains for the more complex hydrocarbons, suggesting that the hydrocarbon production is enhanced in the illuminated front. *We propose that this formation process is related to the photo-erosion of UV-irradiated large carbonaceous compounds at the border of the PDR.* In this scenario, complex hydrocarbons such as C_4H would survive the intense illumination conditions encountered at the molecular cloud interfaces. This also suggests that heavier species (C4, C5) could be observed in UV-irradiated regions.

References

Abergel, A., Bernard, J.-P., et al. 2002, A&A, 389, 239
Allain, T., Leach, S., & Sedlmayr, E. 1996, A&A, 305, 602
Duley, W.W. & Williams, D.A. 1988, MNRAS, 231, 969
Herbst, E. 1991, ApJ, 366, 133
Hollenbach, D.J. & Tielens, A.G.G.M. 1997, ARA&A, 35, 179
Kramer, C., Stutzki, J., & Winnewisser, G., 1996, A&A, 307, 915
Léger, A. & Puget, J.-L. 1984, A&A, 137, L5
McEwan, M.J., Scott, G.B.I., et al. 1999, ApJ, 513, 287
Scott, A., Duley, W.W., & Pinho, G.P. 1997, ApJ, 489, L193

ABUNDANT SOLID CO IN THE EDGE-ON DISK AROUND CRBR 2422.8-3423

W. F. Thi[1,2], K. M. Pontoppidan[2], E. F. van Dishoeck[2], E. Dartois[3], and L. d'Hendecourt[3]

[1]Department of Physics and Astronomy, University College London, Gower Street, London WC1E 6BT, U.K.
[2]Leiden Observatory, P. O. Box 9513, 2300 RA, Leiden, The Netherlands
[3]Astrochimie Expérimentale, Institut d'Astrophysique Spatiale, Université Paris-Sud, Bât. 121, F-91405 Orsay, France

Abstract

We report on the detection of a large quantity of solid CO in the disk around the Class I object CRBR 2422.8-3423. The observations in the M-band window were obtained with ISAAC-VLT at resolving power $R = 10,000$. The spectrum shows the presence of a lower limit of 2.2 $\times$ 10^{18} cm^{-2} of solid CO, the absorption feature being saturated. The abundance of solid CO is the highest observed so far, including high-mass protostars and background field stars. The absorption by foreground material likely accounts for a few percent of the total amount of solid CO, supported by the absence of $C^{18}O$ $J = 2 \rightarrow 1$ emission 30$''$ south. The solid CO is mostly in the apolar form, suggesting a weak alteration of the ice mantle by the stellar radiation coming from a sub-solar luminosity central star, consistent with the non-detection of solid OCN^-, a tracer of thermal/ultraviolet processing. Unresolved gas phase ro-vibrational CO absorption lines are also detected. We derive a gas phase temperature of 50 $\pm$10 K, indicating that the CO gas seen in absorption is located close to the star, presumably in the warm upper layer of the disk, whereas solid CO is located in the cold outer part. The average CO(gas)/CO(ice) along the line of sight is $\sim$1.

Key words: star formation – ISM: dust, extinction – molecules – abundances – ISM: lines and bands

1. Introduction

Jupiter-mass planets and icy planetesimals are likely formed in the circumstellar disk that orbits most pre-main sequence stars (e.g., van Dishoeck & Black 1998; Ehrenfreund & Charnley 2000). The materials in the mid-plane of those disks are so cold ($T_{\rm gas} \simeq T_{\rm dust} < 20$ K) and dense ($n_{\rm H} = 10^4 - 10^9$ cm^{-3}) that most molecules are frozen out onto the grain surface, as predicted by detailed chemical models of disks (e.g., Aikawa et al. 2002). The small abundances of gas-phase molecules observed in disks compared to that found in the quiescent dense interstellar medium providc indirect evidence of this phenomenon (Thi et al. 2001; Dutrey et al. 1997).

Direct observations of gaseous and solid CO have been performed for a few transitional objects from Class I to Class II that are known to posses a disk. Boogert et al. (2002a) observed L 1489 in Taurus—a large 2000 AU rotating disk—but the amount of solid CO is not exceptionally high ($\sim$7% of gaseous CO). This may stem from the fact that these systems are still far from edge-on (inclination $\sim 20° - 30°$) so that the line of sight does not intersect the midplane, the largest reservoir of solid CO. Shuping et al. (2001) found strong CO depletion toward Elias 18 in Taurus, but both the disk structure and its viewing angle are not well constrained.

In the near-infrared, disks seen edge-on show a dark lane surrounded by a bipolar nebulosity. The number of such disks discovered to date is around 20 (Padgett et al. 1999). We observed the CO gas-phase ro-vibrational transitions and solid CO absorption around 4.67 μm in the edge-on disk CRBR 2422.8-3423 located in the ρ Ophiuchus cloud with VLT-ISAAC. The spectral energy distribution of CRBR 2422.8-3423 is consistent with that of a Class I object or a Class II object whose circumstellar disk is seen edge-on and shows a strong silicate band in absorption at 9.6 μm.

In this paper, we first present the spectrum obtained with VLT-ISAAC and derive the column densities. Then a discussion on the location of the gas-phase and solid CO in the disk is provided. We conclude on future possible observations.

2. Results and analysis

We obtained in May 2002 a M-band infrared spectrum, displayed in Figure 1, at $R \simeq 10,000$ of the edge-on disk system CRBR 2422.8-3423 with VLT-ISAAC. The spectrum shows strong gas and solid-phase CO absorption features. The good signal-to-noise spectrum has been ob-

Chemistry as a Diagnostic of Star Formation (2003). *Edited by Charles L. Curry and Michel Fich.* Published in 2003 by NRC Research Press, Ottawa ON, K1A 0R6, Canada. ISBN 0-660-19089-3.

tained after 36 minutes of integration on-source , testifying of the quality of VLT-ISAAC to detect gas and solid phase CO in faint sources. The solid absorption band is saturated, indicating that a large fraction of the molecules may be frozen out onto the grains in the cold midplane. It is the first time that ices are detected toward a unambiguously edge-on disk. The derived lower limit solid CO column of 2.2×10^{18} cm^{-2} is the highest observed so far, including ice-rich high-mass protostars such as NGC 7538 IRS9 (Chiar et al. 1998) and background field stars. The deep M band spectrum does not show the broad absorption attributed to solid OCN^-, a signature of thermal and/or ultraviolet processing.

The VLT spectrum also shows the presence of narrow gas-phase ^{12}CO and ^{13}CO ro-vibrational lines originating from levels up to $J = 9$ (250 K above ground). Synthetic LTE model spectra were fitted to the observed spectrum. The limited number and the high optical depth of the ^{12}CO lines prevent a unique fit, but the best result (see Fig. 1) is obtained for $T_{ex} = 40 - 60$ K, which is probably a mean between the cold and warm components along the line of sight. The optical depth of the ^{12}CO lines is likely underestimated so that the gas-phase CO column density can range from 1 to 6×10^{18} cm^{-2}. Adopting a mean value of 3×10^{18} cm^{-2}, the line of sight average gas/solid CO ratio is ~ 1. Comparing the column of gas-phase CO with that of polar solid CO only, the ratio drops to ~ 0.1. The latter value is comparable to that found for L 1489 (0.07; see Boogert et al. 2002a) but still higher than for Elias 18 (0.01; Shuping et al. 2001). If the lines were significantly wider than 2 km s^{-1} as found for L 1489 ($\sim$20 km s^{-1}), the gaseous CO column density would drop by an order of magnitude.

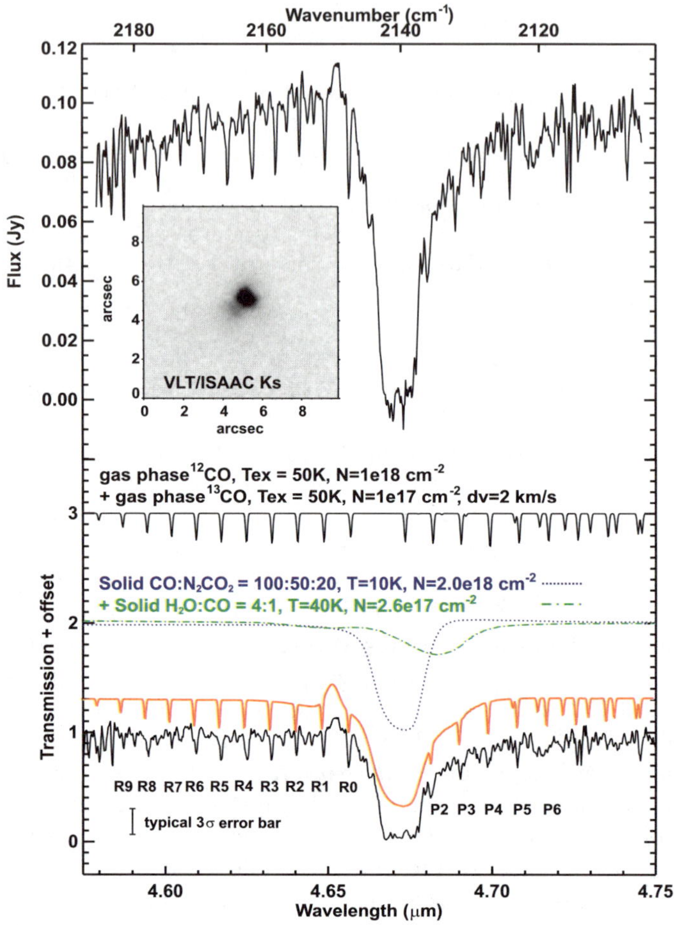

Figure 1. VLT-ISAAC M-band spectrum toward CRBR 2422.8-3423 *(upper panel).* The insert shows the VLT Ks image taken by Brandner et al. (2000). The upper curve of the lower panel shows the gas-phase model, convolved to the observed spectral resolution and shifted in wavelength to account for a source heliocentric velocity of 10 km s^{-1}. The middle curve displays laboratory absorption spectra of apolar *(dotted)* and polar *(dashed)* CO ice mixtures. The sum of the two models including Pfund β emission at 4.65 μm is overlaid on the observed spectrum in the bottom curve.

3. Contamination by foreground material?

Absorption studies toward young stellar objects located in the ρ Oph cloud complex can be dominated by foreground cloud(s) (Boogert et al. 2002b). The SCUBA 850 μm map of ρ Oph by Johnstone et al. (2000) shows that CRBR 2422.8-3423 is indeed at the edge of the ridge which contains both Elias 29 and the nearby Class I object IRS 43. We obtained $C^{18}O$ $J = 2 \rightarrow 1$ observations on-source and at 30″ south off the source with the JCMT to check for any surrounding cloud material that might contribute to the gas-phase CO absorption feature. The lack of CO emission at the offset position indicates that the bulk of the gas-phase CO is located within 3000 AU of CRBR 2422.8-3423. The similarity of the gaseous CO column densities derived from the on-source $C^{18}O$ $2 \rightarrow 1$ emission and the VLT-ISAAC infrared absorption may be fortuitous since the VLT-ISAAC data do not probe very cold CO with small ($\lesssim 1$ km s^{-1}) line widths. We cannot exclude that part of the gas-phase CO seen in the infrared arises in a more extended envelope or cloud, but the fact that the CO excitation temperature is significantly above 10 K indicates that at least some fraction must originate close to the young star in a warm part of the disk.

For solid CO, there are strong arguments that most of the absorption must arise in the disk. The bright nearby

($\sim$ 30$''$ east) source IRS 43 was observed simultaneously with CRBR 2422.8-3423 with VLT-ISAAC and has a CO ice optical depth of only 1.87 $\pm$ 0.02, corresponding to $N(\mathrm{CO_{ice}})$=7 $\times$ 10^{17} $\mathrm{cm^{-2}}$ (Pontoppidan et al., in preparation, and this volume). This is at least a factor of 3 lower than toward CRBR 2422.8-3423 even though its 850 μm flux is a factor of 1.5 higher. Toward Elias 29, solid CO has an optical depth of only 0.25, most of which is believed to be located in the foreground clouds. Compared to the optical depth found toward CRBR 2422.8-3423 (τ_{ice}[CO] $\gtrsim$ 6), this amount of foreground material can account for only an insignificant fraction of the observed solid CO. Finally, if clouds happen to lie in front of CRBR 2422.8-3423, the moderate extinction (A_{V} <10) of those clouds probably prevents them to harbour significant amounts of solid CO (Shuping et al. 2000). Indeed, in a M-band survey of more than 30 young stellar objects, this is the deepest CO ice band observed.

4. Location of gaseous and solid CO in the disk

A simple disk model cf. Chiang & Goldreich (1999) with T_{*}=3500 K and a disk radius of 250 AU has been adopted to investigate the location of the solid and gaseous CO seen in infrared absorption. Because of its higher temperature, the gaseous CO is likely not co-located with the bulk of the apolar solid CO, which evaporates at $\sim$ 20 K. The polar solid CO can however reside in the same region of the disk as the CO gas at 40–60 K. Assuming that CO is frozen out at $<$ 20 K (apolar CO) and $<$ 40 K (polar CO), the best fit to the column densities, gas/solid CO and polar ice/apolar ice ratio is obtained for i=20$\pm$5°. This inclination is consistent with the flux asymmetry seen in the near-infrared image of Brandner et al. (2000). For such line of sight, most of the CO ice is located above the midplane in the outer disk, whereas the CO gas is found in the warm inner disk. Thus, the overall CO depletion could be significantly higher than the ratio of $\sim$ 1 found here.

5. Conclusion

In summary, we detected a large amount of solid CO in the line of sight toward CRBR 2422.8-3423. The majority of this ice is likely located in the flaring outer regions of the edge-on circumstellar disk. Very high resolution ($R > 10^5$) near-infrared spectroscopy is needed to reveal the gaseous CO line profiles and thus their origin and the gas dynamics in the inner disk.

Deep searches for solid features toward other edge-on disks are ongoing. In the L-band, water and methanol ice features will be searched for, while in the long wavelength part of the M-band we will seek to detect the solid ^{13}CO feature centred around 4.78 μm. The results of our observations will help us to better understand the chemistry in the cold midplane of circumstellar disk, the likely birthplace of icy planetesimals.

Future submillimeter interferometer data can probe the velocity pattern and excitation conditions of the gas as functions of disk radius, whereas mid-infrared spectroscopy with, e.g., the *Space Infrared Telescope Facility (SIRTF)* will allow searches for other ice components, in particular solid CO_2.

Acknowledgements

WFT thanks PPARC for a postdoctoral grant to UCL. Astrochemistry in Leiden is supported by a Spinoza grant from the Netherlands Organization for Scientific Research (NWO) and a Ph.D. grant from the Netherlands Research School for Astronomy (NOVA). We thank the ESO staff for their help during the observations, P. Papadopoulos for performing the JCMT observations and D. Johnstone for a blow-up of the SCUBA map of ρ Oph.

References

Aikawa, Y., van Zadelhoff, G. J., Herbst, E., & van Dishoeck, E. F. 2002, A&A, 386, 622
Boogert, A. C. A., Hogerheijde, M. R., & Blake, G. A. 2002a, ApJ, 568, 761
Boogert, A. C. A., Hogerheijde, M. R., Ceccarelli, C., et al. 2002b, ApJ, 570, 708
Brandner, W., Sheppard, S., Zinnecker, H., et al. 2000, A&A, 364, L13
Chiang, E. I., & Goldreich, P. 1999, ApJ, 519, 279
Chiar, J. E., Adamson, A. J., Kerr, T. H., Whittet, D. C. B. 1995, ApJ, 455, 234
Chiar, J. E., Gerakines, P. A., Whittet, D. C. B., et al. 1998, ApJ, 498, 716
Dutrey, A., Guilloteau, S., & Guélin, M. 1997, A&A, 317, L55
Ehrenfreund, P., & Charnley, S. B. 2000, ARA&A, 38, 427
Johnstone, D., Wilson, C. D., Moriarty-Schieven, G., et al. 2000, ApJ, 545, 327
Padgett, D. L., Brandner, W., Stapelfeldt, et al. 1999, AJ, 117, 1490
Shuping, R. Y., Snow, T. P., Chiar, J. E., & Kerr, T. H. 2000, ApJ, 529, 932
Shuping, R. Y., Chiar, J. E., Snow, T. P., & Kerr, T. H. 2001, ApJ, 547, L161
Thi, W. F., van Dishoeck, E. F., Blake, G. A., et al. 2001, ApJ, 561, 1074
van Dishoeck, E. F., & Blake, G. A. 1998, ARA&A, 36, 317

FRAGMENTATION AND COLLAPSE OF TURBULENT MOLECULAR CLOUDS

David Tilley, Ralph Pudritz, and James Wadsley

Department of Physics and Astronomy, McMaster University, Hamilton, ON, Canada L8S 4M1

Abstract

We performed simulations of self-gravitating hydrodynamic turbulence to model the formation of filaments, clumps and cores in molecular clouds. We find that when the mass on the initial computational grid is comparable to the Jeans mass, turbulent pressure is able to prevent gravitational collapse. When the turbulence has damped away sufficiently, gravitational collapse can occur, and the resulting structure closely resembles the pre-singularity collapse of an isothermal sphere of Penston (1969). If several Jeans masses are initially placed on the grid, turbulence may not be sufficient to prevent collapse before turbulence can be significantly damped. In this case, the cores have density structures which are considerably shallower than expected for an isothermal gas, and resemble the solutions for a logatropic equation of state.

Key words: ISM: clouds – ISM: structure

1. Background

There is abundant evidence that stars form in clumps within molecular clouds. In order to understand the mechanism by which stars form, we therefore must understand the environment within a molecular cloud. Observations show that molecular clouds are composed of a hierarchical network of filaments, clumps and cores (Mizuno et al. 1995; Johnstone & Bally 1999; Gahm et al. 2002). The most popular explanation for this substructure is that of turbulent fragmentation (Klessen & Burkert 2000; Ostriker et al. 2001). In this scenario, turbulent fluctuations within the molecular cloud can compress the gas through shocks to high enough densities that the self-gravity of such regions becomes strong enough to induce collapse, resulting in a cloud which has been broken up into several condensing cores. The cores, due to their origins as fragmenting shocks, are arranged along filamentary structures; the intersections of these filaments have many of these smaller cores attracted to their higher average density, resulting in clusters which resemble the observed clumps.

We performed simulations to test whether the structure of cores produced in this turbulent fragmentation scenario are consistent with observation.

2. Simulations

Our simulations were performed using the NCSA ZEUS-MP hydrodynamics code. ZEUS-MP is a mesh-based Eulerian hydromagnetic code capable of performing 3-D simulations. The self-gravity of the fluid is calculated using a FFT gravity solver. Our simulations are stopped when the numerical gravitational instability criterion of Truelove et al. (1997) is violated ($L_J < 4L_{\rm pix}$). We use an isothermal equation of state and periodic boundary conditions.

We start with a uniform initial density and no magnetic field, at a resolution of 128^3 grid cells. Into this box we inject an initial series of plane waves, with their amplitudes described by a Kolmogorov scaling law ($E_k \propto k^{-5/3}$) and with random phases and directions. We ensure that $\nabla \cdot \mathbf{v} = 0$. Our simulations can be described by two parameters, the RMS Mach number used to normalize the velocity fluctuations, and the initial number of Jeans masses on the grid, used to normalize the density. We chose our initial number of Jeans masses from the clump data of Lada et al. (1991).

3. Results

We show the projected density of two of our simulations in Figure 1, each of which is at the point where the Truelove et al. (1997) condition is violated. In the simulation with an initial mass of $1.3M_J$ (where M_J is the initial Jeans mass), the relatively weak gravitational force was not able to induce collapse until much of the turbulent kinetic energy had dissipated. The resulting core is much more homogeneous than might be expected for a turbulent process. The density structure of this core is slightly

Chemistry as a Diagnostic of Star Formation (2003). *Edited by Charles L. Curry and Michel Fich.* Published in 2003 by NRC Research Press, Ottawa ON, K1A 0R6, Canada. ISBN 0-660-19089-3.

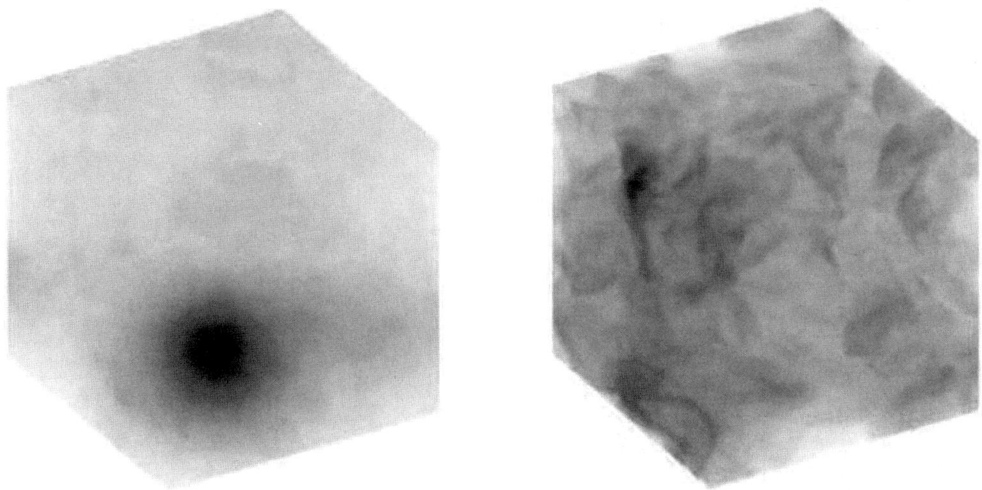

Figure 1. Simulation results at the time when the Truelove et al. condition is violated. Projected density is shown. *Left:* A run with initial $M_{\rm RMS} = 5$, $n_{\rm J} = 1.3$. The simulation has run for 2 crossing times (7.1 free-fall times). *Right:* A run with initial $M_{\rm RMS} = 5.1$, $n_{\rm J} = 5$. The simulation has run for 0.23 crossing times (1.28 free-fall times).

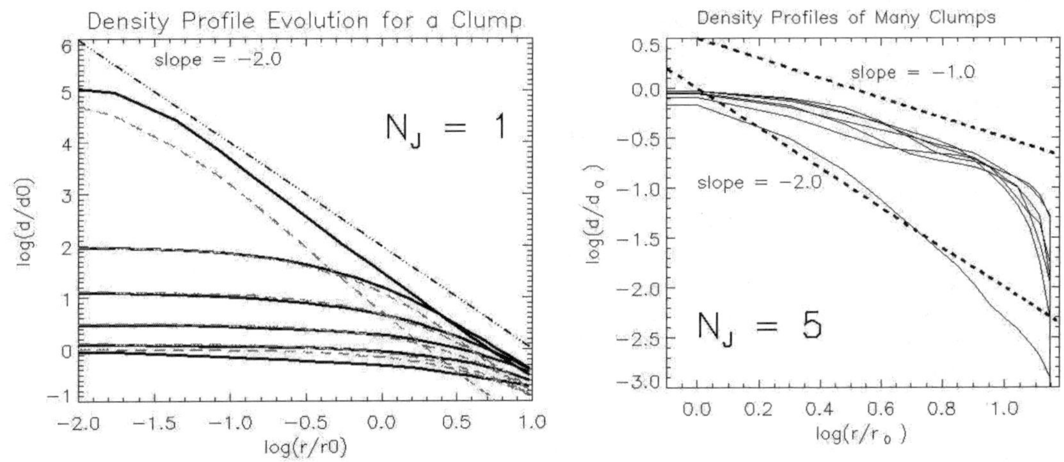

Figure 2. Density structure of cores. *Left:* A run with initial $M_{\rm RMS} = 5$, $n_{\rm J} = 1.3$. The density structure of the largest core is shown, from bottom to top, at 1.6, 1.7, 1.8, 1.9, and 2.0 crossing times. *Right:* Several cores for the run with initial $M_{\rm RMS} = 5$, $n_{\rm J} = 5$, after 0.23 crossing times.

shallower than a Bonnor-Ebert sphere (Fig. 2), suggesting that some small amount of turbulent pressure is providing additional support to the fluid. The evolution is similar to that of Penston (1969) and Foster & Chevalier (1993) for a collapsing pre-protostellar object, prior to what they call "core formation" (when a singularity forms).

For the run with a large number of initial Jeans masses ($5.1 M_{\rm J}$), significant turbulent motions are still present when the simulation is stopped. The fluid has condensed into a massive clump, surrounded by filamentary structures, and many smaller clump-like objects. The individual cores have a shallower density profile than those in the weak-gravity simulation, resembling that of a logatropic sphere ($\rho \propto r^{-[1-1.5]}$) (McLaughlin & Pudritz 1997). This suggests that they are still largely supported by turbulent motions. Similar density structures have been seen in regions of high-mass star formation (van der Tak et al. 1999; van der Tak et al. 2000). In our simulations, the filaments and clumps appear to form simultaneously out of the collision and merger of short shock arcs, rather than undergoing hierarchical collapse. However, the shocks that do not merge do not contain enough mass to undergo gravitational collapse.

We are currently performing and analyzing results from higher resolution simulations. These will be presented in a future paper.

ACKNOWLEDGEMENTS

D. T. is supported by an Ontario Graduate Scholarship. R. E. P. is supported by the Natural Sciences and Engineering Research Council of Canada. J. W. is a SHARCNet Research Associate.

REFERENCES

Bonnor, W.B. 1956, MNRAS 116, 351

Ebert, R., 1955, Z. Ap., 37, 217

Foster, P. N. & Chevalier, R. A., 1993, ApJ 416, 303

Gahm, G. F., Lehtinen, K., Carlqvist, P., Harju, J., Juvela, M., & Mattila, K., A&A 389, 577

Johnstone, D. & Bally, J. 1999, ApJ, 510, L49

Klessen, R. S. & Burkert, A. 2000, ApJS 128, 287

Lada, E. A., Bally, J., & Stark, A. A., 1991, ApJ 368,432

McLaughlin, D. E. & Pudritz, R. E., 1997, ApJ 476, 750

Mizuno, A., Onishi, T., Yonekura, Y., Nagahama, T., Ogawa, H., & Fukui, Y., 1995, ApJ 445, 161

Ostriker, E. C., Stone, J. M. & Gammie, C. F., 2001, ApJ 546, 980

Penston, M. V., 1969, MNRAS 144, 425

Truelove J. K., Klein, R. I., McKee, C. F., Holliman, J. H., Howell, L. H., & Greenoughj, J. A., 1997, ApJ 489, 179

van der Tak, F. F. S., van Dishoeck, E. F., Evans, N. J., Bakker, E. J., & Blake, G. A., 1999, ApJ 522, 991

van der Tak, F. F. S., van Dishoeck, E. F., Evans, N. J., & Blake, G. A., 2000, ApJ 537, 283

FILLING THE *ODIN* BEAM: CS AND SO IN TAURUS CLOUD CORES

N. F. H. Tothill[1,2], E. M. Gregersen[3], C. L. Curry[4], and R. Plume[5]

[1]Department of Astronomy and Physics, St Mary's University, Halifax, Nova Scotia B3H 3C3, Canada
[2]Smithsonian Astrophysical Observatory, 60 Garden Street, Cambridge, Massachusetts 02138, U.S.A.
[3]Department of Physics and Astronomy, McMaster University, Hamilton, Ontario L8S 4M1, Canada
[4]Department of Physics, University of Waterloo, Waterloo, Ontario N2L 3G1, Canada
[5]Department of Physics and Astronomy, University of Calgary, Calgary, Alberta T2N 1N4, Canada

Abstract

We have observed CS and SO emission from a number of dark cloud cores in Taurus, and present maps of CS, SO and their isotopologues towards three of them. On the basis of the maps, there is some evidence that these cores may be good candidates for the detection of molecular oxygen by the *Odin* satellite.

Key words: ISM: molecules – ISM: clouds – ISM: individual (Taurus)

1. Introduction

The detection of emission from the O_2 $1_1{-}1_0$ ground state transition at 119 GHz remains a key objective of the *Odin* mission. However, the interpretation of such a detection or an upper limit is complicated by the large beam (9′) of the *Odin* telescope at this low frequency. By mapping the physical and chemical structure of the molecular gas inside the beam, we hope to facilitate the interpretation of upcoming *Odin* observations of a number of dark cloud sources in Taurus.

2. Transitions and Sources

We selected two transitions to map: CS 2–1 and SO $3_2{-}2_1$. CS is a good tracer of dense gas, and can therefore be used to estimate the density distribution in the *Odin* beam. It has been suggested that SO may be a useful surrogate for O_2, the two molecules being formed and destroyed in similar ways (Nilsson et al 2000).

By observing these two molecules we may also estimate the CS/SO ratio. This is interesting insofar as it should be sensitive to the C/O ratio: although CS/SO need not equal C/O, nor need the ratios be strictly proportional in all physical regimes, the two quantities may well be correlated in cores with very similar physical properties. The C/O ratio is known to be an important constraint in models of O_2 abundance; indeed, a very high C/O ratio has been suggested to explain the lack of O_2 detections to date (Goldsmith et al 2000). However, a direct estimate of C/O from CS/SO requires detailed chemical modelling (Nilsson et al 2000).

Table 1. Source list

Source	R.A. (1950.0)	Dec (1950.0)
Tau Onishi 10	04:14:43.9	27:29:49
TMC-1 NH_3	04:38:14.0	25:43:00
TMC-1 HC_7N	04:38:38.0	25:36:00
L1498	04:07:56.0	25:01:30
L1495	04:11:06.4	28:01:53
B217	04:24:42.4	26:11:13
B18	04:26:23.9	24:24:00
TMC-2A	04:28:53.9	24:24:00
L1536B	04:30:25.9	22:37:09
HCL2-A	04:36:42.7	25:35:44
HCL2-B	04:36:42.7	25:59:14
HCL2-C	04:38:38.0	25:55:15

Our targets are the highest priority dark cloud sources in Taurus from the *Odin* target list (Table 1). These sources are nearby (~ 140 pc), cold (~ 10 K), and relatively dense ($10^4{-}10^5\,\mathrm{cm}^{-3}$).

3. Observations

We carried out our observations at the 14 m telescope of the Five College Radio Astronomy Observatory, using SEQUOIA, their multi-beam receiver. This allowed us to map an area of about $10' \times 10'$ with full Nyquist sampling of the $\sim 50''$ beam of the telescope in a matter of an hour or two.

While all the sources were detected in CS, three were bright in both molecules: Onishi 10, TMC-1 HC_7N and TMC-1 NH_3. We therefore mapped these cores in corresponding transitions of $C^{34}S$ and ^{34}SO. These transitions

Chemistry as a Diagnostic of Star Formation (2003). *Edited by Charles L. Curry and Michel Fich.* Published in 2003 by NRC Research Press, Ottawa ON, K1A 0R6, Canada. ISBN 0-660-19089-3.

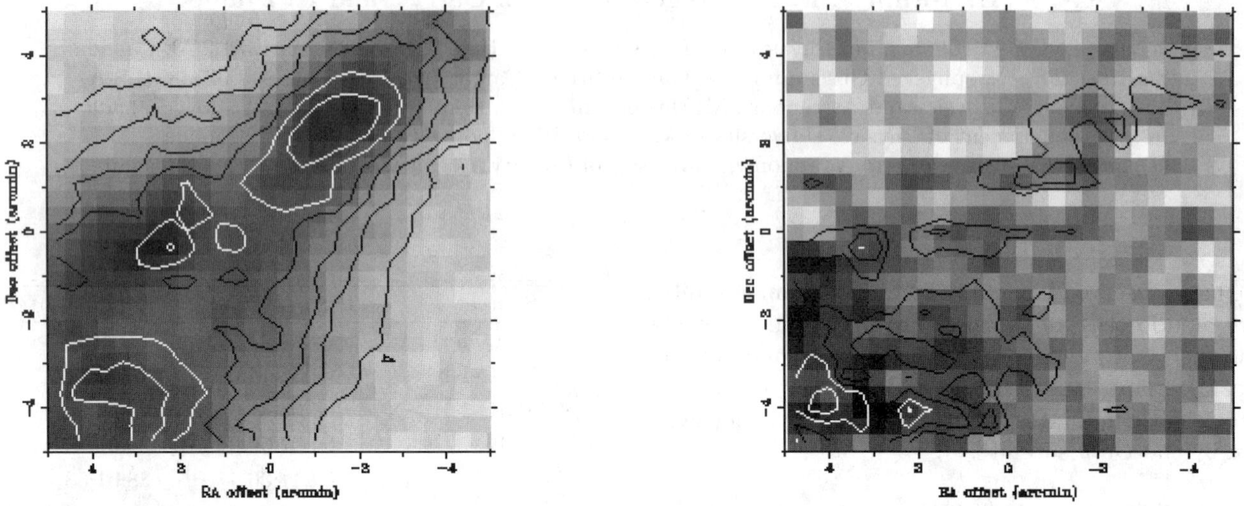

Figure 1. CS (*contours*) and SO (*greyscale*) towards Onishi 10. *Left:* The main isotopologues. Contours are 0.15 to 0.9 K km s^{-1} in steps of 0.15; the greyscale levels are –0.2 K km s^{-1} (*white*) to 1.2 K km s^{-1} (*black*). *Right:* $C^{34}S$ (*contours*) and ^{34}SO (*greyscale*). The contours are 0.1 to 0.25 K km s^{-1}; the greyscale levels are –0.05 K km s^{-1} (*white*) to 0.2 K km s^{-1} (*black*).

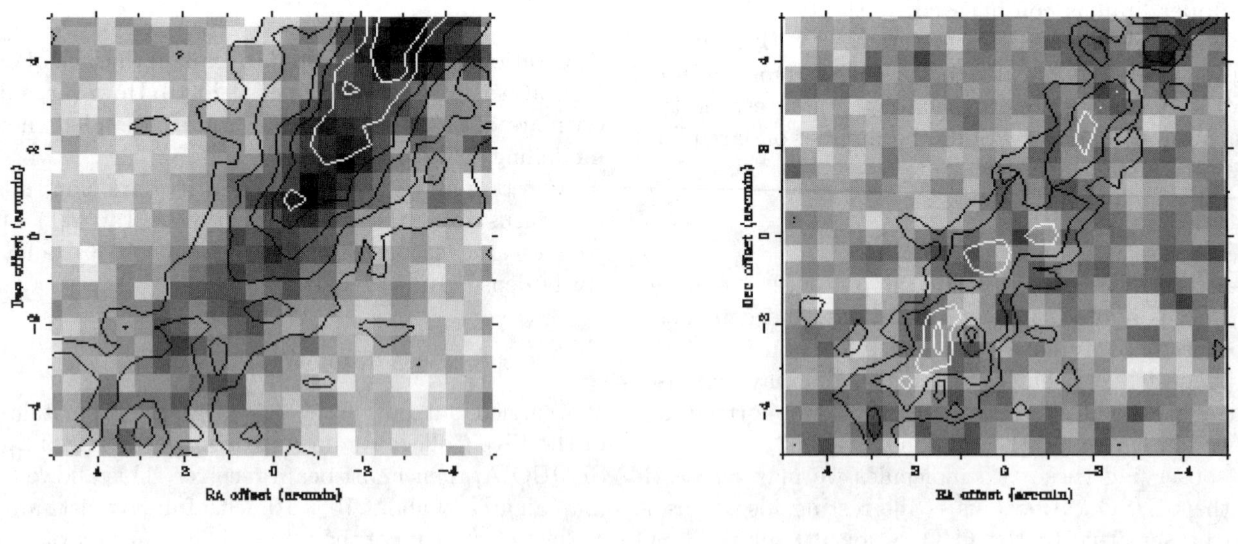

Figure 2. CS (*contours*) and SO (*greyscale*) towards TMC1-HC_7N. *Left:* The main isotopologue. The contours are as in Fig. 1*a*; the greyscale levels are –0.1 K km s^{-1} (*white*) to 0.5 K km s^{-1} (*black*). *Right:* $C^{34}S$ (*contours*) and ^{34}SO (*greyscale*). The contours and greyscale levels are as in Fig. 1*b*.

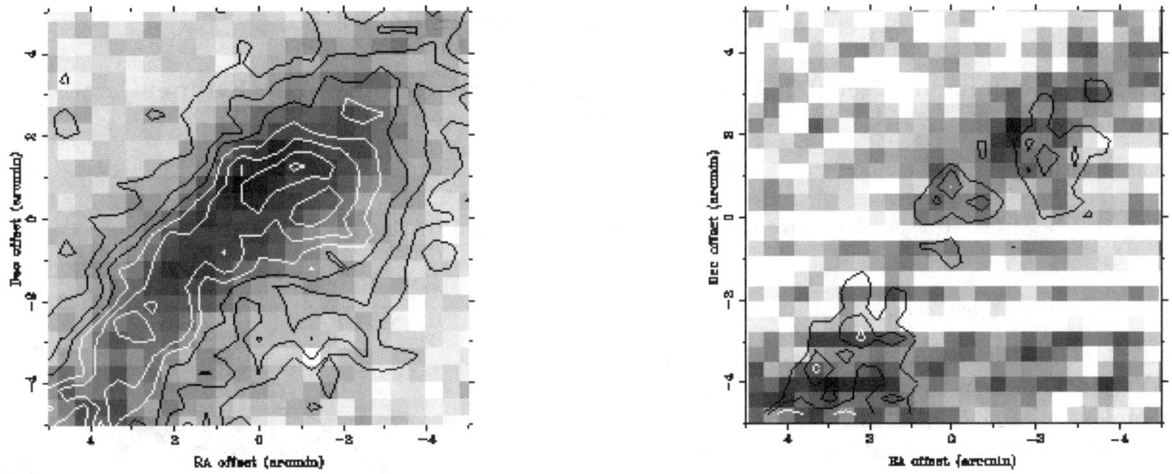

Figure 3. CS (*contours*) and SO (*greyscale*) towards TMC1-NH_3. *Left:* The main isotopologues. The contours are as in Fig. 1*a*; the greyscale levels are −0.1 K km s^{-1} (*white*) to 1.0 K km s^{-1} (*black*). *Right:* $C^{34}S$ (*contours*) and ^{34}SO (*greyscale*). The contours are as in Fig. 1*b*; the greyscale levels are −0.05 K km s^{-1} (*white*) to 0.15 K km s^{-1} (*black*).

should be optically thin, acting as a check on derivations of column density.

Figures 1–3 show the integrated emission of CS, SO and their isotopologues in the three cores mentioned above. The two TMC-1 cores are close enough together that their maps overlap: $(-3', +4')$ with respect to TMC-1 HC_7N is within a few arcseconds of $(+3', -3')$ in the map of TMC-1 NH_3. In general, we see that SO and CS have very similar distributions, with CS brighter (e.g., in TMC-1 HC_7N); however, both TMC-1 NH_3 and Onishi 10 have similar integrated intensities of CS and SO.

Table 2. CS and SO parameters in a 9′ beam

Source	$\frac{I(SO)}{I(CS)}$	$\frac{N(SO)}{N(CS)}$	$\frac{I(^{34}SO)}{I(C^{34}S)}$	$\frac{N(^{34}SO)}{N(C^{34}S)}$
Tau Onishi 10	0.88	3.9	0.42	1.9
TMC-1 NH_3	0.68	3.1	—	—
TMC-1 HC_7N	0.41	1.8	0.46	2.1

4. Results

We integrated the intensities over a 9′ gaussian beam, co-centred with each map, and we report the SO to CS intensity ratios in Table 2, along with estimates of the column density ratios, obtained from the intensity ratios using the relationship given in Nilsson et al (2000) assuming a temperature of 10 K. The optically thin ratios are similar for Onishi 10 and TMC-1 HC_7N, which suggests that the higher SO/CS ratio seen in Onishi 10 may be an optical depth effect. Even so, these ratios are fairly high compared with the results of Nilsson et al (2000), suggesting that these dark cloud cores may be good candidates for the detection of O_2. However, we note that Pagani et al (2003) failed to detect O_2 towards TMC-1 NH_3 using *Odin*. The combination of their upper limit with these maps should help constrain our models of interstellar chemistry.

Acknowledgements

We thank Mark Heyer for showing us how to use the telescope and SEQUOIA, and the NSERC of Canada for financial support. NFHT gratefully acknowledges the physical, moral and culinary support of the staff of the Amundsen-Scott South Pole Station.

References

Goldsmith, P.F., et al. 2000, ApJL 539, L123
Nilsson, A., et al. 2000, A&A 358, 257
Pagani, L., et al. 2003, A&A, 402, L77

A HOT CORE LABORATORY

Fleur A. van Broekhuizen[1], Helen J. Fraser[1], Willem A. Schutte[1], Ewie de Kuijper[2], and Ewine F. van Dishoeck[1]

[1]Raymond and Beverly Sackler laboratory for Astrophysics, Leiden Observatory, P.O. Box 9513, 2300 RA Leiden, The Netherlands

[2]F.M.D., Huygens laboratory, University of Leiden, 2300 RA Leiden, The Netherlands

Abstract

We present a new ultra-high vacuum (UHV) experiment, CRYOPAD, that is specifically designed to improve our understanding of the chemistry occurring in hot cores. CRYOPAD will be used for a detailed study of chemical mechanisms involved in hot core regions, especially the production of volatile complex organics from interstellar ices mediated by UV photons and thermal processes. The results will provide far more information than can be obtained from current high vacuum (HV) experiments, removing errors induced by vacuum contamination, and providing data that can then be used to perform a directed search for organic species in these hot core regions. The empirical data on reaction rates and yields to be obtained from CRYOPAD will be fed into astrochemical models. Second, we address here a study on the formation efficiency of OCN^-. Traditionally this molecule has been used as a signature of UV processing. OCN^- is a plausible carrier of the 4.62 μm feature first observed towards W33A. HV experiments suggest that a thermally mediated acid-base reaction between HNCO and NH_3 is needed to match the strong 4.62 μm feature observed towards W33A. UV mediated production from HNCO cannot be excluded, but the production of OCN^- by UV photo-processing of a classic $H_2O/CO/NH_3$ ice matrix is too inefficient.

Key words: ISM: molecules – H II regions – methods: laboratory – astrochemistry

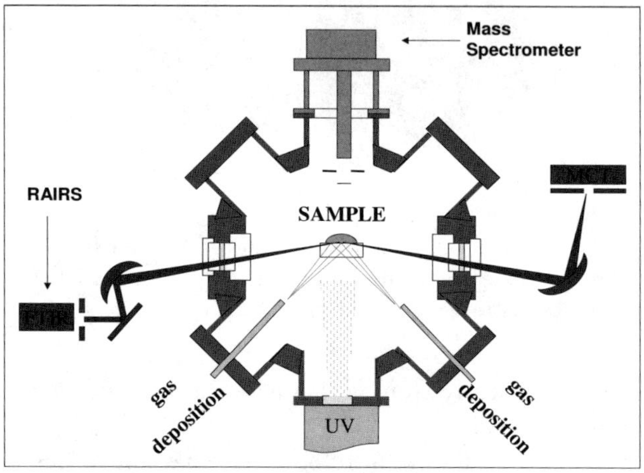

Figure 1. The reaction chamber of CRYOPAD. The ice matrix is grown on top of the sample by gas deposition. Volatile species are detected by a mass spectrometer, also used for TPD product analysis. RAIRS maps the infrared characteristics of the ice matrix. This can be done in parallel with UV photo-processing or during a TPD experiment (van Broekhuizen et al., in preparation).

1. Simulating Hot Core Chemistry

Hot cores (HCs) represent the most chemically rich areas of the interstellar medium. They are characterized as warm (100–200 K), dense ($n[H_2] \sim 10^7\,cm^{-3}$) regions that are predominantly associated with high-mass star formation. Internal heating and ultraviolet radiation inside HCs induce chemical processing in, and evaporation of, the ice mantles on infalling grains. This initiates an extensive gas-phase chemistry that leads to the formation of numerous complex organics (Charnley et al 1995). Gas-phase chemical models, however, cannot explain all the observed molecular abundances. Therefore, chemistry at the surface and in the bulk of the ice mantle is thought to play a key role in the production of many volatile organics that have been observed in hot cores (Gibb et al. 2000; Hatchell et al. 1998).

Previous HV experiments that monitored UV mediated and thermally mediated solid state chemistry of interstellar ice analogs have focused primarily on refractory products (Briggs et al. 1992; Bernstein et al. 2002; Muñoz-Caro et al. 2002). Because of their nature these are unlikely to be present in the gas phase in appreciable quantities. Clues about the chemical species that can

Chemistry as a Diagnostic of Star Formation (2003). *Edited by Charles L. Curry and Michel Fich.* Published in 2003 by NRC Research Press, Ottawa ON, K1A 0R6, Canada. ISBN 0-660-19089-3.

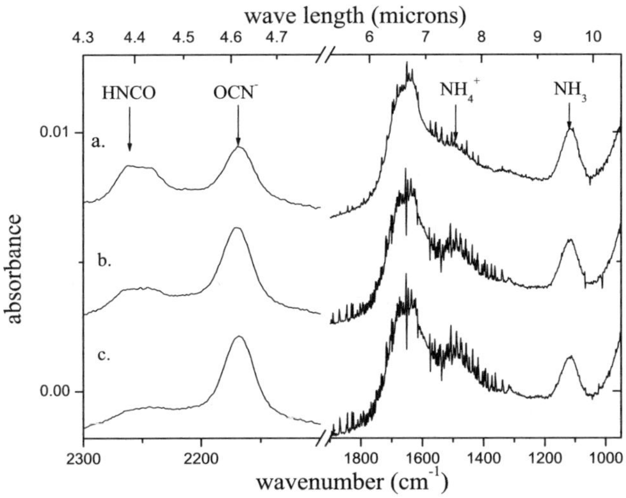

Figure 2. Photon-induced deprotonation of HNCO. Ice composition: $H_2O/HNCO/NH_3$=140/1/8. *(a)* 0 photons cm^{-2}; *(b)* 2.4×10^{17} photons cm^{-2}; *(c)* 7.2×10^{17} photons cm^{-2} (van Broekhuizen et al., in prep.).

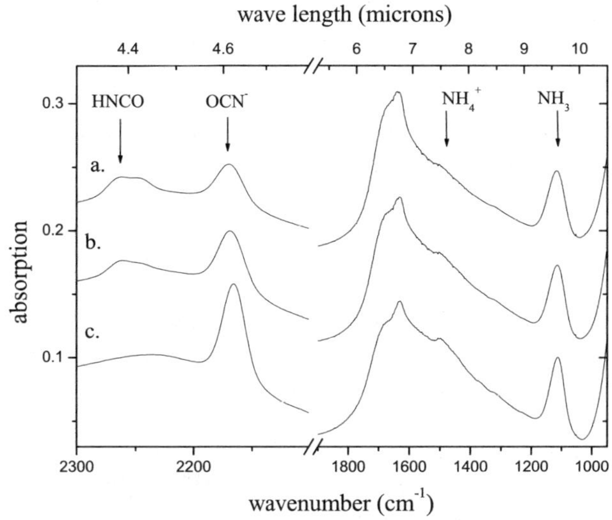

Figure 3. Thermally-induced deprotonation of HNCO. Ice composition: $H_2O/HNCO/NH_3$=120/1/10. *(a)* 15 K; b) 52 K; c) 122 K (van Broekhuizen et al., in prep.).

form and the resulting spectroscopy were obtained, but not much information was provided on the reaction mechanisms and chemistry involved on both refractory and volatile products. Experiments performed under UHV conditions can give this extra information. H_2O accretion onto the ice surface is negligible at UHV during the time scale of the experiments which allows highly sensitive and quantitative detection of the formation of trace volatile species, providing a more complete view of the chemistry.

CRYOPAD has been built to study the production of volatile complex organics from interstellar ice analogs similar to those observed towards HCs. CRYOPAD, whose reaction chamber shown in Figure 1, is capable of detecting the desorption products and surface reaction mechanisms simultaneously, using two traditional surface science techniques, namely TPD and RAIRS. TPD, Temperature Programmed Desorption, gives direct information on the volatiles formed by detecting the molecules with mass spectrometry. Second, their desorption pattern provides information on the structure of the ice matrix. RAIRS, Reflective Absorption InfraRed Spectroscopy, maps the infrared characteristics of the species in the bulk ice and gives additional information about the surface structure. The results may lead to a better understanding of the nature of molecules that sublime into the gas phase in HCs. This will allow for a directed search for molecules in HCs. Furthermore, by giving better constraints on the nature of the subliming species, the results can be used to refine the chemical models that currently exist to describe the gas-phase molecular abundances.

The first experiments with CRYOPAD will focus on the solid state chemistry of methanol and CH_3OH/CO ices mediated by UV photons and thermal warm-up. Infrared observations show that methanol generally has significant abundances in the ice mantles towards YSOs (Allamandola et al. 1992; Dartios et al. 1999 and references therein). This points to solid state chemistry as the possible methanol reaction mechanism. The evaporation of methanol-containing ice mantles from grains in hot cores drives a rich gas-phase chemistry that may explain the abundances of volatile complex organics such as dimethyl ether, methyl formate, ketene and acetaldehyde (Charnley et al. 1995). The contribution of UV and thermally-mediated solid state chemistry to the production of these secondary volatile organic products will be investigated with CRYOPAD.

2. Reproducing the 4.62 μm feature

Laboratory experiments of interstellar ice analogs have assigned the 4.62 μm 'XCN' feature, first observed towards W33A by Soifer et al. (1979), to OCN^- (Grim &

Greenberg 1987; Schutte & Greenberg 1997). Radiative processing of ices containing a mixture of H_2O, CO, and NH_3 produces an excellent fit to the 4.62 μm band towards W33A, leading to the feature being associated with regions exposed to radiative processing. This conclusion is, however, only based on the detection of this band. A careful interpretation of interstellar data requires also a thorough study of the efficiency of the OCN^- formation together with the formation of additional products.

HV studies on the formation efficiency of OCN^- by UV photo-processing have shown that UV processing of H_2O/ CO/NH_3 ice is not efficient enough to fully account for the high abundance of OCN^- towards W33A (van Broekhuizen et al., in prep.). Laboratory HV experiments prove that OCN^- can be produced in sufficient quantities to match the 'XCN' band toward W33A by UV or thermal processing of $H_2O/HNCO/NH_3$ ice mixtures. The similar behaviour of a $H_2O/HNCO/NH_3$ ice upon UV processing and thermal warming is shown in Figures 2 and 3. The figures show that a considerable fraction of HNCO reacts to OCN^- at 15 K prior to any processing. This fraction ranges from ~40 % in a H_2O-dominated ice matrix to ~90 % when no H_2O is present. We conclude that UV processing of $H_2O/CO/NH_3$ ices is unlikely to be the main source of OCN^- towards W33A, but that processing of $HNCO/NH_3$-containing ices, either thermally or by UV, is the dominant mechanism to produce high OCN^- abundances (van Broekhuizen et al., in prep.). HNCO is likely to be produced by grain-surface reactions.

The observed abundance of OCN^- towards W33A is a factor of ~10 higher than that towards other sources, with the exception of AFGL7009S. The relative importance of the above processes for other sources remains to be investigated.

References

Allamandola, L. J., Sandford, S. A., Tielens, A. G. G. M., & Herbst, T. M. 1992, ApJ, 399, 134

Charnley, S. B., Kress, M. E., Tielens, A. G. G. M., & Millar, T. J. 1995, ApJ, 448, 232

Dartois, E., Schutte, W., Geballe, T. R., Demyk, K., Ehrenfreund, P., & d'Hendecourt, L. 1999, A&A, 342, L32-L35

Gibb, E., Nummelin, A., Irvine, W. M., Whittet, D. C. B., & Bergman, P. 2000, ApJ, 545, 309

Grim, R. J. A., & Greenberg, J. M. 1987, ApJ, 321, L91

Hatchell, J., Thompson, M. A., Millar, T. J., & MacDonald, G. H. 1998, A&AS, 133, 29

Schutte, W. A., & Greenberg, J. M. 1997, A&A, 317, L43

Soifer, B. T., Puetter, R. C., Russell, R. W., Willner, S. P., Harvey, P. M., & Gillett, F.C. 1979, ApJ, 232, L53

SULPHUR CHEMISTRY IN THE ENVELOPES OF MASSIVE YOUNG STARS

F.F.S. van der Tak[1], A.M.S. Boonman[2], R. Braakman[2], and E.F. van Dishoeck[2]

[1]Max-Planck-Institut für Radioastronomie, Auf dem Hügel 69, 53121 Bonn, Germany
[2]Sterrewacht, Postbus 9513, 2300 RA Leiden, The Netherlands

Abstract

We present submillimeter observations of SO, SO_2, H_2S, H_2CS, OCS, NS and HCS^+ toward nine massive young stars. The outflow contributes $\approx$50% to the SO and SO_2 emission in $15 - 20''$ beams, more than for CS, where it is 10%. The SO_2 abundance increases from dark cloud levels in the outer envelope ($T < 100$ K) to levels seen in hot cores and shocks in the inner envelope ($T > 100$ K). Molecular abundances are consistent with a model of ice evaporation in an envelope with gradients in temperature and density for a chemical age of $\sim 30,000$ yr. The high observed abundance of OCS, the fact that $T_{\rm ex}$(OCS)$\gg T_{\rm ex}$(H_2S), and the data on solid OCS and H_2S all suggest that the major sulphur carrier in grain mantles is OCS rather than H_2S. For most other sulphur-bearing molecules, the source-to-source abundance variations by factors of up to 10 do not correlate with previously established evolutionary trends in temperature tracers. These species probe the chemically inactive outer envelope. Our data set does not constrain the abundances of H_2S and SO in the inner envelope, which, together with SO_2, are required to use sulphur as a clock.

Key words: ISM: molecules – molecular processes – stars: circumstellar matter – stars: formation

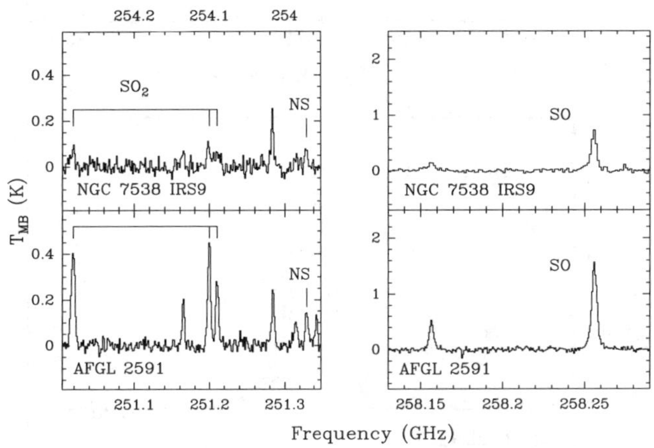

Figure 1. Examples of the spectra taken with the JCMT. The top and bottom frequency scales of the left panel are the two receiver sidebands; the right panel only shows lines from one sideband. Adopted from van der Tak et al. (2002).

1. Introduction

Spectral line surveys at submillimeter wavelengths have revealed considerable chemical differences between star-forming regions (see van Dishoeck 2001 for an overview). While these differences indicate activity, the dependence of molecular abundances on evolutionary state and physical parameters is poorly understood. Better insight into this relation would be valuable for probing the earliest, deeply embedded phases of star formation where diagnostics at optical and near-infrared wavelengths are unavailable. This is especially true for the formation of high-mass stars, for which the order in which phenomena occur is much less well understood than in the low-mass case, and for which the embedded phase is a significant fraction ($\approx 10\%$) of the total lifetime of $\sim 10^6$ yr.

Sulphur-bearing molecules are attractive as candidate tracers of early protostellar evolution (Hatchell et al 1998). Models of "hot core" chemistry (Charnley 1997), as well as of shock chemistry (Pineau des Forêts et al. 1993) predict strong variations in the abundances of sulphur molecules on time scales of $\sim 10^4$ yr. To test these scenarios, this paper discusses submillimeter observations of sulphur-bearing molecules towards nine regions at the earliest stages of high-mass star formation, with luminosities $1 \times 10^4 - 2 \times 10^5$ $L_\odot$ at distances of $1 - 4$ kpc. This paper only summarizes our results; a complete description is given by van der Tak et al (2002).

Chemistry as a Diagnostic of Star Formation (2003). *Edited by Charles L. Curry and Michel Fich.* Published in 2003 by NRC Research Press, Ottawa ON, K1A 0R6, Canada. ISBN 0-660-19089-3.

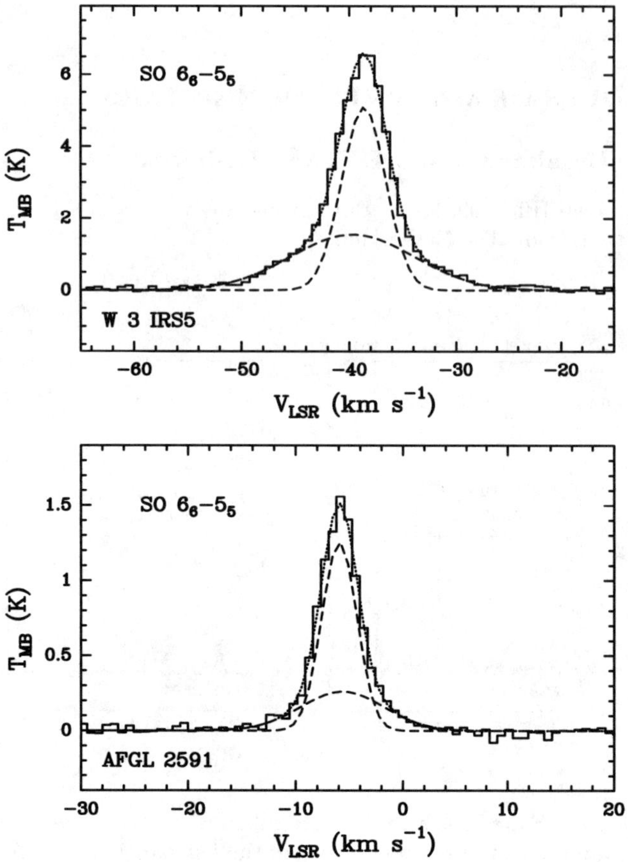

Figure 2. Velocity profiles of the SO 6_6–5_5 line in the sources AFGL 2591 and W3 IRS5.

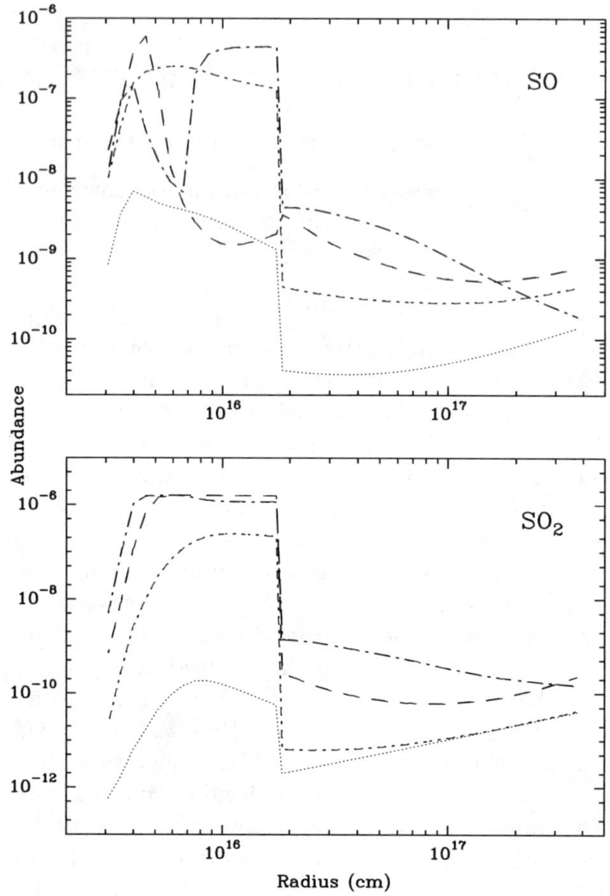

Figure 3. Distributions of the SO and SO_2 abundances in AFGL 2591 for chemical ages of 300, 3000, 30,000 and 300,000 yr (*curves from bottom to top*) after Doty et al. (2002). The $t \sim 3 \times 10^4$ yr model fits our data best.

2. Observations

Spectra in the 230 and 345 GHz bands were taken in 1995–1998 with the JCMT. The overview spectra in Figure 1 show source-to-source differences in the relative and in the absolute strengths of lines of sulphur molecules. The line profiles (Fig. 2) reveal the presence of a narrow component, due to the envelope, and a broad one, due to the outflow. For SO and SO_2, the outflow contributes $\approx 50\%$ of the line flux. This fraction is much larger than for CS, where it is only 10%. Lines of other molecules are too weak to separate the two components.

3. Results

To calculate molecular abundances, we have used the Monte Carlo program by Hogerheijde & van der Tak (2000), and models of the temperature and density structure of the sources (van der Tak et al 2000). Table 1 compares our results with abundances in other types of sources. The best overall match is with "hot cores."

For some sources, the detection of many SO_2 lines from a wide range of energy levels enables us to derive radial abundance profiles. Models with a constant SO_2 abundance of 10^{-9} fit the lines arising $\lesssim 100$ K above ground. However, the higher-excitation lines are underproduced, which indicates a 'jump' in the SO_2 abundance by a factor of $\sim$100 at a temperature of ≈ 100 K. Infrared observations of SO_2 absorption by Keane et al (2001) also indicate abundances of $\sim 10^{-7}$ for the inner envelopes.

Table 1. Molecular abundances

Source	CS $\times 10^{-9}$	H_2CS $\times 10^{-10}$	H_2S $\times 10^{-9}$	HCS^+ $\times 10^{-10}$	NS $\times 10^{-11}$	OCS $\times 10^{-9}$	SO $\times 10^{-9}$	SO_2 $\times 10^{-9}$
W3 IRS5	5	3	9	0.2	–	0.5	10	10
W33A	5	13	8	2	2	20	1	1
AFGL 2136	4	3	<2	1	<0.5	–	1	0.5
AFGL 2591	10	3	20	2	0.5	10	10	2
S140 IRS1	5	7	4	5	1	<2	2	1
NGC 7538 IRS1	10	10	8	2	1	2	5	1
NGC 7538 IRS9	10	10	8	5	1	2	1	0.5
MonR2 IRS3	5	7	<4	–	–	–	5	1
NGC 6334 IRS1	10	7	10	2	10	50	2	2
Hot cores[a]	8	6	9	–	80	5	4	20
PDRs[b]	20	–	6	–	–	–	9	0.1
Dark clouds[c]	1	6	0.8	0.6	–	2	20	4
Shocks[d]	4	8	4000	0.1	–	10	200	100

NOTES: [a] Source average from Hatchell et al (1998). [b] Orion Bar: Jansen et al (1995). [c] L 134N: Ohishi et al (1992). [d] Orion Plateau: Sutton et al (1995); Minh et al (1990).

4. Chemical implications

Most hot core models only consider single values for the temperature and density, while for our sources, these parameters vary strongly along the line of sight. Using the UMIST database, Doty et al (2002) calculated the time-dependent gas-phase chemistry for the temperature and density structure of AFGL 2591. The model has most sulphur initially in S at $T < 100$ K and in H_2S at $T > 100$ K, to mimic the effect of ice evaporation. The resulting 'jump' in the abundances at $r = 2 \times 10^{16}$ cm is consistent with our observations of SO_2. See Figure 3 and Doty (this volume).

The model reproduces most of our observed abundances, and also the increase of the SO_2 abundance at $T \gtrsim 100$ K. However, OCS is under-produced by factors of 10–100, suggesting that most of the sulphur in the grain mantles is in the form of OCS. This idea is supported by the high excitation temperature of OCS (100 K), much higher than the 25 K found for H_2S. In addition, solid OCS has been detected in the infrared towards W33A, while H_2S has not.

The current data set is not suitable to trace chemical evolution within our sample. The CS, H_2CS, HCS^+ and NS molecules mainly probe the chemically inactive outer envelope ($T < 100$ K). While OCS, SO and SO_2 show larger source-to-source variations, only for SO_2, the data allow to determine the abundance in the inner envelope ($T < 100$ K) where ices are evaporating. The situation for H_2S is unclear: models predict strong abundance variations, but the present data do not trace the warm gas. Future observations of high-excitation lines of H_2S, SO and OCS could exploit the full potential of sulphur as a chemical clock.

References

Charnley, S. B. 1997, ApJ, 481, 396

Doty, S. D., van Dishoeck, E. F., van der Tak, F. F. S., & Boonman, A. M. S. 2002, A&A, 389, 446

Hatchell, J., Thompson, M. A., Millar, T. J., & MacDonald, G. H. 1998, A&A, 338, 713

Hogerheijde, M. & van der Tak, F. 2000, A&A, 362, 697

Jansen, D. J., Spaans, M., Hogerheijde, M. R., & van Dishoeck, E. F. 1995, A&A, 303, 541

Keane, J. V., Boonman, A. M. S., Tielens, A. G. G. M., & van Dishoeck, E. F. 2001, A&A, 376, L5

Minh, Y. C., Irvine, W. M., McGonagle, D., & Ziurys, L. M. 1990, ApJ, 360, 136

Ohishi, M., Irvine, W. M., & Kaifu, N. 1992, in *Astrochemistry of Cosmic Phenomena*, IAU Symp. 150, 171

Pineau des Forêts, G., Roueff, E., Schilke, P., & Flower, D. R. 1993, MNRAS, 262, 915

Sutton, E., et al 1995, ApJS, 97, 455

van der Tak, F. F. S., Boonman, A. M. S., Braakman, R., & van Dishoeck, E. F. 2002, A&A, submitted

van der Tak, F. F. S., van Dishoeck, E. F., Evans, N. J., & Blake, G. A. 2000, ApJ, 537, 283

van Dishoeck, E. F. 2001, in *The Promise of the HSO*, ed. G.L. Pilbratt et al. ESA-SP, 460, 185

CIRCUMSTELLAR DISK CHEMISTRY: 2-D UV RADIATIVE TRANSFER AND EFFECTS OF STELLAR UV

G.J. van Zadelhoff[1], Y. Aikawa[2], M.R. Hogerheijde[3], and E.F. van Dishoeck[1]

[1]Leiden Observatory, The Netherlands
[2]Kobe University, Japan
[3]Steward Observatory, University of Arizona, USA

Abstract

Using a new two-dimensional axisymmetric ultraviolet radiative transfer code, the chemistry in circumstellar disks is calculated. Scattering and absorption of photons from the central star and from the interstellar radiation field are taken into account. The molecules are effectively photodissociated in the surface layer of the disk and frozen out onto grains deep in the disk close to the midplane. However, they can exist in the intermediate, moderately warm layers. The lack of molecules in the high-density cold layers makes molecules bad tracers of the total gas mass in these disks. A comparison is made with an approximate 2-D ray-tracing method and it is found that the latter underestimates the ultraviolet field and thus the molecular photodissociation rates in the disks. The full 2-D results show significantly higher abundances of radicals such as CN and C_2H than in previous work, partly due to the fact that CO is dissociated to greater depths. Results for different stellar radiation fields are presented as well. The CN/HCN ratio shows a strong dependence on the stellar spectrum, whereas other ratios such as HCO^+/CO show only little variation.

Key words: Planetary systems: protoplanetary disks – astrochemistry

---

Chemistry as a Diagnostic of Star Formation (2003). *Edited by Charles L. Curry and Michel Fich.* Published in 2003 by NRC Research Press, Ottawa ON, K1A 0R6, Canada. ISBN 0-660-19089-3.

THE [C II] 158 MICRON ABSORPTION LINE TOWARDS THE GALACTIC CENTER: A CONNECTION WITH BRIGHT IR GALAXIES

Charlotte Vastel[1], E. Polehampton[2], J.-P. Baluteau[3], B. Swinyard[2], E. Caux[4], and P. Cox[5]

[1]Caltech 320-47, Pasadena, CA 91125, USA
[2]RAL, Chilton, Didcot, Oxon, UK
[3]LAS, BP 8, F-13004, Marseille, France
[4]CESR, BP 4346, F-31028, Toulouse Cedex 04, France
[5]IAS, F-91405, Orsay Cedex, France

ABSTRACT

Located near the Galactic center, Sgr B2 is one of the largest H II /molecular cloud complexes in the Galaxy. The total mass and compactness of Sgr B2 are comparable to those of the W49N star-forming complex (Vastel et al. 2000; Vastel et al. 2001). We studied the clouds along its line of sight with the help of *ISO*-LWS data combined to ground-based molecular observation. A complete study can be found in Vastel et al. (2002) but we present here the fundamental results.

Key words: ISM: molecules, clouds – line: profiles – infrared: ISM – radio lines: ISM

1. *ISO*-LWS OBSERVATIONS

The [O I] 63.2 μm and 145.5 μm and [C II] 157.7 μm lines have been observed toward Sgr B2 with the LWS instrument using the high resolution mode (L03 prime and non-prime, and L04 prime). Figure 1 shows the spectrum at 63 μm using L03 prime, L03 non-prime and L04 prime observations. In the following we will use the spectrum obtained in the L03 non-prime mode whose systematic and random noise at line center is the lowest. Each absorption feature can be attributed to a cloud along the line of sight adopting a rotation model for the Galaxy.

2. MODELLING

At velocities below 30 km s^{-1}, the [C II] 157.7 μm line is seen in absorption and in emission at larger velocities (see Fig. 2). The C^+ absorption features are fitted by gaussians whose centers and line widths are deduced from the H I 21 cm line parameters (Garwood & Dickey 1989). The optical depth of each absorption feature is adjusted in order to reproduce the observed spectrum when combined with the emission at 60 km s^{-1} parameterized by the [O I]

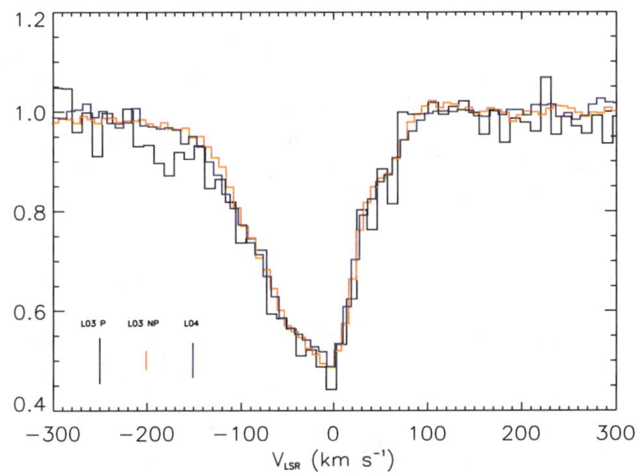

Figure 1. Comparison between the L03 prime *(black)*, L03 non-prime *(red)* and L04 prime data *(blue)* with their combined systematic and random error at line center. The $y-$axis is the line-to-continuum ratio.

145.5 μm line seen in emission. The final fit is shown in Figure 2 after convolution with the instrumental profile.

Molecular observations of Sgr B2 have been carried out in the ^{13}CO and C^{18}O (1–0) and (2–1) lines at the IRAM 30 m telescope. The spectra show a bright emission profile at the velocity of the H II region around 60 km s^{-1}. Many molecular components appear at velocities between -120 km s^{-1} and 35 km s^{-1}. We combined these molecular observation with H I 21 cm observations (Garwood & Dickey 1989) to disentangle the clouds along the line of sight. *Three different layers can be distinguished according to photodissociation region (PDR) models.* Ionized carbon, atomic hydrogen and atomic oxygen coexist in the warm, mostly neutral, layer that lies at the surface on which the UV impinges. In the next (deeper) layer, where molecular hydrogen is able to resist photodissociation, ionized carbon and atomic carbon still coexist. The third layer, which constitutes the cold self-shielded molecular core of the cloud, contains molecular hydro-

Chemistry as a Diagnostic of Star Formation (2003). *Edited by Charles L. Curry and Michel Fich.* Published in 2003 by NRC Research Press, Ottawa ON, K1A 0R6, Canada. ISBN 0-660-19089-3.

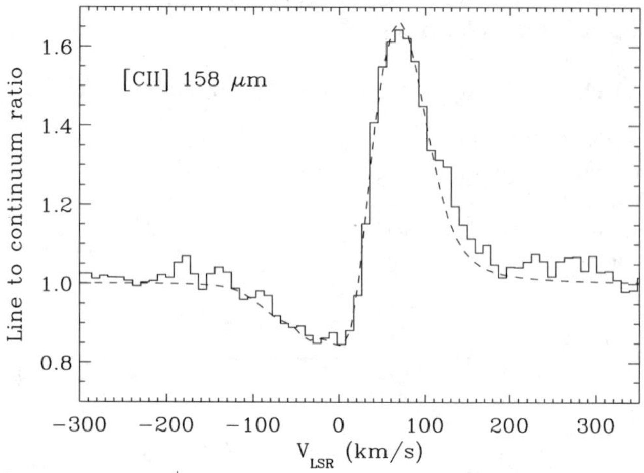

Figure 2. The C^+ line compared to a model profile after convolution with the instrumental profile *(dashed line)*.

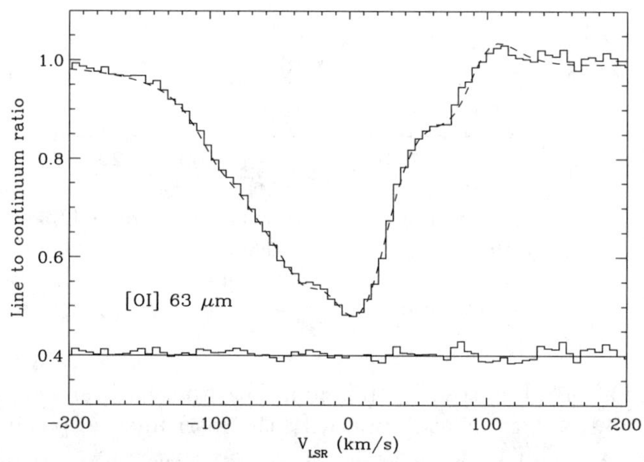

Figure 3. Observed [O I] 63.2 μm absorption profile toward Sgr B2 together with the best fit *(dashed line)* including both the external layers and the molecular cores contributions. The lower plot shows the residuals, $y-$shifted by +0.4 for clarity.

gen, carbon monoxide and atomic oxygen. We present in Table 1 the eight clouds, where the external layers have been traced by H I and the molecular cores by ^{13}CO and $C^{18}O$.

We combined the absorption due to the diffuse layers traced by atomic hydrogen to the absorption due to molecular layers traced by CO and its isotopes for each of the clouds along the line of sight to Sgr B2. The final fit is presented in Figure 3.

Table 1. Properties of the eight clouds found between −110 and +10 km s^{-1}

Cloud label	External layers V_{LSR} (km/s)	Molecular clouds V_{LSR} (km/s)	Galactocentric distance (kpc)
A	−108	−113.6, −104.8	< 1
B	−92		< 1
C	−77	−76	< 1
D	−60.5	−57.8	< 1
E	−51.9		3
F	−44	−46.2, −41	3
G	−21.5	−25.8, −20.5	4
H	−3.5, 5.5	1.7	?

3. Results

3.1. The O^0/CO ratio in the molecular cores

The O^0/CO ratio in the molecular clouds varies between 3 and 37 with a mean value of 18. The O^0 column densities of the 6 clouds exhibiting molecular cores (A, C, D, F, G, H) are plotted versus their CO column densities in Figure 4. There is a clear correlation between N(O) and N(CO), except for cloud H (at 1.7 km s^{-1}) which is believed to result from a blending of clouds with uncertain galactocentric distances. We performed a linear least-squares fit and found that the resulting slope is 2.5 ± 1.8 and the intersection point with the N(O)-axis gives an excess amount of atomic oxygen with an average column density of (5.6 ± 2.0) × 10^{17} cm^{-2}. *This atomic oxygen excess is interpreted as being due to an intermediate layer between the C^+ external layer and the CO core where the atomic oxygen coexists with the atomic carbon, as predicted by PDR models.*

Our derived O^0/CO ratios imply that ~ 70% of gaseous oxygen is in the atomic form and not locked into CO in the molecular clouds along the Sgr B2 line of sight. The upper (respectively, lower) limits of this ratio correspond to ~ 80% (respectively 40%) of the gaseous oxygen in atomic form. Considering the calculated uncertainties, we cannot directly determine the exact oxygen budget in these molecular clouds. However, only the lower limit on the O^0/CO ratio is consistent with the prediction of steady states models and though we cannot exclude values as low as this (O^0/CO ~ 1), the probability of larger ratios is much higher. The oxygen chemistry in dense clouds is still not well known as it is very difficult to reproduce the O_2 and H_2O abundance limits

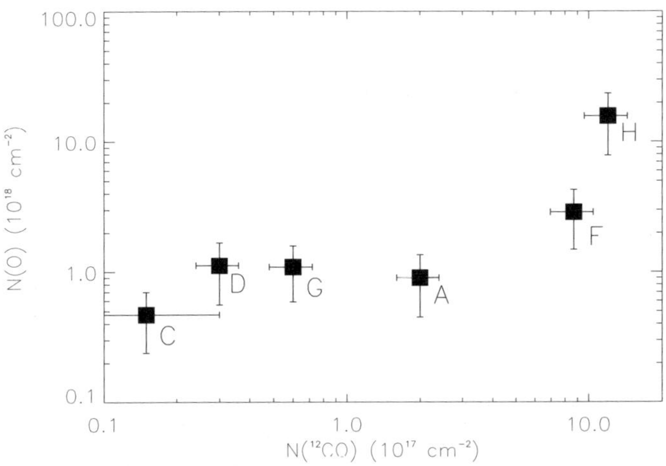

Figure 4. The O^0 column density vs. CO column density for the clouds associated with molecular clouds. Cloud labels identify each data point. Note the logarithmic scale: the fit was carried out on a linear scale.

measured by *SWAS*. Further observations of the O^0, O_2 and H_2O species will provide important clues to constrain the chemistry in these cold molecular clouds.

3.2. The C^0 layer in the clouds

Using the cosmic O/C ratio of 2.3, we deduce an average N(C) in each cloud along the line of sight of $(2.4 \pm 0.9) \times 10^{17}$ cm^{-2}. This value is well within the range of values derived by Usuda, Hasegawa, & White (in preparation) from C^0 observations. Their first detection of the [C I] $^3P_1-^3P_0$ absorption lines in the direction of Sgr B2 in the 10.6″ beam of JCMT leads to an estimate of the C^0 column density in five cloud complexes (clouds labelled here by A, C, D, F and G) between 1 and 13×10^{17} cm^{-2}.

A summary of previous measurements of the C^0/CO ratio in galactic interstellar clouds is illustrated in Fig. 9 of Usuda et al. (in prep.) where this ratio is plotted against the cloud visual extinction. The highest values of this ratio (between 2 and 20) are obtained in diffuse interstellar clouds while the lowest values (ranging from 0.03 to 0.3) are found in dense PDRs. The general slope of the C^0/CO vs. A_V relation indicates that the column density of C^0 is relatively constant in clouds with visual extinction greater than unity. Our derived values of the C^0/CO ratio are consistent with this general relation: clouds C and D could be classified as diffuse clouds where C^0/CO = [16.2, 8.1], clouds A and G as translucent clouds where C^0/CO = [1.2, 4.0] and cloud F as a dense cloud where C^0/CO = 0.3.

Values for the C^+/C^0 ratio in the six clouds associated with a molecular core can be derived from this study. While clouds close to the Galactic center (labelled A, C and D) yield a C^+/C^0 ratio less than 0.7, those at the galactocentric distance of 3-4 kpc (F and G) have ratios close to 2.5. For $G_0/n \leq 3 \times 10^{-3}$ cm^3, the C^+/C/CO transition is drawn to the cloud surface (Kaufman et al. 1999) which could explain the low C^+/C^0 ratio (< 1) found near the Galactic center.

3.3. Optical thickness of the C^+ line

The [C II] fine structure transition at 157.741 μm is the major coolant of the warm interstellar medium and is in the form of C^+ on the surface of far-UV illuminated neutral gas clouds. Depending on the density in this layer, the line may be self-absorbed. Recently, studies of many bright galaxies showed that *the [C II] 157.7 μm line is deficient compared to the total FIR luminosity.* Non-detection or weak detection have been highlighted toward FIR bright galaxies (Malhotra et al. 1997; Luhman et al. 1998). These observations were carried out at a much lower resolution than that used in our study of Sgr B2. Our results indicate that if these galaxies were viewed at high spectral resolution, the spectra could reveal both emission and absorption components present in the C^+ profile. This strengthens the idea that under certain conditions *the C^+ line could be optically thick in the direction of large star-forming complexes or in the nuclei of galaxies.*

References

Garwood, R., & Dickey, J., 1989, ApJ 338, 841
Kaufman, M., Wolfire, M., Hollenbach, D., Luhman, D., 1999, ApJ 527, 795
Luhman, M., Satyapal, S., Fischer, J., Wolfire, M., et al., 1998, ApJ 504, L11
Malhotra, S., Helou, G., Stacey, G., Hollenbach, D., et al., 1997, ApJ 491, L27
Vastel, C., Caux, Ceccarelli, C., et al., 2000, A&A 357, 994
Vastel, C., Spaans, M., Ceccarelli, C., Tielens, X., Caux, E., 2001, A&A 376, 1064
Vastel, C., Polehampton, E., Baluteau, J.P., et al. 2002, ApJ, 581, December 10 issue

THE VELOCITY STRUCTURES OF EMBEDDED CORES IN PERSEUS

N.H. Volgenau[1], L.G. Mundy[1], L.W. Looney[2], and W.J. Welch[3]

[1]University of Maryland
[2]Max-Planck-Institut fuer Extraterrestrische Physik
[3]University of California, Berkeley

Abstract

Recent theoretical studies (e.g., Ostriker et al. 2001; Burkert & Bodenheimer 2000) have presented a scenario for star formation in which dense protostellar cores form rapidly at the collision interfaces of turbulent cloud material. The structure and star-forming capability of these cores are influenced by the conditions of the collision. Such turbulence-induced star formation may help explain both the tendency of young stellar objects to appear in clusters and the plenitude of stellar multiplicity. We seek to determine the extent of turbulence in protostellar cores, with the goal of testing these dynamic star formation theories.

We present maps of the velocity structures of protostellar cores in emission lines from several molecular species. Our observations, made with the BIMA mm-wave interferometer and the FCRAO 14m antenna, are of the $C^{18}O$, N_2H^+, and $H^{13}CO^+$ $J = 1$–0 emission from three embedded cores in the Perseus Molecular Cloud: L1448 N, NGC1333 IRAS2, and NGC1333 IRAS4. The BIMA and FCRAO observations were combined using a maximum entropy method and tuned to resolutions of 10, 5 and 2.5 arcseconds. The resultant maps reveal the gas distributions and velocity structures of the cores on scales from 0.3 pc to 900 AU.

Key words: ISM: kinematics and dynamics – ISM: clouds – turbulence

Chemistry as a Diagnostic of Star Formation (2003). *Edited by Charles L. Curry and Michel Fich.* Published in 2003 by NRC Research Press, Ottawa ON, K1A 0R6, Canada. ISBN 0-660-19089-3.

SULPHUR-BEARING SPECIES IN THE STAR-FORMING REGION L1689N

V.Wakelam[1], A.Castets[1], C.Ceccarelli[1,2], B.Lefloch[2], E.Caux[3] and L.Pagani[4]

[1]Observatoire de Bordeaux, BP 89, 33270 Floirac, France
[2]Laboratoire d'Astrophysique, Observatoire de Grenoble, BP 53, F-38041 Grenoble Cedex 09, France
[3]CESR CNRS-UPS, BP 4346, 31028, Toulouse Cedex 04, France
[4]LERMA and FRE 2460 du CNRS, Observatoire de Paris, 61, Av. de l'Observatoire, 75014 Paris, France

Abstract

We report observations of the expected main S-bearing species (SO, SO_2 and H_2S) in the low-mass star-forming region L1689N. We derived the column density of SO, SO_2 and H_2S, together with SiO and H_2CO—which were observed previously—and their relevant abundance ratios in six interesting regions: the molecular cloud (as reference position), four shocked regions caused by the interaction of the molecular outflows with the cloud, and the protostar IRAS 16293-2422. We find that SiO is the molecule that shows the largest abundance variations in the shocked regions, whereas S-bearing molecules show more moderate variations. We propose a scheme in which the different molecular ratios correspond to different ages of the shocks. Finally, we find that SO, SO_2 and H_2S have significant abundance jumps in the inner hot core of IRAS 16293-2422 and discuss the implications of the measured abundances.

Key words: ISM: abundances – ISM: molecules – Stars: formation – ISM: individual: L1689N, IRAS 16293-2422

1. Introduction

low-mass star-forming regions are formed by at least three main ingredients: the molecular cloud from which protostars are born, the protostars themselves, and the shocked regions at the interface between the cloud and the outflows emanating from the protostars, which have very different physical conditions (van Dishoeck & Blake 1998). We have focused our attention on the abundance changes occurring to the S-bearing molecules and the relevant sulphur chemistry. We present millimeter observations of several transitions of SO, SO_2 and H_2S in the molecular cloud L1689N. L1689N is a molecular cloud located in the ρ Ophiuchi cloud complex at 120 pc from the sun (Knude & Hog 1998). This cloud harbors two young protostellar sources.The first one is IRAS 16293-2422 (hereafter IRAS 16293), a low-luminosity Class 0 protostar (15 $L_\odot$) and the second one is 16293E, a recently discovered low-mass and very young Class 0 source situated southeast of IRAS 16293. The structure of the envelope surrounding IRAS 16293 has been recently modeled by Ceccarelli et al. (2000a,b) showing that a hot core region exists in the inner parts. These protostars induce several bipolar outflows which impact the molecular quiescent gas, creating six shocked regions, named E1, E2, HE1, HE2, W1 and W2.

2. Observations and column densities

Observations of large-scale maps and specific positions inside the L1689N region were performed with the IRAM and SEST telescopes. We measured the emission of the following molecules: SO ($3_2 \rightarrow 2_1$ and $6_5 \rightarrow 5_4$ transitions), SO_2 ($5_{1,5} \rightarrow 4_{0,4}$ transition) and H_2S ($1_{1,0} \rightarrow 1_{0,1}$ and $2_{2,0} \rightarrow 2_{1,1}$ transitions). Combining these observations with a LVG model, we have computed the column densities of the SO, SO_2 and H_2S molecules in the molecular cloud L1689N (as a reference point), in the shocked regions (E1, E2, W1 and W2) and in the hot core of IRAS 16293. To have a global view of the chemistry in these regions, we have compared the sulphur species abundances with the SiO and H_2CO molecules previously observed by Castets et al. (2001). More details on the observations and the density, temperature, and column densities can be found in Wakelam et al. (2002).

3. Discussion

3.1. Shocked regions

Figure 1 shows a graphic representation of the derived abundances ratios in the shocked regions and in the cloud. The first result to note is that, with the exception of the SO_2/H_2CO ratio, all other ratios in the shocked regions differ by more than a factor two when compared to the ratios in the cloud. In particular, the largest variations, by up a factor 100, are shown by the ratios with the SiO,

Chemistry as a Diagnostic of Star Formation (2003). *Edited by Charles L. Curry and Michel Fich.* Published in 2003 by NRC Research Press, Ottawa ON, K1A 0R6, Canada. ISBN 0-660-19089-3.

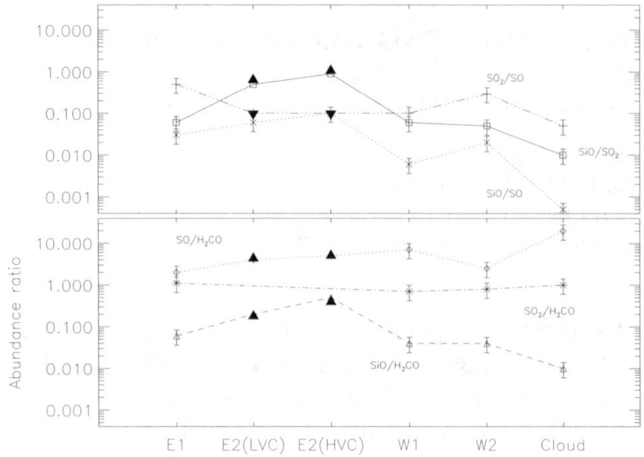

Figure 1. Abundance ratios observed in the different shocked regions and in the molecular cloud. The filled triangles represent upper or lower limits.

Table 1. Proposed scheme of the formation and destruction of SO, SO_2, H_2S, SiO and H_2CO as a function of time. The times indicative and are meant to represent a sequence.

Time (yr)	Formation	Destruction
0^a	H_2S, H_2CO, SiH_2, SiH_4	
10^3	SO, SO_2, H_2S, SiO	
3×10^4		SO_2, H_2CO
10^5		SiO

[a] Injection into the gas phase of the molecules sputtered from the grain mantles.

confirming that SiO is the best molecule to trace shocks (Caselli et al. 1997; Schilke et al. 1997; Garay et al. 2002). Furthermore, the strongest shocks, in E2, show also the largest SiO over the other molecules ratios, in agreement with the idea that as stronger the shock as more abundant is SiO.

A second robust result is that SO_2 is enhanced with respect to SO in the shocks E1, W1 and W2, when compared to the cloud (by about a factor 10). This also confirms a rather general theoretical prediction that SO_2 is over-produced with respect to SO in the shocked gas, because of endothermic reactions that lock most of the gaseous sulphur in SO_2 on timescales relatively small ($\sim 10^3$ yr; e.g., Pineau Des Forêts et al. 1993). Remarkably, SO_2/SO is lowest in the strongest shock of the region, E2. Castets et al. (2001) argued that E2 is an older shock and this would agree with the low SO_2/SO ratio, for SO_2 is expected to be converted into atomic sulphur at later stages ($\geq 3 \times 10^4$ yr; e.g., Hatchell et al. 1998).

The third result is the constant SO_2/H_2CO ratio, within a factor 2, in the shocked regions and in the cloud, which seems a remarkable coincidence for which we do not have an explanation. On the contrary, the constant SO_2/H_2CO ratio in the E1, W1 and W2 shocked regions may be due to the timescales involved in the formation and destruction of the two molecules, even though the mechanisms of SO_2 and H_2CO formation in shocks are expected to be different. On the one hand, formaldehyde is thought to be formed onto the grain mantles, and released into gas phase because of the mantle sputtering in the shock, and finally converted into more complex molecules by gas phase reactions on timescales of 10^4 yr (Charnley et al. 1992). On the other hand, sulphur is supposed to be released into the gas phase mainly as H_2S, following also mantle sputtering, and then H_2S is transformed into SO and SO_2 via neutral-neutral reactions on timescales of $\sim 10^3$ yr (Charnley 1997). Therefore, the constant SO_2/H_2CO ratio in E1, W1 and W2 may tell us that these three shocks have similar ages, around 10^4 yr if models are correct, whereas E2 is a much older shock, $\geq 3 \times 10^4$ yr, and SO_2 and H_2CO both disappear because transformed into S and complex O-bearing molecules, respectively.

The overall emerging picture is that the combination of the variations in the abundance of different molecules can in first instance give an estimate of the age of the shock. Table 1 summarizes the situation. The simultaneous presence of SO, SO_2, H_2S, H_2CO, and SiO would mark relatively young shocks ($\leq 10^4$ yr), whereas the only presence of SiO emission would point towards relatively older shocks ($\geq 3 \times 10^4$ yr).

3.2. The Hot core of IRAS 16293

L1689N has abundances typical of other cold molecular clouds, with the possible exception of formaldehyde, which seems to be underabundant whereas, in the inner core, the sulphuretted molecule abundances exhibit a strong enhancement (see Table 2). The S-bearing molecules are expected to be useful to evaluate the age of hot cores. We have compared the observed ratios H_2S/SO and SO_2/SO in the hot core of IRAS 16293 with the theoretical predictions by Hatchell et al. (1998), and we have found a good agreement between our observations and model predictions for an age of $\sim 3 \times 10^4$ yr which is similar to the independently estimated age of IRAS 16293 (Ceccarelli et al. 2000a).

Following the theoretical expectations that gaseous sulphur in the warm gas is mainly locked into SO, SO_2 and H_2S (e.g. Pineau Des Forêts et al. 1993; Charnley 1997), while silicon is mainly locked into SiO (Herbst et al. 1989), the $(SO+SO_2+H_2S)/SiO$ ratio gives a measure of the gas-phase elemental sulphur over silicon abundance ratio in the IRAS 16293 hot core. This value is about 300, to compare with the solar elemental sulphur over silicon abundance ratio of 0.5. It is unlikely that we overestimate the overall quantity of sulphur. Conversely, it could have been underestimated because of a possible optical thickness of the sulfuretted molecular lines and/or the presence of other important sulphur-bearing molecular species not considered here. Thus, the measured $(SO+SO_2+H_2S)/SiO$ ratio suggests that there is an important deficiency of silicon in the warm gas of the IRAS 16293 hot core. The likely explanation is that silicon is mostly depleted into the non-volatile, refractory cores grains, whereas sulphur is depleted mostly onto the volatile grain mantles. This conclusion supports the thesis by (Ruffle et al. 1999) that sulphur and silicon follow different routes of depletion, and specifically that sulphur is depleted at the time of mantle formation.

Finally, the total abundance of SO, SO_2 and H_2S in the IRAS 16293 hot core is 2.8×10^{-6}, namely more than ten times lower than the solar abundance 3.4×10^{-5}. Even adding up the OCS, whose abundance is about that of SO (Schöier et al. 2002), the overall sulphur abundance is still low. Unfortunately, the other possible reservoir of sulphur, the atomic sulphur (Charnley 1997), is extremely difficult to observe, as it has the ground transition at 57 μm in a wavelength range obscured by the atmospheric atmosphere. Observations obtained with the Long Wavelength Spectrometer on board *ISO* did not detect any signal larger than about 5×10^{-13} erg s^{-1} cm^{-2} (Ceccarelli et al. 1998), corresponding to an upper limit to the S column density of 3×10^{16} cm^{-2}, and an abundance lower than 5×10^{-7}. Even taking into account the factor five uncertainty in the gas density, sulphur seems still deficient (by at least a factor two) in the warm gas of IRAS 16293.

4. Conclusion

Comparing the column densities of the SO, SO_2,H_2S, H_2CO and SiO molecules in the different regions of L1689N, we have found that SiO is the best tracer of shocks and that SO_2 is enhanced with respect to SO in the three regions of shocked gas. Putting together the observed combinations of these molecules ratios, we proposed a scheme in which the different molecular ratios correspond to different ages of the shocks. Finally, we found that SO, SO_2 and H_2S have significant abundance jumps (300, 1300 and 1700 respectively) in the inner hot core of IRAS 16293. We compared the measured abundances with theoretical models and argued that the abundance ratios of S-bearing molecules can indeed give information on the protostar age. Comparing the $SO+SO_2+H_2S/SiO$ ratio in the hot core of IRAS 16293, we found that silicon is largely deficient in the warm gas (by a factor ~ 600), supporting the thesis that silicon is depleted into the grain refractory cores whereas sulphur is depleted into the grain volatile mantles. Nonetheless, sulphur in the IRAS 16293 warm gas is still deficient by more than a factor of two.

Table 2. Abundances of SO, SO_2, H_2S, H_2CO and SiO with respect to H_2 in the hot core and in the molecular cloud

Species	Hot core	Cloud	X_{hc}/X_{cloud}
SO/H_2	1.7×10^{-6}	$(4-12) \times 10^{-9}$	140-420
SO_2/H_2	5.4×10^{-7}	4×10^{-10}	1300
H_2S/H_2	5.3×10^{-7}	3×10^{-10}	1700
H_2CO/H_2	1.0×10^{-7}	4×10^{-10}	250
SiO/H_2	1.5×10^{-8}	4×10^{-12}	3750

Acknowledgements

V. Wakelam is grateful to the organizers of SFCHEM2002 for financial support.

References

Caselli, P., et al., 1997, A&A 322, 296
Castets, A., et al., 2001, A&A 375, 40
Ceccarelli, C., et al., 1998, A&A 331, 17
Ceccarelli, C., et al., 2000a, A&A 355, 1129
Ceccarelli, C., et al., 2000b, A&A 357, L9
Charnley, S. B. 1997, ApJ 481, 396
Charnley, S. B., et al., 1992, ApJ Lett. 399, L71
Garay, G., et al., 2002, ApJ 567, 980
Hatchell, J., et al., 1998, A&A 338, 713
Herbst, E., et al., 1989, A&A 222, 205
Knude, J. & Hog, E. 1998, A&A 338, 897
Pineau Des Forêts, G., et al., 1993, MNRAS 262, 915
Ruffle, D. P., et al., 1999, MNRAS 306, 691
Schilke, P., et al., 1997, A&A 321, 293
Schöier, F. L., et al., 2002, A&A 390, 1001
van Dishoeck, E. F. & Blake, G. A. 1998, ARA&A 36, 317
Wakelam, V., et al., 2002, A&A submitted

A HIGH RESOLUTION STUDY OF STAR-FORMING CLUMPS IN NGC 1333

Andrew J. Walsh[1], Philip C. Myers[1], James Di Francesco[2], David J. Wilner[1], and & Tyler L. Bourke[1]

[1]Harvard-Smithsonian Center for Astrophysics, Mail Stop 42, 60 Garden Street, Cambridge, MA 02138, USA
[2]Radio Astronomy Laboratory, Campbell Hall, University of California at Berkeley, Berkeley, CA 94720, USA

Abstract

We are conducting a large-scale survey of the NGC 1333 star-forming cloud at high resolution with the BIMA interferometer. So far, we have covered the southern half of the cloud in the (1–0) lines of N_2H^+ and HCO^+, as well as 3 mm continuum emission. The N_2H^+ emission not only allows us to study the morphology of emission in the region, but we also obtain precise velocity information, allowing us to study the region dynamically. The HCO^+ line emission is also a good tracer of outflow activity. We find emission throughout most of the field of view in both lines. To identify individual clumps in this busy region, we have used the MIRIAD task CLUMPFIND. We have found over 100 clumps in both the N_2H^+ and HCO^+ lines. We report on the initial findings of our clump census in both lines.

Key words: ISM: kinematics and dynamics – ISM: clouds – ISM: individual (NGC 1333)

Chemistry as a Diagnostic of Star Formation (2003). *Edited by Charles L. Curry and Michel Fich.* Published in 2003 by NRC Research Press, Ottawa ON, K1A 0R6, Canada. ISBN 0-660-19089-3.

ROTATIONAL SPECTROSCOPY AND HOT CORE OBSERVATIONAL ASTRONOMY OF PREBIOTIC MOLECULES

Susanna L. Widicus[1], **Brian J. Drouin**[2], **Kathryn A. Dyl**[3], **and Geoffrey A. Blake**[3]

[1]Division of Chemistry and Chemical Engineering, California Institute of Technology, Pasadena, CA 91125, U.S.A
[2]California Institute of Technology, Jet Propulsion Laboratory, Pasadena, CA 91109, U.S.A.
[3]Division of Geological and Planetary Sciences, California Institute of Technology, Pasadena, CA 91125, U.S.A.

Abstract

Hot core regions have been found to contain a rich array of complex organic molecules. While the presence of many of these species can be explained by models which couple grain surface and gas-phase reactions, several detected compounds are not included in these models, and many species that are predicted to be abundant have not been characterized spectroscopically despite their serving as precursors to important biological species such as amino acids. Therefore, the detection of these organics would offer insight into potential prebiotic chemistry in hot core environments. We present here the preliminary investigation into the spectroscopic characterization of and observational search for the suspected prebiotic interstellar species aminomethanol, aminoethanol, methanediol, and dihydroxyacetone.

Key words: ISM: molecules – Methods: laboratory – – astrobiology – H II regions

1. Introduction

It is now clear that prebiotic chemistry achieves considerable complexity long before planetary systems mature. Indeed, organic molecules associated with the development of life may have been transported to the early earth by comets and meteorites. The complex array of organic material present in carbonaceous chondrites lends support to this theory. The Murchison meteorite was found to contain more than 92 amino acids, only 19 of which exist naturally on Earth (Cronin and Chang 1993). Several sugars and other polyhydroxylated compounds have also recently been detected in the Murchison and Murray meteorites (Cooper et al. 2001). These prebiotic species could have been formed before our planetary system was assembled, integrated into planetesimals, and introduced to the Earth through impact. Alternatively, the planetesimals could have formed from interstellar dust grains whose mantles contained much simpler organic species. More complex prebiotic chemistry could then follow on meteorite parent bodies, with the products ultimately being delivered to the early Earth through impact.

Presently there are over one hundred known interstellar molecules, the majority of which are organic (1991). Theoretical models of grain surface chemistry (Fig. 1) predict precursors to the more complex compounds (Charnley 1999 Charnley 1997). Many potential grain surface reaction pathways are eliminated by the conditions imposed on these models, greatly simplifying the possible products of grain synthesis and eliminating the possibility for much larger organics to form on grain surfaces unless the grain mantles are subjected to repeated thermal cycling (Greenberg et al. 2000).

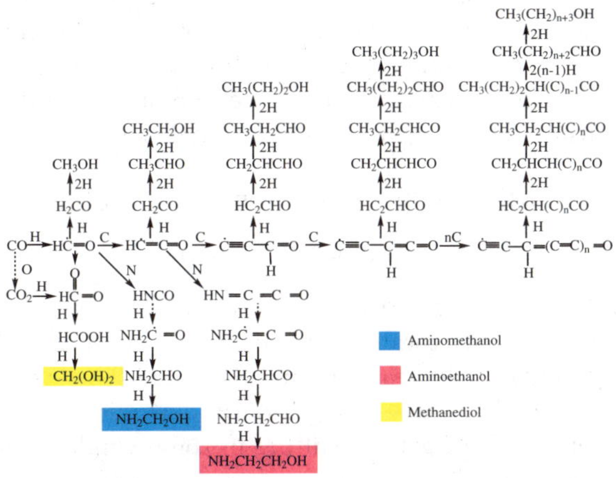

Figure 1. Interstellar grain surface chemistry.

Gas-phase theoretical models of the chemistry in hot cores involving the products of grain surface reactions are therefore required to explain the formation of substantially larger organics. In these models, the temperature of the hot cores leads to thermal evaporation of simple molecules from the grain surface (Fig. 2). These molecu-

Chemistry as a Diagnostic of Star Formation (2003). *Edited by Charles L. Curry and Michel Fich.* Published in 2003 by NRC Research Press, Ottawa ON, K1A 0R6, Canada. ISBN 0-660-19089-3.

les can then undergo gas phase reactions to form more complex species such as amino acids, sugars, and other biologically important molecules.

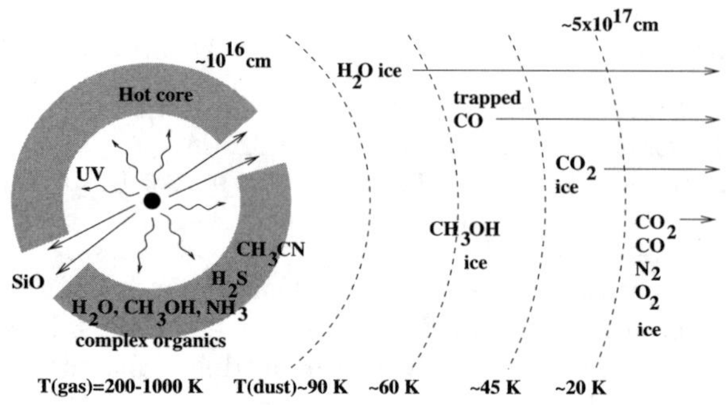

Figure 2. Schematic diagram of a hot protostellar core.

While hot core models predict the formation of many organic species, characterization of and searches for many of even the simplest species have not been undertaken. In addition, these models do not account for the formation of some observed molecules. One such organic is glycolaldehyde ($CHOCH_2OH$), the simplest sugar, which has been detected in Sgr B2(N-LMH) (Hollis et al. 2000). This discovery confirmed the need for further experimental and observational investigation of these reaction pathways. Those species predicted by Charnley et al. (1995, 1997, 1999) to exist in detectable amount and that have limited spectroscopic characterization, namely aminomethanol, aminoethanol, and methanediol, as well as the next most complex sugar, dihydroxyacetone, are the focus of our initial laboratory and observational studies.

2. Laboratory Spectroscopic Studies

2.1. Instrumentation

Prebiotic monomers are expected to have strong torsional transitions in the terahertz (THz) frequency range. Due to experimental limitations in this range, however, the spectral search is more straightforward at microwave and millimeter wavelengths, resulting in these windows being the most natural starting points when spectra are unavailable. Initial laboratory work involved direct absorption millimeter and submillimeter spectroscopy in a flow cell apparatus belonging to the Jet Propulsion Laboratory (JPL) Spectroscopy Group (Widicus et al. 2002 and references therein). A similar apparatus has recently been completed at Caltech and is currently being used simultaneously with the JPL instrument. These initial experiments focused on those molecules available commercially, specifically aminoethanol and dihydroxyacetone.

Those molecules not available commercially will have to be synthesized and may require molecular beam experiments for isolation of the desired species. Spectra for these molecules will be obtained using a Fabry-Perot cavity pulsed Fourier transform microwave (FTMW) spectrometer, also known as a Flygare instrument (Balle and Flygare 1981). In addition to the advantages offered by this approach for unstable species, this setup can be used for high sensitivity studies of the other species being examined. The FTMW spectrometer can obtain large spectral coverage in a short period of time, and is therefore the first approach for those molecules with little to no available spectroscopic information. These data will serve as the basis for further flow cell absorption experiments.

2.2. Molecules of Interest

One proposed hot core pathway involves the protonated forms of two grain surface products, namely aminomethanol (NH_2CH_2OH) and aminoethanol ($NH_2CH_2CH_2OH$). These aminoalcohols react with formic acid (HCOOH) to yield the protonated forms of the amino acids glycine and alanine, respectively (Charnley 1999). Aminomethanol is a transient species and has not been characterized. However, aminoethanol is chemically stable and commercially available. The rotational spectrum of aminoethanol has been characterized up to 305 GHz. We have extended measurements of ground state transitions and assigned transitions for several vibrational states (Widicus et al. 2002). Improved rotational and centrifugal distortion constants have been determined for these states. This work has provided sufficient data to guide observational searches, as predictions of strong spectral features in the submillimeter range are accurate to <1 MHz.

Although aminomethanol is unstable under typical laboratory conditions, recent theoretical studies into its potential loss of water and conversion to methanimine indicate that it is indeed stable under interstellar conditions (Feldmann et al. 2002). This work also indicates that the loss of aminomethanol through this and other pathways can be virtually eliminated by minimizing reactions with other species in a molecular beam experiment. Spectroscopic characterization of aminomethanol should therefore be a straightforward process and is currently underway with the Flygare instrument.

Methanediol, $CH_2(OH)_2$, is also predicted to form on grain surfaces (Charnley 1999). While this molecule does not play a role in further gas-phase prebiotic pathways, it does offer a unique opportunity to place rigid constraints on the expected abundances of grain surface reaction products due to its position early in the proposed reaction schemes. Methanediol's relatively large rotational constants and expected rigid hydrogen-bonded structure lead to a small partition function and strong lines, making this species ideal for observational searches. Its laboratory spectroscopic characterization is underway, with initial millimeter data being analyzed at this time.

The discovery of glycolaldehyde has led to interest in species not yet accounted for by the current hot core models. Recent analysis of carbonaceous chondrites has led to the detection of a suite of sugars and other polyhydroxylated organic species (Cooper et al. 2001). These species were detected in similar concentrations to amino acids, indicating a possible link between the proposed hot core gas-phase reaction schemes and the formation of these compounds. Dihydroxyacetone ($CHO[CH_2OH]_2$), the second most complex sugar, was one of the species detected in this study. Its millimeter and submillimeter spectra have been obtained in our laboratory, and analysis of these spectra is currently underway. Many strong lines suitable for observational studies were observed.

3. Observational Searches

The first searches for aminoethanol were conducted with the Owen's Valley Radio Observatory (OVRO) Millimeter Array at 112.7 GHz in the hot core sources of Orion KL and W51 e2. This molecule should be present in column densities on the order of 10^{13} cm^{-2}. There was no direct detection of aminoethanol in either Orion or W51 with OVRO, but several U-lines were found and column density upper limits of $\sim 10^{15}$ cm^{-2} and $\sim 10^{16}$ cm^{-2} have been calculated, respectively. Initial CSO searches near 250 GHz resulted in upper limits of $\sim 10^{14}$ cm^{-2} for both sources. Additional CSO studies may be necessary to reach the rms levels necessary for a detection at the expected aminoethanol abundances.

Observational searches for the other species of interest are planned. A survey of two hot core sources, namely the low mass young stellar object IRAS 16293-2422 and the hot core region Sagittarius B2(N-LMH), will begin int he summer of 2003, and will include aminoethanol, methanediol, and dihydroxyacetone. Additional surveys of Orion and W51 are planned for the fall of 2003.

4. Conclusion

Laboratory characterization is underway for several prebiotic molecules, namely aminomethanol, aminoethanol, methanediol, and dihydroxyacetone. This study will provide the basis for future studies of the torsional modes of these molecules. Due to the large number of vibrational modes for these molecules, their THz spectra will be quite complex. Therefore, complete characterization of the rotational spectra is necessary before these studies can begin. This work will support future atmospheric and space telescopes such as SOFIA (operations begin in 2004) and *Herschel* (launch in 2007).

In addition, the laboratory aspect of this project is providing the basis for extensive observational searches in hot protostellar cores. The combined results of these studies will offer clues as to the complexity of organic reactions in the interstellar medium, and the role of this chemistry in the formation of planetary systems. If it is found that these species are not present in hot protostellar cores, the pathways currently proposed for the formation of such material should be reconsidered and greater attention paid to the role of meteorite parent bodies in the formation of prebiotic molecules.

Acknowledgements

This work was funded in part by the NASA Exobiology and SARA programs, grant numbers NAG5-8822 and NAG5-11423. We also acknowledge partial support from the CIT Summer Undergraduate Research Program. We thank the JPL Microwave, Millimeter, and Submillimeter Spectroscopy Group for use of the flow cell apparatus and assistance during data collection, and gratefully acknowledge the support and services rendered by the CIT, JPL, OVRO, and CSO staff.

References

Balle T. J. & Flygare W. H. 1981, Rev. Sci. Instr., 52, 33

Charnley S., et al. 1995, ApJ, 448, 232

Charnley S. 1997, in *Astronomical and Biochemical Origins and the Search for Life in the Universe*, ed. C.B. Cosmovici, S. Bowyer, & D. Werthimer (Bologna: Editrice Compositori), 89

Charnley S. 1999, in *The Bridge Between the Big Bang and Biology* (Consiglio Nazionale delle Ricerche, Italy)

Cooper, G., et al. 2001, Nature, 414, 879

Cronin, J.R., & Chang, S. 1993, in *The Chemistry of Life's Origins*, ed. J.M. Greenberg et al. (Dordrecht: Kluwer), 209

Feldmann, M.T., et al. 2002, J. Chem. Phys., in prep.

Greenberg, J.M., et al. 2000, ApJ, 531, L71

Hollis, J.M., Lovas, F.J., & Jewell, P.R. 2000, ApJ, 540, L107

Ohishi, M. 1997, in *Molecules in Astrophysics*, IAU Symposium No. 178, ed. E.F. van Dishoeck (Dordrecht: Kluwer), 61

Penn, R.E., & Curl, R.F. 1971, J. Chem. Phys., 53, 651

Widicus, S.L., et al. 2002, J. Mol. Spec., submitted

THE AUSTRALIA TELESCOPE MILLIMETRE-WAVE UPGRADE

Tony Wong

CSIRO Australia Telescope National Facility, P.O. Box 76, Epping NSW 1710, Australia, Tony.Wong@csiro.au

Abstract

The Australia Telescope Compact Array (ATCA), an array of six 22 m-diameter antennas situated near Narrabri, New South Wales, is the premier aperture synthesis radio telescope in the southern hemisphere. Its frequency coverage is being extended from 1–10 GHz to encompass the 16–26 GHz and 85–105 GHz bands as well. On behalf of the ATNF and many collaborators, I present in this poster paper a selection of recent observational results.

Key words: instrumentation: interferometers

1. The ATCA Millimetre Upgrade

The new 3 mm and 12 mm ATCA receivers use amplifiers based on indium phosphide (InP) monolithic microwave integrated circuit (MMIC) technology, which offers exceptional gain stability and a wide instantaneous bandwidth, and can be operated at front-end temperatures of ~20 K (unlike SIS systems typically used at 3 mm, which must be cooled to ~3 K). The systems receive both linear polarizations and are enclosed in a common dewar. Currently, 3 of the 6 ATCA antennas have millimetre receivers. Typical above-atmosphere system temperatures are around 60 K at 18 GHz and around 300 K at 90 GHz.

A high frequency local oscillator system will provide a reference tone from 11.6–15.2 GHz which can be doubled to enable tuning within the 16–26 GHz band and multiplied by 8 to enable tuning within the 85–105 GHz band, with a possible extension up to 115 GHz. In the current systems, however, this local oscillator is set at a fixed frequency, and only certain frequency bands (16.1–18.9, 20.1–22.5, 84.9–87.3, and 88.6–91.3 GHz) are available for observations.

The ATCA correlator provides up to 128 MHz bandwidth on each of two frequencies. For 3 mm and 12 mm observing, both frequencies must lie within one of the four bands listed above. Since the velocity coverage at 3 mm provided by 128 MHz of bandwidth is relatively poor, a new FX correlator with up to 2 GHz bandwidth is under development.

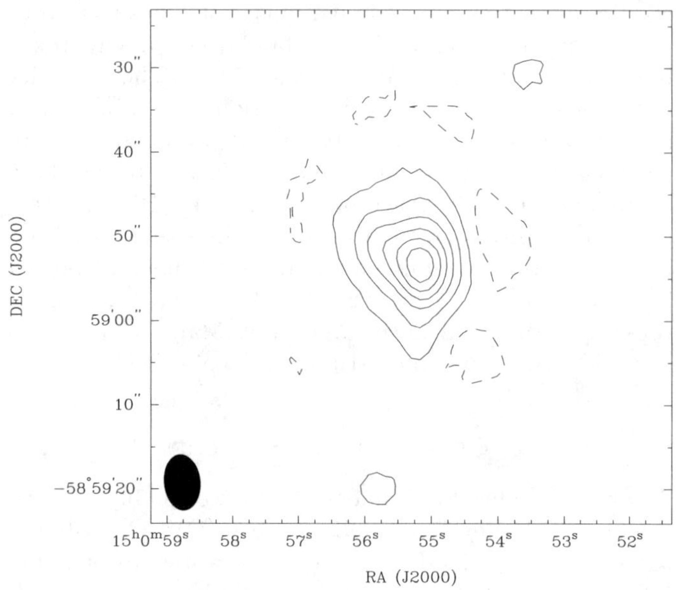

Figure 1. Integrated intensity map for HCO^+ in G318.95-0.17. The contour levels are 1 Jy beam^{-1} km s^{-1}. The synthesized beam is shown at lower left. Credit: V. Minier et al.

As part of the millimetre upgrade, a new north-south track with 5 antenna stations has been constructed to allow the antennas to be placed in a T-configuration. This will avoid the problem of having to observe at low elevations for good coverage of the visibility plane, while improving the imaging of sources north of declination −45°.

2. Selected Results from the 3 mm System

2.1. Observations of Hot Molecular Cores

Recent work (e.g., Walsh et al. 1998) has suggested that the 6.7 and 12.2 GHz methanol maser lines trace the "hot core" stage of massive star formation, prior to the emer-

Chemistry as a Diagnostic of Star Formation (2003). *Edited by Charles L. Curry and Michel Fich.* Published in 2003 by NRC Research Press, Ottawa ON, K1A 0R6, Canada. ISBN 0-660-19089-3.

gence of an ultra-compact H II region. In order to test this hypothesis and investigate the chemistry and kinematics of hot cores, we have conducted observations of the maser source G318.95-0.17, identified as a hot core in a survey with the Mopra single-dish telescope, in several molecular lines. Figure 1 shows the integrated intensity map for HCO^+ emission, with a beam size of 6.″7 × 4.″3. We have also detected and imaged emission from HCN (1–0), CH_3OH (15_3–14_4), and HC_3N (10–9).

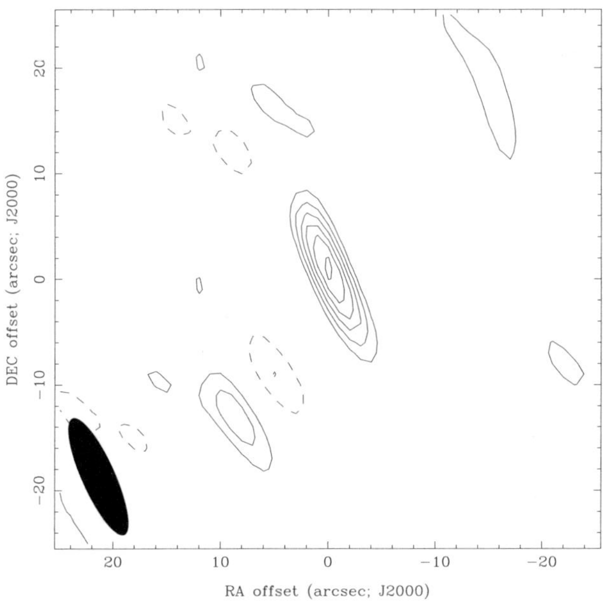

Figure 2. Continuum emission at 89 GHz from TW Hya. The contours are in 5 mJy steps starting at 10 mJy. The flux is roughly 40 mJy. Credit: T. Bourke et al.

2.2. Dust and Gas in Pre-Planetary Disks

A large fraction of young stars are surrounded by dusty disks of material left over from their formation process, which may represent the early stages of planetary system formation. Beyond a few stellar radii, the disks are sufficiently cold that they are best studied at millimetre and sub-millimetre wavelengths. We have undertaken a program to image two such disks, around the T Tauri star TW Hya and the Herbig Be star HD 100546, in the continuum and HCO^+ (1–0) line. In observations during 2002 May and June we detected continuum emission from both sources (Fig. 2) and HCO^+ emission from TW Hya.

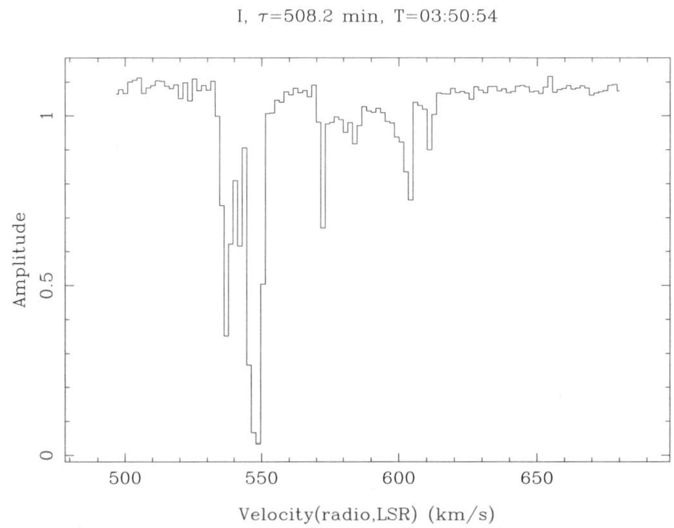

Figure 3. Spectrum of HCO^+ in absorption in front of the radio nucleus of Centaurus A, taken on 6 Oct 2001. The spectrum is an average of 3 baselines (45, 75, and 120m) after 3 hours of integration. The amplitude scale is uncalibrated. Credit: B. Koribalski et al.

2.3. Molecular Gas in Nearby Galaxies

As part of our 3 mm commissioning tests we have observed molecular gas in several nearby galaxies. These include HCO^+ emission in the starburst nuclei of NGC 253 and the Circinus galaxy, as well as absorption by several molecular species in front of the core of the nearby giant elliptical NGC 5128 (Centaurus A). Figure 3 shows the HCO^+ absorption spectrum in front of Centaurus A.

2.4. The Detached Shell Around R Sculptoris

Mass loss in AGB stars is a key process for enriching and replenishing the interstellar medium, yet its mechanisms are poorly understood. Strong variations in the mass loss rate, probably related to He shell flashes, lead to the formation of *detached shells* of molecular material which can be studied with mm-wave interferometry. The carbon star R Scl has been inferred from SEST observations to have a relatively young shell (Olofsson et al. 1996), and is thus an ideal target for ATCA imaging. Our observations in 2002 May and June detected strong line HCN emission but did not resolve the source (Fig. 4); higher resolution observations are planned for October.

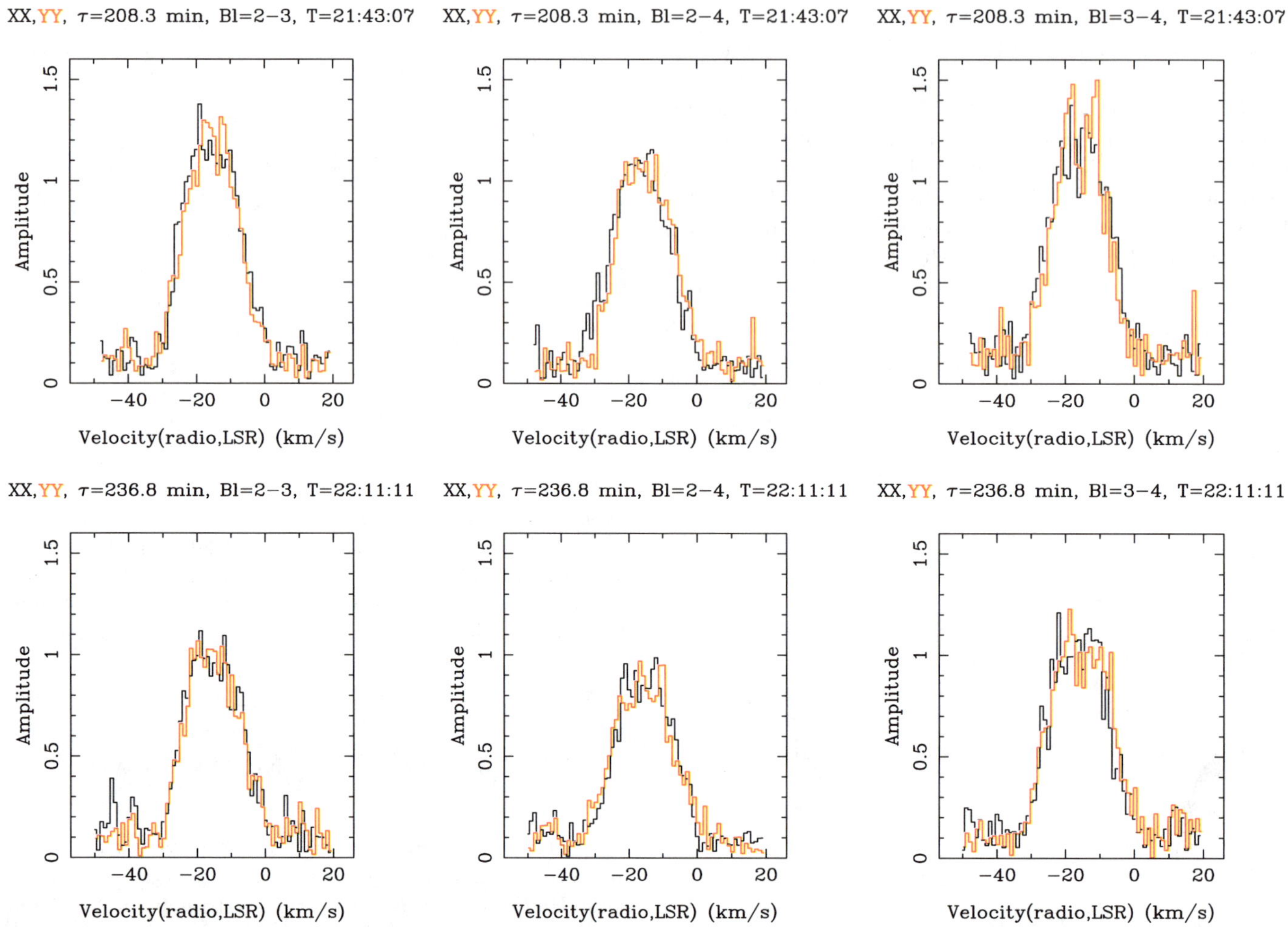

Figure 4. Spectra of HCN emission from R Scl, taken in two different array configurations with east-west baselines of 61, 92, 31, 77, 122, and 46 m (from top left to bottom right). The amplitude scale is in Jy. The two polarizations are shown as black and red lines. Credit: T. Wong et al.

3. Applying for Time

The ATCA is operated by the Australia Telescope National Facility (ATNF), a division of CSIRO Australia. Proposals from Australians and overseas users are evaluated on an equal basis. Successful proposers are expected to travel to Narrabri for their observations and are strongly encouraged to visit ATNF headquarters near Sydney. To allow for some flexibility in dealing with inclement weather, your project may be "swapped" with a centimetre project scheduled a few days later.

The next proposal deadline for which 3 mm and 12 mm proposals will be solicited is the 2003 May term (1 May to 31 August). Proposals must be received by 2003 February 15. For proposal forms and information see the ATNF website at `http://www.atnf.csiro.au/observers`.

Acknowledgements

I am indebted to the ATNF millimetre science and engineering team for the hard work that has made the results presented here possible. I also thank my collaborators for permission to show work in progress.

References

Olofsson, H., Bergman, P., Eriksson, K., & Gustafsson, B. 1996, A&A, 311, 587

Walsh, A. J., Burton, M. G., Hyland, A. R., & Robinson, G. 1998, MNRAS, 301, 640

CHARACTERISTIC DIFFERENCE BETWEEN THE FORMATION PROCESSES OF C_NO AND C_NS MOLECULES (N=2–4) IN INTERSTELLAR SPACE

Masato Yamada and Yoshihiro Osamura

Department of Chemistry, Rikkyo University, 3-34-1 Nishi-ikebukuro, Toshima-ku, Tokyo 171-8501, Japan

Abstract

The mechanism of the formation processes of carbon-chain interstellar molecules C_nO and C_nS (n=2–4) via neutral-neutral reactions has been studied in terms of ab initio molecular orbital theory. The evidence that the sulfur-containing carbon-chain molecules C_nS are more abundant than C_nO except CO may be due to the characteristic difference of chemical processes on the formation of C_nO and C_nS in interstellar conditions. Since the source of carbon chain can be C_nH as abundant and reactive hydrocarbon species, we have considered the reactions between C_nH and oxygen or sulfur atom by employing the B3LYP density functional method in order to obtain the potential energy surfaces of these reactions. Both of the potential energy surfaces indicate that there is no entrance barrier leading to the intermediates HC_nO or HC_nS. Using the excess energies of the bond formation in HC_nO or HC_nS, these molecules can decompose to $C_nO + H$ or $C_nS + H$ without energy barrier. There is another pathway producing $CO + HC_{n-1}$ or $CS + HC_{n-1}$. The energy of the products $C_nS + H$ is calculated to be lower than that of $CS + HC_{n-1}$. The energy of $C_nO + H$ is, however, found to be higher than that of $CO + HC_{n-1}$. This means that the formation of C_nO is less favorable and the reaction between C_nH and O atom mainly produces CO molecules. The potential energy calculation suggests that C_4S being unobserved interstellar molecule can be produced via the reaction of $S + C_4H$.

Key words: ISM: molecules – ISM: abundances – molecular processes

1. Introduction

Series of carbon-chain molecules are most characteristic and abundant species in dark molecular clouds. The cyanopolyyne, $HC_{2n+1}N$, cumulene-type hydrocarbons, C_nH, and sulfur-containing carbon-chain, C_nS, are well surveyed and identified in various molecular clouds, especially in TMS-1 (Hirahara et al. 1992). There are significant difference of the abundances between sulfur-containing molecules, C_nS (n=1–3), and oxygen-containing species, C_nO (n=1–3). Although the cosmic abundance of oxygen atoms is 20 times larger than that of sulfur atoms, C_2O and C_3O are significantly depleted in most of dark molecular clouds. The abundance ratios are 80 and 6 for $n(C_2S)/n(C_2O)$ and $n(C_3S)/n(C_3O)$ in TMC-1, respectively (Suzuki et al. 1988; Petrie 1996). The C_2S molecules show strong intensities and have been used for the characterization of the evolution of molecular clouds (Suzuki et al. 1992).

In order to explore the characteristic difference of the abundance ratio between C_nS and C_nO, we have examined the potential energy surfaces of the formation reactions via neutral-neutral processes. There must be other types of the formation processes such as ion-molecule reactions (Suzuki et al. 1988; Smith et al. 1988; Herbst et al. 1984; Suzuki et al. 1992). Since the ionization potential of oxygen atom is slightly larger than hydrogen atom, oxygen atom is less ionized in the interstellar conditions. Ion-molecule reactions producing oxygen-containing carbon-chain molecules should involve the reaction between neutral oxygen atom and ionized hydrocarbons. On the other hand, sulfur atom is easily ionized due to its small ionization energy. Ionization of sulfur atom gives ion-molecule reaction between S^+ and hydrocarbons.

Theoretically proved reaction is $S^+ + C_2H_2$ which produces $HCCS^+ + H$ (Barrientos et al. 1999). Such ion-molecule reactions between ionized sulfur atom and unsaturated hydrocarbons, however, lead to the intermediates in excited state in most of the cases and cause very complicated chemistry.

Although ion-molecule reactions would be main route on the molecular formation in dark molecular clouds, the neutral-neutral reactions are shown to play important role on the formation of cyanopolyynes (Woon & Herbst 1997; Fukuzawa & Osamura 1997; Fukuzawa et al. 1998). In this respect, we have examined the reaction between

Chemistry as a Diagnostic of Star Formation (2003). *Edited by Charles L. Curry and Michel Fich.* Published in 2003 by NRC Research Press, Ottawa ON, K1A 0R6, Canada. ISBN 0-660-19089-3.

C_nH and S atom as well as the reactions between C_nH and O atom in order to explore the possibilities of C_nS and C_nO formation via neutral-neutral reactions in interstellar space.

2. Theoretical method

The potential energy surfaces of the reactions, O + C_nH and S+C_nH were calculated in terms of ab initio quantum chemical methods. The molecular structures are optimized by employing the hybrid density functional B3LYP method (Becke 1992) with 6-311G(d,p) basis set. All computations were carried out using Gaussian 98 program package (Frisch et al. 1998).

3. Results and discussion

3.1. Neutral-neutral reactions producing C_nS

The neutral-neutral reactions of forming sulfur-containing carbon-chain molecules have been first proposed by Petrie (1996). Following his idea, Flores et al. (2001) have demonstrated that the reaction S + C_2H is feasible to produce C_2S molecule in interstellar conditions. Flores and Gomez (2001) also calculated the potential energy surface of S + C_3H resulting the possible formation of C_3S. The reaction between S atom and C_3H is possible to produce C_3S + H and CS + C_2H. Both of these products are exothermic in energy and can be produced without energy barrier. Since the energy of CS + C_2H is higher than that of C_3S + H, the reaction CS + C_2H can be an alternate route to produce C_3S molecule in interstellar space.

Similar argument might be utilized in order to produce C_4S molecule being undetected species in dark molecular clouds. We have calculated the potential energy surface of the reaction between S atom and C_4H using the B3LYP density functional method as shown in Figure 1. The energetically possible products from S + C_4H are C_4S + H, CS + *linear*-C_3H, and CS + *cyclic*-C_3H. This process involves deep potential minimum of HC_4S intermediate and leads to either hydrogen elimination or C-C bond scission. There is no entrance barrier on the potential between S atom and C_4H species approaching S atom to the terminal carbon atom of C_4H species. There is no exit barrier of the hydrogen elimination from HC_4S intermediate to the C_4S+H products. On the other hand, the CS elimination pathway from HC_4S has a transition state before leading to the products CS + *linear*-C_3H. The energy of this transition state is 13 kJ/mol higher than the products but is below the energy of the reactants S + C_4H. This means that both C_4S + H and CS + C_3H can be formed in the interstellar conditions. There may be a pathway from HC_4S intermediate to CS + *cyclic*-C_3H, but we could not find the transition state due to the convergence problem during the calculation. There are various isomerization pathways from the intermediate from HC_4S, as discussed in the cases of HC_2S and HC_3S (Flores et al. 2001, Flores and Gomez 2001). The isomerization pathways on the potential energy surface of HC_4S involve hydrogen-migration pathways and cyclization pathways. Even if these isomerizations occur on the HC_4S potential energy surface, the final products are only C_4S+H and CS+C_3H, since all other products such as C_2S + C_2H, C_3S + CH, *cyclic*-C_4S + H, etc., are endothermic to be formed. In this respect, we can conclude that the reaction between S atom and C_4H molecule can produce only C_4S+H and CS+C_3H in interstellar space. Since there is an energy barrier to produce CS + C_3H, the reaction between CS and C_3H molecules cannot occur forming C_4S + H in interstellar conditions.

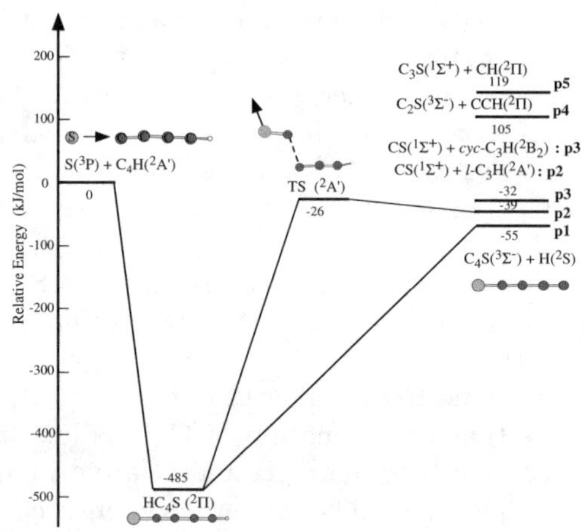

Figure 1. Schematic representation of the potential energy surface of the reaction S + C4H calculated with B3LYP/6-311G(d,p) method.

3.2. Neutral-neutral reaction producing C_nO

The neutral-neutral reaction of forming C_2O via O + C_2H is suggested by Ohishi et al. (1991) at the detection of C_2O molecule in TMC-1. Woon and Herbst (1996) have studied the reaction between O atom and C_3 molecule and show small activation barrier. When Brown et al.(1985) reported the new detection of C_3O in TMC-

1, they have not counted the neutral-neutral reaction of $O + C_3H$. Since the $S + C_nH$ reactions are possible to produce C_nS molecules in interstellar conditions as discussed above, there must be possibility of forming C_nO species via the $O + C_nH$ reactions.

We have examined the potential energy surfaces of $O + C_nH$ (n=2–4) in order to clarify whether C_nO formation is possible or not in interstellar conditions. The reactions between oxygen atom and C_nH molecules are very similar to the case of $S + C_nH$ reactions because of their electronic structures. All reactions go through the deep potential well of HC_nO species following to either hydrogen elimination or C-C bond cleavage. Figure 2 depicts the potential energy surface of the reaction between oxygen atom and C_3H. Similar potential energy surface can be obtained for the $O + C_4H$ reaction. It is noticeable that the energies of the products $C_nO + H$ are always higher than those of $CO + HC_{n-1}$. There are many isomerization pathways from the HC_nO intermediates, but most of these isomerization processes are not important for producing energetically feasible products which are CO and C_nO molecules.

The present result indicates that the carbon monoxide is the most stable product from the reactions between O atom and C_nH species via neutral-neutral reactions. Even if C_nO molecules are produced, C_nO can react with hydrogen atom without energy barrier and yields CO molecule very easily. These evidences rationalize why C_nO molecules are significantly depleted and most of oxygen-containing species end up to CO molecules in interstellar space.

4. Concluding remarks

The present theoretical calculations predicts that the reactions between S atom and C_nH molecules can form C_nS species in interstellar conditions for n=1–4. This means that there is a possibility of detecting C_4S and longer C_nS species in dark molecular clouds. On the other hand, oxygen-containing carbon-chain molecules C_nO should be significantly less abundant than C_nS molecules because the C_nO species are less stable and undergo C-C bond cleavage by hydrogen atom leading to the CO formation. Such difference is due to the large bond energy of CO compared to CS bond. The extremely strong CO bond formation results in the difficulty of the existence of oxygen-containing longer carbon-chain molecules.

Acknowledgements

This research was supported by the Grants-in-Aid for Scientific Research on Priority Areas from the Ministry of Education, Science, and Culture, Japan. Most of the ab initio calculations were carried out at the computer center of the Institute for Molecular Science, Japan.

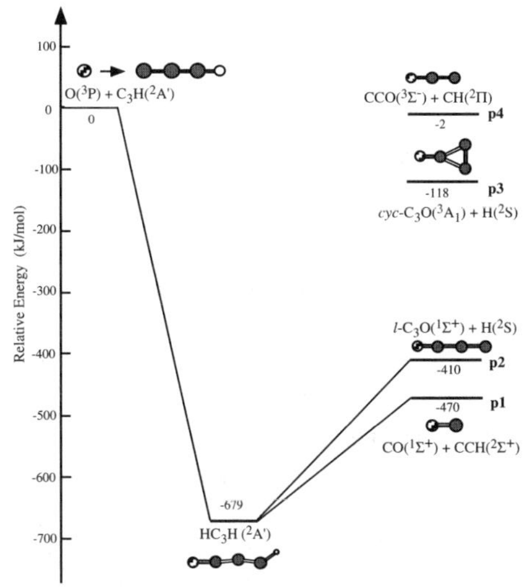

Figure 2. The potential energy surface of the reaction between O atom and C_3H molecule calculated with B3LYP/6-311G(d,p) method.

References

Barrientos, C., Redondo, P. & Largo, A. 1999, Chem. Phys. Lett., 306, 168

Becke, A. D. 1992, J. Chem. Phys., 97, 9173

Brown R. D. et al. 1985, ApJ, 297, 302

Flores J.R. & Gomez, F.J. 2001, J. Phys. Chem. A, 105, 10384

Flores, J.R., Estevez, C. M., Carballeira, L. & Perez Juste, I. 2001, J. Phys. Chem. A, 105, 4716

Frisch, M. J. et al. GAUSSIAN 98, Revision A.9, Gaussian, Inc., Pittsburgh, Pennsylvania, 1998

Fukuzawa, K. & Osamura, Y. 1997, ApJ, 489, 113

Fukuzawa, K., Osamura, Y. & Schaefer III, H. F. 1998, ApJ, 505, 278

Herbst, E., Smith, D. & Adams, N. G. 1984, A&A, 138, L13

Hirahara, Y., Suzuki, H., Yamamoto, S., Kawaguchi, K., Kaifu, N., Ohishi, M., Takano, S., Ishikawa, S. & Masuda, A. 1992, ApJ, 394, 539

Ohishi, M. et al. 1991, ApJ, L39

Petrie, S. 1996, MNRAS, 281, 666

Smith, D., Adams, N. G., Giles, K & Herbst, E. 1988, A&A, 200, 191

Suzuki, H., Ohishi, M., Kaifu, N., Kasuga, T., Ishikawa, S. & Miyaji, T. 1988, Vistas in Astronomy, 31, 459

Suzuki, H., Yamamoto, S., Ohishi, M., Kaifu, N., Ishikawa, S., Hirahara, Y. & Takano, S. 1992, ApJ, 392, 551

Woon, D. E. & Herbst, E. 1996, ApJ, 465, 795

Woon, D. E. & Herbst, E. 1997, ApJ, 477, 204

DIFFUSIVE MIXING AND GRAIN ACCRETION IN DARK CLOUDS

C.J. Yate and T.J. Millar

Department of Physics, UMIST, P.O. Box 88, Manchester M60 1QD, UK

Abstract

We report results of models of dark cloud chemistry incorporating a depth-dependent density distribution with diffusive mixing and adsorption onto grains. The model is based on the approach taken by Xie et al. (1995), with the addition of grain accretion effects. Without diffusion, the central regions of the cloud freeze out in 10^6 years. Freeze-out time is dependent on density, so the diffuse outer region of the cloud remains abundant in gas for about an order of magnitude longer. We find that the gas-phase lifetime of the cloud core can be increased by up to an order of magnitude or more, with fairly small amounts of diffusive mixing. The mixing process delays freeze-out at the centre of the cloud for a time up to an order of magnitude greater than without diffusion, due to material diffusing inward from the edges of the cloud.

Key words: astrochemistry – turbulence – diffusion – ISM: abundances – ISM: dust

1. Introduction

The phenomenon of turbulence is seen on almost all scales in the universe, and is familiar in many situations on Earth. The evidence for turbulence within interstellar clouds lies in widths of molecular emission and absorption lines, which can be broader than would be possible from thermal broadening alone. Modelling turbulence is a mathematically difficult problem and so is neglected in many interstellar chemistry models. However, turbulence within a cloud causes mixing of the gas which affects the chemistry as reactants are redistributed throughout the cloud. Several authors have attempted to model mixing (Chièze & Pineau des Forêts 1987, 1989; Pineau des Forêts et al. 1992; Xie et al. 1995; Rawlings & Hartquist 1997; Willacy et al. 2002). For this work we have used a model similar to that of Xie et al. (1995), with the addition of adsorption onto grains. The mixing effects of turbulence are approximated by a diffusion function, dependent on molecular abundance gradients.

2. Model Details

We use a subset of the UMIST rate file Rate99 (Le Teuff et al. 2000) incorporating 127 species with 1669 reactions. ODEs are solved with the CVODE solver. Physical conditions are represented by a 1-D slice into a plane parallel cloud of depth 1pc. The chemistry and diffusion are calculated at 23 depth points which have density, A_v and temperature taken from Xie et al.. The model cloud is assumed to be isothermal throughout, at 10K. The density at the central point is 3.5×10^4 H_2 cm^{-3}, decreasing by a $1/r$ law to the edge. The turbulent diffusion is calculated by the method described by Xie et al.. In summary, the term included in the ODE for each species is:

$$\phi_i = -Kn(H_2)\frac{df_i}{dz} \quad \mathrm{cm}^{-2}\,\mathrm{s}^{-1}, \tag{1}$$

where f_i is the fractional abundance of each species, $n(H_2)$ is the density of molecular hydrogen, z is the depth coordinate, and K is the diffusion coefficient which sets the speed of diffusion, defined as:

$$K = \langle v_t\, L\rangle \quad \mathrm{cm}^2\ \mathrm{s}^{-1}, \tag{2}$$

where L is the mixing length, and v_t is the turbulent dispersion velocity, which can be estimated from observed linewidths in clouds. The turbulent mixing length is rather more difficult to estimate. Xie et al. suggest a value of 0.1 pc on the grounds that this is 10% of the cloud radius. This assumption is based on work by Kleiner & Dickman (1987). For further reading, Larson (1981) and Leung, Kutner & Mead (1982) also discuss turbulent velocities in more detail. In this paper we show results for models with K up to 10^{23} cm^2 s^{-1} which is around the highest reasonable value for a 1pc cloud, assuming $v_t = 1$ km s^{-1} and $L = 0.5$ pc. The accretion calculation used is the standard formula of d'Hendecourt et al (1985). The ODE for each species i includes an accretion term A_i, defined as:

Chemistry as a Diagnostic of Star Formation (2003). *Edited by Charles L. Curry and Michel Fich.* Published in 2003 by NRC Research Press, Ottawa ON, K1A 0R6, Canada. ISBN 0-660-19089-3.

$$A_i = -S_i \, \sigma_g \, n_g n_i \left(\frac{8 k_b T}{\pi m_i} \right)^{1/2} \text{cm}^{-3}\ \text{s}^{-1}, \qquad (3)$$

where n_g is the dust abundance assumed to be $10^{-12}\, n(H_2)$ cm^{-3}, and n_i is the abundance of species i cm^{-3}. S_i is the sticking coefficient, σ_g is the cross section of the grain (assuming grain radius $= 10^{-5}$ cm), m_i is the molecular mass, and all other symbols have their usual meaning. In these models only H_2 is formed on the grain surface; H is assumed to have a sticking coefficient $S_H = \frac{1}{3}$. All other species have $S_i = 1$, except small species such as H^+, He, He^+ etc., which are assumed not to stick to grains.

3. Results

Figures 1 to 8 show fractional abundance of CH_3OH and C_2H over model shells versus time. The centre is marked with a thick, solid line, and alternate shells are plotted out to the edge. Notice the effect of diffusion on the abundances at the centre of the cloud for times $> 10^5$ years. Graphs are shown for models with $K = 0$, 10^{21}, 10^{22}, and 10^{23} $\text{cm}^2\ \text{s}^{-1}$. We have run models with other values of K, but these are sufficient to show the effect of diffusion when included alongside accretion. The species CH_3OH and C_2H are chosen as methanol peaks strongly toward the centre of the cloud, while the C_2H abundance is fairly constant. Our results show that large enough gradients can occur as a result of grain accretion to affect abundances of most molecules.

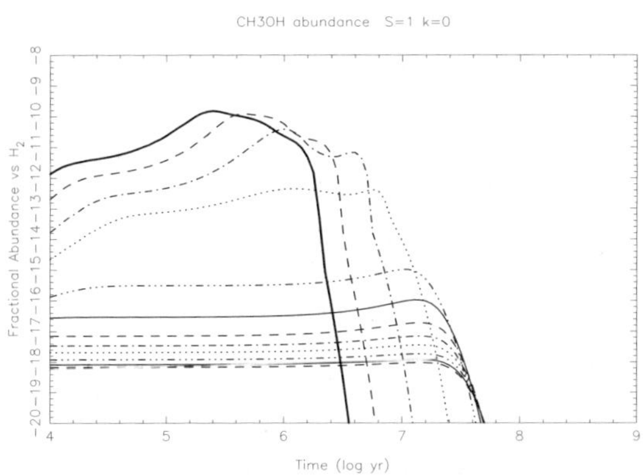

Figure 1. CH_3OH abundances, $K = 0$ $\text{cm}^2\ \text{s}^{-1}$. A heavy solid line marks the centre of cloud, while each line plotted shows each alternate gridpoint.

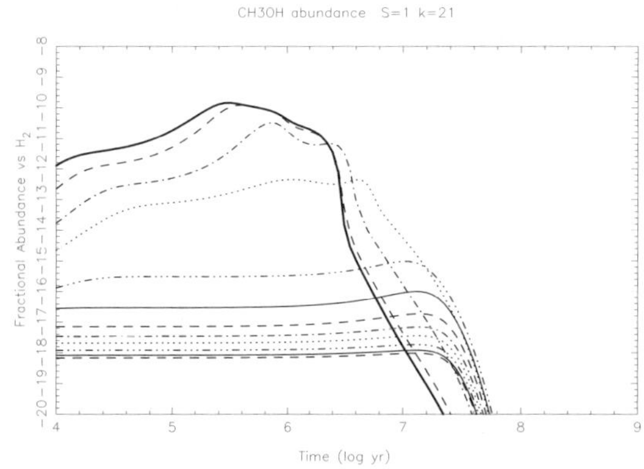

Figure 2. CH_3OH abundances, $K = 10^{21}$ $\text{cm}^2\ \text{s}^{-1}$.

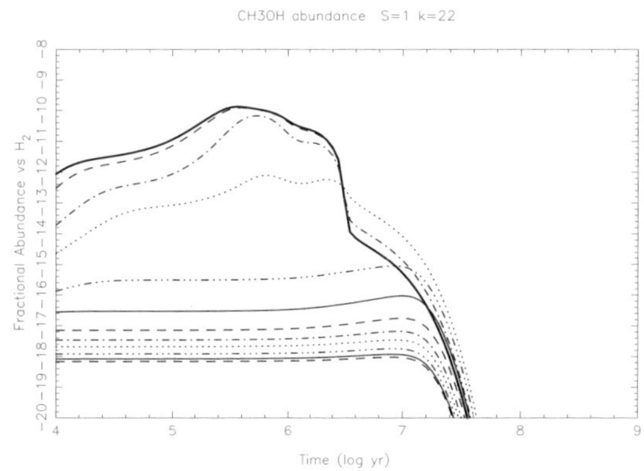

Figure 3. CH_3OH abundances, $K = 10^{22}$ $\text{cm}^2\ \text{s}^{-1}$.

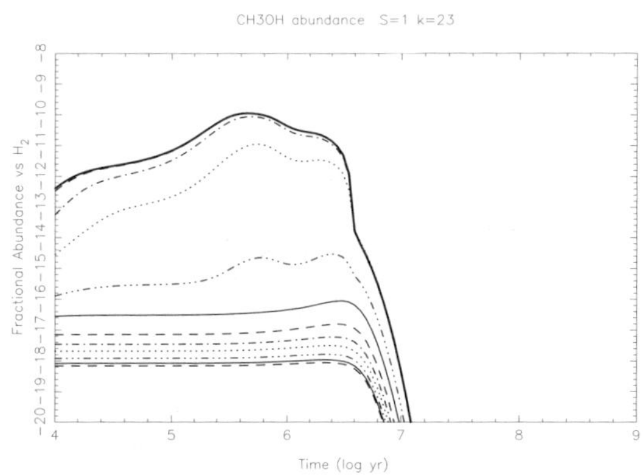

Figure 4. CH_3OH abundances, $K = 10^{23}$ $\text{cm}^2\ \text{s}^{-1}$.

4. Conclusions

The large radial abundance gradients caused by grain accretion help to drive the diffusion process, which tends to homogenize the gas. As a result, the time at which molecules are depleted from the gas is increased when diffusion is included. We find that $K = 10^{21}$ $\text{cm}^2\ \text{s}^{-1}$ is sufficient to increase the abundance of many molecules at

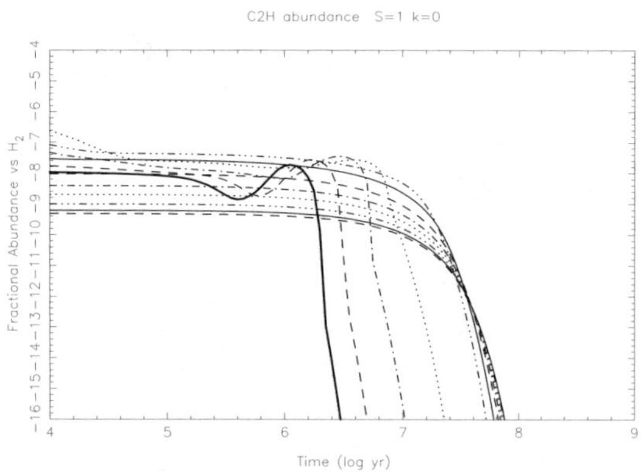

Figure 5. C_2H abundances, $K = 0$ cm^2 s^{-1}.

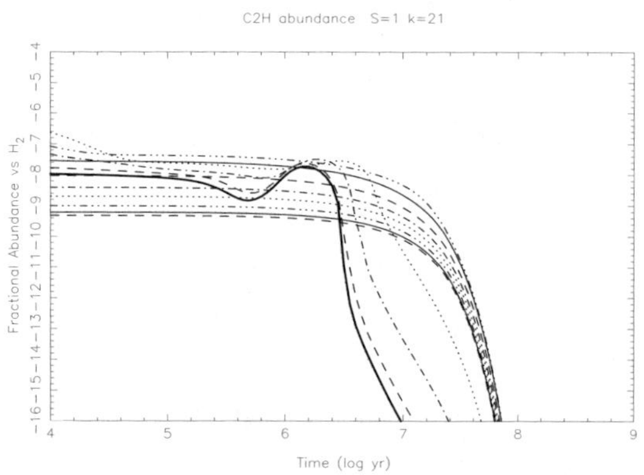

Figure 6. C_2H abundances, $K = 10^{21}$ cm^2 s^{-1}.

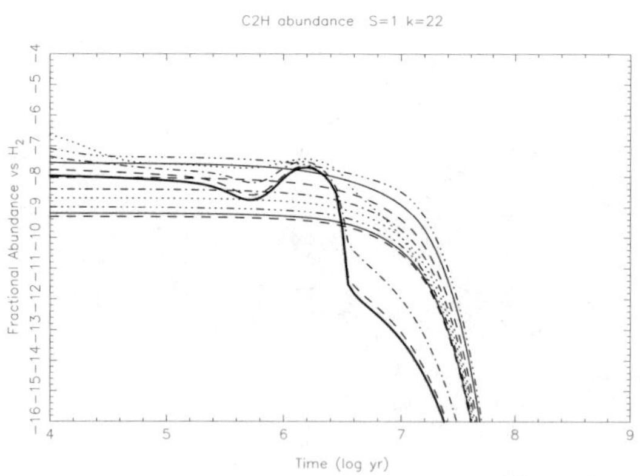

Figure 7. C_2H abundances, $K = 10^{22}$ cm^2 s^{-1}.

times $t > 10^{5.5}$ years, by well over an order of magnitude. This is caused by material at the cloud edge being pulled towards the centre by the diffusion process. Xie et al. required diffusion coefficients of around 10^{23} cm^2 s^{-1} to significantly change species abundances. These values were rejected by Rawlings & Hartquist (1997), on the grounds that such a high diffusion coefficient would be unphysical,

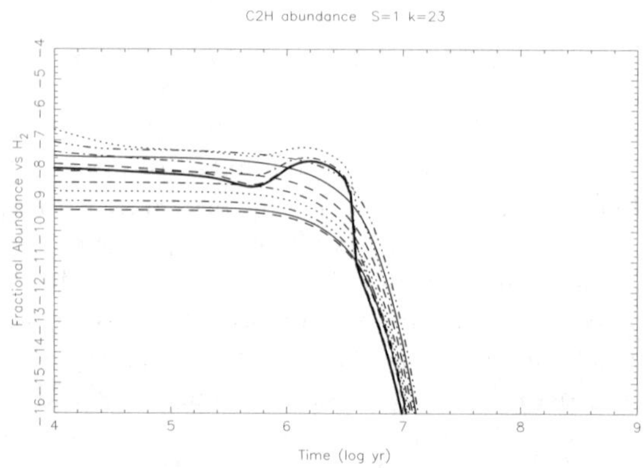

Figure 8. C_2H abundances, $K = 10^{23}$ cm^2 s^{-1}.

particularly in a cloud of size 1pc. Higher values of K mix the gas more thoroughly, but cause the abundances in the outer shells to decrease, as material is transferred to the centre at a larger rate. Willacy et al. (2002) have run similar models to these, but including H_2 photodissociation, and allowing grains to diffuse throughout the cloud. Photodissociation increases H abundances at the edge of the cloud, which leads the H abundance at the centre to be strongly increased by increasing K. We find only small increases in H abundance with K. Our results show that if grain accretion is included in the model it is possible to significantly modify abundances of some molecules, using diffusion coefficients two orders of magnitude or less than those used by Xie et al.. Fuller discussion of the results of this model are given by Yate & Millar (2003).

Acknowledgements

CJY acknowledges receipt of a PPARC studentship. Astrophysics at UMIST is supported by PPARC.

References

Chièze, J.-P. & Pineau des Forêts, G. 1987, A&A, 183, 98
Chièze, J.-P. & Pineau des Forêts, G. 1989, A&A, 221, 89
d'Hendecourt L. B., Allamandola L. J. & Greenberg J. M. 1985, A&A, 152, 130
Kleiner, S. C. & Dickman, R. L. 1987, ApJ, 312, 837
Larson R. B. 1981, MNRAS, 194, 809
Le Teuff Y. H., Millar T. J. & Markwick A. J. 2000, A&AS, 146, 157
Leung C. M., Kutner M. L. & Mead K. N. 1982, ApJ, 262, 583
Pineau des Forêts, G., Flower, D. R. & Chièze, J.-P. 1992, MNRAS, 256, 247
Rawlings J. M. & Hartquist T. W. 1997, ApJ, 487. 672
Willacy, K., Langer, W. D. & Allen, M. 2002, ApJ, 573, 119
Xie, T., Allen, M. & Langer, W. D. 1995, ApJ, 440, 674
Yate, C. J. & Millar, T. J. 2003, A&A, in press

CONFERENCE SUMMARY

CHEMISTRY AS A DIAGNOSTIC OF STAR FORMATION: CONFERENCE SUMMARY

George F. Mitchell

Department of Astronomy and Physics, Saint Mary's University, Halifax, NS, Canada B3H 3C3

SFCHEM 2002 brought together about 130 participants, including many of the leading scientists in the field of astrochemistry. The scientific program featured 37 oral presentations and 86 poster papers. Although the meeting concentrated on chemistry as a diagnostic of star formation, a very broad range of astrochemical topics were discussed. The meeting began with a session on future instrumentation: The capabilities of ALMA, LMT, SOFIA, and *Herschel*, as well as new infrared spectrographs (NIRSPEC, Phoenix, TEXES) and submillimeter bolometer arrays (SCUBA-2), ensure a bright future for astrochemistry in general and for the study of star formation in particular.

The formation of H_2 was the subject of three talks (***Habart, Roser, Joblin***) in a session on fundamental processes. Although it has been recognized for more than thirty years that H_2 must be formed on grains, the mechanism is still poorly understood. Evidence was presented that H_2 emission from PDRs requires a high rate of H_2 formation even at high gas temperatures. Because infrared emission from PAH molecules and from H_2 are spatially correlated, it was suggested that PAH molecules may catalyze H_2 formation. Laboratory studies of this process are in progress. Another molecule of immense interest is H_3^+, the key to ion-molecule chemistry. Its first detection a few years ago, via near-IR absorption lines, was a key validation of our interstellar chemical systems. New observations of H_3^+ were presented (***Geballe***). In addition to providing further confirmation of gas phase ion-molecule chemistry, H_3^+ may soon become a unique tool for the determination of astrophysical parameters such as temperature and cosmic ray flux.

One of the surprises of recent years has been the very high abundances attained by doubly and triply deuterated molecules (e.g. D_2CO, CHD_2OH, and ND_3). Three talks and five posters addressed deuterated molecules (***Phillips, van der Tak, Millar***). As an example of the extreme situation, the abundance of doubly deuterated formaldehyde is eight orders of magnitude higher than would be predicted in the absence of deuteration! Both gas phase deuteration and active grain chemistry have difficulty in explaining the high levels of deuteration.

A large fraction of the results presented at the meeting were naturally concerned with the central issues of star formation, such as the progression from a prestellar core to a deeply embedded protostar to a visible low mass star. If a prestellar core is defined as a gas clump of one to a few solar masses which shows no evidence for the existence of a central point mass, then such cores can be modeled quite well as Bonnor-Ebert spheres. Recent models which relax the isothermal assumption predict slightly cooler central regions (e.g., 8 K instead of 10 K). Although the differences are small, they *are* important for molecular excitation modeling. One talk and four posters addressed molecular depletion (e.g., ***Bergin, Feldman et al., Savva et al.***). It is now very well established that many molecules are strongly depleted due to adsorption onto grains. Even the ion N_2H^+ can be depleted (***Lee et al.***), contrary to some earlier work. Thus, obtaining masses of cores from molecular column densities is difficult (if not impossible). Of course, molecular excitation studies can still yield the gas density (i.e. H_2 density) and thus the mass. Depletion may have one positive application: If we have confidence in depletion rates, the degree of depletion in a core gives the core age. Some very young ages were presented (e.g., ***Bergin***), but these depend on still highly uncertain binding energies and cosmic ray heating rates.

Observations and modeling of embedded protostars was a central topic of the meeting. These are complex systems, consisting of a central object, a disk, a collapsing envelope, an outflow, and surroundings. The temperature and density structure of the envelopes of protostars is still under active study. Work was presented in which molecular brightness profiles and SEDs of a number of Class 0 and Class I envelopes were modeled (using DUSTY) to obtain the density as a function of radius (***Jørgensen et al.***). The density behaviour is usually consistent with a collapse model, although not with any

Chemistry as a Diagnostic of Star Formation (2003). *Edited by Charles L. Curry and Michel Fich.* Published in 2003 by NRC Research Press, Ottawa ON, K1A 0R6, Canada. ISBN 0-660-19089-3.

single model. A model was presented in which chemical abundances were followed in a Larson-Penston collapse (***Aikawa et al.***). In this case, reasonable agreement was found with observations of L1544, including the centrally peaked N_2H^+ emission. Grain mantles in the envelopes of low mass protostars have eluded detection because the embedded sources were too faint for infrared absorption spectroscopy. This has now changed. Impressive first results of a Keck/NIRSPEC study were presented, showing absorption, due to CO ice mixtures, in Elias 29 and L1489IRS (***Boogert***).

It is trite, but true, to say that contributions of astrochemistry to star formation theory will depend on the combination of excellent observations with sophisticated physical and chemical modeling. A number of examples of this kind of effort were presented at the meeting. From a detailed radiative transfer analysis of continuum and line emission from the envelope of the Class 0 object IRAS 16293-2422, jumps in the abundance of some molecules are suggested, providing evidence for evaporation of grain mantles, and thus for a small "hot core" (***Schöier et al.***). There is evidence for infall in this envelope, but the standard inside-out collapse model does not work very well.

A detailed study of the Class 0 object IRAM 04191 was presented (***André***), which made use of new molecular line observations and a high resolution continuum map. The observations were interpreted as showing an extended region infalling at a constant speed, an inner region infalling according to a power law, and differential rotation. The extended region is inconsistent with the inside-out collapse of a singular isothermal sphere and only barely consistent with the isothermal collapse of a Bonnor-Ebert sphere. Angular momentum appears *not* to be conserved in the outer region, suggesting magnetic braking. A suggested scenario was: The rapidly rotating inner envelope is a magnetically supercritical core decoupling from the environment which is still supported by magnetic field (and still undergoing magnetic braking). One possible inconsistency in the scenario is that the continuum image rules out any accretion disk larger than 10 AU. Whether these interpretations hold up, it is exciting to see such a direct confrontation with star formation theory.

Circumstellar disks were the subject of four talks and several posters. In one disk model, 2-dimensional molecular distributions were shown for a disk model with a temperature gradient in the vertical direction (***Aikawa et al.***). The models include 2D radiative transfer and full chemistry, and return molecular line intensities and line profiles. A layered structure is predicted, partly due to grain adsorption in the cold midplane region. Comparison with observations remains at an early stage. A physical model of disks was presented which included transport of gas inwards and outwards (***Henning & Ilgner***). It appears that future chemical models should include horizontal transport. On the observational front, new instrumentation (NIRSPEC, Phoenix, TEXES) is resulting in a post-*ISO* flowering of infrared spectroscopy. Impressive M-band observations were presented showing CO in emission and in absorption (***Blake***). The CO emitting gas is believed to originate in the inner disk, just 0.5 AU from the central source.

The study of massive star formation presents special problems because massive stars are rare, because they form inside massive, dense clumps, and because the formation process is so disruptive. It has not been easy to demonstrate that massive YSOs form via disk accretion (for example). Observations of mm/submm molecular emission lines is relatively easy because the envelopes are warm. Quite a number of embedded massive protostars are bright in the IR, so they can also be observed via IR spectroscopy. Beautiful new IR spectra taken with the TEXES spectrograph were shown (***Lacy***). These included the molecules H_2, SiO, HCN, and C_2H_2 and the sources KL, NGC 7538 IRS1, and AFGL 2591. Models of the high mass YSO AFGL 2591 and the low mass YSO IRAS 16293 point to similarities in their envelopes (***van Dishoeck***). The model of the envelope of AFGL 2591 can fit most, but not all, molecular observations when hot gas chemistry is invoked and when certain species are assumed to be frozen on grains in the outer regions. Results of a BIMA survey of Ultra-Compact H II Regions was presented (***Gibb***): One of the intriguing results is that compact dust cores are often found to be offset from the H II region. This might be evidence for star formation by triggering.

A great deal of research effort is being devoted to numerical simulations of molecular clouds. These simulations now often assume compressible gas and self-gravity. The ensuing supersonic turbulence produces shock compression and fragmentation, resulting in a hierarchical (i.e. scale-free or self-similar) structure. Several of the simulations lead to gravitationally bound cores whose mass function slope is similar to the Salpeter IMF. Although there were three poster papers devoted to turbulence and molecular cloud structure, this meeting did nor deal with the impact of these simulations on our field. In the future, I expect that the observational and modeling expertise of astrochemists will be brought to bear on these simulations. So far, for example, line profiles calculated from simulations seem unable to duplicate the observed line profiles. Simulations show structure while observed lines tend to be smooth. Another point of contact may be pre-stellar cores. Do the cores from simulations look like actual cores?

In summary, this meeting demonstrated that the astrochemical community is deeply involved in the observation and modeling of cores, YSO envelopes and disks. Models are becoming increasingly realistic and are becoming increasingly effective in testing star formation theory, from the structure of cores to the properties of disks.

References

Aikawa, Y., 237

Aikawa, Y., van Zadelhoff, G.-J., & van Dishoeck, E.F., 166

André, P., Bouwman, J., Belloche, A., & Hennebelle, P., 127

Bergin, E.A., 63

Blake, G.A., 178

Boogert, A.C.A., Blake, G.A., & Hogerheijde, M.R., 172

Feldman, P.A., Redman, R.O., Carey, S.J., & Wyrowski, F., 292

Geballe, T.R., McCall, B.J., & Oka, T., 46

Gibb, A.G., Wyrowski, F., & Mundy, L.G., 214

Habart, E., Boulanger, F., Verstraete, L., Walmsley, C.M., & Pineau des Forêts, G., 26

Henning, T., & Ilgner, M., 54

Joblin, C., 45

Jørgensen, J.K., Schöier, F.L., & van Dishoeck, E.F., 318

Lacy, J., Knez, C., Evans, N., & Richter, M., 220

Lee, J.-E., Evans, N.J. II, Shirley, Y.L., & Tatematsu, K., 344

Millar, T.J., 37

Phillips, T.G., & Vastel, C., 3

Roser, J.E., Vidali, G., Manicò, G., & Pirronello, V., 32

Savva, D., Gibb, A.G., Little, L.T., & Phillips, R.R., 404

Schöier, F.L., Jørgensen, J.K., van Dishoeck, E.F., & Blake, G.A., 410

van der Tak, F.F.S., Lis, D.C., Gerin, M., Roueff, E., Schilke, P., & Phillips, T.G., 50

van Dishoeck, E.F., 201

INDEXES

Subject Index

Author Index

Astronomical Object Index

Chemical Species Index (*H_2, CO omitted*)